METAL CUTTING PRINCIPLES

OXFORD SERIES ON ADVANCED MANUFACTURING

Series Editors

J. R. Crookall
Milton C. Shaw
Nam P. Suh

Series Titles

John Benbow and John Bridgewater, *Paste Flow and Extrusion*

John L. Burbidge, *Period Batch Control*

John L. Burbidge, *Production Flow Analysis for Planning Group Technology*

Shiro Kobayashi, S. Oh, and T. Altan, *Metal Forming and the Finite-Element Method*

Milton C. Shaw, *Metal Cutting Principles, 2nd edition*

Milton C. Shaw, *Principles of Abrasive Processing*

Nam P. Suh, *The Principles of Design*

Daniel E. Whitney, *Mechanical Assemblies: Their Design, Manufacture, and Role in Product Development*

METAL CUTTING PRINCIPLES

Second Edition

Milton C. Shaw

Professor Emeritus of Engineering
Arizona State University

New York   Oxford
OXFORD UNIVERSITY PRESS
2005

Oxford University Press

Oxford   New York
Auckland   Bangkok   Buenos Aires   Cape Town   Chennai
Dar es Salaam   Delhi   Hong Kong   Istanbul   Karachi   Kolkata
Kuala Lumpur   Madrid   Melbourne   Mexico City   Mumbai   Nairobi
São Paulo   Shanghai   Taipei   Tokyo   Toronto

Published by Oxford University Press, Inc.
198 Madison Avenue, New York, New York 10016
www.oup.com

Oxford is a registered trademark of Oxford University Press

Library of Congress Cataloging-in-Publication Data
Shaw, Milton Clayton, 1915–
        Metal cutting principles / Milton C. Shaw. — 2nd ed.
          p. cm. — (Oxford series on advanced manufacturing; 5)
        Includes bibliographical references and index.
        ISBN 0-19-514206-3 (cloth)
        1. Metal-cutting.   I. Title.   II. Series.

    TJ1185.S498 2004
    671.5′3—dc22

                                        2003064974

Printing number: 9 8 7 6 5 4 3 2 1

Printed in the United States of America
on acid-free paper

To my wife
Mary Jane

CONTENTS

PREFACE TO THE SECOND EDITION

There were many important developments during the second half of the twentieth century, and these have been incorporated in the appropriate chapters of the first edition. In addition, five new chapters have been added. The first of these (Chapter 20) reviews several new developments in modeling the cutting process. The next two chapters have been added to cover two chip types not fully discussed in the first edition (wavy chips in Chapter 21 and sawtooth chips in Chapter 22). Chapter 23 provides a general discussion of the area of precision engineering, which has played an increasing role in production engineering since the first edition. The final chapter, Chapter 24, discusses a number of noncutting operations where the very unusual conditions associated with metal cutting may be used to advantage in other operations, some of these being as remote from the production of mechanical parts as the production of organic chemicals.

Tempe, Arizona   M. C. S.

PREFACE TO THE FIRST EDITION

Metal cutting is one of the most important methods of removing unwanted material in the production of mechanical components. This treatment identifies the major problem areas and relates observed performance to fundamentals of physics, chemistry, materials behavior, and the engineering sciences of heat transfer, solid mechanics, and surface science (tribology).

The basic two-dimensional (orthogonal) cutting process is first analyzed in detail followed by consideration of representative three dimensional cutting operations. Special attention is directed toward cutting temperatures, tool wear and tool life, and the integrity of finished surfaces. Machining economics and optimization of cutting processes are discussed in terms of representative examples.

Cutting processes are unusually complex largely due to the fact that two basic operations occur simultaneously in close proximity with strong interaction:

1. large strain plastic deformation in a zone of concentrated shear
2. material transport along a heavily loaded region of relative motion between chip and tool

In general, several simplified models which emphasize different aspects of the problem such as thermal, material, and surface considerations are operative simultaneously with varying degrees of importance depending on specific machining conditions. Due to complexities of the problem, a general predictive theory is not possible. Instead of seeking the impossible, a more practical approach is adopted in which a wide variety of experiences are explained in fundamental terms. Thus, the aim is to illustrate how fundamental concepts may be used to explain observed results from carefully planned experiments and how solutions to new machining situations may be achieved by application of scientific principles. Where possible, theoretical discussions are kept simple by use of techniques such as dimensional analysis.

The book should serve as a valuable reference to those engaged in research in metal cutting or as a text in a graduate subject concerned with manufacturing engineering. The background material concerning the plastic flow and fracture of solid materials and the special behavior of surfaces contains many new ideas and should be of interest to a wider group of engineers than those engaged in production engineering.

This monograph began as a set of photo-offset lecture notes first published in 1950 by Technology Press (MIT, Cambridge, Mass.) with the title *Metal Cutting Principles*. Although these notes were issued in three successive editions (the third and last in 1954), they never appeared in complete form. For example, there was no Chapter 8. This was not because of any superstition attached to that number but because Chapter 8 was reserved for what has come to be known as the "shear angle problem" (also Chapter 8 in the present work). In the early 1950s there was considerable interest in this problem and always the hope that a satisfactory general solution was soon to be found. While there have been countless attempts, the ideal solution has never been found largely for reasons discussed in the present Chapter 8.

I have worked continuously on metal removal problems with many very talented graduate students for over thirty-five years (Massachusetts Institute of Technology (MIT), 1946–61; Carnegie Mellon University (CMU), 1961–78; and Arizona State University (ASU), 1978–). This book largely records and integrates their collective ideas and carefully performed experiments. Generous support for this research has been received from a number of industrial companies, the U.S. Government, and the three institutions with which I have been associated.

Metal cutting is a time honored activity having a rich literature much of which goes back well before the work of F. W. Taylor at the turn of the [twentieth] century. No attempt has been made to review all that has been written on a given subject. I have been selective in citing literature that seems to support points of view I consider closest to fact in each problem considered. Ideas and experimental results have been considered from a number of sources, and in each case I have tried to clearly acknowledge them throughout the book.

While I have done my best to eliminate errors of all types, I do not expect complete success in this endeavor and should appreciate being informed of errors which still persist.

I wish to acknowledge with thanks the efforts of Mr. Young Moon Lee of Kyungpook National University, Taegu, Korea, who critically reviewed the entire manuscript and uncovered many errors and inconsistencies. Special thanks are also due to Ms. Elinor M. Lindenberger of ASU for carefully typing most of the manuscript. My wife, Mary Jane, spent many hours editing the text, reading proofs, and generally improving the manuscript, for which I am most grateful.

Tempe, Arizona   M. C. S.
August 1983

SYMBOLS

A	area
	apparent area of contact
	constant in Eq. (9.11)
A_m	factor defined in Eq. (12.11)
A_R	real area of contact
A_S	area of shear plane
$\bar{A}$	area factor (a function of aspect ratio of slider, m/l in Fig. 12.17)
AB	chordal size of "ear" type chip (Fig. 18.8)
$AB*$	nondimensional chip shape = AB/R_c (Fig. 18.8)
B	volume worn away (Chapter 11)
	Bainite
	constant in Eq. (9.11)
C	constant in Taylor tool life equation [Eq. (11.12)]
	stress concentration factor
	constraint factor for an indentor
	volume specific heat
	corner angle of milling cutter
	controlled-contact tool-face length
	concentration of additive in cutting fluid (volume fraction)
	cutting efficiency
C_e	end-cutting edge angle
C_s	side-cutting edge angle
$\bar{C}$	iron carbide (Fe_3C)
CE	chip equivalent (Chapter 17)
D	diameter
E	Young's modulus of elasticity
	output voltage from strain gage
F	force (fundamental dimension in dimensional analysis)
	shear force
	fraction failed in Weibull statistics

F_C	cutting force component parallel to tool face
F_o	probability of failure at given stress level
F_P	cutting force component in power direction (i.e., parallel to V)
F_Q	cutting force component in undeformed chip thickness direction
F_R	cutting force component perpendicular to F_P and F_Q
F_S	cutting force component parallel to shear plane
G	shear modulus of elasticity
H_B	Brinell hardness
H_K	Knoop hardness
H_M	Meyer hardness
H_{RA}	Rockwell A hardness
H_{RB}	Rockwell B hardness
H_{RC}	Rockwell C hardness
H_S	Moh's hardness
H_V	Vickers hardness
I	area moment of inertia
J	polar moment of inertia
	mechanical equivalent of heat
K	bulk modulus
	stress intensity factor in fracture mechanics
	spring constant $[FL^{-1}]$
	probability that a real contact will result in a wear particle
	ratio of (shear stress on shear plane)/(shear stress on tool face)
	incremental speed factor in machine tool drive
	pressure coefficient of ductility in Bridgman fracture equation
L	liquid phase
L	fundamental dimension for length in dimensional analysis
	nondimensional contact length (a'/t)
	helical length of cut in turning
	sliding length (Chapter 11)
	pitch length of helix (drill)
	active length of cutting edge (Fig. 17.32b)
	nondimensional velocity quantity in Chapter 12 = $Vl/2K$
M	drilling torque
	cost machinability ratio = $\mathcal{C}_1/\mathcal{C}_2$
M_f	temperature at which martensitic transformation is complete
	fracture torque in drilling
M_s	temperature at which martensitic transformation begins
M_T	twisting moment
$M^{\#}$	cost machinability ratio under cost optimum conditions
N	r.p.m.
	cycles to fracture
N_C	cutting force component perpendicular to tool face
N_F	nondimensional fracture number = $(\sigma^2 K)/(Eu^2 b)$ (Chapter 11)
N_o	characteristic life in Weibull statistics
N_S	cutting force component perpendicular to shear plane
N_W	wear number = BH/LP (Chapter 11)
P	pearlite

P	force
	principal stress in Griffith crack initiation analysis
	probability
P_f	brittle fracture load
P'_f	brittle fracture load per unit length
Q	energy input per unit time
	principal stress in Griffith crack initiation analysis
	total heat flux
R	resultant force on tool face
	profile radius at neck of tensile specimen
	heat partition coefficient in Chapter 12 = fraction of heat flowing to extensive member of sliding pair (R_1 for shear plane, R_2 for tool face)
	cost ratio = $xT_d + y/x$ in Chapter 19
	risk of rupture in Weibull statistics
	resistance (ohms)
	chip packing ratio = volume of chips/equivalent volume of uncut metal
R'	resultant force on shear plane
R_a	arithmetic average surface roughness
R_n	indenting force due to tool tip radius (Chapter 10)
	radius of principal cutting edge (Fig. 10.14)
R_t	maximum peak-to-valley surface roughness
R_{nP}	power component of R_n (Fig. 10.15)
R_{nQ}	feed component of R_n (Fig. 10.15)
S	Spherodite
S	normal tensile stress (based on original area A_o)
	percent surviving in Weibull statistics
	total saving with stepless machine tool drive
S_d	Steadite
T	tempered martensite
T	fundamental dimension (time) in dimensional analysis
	life in Taylor tool life Eq. (11.12)
	surface energy $[FL^{-1}]$
	absolute temperature
	drilling thrust
T_c	cutting time
T_d	down-time to change and reset tool
T_e	component of drill thrust due to "extrusion" at web
T_f	fracture thrust for drill
T_H	homologous temperature = absolute temperature/absolute melting temperature (both K)
T_m	velocity modified temperature
T_P	surface energy associated with plastic crack growth
$T^\#$	cost optimum tool life (min)
$T^{\#\#}$	production rate optimum tool life (min)
U	total cutting energy per unit time
	elastic energy stored at crack tip per unit length in Griffith analysis (Chapter 6)
V	cutting speed
	fundamental dimension (velocity) in dimensional analysis

	voltage
	volume in Weibull statistics
V_C	chip speed
V_W	velocity of work (Chapter 16)
V'	a volume
$V^\#$	cost optimum cutting speed
$V^{\#\#}$	production rate optimum cutting speed
V_{60}	60-min tool-life
V_{240}	240-min tool-life
W	weight
	limiting wattage for strain-gage
Y	flow (yield) stress in uniaxial tension
Z	nondimensional quantity in Chapter 12 $= (4\pi DK)/(Vt^2)$
a	radius of contact area in indentation hardness
	radius of neck of tensile specimen
	acceleration
	imperfection spacing (Chapter 6)
a'	tool face contact length
a_1, a_2	atomic spacings
b	width of cut (depth of cut in turning)
	slider width (Chapter 10)
	width of beam
	offset distance for chip breaker (Fig. 18.7)
b_C	deformed chip width
c	half-length of crack
	length of chisel edge of drill
d	diameter
	maximum depth of crater on tool face
e	nominal strain based on original gage length (l_o)
	EMF
	secondary cutting edge length (Chapter 14)
f	feed per revolution (drilling)
	feed per tooth (milling)
	width of flat in disc test
f_e	exciting frequency
g	acceleration due to gravity
	cost of cutting fluid additive per gallon (Chapter 19)
h	height of beam
	depth of heated layer (Chapter 10)
	height of scallop left on milled surface
	drop height in drop test
h_c	critical drop height for fracture
i	inclination angle
	electrical current
k	coefficient of thermal conductivity
	plane strain flow stress in shear
	Boltzmann's constant
l	length

	undeformed chip length
	axial length of cut in turning
	half-length of slider (Chapter 12)
l_C	deformed chip length
m	mass
	strain rate sensitivity index
	half-width of slider (Chapter 12)
	Weibull slope
n	strain hardening index
	exponent in Taylor tool-life equation
	number of cutter teeth
	number of revolutions
	number of planes per unit distance (Chapter 9)
p	pressure
	point half angle of drill
	normal stress at center of Mohr's circle in slip line field analysis
$\bar{p}$	mean pressure on punch face (Meyer hardness)
q	heat flux in Chapter 12 $[FL^{-2}T^{-2}]$
r	cutting ratio = t/t_C
	moment arm length
	nose radius of tool
t	undeformed chip thickness (feed in turning)
	time
	axial thickness in disc test
t_C	deformed chip thickness
t_m	maximum undeformed chip thickness (milling)
u	total cutting energy per unit volume
u_A	surface energy per unit volume
u_c	specific elastic tensile energy for fracture in impact
	fracture energy
u_F	friction energy per unit volume
u_M	momentum energy per unit volume
u_o	specific energy of thermal origin (Chapter 9)
u_S	shear energy per unit volume
w	extent of wear land on clearance face of tool
	web thickness of drill
	crack width
w_C	weight of chip
x	cost of machine, operator, and overhead per unit time
y	mean value of single cutting edge
y_o	initial displacement
Δy	thickness of shear plane
α	rake angle
	ferrite
	ship line coordinate direction
	fraction of A_R strongly bonded to mating surface
	angular deformation
α'	decrease in rake angle due to secondary shear (Fig. 3.20)

α_b	back rake angle
α_e	effective rake angle
α_n	normal rake angle
α_r	radial rake angle
α_s	side rake angle
α_v	velocity rake angle
β	friction angle on tool face $= \tan^{-1}(F_C/N_C)$
	slip line coordinate direction
β'	$\tan^{-1}(\tau/\sigma)$ on shear plane
γ	shear strain
γ'	specific weight of work material
$\dot{\gamma}$	time rate of shear strain
δ	linear deformation
	extent of secondary shear zone (Fig. 12.63)
	wedge angle of principal cutting edge (Fig. 10.14)
	inclination of free surface (Chapter 17)
	helix angle (drill)
ε	true (ln) strain
	feed angle of drill
ε_e	effective strain
ε_f	strain at fracture in chip (Chapter 18)
ε_U	strain at point of onset of necking = ultimate strain
ε_y	strain at yield point
$\varepsilon_1, \varepsilon_2, \varepsilon_3$	principal strains
ζ	setting angle
η	angle between tool face and plane of maximum shear stress (Fig. 8.3)
η'	angle between shear plane and plane of maximum shear stress
η_C	chip flow angle (Fig. 16.5)
η_S	shear flow angle [Eq. (16.8)]
θ	temperature (fundamental dimension in dimensional analysis)
	clearance angle
	angular extent of BUE (built-up edge)
	angle of twist in torsion test
	chip breaker face angle (Fig. 18.1a)
	cone half angle (for conical indentor or punch θ_e = effective cone half angle)
	angle in Fig. 8.4a
θ_f	workpiece surface temperature 180° from point of cutting
θ_e	end-relief angle
θ_m	maximum temperature
θ_o	ambient temperature of work (Chapter 12)
θ'_o	ambient temperature of tool (Chapter 12)
θ_s	side relief angle
$\bar{\theta}$	mean cutting temperature
$\bar{\theta}_S$	mean shear plane temperature (Chapter 12)
$\bar{\theta}_T$	mean temperature rise in cutting (Chapter 12)
$\bar{\theta}_t$	mean tool temperature by chip–tool thermocouple technique
$\Delta\bar{\theta}_F$	mean temperature rise on tool face (Chapter 12)
λ	chip compression ratio (reciprocal of cutting ratio = $1/r$)

μ	coefficient of tool-face friction $= \tan \theta = F_C/N_C$
v	Poisson's ratio
Π	nondimensional group in dimensional analysis
ρ	radius at tip of abrasive particle (Chapter 11)
	radius of curvature of tool tip
ρC	volume specific heat
σ	normal stress
σ_C	nominal uniaxial compressive stress at fracture
σ_c	critical tensile stress at fracture
	normal stress corrected for nonuniform stress in neck of tensile specimen
σ_D	nominal tensile stress in disc test at fracture
σ_e	effective normal stress
σ_f	normal stress at fracture
σ_{fo}	constant in Bridgman fracture equation
σ_H	mean principal stress ($\cong$ hydrostatic stress)
σ_0	characteristic stress in Weibull statistics
σ_x	elastic tensile stress at disc center in disc test
σ_y	elastic compressive stress at disc center in disc test
σ_T	nominal uniaxial tensile stress at fracture
σ_U	stress at onset of necking = ultimate stress
σ_{xD}	nominal tensile stress at center of disc at fracture
σ_{xT}	nominal tensile stress in uniaxial tensile test at fracture
$\sigma_1, \sigma_2, \sigma_3$	principal stresses ($\sigma_1 > \sigma_2 > \sigma_3$)
$\sigma'_1, \sigma'_2, \sigma'_3$	principal deviator stresses
$\bar{\sigma}$	mean normal stress
	effective normal stress
τ_0	theoretical shear strength
ϕ	shear angle
	angle of shear coordinate rotation in slip line field analysis
ϕ_n	normal shear angle (Chapter 16)
ψ	some function (in dimensional analysis)
	angle between shear plane and direction of maximum deformation in chip (Fig. 3.12)
ω	angular velocity (rad s^{-1})
	angle in Fig. 8.5
$\cent$	total cost per part
$\cent^*$	optimum cost per part
$\cent^\#$	optimum cost per cut
$^w/_o$	weight percent

INTRODUCTION

There are several methods of changing the geometry of bulk material to produce a mechanical part:

1. by putting material together (+)
2. by moving material from one region to another (0)
3. by removing unnecessary material (−)

These operations may be performed on the atomic, micro, or macro scales. For example, electroplating and electroforming are plus operations at the atomic level while the fabrication of a structure by welding is a plus operation at the other end of the spectrum (macro joining). Rolling, forging, and extrusion are examples of (0) operations performed at the macro level while surface burnishing is at the micro level. Removal operations (−) with which this book is concerned are performed primarily at the macro level (cutting).

Another way of classifying operations is in terms of the temperature pertaining. Mechanical properties are related to the amplitude of vibration of adjacent atoms which varies linearly with absolute temperature. The melting point of a metal represents a critical temperature where the amplitude of atomic vibration is sufficient to cause a structural change from that of a solid to that of a liquid. At equal percentages of melting temperature on the absolute temperature scale, metals have similar properties and this suggests an homologous temperature (T_H) scale which corresponds to the fraction of the melting temperature on the absolute (K or R) temperature scale. Metals deformed below $T_H = 0.5$ behave differently than those deformed above $T_H = 0.5$. Deformation below $T_H = 0.5$ is called cold working and takes place primarily within individual crystals and is relatively strain rate insensitive but strongly strain sensitive (strain hardening). Deformation above $T_H = 0.5$ is called hot working, occurs primarily by grain boundary rearrangement, and is relatively strain rate sensitive but insensitive to strain (negligible strain hardening). Machining that occurs at temperatures above $T_H = 0.5$ is called hot machining. For example, steel has a melting temperature of about 1540 °C (2800 °F) or $1540 + 273 \simeq 1810$ °K (or $2800 + 460 = 3260$ °R). For hot machining, steel should be cut at an homologous temperature of at least $T_H = 0.5$ (or 630 °C or 1160 °F).

There are still other ways of classifying material removal operations that will be developed in subsequent chapters of this book.

IMPORTANCE OF MATERIAL REMOVAL

The importance of material removal operations in the scheme of things may be realized by considering the total cost associated with this activity, including expendable tool cost, labor cost, and cost of capital investment. In the United States, the yearly cost associated with material removal has been estimated at about 10% of the gross national product.

The importance of the cutting process may be further appreciated by the observation that nearly every device in use in our complex society has one or more machined surfaces or holes.

There are several reasons for developing a rational approach to material removal:

1. to improve cutting techniques—even minor improvements in productivity are of major importance in high volume production

2. to produce products of greater precision and of greater useful life

3. to increase the rate of production and produce a greater number and variety of products with the tools available

All basic fields of industrial endeavor have taken similar paths in the course of their development. The earliest work has generally been carried out on a purely empirical basis and in many instances such activities have been highly developed by following the case method. While this method presents a clear picture of each specific job, a great many cases must be considered before sufficient examples have been presented to enable all common situations to be covered. This approach has been extensively used in metal cutting as well as in other fields, such as machine design, hydraulics, metallurgy, and even such nonengineering activities as law and medicine. The weakness of the method lies in its failure to provide a direct means for solving problems which lie beyond the range of current experience. Each new case that is established must be arrived at by a costly procedure of trial and error.

Not too many years ago, steam turbines and power machinery were designed largely in accordance with the judgement of the designer, rather than by following the more rational approach involving stress analysis that is in wide use today. Similarly, the design of hydraulic conduits and machinery that once was done by rule-of-thumb procedures is now being accomplished largely with the art of the principles of fluid mechanics. In the field of metallurgy, steelmaking is being carried out by considering it a special problem in physical chemistry instead of employing the age-old recipe technique. Metal cutting tools and procedures are still largely established by the old case method. This activity has resisted the impact of modern technology and the scientific method, mainly due to the complexity of the operations but also partly due to the attitude held toward metal cutting in engineering schools.

Traditionally, metal cutting has been part of the training of mechanical engineers. However, in the past, a trade-school approach was generally adopted, emphasis being placed entirely upon nomenclature, the mastery of machine manipulation, and the learning of a large number of disconnected empirical rules. In some instances, the major objective has actually been the production of trinkets, thus appealing to the hobby instincts of the student rather than developing the ability to apply fundamental concepts.

APPROACH TO SUBJECT

In this treatment of the subject we will consider the cutting process in fundamental terms. The objective is to explain a number of commonly observed results rather than to present a large mass of empirical constants and a large number of empirical relationships of limited applicability. In this first chapter, we will briefly consider the subject from a qualitative point of view. Basic definitions and concepts will be given and discussed. The wide scope of material removal will be presented and the relation to other fields of science and engineering indicated.

Classification of Material Removal Operations

The entire field of material removal may be divided into the following categories mainly in terms of the size of the individual elements removed:

- cutting
- grinding
- special techniques

Cutting operations involve the removal of macroscopic chips in the form of ribbons or particles having a thickness of from about 10^{-3} in (0.025 mm) to 10^{-1} in (2.5 mm). A wide range of kinematic arrangements briefly discussed in the next chapter are involved in cutting. Grinding operations usually involve subdivision of the material removed into smaller particles than in cutting. Grinding chips will usually range in thickness from 10^{-4} in (0.0025 mm) to 10^{-2} in (0.25 mm). Other removal techniques such as electrochemical machining (ECM), electrodischarge machining (EDM), ultrasonic machining (UM), or electron beam machining (EBM) involve chips of atomic or submicroscopic size.

Overview of the Cutting Process

The chisel (Fig. 1.1) was probably one of the first cutting tools used by man. The earliest stone implements were undoubtedly blunt as shown in Fig. 1.1a but as experience was gained, the importance of three basic angles became apparent—the rake angle (α), the clearance angle (θ), and the setting angle (ζ) (Fig. 1.1b). Modern tools have a wide variety of forms and their geometry and kinematics differ considerably from those of the chisel. However, all of our modern tools have effective rake, clearance, and setting angles.

Certain superficial observations may be made by merely observing a metal cutting tool in operation. These include

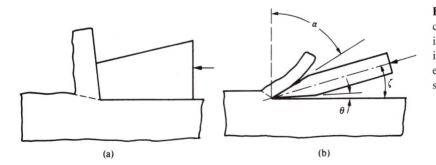

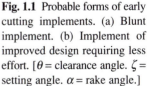

Fig. 1.1 Probable forms of early cutting implements. (a) Blunt implement. (b) Implement of improved design requiring less effort. [θ = clearance angle. ζ = setting angle. α = rake angle.]

(a)   (b)

1. The basic difference between the cutting of wood and metal. Formerly it was believed that when metal was cut, the material merely split off in front of the tool as the tool advanced—like the chip formed when an axe splits a log. When thickness of a metal chip is measured and compared with the depth of layer removed, it is found that the chip is thicker than the actual depth of layer removed and the chip correspondingly shortened.

2. There is essentially no flow of metal at right angles to the direction of chip flow. For the purpose of simplifying the geometry involved in cutting, it is advantageous to start with a two-dimensional process. We then need to consider what happens in but one representative plane. Although most cutting operations involve tools and processes which are not strictly two dimensional, many processes such as planing, sawing, and certain turning operations are essentially two dimensional. Later we will briefly discuss the complications introduced by the three-dimensional aspects of cutting tools.

3. Flow lines are evident on the side and back of a chip. These lines suggest that cutting involves a shearing mechanism.

4. Some chips are in the form of a continuous ribbon while others are cyclic.

5. The chip, tool, and workpiece are hot to the touch. Considerable thermal energy is associated with the cutting process.

The significance of these observations will be discussed in subsequent chapters.

A photomicrograph of a partially formed chip reveals much concerning the cutting process (Fig. 1.2). Such photomicrographs are obtained in the following manner. During the course of a cutting operation, the tool is brought to a sudden stop. Then the tool is carefully removed leaving a partially formed chip attached to the workpiece. The section of the metal in the vicinity of the partially formed chip is cut from the workpiece and mounted in plastic for convenience in handling. The mounted specimen is ground and polished to produce a very smooth flat surface. Next, the polished surface is etched with a fluid such as a 1% mixture of nitric acid in alcohol which reacts at different rates with the different components of the metal. The etched surface of the specimen is photographed through a microscope. Examination of a photomicrograph of a partially formed chip such as that of Fig. 1.2 reveals

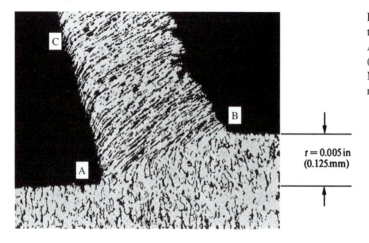

$t = 0.005$ in
(0.125 mm)

Fig. 1.2 Photomicrograph of partially formed chip. Work material, AISI 1015 steel cut at 24 f.p.m. (0.13 m s^{-1}) using water + 0.1 $^{w}/_{o}$ NaNO$_2$. Undeformed chip thickness (t) = 0.005 in (0.125 mm).

1. There is generally no crack extending in front of the tool point.

2. A line separates the deformed and undeformed regions. Line AB in Fig. 1.2 divides the work from the chip. The material below this line is undeformed. The chip or material above the line has been deformed by a concentrated shearing process. When the line AB is projected perpendicular to the paper and parallel to itself, it describes what is known as the shear plane (Fig. 1.3), making an angle ϕ with the direction of cut.

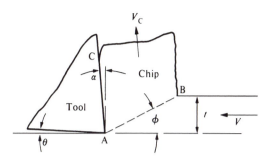

Fig. 1.3 Orthogonal (two-dimensional) cutting process.

3. The chip is in intimate contact with the rake face of the tool from A to C and is subject to a substantial shear stress sufficient to cause the secondary subsurface to shear (as evident in Fig. 1.2 along the tool face).

4. The rate at which metal is deformed along shear plane AB is high as a consequence of the thinness of the region in which shear occurs.

OTHER CHIP TYPES

Not all chip formation involves a steady state where a chip in the form of a continuous ribbon of uniform thickness is involved as in Fig. 1.3. In some cases chip formation is cyclic.

A stationary nose or built-up edge (BUE) such as that shown in Fig. 1.4 sometimes forms at the tip of a tool and significantly alters the cutting process. The BUE is one of the major sources of surface roughness and also plays an important role in tool wear. Data regarding BUE formation are in Chapter 9.

Other types of cyclic chip formation are

- discontinuous (Fig. 1.5a)
- wavy (Fig. 1.5b)
- sawtooth (Fig. 1.5c)

Figure 1.5a shows a sequence of high-speed motion pictures when machining a very brittle material. Periodically a crack extending across the chip width initiates at the tool tip and rapidly progresses from A to B. This causes the segment above AB to eject as stored energy is released. As the tool progresses, material below AB is extruded upward as a new chip is formed. Frame 40 is identical to frame 1 and a new cycle begins. Material containing points of stress concentration such as graphite flakes in cast iron or manganese sulfide inclusions in a free machining steel frequently produce chips in the form of discrete particles. Further discussion of discontinuous chip formation is given in Chapter 15.

Figure 1.5b shows a wavy chip produced when the shear angle (ϕ) fluctuates cyclically between high and low

Fig. 1.4 Photomicrograph of partially formed chip showing large built-up edge (BUE) and portions of BUE along finished surface and face of chip. Work material, AISI 1020 steel; cutting speed, 90 f.p.m. (4.84 m s^{-1}); fluid, air; $t = 0.008$ in (0.2 mm).

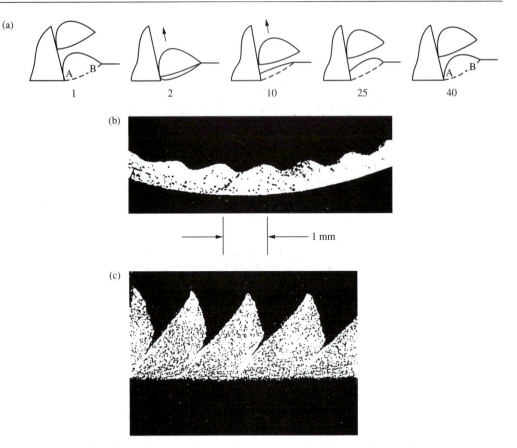

Fig. 1.5 Cyclic chip types. (a) Discontinuous. (b) Wavy. (c) Sawtooth.

values for any one of a number of reasons. This type of chip formation is considered in detail in Chapter 21.

Figure 1.5c is a photomicrograph of a sawtooth chip, which derives its name from its resemblance to a saw blade. This type of chip formation is far more difficult to analyze than steady state chip formation and will be discussed in detail in Chapter 22.

CHIP FORMATION ANALYSIS

The first step in a consideration of the mechanics of chip formation is to identify correctly the type of chip involved (steady state, BUE, discontinuous, wavy, or sawtooth). If, at the outset of a metal cutting analysis, an incorrect chip type is assumed, the results obtained may be misleading. It is extremely difficult to predict the type of chip that will form and the best way of determining this is experimentally. A very few cuts under the conditions of interest will clearly indicate the type of chip involved and hence the type of analysis to be applied.

The friction between the chip and the tool plays a significant role in the cutting process. This friction may be reduced by

1. improved tool finish and sharpness of the cutting edge

2. use of low-friction work or tool materials

3. increased cutting speed (V)

4. increased rake angle (α)

5. use of a cutting fluid

When tool face friction is decreased there is a corresponding increase in shear angle and an accompanying decrease in the thickness of the chip. As shown in Fig. 1.6, the plastic strain in the chip decreases as the shear angle increases. The length of the shear plane is seen to be significantly decreased as the shear angle increases. The force along the shear plane will increase as the area of the shear plane increases, assuming the shear stress on the shear plane remains constant.

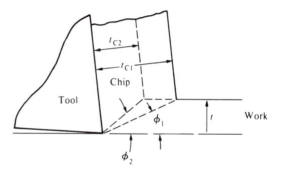

Fig. 1.6 Effect of small (ϕ_1) and large (ϕ_2) shear angle on chip thickness (t_C) and length of shear plane for a given tool and undeformed chip thickness (t).

The temperature of a cutting tool may reach a high value particularly when a heavy cut is taken at high speed. This is evident when the work or tool is touched; by the presence of temper colors on the chip, work, or tool; or may even be evident due to the loss of hardness of the tool point with an attendant loss of tool geometry and failure by excessive flow at the cutting edge.

The operational characteristics of a cutting tool are generally described by a single word—*machinability*. There are three main aspects of machinability:

1. tool life

2. surface finish

3. power required

These important items will be considered in detail later.

Regions of Interest

There are three regions of interest in the cutting process. The first area shown in Fig. 1.7 extends along the shear plane and is the boundary between the deformed and undeformed material or the chip and the work. The second area includes the interface between the chip and the tool face, while the third area includes the finished or machined surface and the material adjacent to that surface. We are primarily interested in the plastic deformation characteristics of the material cut in the first area, the friction and wear characteristics of the tool–work combination in the second area, and the surface roughness and integrity of the finished surface constituting the third area.

The understanding of what happens in each of the three major regions of the cutting process involves a knowledge of several fields of engineering and science including

1. solid state physics

2. engineering mechanics

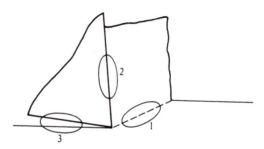

Fig. 1.7 Principal areas of interest in machining.

3. materials science
4. engineering plasticity and fracture mechanics
5. fundamentals of lubrication, friction, and wear
6. basic concepts of chemistry and physics
7. principles of physical metallurgy
8. thermodynamics and heat transfer

In subsequent chapters some of the fundamental concepts of these many fields of endeavor will be reviewed and applied to the cutting process.

PROCESS OPTIMIZATION

Since material removal is a workshop related activity where very strong economic or production rate constraints pertain, it is important that the selection of tools, fluids, operating conditions, etc., be based upon a rational optimization procedure. Therefore, the basic principles of engineering economics will be presented and applied to optimize typical removal operations at appropriate points in the text. While the most important optimization variable is usually total cost per part (including tool, operator, and overhead costs), there are occasions when maximum rate of production must take precedence (as in the case of a production bottleneck or a national emergency); consequently, both of these points of view are covered by optimization procedures discussed at several points throughout the text.

HISTORY

Early attempts were made to explain the mechanics of cutting in fundamental terms in the nineteenth century and Finnie (1956) has reviewed some of this early work as has Zorev (1965). Shaw (1968) has reviewed some of the major metal cutting studies of the first half of the twentieth century, as has Komanduri (1993).

REFERENCES

Finnie, I. (1956). *Mech. Engng.* **78**, 715.
Komanduri, R. (1993). *Appl. Mech. Rev.* **46/3**, 80–132.
Shaw, M. C. (1968). In *Metal Transformations*. Gordon and Breach, New York, p. 211.
Zorev, N. N. (1965). *Metal Cutting Mechanics*. Pergamon Press, Oxford.

2 TYPICAL CUTTING OPERATIONS

PRINCIPAL CUTTING OPERATIONS

The three most widely used cutting operations are

- turning
- milling
- drilling

Turning (Fig. 2.1) is a process using a single point tool that removes unwanted material to produce a surface of revolution. The machine tool on which this is accomplished is a lathe (Fig. 2.2). The variables adjusted by the operator are the cutting speed V (f.p.m. or m s^{-1}), the feed f (i.p.r. or mm rev^{-1}), and the depth of cut d (in or mm). Since the depth of cut (d) is usually at least five times the feed (f), the chip is produced in plane strain and hence the width of chip is equal to the undeformed chip width, to a good approximation.

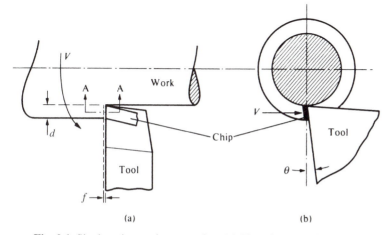

(a)   (b)

Fig. 2.1 Single point turning operation. (a) Plan view. (b) Side view.

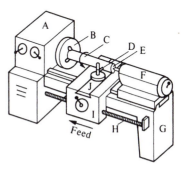

Fig. 2.2 Engine lathe. [A = head stock and gear box. B = spindle with chuck. C = workpiece. D = tool post. E = dead center. F = tail stock. G = base. H = lead screw to provide feed. I = apron. J = cross slide.]

9

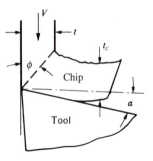

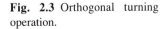

Section AA of Fig. 2.1 ($i = 0$)

Fig. 2.3 Orthogonal turning operation.

The cutting speed and feed in Fig. 2.1 actually vary along the cutting edge. This causes the chip to flow away from the new surface on the workpiece as shown in Fig. 2.1a. There also will be a small change in chip thickness across the width of the chip. These small variations along the cutting edge are usually ignored, and maximum values of cutting speed, and mean values of feed, and chip thickness are used as shown in Fig. 2.3, which is a two dimensional representation of the cutting process. In this figure the depth of cut and width of cut (perpendicular to the paper in Fig. 2.3) are constant across the chip width. This is called an *orthogonal* cutting representation, where the direction of chip flow is perpendicular to the cutting edge at all points along the cutting edge. In this two dimensional approximation, the undeformed chip thickness is no longer the actual feed (f) and the chip width is no longer the depth of cut (d) shown in Fig. 2.1. Therefore, these quantities in the two dimensional representation shown in Fig. 2.3 are designated t for the undeformed chip thickness and b for the width of cut (perpendicular to the paper in Fig. 2.3).

When the cutting edge is not perpendicular to the axis of the work, as it is in Fig. 2.1, the angle between these two directions is called the inclination angle (i). The inclination angle in Fig. 2.1 is 0°. When i is not zero, this causes the chip to further change its direction of flow across the tool face beyond that shown in Fig. 2.1a that is due to variation of cutting speed and feed across the chip width. When the direction of chip flow is not perpendicular to the cutting edge, as in Fig. 2.1, this gives rise to a change in the effective rake angle from that provided on the tool as discussed in Chapter 16.

Milling is a process for producing flat and curved surfaces using multipoint cutting tools. The machine tool involved is a milling machine (Fig. 2.4). There are three basic types of milling cutters (Fig. 2.5).

The plane-milling cutter is used to produce flat surfaces in the manner shown in Fig. 2.6. The cutting edge may be parallel to or inclined to the axis of the cutter. If the cutting edge is inclined to the cutter axis, this inclination angle is referred to as a helix angle. A plane-milling cutter produces a cut

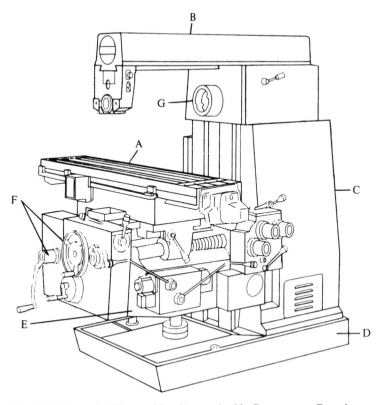

Fig. 2.4 Horizontal milling machine. [A = work table. B = overarm. C = column. D = base. E = knee. F = manual table controls. G = spindle.]

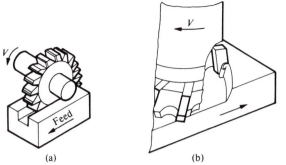

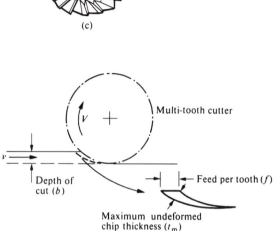

Fig. 2.5 Principal types of milling cutters. (a) Plane. (b) Face. (c) End.

of variable undeformed chip thickness (t) and the simple orthogonal cutting situation (Fig. 2.3) is only a crude approximation because of the variable undeformed thickness (t) involved (Fig. 2.6). In the plane-milling operation, the undeformed chip width is measured parallel to the cutting edge.

A face-milling cutter (Fig. 2.5b) is also used to machine flat surfaces as shown in Fig. 2.7. However, in this case, the axis of the cutter is perpendicular to the finished surface instead of being parallel to it as in the case of a plane-milling cutter. Also, the undeformed chip thickness (t) is constant throughout the cut and corresponds to the feed per tooth, while the depth of cut corresponds to the undeformed chip width (b). The inclination angle may be 0° or some value up to about 45°.

An end mill is used to produce slots and surface profiles and operates as shown in Fig. 2.8. It closely resembles a face mill in the manner it is presented to the work but is a very much smaller cutter. As in the case of the face-milling cutter, there is a major cutting edge parallel to the cutter axis, a secondary cutting edge perpendicular to the cutting axis, and a nose radius connecting the two. However, the undeformed chip thickness is variable throughout a cut as in the case of the plane-milling cutter, but the undeformed chip width corresponds to the depth of cut.

A twist drill (Fig. 2.9) is a complex cutting tool used to produce rough holes. The machine tool involved in this case is a drill press (Fig. 2.10). There are two cutting edges that produce chips similar to those produced in turning on a lathe and a central web region that extrudes the metal at

Fig. 2.6 Individual chips produced in plane-milling.

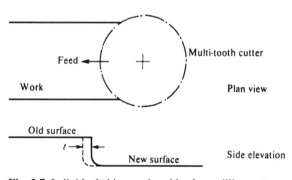

Fig. 2.7 Individual chips produced by face-milling cutter.

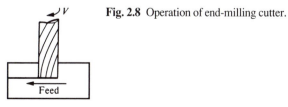

Fig. 2.8 Operation of end-milling cutter.

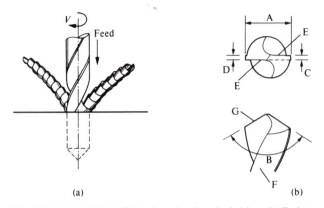

Fig. 2.9 Twist drill. (a) Side view showing dual chips. (b) Twist drill nomenclature. [A = diameter. B = point angle. C = web thickness. D = margin. E = cutting edges. F = flute. G = drill point.]

Fig. 2.10 Drill press. [A = base. B = column. C = head. D = feed box. E = spindle. F = drill. G = work table.]

the center of the hole not removed by the cutting edges. Both lips (cutting edges) of a drill operate with variable rake angle, inclination angle, and clearance angle along the cutting edge. The flutes of a drill play the important role of conveying the chips out of the hole and the helix angle of the drill is important in this connection. The fundamental quantities of importance (rake angle (α), undeformed chip thickness (t), and inclination angle (i)) are related in a complex way to be discussed in Chapter 16 to the quantities specified on a drill drawing (helix angle, point angle, and web thickness) as well as the principal operating variables (speed and feed).

OTHER CUTTING OPERATIONS

In addition to the three most widely used operations discussed above, there are many others including

1. sawing (Fig. 2.11a)
2. reaming (Fig. 2.11b)
3. tapping (Fig. 2.11c)
4. planing (Fig. 2.11d)
5. broaching (Fig. 2.11e)
6. boring (Fig. 2.11f)
7. threading (Fig. 2.11g)

These operations cover a wide range of complexity similar to that encountered in turning and drilling. While some of these operations involve simultaneous cutting on two edges connected by a nose radius as in the case of a straight turning operation, the length of cut on one edge is usually minor relative to the other.

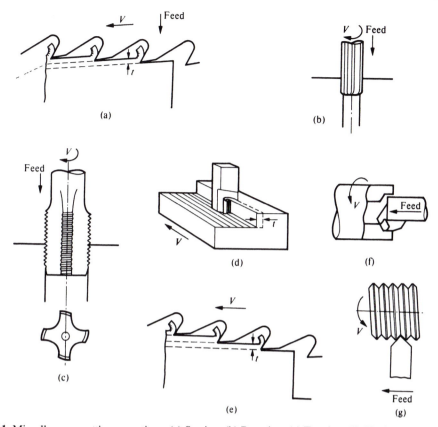

Fig. 2.11 Miscellaneous cutting operations. (a) Sawing. (b) Reaming. (c) Tapping. (d) Planing. (e) Broaching. (f) Boring. (g) Threading.

ORTHOGONAL CUTTING

In practically all cases, the simple orthogonal cutting operation represents a reasonably good approximation to performance on the major cutting edge and for that reason has been extensively studied. Two widely used orthogonal cutting arrangements are shown in Fig. 2.12. The arrangement shown in Fig. 2.12a has the tool attached to the stationary overarm of a milling machine while the work is moved against the tool by the table feed, planer fashion, and is convenient for studies at low and moderate cutting speed (V). For higher cutting speeds the arrangement

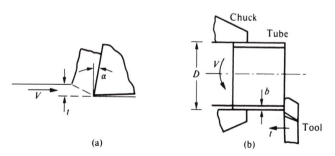

Fig. 2.12 Orthogonal cutting operations. (a) Planing. (b) Turning end of tube.

shown in Fig. 2.12b is convenient where the end of a tube ($b \ll D$) is cut in a lathe by a tool with zero inclination angle. In both cases (Figs. 2.12a and 2.12b), the width of the tool is made greater than the width of the workpiece.

Concluding Remarks

This chapter represents a very brief introduction to machine tool and cutting tool nomenclature. It is beyond the intended scope of this book to go into more detail. The reader is directed to one of the many texts covering the more descriptive aspects of manufacturing engineering for more details. One of the better texts of this type is S. Kalpakjian's *Manufacturing Engineering and Technology*, 3rd ed. (1995, Addison Wesley, Reading, Mass.).

3 MECHANICS OF ORTHOGONAL STEADY STATE CUTTING

As is evident in Fig. 1.2, metal cutting involves concentrated shear along a rather distinct shear plane. As metal approaches the shear plane, it does not deform until the shear plane is reached. It then undergoes a substantial amount of simple shear as it crosses a thin primary shear zone (Fig. 3.1). There is essentially no further plastic flow as the chip proceeds up the face of the tool. The small amount of secondary shear along the tool face is generally ignored in a first treatment of the cutting process, and the motion of the chip along the tool face is considered to be similar to that of a friction slider of constant coefficient from A to C.

The back of a chip is rough due to the strain being inhomogeneous. This is due to the presence of points of weakness or stress concentration present in the metal being cut. A shear plane passing through a point of stress concentration will deform at a lower value of stress than one that does not include a point of stress concentration. Once deformation begins on a given shear plane, it tends to continue there as though the material exhibited negative strain hardening. Reasons for this apparent negative strain hardening will be discussed in Chapter 9. Thus, some metal in the chip strains more than other metal resulting in a wavy surface on the back of the chip. These waves however are not continuous clear across the back of the chip but are of limited extent in the width direction as illustrated in the insert in Fig. 3.1 which represents the back of a chip as viewed in a normal direction.

In the early 1940s, Merchant (Ernst and Merchant, 1941; Merchant, 1945) published his world famous model of chip formation based on the concentrated shear process to be discussed in detail in this chapter. However, soon after this, it was suggested by several authors that all steady state chips do not behave in accordance with the model of Fig. 3.2a. If the work is relatively soft and not prestrain hardened before cutting, chip formation will involve a pie shaped zone (Fig. 3.2b) and an even more extensive shear zone if the radius at the tool tip (ρ) is significant relative to the undeformed chip thickness, t

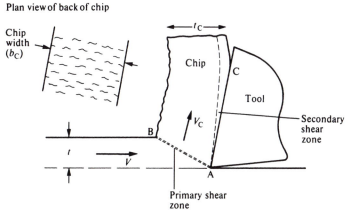

Fig. 3.1 Schematic of orthogonal steady state cutting operation.

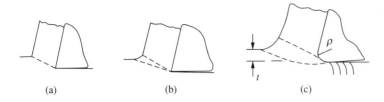

Fig. 3.2 Chip formation for "flow" type chips. (a) Concentrated shear model for precoldworked soft material. (b) Pie shaped shear zone for soft annealed material. (c) More extensive shear zone with subsurface plastic flow for tool with rounded tip.

(Fig. 3.2c). Since it is next to impossible to predict which of the three steady state modes in Fig. 3.2 will pertain, Fig. 3.2a is usually assumed as a useful approximation.

CARD MODEL

Despite its limitations, the so-called card model of the cutting process (Fig. 3.3) due to Piispanen (1937) is very useful. This model depicts the material cut as a deck of cards inclined to the free surface at an angle corresponding to the shear angle (ϕ). As the tool moves relative to the work, it engages one card at a time and causes it to slide over its neighbor. While this model

1. exaggerates the inhomogeneity of strain,
2. depicts tool face friction as elastic rather than plastic,
3. assumes shear to occur on a perfectly plane surface,
4. ignores any BUE that may be present,
5. involves an arbitrarily assumed shear angle (ϕ), and
6. does not explain chip curl or predict chip–tool contact length,

it does, however, contain the main concepts in the steady state chip-forming process and is easy to understand.

In the card model all atomic planes are not active shear planes, but only those associated with a structural defect (second phase particle, missing atom, an impurity, a grain boundary, etc.) (Shaw, 1950). This results in inhomogeneous strain and a series of sharp points on both surfaces of the chip. The points on the free surface remain and serve as an indication of the degree of inhomogenity pertaining. Those on the tool face side of the chip are removed by burnishing. If the structural defects are widely spaced, the free surface of the chip will have a sawtoothed appearance. For less intense and less widely spaced defects, the free surface of the chip will be relatively smooth. The "teeth" will not extend continuously across the chip but will be localized and staggered (as in the insert in Fig. 3.1).

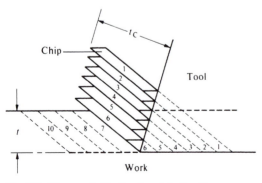

Fig. 3.3 Piispanen's idealized model of cutting process. (Piispanen, 1937)

ORTHOGONAL CUTTING

Figure 3.4 is a related model of the "ideal" cutting process that completely suppresses the concept of inhomogeneous strain by assuming the material to behave in a completely homogeneous fashion. The assumptions on which this two dimensional model is based include

1. The tool is perfectly sharp and there is no contact along the clearance face.

2. The shear surface is a *plane* extending upward from the cutting edge.

3. The cutting edge is a straight line extending perpendicular to the direction of motion and generates a plane surface as the work moves past it.

4. The chip does not flow to either side (plane strain).

5. The depth of cut is constant.

6. The width of the tool is greater than that of the workpiece.

7. The work moves relative to the tool with uniform velocity.

8. A continuous chip is produced with no built-up edge.

9. The shear and normal stresses along shear plane and tool are uniform (strength of materials approach). Such an ideal two-dimensional cutting operation is referred to as orthogonal cutting.

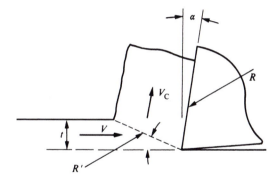

Fig. 3.4 Resultant forces R and R' in orthogonal cutting.

FORCES

When the chip is isolated as a free body (Fig. 3.5), we need to consider only two forces—the force between the tool face and the chip (R) and the force between the workpiece and the chip along the shear plane (R'). For equilibrium these must be equal

$$R = R' \qquad (3.1)$$

The forces R and R' are conveniently resolved into three sets of components as indicated in Fig. 3.5:

1. in the horizontal and vertical direction, F_P and F_Q
2. along and perpendicular to the shear plane, F_S and N_S
3. along and perpendicular to the tool face, F_C and N_C

If the forces R' and R are plotted at the tool point instead of at their actual points of application along the shear plane and tool face, we obtain a convenient and compact diagram (Fig. 3.6). Here R and R' (which are equal and parallel) are coincident and are made the diameter of the dotted reference circle shown. Then, by virtue of the geometrical fact that lines that terminate at the ends of diameter R and intersect at a point on the circle will intersect at right angles, we have a convenient means for graphically resolving R into orthogonal components in any direction. This type of plot was suggested by Merchant (1945).

Analytical relationships may be obtained for the shear and friction components of force in terms of the horizontal and vertical components (F_P and F_Q) which are the components normally determined experimentally by means of a dynamometer (Chapter 7). From

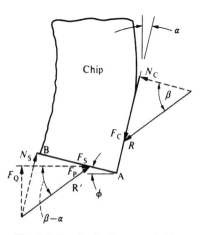

Fig. 3.5 Free body diagram of chip.

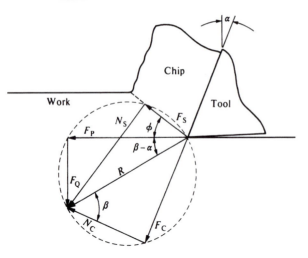

Fig. 3.6 Composite cutting force circle.

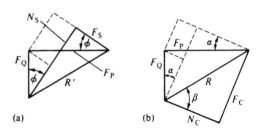

Fig. 3.7 Resolution of resultant cutting forces. (a) Shear plane. (b) Tool face.

Fig. 3.7a, it is evident that

$$F_S = F_P \cos \phi - F_Q \sin \phi \tag{3.2}$$

$$N_S = F_Q \cos \phi + F_P \sin \phi = F_S \tan (\phi + \beta - \alpha) \tag{3.3}$$

Similarly from Fig. 3.7b,

$$F_C = F_P \sin \alpha + F_Q \cos \alpha \tag{3.4}$$

$$N_C = F_P \cos \alpha - F_Q \sin \alpha \tag{3.5}$$

The tool face components are of importance since they enable the coefficient of friction on the tool face ($\mu = \tan \beta$ where β is the friction angle shown in Fig. 3.5) to be determined:

$$\mu = \frac{F_C}{N_C} = \frac{F_P \sin \alpha + F_Q \cos \alpha}{F_P \cos \alpha - F_Q \sin \alpha} = \frac{F_Q + F_P \tan \alpha}{F_P - F_Q \tan \alpha} \tag{3.6}$$

The nomographs given in Figs. 3.8 and 3.9 due to Merchant and Zlatin (1945) are useful for computing the friction force and the coefficient of friction. Alternatively, programs for a personal computer (PC) may be readily written.

STRESSES

The shear plane components of force [Eqs. (3.2) and (3.3)] are of importance since they enable the mean shear and normal stresses on the shear plane to be determined. Thus,

$$\tau = \frac{F_S}{A_S} \tag{3.7}$$

where A_S is the area of the shear plane, and

$$A_S = \frac{bt}{\sin \phi} \tag{3.8}$$

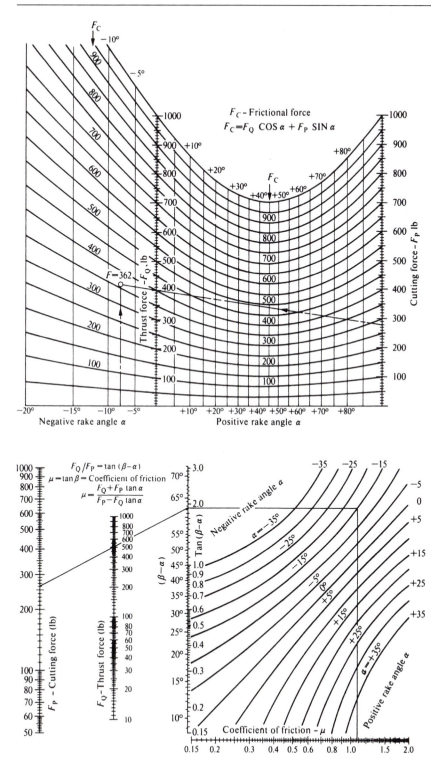

Fig. 3.8 Nomograph for friction force. (after Merchant and Zlatin, 1945) [Example: $F_P = 280$ lb. (1232 N), $F_Q = 400$ lb. (1760 N), $\alpha = -7.5°$, $F_C = 362$ lb. (1593 N).]

Fig. 3.9 Nomograph for coefficient of friction. (after Merchant and Zlatin, 1945) [Example: $F_P = 260$ lb. (1144 N), $F_Q = 500$ lb. (2200 N), $\alpha = -15°$, $\mu = 1.09$.]

where b is the width of cut and t is the depth of cut. From Eqs. (3.2), (3.3), (3.7), and (3.8),

$$\tau = \frac{(F_P \cos \phi - F_Q \sin \phi) \sin \phi}{bt} \tag{3.9}$$

Similarly,

$$\sigma = \frac{N_S}{A_S} = \frac{(F_P \sin \phi + F_Q \cos \phi) \sin \phi}{bt} \tag{3.10}$$

SHEAR ANGLE

In addition to F_P and F_Q, which are obtained by methods to be described in Chapter 7, we must know the shear angle (ϕ) before Eqs. (3.9) and (3.10) may be applied. The shear angle could be obtained by direct measurement of a photomicrograph, but this is not very convenient. Another method of obtaining the shear angle is by means of the cutting ratio (r) which is the ratio of the depth of cut (t) to the chip thickness (t_C). This ratio may be determined by directly measuring t and t_C.

When metal is cut, it is experimentally found that there is no change in density and hence,

$$tbl = t_C b_C l_C \tag{3.11}$$

where t, b, and l are the depth of cut, width of cut, and length of cut, and the subscript C refers to the corresponding measurements on the chip. If $b/t \gtrless 5$, it is found experimentally that the width of the chip is the same as that of the workpiece and hence, from Eq. (3.11),

$$\frac{t}{t_C} = \frac{l_C}{l} = r \tag{3.12}$$

The relationship between cutting ratio (r) and the shear angle (ϕ) may be obtained with the aid of Fig. 3.10. Here it is evident that

$$r = \frac{t}{t_C} = \frac{AB \sin \phi}{AB \cos (\phi - \alpha)} \tag{3.13}$$

or solving for ϕ:

$$\tan \phi = \frac{r \cos \alpha}{1 - r \sin \alpha} \tag{3.14}$$

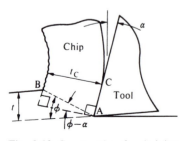

Fig. 3.10 Construction for deriving relation between shear angle (ϕ) and cutting ratio (r).

This equation is most conveniently solved with the aid of the nomograph given in Fig. 3.11.

While the cutting ratio can be obtained by measuring chip thickness (t_C) and depth of cut (t), this is not the most precise procedure. Due to the roughness on the back surface of a chip, it is difficult to obtain a representative chip thickness. A better procedure is to measure the chip length (l_C) and corresponding work length (l). From Eq. (3.12) it can be seen that r is also equal to the chip length ratio (l_C/l). The work length l is easily determined in the case of turning. A useful technique in this type of operation is to provide the workpiece with a deep scratch which will cause a notch to be formed in the chip for each revolution. Then, the work length may be computed from the work diameter.

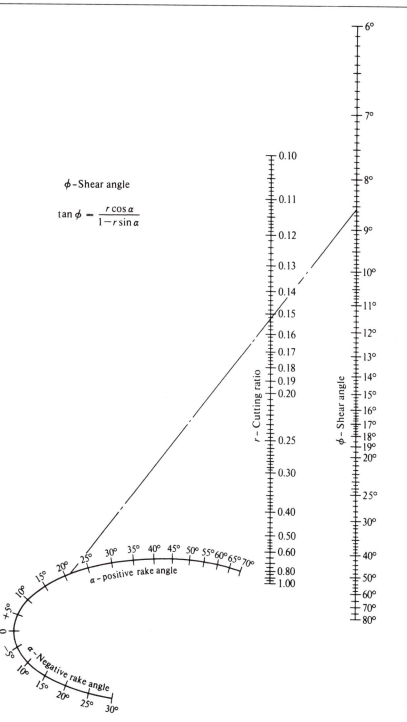

Fig. 3.11 Nomograph for shear angle. (after Merchant and Zlatin, 1945) [Example: Rake angle = 21°, cutting ratio = 0.152, shear angle = 8.55°.]

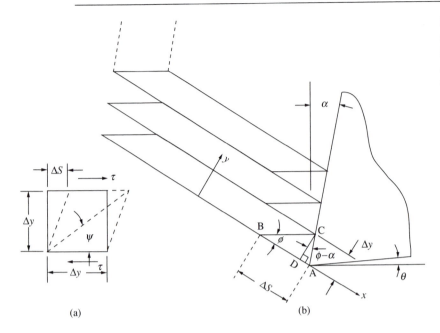

Fig. 3.12 Determination of shear strain in cutting. (a) Shear strain in general ($\gamma = \Delta S/\Delta y$). (b) Shear strain in cutting ($\gamma = \Delta S/\Delta y$).

(a)   (b)

The work length may also be determined by weighing a chip. If the weight of a chip is w_C, then

$$1 = \frac{w_C}{tb\gamma'} \tag{3.15}$$

where γ' is the specific weight of the metal cut in pounds per cubic inch (N m^{-3}).

SHEAR STRAIN

The shear strain is another item of interest associated with the cutting process. The shear strain (γ) is defined as $\Delta S/\Delta Y$ (Fig. 3.12) and hence in cutting,

$$\gamma = \frac{\Delta S}{\Delta Y} = \frac{AB'}{CD} = \frac{AD}{CD} + \frac{DB'}{CD} = \tan(\phi - \alpha) + \cot\phi \tag{3.16}$$

A nomograph due to Merchant and Zlatin (1945) for use in determining shear strain in cutting (γ) is given in Fig. 3.13. It may also be readily shown that

$$\gamma = \frac{\cos\alpha}{\sin\phi\cos(\phi - \alpha)} \tag{3.17}$$

VELOCITY RELATIONS

There are three velocities of interest in the cutting process (Fig. 3.14):

1. the cutting velocity (V), which is the velocity of the tool relative to the work and directed parallel to F_P

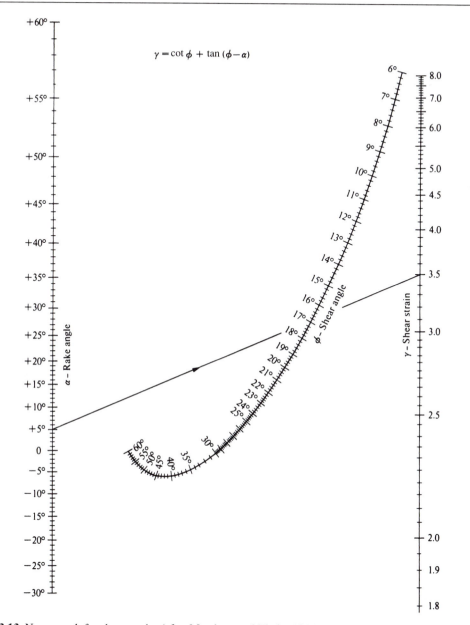

Fig. 3.13 Nomograph for shear strain. (after Merchant and Zlatin, 1945) [Example: Rake angle = 5°, shear angle = 17°, shear strain = 3.5.]

$$\gamma = \cot \phi + \tan (\phi - \alpha)$$

2. the chip velocity (V_C), which is the velocity of the chip relative to the tool and directed along the tool face

3. the shear velocity (V_S), which is the velocity of the chip relative to the workpiece and directed along the shear plane

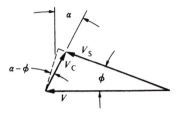

Fig. 3.14 Hodograph for cutting.

In accordance with the principles of kinematics, these three velocity vectors must form a closed velocity diagram as shown by the hodograph of Fig. 3.14. The vector sum of the cutting velocity and the chip velocity is equal to the shear velocity vector. Analytically, we have

$$V_C = \frac{\sin \phi}{\cos (\phi - \alpha)} V \tag{3.18}$$

$$V_S = \frac{\cos \alpha}{\cos (\phi - \alpha)} V \tag{3.19}$$

It can be readily shown that these equations may also be written

$$V_C = rV \tag{3.20}$$

$$V_S = \gamma \sin \phi\, V \tag{3.21}$$

RATE OF STRAIN

The rate of strain in cutting is

$$\dot{\gamma} = \frac{\Delta S}{\Delta y \Delta t} = \frac{V_S}{\Delta y} \tag{3.22}$$

where Δt is the time elapsed for the metal to travel a distance ΔS along the shear plane, and Δy is the thickness of the shear zone. From Eqs. (3.17), (3.21), and (3.22),

$$\dot{\gamma} = \frac{\cos \alpha}{\cos (\phi - \alpha)} \frac{V}{\Delta y} \tag{3.23}$$

Since the shear "plane" is often very thin (Kececioglu, 1958) as shown by photomicrographs such as that of Fig. 1.2, it is to be expected that very few slip planes will be active at any one time. A reasonable mean value for the spacing of successive slip planes (Δy) would appear to be about 10^{-3} in ($\sim 25\ \mu$m).

An idea of the rate of strain to be expected in cutting may be had by substituting representative values into Eq. (3.23) (Drucker, 1949):

$$V = 100 \text{ f.p.m.}, \ \alpha = 0°, \text{ and } \phi = 20°$$

then

$$\Delta y = 10^{-3} \text{ in}$$

The mean rate of strain corresponding to these values is found to be 21,000 s^{-1}. We may thus say that the rate of strain in cutting will be of the order of 10^4 s^{-1} or higher. This is recognized as a very high value when it is realized that the rate of strain in an ordinary tensile test is only about 10^{-3} s^{-1} and 10^3 s^{-1} in the most rapid impact test.

ENERGY CONSIDERATIONS

When metal is cut in a two-dimensional cutting operation, the total energy consumed per unit time is

$$U = F_p V \tag{3.24}$$

The *total energy per unit volume* (specific energy, u) of metal removed will therefore be

$$u = \frac{U}{Vbt} = \frac{F_P}{bt}$$ (3.25)

where b and t are the width and depth of cut respectively.

This total energy per unit volume will be consumed in several ways (Shaw, 1954):

1. as shear energy per unit volume (u_S) on the shear plane
2. as friction energy per unit volume (u_F) on the tool face
3. as surface energy per unit volume (u_A) due to the formation of new surface area in cutting
4. as momentum energy per unit volume (u_M) due to the momentum change associated with the metal as it crosses the shear plane

Specific energy is an intensive quantity that characterizes the cutting resistance offered by a material just as tensile stress and hardness characterize the strength and plastic deformation resistance of a material, respectively. In metal cutting the specific energy is essentially independent of the cutting speed V, varies slightly with rake angle α (increases about 1% per degree decrease in rake angle), but varies inversely with undefomed chip thickness t to an appreciable degree. For practical values of undeformed chip thickness, $t > 0.001$ in.

For a sharp tool practically all of the energy is consumed along the shear plane and along the tool face. A worn tool will generally have a small radius at the tool tip, a region of zero clearance on the clearance surface (wear land), and a crater along the chip-tool contact area on the tool face. The wear land and tool tip radius will cause additional energy to be dissipated in the finished surface and slightly below the finished surface, respectively, thus causing an increase in specific cutting energy as wear progresses. The crater will induce chip curl which in turn will decrease the length of chip tool contact and hence will result in a decrease in specific energy.

The bulk of the energy involved in metal cutting is consumed along the shear plane and essentially all of this energy is convected away by the chip. The energy consumed along the tool face is divided between chip and tool with most of it going to the extensive member (the chip). The energy consumed on the finished surface with a worn tool is divided between the tool and the work with most of the energy again going to the extensive member. The net result for a typical cutting operation is an energy distribution as follows:

- 90% of total energy to chip
- 5% of total energy to tool
- 5% of total energy to work

The shear energy per unit volume may be obtained as follows:

$$u_S = \frac{F_S V_S}{Vbt} = \tau \left(\frac{V_S}{V \sin \phi} \right)$$ (3.26)

However, from Eq. (3.21)

$$u_S = \tau \gamma$$ (3.27)

The friction energy per unit volume may be similarly found

$$u_F = \frac{F_C V_C}{Vbt}$$ (3.28)

or from Eq. (3.20)

$$u_F = \frac{F_C r}{bt} \tag{3.29}$$

When a new surface is generated in a solid substance, sufficient energy must be supplied to separate the ions at the interface. A certain energy is thus associated with the formation of new surface and this is usually referred to as the surface energy of the substance (T). The surface energy of a solid is analogous to the surface tension of a liquid. The surface energy per unit volume in cutting will be

$$u_A = \frac{(T)(2Vb)}{Vbt} = \frac{2T}{t} \tag{3.30}$$

The quantity 2 in Eq. (3.30) comes from the fact that two surfaces are generated simultaneously when a cut is made. The value of T for most metals is about 0.006 in lb in^{-2} (1000 dyne cm^{-1}).

The resultant momentum force associated with the change in momentum of the metal as it crosses the shear plane will obviously be in the direction of slip or along the shear plane. This resultant momentum force (F_M) may, therefore, be obtained by applying the linear momentum law of applied mechanics along the shear plane:

$$F_M = \frac{\gamma'}{g}(Vbt)[V_C \sin(\phi - \alpha) + V \cos \phi] \tag{3.31}$$

where γ'/g is the mass density of the metal. By use of Eqs. (3.13), (3.17), and (3.18) this may be written

$$F_M = (\gamma'/g)V^2 \, bt \, \gamma \sin \phi \tag{3.32}$$

Hence, the momentum energy per unit volume will be

$$u_M = \frac{F_M V_S}{Vbt} = \gamma'/gV \, \gamma \sin \phi \, V_S \tag{3.33}$$

From Eq. (3.21) this becomes

$$u_M = \gamma'/gV^2 \, \gamma^2 \sin^2\phi \tag{3.34}$$

As will be shown in the next section, the surface energy per unit volume (u_A) and momentum energy per unit volume (u_M) are negligible relative to the other two components and hence to a very good approximation we have

$$u = u_S + u_F \tag{3.35}$$

Practically all of the energy associated with a cutting operation is consumed in either plastic deformation or friction, and essentially all of this ends up as thermal energy. Further details concerning the mechanics of orthogonal cutting may be found in the literature (Merchant, 1945; Piispanen, 1948; Drucker, 1949).

EXAMPLE

As an example Fig. 3.15 gives an analysis for the photomicrograph of Fig. 1.2. Figure 3.15a is a reproduction of Fig. 1.2. Figure 3.15b shows the chip as a free body with resultant forces R and R'

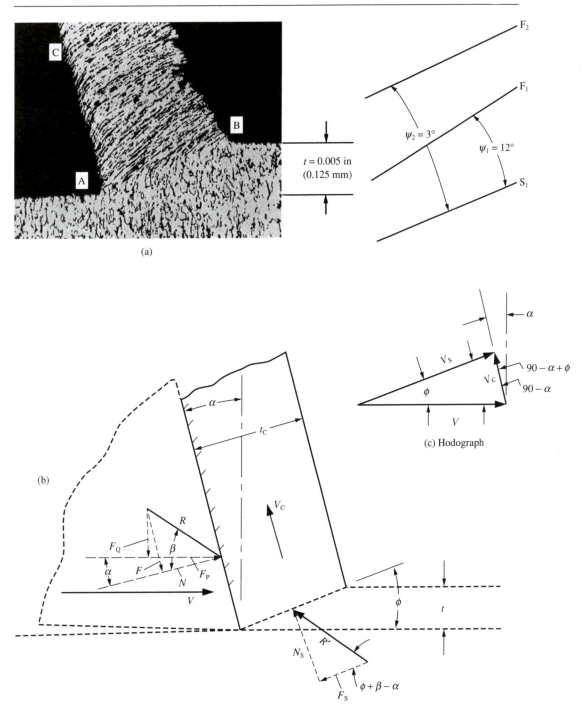

Fig. 3.15 Steady state chip formation. (a) Photomicrograph of partially formed two dimensional (orthogonal) chip of AISI 1041 steel cut at 24 f.p.m. (7.3 m/min) with a rake angle (α) of +15° and a shear angle (ϕ) of 20°. (b) Free body diagram for the chip. (c) Hodograph relating the three steady state velocities involved.

acting on the tool face and shear plane respectively. These are shown resolved into components F_Q and F_P and F and N in the tool face and N_S and F_S on the shear plane. Figure 3.15c shows the velocity diagram (hodograph) to scale. Figure 3.15a is seen to be well represented by Fig. 3.15b where the shear angle is about 20°, $\alpha = 15°$, and the width of the shear plane is less than 500 μm (12.5 μin). The lines S_1, F_1 and F_2 in Fig. 3.15a will be discussed later in this chapter.

From Fig. 3.15c:

$$\frac{V_C}{V} = 0.34$$

$$\frac{V_S}{V} = 0.97$$

The quantity V_C/V is of course the cutting ratio (r). For steady state chip formation this will always be less than one. Due to the constancy of volume,

$$r = \frac{t}{t_C} = 0.37$$

From Eqs. (3.25), (3.26), and (3.27), it may be shown that

$$\frac{u_S}{u} = \left[\frac{\cos\,(\phi + \beta - \alpha)}{\cos\,(\beta - \alpha)} \right] \left[\frac{V_S}{V} \right] \tag{3.36}$$

$$\frac{u_F}{u} = \left[\frac{\sin\,\beta}{\cos\,(\beta - \alpha)} \right] \left[\frac{V_C}{V} \right] \tag{3.37}$$

For Fig. 3.15a, $\phi = 20°$, $\beta = 45°$, $\alpha = 15°$, $V_C/V = 0.34$, and $V_S/V = 0.97$, so substituting these values into Eqs. (3.36) and (3.37) yields

$$\frac{u_S}{u} = 0.72$$

$$\frac{u_F}{u} = 0.28$$

All of the above values are typical for steady state chip formation of steel.

REPRESENTATIVE DATA

Representative orthogonal cutting data are given in Tables 3.1 and 3.2 to illustrate the nature of the results obtained when cutting dry. The values of strain (γ) are found to be much larger in cutting than corresponding values in ordinary materials tests. The reason for this lies in the fact that gross fracture is postponed by the presence of a large normal stress on the shear plane. This and other materials aspects will be discussed in detail in Chapter 6.

The magnitudes of the several energy components discussed in the foregoing section may be compared by use of the data in the first line of Table 3.1. From Eqs. (3.25), (3.27), (3.29), (3.30), and (3.34), respectively,

TABLE 3.1 Representative Metal Cutting Data

(a) English Units

t in ($\times 10^3$)	V ft min^{-1}	α deg	Cutting Ratio r	ϕ deg	γ	F_P lb	F_Q lb	μ	τ p.s.i. ($\times 10^{-3}$)	σ p.s.i. ($\times 10^{-3}$)	u in lb in^{-3} ($\times 10^{-3}$)	u_S in lb in^{-3} ($\times 10^{-3}$)	$u_{S/u}$
3.70	197	+10	0.29	17.0	3.4	370	273	1.05	85	115	400	292	0.73
3.70	400	+10	0.33	19.0	3.1	360	283	1.11	88	137	390	266	0.68
3.70	642	+10	0.37	21.5	2.7	329	217	0.95	90	129	356	249	0.70
3.70	1186	+10	0.44	25.0	2.4	303	168	0.81	93	130	328	225	0.69
3.70	400	−10	0.32	16.5	3.9	416	385	0.64	89	153	450	342	0.76
3.70	637	−10	0.37	19.0	3.5	384	326	0.58	90	152	415	312	0.75
3.70	1160	−10	0.44	22.0	3.1	356	263	0.51	94	154	385	289	0.75
1.09	542	+10	0.33	19.0	3.1	127	101	1.12	103	164	465	308	0.66
2.34	542	+10	0.32	18.5	3.1	242	186	1.08	93	140	414	287	0.69
3.70	542	+10	0.37	21.5	2.7	336	226	0.96	91	131	363	249	0.69
7.88	542	+10	0.44	25.0	2.4	605	315	0.76	89	116	307	214	0.70
1.09	542	−10	0.23	12.5	5.0	181	198	0.78	103	179	664	523	0.79
2.34	542	−10	0.30	16.0	4.0	295	291	0.70	96	172	505	385	0.76
3.70	542	−10	0.37	19.0	3.5	401	350	0.60	94	164	434	306	0.71
7.88	542	−10	0.46	22.5	3.1	698	472	0.46	90	138	354	270	0.76

(b) Metric Units

t mm	V m min^{-1}	α deg	Cutting Ratio r	ϕ deg	γ	F_P N	F_Q N	μ	τ MPa	σ MPa	u Nm m^{-3}	u_S Nm m^{-3}	$u_{S/u}$
0.094	60	+10	0.29	17.0	3.4	1646	1214	1.05	586	793	2758	2013	0.73
0.094	122	+10	0.33	19.0	3.1	1601	1259	1.11	607	945	2689	1834	0.68
0.094	196	+10	0.37	21.5	2.7	1463	965	0.95	621	889	2455	1717	0.70
0.094	362	+10	0.44	25.0	2.4	1348	747	0.81	641	896	2262	1551	0.69
0.094	122	−10	0.32	16.5	3.9	1850	1712	0.64	614	1055	3103	2358	0.76
0.094	194	−10	0.37	19.0	3.5	1708	1450	0.58	621	1048	2861	2151	0.75
0.094	354	−10	0.44	22.0	3.1	1583	1170	0.51	648	1062	2655	1993	0.75
0.001	165	+10	0.33	19.0	3.1	565	449	1.12	710	1131	3206	2124	0.66
0.059	165	+10	0.32	18.5	3.1	1076	827	1.08	638	965	2855	1979	0.69
0.094	165	+10	0.37	21.5	2.7	1495	1005	0.96	627	903	2503	1717	0.69
0.200	165	+10	0.44	25.0	2.4	2691	1401	0.76	614	800	2117	1476	0.70
0.001	165	+10	0.73	12.5	5.0	805	881	0.78	710	1234	4578	3606	0.79
0.059	165	+10	0.30	16.0	4.0	1312	1294	0.70	662	1186	3482	2655	0.76
0.094	165	+10	0.37	19.0	3.5	1784	1557	0.60	648	1131	2992	2110	0.71
0.200	165	+10	0.46	22.5	3.1	3105	2099	0.46	621	952	2441	1862	0.76

The data were obtained by Merchant (1945), using two-dimensional tools. The work material was NE 9445 steel having a Brinell hardness of 187. The tool material was sintered carbide. The width of cut was 0.25 in (6.35 mm). All chips were of the continuous type and no cutting fluids were used.

TABLE 3.2 Representative Metal Cutting Data

(a) English Units

t in ($\times 10^3$)	α deg	Cutting Ratio r	ϕ deg	γ	F_P lb	F_Q lb	μ	τ p.s.i. ($\times 10^{-3}$)	σ p.s.i. ($\times 10^{-3}$)	u in lb in^{-3} ($\times 10^{-3}$)	u_S in lb in^{-3} ($\times 10^{-3}$)	$u_{S/u}$
2.5	25	0.358	20.9	2.55	380	224	1.46	82.8	104.0	320	209.0	0.65
3.5	25	0.366	21.5	2.48	475	281	1.46	74.6	96.0	286	180.0	0.63
5	25	0.407	24.0	2.23	643	357	1.38	75.5	100.0	270	169.0	0.63
6	25	0.345	20.1	2.65	728	398	1.36	65.8	86.5	255	175.0	0.69
8.5	25	0.383	22.4	2.38	992	551	1.38	66.5	83.6	246	159.0	0.65
2.5	35	0.527	31.6	1.56	254	102	1.53	71.4	96.0	214	112.0	0.52
3.5	35	0.528	31.9	1.55	306	122	1.52	61.2	84.0	184	96.4	0.52
5	35	0.529	32.0	1.55	433	166	1.48	62.4	82.9	182	96.5	0.53
6	35	0.533	32.2	1.54	507	184	1.43	61.8	88.6	178	95.5	0.54
8.5	35	0.532	32.0	1.55	675	234	1.38	58.5	80.9	167	91.3	0.55
2.5	40	0.585	35.7	1.32	232	71	1.54	72.0	94.5	195	94.0	0.48
3.5	40	0.580	35.4	1.33	296	87	1.50	66.5	84.3	178	87.8	0.49
5	40	0.611	37.5	1.26	412	112	1.44	66.1	87.0	173	83.1	0.48
6	40	0.606	37.2	1.27	475	127	1.42	63.5	90.0	167	81.1	0.49
8.5	40	0.606	37.2	1.27	634	153	1.35	61.5	83.1	157	78.1	0.50
2.5	45	0.670	41.9	1.06	232	68	1.83	71.1	114.6	195	74.8	0.38
3.5	45	0.670	41.9	1.06	285	77	1.74	64.2	98.8	171	68.5	0.40
5	45	0.649	40.2	1.10	386	94	1.64	63.8	87.2	162	69.6	0.43
6	45	0.642	39.6	1.01	443	102	1.60	61.7	80.8	155	62.7	0.40
8.5	45	0.646	39.9	1.11	581	117	1.51	58.7	73.6	144	65.0	0.45

(b) Metric Units

t mm	α deg	Cutting Ratio r	ϕ deg	γ	F_P N	F_Q N	μ	τ MPa	σ MPa	u Nm m^{-3}	u_S Nm m^{-3}	$u_{S/u}$
0.064	25	0.358	20.9	2.55	1690	996	1.46	571	717	2206	1441	0.65
0.089	25	0.366	21.5	2.48	2113	1250	1.46	514	662	1972	1241	0.63
0.127	25	0.407	24.0	2.23	2860	1588	1.38	521	690	1862	1165	0.63
0.152	25	0.345	20.1	2.65	3238	1770	1.36	454	596	1758	1207	0.69
0.216	25	0.383	22.4	2.38	4412	2451	1.38	459	576	1696	1096	0.65
0.064	35	0.527	31.6	1.56	1130	454	1.53	492	662	1476	772	0.52
0.089	35	0.528	31.9	1.55	1361	543	1.52	422	579	1269	665	0.52
0.127	35	0.529	32.0	1.55	1926	738	1.48	430	572	1255	665	0.53
0.152	35	0.533	32.2	1.54	2255	818	1.43	426	611	1227	658	0.54
0.216	35	0.532	32.0	1.55	3002	1041	1.38	403	558	1151	630	0.55
0.064	40	0.585	35.7	1.32	1032	316	1.54	496	652	1345	648	0.48
0.089	40	0.580	35.4	1.33	1317	387	1.50	459	581	1227	605	0.49
0.127	40	0.611	37.5	1.26	1833	498	1.44	456	600	1193	573	0.48
0.152	40	0.606	37.2	1.27	2113	565	1.42	438	621	1151	559	0.49
0.216	40	0.606	37.2	1.27	2820	681	1.35	424	573	1083	539	0.50
0.064	45	0.670	41.9	1.06	1032	302	1.83	490	790	1345	516	0.38
0.089	45	0.670	41.9	1.06	1268	343	1.74	443	681	1179	472	0.40
0.127	45	0.649	40.2	1.10	1717	418	1.64	440	601	1117	480	0.43
0.152	45	0.642	39.6	1.01	1970	454	1.60	425	557	1069	432	0.40
0.216	45	0.646	39.9	1.11	2584	520	1.51	405	507	993	448	0.45

The data were obtained by Lapsley, Grassi, and Thomsen (1950), using a two-dimensional tool. The work material was pierced seamless tubing (probably SAE 4130). The tool material was high-speed steel. The width of cut was 0.475 in (12.1 mm) and the cutting speed in all tests was 90 ft min^{-1} (27 m min^{-1}). All chips were continuous and no cutting fluids were used.

$$u = \frac{F_P}{bt} = 400,000 \text{ p.s.i.}$$

$$u_S = 292,000 \text{ p.s.i.}$$

$$u_F = \frac{F_C r}{bt} = 108,000 \text{ p.s.i.}$$

$$u_A = \frac{2T}{t} = \frac{2(0.006)}{0.0037} = 3.2 \text{ p.s.i.}$$

$$u_M = \gamma'/gV^2 \, \gamma^2 \, \sin^2\phi = \frac{0.28}{386} \frac{197(12)^2}{60} (3.4)^2 (\sin 17°)^2 = 1.1 \text{ p.s.i.}$$

It is evident from this example that u_A and u_M are negligible compared with u and that the sum of u_F and u_S is therefore equal to the total energy per unit volume (u). In the foregoing example shear energy is seen to account for 73% of the energy, while 27% is due to friction. In general the friction energy is found to be about $\frac{1}{3}$ the shear energy in a turning operation.

While the treatment of cutting mechanics presented in this chapter is highly idealized and based on many simplifying assumptions, it does represent a useful first approximation to what is taking place when metal is cut so as to produce a continuous chip. The dominant action occurs on the shear "plane" and this may be characterized as concentrated simple plastic shear. Since the plastic behavior of the material in front of the tool plays such an important role in metal cutting, this topic will be considered in some detail in Chapter 5. In subsequent chapters, frictional behavior under conditions pertaining at the chip–tool interface will be considered (Chapter 10) as well as the character of the finished surface (Chapter 17).

FLOW LINES IN A CHIP

The flow lines in a chip are usually not in the direction of the shear plane. This was demonstrated by Ernst and Merchant (1941) by means of a card model (Fig. 3.16). This is an end view of a stack of cards on which small circles have been drawn before orthogonal chip formation. The circles represent grain boundaries in an annealed metal. These grains are seen to be elongated in the chip in a direction that is quite different from the direction of maximum shear stress (i.e., at an angle ψ to the direction of the shear plane).

Townend (1947) has derived an expression that relates angles ψ, ϕ, and α for orthogonal steady state cutting with no secondary shear strain along the tool face. In Fig. 3.17, section ABCD in the work becomes BCEF in the chip. BD and BH will be equal (to a) if BD and BH are perpendicular to the shear plane. Line CG in the chip represents the direction of maximum crystal elongation making an angle ψ to the shear plane. The line in the workpiece corresponding to CG in the chip will be DJ. Angle ψ will correspond to the direction of maximum crystal elongation if GC/DJ is a maximum. Townend showed by trigonometrical analysis that GC/DJ will be a maximum when

$$2\cot 2\psi = \cot \phi + \tan (\phi - \alpha) \qquad (3.38)$$

As an example, consider the model of Fig. 3.15, where $\alpha = 15°$, and $\phi = 30°$. From Eq. (3.38),

$$2\cot 2\psi = \cot 30 + \tan 15 = 2.00$$

and hence, $\psi = 22.5°$. This is found to be in good agreement with ψ shown in the model of Fig. 3.15.

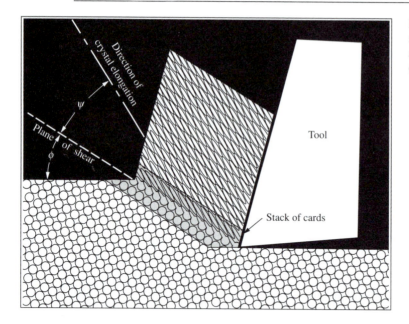

Fig. 3.16 Stack of cards model showing difference in directions of shear plane and flow lines in a chip. (after Ernst and Merchant, 1941)

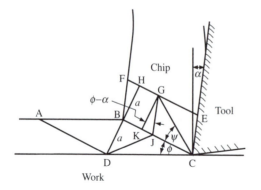

Fig. 3.17 Diagram for obtaining the relationship between ψ, ϕ, and α for orthogonal steady state cutting in the absence of secondary shear in a chip. (after Townend, 1947)

For the quick stop photomicrograph of Fig. 3.15a, $\alpha = 15°$, $\phi = 20°$, and, from Eq. (3.38), $\psi = 17.6°$. The line associated with Fig. 3.15a marked S_1 indicates the direction of the shear plane, while F_1 shows the flow direction in the chip near the shear plane and F_2 shows the chip flow direction where the chip leaves contact with the tool. The analytical value $\psi = 17.6$ from Eq. (3.38) is in poor agreement with measured value $\psi_1 = 12°$. The value of ψ_2 where the chip leaves contact with the tool is still much lower. This is due to secondary shear extending the length of the chip along the tool face causing flow lines to be more nearly parallel to the shear plane as the chip proceeds up the face of the tool. It would thus appear that Eq. (3.38) gives a good approximation for ψ only when secondary shear is not present. When secondary shear is present, Eq. (3.38) will give a value of ψ that is much too high for a fully developed chip that has left contact with the tool face. Another possible difficulty is that the uncut material may not be fully annealed and have an initial set of flow lines.

OTHER CHIP TYPES

Up to this point, it has been assumed that the chip produced is a continuous ribbon of uniform thickness (t_C) with a well-defined thin shear plane (Figs. 1.2 and 3.1). However, in practice, this is not always the case (Shaw, 1967).

At low cutting speeds or when cutting a material containing points of stress concentration (such as the graphite flakes in cast iron or the manganese sulfide inclusions in a free-machining steel) discontinuous chips may form. In this case, cutting begins at a relatively high shear angle which decreases as the cutting proceeds (Field and Merchant, 1949). When the strain in the chip reaches a critical value, the chip fractures and the process begins over again. The result is a series of discrete chip segments that have been produced by a process that resembles an extrusion process more than that of the concentrated shear obtained in orthogonal cutting (Cook, Finnie, and Shaw, 1954). The orthogonal model and subsequent calculations represent a very poor approximation for cutting with discontinuous chips. The mechanics of discontinuous chip formation are discussed in more detail in Chapters 15 and 17.

Another deviation from steady state orthogonal cutting is chip formation with a built-up edge. At speeds where the temperature at the chip–tool interface is relatively low, fracture may occur within the chip along a plane approximately at right angles to the shear plane (Nakayama, 1957), leaving behind a portion of the chip attached to the tool face (Figs. 1.4 and 3.18). This detached material then acts as the cutting edge and is called a built-up edge (BUE). A BUE tends to grow until it reaches a critical size and then passes off with the chip. This gives rise to a cyclic variation in size of BUE. Since the BUE grows outward and downward, this gives rise to a variation in depth of the cut surface which represents a major component of surface roughness when cutting with a BUE (Fig. 3.19). As cutting speed (and hence chip–tool interface temperature) increases, the size of the BUE decreases. The BUE disappears when thermal softening at the interface causes a lower flow stress at the interface than in the main body of the chip. Under ordinary turning conditions, for steel the BUE will usually have its largest value at a cutting speed of about 50 f.p.m. (0.3 m s^{-1}) and will disappear at about 200 f.p.m. (1.2 m s^{-1}).

The simple orthogonal steady state model represents only a fair approximation when cutting with a continuous chip and BUE. The main influences of the BUE are to give an unusually high friction stress (fracture versus sliding stress) at the tool point but an increase in the effective rake angle. These two effects influence cutting forces in opposite directions and hence tend to cancel each other. Similarly, the BUE has opposing actions relative to tool wear. The rough particles of hard highly worked BUE passing off with the chip tend to increase abrasive wear while the BUE actually protects the cutting edge from wear. On the other hand, cutting with a large variable BUE always gives rise to a substantial increase in surface roughness. Further discussion of the BUE is presented in Chapter 17.

Still another deviation from the simple orthogonal model involves secondary shear on the tool face. When the friction stress on the tool face reaches a value equal to the shear flow stress of the chip material, flow occurs internally within the chip adjacent to the tool face. This is called secondary shear flow to distinguish it from shear flow occurring on the shear plane (primary shear). Fig. 1.2 is a partially formed chip where secondary shear flow is clearly evident. Fig. 3.20 is a diagrammatic sketch corresponding to Fig. 1.2. The secondary shear zone is seen to increase in thickness as the chip travels from A to C. Since secondary shear is due to high shear stress on the tool face, it is accompanied by an increase in cutting forces. One

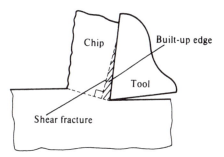

Fig. 3.18 Formation of built-up edge (BUE) by shear fracture within chip.

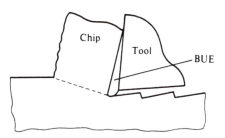

Fig. 3.19 Surface roughness due to built-up edge (BUE).

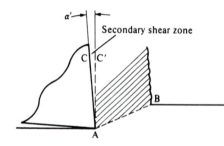

Fig. 3.20 Secondary shear.

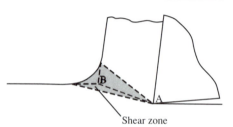

Fig. 3.21 Pie-shaped shear zone.

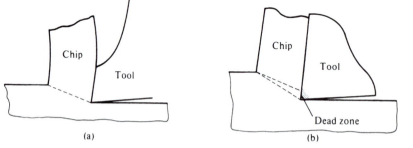

(a)   (b)

Fig. 3.22 (a) Reduced contact tool. (b) Tool with chamfered edge.

way of considering the influence of secondary shear is in terms of a decrease in rake angle by an amount α' (Fig. 3.20). AC′ in Fig. 3.20 corresponds to an equivalent rake face when cutting with sticking friction (i.e., with shear stress on the tool face = flow stress of the chip material in shear). Any action that tends to increase shear angle ϕ will tend to decrease the magnitude of secondary shear (α'). Since angle (α') is usually small, the orthogonal steady state model represents a good approximation for cutting with secondary shear flow provided the chip is continuous.

When the material cut is soft and unstrain hardened, the shear zone will tend to be pie-shaped as shown in Fig. 3.21 instead of being uniformly thin as in Fig. 3.1. An increase in feed (t) will tend to favor a pie-shaped shear zone. When cutting with a pie-shaped shear zone, there is no definite shear angle. Use of the dotted line (AB) in Fig. 3.21 for the equivalent shear angle in the steady state analysis represents a good first approximation.

Controlled contact cutting (Fig. 3.22a) and cutting with a small negative rake honed on the cutting edge (Fig. 3.22b) represent further departures from the ideal orthogonal model. The dead zone shown in Fig. 3.22b is a trapped volume of stationary metal. In these cases there is a composite rake angle and the only reasonable approximation appears to be the use of an equivalent rake angle that represents a compromise between the extremes.

Cutting with chip curlers (Fig. 3.23) of different types designed to break the chip periodically represents still another departure from the idealized model (Fig. 3.1). Cutting under such conditions is very complex and it is difficult to suggest a simple equivalent that approximates the idealized model discussed in this chapter.

Under certain conditions, strain in the chip is very inhomogeneous even though the chip is continuous. (Situations of this type are considered in Chapters 21 and 22.)

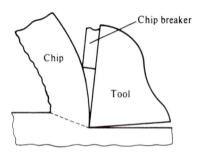

Fig. 3.23 Clamped chip breaker.

ESTIMATION OF CUTTING FORCES

It is frequently important to estimate cutting forces F_P and F_Q in practice. This may best be done in terms of total specific energy (u) since this tends to remain approximately constant for a given work material operating under different cutting conditions. The specific cutting energy will be essentially independent of cutting speed (V) over a wide range of values, provided a large BUE is not obtained. The following items do have the influence indicated on specific energy:

1. workpiece chemistry and structure (specific energy is approximately proportional to workpiece hardness and Table 3.3 gives approximate average values for a few common classes of work materials cut at representative hardness levels)
2. effective rake angle of tool α_e (u decreases about 1% per degree increase in α_e)
3. undeformed chip thickness t (the specific cutting energy varies approximately as follows with undeformed chip thickness, t in the usual range of chip thickness; this inverse relationship is sometimes referred to as the size effect):

$$u \sim 1/t^{0.2} \tag{3.39}$$

Item 2 may be illustrated by comparing data in rows 2 and 5 of Table 3.1a. The difference in u is 60,000 p.s.i. for a mean value of u of 420,000 p.s.i. This corresponds to a $(60,000/420,000)(100)$ = 14.3% difference in u for a 20° difference in α, or a 0.7% difference in u for a 1° change in α. Within the accuracy to be expected for this approximation, this should be rounded to a 1% decrease in u per 1° increase in rake angle.

Item 3 may be estimated with data from Table 3.2a. Figure 3.24 shows values of u versus values of t for rake angles of 25°, 35°, and 45°. The straight lines shown correspond to Eq. (3.39) and are seen to represent good approximations relative to the experimental data.

The principal cutting force responsible for the power consumed (F_P) may be estimated as follows:

$$F_P = \frac{u\,Vbt}{V} = ubt \tag{3.40}$$

The cutting force component in the feed (t) direction is not easily computed but to a first approximation may be assumed to be one half of F_P.

TABLE 3.3 Approximate Values of Specific Energy for Different Materials Cut with $\alpha_e = 0°$ and $t = 0.010$ in (0.25 mm) for Cutting with Continuous Chip and No BUE

Material	u_0 in lb in^{-3}[†]	u_0 J m^{-3}[‡]
Aluminum alloy	100,000	7.02×10^8
Gray cast iron	150,000	10.53×10^8
Free-machining brass	150,000	10.53×10^8
Free-machining steel (AISI 1213)	250,000	17.55×10^8
Mild steel (AISI 1018)	300,000	21.06×10^8
Titanium alloys	500,000	35.10×10^8
Stainless steel (18-8)	700,000	49.14×10^8
High-temperature alloys (Ni or Co base materials)	700,000	49.14×10^8

[†] In the workshop specific cutting energy is sometimes expressed in terms of the units h.p./(in^3/min), where 1 h.p./(in^3/min) = 396,000 in lb/cu in.

[‡] In converting from English to SI units it is convenient to know that 1 in lb in^{-3} = 6894 J m^{-3}.

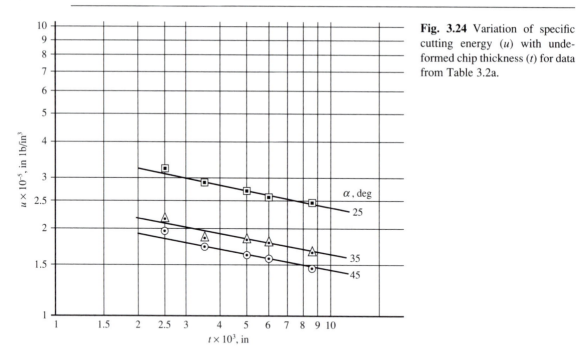

Fig. 3.24 Variation of specific cutting energy (u) with undeformed chip thickness (t) for data from Table 3.2a.

That is,

$$F_Q = \frac{F_P}{2} \tag{3.41}$$

The horsepower consumed at the tip of the tool will be

$$\text{h.p.} = \frac{F_P V}{33,000} \quad (F_P \text{ in lb, } V \text{ in f.p.m.}) \tag{3.42}$$

or

$$\text{kW} = \frac{F_P V}{1000} \quad (F_P \text{ in N, } V \text{ in m s}^{-1}) \tag{3.43}$$

The foregoing method of estimating cutting forces is only very approximate and is for cutting with a continuous chip and no BUE. With more complex chip formation, cutting forces and power may be quite different as follows:

1. Forces will generally be less with discontinuous chip formation.

2. Forces will in general be less with a BUE, the amount of decrease depending on the size of the BUE. Thus, cutting forces will in general decrease with decrease in cutting speed in the range of cutting speeds where the size of BUE increases with decrease in speed.

3. Cutting forces will tend to increase with the extent of secondary shear (conveniently measured by α').

4. Cutting forces will increase with increased strain hardening during cutting. When cutting an annealed material that strain hardens substantially during cutting, a fan-shaped shear zone will tend to be produced as already discussed in connection with Fig. 3.21.

5. Cutting forces will usually decrease with a controlled contact length tool design (but not always—this depends on whether optimum contact length is larger or smaller than that obtained with a conventional tool).

6. Cutting forces will be greater when cutting with a tool tip protecting chamfered cutting edge (Fig. 3.22b).

The influence of prior cold working of the work material on cutting forces is quite complex. On the one hand, forces will tend to be higher due to an increase in initial hardness of the work. On the other hand, there will be less tendency for a pie-shaped shear zone to develop. While these two tendencies tend to cancel each other, either may predominate.

EXAMPLE: SINGLE-POINT TURNING

Figure 3.25 shows a cutting tool having zero side-cutting edge angle in a straight turning situation. The depth of cut is b and the feed rate t (i.p.r. or mm rev^{-1}). There are two cutting edges (of extent b and t, respectively). In most cases $b/t \sim 10$; therefore, cutting on the end of the tool will be negligible compared with cutting along the main cutting edge of extent b. As a first approximation, therefore, this three-dimensional chip formation situation may be approximated by an equivalent two-dimensional orthogonal case where

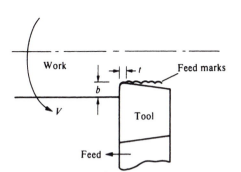

Fig. 3.25 Single-point turning tool showing depth of cut (b) and feed (t).

α = side rake angle of tool

t = feed per revolution (undeformed chip thickness)

b = depth of cut (conventional workshop nomenclature)

V = surface speed of work at outside diameter (o.d.).

The cutting ratio may be obtained by taking a chip of any convenient length (l_C) and calculating the corresponding undeformed length (l) from the weight of the chip, the specific weight of the work material, and the known values of b and t.

Consider the following specific turning case:

- back rake angle (α_b) = 15°
- side rake angle (α_S) = 10°
- side cutting edge angle (C_s) = 10°
- work material: AISI 1020 steel
- workpiece r.p.m. (N) = 200
- cutting speed (V) = 200 f.p.m. (61 m min^{-1})
- feed rate (t) = 0.015 i.p.r. (0.38 mm rev^{-1})
- depth of cut (b) = 0.150 in (3.8 mm)

From Table 3.3, Eq. (3.39), and assuming a continuous chip without BUE and small secondary shear

$$u = (300,000) \left(1 - \frac{15 - 0}{100} \right) \left(\frac{0.010}{0.015} \right)^{0.2} = 235,100 \text{ p.s.i. } (34.1 \text{ MPa})$$

From Eq. (3.40)

$$F_P = ubt = (235{,}100)\ (0.150)\ (0.015) = 529\ \text{lb}\ (2328\ \text{N})$$

and Eq. (3.41)

$$F_Q = \frac{F_P}{2} = 265\ \text{lb}\ (1166\ \text{N})$$

and Eq. (3.42)

$$\text{h.p.} = \frac{F_P V}{33{,}000} = \frac{(529)(200)}{33{,}000} = 3.2\ \text{h.p.}\ (2.39\ \text{kW})$$

The horsepower (h.p.) required to feed the tool would be

$$(\text{h.p.})_t = \frac{(F_Q)(t/12\ N)}{33{,}000} = \frac{(529)(0.015/12)(200)}{33{,}000} = 0.004\ \text{h.p.}\ (0.003\ \text{kW})$$

This is seen to be negligible compared with the power required to turn the work (3.2 h.p. or 2.39 kW).

It should be kept in mind that the foregoing method of estimating cutting forces and power is only approximate. Nevertheless, the method is useful in estimating the capacity of a machine tool required to do a given job or in estimating the order of magnitude of the forces that are apt to be encountered. While a more elaborate table of specific energy values than that of Table 3.3 could be developed, this does not appear justified in view of the other approximations involved. In fact, one might even question whether the corrections for change in rake angle and feed are justified. The only reason for including these in the discussion is to indicate the direction and relative magnitudes of these effects, other things being equal.

REFERENCES

Cook, N. H., Finnie, I., and Shaw, M. C. (1954). *Trans. Am. Soc. Mech. Engrs.* **76**, 153.

Cooke, W. B. H., and Rice, W. B. (1973). *J. Engng. Ind.* **95**, 844.

Drucker, D. C. (1949). *J. Appl. Phys.* **20**, 1013.

Ernst, H., and Merchant, M. E. (1941). Chapter in *Surface Treatment of Metals*. Am. Soc. for Metals, pp. 299–378.

Field, M., and Merchant, M. E. (1949). *Trans. Am. Soc. Mech. Engrs.* **71**, 421.

Kececioglu, D. (1958). *Trans. Am. Soc. Mech. Engrs.* **80**, 158.

Lapsley, J. T., Grassi, R. C., and Thomsen, E. G. (1950). *Trans. Am. Soc. Mech. Engrs.* **72**, 979.

Merchant, M. E. (1944). *Trans. Am. Soc. Mech. Engrs.* **66**, A-168.

Merchant, M. E. (1945). *J. Appl. Phys.* **16**, 267(a) and 318(b).

Merchant, M. E., and Zlatin, N. (1945). *Mech. Engng.* **67**, 737.

Nakayama, K. (1957). *Bull. Fac. Engng. Yokohama Natn. Univ.* **6**, 1.

Piispanen, V. (1937). *Eripaines Teknilliseslä Aikakauslehdeslä* **27**, 315.

Piispanen, V. (1948). *J. Appl. Phys.* **19**, 876.

Shaw, M. C. (1950). *J. Appl. Phys.* **21**, 599–606.

Shaw, M. C. (1954). *Metal Cutting Principles*, 3rd ed. MIT Press, Cambridge, Mass.

Shaw, M. C. (1967). Report 94, Iron and Steel Institute, London.

Townend, G. H. (1947). *J. Appl. Phys.* **18**, 784.

4 ELASTIC BEHAVIOR

The mechanical behavior of a material being cut plays a very important role, and it is therefore useful at the outset to review briefly the behavior of materials relative to elastic and plastic behavior and fracture. This will be done in the next three chapters.

When a metal is loaded in uniaxial tension, uniaxial compression, or simple shear (Fig. 4.1), it will behave elastically until a critical value of normal stress (S) or shear stress (τ) is reached and then it will deform plastically. In the elastic region, the atoms are temporarily displaced but return to their equilibrium positions when the load is removed. Stress (S or τ) and strain (e or γ) in the elastic region are defined as indicated in Fig. 4.2: the usual convention of considering tensile stress and strain positive and compressive stress and strain negative is followed here. Poisson's ratio (v) is the ratio of transverse to direct strain in tension or compression:

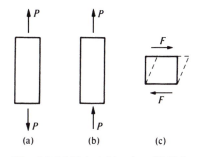

Fig. 4.1 (a) Uniaxial tension. (b) Uniaxial compression. (c) Simple shear.

$$v = -\frac{e_2}{e_1} \qquad (4.1)$$

where e_2 is the transverse strain that accompanies strain in the direction of the applied load (e_1 = direct strain). In the elastic region v is between $\frac{1}{4}$ and $\frac{1}{3}$ for metals. The relation between stress and strain in the elastic region is given by Hooke's law:

$$S = Ee \text{ (tension or compression)} \qquad (4.2)$$

$$\tau = G\gamma \text{(simple shear)} \qquad (4.3)$$

where E is Young's modulus of elasticity and G is the shear modulus of elasticity.

A small change in specific volume (change in volume per unit volume) is associated with elastic deformation which may readily be shown to be as follows for an isotropic (same properties in all directions) material:

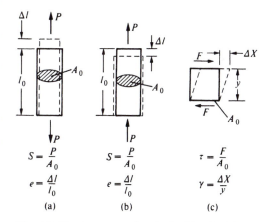

Fig. 4.2 Elastic stress and strain for (a) uniaxial tension, (b) uniaxial compression, and (c) simple shear.

39

$$\frac{\Delta\text{Vol}}{\text{Vol}} = e_1(1 - 2v) \tag{4.4}$$

The bulk modulus (K = reciprocal of compressibility) is defined as follows:

$$K = \frac{\Delta p}{\dfrac{\Delta\text{Vol}}{\text{Vol}}} \tag{4.5}$$

where Δp is the pressure acting at a particular point. For an elastic solid loaded in uniaxial compression (S),

$$K = \frac{S}{\dfrac{\Delta\text{Vol}}{\text{Vol}}} = \frac{S}{e_1(1 - 2v)} = \frac{E}{1 - 2v} \tag{4.6}$$

Thus, an elastic solid is compressible as long as v is less than $\frac{1}{2}$ which is normally the case for metals.

STRESS VERSUS DIRECTION

A distinguishing feature of solids from fluids is that stresses and strains at a point are different in different directions. Fig. 4.3 shows stresses acting on a small cube having an arbitrary orientation relative to external loads. Normal stresses are designated S and shear stresses τ. The subscript xy designates a stress in a plane normal to the x-axis but in the y-direction (a shear stress—hence τ_{xy}). Fig. 4.4a is the two dimensional counterpart of Fig. 4.3, and Fig. 4.4b is a free-body diagram of the part of the square of Fig. 4.4a below oblique plane AB. The normal to plane AB makes an angle θ with the x-axis. The normal (S') and shear (τ') stresses on plane AB may be found relative to stresses on the faces normal to the x- and y-axes and angle θ by writing the three equations for static equilibrium for the free body of Fig. 4.4b (for static equilibrium relative to rotation ($|\tau_{xy}| = |\tau_{yx}|$). After simplification, the transformation stress equations become

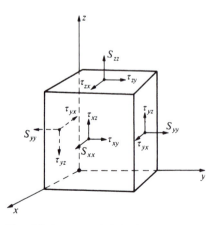

Fig. 4.3 Stress on infinitesimal cube at a point.

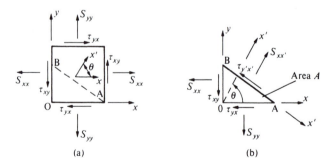

(a)   (b)

Fig. 4.4 Two-dimensional stresses at a point. (a) Oblique plane AB oriented with its normal angle θ to x-axis. (b) Free-body diagram corresponding to (a).

$$S_{x'x'} = \frac{S_x + S_y}{2} + \frac{S_x - S_y}{2} \cos 2\theta + \tau_{xy} \sin 2\theta$$

$$S_{y'y'} = \frac{S_x + S_y}{2} - \frac{S_x - S_y}{2} \cos 2\theta - \tau_{xy} \sin 2\theta \qquad (4.7)$$

$$\tau_{x'y'} = \frac{S_y - S_x}{2} \sin 2\theta + \tau_{xy} \cos 2\theta$$

Mohr Stress Diagram

The Mohr stress diagram is a convenient technique for applying these equations. This is a diagram in stress space (S, τ) instead of real space (x, y). Figure 4.5b is the Mohr's circle corresponding to stresses on the elemental square of Fig. 4.5a. The Mohr's stress diagram is started by plotting normal stress (S_x) and shear stress (τ_{xy}) for the X face to scale. It is convenient to consider a clockwise stress to be positive and a counterclockwise shear stress to be minus. A tensile stress is plotted plus and a compressive stress minus. Next, the stresses on the Y-face are plotted and a circle is drawn with XY as diameter. Points 1 and 2 in Fig. 4.5b are the principal stresses (normal stresses at A on a square oriented such that shear stresses τ are zero). The orientation of the planes of principal stress may be found by noting that angles turned through in the stress space diagram are twice as large as those turned through in the real space diagram and in the same direction provided the above sign convention is adopted for shear stresses. Figure 4.5c shows the orientation of principal planes (1, 2) corresponding to the Mohr's circle of Fig. 4.5b. The double angle relationship in stress space arises from the fact that only (2θ) appears in Eqs. (4.7).

The Mohr's circle technique is widely used to convert stresses at a point with one orientation to another. However, it is not usually used by plotting to scale but simply as a nemonic (memory aid) device for conveniently applying Eqs. (4.7).

The Mohr's circle concept may be readily extended to three dimensions. Figure 4.6b shows the three-dimensional Mohr's circle representation for the elemental cube shown in Fig. 4.6a which is oriented so that the cube faces correspond to the three principal directions (1, 2, 3). All possible combinations of stress lie in the shaded region. The three circles give values of S and τ that lie in the 1–2, 2–3, and 3–1 planes. Of particular interest is the overall maximum shear stress which is $(S_1 - S_3)/2$.

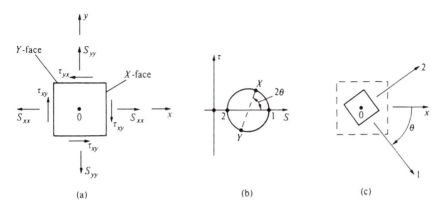

(a)   (b)   (c)

Fig. 4.5 (a) Two-dimensional stresses at point O. (b) Corresponding Mohr's circle diagram. (c) Orientation of planes of principal stress.

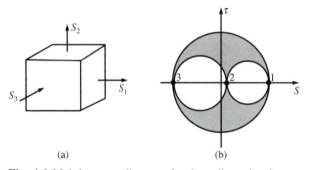

(a)   (b)

Fig. 4.6 Mohr's stress diagrams for three-dimensional case. (a) Principal stresses at point 0. (b) Corresponding Mohr's circle diagrams.

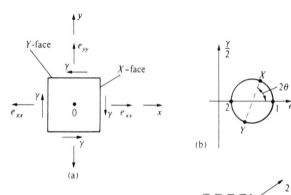

(a)

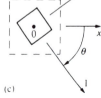

(c)

Fig. 4.7 (a) Two-dimensional strains at point O. (b) Corresponding Mohr's circle diagram. (c) Orientation of planes of principal strain.

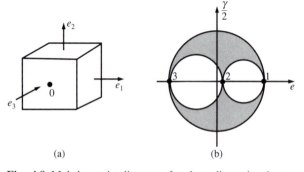

(a)   (b)

Fig. 4.8 Mohr's strain diagrams for three-dimensional case. (a) Principal strains at point O. (b) Corresponding Mohr's circle diagrams.

MOHR STRAIN DIAGRAM

A parallel treatment of strain at a point leads to a Mohr's circle of strain. In this case the strain coordinates are e and $\gamma/2$ in place of S and τ, otherwise all details of Mohr's circle for stress have a direct counterpart in Mohr's circle for strain. Figure 4.7a shows the strains at a point A for the X and Y planes while Fig. 4.7b is the corresponding Mohr's strain diagram and Fig. 4.7c shows the corresponding directions of principal strain ($\gamma = 0$). Angles measured about 0′ in the Mohr diagram again correspond to twice the angle turned through in the same direction in real space, provided a clockwise shear strain is considered positive.

The Mohr's strain circle concept may be extended to three dimensions, and Fig. 4.8 is the strain counterpart of Fig. 4.6 for three-dimensional stress.

GENERALIZATION OF HOOKE'S LAW

Hooke's law [Eq. (4.2)] for uniaxial tension may be generalized for a three-dimensional elastic situation as follows:

$$e_1 = 1/E\,[S_1 - v(S_2 + S_3)]$$
$$e_2 = 1/E\,[S_2 - v(S_3 + S_1)] \qquad (4.8)$$
$$e_3 = 1/E\,[S_3 - v(S_1 + S_2)]$$

where e_1, e_2, e_3, and S_1, S_2, S_3 are the principal strains and stresses for an isotropic homogeneous material that is elastically loaded.

EXAMPLES

Several special states of stress and strain are of particular interest because of their importance in materials testing in the laboratory, and Fig. 4.9 shows the Mohr diagrams for a few of these special cases. It is convenient to number the principal stresses and strains taking 1 as the algebraically largest value (one furthest to the right) and 3 as the algebraically lowest value,

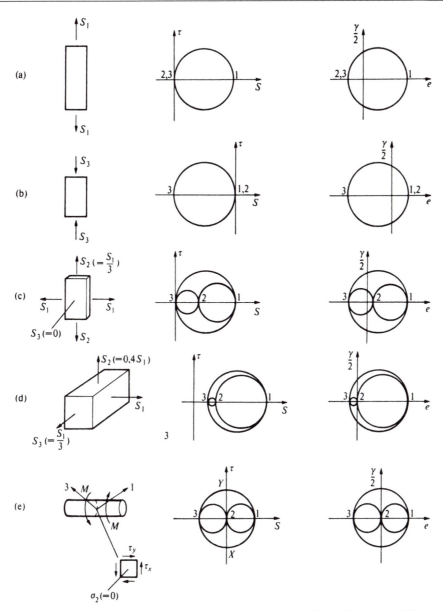

Fig. 4.9 Mohr's circle diagrams for special states of stress and strain. (a) Uniaxial tension. (b) Uniaxial compression. (c) Biaxial stress ($S_2 = S_1/3$) for thin specimen (plane stress, $S_3 = 0$). (d) Biaxial stress ($S_3 = S_1/3$) for wide specimen (plane strain, $e_2 = 0$). (e) Torsion (stress and strain in outer surface of round bar). Poisson's ratio has been assumed to be 0.30 in all cases.

and this has been done in Fig. 4.9. Equations (4.8) may then be used to find missing values of stress or strain. Cases (c) (plane stress) and (d) (plane strain) in Fig. 4.9 are very important in the plastic regime where they have somewhat different Mohr's circle diagrams due to Poisson's ratio being one half in the plastic region (see Fig. 5.39 in the next chapter).

REFERENCES

The material in this chapter is to be found in many texts. If additional discussion of elastic behavior is desired, the following books may be consulted.

Alexander, J. M., and Brewer, R. C. (1963). *Manufacturing Properties of Materials.* D. Van Nostrand, London.

Backofen, W. A. (1972). *Deformation Processing.* Addison Wesley, Reading, Mass.

Crandall, S. H., and Dahl, N. C. (1959). *An Introduction to the Mechanics of Solids.* McGraw-Hill, New York.

Dieter, G. E. (1976). *Mechanical Metallurgy,* 2nd ed. McGraw-Hill, New York.

Flinn, R. A., and Trojan, P. K. (1975). *Engineering Materials and Their Applications.* Houghton Mifflin, Boston.

McClintock, F. A., and Argon, A. S. (1966). *Mechanical Behavior of Materials.* Addison Wesley, Reading, Mass.

Polakowski, N., and Ripling, E. T. (1964). *Strength and Structure of Engineering Materials.* Prentice-Hall, New York.

Suh, N. P., and Turner, A. P. L. (1974). *Elements of the Mechanical Behavior of Solids.* McGraw-Hill, New York.

Van Vlack, L. H. (1970). *Materials Science for Engineers.* Addison Wesley, Reading, Mass.

5 PLASTIC BEHAVIOR

At a certain value of strain ($e \cong 10^{-3}$) called the yield point, a material ceases to behave elastically —does not return to its original condition when the load is removed but undergoes permanent deformation. This is called plastic behavior. It is found experimentally that there is negligible change in specific volume during plastic flow, and this suggests that more than the stretching of atoms is involved. A suitable model to explain plastic flow and its negligible change in density is slip in which one layer of atoms slides over another in shear. However, when the shear stress required to cause one layer of atoms in a perfect lattice to slide over an adjacent layer is estimated, the required value far exceeds the flow stress of ordinary materials.

THEORETICAL SHEAR STRENGTH

Figure 5.1a shows two layers of atoms in a perfect crystal under no-load equilibrium conditions. The atoms represented by spheres are all vertically aligned. Figure 5.1b shows the atoms displaced

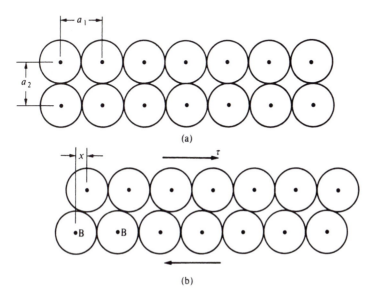

(a)

(b)

Fig. 5.1 Slip of one layer of atoms over another layer of atoms in a perfect crystal. (a) Before deformation. (b) After shear displacement x.

relative to each other due to the application of shear stress (τ). The resisting stress (τ) may be assumed to vary linearly with displacement (x). When $x = a_1$ (the horizontal atom spacing), the atoms will be realigned and $\tau = 0$. Also, when $x = a_1/2$, a displaced atom will be equally attracted to its nearest neighbors B in Fig. 5.1b and τ should again be zero. As a first approximation, the maximum resisting shear stress (τ_0 = shear strength) might be expected to occur when $x = a_1/4$. For an assumed linear relation between stress and displacement, it follows that

$$\frac{\tau}{\tau_0} = \frac{x}{a_1/4} \tag{5.1}$$

From the definition of simple shear strain (γ),

$$\gamma = \frac{x}{a_2} \tag{5.2}$$

from Eqs. (5.1) and (5.2),

$$\tau_0 = \frac{\tau a_1}{4 a_2 \gamma} \tag{5.3}$$

but from Eq. (4.3),

$$\tau_0 = \frac{G}{4} \frac{a_1}{a_2} \cong \frac{G}{4} \tag{5.4}$$

since $a_1 \cong a_2$.

When this value of theoretical shear strength (τ_0) is compared with values for ordinary materials, a very large discrepancy is found. For example, mild steel has a value of G of about 11.6×10^6 p.s.i. (80,093 MN m^{-2}). Thus from Eq. (5.4)

$$\tau_0 = \frac{G}{4} = \frac{11.6 \times 10^6}{4} = 2.9 \times 10^6 \text{ p.s.i. (19,651 MPa)}$$

However, the actual shear strength of such a material is only about 30,000 p.s.i. (207 MPa). This represents a discrepancy of $2.9 \times 10^6/0.03 \times 10^6 \cong 97$ or two orders of magnitude.

While the assumed linear variation of stress with displacement to such a large strain ($\gamma = \frac{1}{4}$) might be questioned, this is not the problem. This may be shown by substituting a sinusoidal variation of stress with displacement for a linear one. That is, when $\tau = \tau_0 \sin x/(a_1/4) (2\pi)$ instead of $\tau = \tau_0 x/(a_1/4)$ [Eq. (5.1)], the value of τ_0 is found to be $G/2\pi$ instead of $G/4$. This changes the discrepancy to $1.85 \times 10^6/0.03 \times 10^6 = 62$ which is still a very large value. Thus, the assumed relation between stress and displacement does not appear to represent the basic difficulty.

DISLOCATION THEORY

In 1934 a linear crystal imperfection called a dislocation was independently and simultaneously proposed by three materials scientists (Orowan, Polanyi, and Taylor). Figure 5.2 shows one type of dislocation known as an edge dislocation in which there is one more atom above the plane of slip than below. This results in some atoms above the slip plane being attracted to their nearest neighbors below the plane while others are resisting separation from their nearest neighbors. The net effect of such a configuration is to require far less energy to cause displacement of the upper row relative to the lower row than if all atoms acted in the same way as in Fig. 5.1.

A couple of useful analogies explain the difference in flow stress with and without the presence of dislocations. Figure 5.3 shows the mode of transport of an inchworm over the ground. The progressive movement of a small section of its body relative to the ground, instead of sliding all at the same time, requires much less energy. Similarly, if a large rug is to be slid across the floor, the easiest way of doing this is to form a ruck in the rug and then kick this over the length of the rug instead of sliding the entire rug over the floor at the same time (Fig. 5.3b).

When a dislocation runs across a crystal, this results in a displacement of one atom spacing (a_1). If dislocations were found in every plane of atoms in the a_2 direction of Fig. 5.1, the passage of one

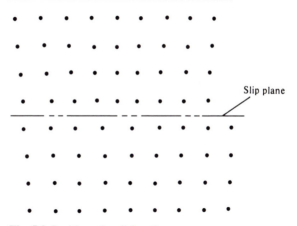

Fig. 5.2 Positive edge dislocation.

dislocation across a crystal would correspond to a local shear strain of unity $= a_1/a_2$. However, it is found that dislocations are located on planes that are very widely separated relative to atomic spacing. This means that the passage of a large number of dislocations is required on the active slip planes to account for observed plastic strains.

While several complex mechanisms of dislocation formation have been proposed, only one will be considered here. It is well established that real materials contain imperfections at a spacing of about one micrometer which corresponds roughly to the maximum spacing of active shear planes. The main effect of such imperfections is to cause an intensification of stress (and strain) above that obtained with a homogeneous material (same properties at all points). While all imperfections are not microcracks, a microcrack offers a useful model to explain the significance of such a stress intensifier. Figure 5.4a shows the manner in which the stress (σ_t) at the tip of a microcrack rises above the homogeneous value (σ_0) as a crack tip is approached. Figure 5.4b shows atoms along a line 45° to the direction of homogeneous stress (σ_0). The closed circles represent the unloaded positions of the atoms while the open dots represent the deformed positions. The open dots constitute a dislocation that once formed will run across the crystal at a fraction of the energy required to move all atoms simultaneously. Each time a dislocation is formed and moves off under the action of a tensile stress, the crack will grow and when the crack becomes sufficiently large, it will grow spontaneously as discussed later.

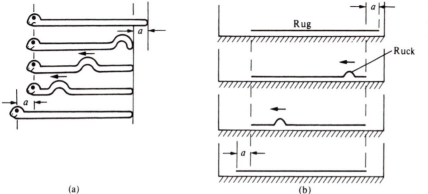

(a)

(b)

Fig. 5.3 Dislocation analogies. (a) Inch worm (after Orowan). (b) Displacement of ruck in rug.

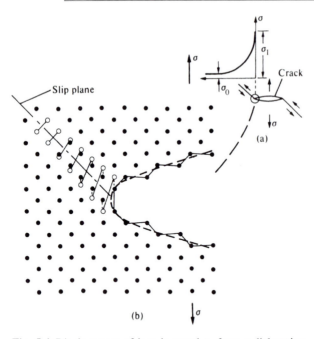

Fig. 5.4 Displacement of ions in metal to form a dislocation at the tip of a microcrack. (a) Stress intensification at crack tip. (b) Displacement of ions at crack tip. Solid points represent unloaded ions, open points represent ion centers after stressing.

An important characteristic of plastic flow is strain hardening which is the increase in plastic flow stress with strain. Dislocations will move across a crystal until a crystal boundary or second phase or impurity atom is encountered. Stuck dislocations project a back stress to the point of dislocation formation or interfere with the motion of dislocations and these actions constitute the major sources of strain hardening.

The increase in flow stress with plastic strain can be removed by heating a material to a sufficiently high temperature so that the increased amplitude of vibration of atoms is sufficient to shake dislocations from the structure. Such a softening operation is accompanied by a rearrangement of crystal boundaries and a decrease in grain size. When a metal is plastically deformed at a temperature above the strain recrystallization temperature ($\cong$ absolute melting point/2), this is called hot working and strain hardening is avoided because dislocations are removed by grain boundary rearrangement as fast as they are formed.

While dislocations are too small to be photographed directly, their presence may be revealed by indirect evidence such as the formation of etch pits or precipitation (decoration) at dislocation pile-ups. X-ray diffraction and transmission electron microscopy (TEM) of very thin sections of plastically deformed material have also proved useful in verifying the existence of dislocations.

Probably the most convincing evidence for dislocations is provided by the soap bubble analogy devised by Bragg and Nye (1947). Small air bubbles of uniform size are attracted to form a close-packed array (Fig. 5.5) due to surface tension but resist approaching beyond an equilibrium spacing due to internal pressure. There are two opposing forces as in the case of atoms in a metal that similarly give rise to an equilibrium bubble spacing. In addition to crystal boundaries, dislocations also appear in a bubble raft as accidents of growth. When such a two-dimensional analogy of a metal structure is sheared, dislocations are observed to run across single "grains" and to be generated at points of stress concentration represented by a microcrack tip. In addition to dislocations formation and movement, the Bragg–Nye analogy also clearly demonstrates strain recrystallization. An excellent motion picture that illustrates many of the details of dislocation mechanics in terms of the bubble model has been produced by Bragg, Lomer, and Nye (1954).

In metal cutting the strains (γ) in steady state chip formation are unusually high (frequently as high as 3 and occasionally as high as 5, as may be seen in Table 3.1). There is considerable experimental evidence that microcracks are formed on the shear plane at points of stress concentration and that these are the principal cause for the size effect (increase in specific energy with decrease in the volume deformed in chip formation). It is also found experimentally that normal stress on the shear plane has a substantial influence on the magnitude of shear stress on the shear plane. This further suggests that microcracks together with normal stress on the shear plane give rise to the migration of microcracks across the shear plane as they form, weld, reform, and reweld. It is suggested that this migration of microcracks joins the migration of dislocations in accounting for the unusually high strains associated with steady state chip formation. All of this is discussed in detail in Chapter 20.

METAL WHISKERS

It was found quite by accident (Herring and Galt, 1952) that very small samples of metals free of dislocations may be produced by electrodeposition at very low values of potential. These "whiskers" are produced so slowly there is plenty of time for each atom to find its proper place and there are no accidents of growth (dislocations). When such whiskers are tested in bending, they

1. are elastic to the point of fracture (no plastic flow)

2. are nonlinearly elastic, the elastic stress–strain curve resembling the first quarter cycle of a sine wave

3. have a very large strain at fracture approaching $\gamma = \frac{1}{4}$ suggested by the derivation for theoretical strenth (τ_0)

From the foregoing discussion, it is evident that metals are ductile and flow plastically due to the presence of structural defects called dislocations. Plastic behavior is ordinarily possible only due to dislocations. In the absence of dislocations, metals are perfectly brittle (yield strain $\geq$ fracture strain). The ductility characteristic of metals is due to the tendency for metals to generate large numbers of dislocations.

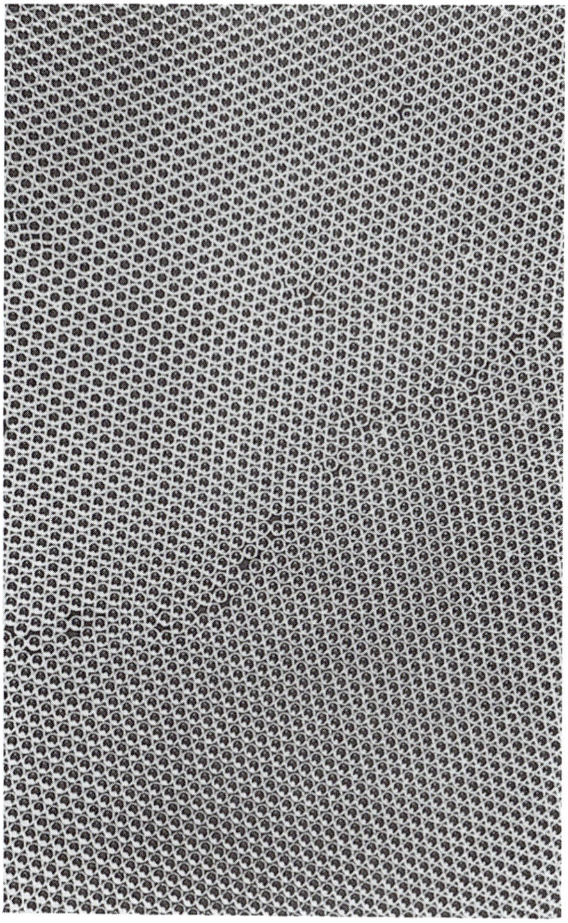

Fig. 5.5 Bubble model showing grain boundary running diagonally across center of field and dislocation just above center of field.

CONSTANCY OF VOLUME AND POISSON'S RATIO

It was previously stated that there is essentially no change in specific volume with plastic flow. From Eq. (4.4), it follows that in order that $\Delta \text{Vol}/\text{Vol} = 0$, the value of Poisson's ratio (ν) must be $\frac{1}{2}$. Thus, Poisson's ratio is always $\frac{1}{2}$ for all metals in the plastic region.

UNIAXIAL TENSILE TEST

Figure 5.6 shows a stress–strain curve for a ductile metal as frequently determined in acceptance testing. Here stress (S) is the applied load (P) divided by the original area (A_0), and strain (e) is the change in length (Δl) divided by the original gage length (l_0). Also, S_Y is the yield stress (limit of linear elasticity) and S_U is called the ultimate stress (P = a maximum) while S_F is the fracture stress. What appears to be negative strain hardening beyond point U is really due to a degree of confusion

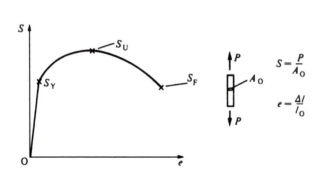

Fig. 5.6 Engineering stress–strain uniaxial tensile curve. [S_Y = Yield stress. S_U = Ultimate stress. S_F = Fracture stress.]

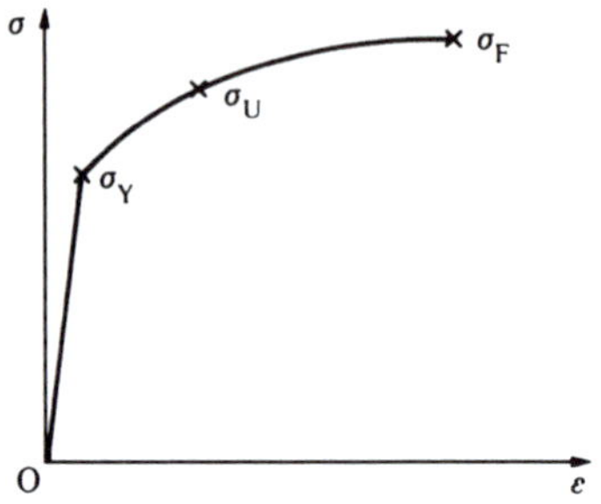

Fig. 5.7 True-stress–true-strain uniaxial tensile curve. [σ_Y = Yield stress. σ_U = Ultimate stress. σ_F = Fracture stress.]

introduced by defining the stress in terms of the original area instead of the actual instantaneous area. Figure 5.7 shows the curve of Fig. 5.6 replotted in terms of true stress (σ) and true strain (ε) where stress and strain are now defined in terms of the instantaneous area and gage length instead of initial values (A_0 and l_0). A metal that is highly strain hardened before testing will yield a much flatter curve that approaches the "ideal plastic" which has a perfectly flat σ–ε curve in the plastic region. Since there is no change of volume in the plastic regime,

$$Al = A_0 l_0 \tag{5.5}$$

and by definition

$$\sigma = \frac{P}{A} = \frac{Pl}{A_0 l_0} = \frac{S(l_0 + \Delta l)}{l_0} = S(1 + e) \tag{5.6}$$

Also,

$$d\varepsilon = \frac{dl}{l}$$

Integrating and evaluating the constant of integration:

$$\varepsilon = \ln \frac{l}{l_0} = \ln \left[\frac{l_0(1 + e)}{l_0} \right] = \ln (1 + e) \tag{5.7}$$

Equations 5.6 and 5.7 are useful in converting S and e into σ and ε. From Eq. (5.7) it is evident that the difference in ε and e is negligible for values of e less than 0.1. The strain at the yield point S_Y is $\varepsilon_Y \cong 10^{-3}$.

If Fig. 5.7 is replotted on log–log coordinates (Fig. 5.8), a straight line is obtained having a slope n and hence beyond the yield point

$$\sigma = \sigma_1 \varepsilon^n \tag{5.8}$$

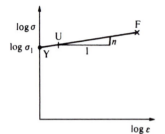

Fig. 5.8 Log–log true-stress–true-strain uniaxial tensile curve. [Y = Yield point. U = Ultimate point. F = Fracture point.]

where σ_1 is the stress corresponding to $\varepsilon = 1$ and n is called the strain hardening index. Equation (5.8) is widely used in analytical studies involving plastic flow in the cold working regime.

NECKING

In the initial plastic region (between Y and U in Figs. 5.6 and 5.7) the area decreases uniformly along the gage length and the increase in load (P) required due to strain hardening exceeds the decrease in P required due to reduction in area. However, at point U, a localized neck begins to form. Then the change in P required due to area reduction exceeds the change in P due to strain hardening, and what appears to be negative strain hardening occurs if stress is measured in terms of original area A_o instead of actual area A. The necking strain may be obtained as follows:

$$P = \sigma A \qquad \text{(by definition)} \tag{5.9}$$

Differentiating both sides

$$dP = \sigma \, dA + A \, d\sigma \tag{5.10}$$

At point U, $dP = 0$ and hence

$$\frac{d\sigma}{\sigma} = -\frac{dA}{A} \tag{5.11}$$

Since

$$Al = \text{constant} \tag{5.12}$$

$$A \, dl + l \, dA = 0 \tag{5.13}$$

or

$$-\frac{dA}{A} = \frac{dl}{l} = d\varepsilon \tag{5.14}$$

Hence from Eqs. (5.11) and (5.14)

$$\frac{d\sigma}{d\varepsilon} = \sigma \tag{5.15}$$

and from Eq. (5.8)

$$\frac{d\sigma}{d\varepsilon} = \sigma_1 n \varepsilon^{n-1} = \sigma \tag{5.16a}$$

and

$$\sigma_1 = \sigma / \varepsilon^n \tag{5.16b}$$

then

$$\frac{n\sigma}{\varepsilon} = \sigma \tag{5.17}$$

or at point U

$$\varepsilon_U = n \tag{5.18}$$

Since the value of n is about 0.2 for a ductile metal (annealed), the natural strain at necking (ε_U) is about 0.2.

When a concentrated neck forms in a tensile specimen, added complications are involved. First of all, the strain cannot be measured in terms of length since the strain is nonuniform along the gage length. However, due to the constance of volume,

$$ld^2 = l_o d_o^2 \tag{5.19}$$

$$\varepsilon = \ln \frac{l}{l_o} = \ln \left(\frac{d_o}{d} \right)^2 = 2 \ln \left(\frac{d_o}{d} \right) \tag{5.20}$$

Thus, after necking, the strain must be measured in terms of the minimum diameter at the neck (d).

STRESS DISTRIBUTION IN NECK

The stress is no longer uniformly distributed across the minimum area once a neck forms. The reason for this becomes evident when we consider two fibers A and B in the necked tensile specimen shown in Fig. 5.9. While the stress is axially directed in the region of uniform area, a considerable radial component of stress will be present in the neck which will increase as the center of the specimen is approached. An increase in radial stress must be accompanied by an equal increase in axial stress as a result of a flow criterion (Tresca criterion) to be discussed presently. Thus, the state of stress across the minimum section of a necked specimen will vary from a uniaxial stress (σ_c) at the periphery of the neck to ($\sigma_c + \sigma_H$) at the axis where σ_H is a hydrostatic component of stress due to the shape of the neck.

Bridgman has found the variation of axial stress (σ_a) with radial distance (r) from the specimen centerline to be

$$\sigma_a = \sigma_c \left[1 + \ln \left(\frac{a^2 + 2aR - r^2}{2aR} \right) \right] \tag{5.21}$$

where a is the section radius and R is the profile radius at the neck (Fig. 5.9). Thus, the maximum tensile stress occurs at the center of the specimen and equals

$$\sigma_{a,max} = \sigma_c \left[1 + \ln \left(1 + \frac{a}{2R} \right) \right] \tag{5.22}$$

The corrected uniaxial tensile stress (σ_c) is equal to the stress that would obtain in the absence of a neck and in turn is equal to the stress at the outer periphery of the neck:

$$\sigma_c = \frac{\bar{\sigma}}{\left(1 + 2\frac{R}{a} \right) \ln \left(1 + \frac{a}{2R} \right)} \tag{5.23}$$

where $\bar{\sigma}$ is the mean stress

$$\bar{\sigma} = \frac{P}{\pi a^2}$$

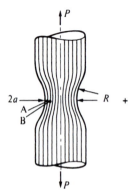

Fig. 5.9 Neck in tensile specimen.

The quantity in the denominator of Eq. (5.23) is known as the Bridgman correction factor and is shown plotted against a/R in Fig. 5.10. Figure 5.11 illustrates the manner in which the stress varies across the neck and the relation between

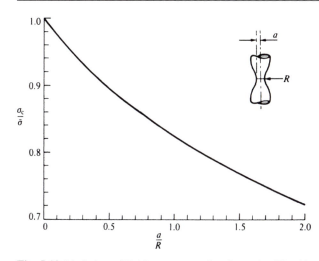

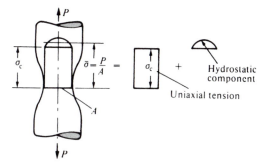

Fig. 5.11 Stresses in the neck of a tensile specimen.

Fig. 5.10 Variation of Bridgman correction factor ($\sigma_c/\bar\sigma$) with a/R ratio.

$\bar\sigma$, σ_c, and $\sigma_{c,max}$. Figure 5.12 is an empirical curve due to Bridgman (1944) for estimating $\sigma_c/\bar\sigma$ for different values of strain (ε) as defined in Eq. (5.20). Also shown on this curve are representative experimental values for mild steel and copper. The Bridgman approximation is found to be much better for steel than for nonferrous metals such as copper. The validity of the Bridgman correction has been verified by Marshall and Shaw (1952) by use of tensile specimens into which necks of known curvature had been machined.

Figure 5.13 shows the relationship between the so-called engineering stress–strain curve based on the original area, the true stress–strain curve and the corrected true stress–strain curve where the stress plotted (σ_c) is the uniaxial tensile stress in the absence of the hydrostatic component introduced by curvature of the neck.

From the foregoing discussion, it is evident that interpretation of tensile test results is really quite involved despite the apparent simplicity of the test.

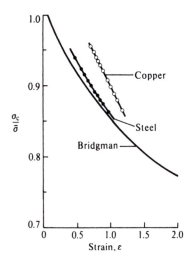

Fig. 5.12 Bridgman's empirical correction factor versus true strain (ε).

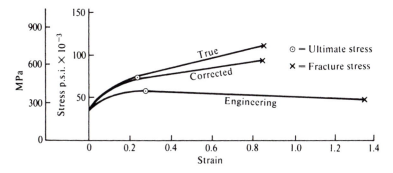

Fig. 5.13 Relationship between engineering, true, and corrected tensile stress–strain curves for AISI 1112 steel.

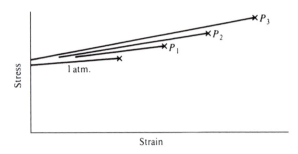

Stress (y-axis), Strain (x-axis), 1 atm., $\times P_1$, $\times P_2$, $\times P_3$

Fig. 5.14 Influence of hydrostatic pressure on flow stress–strain curve and fracture stress (x) $P_1 < P_2 < P_3$.

HYDROSTATIC STRESS

When tensile specimens are tested under different values of hydrostatic pressure, a negligible effect is found upon the yield stress and the post yield shape of the true stress–strain curve. For example, Bridgman (1952) found an increase of less than 2% in the ultimate stress ($\varepsilon = 0.2$) of AISI 1045 steel when subjected to a hydrostatic pressure of 100,000 p.s.i. (690 MPa). He similarly found an increase in Brinell hardness of less than 4% with a hydrostatic pressure of 100,000 p.s.i. However, hydrostatic stress has an appreciable influence on fracture stress, a compressive hydrostatic stress increasing fracture stress and a tensile hydrostatic stress decreasing fracture stress. Figure 5.14 illustrates these effects. Bridgman expresses the increased fracture strength (σ_f) with pressure in terms of a pressure coefficient of ductility (K):

$$\sigma_f = \sigma_{f0} + K\sigma_H \qquad (5.24)$$

where σ_H is the hydrostatic pressure (+ for compression).

TEMPERATURE AND STRAIN RATE EFFECTS

When specimens are tested in tension at different temperatures, they are found to be more brittle (low strain at fracture) at low temperature than at elevated temperatures. However, in the vicinity of 550 °F (288 °C), steels tend to have a higher yield stress and lower strain at fracture than at room temperature. This anomalous behavior, termed blue brittleness, is due to the migration of intersticial carbon and nitrogen to dislocations resulting in their immobilization. Figure 5.15 shows some typical tensile results for AISI 1112 steel tested at different temperatures but ordinary strain rate (10^{-3} s^{-1}). As a first approximation, the blue brittle anomaly is ignored and the yield stress is considered to decrease with increase in temperature.

When metals are tested in tension at different strain rates, the flow stress corresponding to a given strain is found to increase with strain rate. Figure 5.16 illustrates

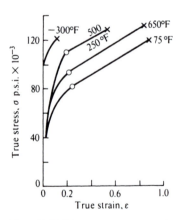

Fig. 5.15 True-stress–true-strain tensile curves for AISI 1112 steel at different temperatures. Strain rate = 10^{-3} s^{-1}. (after MacGregor, 1944)

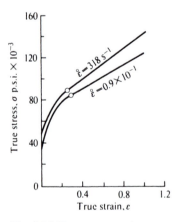

Fig. 5.16 True stress–strain curves for AISI 1112 steel at different strain rates. Temperature = 70 °F (21 °C). (after MacGregor, 1944)

this behavior. The following equation resembling Eq. (5.8) for strain hardening is frequently used to relate flow stress and strain rate at a given strain and temperature

$$\sigma = \sigma_1 \dot{\varepsilon}^m \tag{5.25}$$

where $\dot{\varepsilon} = d\varepsilon/dt$ and σ_1 and m are material constants. The exponent m (strain rate sensitivity) is found to increase with temperature, especially above the strain recrystallization temperature. In the hot working region, metals tend to approach the behavior of a Newtonian liquid for which $m = 1$, particularly metals of very small crystal size ($\sim 1\ \mu m$) which are superplastic (capable of deformation to very large strains at high temperature and low strain rate).

In general, increased temperature and strain rate have opposite effects on the flow stress of a material.

VON MISES FLOW CRITERION

A basic problem in design is to predict the yield stress (or flow stress corresponding to a given plastic strain) under a complex state of stress given the yield stress (or flow stress curve) in uniaxial tension or other standardized materials test. There is considerable experimental evidence that plastic flow is not influenced by a purely hydrostatic component of stress. Since the mean principal stress given by

$$\sigma_H = \frac{\sigma_1 + \sigma_2 + \sigma_3}{3} \tag{5.26}$$

approximates hydrostatic stress, it is convenient to use this in formulating a flow criterion. The first step in formulating a flow criterion is to subtract (σ_H) from the principal stresses pertaining and then to formulate a criterion based on these altered principal stresses which are termed deviator stresses. The principal deviator stresses are

$$\sigma_1' = \sigma_1 - \sigma_H$$
$$\sigma_2' = \sigma_2 - \sigma_H \tag{5.27}$$
$$\sigma_3' = \sigma_3 - \sigma_H$$

Since, for an annealed material, the yield stress in uniaxial tension is the same as for uniaxial compression, the function of the deviator stresses constituting the yield criterion should be an even function—i.e., give the same result for plus or minus values of σ'. Von Mises (1913) took the simplest even function of the stress deviators thus

$$\sigma_1'^2 + \sigma_2'^2 + \sigma_3'^2 = \text{constant} \tag{5.28}$$

for his yield criterion. When this is rewritten in terms of the actual principal stresses, we obtain

$$[(\sigma_1 - \sigma_2)^2 + (\sigma_2 - \sigma_3)^2 + (\sigma_3 - \sigma_1)^2] = \text{constant} \tag{5.29}$$

If a uniaxial tensile test is to be the basis of comparison, the following values are substituted to evaluate the constant:

$$\sigma_1 = \sigma_y$$
$$\sigma_2 = 0$$
$$\sigma_3 = 0$$

When these values are substituted into Eq. (5.29), we obtain

$$\sigma_y = \frac{1}{\sqrt{2}} [(\sigma_1 - \sigma_2)^2 + (\sigma_2 - \sigma_3)^2 + (\sigma_3 - \sigma_1)^2]^{1/2} = \text{constant} \tag{5.30}$$

This is known as the von Mises yield criterion. The stress σ_y is the equivalent of any complex state of stress for which the principal stresses are σ_1, σ_2, σ_3 as far as yield or flow at the same strain is concerned.

This equation which is in good agreement with experiment has also been obtained by Nadai (1950) by a more physical approach. He reasoned that flow would begin when the shear stress on a plane oriented so that its normal stress was equal to σ_H reached a particular value. The so-called octahedral plane is one that makes equal angles with the three principal stress coordinates, and Nadai reasoned that flow would occur when the shear stress on this plane reached a critical value. The result is the same as Eq. (5.30). Maxwell (1856) and later Hencky (1924) arrived at Eq. (5.30) from still another point of view. They assumed that flow will occur when a material absorbs a certain shear strain energy per unit volume. The strain energy per unit volume associated with an elastic body subjected to principal stresses σ_1, σ_2, σ_3 is

$$u_S = \frac{1 + v}{3E} [(\sigma_1 - \sigma_2)^2 + (\sigma_2 + \sigma_3)^2 + (\sigma_3 - \sigma_1)^2] \tag{5.31}$$

For a tensile specimen

$$u_S = \frac{1 + v}{3E} (2\varepsilon_y^2) \tag{5.32}$$

Equating Eqs. (5.31) and (5.32) again leads to Eq. (5.30). The Maxwell–Hencky criterion may also be derived in terms of principal strains and when this is done, the flow criterion becomes

$$\varepsilon_y = \frac{1}{\sqrt{2}(1 + v)} [(\varepsilon_1 - \varepsilon_2)^2 + (\varepsilon_2 - \varepsilon_3)^2 + (\varepsilon_3 - \varepsilon_1)^2]^{1/2} \tag{5.33}$$

where ε_y is the principal yield strain in a tensile test.

TRESCA FLOW CRITERION

Tresca (1864) and Mohr (1914) took a different approach to the flow criterion problem. They assumed that flow in a polycrystalline material will occur when the greatest of the principal shear stresses reaches a critical value. For plane stress

$$\frac{\sigma_1 - \sigma_2}{2} = \frac{\sigma_y}{2} \tag{5.34}$$

where $\sigma_y/2$ is the maximum shear stress in a tensile test at the onset of plastic flow.

Equation (5.34) may be written

$$(\sigma_1 - \sigma_2)^2 - \sigma_y^2 = 0 \tag{5.35}$$

or when generalized for the three-dimensional case

$$[(\sigma_1 - \sigma_2)^2 - \sigma_y^2] [(\sigma_2 - \sigma_3)^2 - \sigma_y^2] [(\sigma_3 - \sigma_1)^2 - \sigma_y^2] = 0 \tag{5.36}$$

or when written in terms of deviator stresses

$$[(\sigma_1' - \sigma_2')^2 - \sigma_y^2][(\sigma_2' - \sigma_3')^2 - \sigma_y^2][(\sigma_3' - \sigma_1')^2 - \sigma_y^2] = 0 \quad (5.37)$$

It is thus seen that the Tresca criterion leads to an even function of the stress deviator components just as the von Mises relationship did. While Eq. (5.37) is more complex than that arbitrarily assumed by von Mises, there is no theoretical reason why it is any more or less correct.

The results predicted by Eqs. (5.30) and (5.37) are usually in close agreement (Fig. 5.17). While the Tresca criterion is more appealing on physical grounds, those analytically inclined seem to favor the von Mises approach. However, the two criteria appear to be equally useful and the one to use is that which is more convenient for the problem involved.

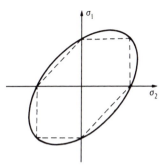

Fig. 5.17 Yield loci for plane stress. [Solid line = von Mises criterion. Dotted line = Tresca criterion. Maximum difference = 15.5%.]

EFFECTIVE STRESS AND STRAIN

The yield criterion is sometimes extended to cover the entire flow curve. When this is done, equivalent stress (σ_e) and equivalent strain (ε_e) are defined as follows based on the von Mises approach:

$$\sigma_e = \frac{1}{\sqrt{2}} [(\sigma_1 - \sigma_2)^2 + (\sigma_2 - \sigma_3)^2 + (\sigma_3 - \sigma_1)^2]^{1/2} \quad (5.38)$$

$$\varepsilon_e = \frac{\sqrt{2}}{3} [(\varepsilon_1 - \varepsilon_2)^2 + (\varepsilon_2 - \varepsilon_3)^2 + (\varepsilon_3 - \varepsilon_1)^2]^{1/2} \quad (5.39)$$

The numerical multipliers in each case have been chosen to yield the uniaxial tensile values for σ_e and ε_e when values for principal stress and strain corresponding to uniaxial tension are substituted:

$$(\sigma_1 = \sigma_y, \sigma_2 = \sigma_3 = 0; \varepsilon_1 = \varepsilon_y, \varepsilon_2 = \varepsilon_3 = -\varepsilon_y/2)$$

Similar equations based on the Tresca approach are

$$\sigma_e = \sigma_1 - \sigma_3 \quad (5.40)$$

$$\varepsilon_e = \frac{2}{3} (\varepsilon_1 - \varepsilon_3) \quad (5.41)$$

where σ_1, ε_1 and σ_3, ε_3 are the algebraically largest and smallest principal stresses and strains respectively.

UNIAXIAL COMPRESSION TEST

The simple uniaxial compression test (Fig. 5.18) offers another way of obtaining the plastic flow curve of a material. In performing such a test, the height of the specimen (H_o) should be about 1.5 times the diameter (D_o), and friction on the loading surface should approach zero. The role of friction in this test is to cause barrelling (Fig. 5.19) by making it more difficult for the material near the loading surfaces to flow outward than that in the central region. Barrelling plays a similar role in the compression test to necking in the tensile test—that is, it causes the radial stress distribution to be nonuniform (Fig. 5.20) and the Bridgman correction may be applied by use of a negative value

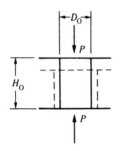

Fig. 5.18 Uniaxial compression test with zero friction on loading surfaces.

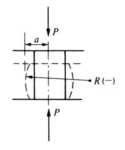

Fig. 5.19 Compression test with appreciable friction on loading surfaces leading to barrelling.

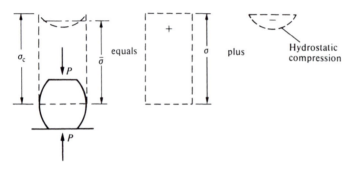

Fig. 5.20 Stress distribution across a compression specimen after barrelling.

for R (where R = the curvature of the barrel). However, the effect of barrelling is usually avoided and the corrected flow curve is directly obtained. This is done by either arranging for negligible friction on the die faces or by periodically returning the specimen to cylindrical shape by machining the outer surface of the specimen in a lathe.

Friction on the die surface may be reduced to a very low value by providing a helical scratch in the die face filled with a grease containing a solid lubricant (MoS_2 or graphite) or by use of a thin (~ .004 in or 0.1 mm) film of Teflon. Where the specimen must be deformed to a very large strain before fracture, it is advisable to eliminate friction as much as possible and then remove the residual bulge by machining the specimen periodically.

The corrected flow curves for uniaxial tension and compression will be the same for an annealed material, except that the strain at fracture will be greater in compression than in tension. This is in accordance with Bridgman's equation [Eq. (5.24)] since for tension $\sigma_H = +\sigma_f/3$ while in compression $\sigma_H = -\sigma_f/3$ where σ_f is the uniaxial stress at fracture. While the strain at fracture in compression is somewhat greater than that in tension, both strains at fracture are considerably smaller than in cutting, and both must be extrapolated considerably in order to be able to predict the flow stress in cutting at appropriate values of strain.

However, need for extrapolation is not the main problem with using tensile or compression data to predict the flow stress in cutting. In cutting there is a steep stress gradient in front of the tool and a strong stress concentration in the form of the relatively sharp cutting edge. Both of these require the material to flow in a very small volume (the shear band) at any one time instead of flowing over a larger volume of specimen as in a compression or tension test. In cutting, the probability of finding a point of weakness in the loaded zone is substantially less than in compression. This "size effect" is but one of several reasons why extrapolation of tensile or compression data to the strain pertaining in cutting will yield a flow stress that in general represents a poor approximation to that involved in cutting. Other reasons will be discussed later.

HARDNESS

Hardness is a term having different meaning to different people. It is resistance to *penetration* to a metallurgist, resistance to *wear* to a lubrication engineer, a measure of *flow stress* to a design engineer, resistance to *scratching* to a mineralogist, and resistance to *cutting* to a machinist. While these several actions appear to differ greatly in character, they are all related to the plastic flow stress of the material (Y).

The wide variety of hardness test procedures that have been used may be classified as follows:

1. *Static indentation* tests in which a ball, cone, or pyramid is forced into a surface and the load per unit area of impression is taken as the measure of hardness. The Brinell, Vickers, Rockwell, Monotron, and Knoop tests are of this type.

2. *Scratch* tests in which we merely observe whether one material is capable of scratching another. The Mohs and file hardness tests are of this type.

3. *Plowing* tests in which a blunt element (usually diamond) is moved across a surface under controlled conditions of load and geometry and the width of the groove is the measure of hardness. The Bierbaum test is of this type.

4. *Rebound* tests in which an object of standard mass and dimensions is bounced from the test surface and the height of rebound is taken as the measure of hardness. The Shore Scleroscope is an instrument of this type.

5. *Damping* tests in which the change in amplitude of a pendulum having a hard pivot resting on the test surface is the measure of hardness. The Herbert pendulum test is of this type.

6. *Cutting* tests in which a sharp tool of given geometry is caused to remove a chip of standard dimensions.

7. *Abrasion* tests in which a specimen is loaded against a rotating disc and the rate of wear is taken as a measure of hardness.

8. *Erosion* tests in which sand or abrasive grain is caused to impinge upon the test surface under standard conditions and the loss of material in a given time is taken as the measure of hardness. The hardness of grinding wheels is measured in this way.

The equipment and detailed test conditions for most of the hardness tests in use today may be found in texts by O'Neill (1934), Williams (1942), Tabor (1951), and von Weingraber (1952).

Scratch Hardness

One of the oldest scales of hardness, and one that is still in use by geologists, is that due to Mohs (1822) in which the scratch resistance of a material is compared with the scratch resistances of a standard series of ten minerals. Each of these minerals is assigned a number from 1 to 10. A Mohs hardness of 1 corresponds to a very soft material while 10 refers to the most scratch resistant of all known materials—the diamond. Mohs' scale of hardness is given in Table 5.1.

In applying the Mohs scale practically, the standard minerals are not employed, but the simple tests given in Table 5.1 under the heading *working scale* are used instead. The Mohs hardness of engineering metals falls within the narrow range from 4 to 7.

Tabor (1956) has shown that whereas a stylus must be 20% harder than a flat surface in order to scratch the flat surface, the minerals constituting the Mohs' scale increase in hardness in geometric ratio by a factor of approximately 1.4. Diamond is an exception since it is approximately three times as hard as corundum.

As can be seen, Mohs' method of describing hardness is little more than qualitative. While attempts have been made to put the scratch technique on a firmer quantitative basis by measuring the width of a scratch made with a diamond under standard conditions (Bierbaum, 1930), such tests are difficult to carry out with the necessary precision. The indentation method is by far the most practical and useful method of measuring hardness.

TABLE 5.1 Mohs' Scale of Hardness

Mineral	Mohs Hardness Number	Working Scale
Talc	1	Very easily scratched by fingernail
Gypsum	2	Easily scratched by fingernail
Calcite	3	Scratched by copper coin
Fluorite	4	Easily scratched by knife
Apatite	5	Scratched with difficulty by knife
Orthoclase	6	Easily cut by file
Quartz	7	Hardly cut by file—scratches glass
Topaz	8	Scratches glass
Corundum	9	Scratches glass
Diamond	10	Scratches glass

INDENTATION HARDNESS TESTS

Two important types of indentation tests are in use. In one type (Brinell, Vickers, Knoop), the size of the impression left after a hard indentor is pressed into the surface is measured. In the other type (Rockwell), the depth to which the indentor penetrates the specimen is the measure of hardness.

The indentation hardness test measures the resistance material offers to plastic indentation. The quantity measured is the flow stress at relatively small strain ($\varepsilon < 0.1$), low strain rate ($\sim 10^{-3}$ s^{-1}), and negligible friction. The main distinguishing feature is the presence of a large constraint relative to plastic deformation which causes the flow stress to be several times that involved in uniaxial compression.

Indentation hardness is usually expressed in units of pressure (p.s.i. or kg mm^{-2}), obtained by dividing the maximum applied load (P) by the area of the indentation measured either over the surface of the indentor (Brinell hardness) or in the plane of the surface indented (Meyer hardness). In both the Brinell and Meyer tests, the indentor is a sphere (Fig. 5.21a) and

$$H_B = \text{Brinell hardness} = \frac{2P}{\pi D^2 \{1 - \sqrt{[1 - (2a/D)^2]}\}} \tag{5.42}$$

$$H_M = \text{Meyer hardness} = \frac{P}{\pi a^2} \tag{5.43}$$

Brinell (1901) and Meyer (1908) hardness values do not differ greatly, but the Meyer value is often preferred because of its simplicity and correspondence to the true mean stress over the area of contact.

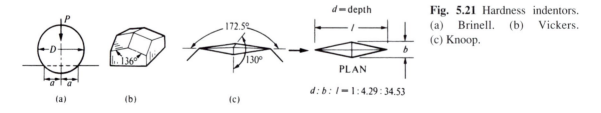

$d : b : l = 1 : 4.29 : 34.53$

Fig. 5.21 Hardness indentors. (a) Brinell. (b) Vickers. (c) Knoop.

(a)   (b)   (c)

Vickers (Fig. 5.21b) and Knoop (Fig. 5.21c) indentors are blunt pyramids. Faces of the Vickers indentors are inclined at an angle of 136°, while the ridges of the Knoop indentor have angles of 130° and $172\frac{1}{2}$°, respectively. Because the Knoop indentor penetrates only about half as deeply as the Vickers for the same load, it is frequently preferred for studies of superficial hardness. The Vickers hardness is the load divided by the contacting surface area, while the Knoop hardness is the load divided by the projected area, and hence corresponds to the Meyer value.

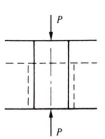

Fig. 5.22 Plane strain uniaxial compression test.

The hardness test is very easily conducted, but not so easily interpreted. Action beneath the indentor is complex and must be understood if full use is to be made of hardness values.

The simple compression test, Fig. 5.22, provides another measure of resistance to plastic flow that is more widely used in design analysis. If friction is kept to a low value on the die faces, a compression specimen will deform as shown by the dotted lines in Fig. 5.22 without barreling, and the uniaxial flow stress will be

$$\sigma_1 = \frac{P}{A} \tag{5.44}$$

where A is the cross-sectional area of the specimen.

It is important to relate Meyer hardness to uniaxial flow stress, a term with which most engineers are accustomed. The plastic zone beneath a hardness indentation is surrounded by elastic material which acts to hinder plastic flow in a manner similar to the die surfaces in a closed die forging. In the simple compression test the entire specimen goes plastic, and there is no resistance to side flow because the specimen is surrounded by air. Therefore, a greater mean stress is required to cause plastic flow in the hardness test than in the simple compression test.

The relation between the Meyer hardness and the uniaxial flow stress may be expressed as follows:

$$H_M = C\sigma_1 \tag{5.45}$$

where C is called the constraint factor for the hardness test. Experimentally, C approximates three for the Brinell, Vickers, and Knoop hardness tests. A central problem in the theory of hardness is to explain the origin of constraint factor, C.

SLIP LINE FIELD THEORY

The generally adopted explanation of indentation hardness is given in terms of the slip line field (SLF) theory. According to this theory, the material beneath a punch flows plastically in plane strain over a region consistent with the material displaced by the punch. At all other points, the specimen is considered rigid. The SLF is a network of curves along which the shear stress or shear strain rate is maximum. A suitable flow pattern need only be consistent from the point of view of velocities; when this is so, it is said to be a kinematically admissible solution.

The first kinematically admissible solution for a flat two-dimensional punch is that shown in Fig. 5.23a, due to Prandtl (1920). The solid lines are directions of constant maximum shear stress. They constitute a set of orthogonal shear stress coordinates designated α and β. The normal stress on shear planes α and β will be constant as long as the slip lines are straight but will change when traversing a curved slip line. The normal stress on surface AB (σ_1) will be zero since this communicates with air at zero gage pressure.

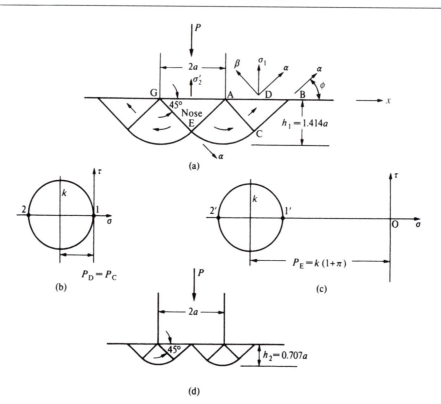

Fig. 5.23 Slip line field solutions for flat two-dimensional punch. (a) Prandtl (1920) SLF for punch face with high friction. (b) Mohr's circle for field ABC. (c) Mohr's circle for field AEG. (d) Hill (1950) for frictionless punch face.

The Mohr's stress circle (Chapter 3) for all points in region ABC is shown in Fig. 5.23b. The convention adopted in labeling the α and β directions is such that the directions of the algebraically greatest stress (σ_1) lies in the first quadrant of the α–β coordinate system. Thus the positive α and β directions at point D are directed as shown in Fig. 5.23a according to the usual convention. The Mohr's circle of Fig. 5.23b will hold for all points within region ABC. The normal stress on the shear plane for any point within region ABC will be $p_D = k$.

In going from C to E along the curved α line there will be no change in shear stress but a change in normal stress according to the following Hencky (1924) equation:

$$p + 2\phi k = \text{constant} \tag{5.46}$$

where ϕ is the angle between the α and x axes. At point C (Fig. 5.23a) $\phi = \pi/4$ and at point E, $\phi = -\pi/4$. From Eq. (5.46)

$$p_E = p_c + 2k\frac{\pi}{2} = k(1 + \pi) \tag{5.47}$$

The Mohr's circle diagram for point E and all points in region AEG is shown in Fig. 5.23c. The normal stress on the face of the punch (σ_2') in Fig. 5.23c will be

$$\sigma_2' = k + p_E = 2k\left(1 + \frac{\pi}{2}\right) = 2k(2.57) \tag{5.48}$$

It should be noted that the foregoing slip line field analysis

1. is for plane strain
2. ignores strain hardening (k constant) and strain history (Bauschinger effect) (see the end of this chapter for a definition)
3. involves a kinematically admissible flow pattern but ignores equilibrium except as approximately involved in the Hencky relationship
4. assumes the material beyond the zone of plastic flow to be rigid instead of elastic
5. assumes sticking friction on punch face and formation of stationary nose

The corresponding mean stress on the punch for plane strain uniaxial compression (Fig. 5.22) is $2k$. Thus the constraint factor C [Eq. (5.45)] for the *plane strain* hardness test is

$$C = \frac{(1 + \pi/2)2k}{2k} = 2.57 \tag{5.49}$$

According to the von Mises flow criterion, the relation between the plane strain flow stress ($2k$) and the axisymmetric (3D) flow stress (Y) is

$$2k = 2Y/\sqrt{3} \tag{5.50}$$

This may be shown by substituting $\sigma_1 = Y$ and $\sigma_2 = \sigma_3 = 0$ into the von Mises equation [Eq. (5.30)] for the axisymmetric test and $\sigma_1 = 2k$, $\sigma_2 = k$, and $\sigma_3 = 0$ for the plane strain test. Therefore the axisymmetric constraint factor assuming two-dimensional (plane strain) and three-dimensional (axisymmetric) constraints to be comparable will be

$$C = \frac{\sigma_2'}{Y} = \frac{(1 + \pi/2)2k}{(\sqrt{3}/2)2k} = 2.97 \tag{5.51}$$

Experimentally, the three-dimensional constraint factor is found to be about three, which is in excellent agreement with the above analytical value.

A solution that satisfies flow kinematics but not equilibrium will lead to results on the high side of truth and is termed an upper bound solution. One which ignores flow considerations but satisfies equilibrium leads to a lower bound solution. Thus, a SLF solution should yield an upper bound since it does not take equilibrium *fully* into account. According to the SLF approach, upward flow accounts for the material displaced by the punch and therefore constraint factor C may be termed a flow constraint from this point of view.

The SLF of Fig. 5.23a assumes a dead zone of material on the punch face and hence corresponds to maximum (sticking) friction. Hill (1950) has suggested a second SLF solution (Fig. 5.23d) that corresponds to zero punch face friction. When a slip line field analysis is performed on Fig. 5.23d, the same result as that for Fig. 5.23a is obtained. It is found experimentally that friction has a negligible influence on the constraint factor C associated with indentation by a punch. When applying the SLF technique, it is important to check that the flow pattern assumed is reasonably close to that actually obtained in practice. More details concerning the SLF approach to plasticity problems may be found in Johnson and Mellor (1962) and in Rowe (1977).

UPPER BOUND APPROACH

The SLF technique is a quasi upper bound approach since equilibrium is taken into account in part by the Hencky equation but not completely on a point-to-point basis. Solutions that ignore

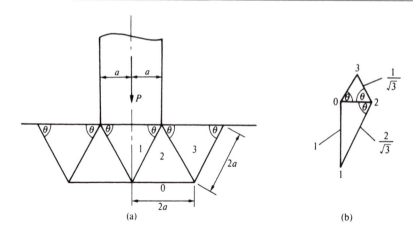

Fig. 5.24 Upper bound solution for plane strain punch. (a) Slip lines for $\theta = 60°$. (b) Corresponding hodograph for flow to right for unit punch velocity ($v = 1$).

equilibrium completely are simpler but more approximate than SLF solutions, and these are generally referred to as upper bound solutions. An example of an upper bound solution for the indentation hardness problem is given in Fig. 5.24. In this solution, deformation is assumed to occur by slip only along the solid lines shown in Fig. 5.24a. At all other points, the material is considered to be rigid. The shear flow stress at all points is assumed to be k (uniaxial plane strain flow stress in shear) as in the SLF approach. The external work per unit time is then equated to the internal work per unit time. The external work per unit time is the product of the punch force P and the punch velocity V, while the internal work per unit time is the sum of the products of shear force on each flow surface and the appropriate velocity. The appropriate velocity is obtained from a hodograph (velocity diagram) that satisfies velocity continuity. Figure 5.24b shows the synthesis of the partial hodograph for the right half of Fig. 5.24a. Here, distance 0–1 corresponds to the absolute velocity of the punch (taken to be unity), 1–2 to the relative velocity of region 2 of Fig. 5.24a relative to region 1, and 2–3 to the velocity of region 3 relative to region 2. Velocity vectors 0–2 and 0–3 correspond to the absolute velocities of blocks 2 and 3. Fig. 5.24b satisfies continuity of flow to the right since the rate of flow into the deformation zone equals the rate of flow from the deformation zone ($a \times 1 = 2a \times \frac{1}{2}$).

Equating the external and internal work per unit time per unit distance perpendicular to the paper for flow to the right,

$$\frac{P}{2}(1) = k(2a \times 1)\frac{2}{\sqrt{3}} + 3k(2a \times 1)\frac{1}{\sqrt{3}} = \frac{5k(2a)}{\sqrt{3}}$$

or

$$\frac{P}{(2a)(2k)} = \frac{p}{2k} = \frac{5}{\sqrt{3}} = 2.89$$

This value is larger than that for the SLF solution [Eq. (5.49)] as it should be since it represents a more approximate upper bound approach where force equilibrium is completely ignored. Other upper bound solutions may be obtained by making θ in Fig. 5.24a other than 60°. The lowest (and best) value of $p/2k = 2.83$ is obtained when $\theta = 54.7°$.

ELASTIC THEORY

When a large block of material having a grid applied to a central plane is loaded by a spherical indentor, flow patterns such as those shown in Fig. 5.25 are obtained. Study of these patterns reveals a plastic zone that passes through the edges of the punch (Fig. 5.25c). There is no evidence of upward flow, and little resemblance to the plastic zones of Figs. 5.23 and 5.24.

The deformed grids of Fig. 5.25 clearly indicate an elastic–plastic boundary, which has a shape resembling that of a line of constant maximum shear stress beneath a sphere pressed against a flat surface. This elastic problem was first studied by Hertz (1895). He found that the contact stress was distributed in a hemispherical pattern over the surface, and that lines of constant maximum shear stress were as shown in Fig. 5.26, where

$$M' = \frac{\tau_{max}}{\bar{p}} \tag{5.52}$$

and

$$\tau_{max} = \text{maximum shear stress}$$

$$\bar{p} = \text{mean pressure on punch face (Meyer hardness)}$$

In Fig. 5.26 the punch face is shown flat for simplicity, whereas in reality it is the surface of a large-radius sphere. The elastic–plastic boundary of Fig. 5.25c closely resembles a line between $M' = 0.15$ and 0.20.

An alternative approach to hardness has been presented by Shaw and DeSalvo (1970) in which the material is assumed to be plastic–elastic instead of plastic–rigid.

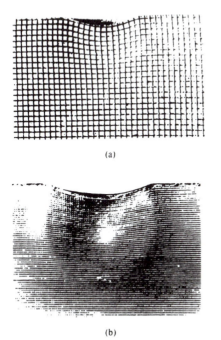

(a)

(b)

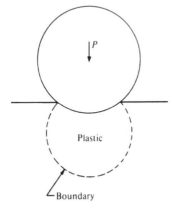

(c)

Fig. 5.25 Deformation of grid on meridional plane in a Brinell test. (a) Plasticene. (b) Mild steel. (c) Interpretation of (a) and (b).

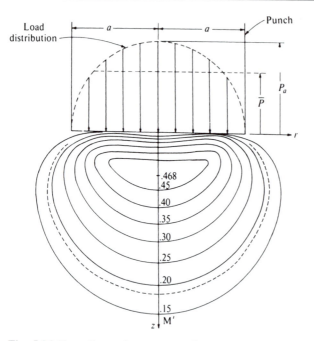

Fig. 5.26 Hertz lines of constant maximum shear stress for a frictionless spherical indentor. The dotted line of constant maximum shear stress is the elastic–plastic boundary ($M' = 0.17$).

When applied to the indentation hardness problem, the alternative theory suggests that, if there is sufficient material beneath an indentor, the displaced material may be completely accounted for by the decrease in volume of the material elastically loaded in compression. There is then no upward flow, as called for in all slip line field solutions. The constraint factor that arises in this way is termed an elastic constraint since the displaced volume is accommodated by an elastic decrease in volume (instead of by upward flow, as in the slip-line field approach).

Since there is no evidence of upward flow in Fig. 5.25, the constraint involved is of the elastic variety in both instances.

By maximum shear theory of plasticity, one of the lines of Fig. 5.26 should correspond to the elastic–plastic boundary. The shear stress on this particular line should be $Y/2$ (where Y is the uniaxial flow stress of the material). Analysis reveals that the curve for $M' = 0.177$ corresponds to the elastic–plastic boundary and

$$M' = \frac{\tau_{\max}}{\bar{p}} = \frac{Y}{2\bar{p}} = 0.177$$

Or the corresponding value of constraint factor C is

$$C = \frac{\bar{p}}{Y} = \frac{1}{2(0.177)} = 2.82$$

However, this constraint factor is referred to the area of the punch in actual contact during indentation (radius a in Fig. 5.26) instead of the area of the plastic impression that remains after the test. These two areas will differ because the edge of the indentor is elastically loaded. When an adjustment is made for the elastically loaded area, the constraint factor based on the plastic impression is found to be $2.82/0.94 = 3.0$, which is in excellent agreement with experiment.

The amount of material required to enable an elastic constraint to be fully developed in a Brinell test of mild steel corresponds to a hemisphere of radius $10(2a) = 20a$. The coefficient 10 is directly proportional to the ratio Young's modulus/Meyer hardness = $E/\bar{p}$. For the standard Brinell test of mild steel (3000 kg load on a 10 mm ball with $2a/D = 0.4$), the impression should be surrounded by a sphere of material of radius 40 mm (1.58 in). If less material surrounds the impression, there must be some upward flow. In fact, complete upward flow may be demonstrated by use of a thin layer of modelling clay (low E) on steel (very high E). The material beyond the plastic zone (steel) then has such a high Young's modulus relative to that of the clay that the plastic-rigid assumption holds, producing flow patterns almost identical in appearance to those in Figs. 5.23a and 5.23d as shown in the plane strain flow patterns of Fig. 5.27.

In order that there be no upward flow, an indentor must be blunt (large cone angle). Increased friction will postpone upward flow as will an increased tendency for strain hardening. The effective cone angle for most indentors, however, is such that some upward flow results even when there is

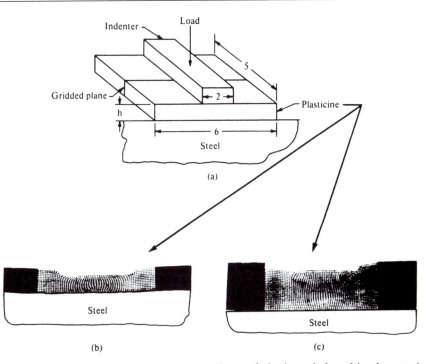

Fig. 5.27 Photographs of grid patterns on transverse planes of plasticene indented in plane strain by flat punch. (a) Test arrangement. Pattern (b) when $h = 1.414a$ as in Fig. 5.23a and (c) when $h = 0.707a$ as in Fig. 5.23d. In both cases the specimens were supported on a steel plate (rigid relative to plasticene).

sufficient material surrounding the indentor to provide a full elastic constraint. Thus, most hardness tests correspond to a constraint that is predominantly elastic, but with a small flow component.

CONE ANGLE EFFECT

The effective cone angle is a very important variable in hardness testing. Dugdale (1954, 1955) has extensively studied the constraint factor of indentors of different cone angle; his experimental results are summarized in Fig. 5.28. The constraint value was found to decrease markedly with decrease in cone semiangle (θ) when a fully worked specimen was indented with very low friction (line AB). However, the constraint increased with decrease in θ when an annealed material was indented with high friction (line BC) in Fig. 5.28, which clearly reveals the influence of friction and strain-hardening tendency, the horizontal dotted line shows the fully elastic constraint value from the elastic constraint point of view.

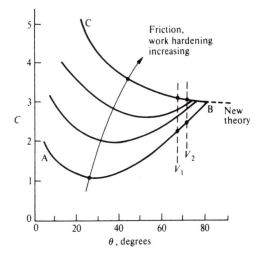

Fig. 5.28 Variation in constraint factor $C = \bar{p}/Y$ for cone semiangles (θ) for metals work-hardened to different degress and with different amounts of friction on the surface of the indentor. (after Dugdale, 1954, 1955)

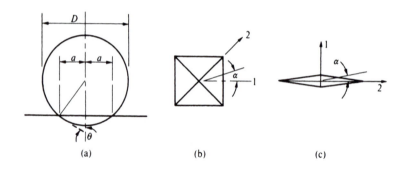

Fig. 5.29 Effective cone angles. (a) Brinell test, side elevation. (b) Vickers indentation, plan view. (c) Knoop indentation, plan view.

(a)   (b)   (c)

While most of the practical indentors are not conical, they may be assigned an effective cone half angle (θ_e). The effective cone half angle for the sphere is the angle that the tangent to the sphere makes with the vertical at the edge of the indentation (Fig. 5.29a). For the sphere,

$$\theta = \cos^{-1} \frac{2a}{D} \tag{5.53}$$

When $2a/D = 0.4$, $\theta_e = 66.4°$.

For the Vickers indentor, the cone half angle θ is $136/2 = 68°$ for flow in direction 1 in Fig. 5.29b. For flow in a direction inclined at an angle α to direction 1,

$$\theta = \tan^{-1} \left(\frac{\tan 68°}{\cos \alpha} \right) \tag{5.54}$$

For direction 2, this gives $\theta = 72.18°$. The mean value of θ may be obtained as follows:

$$\theta_e = \frac{4}{\pi} \int_0^{\pi/4} \tan^{-1} \left(\frac{\tan 68°}{\cos \alpha} \right) d\alpha = 70.3° \tag{5.55}$$

For the Knoop indentor (Fig. 5.27c) the cone angle (θ) is $130/2 = 65°$ in direction 1, and $172.5/2 = 86.25°$ in direction 2. The mean value $\bar{\theta}$, found as above, is $72.12°$.

These values are summarized as follows:

Indentor	$2\bar{\theta}$	$2\theta_{min}$	$2\theta_{max}$
Sphere ($2a/D = 0.4$)	132.8	—	—
Vickers	140.6	136	144.4
Knoop	144.2	130	172.5

A sphere loaded so that $2a/D = 0.4$ is seen to be less blunt than either the Vickers or Knoop indentors.

STRAIN IN HARDNESS TEST

Tabor (1951) has empirically estimated the strain at the outer edge of a Brinell impression to be

$$\varepsilon \cong 0.2 \frac{2a}{D} \tag{5.56}$$

From Eq. (5.53) this will be

$$\varepsilon \cong 0.2 \cos \theta \tag{5.57}$$

For a value of θ of 70°, ε corresponds to 0.068 (a relatively low value for plasticity).

PILING UP AND SINKING IN

When a Vickers indentor is used on fully work-hardened material, the impression will be barrel shaped, as at A in Fig. 5.30, but may have a pin-cushion shape, as at B in Fig. 5.30, when an annealed material is indented. The explanation for this phenomenon is found in Fig. 5.28. At the ridge of the Vickers indentor, the effective cone angle is 144.4°, but is only 136° midway between the ridges. In Fig. 5.28, these values are indicated as V_2 and V_1, respectively. With annealed material, it is evident that it will take less force to displace material along the ridge (V_2) than between the ridges (V_1). As a consequence, this effect will lead to a flow pattern that is more extensive at the corners than between the corners (the pin-cushion pattern). Similarly, if the material is fully cold-worked, line AB in Fig. 5.28 will pertain instead of BC. Then it will take more force to displace material along the ridge (V_2) than between ridges (V_1). This effect, of course, leads to the barrel-shaped pattern.

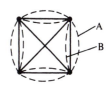

Fig. 5.30 Plan view of Vickers indentation showing barrel-shaped pattern (A) and pin-cushion pattern (B).

Due to the cone angle being about 140° for most practical indentors, there will be a small upward flow. This upward flow is accentuated if the elastic constraint is kept from developing due to insufficient material being present beneath the indentor. The upward flow will lie close to the surface of the indentor and will "pile up" if the material is fully work-hardened and friction is low. For an annealed material, the upward flow will extend farther from the indentor and will give the appearance of "sinking in" at the indentor if friction is high.

If there were no elastic action, residual stresses would not arise. The large residual compressive stresses produced by shot peening to extend fatigue life are unexplained by the SLF theory of indentation hardness, which assumes the material to act in a plastic-rigid manner. The elastic constraint enables the magnitude and extent of residual stresses to be estimated.

A spherical indentor develops the full constraint of three gradually with load and should not be used until $2a/D$ has reached a value of about 0.4 for a spherical indentor made of steel. Since Vickers and Knoop indentors show no such gradual approach to full plasticity, they can be used over a wider range of loads.

RELATIVE "BLUNTNESS" OF INDENTOR

The more blunt the indentor, the smaller will be the amount of upward flow. In the absence of upward flow, the hardness will be the same for an annealed and fully strain-hardened specimen, and friction will play no role. Under such conditions, the hardness measured corresponds to the initial yield point of the material.

An indentor that is blunt enough to prevent upward flow produces an impression that is difficult to see and measure accurately. The effective cone angle must be decreased to a value that approximates that for the Vickers and Knoop indentors for accurate measurement. Presence of the ridges in Vickers and Knoop indentations improves visibility and hence accuracy of measurement.

With indentors having effective cone angles as small as those in the Vickers and Knoop indentors, there will be a small amount of upward flow. The resulting constraint will still be essentially of elastic origin with a small flow component included. This flow component causes the hardness value to correspond to the flow stress at a small plastic strain (approximately 0.05). It also alters the constraint factor upward or downward, depending upon whether the metal is fully strain-hardened, or annealed, and whether the indentor friction is high or low.

FLOW ON UNLOADING INDENTOR

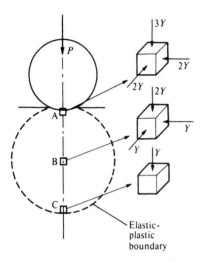

Fig. 5.31 States of stress at different points beneath a hardness indentor when load is still applied.

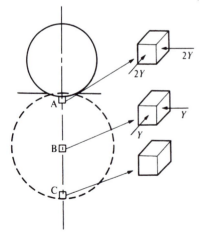

Fig. 5.32 States of stress at different points beneath a hardness indentor just after the load is removed but before the second plastic flow takes place.

Initial plastic flow occurs when load is applied to a hardness indentor and a second flow occurs on unloading. Removal of the load may be approximated by superimposing the Hertz solution to an upward force on the ball equal to the downward force in the hardness test. This will correspond to a mean tensile stress on the contact area = $3Y$ (point A in Fig. 5.31) where Y equals the uniaxial flow stress of the material. The approximate vertical tensile stresses at points B and C will be $2Y$ and Y, respectively. Thus after the load has been removed, the state of stress beneath the indentor will be approximately as shown in Fig. 5.32.

However, this is not consistent with the Tresca flow criterion, and hence plastic flow will occur until the difference between principal stresses at A, B, and C are each equal to Y. The material at A will flow upward until the lateral stresses drop from $2Y$ to Y. Flow will then cease and the residual biaxial stresses at A will be equal to Y. At B, plastic flow will take place until the difference between the vertical stress that develops and the lateral biaxial stress equals Y. The residual stress at B will be triaxial compression, and the transverse residual stresses at B will be greater than those at A. There will be no secondary plastic flow on unloading at point C. Hence, the second plastic flow volume associated with unloading will be smaller than the initial flow volume associated with loading.

The concept of a second plastic flow associated with unloading has several important practical implications.

1. In fundamental studies of indentation hardness, the zone of plastic flow after unloading is often observed and related to the applied load. In interpreting such results, it is important to realize that a second plastic flow will have taken place on unloading; hence care must be taken in inferring the shape of the free surface when the load is still applied from that observed after the load has been relaxed.

2. Ball bearings and gears usually involve a small amount of plastic flow near the surface. In interpreting surface fatigue characteristics of such items, it is important to realize that the plastic flow that occurs on unloading is far more likely to cause local brittle fracture and fatigue crack growth than the plastic flow that occurs on loading. This is a point of view that apparently has not been pursued previously in ball-bearing life analysis.

3. The work material at the tip of a cutting tool is plastically deformed, and the adverse flow that takes place on unloading is apt to result in surface cracks in the finished surface. Thus, plastic flow associated with unloading may play an important role relative to the integrity of machined surfaces, particularly when the metal machined has a relatively low fracture strain.

4. In forming operations, such as extrusion, drawing, rolling, forging, thread-rolling, peening, swaging, etc., the second plastic flow associated with unloading may give rise to surface cracks.

5. In shot peening, to increase fatigue life, the surface is bombarded by hard spheres which make a large number of small Brinell impressions in the surface and leave behind residual biaxial and triaxial compressive stresses. The second plastic flow that accompanies unloading will obviously be involved in any rational approach to shot peening that may be developed in the future. The simple qualitative approach presented above suggests the origin and character of these useful residual compressive stresses. At the same time, it becomes evident why shot peening may not always be useful and why, with somewhat brittle materials, shot peening may have a negative effect on fatigue life. This will be the case when the plastic flow that accompanies unloading gives rise to surface imperfections (cracks) that are more detrimental to fatigue life than the residual compressive stresses are beneficial.

6. In burnishing and ball sizing operations, care must be taken that the plastic flow associated with unloading does not result in surface cracks. However, a recent Russian development takes advantage of cracks thus developed. In this case, a carbide roller is loaded against the workpiece just ahead of the cutting tool. This roller provides sufficient pressure to produce cracks in the material just in front of the cutting tool, substantially reduces cutting forces, and increase tool life by as much as 50% when chilled cast-iron rolls are machined.

The best way of experimentally proving the existence of a second plastic flow on unloading is to trace the shape of the free surface with a sensitive stylus instrument first with the load applied and then after the load has been removed. However, this cannot be done with existing tracing instruments since it is not possible to get the tracing point close enough to the ball when the load is still applied.

It was therefore decided to make a replica of the surface under load and to compare the shape of the free surface of the replica with that of the metal surface after the load had been removed.

A $\frac{1}{4}$ inch (6.35 mm) diameter steel sphere was loaded into a flat steel surface as shown in Fig. 5.33 at loads varying from 1000 to 4000 kg. In each case, a casting was made with the load on the ball held constant throughout the period of solidification. The load was then removed and the surfaces of the steel and plastic were traced in four directions oriented 90° from each other. In all cases, there was a substantial difference between the loaded and unloaded surfaces.

Figure 5.34 is a representative pair of tracings showing the difference between the loaded and unloaded surface shapes.

It appears that a substantial reverse plastic flow occurs when a Brinell ball is unloaded.

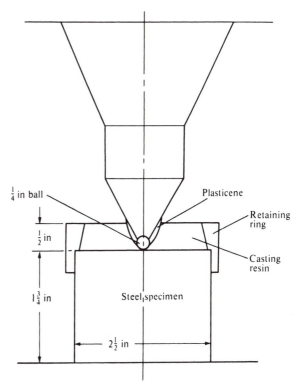

Fig. 5.33 Test arrangement used to obtain a replica of the shape of a loaded surface. (after Shaw, Hoshi and Henry, 1979)

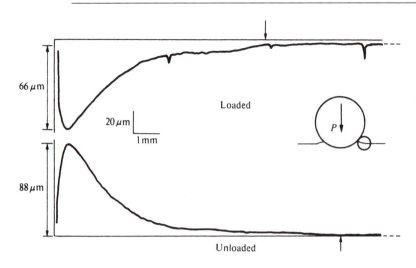

Fig. 5.34 Representative Tallysurf tracings of loaded (upper—plastic replica) and unloaded (lower—steel surfaces in region indicated in insert). (after Shaw, Hoshi, and Henry, 1979)

ROCKWELL HARDNESS

Rockwell hardness values (Rockwell, 1922) are also in common use. In this case the hardness value is an arbitrary number that varies inversely with the depth of penetration under standard loading conditions. Indentors of different geometries are used for materials of different ranges of hardness. The B-scale (B for ball) is used for relatively soft materials while the C-scale (C for cone) and A-scales are used for materials of increasing ranges of hardness. Rockwell hardness determinations leave a small dent and hence represent relatively nondestructive tests (along with Vickers and Knoop) compared with the standard Brinell test.

HARD BRITTLE MATERIALS

By use of small loads it is possible to obtain hardness readings of extremely hard and brittle materials. Indentation hardness values may be obtained for materials such as glass, which is normally considered to be completely brittle, if the size of the impression (load) is sufficiently small.

Brittle materials such as glass contain many surface cracks, and if the indentor engages one or more of these, the specimen will be broken out before a fully developed impression can be made. By using an impression that is smaller than the mean crack spacing, it is possible to produce a perfect indentation in a brittle material. Brittleness is a function of specimen size. Marble, which is so brittle that it will break into many pieces if a Brinell test is attempted, may be indented successfully if the load is sufficiently small. The three impressions shown in Fig. 5.35 were made at loads 15, 10, and 5 grams respectively. While impressions (a) and (b) show cracks, that at (c) is free of cracks. There is some evidence that the cracks in specimens such as (b) are not present while hydrostatically loaded, but appear when the indentor is unloaded.

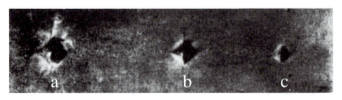

Fig. 5.35 Impressions made in marble with a diamond indentor under loads of (a) 15 gm, (b) 10 gm, (c) 5 gm. Magnification = 1200×. (after Shaw, 1954)

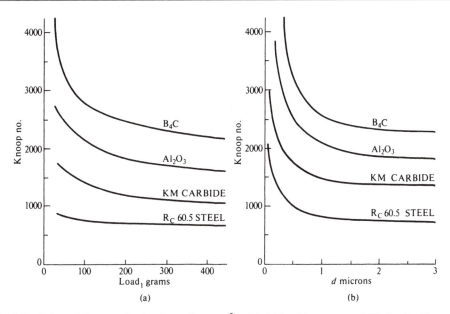

Fig. 5.36 Variation of Knoop microhardness (kg mm^{-2}) with (a) load in grams and (b) depth of impression (d) in microns produced by loads on indentor of varying magnitude. (after Thibault and Nyquist, 1947)

More ductile hard materials also contain many closely spaced imperfections and likewise exhibit a *size effect*. As long as the size of the impression is large compared with the mean imperfection spacing, the normal hardness vale will be obtained. However, as the load is decreased to the point where the impression size approaches imperfection spacing, hardness values begin to increase. Values of Knoop hardness versus load and depth are shown in Fig. 5.36 for several hard materials. The hardest material (B$_4$C) is seen to give values which begin to increase with decreased load at a larger value of load than the softer material (tool steel). This is because it takes a higher load to produce the same size impression in B$_4$C than it does in tool steel. When hardness is plotted against depth of impression in microns (Fig. 5.36b), all curves are found to be of similar shape and to turn upward for depths of impression less than about 2 microns.

Representative microhardness values for a variety of hard materials and metallographic constituents are given in Table 5.2.

CONVERSION CURVES

The curves of Fig. 5.37 enable relative magnitudes of hardness, expressed in several systems, to be compared. Brinell hardness has been used as the standard of comparison. The Knoop hardness is seen to be very nearly equivalent to Brinell hardness over the entire range while the other hardness scales are significantly nonlinear relative to the Brinell scale. The Mohs scale of hardness is seen to be very nonlinear, particularly at the upper end of the scale. The equivalent Brinell hardness values for Mohs hardness values of 8, 9, and 10 are as follows: 1150, 1900, and 7000. From this it is evident that the range from 9 to 10 on the Mohs scale covers more than four times the range of hardness corresponding to the range from 1 to 9.

TABLE 5.2 Vickers Hardness Values for a Variety of Materials

Material	kg mm^{-2}	Material	kg mm^{-2}
Structural constituents of steel:		Chromium-tungsten carbide	2000
Austenite	400	Molybdenum-tungsten carbide	2100
Cementite	1100	SiC	2400
Pure iron	70	MO_2C	2000
Ferrite	80	TaC	1800
Graphite	10	TiC	2400
Martensite	800	WC	1600
Tempered martensite	250–800	WC + 6% Co	1400
Pearlite (eutectoid)	250	WC + 13% Co	1300
Spheroidite	175	VC	2800
Sorbite	275	Al_2O_3	2100
Bainite	485	Fe_2O_3	1100
Troostite	550	Fe_3O_4	650
Fully hardened tool steel	350–700	Diamond	8000
Carbides and oxides:		Quartz	1100
B_4C	2800	Glass	400–600
Cr_3C	1200	Hard Chromium plate	1000
Chrom-vanadium carbide	2700	Nickel plate	340

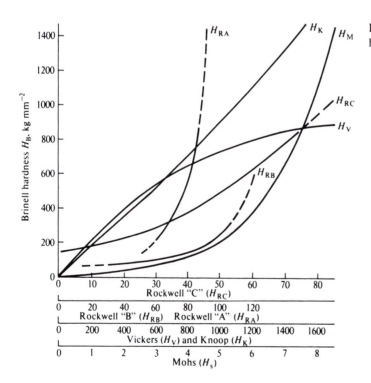

Fig. 5.37 Conversion curves for several hardness scales.

TORSION

The torsion test is an additional method of characterizing the plastic performance of materials. This test is particularly useful for studies involving large plastic strains since neither necking (with all its complications in the tensile test) nor barrelling (compression test) occurs. However, there is a large strain gradient from the center to the outside diameter of the specimen which makes it somewhat awkward to convert measured quantities (torque, θ, and angle of twist, M_T) into shear stress (τ) and shear strain (γ).

Three of the important applications of the torsion test are

1. to directly measure the shear modulus G in the elastic region

2. to evaluate the strain rate sensitivity of materials in the hot working region (homologous temperature > 0.5) at high rates of strain (Gleeble test)

3. to evaluate the fracture stress of quasi-brittle materials such as quenched and tempered tool steels. Since the mean principal stress (σ_H) in the torsion test is zero (Fig. 4.9e) instead of being tensile ($\sigma_H = \sigma_1/3$) as in the uniaxial tensile test, there is a greater spread of fracture values in torsion for specimens of different quality than in tension

The stress–strain curve in torsion has the same appearance as a true-stress–true-strain tensile curve (with yield point, strain hardening, and finally fracture).

In both the elastic and plastic regimes, the shear strain increases linearly with radius, being zero at the center of the specimen and a maximum (γ_m) at the outside diameter (o.d.) where

$$\gamma_m = \frac{r\theta}{L} \tag{5.58}$$

where

r = radius at o.d.

θ = angle of twist of one reference plane relative to the other

L = axial distance between reference planes

In the elastic regime the shear stress at the o.d. is

$$\tau = \frac{M_T r}{J} \tag{5.59}$$

where

M_T = twisting moment

r = radius at o.d.

J = polar moment of inertia

($J = 16\pi r^4/32$ for cylindrical specimen)

In the plastic regime it may be readily shown as Nadai (1950) has done that the shear stress at the o.d. (τ_m) may be obtained from the moment of twist (M_T) versus angle of twist (θ) curve as follows (Fig. 5.38):

$$\tau_m = \frac{1}{2\pi r^3}(AB + 3AD) \tag{5.60}$$

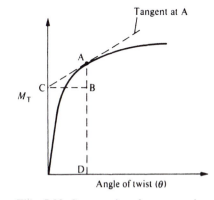

Fig. 5.38 Construction for converting torsional moment vs angle of twist curve to shear-stress–shear-strain curve. (after Nadai, 1950)

The effective stress σ_e and effective strain ε_e for a torsion test may be readily shown to be

$$\sigma_e = \sqrt{3}\,\tau_m \tag{5.61}$$

$$\varepsilon_e = \frac{\gamma_m}{\sqrt{3}} \tag{5.62}$$

When such values are compared with σ_e and ε_e values for a uniaxial tensile test on the same material ($\sigma_e = \sigma_1$ and $\varepsilon_e = \varepsilon_1$ in uniaxial tension), the same curve is obtained to a good approximation. While this is frequently cited as evidence for the general applicability of the concept of equivalent stress and strain, it should be kept in mind that there is no support for the equivalent strain concept in dislocation theory.

MOHR'S CIRCLE IN PLASTIC REGION

Strictly speaking, only Mohr's circle for stress may be applied in the plastic region. Mohr's strain circle only holds exactly for infinitesimal strain increments. However, as Backofen (1972) has shown, Mohr's strain circle also holds providing proportional straining obtains which is often the case. Mohr's circles for stress and strain are frequently employed in the plastic region (first approximation for strain). Equations (5.63) which are the plastic equivalent of elastic equations [Eqs. (4.8)] may also be employed as a first approximation where plastic normal stress is designated σ instead of S and plastic normal strain is designated ε instead of e:

$$\varepsilon_1 \sim \sigma_1 - v(\sigma_2 + \sigma_3)$$

$$\varepsilon_2 \sim \sigma_2 - v(\sigma_3 + \sigma_1) \tag{5.63}$$

$$\varepsilon_3 \sim \sigma_3 - v(\sigma_1 + \sigma_2)$$

As previously mentioned, it is found experimentally that the change in volume in the plastic region is essentially zero and therefore

$$\varepsilon_1 + \varepsilon_2 + \varepsilon_3 = 0 \tag{5.64}$$

When Eqs. (5.63) are substituted into Eq. (5.64), it is found that

$$v = \tfrac{1}{2} \tag{5.65}$$

Plane stress and plane strain are important states of stress in plasticity and Fig. 5.39 gives the Mohr's circle diagrams for these two cases. In the example of Fig. 5.39, σ_2 has been taken to be $\sigma_1/3$ and Eqs. (5.63) and (5.65) have been used to determine the unknown strain and stress, respectively.

As mentioned in the previous section, another relatively crude approximation in plasticity is to use an effective flow strain [Eq. (5.39)] as well as an effective flow stress [Eq. (5.38)]. When this is done, it is equivalent to assuming that strain hardening is a scalar type effect. In reality, strain hardening is strongly dependent on the direction of the strain. The Bauschinger (1881) effect in which the flow stress in compression following plastic flow in tension is found to be substantially reduced (and vice versa) is direct evidence that strain hardening not only depends upon the magnitude of the strain but also the direction of the strain. Considering all strain to be equivalent relative to strain hardening is equivalent to assuming the Bauschinger effect does not exist.

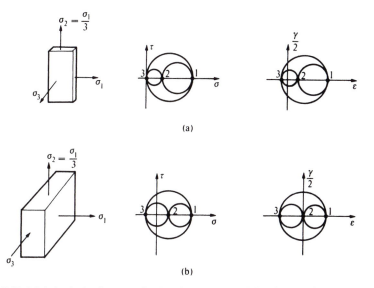

Fig. 5.39 Mohr's circle diagrams for (a) plane stress and (b) plane strain ($\sigma_2 = \sigma_1/3$).

REFERENCES

Backofen, W. A. (1972). *Deformation Processing*. Addison Wesley, Reading, Mass.

Bauschinger, J. (1881). *Zivilingur* **27**, 289.

Bierbaum, C. H. (1930). *Trans. Am. Soc. Steel Treat.* **18**, 1009.

Bragg, Sir Lawrence, Lomer, Wm., and Nye, J. F. (1954). *Experiments with the Bubble Model of a Metal Structure* (16-mm silent film). Brent Laboratories Ltd., London.

Bragg, Sir Lawrence, and Nye, J. F. (1947). *Proc. R. Soc.* **A190**, 474.

Bridgman, P. (1944). *Trans. Am. Soc. Metals* **32**, 553.

Bridgman, P. (1952). *Studies in Large Plastic Flow and Fracture*. McGraw-Hill, New York.

Brinell, J. (1901). *J. Iron Steel Inst.* **51**, 243.

Cottrell, A. H. (1953). *Dislocations and Plastic Flow in Crystals*. Oxford University Press, New York.

Dugdale, D. S. (1954). *J. Mech. Phys. Solids* **2**, 267.

Dugdale, D. S. (1955). *J. Mech. Phys. Solids* **3**, 197 and 206.

Hencky, H. (1924). *Z. für Angew. Math. u Mech.* **4**, 323.

Herring, C., and Galt, J. (1952). *Phys. Rev.* **85**, 1060.

Hertz, H. (1895). *Gesammelte Werke* (Leipzig) **1**, 156.

Hill, R. (1950). *The Mathematical Theory of Plasticity*. Clarendon Press, Oxford.

Johnson, W., and Mellor, P. B. (1962). *Plasticity for Mechanical Engineers*. D. Van Nostrand Co., Inc., Princeton, N.J.

Knoop, F., Peters, G., and Emerson, W. B. (1939). *J. Res. Natn. Bur. Stands.* **23**, 39.

MacGregor, C. W. (1944). *J. Franklin Inst.* **238**, 111.

Marshall, E. R., and Shaw, M. C. (1952). *Trans. Am. Soc. Metals* **44**, 7.

Maxwell, C. (1856). Letter to W. Thomson, 18 Dec. 1856.

Meyer, E. (1908). *Z. Ver. Dtsch. Ing.* **52**, 645.

Mohr, O. (1914). *Abhandlung aus dem Gebiet der Technischen Mechnik*, 2nd ed. W. Ernst & Sohn, Berlin.

Mohs, F. (1822). *Grundriss der Mineralogie*. Dresden.

Nabarro, F. R. N. (1967). *Theory of Crystal Dislocations*. Oxford University Press, New York.

Nadai, A. (1950). *Theory of Flow and Fracture of Solids*, Vol. I. McGraw-Hill, New York.

Nadai, A. (1963). *Theory of Flow and Fracture of Solids*, Vol. II. McGraw-Hill, New York.

O'Neill, H. (1934). *The Hardness of Metals and Its Measurement*. Chapman and Hall, London.

Orowan, E. (1934). *Z. Phys.* **89**, 605.

Polanyi, M. (1934). *Z. Phys.* **89**, 660.

Prandtl, L. (1920). *Nachr. Ges. Wiss. Gottingen, Math. Phys. K1.*, 74.

Rockwell, S. R. (1922). *Trans. Am. Soc. Steel Treat.* **2**, 1013.

Rowe, G. W. (1977). *Principles of Industrial Metal Working*, 2nd ed. E. Arnold, London.

Shaw, M. C. (1954). *Proc. Natn. Acad. Sci. U.S.A.* **46**, 394.

Shaw, M. C., and DeSalvo, G. J. (1970). *J. Engng. Ind.* **92**, 469 (Part I) and 480 (Part II).

Shaw, M. C., Hoshi, T., and Henry, D. (1979). *J. Engng. Ind.* **101**, 104.

Smith, R., and Sandland, G. (1922). *Proc. Inst. Mech. Engrs.* (London) **623**.

Smith, R., and Sandland, G. (1925). *J. Iron Steel Inst.* **111**, 285.

Tabor, D. (1951). *The Hardness of Metals*. Clarendon Press, Oxford.

Tabor, D. (1956). *Br. J. Appl. Phys.* **7**, 159.

Taylor, G. I. (1934). *Proc. R. Soc.* **A145**, 362.

Thibault, N. W., and Nyquist, H. S. (1947). *Trans. Am. Soc. Metals* **38**, 271.

Tresca, H. (1864). *C. r. hebd. Séanc. A cad. Sci. Paris* **59**, 754.

Tresca, H. (1867). *C. r. hebd. Séanc. A cad. Sci. Paris* **64**, 809.

von Mises, R. (1913). *Nachr. Ges. Wiss. Göttingen, Math. Phys. K1.*, 582.

von Weingraber, N. (1952). *Technische Hartemessung*. Carl Hauser Verlag, Munich.

Williams, S. R. (1942). *Hardness and Hardness Measurement*. Am. Soc. Met., Cleveland.

6 FRACTURE

There are in general two types of fracture—ductile fracture and brittle fracture. Ductile fracture occurs after appreciable plastic strain, and the criterion usually employed is the same as for yield (Tresca or von Mises). Brittle fracture occurs at a strain that is below the yield stress (perfectly brittle material) or relatively close to the yield point (quasi-brittle material), and the criterion usually employed in such a case is the maximum tensile stress criterion. This states that fracture occurs on a plane perpendicular to the direction of maximum tensile stress (σ_1) when the maximum tensile stress reaches a critical value ($\sigma_1 = $ constant). The fracture surface is usually in the direction of maximum shear stress in the case of ductile fracture. Brittle or quasi-brittle materials exhibit a greater "pressure coefficient of ductility"—K in Eq. (5.24)—than do ductile materials.

In some cases both types of fracture are involved. For example, in the tensile test of a ductile metal, initial fracture occurs at the center of the neck where the hydrostatic tensile stress is a maximum. As load is further increased, this initial tensile crack grows radially outward until the remaining area is insufficient to support the shear stress pertaining at which time final fracture occurs along a conical surface inclined approximately 45° to the axis of the specimen. This is called a cup-cone fracture due to the appearance of the fracture surfaces.

Figure 6.1 shows the fracture obtained in tension for (a) a perfectly brittle material such as hard high-carbon steel, (b) a relatively ductile material such as low-carbon steel that necks before fracture, and (c) a perfectly ductile material such as poly-crystalline lead.

GRIFFITH'S THEORY

The first rational approach to fracture was due to Griffith (1924) for perfectly brittle materials. Griffith reasoned that all real materials contain microcracks or equivalent regions of stress concentration that will grow with tensile strain until the elastic strain energy stored at the tip of the crack becomes sufficient to satisfy the surface energy associated with the generation of the new surface area accompanying crack growth, when the crack will spread spontaneously. The microcrack visualized by Griffith was elliptical in shape as shown in Fig. 6.2a for a surface crack or in Fig. 6.2b for

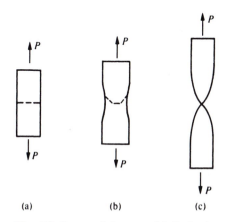

Fig. 6.1 Types of fracture. (a) Perfectly brittle material. (b) Partially ductile material showing cup-cone fracture pattern. (c) Perfectly ductile material.

79

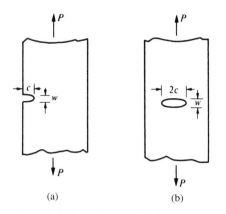

(a)   (b)

Fig. 6.2 Elliptical crack in plane stress tensile specimen. (a) Crack at surface. (b) Internal crack.

an internal crack in a thin plate (plane stress). The stress concentration factor at the tip of such a crack is $(1 + 2c/w) \cong 2c/w$. The elastic energy per unit width stored at the tip of the crack of Fig. 6.2b will be $\pi c^2 \sigma^2 / E$ while the surface energy associated with the two areas of the crack will be $2(2c \times 1)(T)$ where T is the surface energy of the metal. The change in energy ΔU due to growth of the crack will be

$$\Delta U = \left(4cT - \frac{\pi c^2 \sigma^2}{E} \right) \qquad (6.1)$$

where energy absorbed is considered to be (+) and energy released during crack growth is (−). For the crack to grow according to Griffith's thesis

$$\frac{d\Delta U}{dc} = 0 = \frac{d\left(4cT - \dfrac{\pi c^2 \sigma^2}{E} \right)}{dc} = 4T - \frac{2\pi c \sigma^2}{E} \qquad (6.2)$$

or solving for σ at fracture

$$\sigma = \sqrt{\left(\frac{2TE}{\pi c} \right)} \text{ (for plane stress)} \qquad (6.3)$$

This equation is in excellent agreement for perfectly brittle materials such as glass, but only in qualitative agreement with experiment for quasi-brittle materials, for example, $\sigma \sim c^{-1/2}$ for a quasi-brittle material.

The equivalent plane strain relation is found to be

$$\sigma = \sqrt{\left[\frac{2TE}{(1 - v)^2 \pi c} \right]} \qquad (6.4)$$

For glass, the value of $2c$ is estimated to be about 1 μm and T is about 1000 erg cm^{-2} in dry air. However, the fracture strength of glass is significantly reduced by coating a surface crack with water or kerosene, as every glass worker knows. This is because the fluid lowers T which in turn lowers σ [Eq. (6.3)]. The sensitivity of a brittle material to surface environment is known as the Joffe (1928) effect.

Orowan (1950) has pointed out that the material at the tip of a crack in a quasi-brittle material will be subjected to plastic deformation and the energy thus associated with crack growth (T_p) will be very much greater than the surface energy (T). Therefore, according to Orowan, Eq. (6.3) should be written as follows for a quasi-brittle material ($T_p \gg T$):

$$\sigma = \sqrt{\left(\frac{2T_p E}{\pi c} \right)} \text{ (plane stress)} \qquad (6.5)$$

The units of T_p are the same as those for T (erg cm^{-2}).

FRACTURE MECHANICS

In the mid-1950s, Irwin and coworkers laid the foundations for what has since become known as linèar elastic fracture mechanics (LEFM) (Irwin, 1957). The most important contribution of this development has been to introduce an experimentally determined material constant called the stress intensity factor (K) that characterizes the significance of the defects present in a material from the point of view of brittle crack growth. The stress intensity factor is defined as follows:

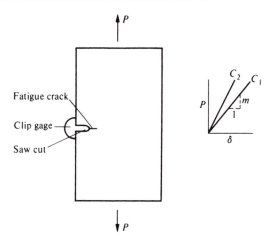

$$K = \sigma \sqrt{(\pi c)}, \text{(p.s.i.)(in)}^{1/2} \quad \text{or} \quad (N/m^2)m^{1/2} \quad (6.6)$$

where σ is the fracture stress and c is the half length of an internal flaw. From Eqs. (6.5) and (6.6), it is evident that

$$K = \sqrt{(2T_pE)} \quad (6.7)$$

Fig. 6.3 Fracture mechanics test for determining mode I stress intensity factor (K_{IC}).

Thus, in effect, K is simply a measure of T_p in the Griffith-Orowan approach to brittle fracture. However, fracture mechanics has contributed a standardized procedure for measuring K for a given material whereas previously T_p was only a qualitative concept. Determination of K involves finding the slopes (m) of the load (P)–deflection (δ) curves for several values of crack half length (c) in the experiment shown in Fig. 6.3. The crack is started by making a saw cut in the thin specimen which is then sharpened by growth in fatigue. The strain clip gage is used to measure the deflection (δ) that results for different values of P.

As may be seen from Eq. (6.6), the variation in fracture stress with crack length may be determined as soon as the stress intensity factor (K) has been determined for a given material.

The stress intensity factor described above involves a crack opening mode of loading (mode I) and is therefore frequently designated K_{IC} (C for critical). This is to distinguish this most widely used value of K from those obtained by other modes of loading (II = shear perpendicular to crack front, and III = shear parallel to crack front as shown in Fig. 6.4).

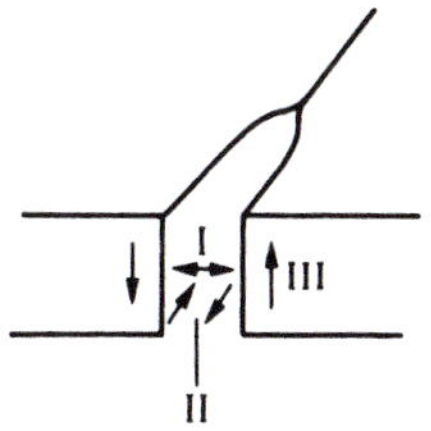

Fig. 6.4 Modes of loading for different critical stress intensity factors K_{IC}, K_{IIC}, K_{IIIC}.

CRACK INITIATION IN COMPRESSION

If brittle materials fail according to a maximum tensile stress criterion, it is a fair question to ask how it is that material such as rock, concrete, plaster, and glass can fail brittly when subjected to uniaxial compression. Griffith (1924) provided the answer to this many years ago. Griffith reasoned that all materials, particularly very brittle materials, contain micro-cracks and that tensile stresses can develop at the tips of such cracks even when the nominal stress is purely compressive. Griffith's analysis was two dimensional. He assumed very thin two-dimensional cracks, plane stress, and principal stresses P and Q oriented relative to the crack as shown in Fig. 6.5. (P is algebraically

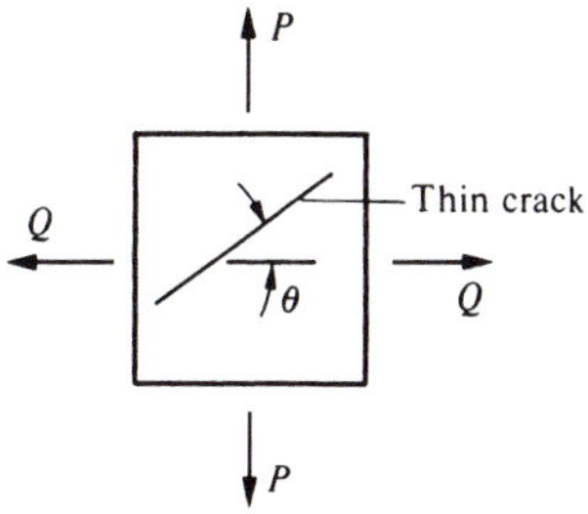

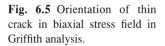

Fig. 6.5 Orientation of thin crack in biaxial stress field in Griffith analysis.

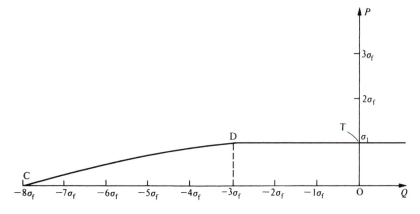

Fig. 6.6 Fracture locus according to Griffith (1924) in tension–compression regime. [Fracture stress in uniaxial tension (T) = σ_f. Fracture stress in uniaxial compression (C) = $8\sigma_f$.]

greater than Q.) The crack in Fig. 6.5 extends all the way through the specimen. Using the elastic solutions for the stresses at a crack tip developed earlier by Inglis (1913), Griffith computed the maximum tensile stress at the crack tip and found that

$$\sigma_{max} = P \text{ if } (3P + Q) \text{ is plus and } \theta = 0$$

$$\sigma_{max} = \frac{(P - Q)^2}{8(P + Q)} \text{ if } (3P + Q) \text{ is minus and } \cos 2\theta = \frac{P - Q}{2(P + Q)}$$

(6.8)

The fracture locus according to this elastic two-dimensional analysis is shown in Fig. 6.6. It is seen that the fracture stress in uniaxial compression (Point C) is 8 times the fracture stress in uniaxial tension (Point T).

Paul and Mirandy (1975) have extended the Griffith approach to three dimensions assuming very thin three-dimensional microcracks having an elliptical plan form. Like Griffith, they assumed a linearly elastic material and a maximum tensile stress fracture criterion. The surfaces of the thin flaws were assumed not to touch and the microcracks were assumed to be far enough apart so that there was no interaction. Figure 6.7 shows fracture loci for very thin defects of different plan shape (b/a), where a and b are the semi major and minor diameters of the assumed elliptical defect and v is Poisson's ratio. In these plots S_t is the uniaxial elastic tensile stress at fracture. It is evident that the ratio of uniaxial compressive strength to uniaxial tensile strength is between about 4.5 and 8 depending upon the values of b/a and v. Paul and Mirandy stress the fact that the fracture loci presented are for initial fracture and that initial cracks which form may not be self propagating.[†] This analysis, like Griffith's, is based on a maximum tensile fracture stress criterion and hence the results are independent of the intermediate stress and should hold for either plane stress or plane strain.

Usui, Ihara, and Shirakashi (1979) have applied the analysis of Paul and Mirandy to the fracture of cutting tool materials, incorporating a Weibull probability distribution into the analysis. In order to evaluate the Weibull constants, four-point bending and uniaxial compression tests were performed on K10 (5.5 $^w/_o$ Co + 94.5 $^w/_o$ WC) and P20 (8 $^w/_o$ Co and 92 $^w/_o$ mixed carbides)

[†] For brittle materials there is a relatively small difference between crack initiation and propagation, and the size of the crack plays a minor role. For ductile materials, however, crack initiation does not ensure crack propagation since sufficient energy must be stored elastically in the vicinity of the crack tip to satisfy the substantial plastic energy involved in crack growth. The size of the crack plays a major role in crack propagation in ductile materials since crack size is directly related to the amount of stored energy available for crack propagation.

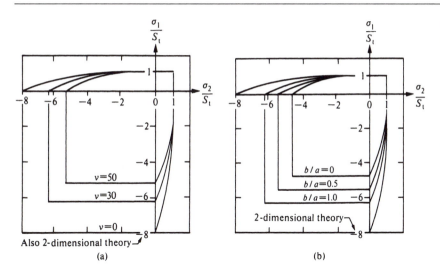

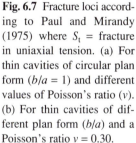

Fig. 6.7 Fracture loci according to Paul and Mirandy (1975) where S_t = fracture in uniaxial tension. (a) For thin cavities of circular plan form ($b/a = 1$) and different values of Poisson's ratio (v). (b) For thin cavities of different plan form (b/a) and a Poisson's ratio $v = 0.30$.

sintered tungsten carbide tool materials. The ratio of median values of uniaxial compressive strength to uniaxial tensile strength $(\sigma_C/\sigma_T)_f$ obtained were approximately as follows:

Material	$(\sigma_C/\sigma_T)_f$
K10	1.95
P20	2.64

Fisher (1953) has suggested a somewhat different approach to brittle fracture. He assumed that brittle materials yield just before fracture; therefore, the initial flow stress is a good approximation for the fracture stress of a brittle material such as cast iron. Like Griffith, Fisher assumed a random distribution of defects. However, his defect was not a void but a graphite platelet which he assumed would intensify tensile stresses but that compressive stress would be transmitted across the defect without intensification. The flow criterion used was that due to von Mises (distortion energy criterion). In applying his theory, the intensification factor for tensile stress (K) due to the defect and the flow stress of the material were quantities to be adjusted to give the best fit between theory and experiment. An example from Fisher's paper (Fig. 6.8) shows experimental fracture data due to Grassi and Cornet (1948) who tested gray cast-iron tubes to fracture that were subjected to internal pressure and end loading. The curve shown is based on Fisher's theory with $K = 3.2$ and the yield stress = 90,000 p.s.i. (622 MPa), which were the values found to give the best fit. The agreement between theory and experiment is excellent but in saying this, it must be kept in mind that the theory contains two floating constants.

When the experimental values of σ_C/σ_T of Usui et al. (1979) for tungsten carbide and Grassi and Cornet (1948) for cast iron are compared with those from the Griffith/Paul and Mirandy analysis (4.5 to 8), a very large discrepancy is observed. This suggests that some of the assumptions in the analysis do not hold for tungsten carbide or cast iron. In examining these assumptions, it is found that the value of $(\sigma_C/\sigma_T)_f$ depends on the shape of the defect assumed. By assuming a thin, hair-line defect, Griffith obtained a predicted value of $(\sigma_C/\sigma_T)_f$ of 8 for the two-dimensional case, and Paul and Mirandy a range of values from 4.5 to 8 for three-dimensional defects. As will be shown below,

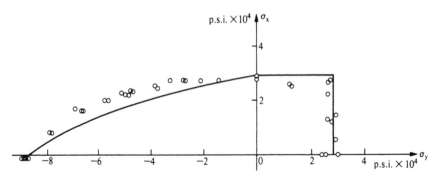

Fig. 6.8 Fracture locus according to Fisher (1953) with data points of Grassi and Cornet (1948). The fracture stress in compression was assumed to be 90,000 p.s.i. (622 MN m^{-2}) and the stress concentration factor K in the Fisher analysis = 3.2 in fitting the theoretical curve to the experimental data. (1 p.s.i. = 6905 N m^{-2})

a circular defect leads to a predicted value of $(\sigma_C/\sigma_T)_f$ of 3 which is in very much better agreement with experimental values for quasi-brittle materials such as cemented tungsten carbide and cast iron than the values of Griffith or Paul and Mirandy.

THEORY ASSUMING SPHERICAL VOIDS

The theory of brittle fracture for materials subjected to biaxial stresses presented here is based on the following assumptions:

1. Following Griffith and Paul and Mirandy (but not Fisher), fracture is assumed to occur when the maximum tensile stress reaches a critical value (maximum tensile stress criterion for fracture).

2. Like Griffith and Paul and Mirandy, defects are assumed to play an important role in intensifying local stresses. However, the defects are assumed to be small, circular voids instead of the thin, hair-line cracks assumed by Griffith and Paul and Mirandy (Fig. 6.9).

3. As in the case of Griffith, a two-dimensional defect is adopted in the interest of simplicity.

4. Voids are considered to be far enough apart so they may be considered in isolation.

Figure 6.10a shows the tangential stress at point A on a circular void when the nominal applied stress is σ_x in tension. The intensified stress at A will be $3\sigma_x$. Figure 6.10b shows the tangential stress at A when the nominal applied stress is σ_y (compressive). The stress at A will then be σ_y (tension). Thus, in uniaxial tension (Fig. 6.10a), the fracture stress at A should be reached when the nominal stress is one third of that for uniaxial compression (Fig. 6.10b).

Figure 6.11 shows the fracture locus to be expected from the maximum tensile stress criterion when nominal stresses are intensified by a two-dimensional defect (void). Point T is for uniaxial tension and point C for uniaxial compression. The fracture stress for uniaxial compression is three times that for uniaxial tension.

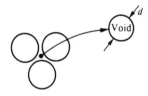

Fig. 6.9 Diagramatic representation of small spherical void located between three solid particles.

The complete fracture locus is shown by the solid line in Fig. 6.12. In the third quadrant, CB' is drawn parallel to σ_x because the third stress σ_z is zero (plane stress) and the stress combination σ_y, σ_z will lead to fracture at a lower value of applied load than the combination σ_x, σ_y. In other words, in the

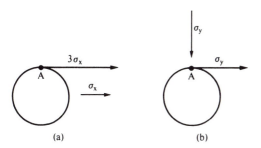

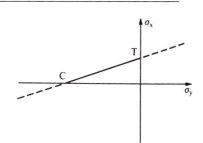

Fig. 6.10 Tangential tensile stress at surface of circular void. (a) Due to nominal tensile stress σ_x. (b) Due to nominal compressive stress σ_y.

Fig. 6.11 Fracture locus in second quadrant for specimen subjected to biaxial plane stress.

second quadrant, the critical defect has its axis parallel to the z-axis but in the third quadrant, the critical defect has its axis parallel to the y-axis.

In the first quadrant line TB is similarly drawn parallel to σ_y instead of being directed along CT extended. This is because the third stress σ_z is again zero (plane stress) and the stress combination σ_x, σ_z will lead to fracture at a lower value of applied load than the combination σ_x, σ_y.

It is well established experimentally that for a biaxial tensile loading in plane stress, brittle fracture will occur when the major nominal tensile stress reaches a critical value regardless of the magnitude of the minor tensile stress. The explanation for this lies in the fact that for plane stress, the third stress $\sigma_3 = 0$ and hence values in the first quadrant are equivalent to uniaxial tension (point T in Fig. 6.12). This ignores the secondary effect sometimes termed triaxiality which reflects the fact that the critical maximum tensile stress (a function of the mean principal stress $\sigma_H = (\sigma_1 + \sigma_2 + \sigma_3)/3$) is somewhat less when σ_H is positive (tensile) than when σ_H is negative (compressive).

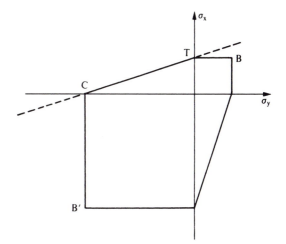

Fig. 6.12 Complete fracture locus for quasi-brittle material such as tungsten carbide. [T = uniaxial tensile strength. C = uniaxial compressive strength.]

In the case of plane strain loading in the third quadrant with Poisson's ratio $= \frac{1}{2}$ (slight plastic flow before fracture), the fracture locus would be along the dotted line (line CT extended) and not along line CB′. In the case of a true hydrostatic loading in compression such as the submergence of a material to a great depth in the ocean point B′ would be at infinity. Glass which contains numerous defects will not fracture when subjected to purely hydrostatic compression no matter how high the stress becomes. This is a result in excellent agreement with experiment and the success of glass for deep submergence vessels.

The fracture criterion of Fig. 6.12 is in excellent agreement with experimental data for tungsten carbide. In fact, one large carbide manufacturer suggests the locus of Fig. 6.12 for use in structural design applications of carbide. However, the reason cited for doing this is the so-called internal friction flow theory which has been proposed in the past on many occasions but always rejected on theoretical grounds. It appears that Fig. 6.12 is a fracture locus in excellent agreement with experiment but previously explained by an incorrect theory.

While the simple theory for plane stress fracture presented here is in excellent general agreement with a wide range of experiments, there is one detail that is not predicted exactly. This is the ratio of nominal fracture stress in uniaxial compression to that in uniaxial tension $(\sigma_C/\sigma_T)_f$. The theory presented predicts a ratio of 3 whereas in practice the observed value is somewhat greater or less than 3 depending on the material tested. The main reason for this deviation probably lies in the assumed shape of the void. While a circular void appears to be a reasonable one for the very small ($\gtrsim 1$ μm) sintering defects for a tungsten carbide tool material, other materials may have defects of different characteristic shape.

Other reasons for the small variation from 3 are

1. The circular void assumed has been considered to be two dimensional whereas in reality, it will be three dimensional with implications similar to those observed when comparing the Griffith (two-dimensional) and Paul and Mirandy (three-dimensional) solutions. This should tend to decrease the predicted value of $(\sigma_C/\sigma_T)_f$ to a value slightly below 3.

2. Strain hardening and deviation from the elastic stress distribution assumed in the analysis due to the small amount of plastic flow that usually precedes fracture in quadrants 2 and 3.

3. The possibility of a high pressure or an equivalent residual stress existing in or around sintering voids. Due to surface tension, a sintering void at equilibrium at high temperature must have a high internal pressure. As the body cools, this pressure will remain or be converted to an equivalent residual stress pattern in the vicinity of the void. The influence of a high void pressure or its residual stress equivalent has been ignored in the analysis presented. While this is believed to be a secondary-order effect, it could account in part for the observed deviation of $(\sigma_C/\sigma_T)_f$ from 3.

4. Plane stress has been assumed but in reality there may be a small third stress in situations that approximate plane stress.

5. Triaxiality which acknowledges the fact that tensile stress at fracture is less when the mean principal stress ($\sigma_H = (\sigma_1 + \sigma_2 + \sigma_3)/3$) is positive than when it is negative has been ignored.

The last two of these are relatively unlikely. While considerable emphasis has been attributed to triaxiality in the past, if this were a strong influence, TB in Fig. 6.12 would be inclined downward which is not in agreement with experiment. This suggests that a large degree of what has been attributed to triaxiality is in reality due to a difference in stress intensification with different stress states in the presence of defects.

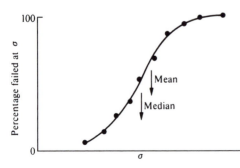

Fig. 6.13 Variation of percentage of specimens failed versus stress level σ.

WEIBULL STATISTICS

In any operation that involves fracture there is considerable unavoidable test-to-test variation and a number of samples must be tested in order to gain a picture of the performance of a given material.

While one could average such results and compare averages, this would not be making full use of the data collected. A better method is to plot the percent of a batch failed at a stress σ against this stress (σ) (Fig. 6.13). The range of breaking stresses and the distribution of values is then evident at a glance as well as the mean and median (equal number of test values above and below) values.

Probability paper has coordinate scales designed to make data such as that of Fig. 6.13 plot as a straight line.

Weibull paper is a special kind of probability paper that is very convenient for analyzing a wide variety of industrial data, particularly that involving fracture and fatigue. This is based on some special statistics developed by Weibull (1939, 1951) to analyze fatigue data and the life of rolling contact bearings.

Weibull found empirically that the percentage (S) of a group of bearings that would survive N cycles at a given load and set of operating conditions was given by

$$S = e^{-(N/N_0)^m} \tag{6.9}$$

where N_0 is a constant for the test series called the characteristic life and m is a constant for the test series called the Weibull slope.

It has subsequently been found more convenient to work with the fraction failed (F) instead of the fraction surviving (S). Thus,

$$F = 1 - S = 1 - e^{-(N/N_0)^m} \tag{6.10}$$

Rearranging Eq. (6.10)

$$e^{(N/N_0)^m} = \frac{1}{1 - F} \tag{6.11}$$

and, therefore,

$$\log \ln \frac{1}{1 - F} = m \log N - m \log N_0 \tag{6.12}$$

Weibull paper is ruled with $\log \ln (1/1 - F)$ as ordinate and $\log N$ as abscissa (Fig. 6.14). When experimental data (F, N) is plotted on this paper, a straight line is usually found having a slope m and a characteristic value N_0. From Eq. (6.12) it is evident that when $N = N_0$, $F = 63.2\%$.

In using Weibull paper the data are first ordered (arranged in increasing order of main dependent variable = fracture load or number of cycles to failure, depending on the problem) and the values of F determined for each rank order. For example, if five specimens are broken in a fracture test and the results for breaking load P given in Table 6.1 are obtained, then after ordering the data we have: For F_1 we want the fraction failing below the first value (56). For this we could use $1/5 = 20\%$ since we might assume that 20% of a large batch would fail below 56 lb based on our small group of 5 tests. Similarly for F_2 we could use $2/5 = 40\%$. However, in doing this we would be assuming a constant frequency distribution. In reality a group of test results are generally normally distributed

TABLE 6.1 Data Ordered According to Rank

Rank Order	P lb	Rank (F)
1	56	F_1
2	70	F_2
3	75	.
4	90	.
5	105	.

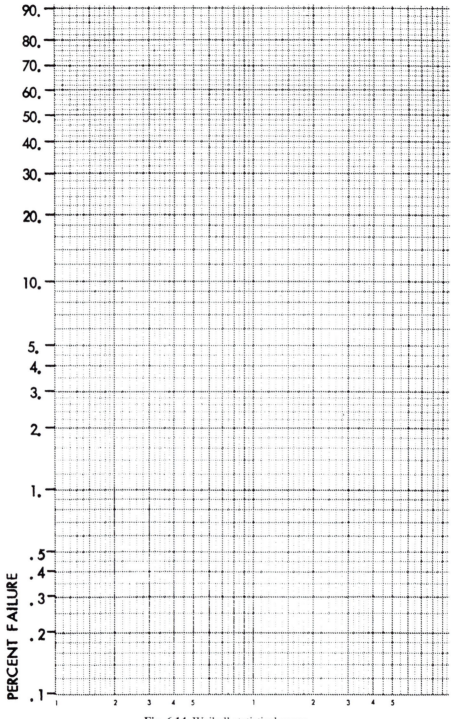

Fig. 6.14 Weibull statistical paper.

and values of F based on this type of distribution are termed median ranks. Tables of median ranks for different sample sizes (number of tests) are available in the literature.

It may be shown that the jth rank (F_j) for a case where there were n observations will be

$$F_j = \frac{j - 0.3}{n + 0.4} \times 100$$

For example, with a 5-test sample ($n = 5$), then the first median rank ($j = 1$) would be

$$F_1 = \frac{1 - 0.3}{5 + 0.4} \times 100 = 12.96$$

and the second median rank would be

$$F_2 = \frac{2 - 0.3}{5 + 0.4} \times 100 = 31.48$$

This is the basis for the values in Table 6.2.

The Weibull procedure is merely an effective way of estimating the distribution of results that would be obtained in a very large test series from a relatively small sample.

F and P may now be plotted on Weibull paper (Fig. 6.15).

The P_{10} point may be of special interest. This is the load P at which only 10% of a large test series will have broken. For this example $P_{10} = 53$ lb.

When the Weibull plot has a low slope (approaching $45°$ or $m = 1$), this indicates a very large variability in the test results. For ductile metals the value of m for breaking strength will be 40 or over which means that there is little scatter in the data and design may be based on the average strength and a factor of safety. For a brittle material the value of m will usually be 20 or less corresponding to substantial scatter. In such cases it is often inadvisable to base a design on the mean strength, since the early failures are usually far more significant. The Weibull plot offers a convenient way of predicting the early failure stress from a relatively small number of tests. The lower the slope of the Weibull curve the more uncertain it will be working with a small sample. The value of m (slope) for Fig. 6.15 is 4.53.

The reason that brittle materials exhibit such a large range of fracture values is that they contain flaws covering a large range of intensity (holes, grain boundaries, impurities, etc.). Brittle materials are incapable of flowing plastically to relieve the stress at a point of stress concentration and hence fracture over a wide range of values as a result of the variable probability of finding a large or small stress concentration

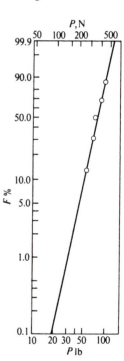

Fig. 6.15 Weibull plot for example. (after Shaw, Braiden, and DeSalvo, 1975)

TABLE 6.2 Rank Values

Rank No.	P lb	Rank (F)
1	56	12.96
2	70	31.48
3	75	50.00
4	90	68.51
5	105	87.03

in a given specimen. Brittle materials also break at higher values of stress when small (size effect) or when loaded to have a stress gradient across the specimen. This is because the probability of finding a large imperfection decreases as the volume of specimen under high stress decreases.

Ductile materials (metals) flow plastically to reduce the stress at points of stress concentration and hence do not exhibit nearly as much scatter in breaking stress as brittle materials. They also do not tend to give different results for different specimen sizes or for loading conditions involving a stress gradient.

Weibull statistics may also be used to estimate the effects of specimen size and stress gradient on fracture stress. In this treatment it is assumed that a specimen fails when a critical tensile stress is reached. It is also assumed that the flaws causing variation in the fracture stress are uniformly distributed.

If F_0 is the probability of a failure occurring at a given stress level per unit volume, then the probability of survival per unit volume will be

$$S_0 = 1 - F_0 \tag{6.13}$$

For V units of volume the probability of survival will be

$$S = (1 - F_0)^V \tag{6.14}$$

or

$$\log S = V \log (1 - F_0) \tag{6.15}$$

Weibull defines the risk of rupture R as

$$R = -\log S \tag{6.16}$$

and, therefore,

$$dR = -\log (1 - F_0)dV \tag{6.17}$$

since $-\log (1 - F_0)$ is assumed to be a function of stress alone, Weibull experimentally found that

$$-\log (1 - F_0) = \left(\frac{\sigma}{\sigma_0}\right)^m \tag{6.18}$$

where σ_0 is the characteristic stress and m the Weibull slope. Thus, from Eqs. (6.17) and (6.18)

$$dR = \left(\frac{\sigma}{\sigma_0}\right)^m dV \tag{6.19}$$

$$R = \int_V \left(\frac{\sigma}{\sigma_0}\right)^m dV \tag{6.20}$$

From Eq. (6.16)

$$S = e^{-\int(\sigma/\sigma_0)^m dV} \tag{6.21}$$

It may be shown that in general R [Eq. (6.20)] will integrate as follows:

$$R = KV \left(\frac{\sigma}{\sigma_0}\right)^m \tag{6.22}$$

where K is a factor depending on the stress gradient involved and V is the volume of the specimen subjected to tensile stress. From Eqs. (6.21) and (6.22)

$$F = 1 - S = 1 - e^{-KV(\sigma/\sigma_0)^m} \qquad (6.23)$$

which may be rewritten as

$$\log \ln (1/1 - F) = m \log \sigma - m \log \sigma_0 + \log KV \qquad (6.24)$$

This is seen to be equivalent to Eq. (6.12).

In order to see how Eq. (6.20) may be used to show the influence of specimen size consider two tensile specimens of volume V_1 and V_2 with rupture stress values of σ_1 and σ_2, respectively. Then, since σ is constant for a tensile specimen, Eq. (6.24) gives

$$R_1 = \left(\frac{\sigma_1}{\sigma_0}\right)^m V_1 = R_2 = \left(\frac{\sigma_2}{\sigma_0}\right)^m V_2 \qquad (6.25)$$

or

$$\frac{\sigma_1}{\sigma_2} = \left(\frac{V_2}{V_1}\right)^{1/m} \qquad (6.26)$$

For example, if $V_2/V_1 = 10$ and $m = 10$, $\sigma_1/\sigma_2 = 1.26$.

It should be noted that the size effect depends on the Weibull slope (m).

Also, the value of K in Eq. (6.21) is seen to be unity for the tensile test.

Consider next a transverse rupture test in three point loading (Fig. 6.16).

In this case Eq. (6.20) is evaluated only in the tensile region shaded in Fig. 6.16. This is because brittle materials are much stronger in compression than tension. In fact, a brittle material will be about three times as strong in compression as in tension. The tensile stress at any point x is

$$\sigma = \frac{MC}{I} = \frac{(Px/2)y}{bh^3/12} = \frac{6Pxy}{bh^3} \qquad (6.27)$$

Substituting into Eq. (6.20),

$$R_B = 2 \int_0^{1/2} \int_0^{h/2} \left(\frac{6Pxy}{bh^3\sigma_0}\right)^m b\,dx\,dy \qquad (6.28)$$

$$= \frac{1}{(m+1)^2} V \left(\frac{\sigma}{\sigma_0}\right)^m \qquad (6.29)$$

where

$$V = \text{volume subjected to tension} = \frac{lbh}{2} \qquad (6.30)$$

$$\sigma_m = \text{maximum tensile stress} = \frac{3Pl}{2bh^2} \qquad (6.31)$$

Equation (6.29) is also seen to correspond to Eq. (6.22) where

$$K = \frac{1}{(m+1)^2}$$

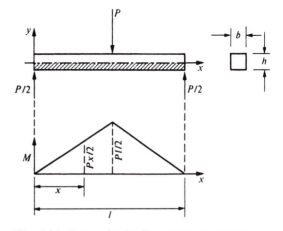

Fig. 6.16 Three-point loading transverse rupture test and resultant bending moment (M).

If we write R_T for a tensile specimen, then from Eq. (6.20)

$$R_T = (1) \, V_T \left(\frac{\sigma}{\sigma_0} \right)^m \tag{6.32}$$

For equal risk of rupture $(R_T) = (R_B)$

$$\frac{\sigma_B}{\sigma_T} = \left[(m+1)^2 \, \frac{V_T}{V_B} \right]^{1/m} \tag{6.33}$$

For example, if the volume of the tensile specimen (V_T) is equal to the volume of the transverse rupture specimen in tension (V_B), and $m = 10$,

$$\frac{\sigma_B}{\sigma_T} = 1.614$$

This explains why transverse rupture experiments give higher values of tensile stress at rupture than tensile tests even when great pains are taken in the alignment of the tensile specimen.

The ratio of σ_B to σ_T for this case is seen to consist of two terms $(V_T/V_B)^{1/m}$ which is the size effect and $(m+1)^{2/m}$ which is the stress gradient effect.

Most of the fracture problems to which Weibull statistics have been applied have involved tensile stresses only (tensile specimens and slender beams). Such problems are readily handled by the methods previously outlined. The risk of rupture (R) is obtained by integrating Eq. (6.20) for two cases to be compared and the values of R made equal so that a comparison may be made at the same risk of rupture. In this way the effects of specimen size and stress gradient are taken care of at the same time. However, this procedure becomes very difficult and involved for more complex situations involving transverse stresses as well as direct stresses.

A simpler procedure that appears to work well is the following. The most vulnerable region of a specimen to fracture is identified following an elastic stress analysis. It is then assumed that the earliest failures occur by cracks which propagate from this region. The volume of this critical region is estimated by considering the space over which the effective stress causing the first fractures is within 10% of the maximum value.

The effective stress for one type of specimen need not be adjusted for a stress gradient when determining the effective stress for another type of loading, since the analysis is confined to fracture that occurs in a small critical volume over which the stress may be considered constant. It is necessary however to make an adjustment for volume difference, and this is done by use of Eq. (6.26).

THE DISC TEST

The disc test (Fig. 6.17) introduced for testing concrete (Carniero and Barcellos, 1953) involves a thin disc that is diametrically loaded to failure. It is often used to measure the tensile strength of brittle materials, particularly concrete, rock, and coal. This test has been used and improved by many research workers (for example: Peltier, 1954; Mitchell,

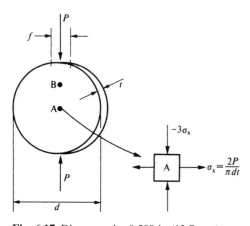

Fig. 6.17 Disc test: $d = 0.500$ in (12.7 mm), $t = 0.010$ in (0.25 mm), $f = 0.100$ in $+ 0.010$ in thick brass shim (2.5 mm $+ 0.25$ mm thick brass shim).

1961; Rudnick, Hunter, and Holden, 1963; Spriggs, Brisette, and Vasilos, 1964; and Berenbaum and Brodie, 1959).

The stresses pertaining at all points will correspond to the plane stress elastic solution given in Fig. 6.18. This is a finite element solution provided by Usui of Tokyo Institute of Technology and his colleagues for a thin (plane stress) disc uniformly loaded over flat surfaces extending 20% of the diameter of the disc. This solution is in excellent agreement over the critical central region with that of Frocht (1948) for a plane stress disc subjected to concentrated diametrically opposed loads. If the load is well distributed at the top and bottom of the specimen, as by use of flats extending about 20% across the specimen and thin brass shims, no fracture will occur in the vicinity of the loading points and instead, fracture will occur at the center of the specimen (A in Fig. 6.17). At a critical value of load, a crack will run from A upward and downward diametrically across the specimen. The tensile stress at A perpendicular to the diametral crack at fracture will be

$$\sigma_x = \frac{2P}{\pi dt} \text{ (tension)} \qquad (6.34)$$

while the radial stress at fracture will be

$$\sigma_y = -3\sigma_x \text{ (compression)} \qquad (6.35)$$

where P is the load at fracture.

According to the maximum tensile stress criterion, the value of nominal tensile stress at fracture (σ_x) should correspond to that for a uniaxial tensile test (four-point bending (Fig. 6.19) or ring test[†]). However, the value of σ_x at fracture in the disc test is found to be appreciably lower (by a factor of about two) than the fracture stress in a uniaxial tensile test. This discrepancy has baffled

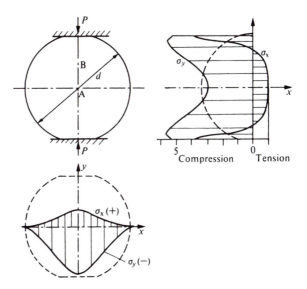

Fig. 6.18 Stress distribution for transversely loaded disc ($v = 0.3$). (Courtesy of E. Usui, Tokyo Institute of Technology)

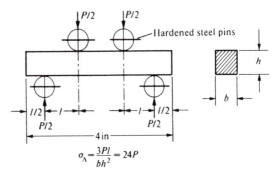

$$\sigma_A = \frac{3Pl}{bh^2} = 24P$$

Fig. 6.19 Four-point bending test; pin dia = 0.500 in (12.7 mm), l = 1.000 in (25.4 mm), $h = b = 0.500$ in (12.7 mm).

[†] The ring test (Sedlacek and Holden, 1962) is one of the best tests for determining the uniaxial tensile strength of a brittle material. The main disadvantage of the test lies in the high cost of test specimens. The test consists of a thin ring typically 2 in (50 mm) o.d. (d), 0.1 in (2.5 mm) wall thickness (t); and 0.5 in axial width (12.5 mm) (b). The thin ring is loaded internally by pressure acting through a rubber bag until fracture occurs. The uniaxial tensile stress is then

$$\sigma_T = \frac{pd}{2t} \qquad (6.36)$$

where p is the internal pressure at fracture.

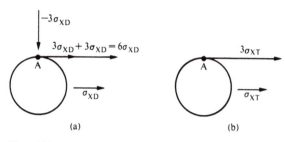

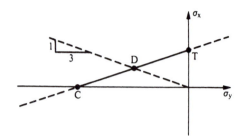

Fig. 6.20 Intensified tensile stress at top of circular void (A) in (a) disc test where σ_{xD} is the nominal tensile stress at center of the disc at fracture and in (b) a uniaxial tensile test where σ_{xT} is the nominal uniaxial tensile stress at fracture.

Fig. 6.21 Fracture point D for disc test on fracture locus of Fig. 6.11.

users of the disc test for many years (Berenbaum and Brodie, 1959; Wright, 1955; Snyder et al., 1962; Akazawa, 1953; Lecrivain and Lambert, 1961). Just as an assumed circular defect explains why a quasi-brittle material such as tungsten carbide can fail brittly in uniaxial compression it also explains why the nominal tensile stress in the disc test at fracture is about half of the fracture stress in a uniaxial tensile test. Figure 6.17 shows a circular defect at the critical point of crack initiation in the disc test (point A in Fig. 6.17). The plastic strain at fracture will be very small and an elastic analysis which leads to the following stresses at A represents a good approximation:

$$\sigma_x \text{ (tensile)}$$

$$\sigma_y = -3\sigma_x \text{ (compressive)}$$

The tensile stress at the top of the circular defect will thus be $6\sigma_{xD}$ for the disc test (Fig. 6.20a) where σ_{xD} is the nominal tensile stress at the center of the disc at fracture. Figure 6.20b shows the nominal tensile stress at fracture (σ_{xT}) for a uniaxial tensile test together with the intensified stress at the top of a circular defect ($3\sigma_{xT}$). When the intensified tensile stresses at the tops of the circular defects are equated for the disc and uniaxial tensile tests it is evident that the nominal stress in the tensile test at fracture $(\sigma_{xT})_f$ should be twice that for the disc test $(\sigma_{xD})_f$.

Figure 6.21 shows the plane stress fracture locus with the uniaxial tensile value (T) disc test value (D) and uniaxial compression (C) values indicated. Point D is seen to lie at the point of intersection of line CT and a line having a 3:1 slope corresponding to the ratio of $|\sigma_y|$ to $|\sigma_x|$ at the center of the disc.

The assumption of a circular defect is thus seen to represent an excellent approximation in that it not only explains why $(\sigma_C/\sigma_T)_f$ is about 3 but also why $(\sigma_T/\sigma_D)_f$ is about 2. The latter ratio is not found to be exactly 2 for the same reasons $(\sigma_C/\sigma_T)_f$ is not exactly 3 as discussed above.

The reason that fracture occurs first at the center of the disc (A in Fig. 6.17) and not at points closer to the applied load (for example B in Fig. 6.17) is due to the earlier onset of plastic flow at points closer to the applied load.

It has been found by use of strain gages that whereas tungsten carbide remains elastic to the point of fracture in uniaxial tension, it undergoes plastic flow before fracture in the disc test. For example, at the center of the disc (A in Fig. 6.17), the elastic stresses are given by Eqs. (6.34) and (6.35) and by the Tresca criterion, the material at A should go plastic when $\sigma_x = Y/4$ where Y is the uniaxial flow stress in tension. However, for point B in Fig. 6.17 which is about half-way from A to the applied load, the values of σ_x and σ_y will be approximately as follows:

$$\sigma_x \cong \frac{2P}{\pi dt} \text{ (tension)}$$

(6.37)

$$\sigma_y \cong -4.5\sigma_x \text{ (compression)}$$

By the Tresca criterion, plastic flow should occur when $\sigma_x = Y/5.5$. Thus, plastic flow will occur at B at a lower value of P than that required at A and thus limit the stress that develops at B. The nominal and intensified tensile stresses at fracture at A will be greater than those at B and hence fracture should originate close to point A.

FRACTURE OF CEMENTED TUNGSTEN CARBIDE

Brittle fracture is an important mode of tool failure particularly for tungsten carbide and other hardmetals and ceramics. Materials are generally considered to fail by brittle fracture when the maximum principal tensile stress reaches a critical value. The most widely employed test to characterize the fracture strength of tungsten carbide tool materials is the tensile stress as measured in a three- or four-point bending test at fracture (Fig. 6.19). The tensile strength of most tungsten carbide tools will be about 150,000 p.s.i. (1030 MPa) and the tensile strength is found to increase as the amount of binder employed is increased. For example, the mean tensile strength of a tungsten carbide containing 6 $^w/_o$ cobalt is about 115,000 p.s.i. (790 MPa), while one containing 12 $^w/_o$ cobalt is about 165,000 p.s.i. (1140 MPa).

In order to verify some of the observations made above, tests were run on specially produced tungsten carbide specimens made in the form of bars ($\frac{1}{2}$ in $\times$ $\frac{1}{2}$ in $\times$ 4 in (12 $\times$ 12 $\times$ 100 mm)), discs ($\frac{1}{2}$ in D $\times$ 0.10 in (12.50 $\times$ 2.5 mm)), and $\frac{1}{4}$ in (6.4 mm) square $\times$ $\frac{1}{2}$ in (12.5 mm) high compression specimens. These samples were made from the same batch of powder that was compacted and sintered under identical conditions. Two compositions were involved—6 $^w/_o$ Co, 94 $^w/_o$ WC and 12 $^w/_o$ Co, 88 $^w/_o$ WC. The specimens were diamond ground all over before testing and flats extending 20% of the diameter were provided on the discs. Figure 6.22 shows Weibull plots of

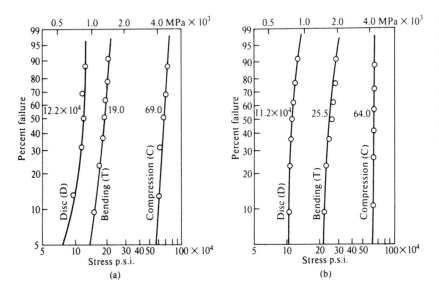

Fig. 6.22 Weibull plots of fracture stresses for disc, bending, and compression fracture stresses for (a) 6 $^w/_o$ Co and (b) 12 $^w/_o$ Co. The numbers indicated next to each curve are the median values. (1 p.s.i. = 6905 N m^{-2}) (after Takagi and Shaw, 1981, 1983)

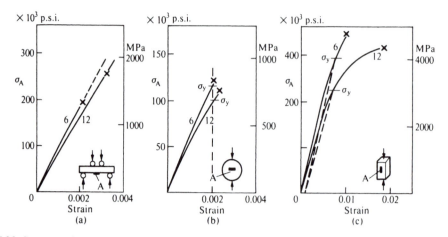

Fig. 6.23 Stress–strain curves. (a) Bending (uniaxial tensile stress). (b) Disc test with strain gage mounted transverse to load at center of disc. (c) Compression test with strain gage mounted parallel to load. X marks the point of fracture and σ_y the yield stress. (after Takagi and Shaw, 1981)

tensile stresses at fracture for the bending tests, disc tests and compression tests. Elastic behavior was assumed to fracture in all cases. It is found that

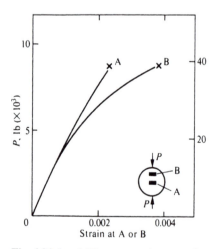

Fig. 6.24 Load (P) versus strain curves for 12 $^w/_o$ Co cemented tungsten carbide with strain gage mounted transversely at center of disc (curve A) and approximately at midpoint between point A and the load (curve B). The material at B goes plastic before fracture occurs at A while fracture at A occurs very near the yield point. (after Takagi and Shaw, 1981)

1. A considerable difference in the tensile median fracture stresses at fracture exists for the disc ($\sigma_{50} = 122,000$ p.s.i., 840 MPa) and bending tests ($\sigma_{50} = 190,000$ p.s.i., 1310 MPa) for 6 $^w/_o$ Co.

2. The high cobalt material is strongest in bending but weakest in the discs and compression tests.

3. The compressive stress at fracture is considerably higher than the uniaxial tensile stress at fracture.

Figure 6.23 shows stress–strain results for typical bending, disc, and compression tests. The strain readings are from very small gage length strain gages made from foil having an unusually high yield strain since they were to be used to measure strains extending slightly beyond the yield strain of the tungsten carbides employed. The gages were mounted as shown by the insets in Fig. 6.23. The curves of Eq. (6.36) are slightly nonlinear all the way to the origin (zero applied load) and it is difficult to identify an exact yield point. An attempt to do this was made by loading and unloading and looking for the onset of a permanent set after unloading. This did not prove to be useful since before the material at A (center of disc) went plastic, that nearer the load at B went plastic (see Fig. 6.24). Upon loading, the plastic zone gave rise to a residual compressive stress which results in a slight tensile permanent set at A. Since there is no apparent way of distinguishing a small permanent set due to the development of a residual compressive stress near the applied load and actual plastic flow at point A, it was finally decided to arbitrarily consider the yield point at point A in Fig. 6.23b to be the point where the observed strain under load was 2×10^{-3}. The stresses in Fig. 6.23 corresponding to this yield point are designated σ_y. From Figs. 6.23 and 6.24 it is evident that

1. The yield stress is greater than the fracture stress for both materials in uniaxial tension (Fig. 6.23a).

2. The yield stress is below the fracture stress for both materials in the disc test and in the uniaxial compression test (Figs. 6.23b and 6.23c).

3. The high cobalt material yields at a lower value of stress than the low cobalt material which is consistent with the high cobalt material being weaker in the disc test (Fig. 6.23b).

4. The material at B (Fig. 6.24) goes plastic before that at A which means that the stress at A should be greater than that at B at fracture. As already mentioned, this explains why the critical crack should initiate at A and not at some point a greater distance from the center (such as B) if the material is reasonably homogeneous.

5. The material in the uniaxial compression test undergoes appreciable plastic strain before fracture, particularly in the case of material B in Fig. 6.23c.

While the exact values of strain pertaining are in question since strain gages have been used to measure strains that extend into the plastic region, the above observations appear to be well justified.

The experimental fracture loci for the two carbides are shown in Fig. 6.25 where it is evident that

Carbide, $^w/_o$	$(\sigma_C/\sigma_T)_f$
6	3.63
12	2.51

The experimental loci of Fig. 6.25 are seen to be in excellent agreement with the analytical fracture locus of Fig. 6.12.

It is of interest to note that whereas the 12 $^w/_o$ Co tungsten carbide is stronger than 6 $^w/_o$ tungsten carbide in uniaxial tension, the reverse is true in the disc and uniaxial compression tests. This is due to the fact that the more brittle 6 $^w/_o$ Co material undergoes a greater increase in strength with increase in mean (hydrostatic) compressive stress than the more ductile 12 $^w/_o$ material does.

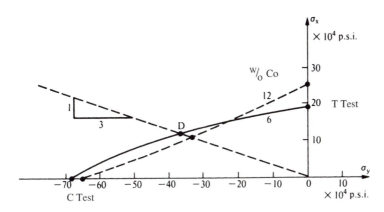

Fig. 6.25 Fracture loci for 6 $^w/_o$ and 12 $^w/_o$ tungsten carbide in the tension–compression regime. The solid dots are experimental median values from Fig. 6.22. (1 p.s.i. = 6905 MPa) (after Takagi and Shaw, 1981, 1983)

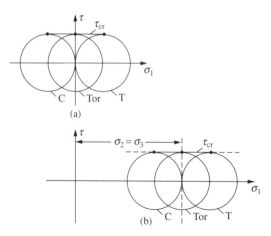

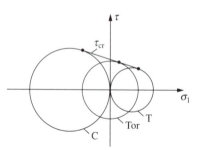

Fig. 6.27 Critical fracture stress for a brittle material with zero transverse stress. The corresponding flow stress diagram will be similar to Fig. 6.26a. [Mohr's fracture criterion for tensile test = T. Mohr's fracture criterion for torsion test = Tor. Mohr's fracture criterion for compression test = C.]

Fig. 6.26 Critical flow or fracture stress (τ_{cr}) for a perfectly ductile material. (a) For zero transverse stress. (b) With transverse stress $\sigma_2 = \sigma_3$. [Mohr's circle for tensile test = T. Mohr's circle for torsion test = Tor. Mohr's circle for compression test = C.]

MOHR FRACTURE CRITERION

In design, failure is often defined in terms of yielding. For completely ductile materials, the yield point in tension equals the yield point in compression. Figure 6.26a shows two-dimensional Mohr's circle yield diagrams for tension, compression, and torsion for a ductile material. The lines tangent to the tops of these circles give the maximum shear stress for yielding for all types of loading (a constant). The von Mises criterion is an appropriate flow and fracture criterion for perfectly ductile materials. If a transverse compressive stress $\sigma_2 = \sigma_3$ (hydrostatic compressive stress in a three-dimensional situation) is applied, Fig. 6.26a changes to Fig. 6.26b. The critical shear stress (τ_{cr}) does not change and the von Mises criterion gives the same result with or without the transverse stress. Applying this to steady state cutting, the flow stress on the shear plane should be independent of the normal stress on the shear plane for a perfectly ductile material.

In the case of brittle materials, the fracture strength in compression is greater than in tension (Fig. 6.27). While the von Mises criterion still holds for flow, it does not hold for fracture. An increase in

Fig. 6.28 Back (free surface) of chip showing disconnected regions of concentrated microfracture. Material: titanium alloy.

transverse compressive stress causes an increase in fracture stress for all types of loading. In the case of steady state chip formation, strains are unusually high and there is evidence that strain hardening materials tend to undergo some fracture even when chips are in the form of continuous ribbons. In this case, fracture is localized and discontinuous across the chip width at points of stress concentration. Such microcracks are most likely to occur on the shear plane near the back of the chip where normal stress on the shear plane is a minimum. Figure 6.28 is a photomicrograph of the back of a chip showing roughness associated with microcrack formation. When some microcrack formation is involved in steady state chip formation, the mean shear stress on the shear plane should increase with increase in normal stress on the shear plane. If this is the case, the von Mises criterion should not be used in any analysis, but a von Mises–Mohr criterion should be used instead. This will be discussed further in Chapters 9 and 20.

REFERENCES

Akazawa, T. (1953). International Union of Testing and Research Laboratories for Materials and Construction, RILEM, Paris, Bull. No. 16.

Berenbaum, R., and Brodie, L. (1959). *Br. J. Appl. Phys.* **10**, 281.

Carnerio, F. L. L. B., and Barcellos, A. (1953). Union of Testing and Research Laboratories for Materials and Structures (Brazil) Bull. No. 13.

Fisher, J. C. (1953). *Bull. Am. Soc. Test. Mater.* **TP76**, 74.

Frocht, M. M. (1948). *Photoelasticity*, Vol. II. Wiley, New York.

Grassi, R. C., and Cornet, I. (1948). *Mech. Engng.* **70**, 918.

Griffith (1924). *Proc. First Int. Cong. Appl. Mech.* Delft, 55 (reprinted 1968) *Trans. Am. Soc. Metals* **61**/87.

Inglis, C. E. (1913). *Trans. Inst. Nav. Architects* **55–1**, 219.

Irwin, G. R. (1957). *J. Appl. Mech.* **24**, 361.

Joffe, A. F. (1928). *The Physics of Crystals*. McGraw-Hill, New York.

Lecrivain, L., and Lambert, B. (1961). *L'Industrie Ceramique*, No. **534**, 367.

Mitchell, N. B. (1961). *Mater. Res. Stand.* **1**, 780.

Orowan, E. (1950). In *Fatigue and Fracture of Metals*. Wiley, New York.

Paul, B., and Mirandy, L. (1975). *J. Engng. Mater. Tech.* **98**, 153.

Peltier, H. (1954). Union of Testing and Research Laboratories for Materials and Structures (Brazil), Bull. No. 19.

Rudnick, A., Hunter, A. R., and Holden, F. C. (1963). *Mater. Res. Stand.* **3**, 283.

Sedlacek, R., and Holden, F. C. (1962). *Rev. Sci. Instrum.* **33**, 298.

Shaw, M. C., Braiden, P. M., and Desalvo, G. J. (1975). *J. Engng. Indus.* **96**, 77.

Snyder, M. J., Rudnick, A., Duckworth, W. H., and Hyde, C. (1962). Battelle Tech. Doc. Rep. ASD–TDR–62–7.

Spriggs, R. M., Brisette, L. A., and Vasilos, T. (1964). *Mat. Res. and Stand.* **4**, 218.

Takagi, J., and Shaw, M. C. (1981). *Ann. of CIRP* **30/1**, 53.

Takagi, J., and Shaw, M. C. (1983). *J. Eng. Industry* **105**, 143.

Usui, E., Ihara, T., and Shirakashi, T. (1979). *Jap. Soc. Precis. Engrs.* **13**, 189.

Weibull, W. (1939). *Ing. Vetenskaps Adakmien-Handlinger* (Stockholm), No. **151**, 1, and No. **153**, 1.

Weibull, W. (1951). *J. Appl. Mech.* **73**, 293.

Wright, P. J. F. (1955). *Mag. Concr. Res.* (July), 87.

Zener, C. (1948). *The Micromechanism of Fracture*. Am. Soc. Met., Cleveland.

7 DYNAMOMETRY

In order to put the analysis of the metal cutting operation on a quantitative basis, certain observations must be made before, during, and after a cut. The number of observations that can be made during the cutting process is rather limited, one of the more important measurements of this type being the determination of cutting force components. In this chapter tool force-measuring apparatus will be considered. Design criteria will be discussed first, followed by a review of several types of force measuring techniques. Finally, a variety of useful dynamometers will be described. Although each dynamometer must be specially designed to meet specific requirements, the examples presented should serve to illustrate general considerations.

DYNAMOMETER REQUIREMENTS

Requirements of importance in dynamometer design are rigidity, sensitivity, and accuracy. The sensitivity of a good research dynamometer should be such that determinations are sensitive and accurate to within ±1%. That is, if a dynamometer is designed for a mean force of 100 lb, one pound increments should be easily readable and accurately indicated.

Some deformation is associated with the operation of every dynamometer. However, a dynamometer should be rigid enough so that the cutting operation is not influenced by the accompanying deflections. In general, the static stiffness of a metal cutting dynamometer should be about 10^6 lb in^{-1} (10^8 N m^{-1}). Frequently, the dominating stiffness criterion is the natural frequency of the dynamometer. All machine tools operate with some vibrations, and in certain cutting operations these vibrations may have large amplitudes (i.e., milling, grinding, shaping). In order that the recorded force (or the actual cut) be not influenced by any vibrating motion of the dynamometer, its natural frequency must be large (at least four times as large) compared to the frequency of the exciting vibration. For purpose of analysis, any dynamometer can be reduced approximately to a mass supported by a spring. The natural frequency (f_n) of such a system is equal to

$$f_n = \frac{1}{2\pi} \sqrt{\frac{K}{m}} \text{ c.p.s.} \tag{7.1}$$

where K is the spring constant in lb in^{-1} (MN m^{-1}) and m is the mass in lb s^2 in^{-1} (kg). In terms of the supported weight of the dynamometer (W),

$$f_n = \frac{1}{2\pi} \sqrt{\frac{386K}{W}} \text{c.p.s.} \qquad (7.2)$$

where W is in lb (N). If, for example, a grinding machine is being run at $N = 3600$ r.p.m., machine vibrations are apt to be present with an exciting frequency of

$$f_e = \frac{N}{60} \text{c.p.s. (Hz)} \qquad (7.3)$$

In this case, the natural frequency of the dynamometer (f_n) should be at least 240 c.p.s. (Hz).

In general, a dynamometer must measure at least two force components. In a three-dimensional cutting operation, three force components are necessary, while in drilling or tapping, only a torque and a thrust are required. It is usually most convenient to measure force relative to a set of rectangular coordinates (x, y, z), and it is advisable that there be no *cross sensitivity* between these components. That is, an applied force in the x-direction should give no reading in the y- or z-directions. If mutual interference of the force-measuring elements exists, determination of the force components requires the solution of simultaneous equations. This prevents the immediate interpretation of the data.

When certain of the electric transducers are suitably located and connected, unwanted strain components can often be cancelled electrically. How this may be done will be subsequently illustrated.

It is convenient to use a system having a linear calibration. In such a case, the force is determined with the precision with which a strain increment can be measured relative to an arbitrary datum. If the system is not linear, it is then necessary that the zero load point be accurately known as well as the strain increment. This introduces an additional quantity that must be carefully measured.

A dynamometer should be stable with respect to time, temperature, and humidity. Once a calibration is made, it should only have to be checked occasionally. Many existing dynamometers use devices for the separation of force components which involve friction, (i.e., rollers, balls, and sliding surfaces). As friction conditions are usually variable due to dirt or environmental changes, such instruments are of limited usefulness.

Force readings can be obtained from recording or indicating type dynamometers. Although recording dynamometers cost more than a direct-reading type, the added convenience and other obvious advantages of a recording system usually offset the added expense.

There are many other special dynamometer requirements that must be frequently met, such as size, ruggedness, and adaptability to several jobs. Such special requirements assume different degrees of importance in different applications. For example, dynamometer requirements are quite different for an adaptive control sensor used under production conditions than they are for a unit used in the laboratory for research purposes.

FORCE MEASUREMENT

In most dynamometers the force is applied to some sort of spring and the deflection thus produced is measured. Since dynamometers must be stiff, only deflection measuring devices that are capable of measuring small deflections (10^{-6} to 10^{-4} in or ~ 1 μm) are of interest here. Some of the devices that have been used to measure such small deflections in dynamometers will now be briefly described.

The dial indicator is capable of reading deflections to about 10^{-4} in (~ 2 μm) when functioning properly. However, dial indicators are subject to sticking and are not to be relied upon for absolutely static readings. A lever system may be used in conjunction with a dial indicator as shown in Fig. 7.1. This is a relatively crude planing dynamometer designed and used by the author about 1940.

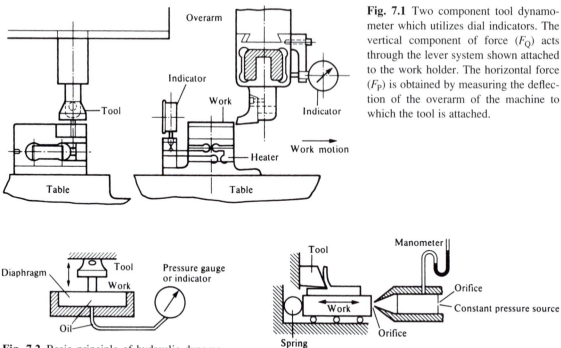

Fig. 7.1 Two component tool dynamometer which utilizes dial indicators. The vertical component of force (F_Q) acts through the lever system shown attached to the work holder. The horizontal force (F_P) is obtained by measuring the deflection of the overarm of the machine to which the tool is attached.

Fig. 7.2 Basic principle of hydraulic dynamometer.

Fig. 7.3 Basis principle of pneumatic type dynamometer.

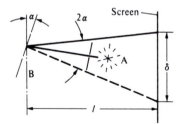

Fig. 7.4 Principle of optical lever shown diagrammatically.

Hydraulic pressure cells have also been used in conjunction with pressure gages to measure or record the force on a tool. The manner in which this may be done is illustrated diagramatically in Fig. 7.2. When this method is used, the force may be read at a distance from the pressure cell.

Pneumatic devices such as the Solex micrometer in which the change in back pressure that occurs when a flat surface is brought into closer contact with a sharp-edged orifice are sometimes used to measure deflections in tool dynamometers (Fig. 7.3). Such systems are simple and reliable if carefully supplied with clean, constant pressure air. There is, however, a limited region over which they are linear, and dynamometers of this type tend to be bulky.

Optical devices of several types have also been employed. Interferometric methods can be used to give very precise measurements using the wavelength of light as a unit. However, this principle is not easily adapted to the measurement of the dynamic deflections in metal cutting. Very small angular deflections can be readily measured by reflecting a beam of light from the moving surface. In Fig. 7.4 a narrow beam of light travels from A to B where it is reflected and hence travels to screen C. When the surface B rotates through an angle α, the spot of light moves through a distance δ on the screen. If α is small, then

$$\delta = 2l\alpha \tag{7.4}$$

This device acts as an optical lever, and may be used to measure small angular displacements if l is large. However, even though such a device is simple in principle, it can be applied to a metal-cutting dynamometer only with considerable difficulty.

Piezoelectric crystals are also widely used as force-measuring units in metal cutting dynamometers. However, since piezoelectric crystals produce an electric charge rather than a current, leakage effects can be troublesome. However, piezoelectric dynamometers are well suited for making dynamic measurements since in general they can be designed to have a higher natural frequency of vibration than other types of dynamometers.

ELECTRIC TRANSDUCERS

Several transducers that transform a physical displacement into an electrical signal are commercially available:

1. the electronic transducer tube
2. the differential transformer
3. the magnetic strain gage
4. the unbonded wire resistance strain gage
5. the bonded wire resistance strain gage

An electronic transducer tube is shown schematically in Fig. 7.5. It is essentially a very small ($\frac{1}{4}$ in $\times \frac{1}{2}$ in or 6 $\times$ 12 mm) triode vacuum tube with a movable plate, the tube characteristics being changed as the plate moves. The pin at A can be rotated approximately $\pm\frac{1}{2}$ degree about the diaphragm and the minimum motion of the end of the pin that can be accurately measured is of the order of one-millionth of an inch (0.025 μm). The electrical system associated with this tube is relatively simple and there are no serious frequency limitations. A dynamometer employing this device will be described later in this chapter.

The differential transformer (Fig. 7.6) consists essentially of three transformer coils on a common axis with a common movable core. AC current is supplied to the center primary coil which induces an electromotive force (EMF) in the two secondary coils. The outputs of the two secondary coils are wired to oppose each other so that when the core is in the center, there is zero net output. When the core is displaced, an output is obtained which is proportional to the displacement, with a phase depending upon the direction of motion. The electrical system associated with this instrument is rather complex because of the necessary high amplification and phase sensitivity. However, AC carrier wave type strain gage amplifiers and recorders are commercially available and are quite satisfactory, giving a sensitivity of about 1 μin (0.025 μm).

The magnetic type strain gage shown in Fig. 7.7 allows point A to move with respect to B thus changing the inductance of coil B. If initially the balancing unit C is adjusted to give zero current at G, any motion of A will cause a current flow through G. This flow can be measured directly, or the balancing unit (at C) can be adjusted to give zero flow at G. The sensitivity of such a gage is about 10 μin (0.25 μm). This lack of sensitivity is somewhat offset by the simplicity of the associated electrical system. Since the input can safely be made quite high, the output can be measured directly without amplification.

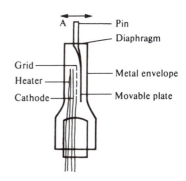

Fig. 7.5 Schematic diagram of electronic transducer tube.

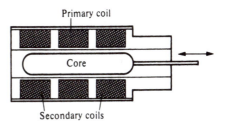

Fig. 7.6 Schematic diagram of differential transformer.

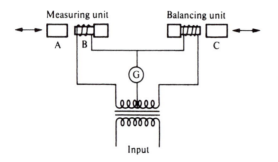

Fig. 7.7 Schematic diagram of magnetic type strain gage.

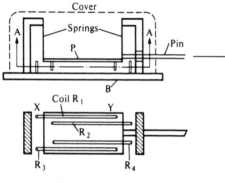

Section A-A

Fig. 7.8 Schematic diagram of unbonded wire resistance strain gage.

The fact that the electrical resistance of a thin conductor is changed when it is stretched led to the development of extremely useful resistance strain gages about 1940. These may be of two types. In the unbonded type of gage (Fig. 7.8), a movable plate P is hung from two spring members. Four coils, R_1, R_2, R_3, and R_4, preloaded in tension are supported by small pins projecting from base B and plate P.

When the plate is moved relative to the base, two of the coils will undergo an increase in tension (and hence an increase in resistance) while the other two coils undergo a decrease in tension (and hence a decrease in resistance). The maximum deflection of an unbonded gage is usually limited by stops to about 0.0015 in ($\frac{1}{3}$ mm). When this type of gage is connected in the form of a Wheatstone bridge (to be presently described), it can measure deflections of the order of 1 μin (0.04 μm). One advantage of a unit of this type over that to be presently described (bonded type) is that it can be detached from the part under test and used over again. There are disadvantages however, not the least of which is cost.

BONDED STRAIN GAGES

Three types of bonded wire resistance strain gages have been commercially available as shown schematically in Fig. 7.9. In each case the necessary length of gage wire (i.e. that to produce the desired resistance which is usually about 120 ohms) is present in the form of a flat coil which is cemented (bonded) between two thin insulating sheets of paper or plastic. Such a gage cannot be used directly to measure deflection, but must be first cemented to the member to be strained. A strain gage is applied to a structure by first cleaning the surface thoroughly and then cementing the gage in place. After cementing, the unit is usually baked at about 180 °F (90 °C) to exclude moisture and then coated with wax or resin to provide some mechanical protection and also to prevent atmospheric moisture from causing difficulty. When the resistance between the metal part under test and the gage itself is tested, it should be at least 50 megaohms if the gage has been properly applied. Because the total area of all conductors

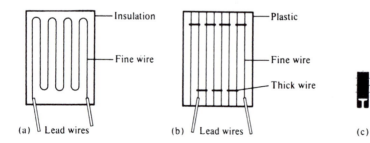

Fig. 7.9 Bonded type resistance strain gages. Those shown in (a) and (b) are wire type gages while type (c) is produced by etching a thin constantan foil (actual size, $\frac{1}{4}$ in [6.35 mm] gage length).

is small compared with the area of the backing material, the cement can easily transmit the force necessary to deform the wire. Strain gages of this type are usually somewhat smaller than a common postage stamp.

Since several of the dynamometers to be described in detail later employ bonded wire or foil gages, it would appear justified to discuss their use in some detail. The manufacturer of a strain gage supplies two important pieces of information with every gage: the resistance (R) and the gage factor (F). The gage factor is a measure of the sensitivity of the gage as defined below

$$F = \frac{\Delta R/R}{\Delta l/l} = \frac{\Delta R}{eR} \qquad (7.5)$$

where e is the normal elastic strain $\Delta l/l$. The gage factor (F) varies from about 1.75 to 3.5 for most gages. Bonded gages can measure strains as low as 10^{-7}, however, commercial strain gage amplifiers limit the practical range to about 10^{-6}. Due to fatigue, the maximum useful strain is of the order of 0.002. Thus, the practical range of operation is about 2000 to 1.

WHEATSTONE BRIDGE

From Eq. (7.5) it is evident that in order to measure strains of the order of 1 μin (0.025 μm) it is necessary to measure small changes of resistance per unit resistance. The change in resistance of a bonded strain gage is usually less than 0.5%. Measurements of this order of magnitude may be made by means of the Wheatstone bridge shown in Fig. 7.10. No current will flow through the galvonometer (G) if the four resistances satisfy the equation

$$\frac{R_1}{R_4} = \frac{R_2}{R_3} \qquad (7.6)$$

In order to demonstrate how a Wheatstone bridge operates, a voltage scale has been drawn at points C and D of Fig. 7.10. Assume that R_1 is a bonded gage and that initially Eq. (7.6) is satisfied. If R_1 is now stretched so that its resistance increases by one unit $(+\Delta R)$, the voltage at point D will be increased from zero to plus one unit of voltage $(+\Delta V)$, and there will be a voltage difference of one unit between C and D which will give rise to a current through G. If R_4 is also a bonded gage, and at the same time that R_1 changes by $+\Delta R$, R_4 changes by $-\Delta R$, the voltage at D will move to $+2 \Delta V$. Also, if at the same time R_2 changes by $-\Delta R$, and R_3 changes by $+\Delta R$, then, the voltage of point C will move to $-2 \Delta V$, and the voltage difference between C and D will now be 4 ΔV. It is then apparent that although a single gage can be used, the sensitivity can be increased fourfold if two gages are used in tension while two others are used in compression.

It should be noted that the four active arm bridge is fully compensating for any change in resistance due to temperature providing all gages experience the same temperature change. This is important, since a single gage is frequently so sensitive to temperature changes as to make it useless for strain measurements in a region of steep temperature gradients. The output voltage (E) from a strain gage bridge is given by the following expression

$$E = \frac{a}{4}(F)(V)(e) \qquad (7.7)$$

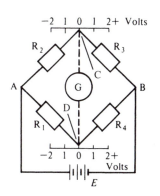

Fig. 7.10 The Wheatstone bridge.

where

a = the number of active arms in the bridge (from 1 to 4)

F = the gage factor of the gages

V = the input voltage to the bridge

e = the longitudinal strain on the gages (neglecting transverse effects which are very small)

This equation is based on the assumption that the instrument used to measure E has a high impedance and hence does not draw an appreciable current from the bridge. If this is not so, the output (E) will no longer be a linear function of the strain e. It is again evident from Eq. (7.7) that when maximum sensitivity (E/e) is required, (the usual case with metal cutting dynamometers) it is desirable to make all four arms of the bridge active. It is further advantageous to use gages having a high gage factor.[†]

The impressed bridge voltage V is another quantity which may be varied to alter the sensitivity of a strain gage bridge circuit. With some commercial instruments this voltage is fixed at from 3 to 6 V for reasons of stability. However, higher voltages may often be used, especially with the larger gages whose greater area allows better heat dissipation. In this respect aluminum makes a better material to which to cement gages than steel because of its higher rate of heat conductivity. The use of two strain gages in each arm of the bridge in place of one enables the impressed voltage to be doubled without overheating the gages. The maximum safe voltage in each case is determined by the desired stability. Too high a voltage will cause zero drift. When a very stable zero setting is required over a long time, the voltage must be reduced. The following expression may be used as a rough guide in determining the maximum permissible impressed voltage per gage when the gage is cemented to steel.

$$V = 2(WR)^{1/2} \tag{7.8}$$

where W is the limiting wattage which a gage may dissipate as heat, an average value being on the order of 0.075 W; R is the resistance of a single gage in ohms. As previously mentioned the strain e should be kept below 0.002 to insure an adequate fatigue life for the strain gages making up the bridge.

The unbalance of a Wheatstone bridge can be determined by several methods. A variable resistance can be used in parallel with one arm of the bridge and adjusted to give zero current at G, thus providing a null instrument. This is the usual potentiometer set-up. Again a null instrument can be provided by use of a calibrated "bucking" voltage applied across CD to give zero current flow through G. On the other hand, the voltage across CD can be measured directly. However, it may be readily demonstrated that if the bridge circuit is used in an unbalanced condition a microammeter without amplification cannot give the sensitivity that is directly available from a good potentiometer.

DESIGN CONSIDERATIONS

In order to use any of the foregoing deflection measuring devices for force measurement, it is necessary to construct a dynamometer which will deflect the proper amount under load; and then mount the deflection detectors in such a manner that they will be unaffected by forces or moments

---

[†] While solid state (semi-conductor) strain gages having a gage factor an order of magnitude greater than metallic gages are available, this increase in gage factor is more than offset by the much greater adverse temperature sensitivity of the solid state gages.

other than the ones which they are supposed to detect. In general, dynamometers can be divided into two types shown schematically in Fig. 7.11. In the first type (Fig. 7.11a) the springs which support the external load (P) are independent of the deflection meter (M); while in the second type (Fig. 7.11b) the spring and the meters are integral. Any of the aforementioned deflection meters are suitable for use in the first way while only the bonded gages are effective in the second.

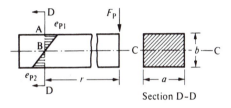

(a)   (b)

Fig. 7.11 Schematic views showing two types of dynamometers.

In many cases the force measuring element can be attached to either a stationary or moving member. Wherever possible the dynamometer should be placed on the stationary element inasmuch as this makes it unnecessary to use moving electrical contacts which may exhibit variable contact resistance. If slip rings have to be used it has been found advisable to employ mercury cups in which amalgamated copper discs rotate. The contact resistance is then practically zero and absolutely constant, as it must be in any successful strain gage circuit.

The design of a dynamometer may be influenced by a variety of special operating conditions that must be met. For example, if a cutting fluid is to be used, provisions must be made to keep the unit watertight. Many other special problems frequently arise in dynamometry. The dynamometers that are described in the following sections should be useful in illustrating designs that have been successful.

STRAIN GAGE LATHE DYNAMOMETER

For simplicity of analysis, it is frequently convenient to reduce the lathe operation to a two-dimensional process (orthogonal cutting). In this case, the resultant force will act in a known plane and only two force components are required (F_P and F_Q) as shown in Fig. 7.12.

It is apparent that the force F_P will cause a bending moment M_P at a distance (r) from the cutting edge and that F_Q will cause a corresponding moment M_Q where

$$M_P = F_P r \qquad (7.9)$$

$$M_Q = F_Q r \qquad (7.10)$$

The distribution of strain at a section through A and B, caused by bending moment M_P is as shown in Fig. 7.13. At the surface the strain is a maximum and equal to

$$e_{P1,2} = \frac{M_P(b/2)}{EI_P} \qquad (7.11)$$

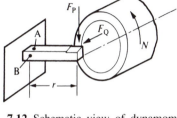

Fig. 7.12 Schematic view of dynamometer for measuring cutting forces in two-dimensional turning operation.

Fig. 7.13 Distribution of strain due to bending moment $M_P = rF_P$.

where e_{P1} is a tensile strain (+), e_{P2} is a compressive strain (−), E is Young's modulus of elasticity and I_P is the area moment of inertia of the section about axis C–C. For a rectangular section such as that shown in Fig. 7.13.

$$I_P = \frac{ab^3}{12} \tag{7.12}$$

and Eq. (7.11) becomes

$$e_{P1,2} = \frac{6M_P}{ab^2E} \tag{7.13}$$

$$e_{Q1,2} = \frac{6M_Q}{ba^2E} \tag{7.14}$$

In order to measure the moment M_P, two strain gages are applied to the top of the measuring section (T_1 and T_2 in Fig. 7.14a) and two gages are applied directly below (C_1 and C_2) thus, when M_P is applied, two gages are put in tension, while two others receive an equal amount of compression, thereby satisfying the requirements for a complete Wheatstone bridge (Fig. 7.14b). In a similar manner M_Q is measured by the four gages T_3, T_4, C_3, C_4, which are connected as shown in Fig. 7.14c. It will be recalled that one of the important design criteria was that there be no cross sensitivity. When a moment M_P is applied, gages T_3 and C_4 will be subjected to tension while T_4 and C_3 are subjected to compression (Fig. 7.14a). However, if the gages are symmetrically placed with respect to the axis of the section, the strains will be of equal magnitude, and there will be no net output from the Wheatstone bridge (see Fig. 7.14c).

The foregoing discussion indicates how turning forces may be measured; we shall now consider a practical design employing such principles. Since it is inconvenient to place the gages directly on the tool, a tool-holder with a built-in measuring section is employed. A dynamometer of this type is shown in Fig. 7.15. It is important that the cutting edge be kept on the axis of the measuring section at a known distance from the gages. When the particular unit is connected to commercial strain recording equipment, a force of 3 lb (13.4 N) gives a 1 mm pen deflection, and this represents adequate sensitivity for turning studies. The stiffness of the unit shown is actually greater than the customary tool holder for the $\frac{5}{8}$ in square tool bit that is used.

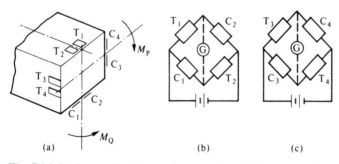

(a)   (b)   (c)

Fig. 7.14 Strain gage locations and connections. (a) Gage locations. (b) Gage circuit for measuring F_P. (c) Gage circuit for measuring F_Q.

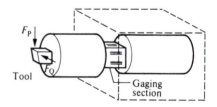

Fig. 7.15 Two-dimensional turning dynamometer. [Force range: 2 to 2000 lb (10 to 10^4 N). Stiffness: 500,000 lb in^{-1} (87,600 N mm^{-1}). Tool size: $\frac{5}{8} \times \frac{5}{8}$ in (15.9 × 15.9 mm).] (after Cook, Loewen, and Shaw, 1954)

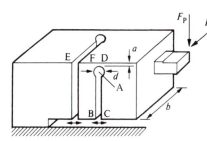

Fig. 7.16 Diagrammatic sketch of transducer tube lathe dynamometer.

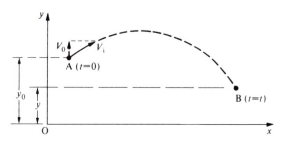

Fig. 7.17 Trajectory of projectile acted on by gravity when released from point A with initial velocity V_i.

TRANSDUCER TUBE LATHE DYNAMOMETER

In the foregoing discussion we have seen that it is only necessary to measure and separate moments about orthogonal axes that lie in a single plane in the case of a two-dimensional lathe dynamometer. One way of doing this has already been presented. The device shown in Fig. 7.16 will also measure the moments associated with forces F_P and F_Q, without interference. The block of metal has a hole of diameter d at A that is connected to the free surface at B and C by a long slot so that stress will be transmitted only across section AD. Force F_P gives rise to a bending moment $M_P = rF_P$ at the minimum section that will cause points B and C to approach each other. However, the distance BC will be unchanged by force F_Q due to the very large difference in the section moduli at section AD about horizontal and vertical axes. By use of the second slot and hole, distance EF may be used to measure force F_Q and this distance will be uninfluenced by force F_P. By placing a sensitive transducer across BC and another across EF, we have a convenient two-component dynamometer. The previously described electronic transducer tube offers a suitable means for measuring changes in BC and EF.

The choice of the slot and hole sizes in the design of the foregoing dynamometer is not readily obtained from elementary elasticity theory, but an analysis of the slotted member may be made by a combination of dimensional analysis, simple logic, and the results of a few easily performed experiments.

This is a convenient point to stop and review the principles of dimensional analysis which will be used to complete the design of the dynamometer of Figs. 7.16 and 7.17 as well as for other purposes in subsequent chapters.

DIMENSIONAL ANALYSIS

Before discussing the details of dimensional analysis, we shall examine the nature of the end result. This will be done by considering a familiar elementary problem—the vertical displacement of a freely falling body. If we ignore air resistance, the only force acting upon a freely falling body is that due to gravity and the body will experience a downward acceleration (g = 32.2 ft s^{-2} or 9.81 m s^{-2}). Thus, if a body is projected from point A with an initial velocity V_0; (Fig. 7.17), its acceleration immediately after release will be

$$\frac{d^2y}{dt^2} = -g \tag{7.15}$$

(where t designates time in seconds) and will remain at this constant value until it strikes another object or until some additional force acts on the body.

Integrating Eq. (7.15)

$$\frac{dy}{dt} = -gt + C_1 \tag{7.16}$$

and again

$$y = \frac{-gt^2}{2} + C_1 t + C_2 \tag{7.17}$$

where C_1 and C_2 are constants of integration which may be evaluated by the following *boundary conditions*:

$$\frac{dy}{dt} = V_0 \qquad \text{when } t = 0$$

$$y = y_0 \qquad \text{when } t = 0$$

After the constants are evaluated, Eq. (7.17) becomes

$$y = y_0 + V_0 t - \frac{gt^2}{2} \tag{7.18}$$

This is the complete solution in closed form.

In considering this problem by the method of dimensions, we would first observe that the main dependent variable (y) will be some function (ψ) of the variables y_0, V_0, t, and g

$$\text{i.e.,} \quad y = \psi(y_0, V_0, t, g) \tag{7.19}$$

After performing the dimensional analysis, we would write

$$\frac{y}{gt^2} = \psi_1\left(\frac{y_0}{gt^2}, \frac{V_0}{gt}\right) \tag{7.20}$$

where ψ_1 is some function of the two quantities in parentheses. The number of variables has been reduced from 5 to 3 but ψ_1 must still be evaluated either experimentally or by further analysis.

Dimensional analysis thus represents a partial solution to a problem, but one which is very easily obtained. In the solution of complex problems it can play a very valuable role.

Before considering the method of proceeding from Eq. (7.19) to Eq. (7.20), a few quantities will be defined.

Fundamental (or primary) dimensions are properties of a system under study that may be considered independent of the other properties of interest. For example, there is but one fundamental dimension in any geometry problem and this is length [L]. The fundamental dimensions involved in different classes of mechanics problems are outlined in Table 7.1, where L stands for length, F for force, and T for time. Dimensions (other than F, L, and T) which are considered fundamental in areas other than mechanics include: temperature (θ), heat (H), electrical charge (Q), magnetic pole strength (P), chemical yield (Y), and unit cost (¢).

Dimensional equations relate the dimensions of the fundamental quantities entering a problem to nonfundamental or secondary quantities. For example, acceleration (a) is a quantity of importance in many kinematic problems, and the dimensions of acceleration are related to fundamental dimensions L and T by the following dimensional equation

$$a = [LT^{-2}] \tag{7.21}$$

TABLE 7.1 Fundamental Dimensions for Different Classes of Problems

Type of Problem	Fundamental Dimensions
Geometry	L
Statics	F
Temporal[†]	T
Kinematics	L, T
Work or Energy	F, L
Momentum or Impulse	F, T
Dynamics	F, L, T

[†] Temporal problems are those involved in time study, frequency analysis, time tables.

It is customary to enclose the combination of fundamental dimensions employed in writing a dimensional equation in brackets.

Dimensional units are the basic magnitudes used to specify the size of a fundamental quantity. In engineering problems, lengths are frequently measured in feet (meters) and times in seconds. In this system of dimensional units, an acceleration would be expressed in units of feet per (second)^2 [meters per (second)^2] which may be verified by reference to Eq. (7.21).

A nondimensional quantity is one whose dimensional equation has unity or $F^0L^0T^0$ on the right-hand side. The quantities (y/gt^2) and (V_0/gt) in Eq. (7.20) are clearly nondimensional. A group of variables that cannot be combined to form a nondimensional group are said to be dimensionally independent.

The choice of fundamental quantities is somewhat arbitrary as the following discussion will reveal. In Table 7.1 the fundamental quantities for the general dynamics problem were stated to be F, L, and T. The quantities M (mass), L, and T could also have been used and then F would be a secondary variable related to fundamental variables M, L, and T by a dimensional equation based on Newton's law:

$$F = ma \tag{7.22}$$

$$\text{i.e.,} \quad F = [MLT^{-2}] \tag{7.23}$$

While any group of dimensionally independent quantities entering a dynamics problem may in fact be chosen as the fundamental quantities (such as mass, velocity, and acceleration), it is customary to use either F, L, and T or M, L, and T and we will use the former since most engineering dimensions are expressed in units of force and not in units of mass.

It is not permissible to consider both mass (M) and force (F) as fundamental quantities in dynamics problems as long as Newton's law holds.

At this point we will introduce the important properties of dimensional and physical equations on which dimensional analysis is based.

The ratio characteristic that most systems of measurement satisfy requires that the ratio of two measurements be independent of the units adopted. Thus, the ratio of two lengths will be the same whether the individual measurements be made in terms of inches, centimeters, cubits, or Swedish miles. Some common systems of measurement that do not satisfy this ratio requirement are the pH scale in chemistry, the Mohs' scale of hardness in minerology, and the decibel scale of acoustical engineering.

Bridgman (1931) has observed that a restriction of form for a dimensional equation exists when the fundamental quantities satisfy the ratio requirement.

This is best illustrated by an example. Let A, B, and C be three dimensional quantities that are related by some function (ψ_1) as follows:

$$A = \psi_1(B, C) \tag{7.24}$$

For example, Eq. (7.24) could be the following specific function which is in agreement with Eq. (7.24)

$$A^2 = 3B + BC \tag{7.25}$$

If Eq. (7.24) is a physically correct formulation then according to Bridgman the three variables A, B, and C may be replaced by three other variables which consist of products of A, B, and C raised to powers. One of the infinite number of such possibilities is

$$A^2B^2 = \psi_2(BC, C^2) \tag{7.26}$$

That Eq. (7.26) is valid for the specific example of Eq. (7.25) may be seen by multiplying both sides of this equation by B^2:

$$A^2B^2 = 3B^3 + B^3C \tag{7.27}$$

which may obviously be written as follows, according to Bridgman's observation.

$$A^2B^2 = \psi_3\left[\frac{B^3C}{(3B^3)^{2/3}}, \frac{(B^3C)^2}{(3B^3)^2}\right] \tag{7.28}$$

or

$$A^2B^2 = \psi_4[BC, C^2] \tag{7.29}$$

which is identical to Eq. (7.26).

The principle of dimensional homogeneity first expressed by Fourier states that the dimensions of each term of a physically correct equation must reduce to the same value. For example, Eq. (7.18) is readily found to meet this test

$$[L] = [L] + [LT^{-1}T] - [LT^{-2}T^2] \tag{7.30}$$

A homogeneous equation is one that contains no dimensional constants and holds true for all systems of units.

We are now ready to return to the displacement problem and to consider the steps involved between Eqs. (7.19) and (7.20).

Since this is a kinematic problem there are but two fundamental dimensions (L, T). Table 7.2 gives the dimensional equations for all quantities.

The quantities g and t are dimensionally independent (may not be combined by raising to powers and multiplying together to form a nondimensional group). If a function exists as stated in Eq. (7.19), then according to Bridgman we may also write

$$yg^{-1}t^{-2} = \psi_2(g, t, y_0 g^{-1}t^{-2}, V_0 g^{-1}t^{-1}) \tag{7.31}$$

TABLE 7.2 Dimensional Equations for Illustrative Dimensional
Analysis Problem

Quantity	Dimensional Equation
Vertical displacement, y	$[L]$
Initial vertical displacement, y_0	$[L]$
Acceleration due to gravity, g	$[LT^{-2}]$
Time, t	$[T]$
Initial vertical velocity, V_0	$[LT^{-1}]$

The left-hand side of this equation is nondimensional and by the principle of dimensional homogeneity, all terms of ψ_2 must also be nondimensional. The quantities g and t therefore cannot appear in ψ_2 except where combined with other quantities to form a nondimensional group. Thus,

$$\frac{y}{gt^2} = \psi_2\left(\frac{y_0}{gt^2}, \frac{V_0}{gt}\right)$$

which is Eq. (7.20).

The foregoing procedure may be followed in any case where all units satisfy the ratio requirement.

Exponents a and b, required to make $yg^a t^b$ nondimensional, may be found by writing simultaneous equations as follows, although usually these exponents may be written directly by inspection:

$$yg^a t^b = [L(LT^{-2})^a T^b] = [L^0 T^0]$$
$$\text{I}\phantom{(LT^{-2})^a T^b] = [}\text{II}$$

Equating exponents of L in I and II yields

$$1 + a = 0$$

while equating the exponents for T in I and II gives

$$-2a + b = 0$$

When these equations are solved simultaneously, a and b are found to be -1 and -2 respectively, in agreement with the left-hand side of Eq. (7.20).

The result of a dimensional analysis is sometimes written in symbolic form in terms of pi quantities as follows:

$$\pi_1 = \psi(\pi_2, \pi_3, \text{etc.}) \tag{7.32}$$

where π_1 is a nondimensional group involving the main dependent variable (y in the displacement problem) and the other pi quantities represent the remaining nondimensional quantities entering the problem.

The dimensionally independent quantities can usually be chosen in more than one way and hence there will often be several correct answers to a dimensional analysis. One of these answers may prove to be more convenient for a given purpose than the others. In the displacement problem all of the pairs of dimensionally independent quantities listed in Table 7.3 could have been used to obtain the dimensionless equations listed opposite each pair.

TABLE 7.3 Different Dimensional Analysis Solutions for Different Choices of Dimensionally Independent Quantities

Dimensionally Independent Quantities	Resulting Equation
y_0, g	$\dfrac{y}{y_0} = \psi\left(\dfrac{gt^2}{y_0}, \dfrac{V_0^2}{y_0 g}\right)$
y_0, t	$\dfrac{y}{y_0} = \psi\left(\dfrac{gt^2}{y_0}, \dfrac{V_0 t}{y_0}\right)$
y_0, V_0	$\dfrac{y}{y_0} = \psi\left(\dfrac{g y_0}{V_0^2}, \dfrac{V_0 t}{y_0}\right)$
g, t	$\dfrac{y}{gt^2} = \psi\left(\dfrac{y_0}{gt^2}, \dfrac{V_0}{gt}\right)$
g, V_0	$\dfrac{yg}{V_0^2} = \psi\left(\dfrac{gt}{V_0}, \dfrac{y_0 g}{V_0^2}\right)$
t, V_0	$\dfrac{y}{V_0 t} = \psi\left(\dfrac{gt}{V_0}, \dfrac{y_0}{V_0 t}\right)$

While there are no hard and fast rules concerning choice of dimensionally independent quantities, the following considerations serve as a useful guide:

1. The main dependent variable should not be chosen.
2. Variables having the greatest significance should be chosen.
3. The variables chosen should represent as many physically different aspects of the problem as possible.

In performing a dimensional analysis, it is important to include all quantities of importance to the problem. Otherwise, an incorrect and misleading result will be obtained. It is less important to include a variable about which doubt exists than to omit one which proves to be significant. Frequently, combinations of variables that are known to appear in a given class of problems in a unique association can be treated as a single quantity. This will result in fewer pi quantities in the final result. Another method of reducing the resulting pi quantities to a minimum is to limit the scope of the problem. Results of auxiliary or approximate analysis can sometimes be combined with conventional dimensional reasoning to greatly increase the power of dimensional analysis. Such auxiliary analysis includes

1. arguments involving symmetry
2. assumptions of linearity or known behavior
3. special solutions for large or small values of an important variable

The meaning of some of these more or less abstract statements will become clearer when specific applications are discussed in later chapters. Additional details concerning dimensional analysis may be found in Van Driest (1946) and Taylor (1974).

DIMENSIONAL ANALYSIS OF TRANSDUCER TUBE LATHE DYNAMOMETER

Returning now to the transducer tube dynamometer design, the angular deflection α between points B and C of Fig. 7.16 will be a function of the variables indicated in Eq. (7.33):

$$\alpha = \psi_1(a, b, d, M_P, E) \qquad (7.33)$$

where M_P is the bending moment at the minimum section. Table 7.4 gives the dimensions of each of the quantities in Eq. (7.33) in terms of F, L, T.

The quantities E and a are dimensionally independent and hence we may write

$$\alpha = \psi_2\left(\frac{b}{a}, \frac{d}{a}, \frac{M_P}{Ea^3}\right) \qquad (7.34)$$

Experience with elastic structures teaches that α should vary directly with (M_P) and inversely with b. When this constraint is applied to Eq. (7.34), we obtain

$$\alpha = \left(\frac{M_P}{Ea^3}\right)\left(\frac{a}{b}\right)\psi_3\left(\frac{d}{a}\right) \qquad (7.35)$$

In order to determine the function ψ_2 it is necessary to perform several experiments on simple bars (Fig. 7.18a) with various d/a ratios obtained by gradually increasing d. The results of such a series of tests is shown plotted in Fig. 7.18b. The fact that all points lie on a smooth curve is consistent with a correct dimensional analysis. The experimental curve can be well approximated by the following exponential function:

$$\alpha \frac{Ea^2b}{M} = 17\left(\frac{d}{a}\right)^{0.395} \qquad (7.36)$$

By use of this equation, the slotted member may be designed to give the proper stiffness for the force range expected and the type of transducer available. The above example illustrates how dimensional reasoning may sometimes be combined with other known facts to yield a solution to a problem that is otherwise not readily tractable.

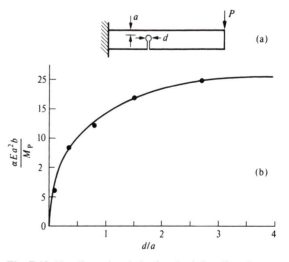

Fig. 7.18 Nondimensional plot for elastic bending characteristics of slotted bar shown at (a).

TABLE 7.4 Dimensional Equations for Transducer Tube Lathe Dynamometer

Quantity	Dimensional Equation
Angular displacement, α	
Linear dimension, a	$[L]$
Linear dimension, b	$[L]$
Linear dimension, d	$[L]$
Bending moment, M_P	$[FL]$
Young's modulus, E	$[FL^{-2}]$

STRAIN RINGS

In the foregoing lathe dynamometers, the force must always be applied at the same point since such an instrument is essentially a moment-measuring device. For many purposes, it is necessary to measure the force applied to the work piece as this force moves with the progress of a cut. In order to measure such an arbitrarily placed three-dimensional force or torque, the strain ring has proven to be a useful device. We will first consider the characteristics of several strain rings and then their application to dynamometry.

Strain rings (Fig. 7.19a) provide a high ratio of sensitivity to stiffness at the same time having adequate stability against buckling. A ring is easily produced and simple to mount. The fact that the inside is always in an opposite state of strain from the outside allows four active arms to be effectively used in a bridge circuit as already described. The symmetry of a ring provides parallel paths for heat flow and hence it is to be expected that equivalent points on opposite sides of a ring will be at the same temperature. This enables drift due to temperature gradient in the vicinity of the dynamometer to be eliminated by connecting the gages to form a complete bridge circuit. Aluminum is a good material from which rings may be made since it will not corrode as readily as steel, is light in weight, easily machined, and has excellent heat conductivity. A thin metal ring of radius r, thickness t, and axial width b is shown in Fig. 7.19a. The ring is fixed at the bottom while a radial force F_r and a tangential force F_t are applied at the top. The point of load application is maintained horizontal by a suitable moment M. When only F_r is applied, the ring will deform as in Fig. 7.19b. Thin ring elastic theory shows the strain at the inside and outside surfaces of the ring at points A to be

$$e_A = \pm \frac{1.09 F_r r}{E b t^2} \tag{7.37}$$

while the strain at the point B ($39.6°$ from the vertical axis) is zero. When only F_t is applied (Fig. 7.19c) the strain at A is zero while the strain at B is

$$e_B = \pm \frac{2.18 F_t r}{E b t^2} \tag{7.38}$$

By placing strain gages at the inside and outside surfaces of a ring at points A and B (Fig. 7.19a), it is possible to separate and measure the F_r and F_t components of force. Only force component F_r will cause a change in resistance in the gages at A with no change in resistance (or strain) at B. Thus, if the inside and outside gages at A are mounted in the opposite arms of a Wheatstone bridge circuit we have a sensitive means for detecting changes in F_r. In a similar way the gages at B are only sensitive to changes in F_t. As long as the strains at A and B are within the elastic limit of the ring material, it is evident from Eqs. (7.37) and (7.38) that the strains will vary linearly with the force, and hence the strain ring satisfies the desired requirement of linearity.

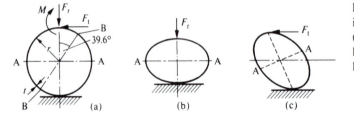

Fig. 7.19 Circular strain ring showing the type of deformation produced (a) by combined load, (b) by radial load (F_r) only, (c) by a tangential load (F_t) only.

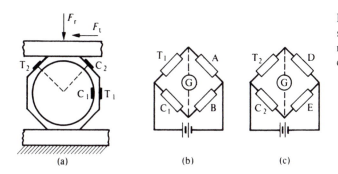

Fig. 7.20 (a) Octagonal strain ring showing gage positions. (b) Bridge connections for measuring F_r. (c) Bridge connections for measuring F_t.

The values of stiffness of a thin circular ring in the radial and tangential directions are as follows:

$$K_r = \frac{F_r}{\delta_r} = \frac{Ebt^3}{1.8r^3} \tag{7.39}$$

$$K_t = \frac{F_t}{\delta_t} = \frac{Ebt^3}{3.6r^3} \tag{7.40}$$

While Eqs. (7.37) and (7.38) hold quite well even for fairly thick rings, the stiffness values given in Eqs. (7.39) and (7.40) are usually found to be much less reliable. This is due in part to the difficulty of preventing rolling at the points of attachment, which changes the end conditions from those assumed. Rings are usually mounted by bolts extending through holes in the top and bottom surfaces.

End conditions can be readily maintained by making the outside surface of the ring octagonal instead of circular (Fig. 7.20).[†] Use of octagonal rings provides points of stress concentration at positions on the horizontal axis and at 45° to the vertical axis. If gages are placed at these points of stress concentration and the minimum value of t is used in Eqs. (7.37) and (7.38), good results are obtained even though the inclined gages are at points 45° from the vertical instead of the 39.5° called for by the circular ring theory. The octagonal rings are stiffer than circular ones of the same minimum section.

Proper strain gage positions are shown in Fig. 7.20a while the manner of connecting the gages to independently measure F_r and F_t is shown in Figs. 7.20b and 7.20c. The resistances A and B in Fig. 7.20b might be a pair of fixed resistances or, if two or more rings are used simultaneously, they would be active strain gages at the horizontal section of another ring on the compressive and tensile sides respectively. In the latter case, the sum of the vertical loads on the two rings would be recorded. Similar remarks might be made for resistances D and E of Fig. 7.20c. In order that there be no cross sensitivity it is important that gages T_2 and C_2 be mounted exactly symmetrically with respect to the ring axes.

If both sides of Eq. (7.37) are divided by force F_r we obtain

$$\frac{e_A}{F_r} = \frac{1.09r}{Ebt^2} \tag{7.41}$$

[†] It has also been found important to make the top and bottom surfaces of the octagon (surfaces of attachment) slightly concave to avoid a change in end conditions under load. If this is not done, a troublesome hysteresis effect is apt to occur.

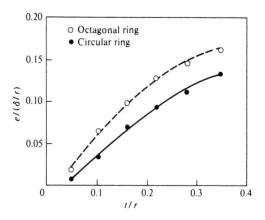

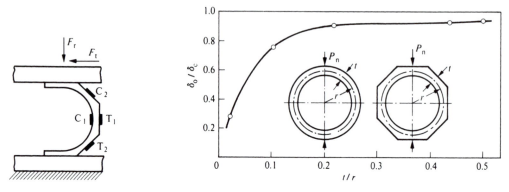

Fig. 7.21 The octogonal half-ring.

Fig. 7.22 Difference in displacement of circular ring (δ_c) and octogonal ring (δ_0). (after Ito, Sakai, and Ishikawa, 1980)

If this equation and Eq. (7.39) are multiplied together we have an expression for the *strain per unit deflection*

$$\frac{e_A}{\delta_r/r} = \frac{1.09}{1.8}\frac{t}{r} = 0.61\frac{t}{r} \qquad (7.42)$$

where δ_r is the deflection in a radial direction due to load F_r. Neither the ring width (b) nor modulus of elasticity (E) is of importance with regard to the strain per unit deflection. It is evident that we should like e_A/δ_r to be as large as possible in the interest of maximum sensitivity and rigidity. This calls for r being as small as possible and t as large as possible. However, r can be reduced only to the point where it becomes too difficult to mount internal gages accurately. Thus, for a ring of given size r, we should like t to be as large as it can be consistent with the desired sensitivity. Therefore, it is clear that the total load should be divided between the smallest number of rings possible. This suggests the half octagonal ring shown in Fig. 7.21. Here strain gage T_2 must be mounted on the lower half of the ring instead of on the upper half of the ring as in Fig. 7.20a.

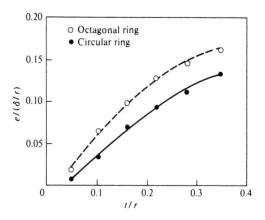

Fig. 7.23 Relation between t/r and strain output for circular and octagonal rings. (after Ito, Sakai, and Ishikawa, 1980)

Ito, Sakai, and Ishikawa (1980) have performed a finite element analysis for the elastic behavior of octagonal rings. Figure 7.22 shows the relative radial deflection of octagonal (δ_0) and circular (δ_c) rings of different thickness (t) to mean radius (r) ratio when subjected to the same vertical load. In this analysis, E was 21,000 kg mm^{-2} ($\sim 30 \times 10^6$ p.s.i.) and Poisson's ratio was 0.28. Figure 7.22 shows that whereas the octagonal ring is substantially stiffer than the circular ring when t/r is 0.05 or less, the difference is less than 10% if t/r is 0.25 or greater. Figure 7.23 shows the corresponding sensitivity to stiffness ratio ($e/\delta/r$) for octagonal and circular rings of different t/r ratio. Figure 7.24 shows the angle θ measured from the direction of the radial load for the null point in strain for circular and octagonal rings for different t/r ratios. For a thin ($t/r = 0.1$) circular ring, $\theta = 39.5°$. This figure shows that the null point in strain for a radial load on an octagonal ring occurs close to 45°. These results are generally consistent with the thin ring theory

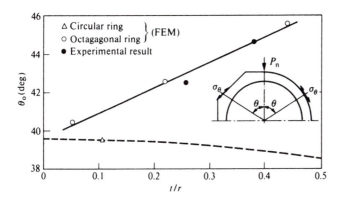

Fig. 7.24 Relation between the position
where the tangential stress is zero and
t/r. (after Ito, Sakai, and Ishikawa, 1980)

considered initially and suggest that the thin ring theory is sufficiently accurate for design purposes. An exact prediction of dynamometer sensitivity is not essential since the final dynamometer must be carefully calibrated in any case.

MILLING DYNAMOMETER

In a plane-milling operation using a helical-milling cutter, three orthogonal components of force must be recorded simultaneously. The dynamometer shown in Fig. 7.25 employs four octagonal rings mounted between two plane rigid surfaces oriented as shown in the plan view of Fig. 7.25. The axes of rings B and C are in the y-direction while rings A and D have their axes in the x-direction.

When an arbitrarily placed vertical load (F_V) is applied it is carried by all four rings, and hence the two vertical measuring gages (T_1 and C_1) in Fig. 7.20 of each ring are wired into one eight-gage bridge circuit in such a way that all outputs add. A force in the x-direction (F_x) is measured by the gages in the inclined positions. T_2 and C_2 in Fig. 7.20 are connected in a four-gage bridge such that their deflections add. A force in the x-direction produces no bending in rings A and D since their axes are in the direction of the force (a very stiff direction). The gages on rings A and D are used to measure forces in the y-direction just as those on rings B and C are used for forces in the x-direction.

In designing a dynamometer such as that in Fig. 7.25, we have need for the stiffness of a ring in the axial direction. From thin ring theory we can obtain an expression for the stiffness (K_a) in the axial direction K, just as we did previously for the radial and tangential directions [Eqs. (7.39) and (7.40)]. However, due to the uncertainty of the end conditions this equation in the K direction does not hold too well. The following semiempirical expression for K_a has been found to give a better estimate than that derived from thin ring theory.

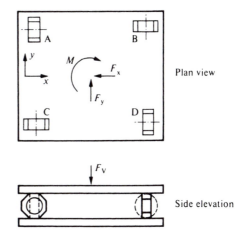

Fig. 7.25 Four-dimensional dynamometer for milling, planing, shaping, drilling studies utilizing four octagonal rings as measuring elements. The dynamometer measures F_x, F_y with rings A and D, F_V with all four rings, moment about vertical axis (M) by rings A and B. [Range: 5 to 5000 lb (25 to 2.5×10^4 N). Stiffness: 25,000 lb in^{-1} (4350 n mm^{-1}).] (after Cook, Loewen, and Shaw, 1954)

$$K_a = \frac{Ebt}{40r} \qquad (7.43)$$

The dynamometer shown in Fig. 7.25 can measure vertical (z-direction) loads up to 5000 lb (22 kN) with a sensitivity of 5 lb (22 N). The measurements are independent of the point of application of the load within the square outlined by the rings.

DRILLING DYNAMOMETER

In drilling and tapping studies, it is only necessary to measure the axial thrust force and torque on the tool.

Reversing the positions in the foregoing bridge circuit of the inclined gages on ring 1 while those on ring 3 are kept the same, enables a torque to be measured about the z-axis. Thus, the device shown in Fig. 7.25 can measure not only force components in three orthogonal directions without interference, but can also be used to measure torques by merely throwing a selector switch that reconnects the gages.

Dynamometers of the basic type shown in Fig. 7.25 may be used to study milling, planing, shaping, broaching, and drilling operations.

THREE-COMPONENT TURNING DYNAMOMETER

A three-component turning dynamometer was constructed using four extended half-rings machined from a single block of steel to avoid hysteresis effects (Fig. 7.26). Force component F_R puts all four half-rings in compression, F_Q subjects all elements to a tangential force while F_P puts the lower elements in compression and the upper ones in tension. By properly connecting the gages, all three components of force may be simultaneously recorded with negligible cross sensitivity. The characteristics of this dynamometer were as follows: range, 4 to 2000 lb (18 N to 8.9 kN); stiffness, 5×10^6 lb in^{-1} (566 kN m^{-1}).

SURFACE GRINDING DYNAMOMETER

Figure 7.27 shows the manner in which the foregoing principles have been applied to produce a dynamometer for use in measuring radial and tangential forces between a grinding wheel and workpiece in a plunge (no crossfeed) surface grinding operation. Here, we have in effect two octagonal half-rings which are machined from a single piece of aluminum to provide a sensitive element of low mass and high natural frequency. Four strain gages are mounted on each half ring and connected

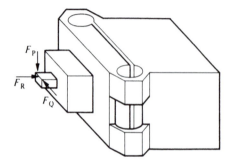

Fig. 7.26 Three-dimensional lathe dynamometer.

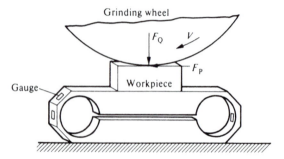

Fig. 7.27 Two-dimensional grinding dynamometer.

to form two complete Wheatstone bridge circuits; one for measuring the power component of force (F_P) and one for measuring the feed component (F_Q). By using two gages from each ring in each of the bridge circuits, the instrument becomes independent of the point of application of the load between the rings.

The characteristics of this dynamometer are as follows: range, 0.1 to 100 lb (0.45 N to 445 N); stiffness, 4000 lb in^{-1} (453 N m^{-1}).

A two-component milling dynamometer having the same design but stiffer rings had the following characteristics: range, 2 to 2000 lb (8.9 to 8910 N); stiffness, 10^6 lb in^{-1} (113 kN m^{-1}).

Further discussion of inexpensive strain gage dynamometers for use in grinding and super-finishing studies may be found in Shaw (1996).

INSTRUMENTATION

It need hardly be mentioned that even the most efficiently designed dynamometer will be of little value if the measuring instrument used with it is not accurate or sensitive enough, or does not have a frequency response that is sufficiently high to follow significant fluctuations of force. Several instruments are available for use with strain gage bridge circuits. The one item all of these instruments have in common is an amplifier which is a necessity since strain gage output voltages are so minute. In other respects the several instruments differ widely.

A most important distinction may be made between those instruments which yield a permanent record and those which are merely indicators. Aside from the inconvenience associated with reading two or more indicators simultaneously there is the more important matter of recording an average or characteristic value of a fluctuating force. It is possible for different operators to record widely differing "average" values for a fluctuating characteristic when an indicator is used. However, when the force curve is continuously recorded, it is then a matter of correctly interpreting the recorded data.

Another important decision which must be made concerns the choice between an AC and DC bridge circuit. The AC circuit has the advantage of eliminating the possibility of thermoelectric effects in the circuit wiring and generally offers a greater range of stable amplication than the DC circuit. On the other hand the DC circuit requires no shielding and offers no problem from 60-cycle mains pick-up. In either case the instrument may be of either the deflection type or the null-balance potentiometric type that requires manual or automatic balancing.

A useful instrument for recording low DC voltages is the photoelectric potentiometer recorder. This device operates essentially as an ordinary potentiometer except that the galvanometer is "read" by two phototubes and instead of the back voltage having to be adjusted manually, it is automatically controlled by an electronic circuit. The result is a high gain without the usual troubles associated with DC amplifiers. With four active strain gages, this instrument is capable of measuring strains as small as 0.1 μin in^{-1}. Despite this sensitivity, it can withstand considerable overloads without damage.

If a permanent record is not required, then any DC voltmeter can be used provided it is sensitive enough and has a high input impedance. A sensitive DC vacuum tube millivoltmeter has proved quite useful. If speed is not essential any of the better thermocouple potentiometers can be used, especially the self-balancing types. These instruments are distinguished by their high sensitivity, (up to 0.003 mV per division) combined with a large scale, (up to 30 in or 0.76 m). However, the speed of response of these devices is limited (about 2 s for full scale). The aforementioned photoelectric potentiometer recorder can follow small variations up to about 4 cycles s^{-1} ($\frac{1}{2}$ s full scale), but like all automatic feedback potentiometers it becomes either unstable or sluggish as the bridge resistance increases. Vacuum tube voltmeters do not suffer from this restriction.

Probably the most familiar of the AC strain gage bridge instruments is the portable strain indicator, which is a manually balanced, null-type instrument. The fact that the strain indicator involves both manual balancing and reading makes it rather slow to use, which is awkward under fluctuating conditions.

PIEZOELECTRIC DYNAMOMETRY

Certain crystalline materials, notably quartz (SiO_2) and barium titanate ($BaTiO_3$), have unit cells with an identifiable dipole. In a strong electrical field, all cells tend to align, producing positive and negative ends. If electrodes attached to these ends are subjected to tensile or compressive stresses, they separate or approach each other and a voltage change results. Or, if these electrodes are subjected to a voltage, a change in the distance between electrodes results. Crystals that exhibit such an electromechanical action are referred to as *electromechanical transducers* and are called *piezoelectric* (pressure-electric) devices. Depending on the location of the surfaces of a piezoelectric crystal relative to the crystal axes, a crystal element is capable of reacting to normal or shear stresses. When the input is a stress, the output is a small electrical charge that may be used to measure dynamometer forces. When the input is an EMF, the output is a displacement often used in ultrasonic generators (Iketa, 1990).

Piezoelectricity was discovered by Pierre and Jacques Curie in 1880 but was not widely used in dynamometers until the 1960s, after the charge amplifier and high resistance materials such as Teflon and kapton were developed.

Quartz is the piezoelectric material used in most modern dynomometers. Most force sensors have an elastic element, the deflection of which is a measure of force. With quartz piezoelectric sensors, the deflections involved are of atomic dimensions rather than microns as with strain gages. This provides a transducer of very great stiffness, high natural frequency, low rise time (time for the output to go from 10 to 90% of the final value), and a high time constant (time for a DC signal to decay to 37% of the initial value). A quartz transducer also has a very high resistance (~ 10^{13} ohms) and useful anisotropic properties. Depending on the orientation of a platelet cut from a single crystal, it may be sensitive to a compressive stress or a shear stress as shown in Fig. 7.28. Differently orientated platelets may be stacked and preloaded (to prevent slip with shear loading) with insulation between the elements, to produce a compact three-force component washer. Figure 7.29 shows such a sensor, where quartz

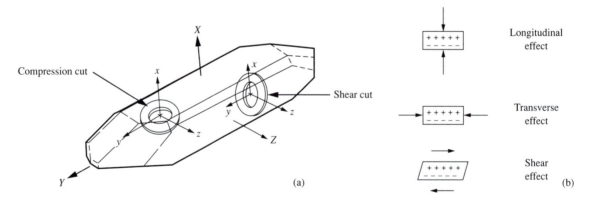

Fig. 7.28 (a) Quartz crystal showing orientation of compression and shear platelets relative to the crystal axes. (b) Loading orientations for producing longitudinal, transverse, and shear piezoelectric effects on compression and shear platelets. (after Kistler Instrument Corp.)

elements A, B, and C are oriented to be sensitive to forces in the x, y, and z directions, respectively. Load washers are available in diameters from 0.4 to 1.4 in (10 to 35 mm), for maximum loads from 1600 to 20,000 lb (7300 to 90,000 N) and natural frequencies from 200 to 65 kHz.

Figure 7.30a shows a platform with four washers (each a three-component force sensor) mounted between upper and lower plates and interconnected to produce a three-component platform dynamometer. This type of dynamometer is convenient for turning, milling, and a variety of other metal cutting studies. These platforms come in a variety of sizes and are sealed against contact of the sensors with cutting fluids. Figure 7.30b shows this dynamometer with a tool-holding attachment mounted on the top plate. Such a dynamometer would be used in turning studies with tools of moderate size. The tool holder shown in Fig. 7.30b will accommodate tool shanks up to 1×1 in (26×26 mm) in size. For heavier turning operations, larger platforms with internal water cooling are available as well as tool holders accommodating tool shanks of larger size. The dynamometer of Fig. 7.30a could also be used for a variety of milling studies. The four three-dimensional sensors near the corners of the dynamometer of Fig. 7.30a play a similar role as the rings in Fig. 7.25 and when properly connected can provide a four-dimensional platform (F_x, F_y, F_z, and M_z).

Figure 7.31a shows a stack of four elements, the top one consisting of a number of F_x platelets arranged in a circle and connected to measure torque M_z about the z-axis. The other three platelets are

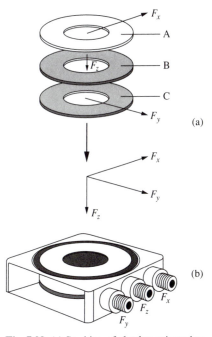

Fig. 7.29 (a) Stacking of platelets oriented to be sensitive to force components in the x, y, and z directions. (b) Stack of three insulated platelets interconnected to produce a washer sensitive to three orthogonal force directions. (after Kistler Instrument Corp.)

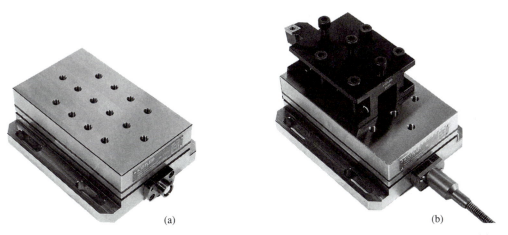

Fig. 7.30 (a) Three-component quartz dynamometer for measuring three orthogonal components of force. [Specifications: range for F_x, F_y, F_z, -1.12 to 1.12 klb (-5 to $+5$ kN); overload for F_x, F_y, F_z, -1.69 to 1.69 klb ($-7.5/7.5$ kN); threshold, <0.002 lb (<0.01 N); sensitivity for F_x, F_y -1.69 pC/lb (-7.5 pC/N); sensitivity for F_z, -0.84 pC/lb (-3.7 pC/N); linearity (all ranges), $<1\%$ FSO; cross talk (influence of F_x on F_y or F_z), $<2\%$; rigidity in x dir., <111 lb/μin (<1 kN/μm); weight, 16 lb (7.3 kg).] (b) Fig. 7.30a with a turning tool mounted on top of the platform. (after Kistler Instrument Corp.)

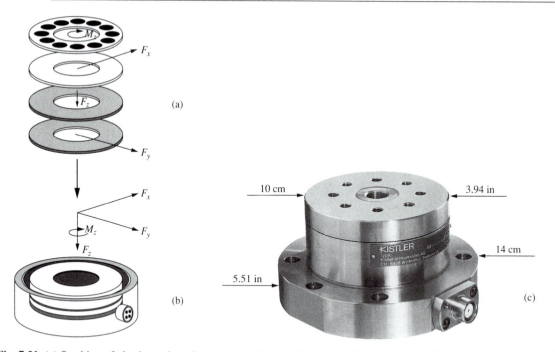

Fig. 7.31 (a) Stacking of platelets oriented to measure F_x, F_y, or F_z and M_z. (b) Assembled finished sensor. (c) Sensor of Fig. 7.31b mounted between concentric plates to produce a dynamometer for measuring four components: F_x, F_y, F_z, and M_z (all dimensions in inches). [Specifications: range for F_z, −1.1 to 4.5 klb (−4.9 to 20 kN); range for F_x, F_y, ±1.1 klb (±4.9 kN); range for M_z, ±148 ft-lb (200 mN); maximum force for F_z, −1.3 to 5.4 klb (−5.8 kN to 24 kN); maximum force for F_x, F_y, ±1.3 klb (±5.8 kN); maximum force for M_z, ±177 ft-lb (240 mN); threshold for F_x, F_y, F_z, 0.004 lb (0.018 N); threshold for M_z, 0.002 in-lb (0.0027 m-N); sensitivy for F_z and F_x, F_y, −17 and −34 pC/lb (−3.82 and −7.64 pC/N); rigidity for F_z and F_x, F_y, 11.4 and 2.3 lb/μin (103 and 20.7 N/μm); nat. freq. for z and x, y, 6 and 3 kHz; cross talk (force components), <±2%; weight, 9.3 lb (4.2 kg).] (after Kistler Instrument Corp.)

connected to measure force components F_x, F_y, and F_z. Figure 7.31b shows the resulting washer, while Fig. 7.31c shows a washer mounted between concentric plates that yields a dynamometer useful for drilling, tapping, reaming, and even face-turning studies.

The dynamometer shown in Fig. 7.32 is designed for applications such as end milling where a rotating cutting force is to be measured as well as three components of force (F_x, F_y, F_z). The signals are transmitted from a rotating dynamometer by telemetry. A dynamometer of this type is useful for multi-edge tool studies where forces on individual cutting edges are of interest.

Piezoelectric crystals are active elements in that they produce an output (charge) only when subjected to a change in force. However, their very high impedance when combined with high-impedance circuitry enables them to retain a charge for an extended period of time providing a quasi-static character. The output of a quartz piezoelectric sensor is a very small charge measured in pico coulombs (pC, where $1 \text{ pC} = 10^{-12} \text{ C} = 10^{-12} \text{ A s}$). This very small charge must be converted to an equivalent change in voltage in millivolts. This is the function of the charge amplifier.

Figure 7.33 shows the typical measuring chain for a drilling operation. Dynomometers that require a charge amplifier are called high-impedance units. Low-impedance units introduced in the mid-1960s have built-in charge amplifiers for each element, eliminating need for an external multichannel charge amplifier.

SINGLE ABRASIVE GRAIN DYNAMOMETER

The characteristics of a single abrasive particle mounted in the periphery of a metal disc are frequently studied to evaluate different types of abrasive materials. At a wheel speed of 6000 f.p.m. (30 m s^{-1}), a maximum depth of cut of 0.001 in (2.5 μm), the cutting time per cut would be about 168 μs. The natural frequency of the dynamometer required for such a case should be about $4 \times (10^6/168) = 24,000$ Hz. It is not easy to design a dynamometer with such a high natural frequency. This calls for a very stiff transducer and very low total mass (including that of the workpiece).

Eiss (1967) and coworkers first made a mini-dynamometer for single abrasive grain studies using two ferro-electric ceramic elements. One piezoelectric transducer was oriented to measure the normal component of force and the other to measure the tangential component. Crisp, Seidel and Stokey (1968) built the improved two component piezoelectric dynamometer shown diagrammatically in Fig. 7.34. By substituting glass for plastic insulation between the crystals and cementing the workpiece to the dynamometer, it was possible to obtain natural frequencies of 120,000 Hz (normal direction) and 60,000 Hz (tangential direction) with essentially no hysteresis. Signals for the dynamometer were displayed on a dual beam oscilloscope and recorded

Fig. 7.32 Rotating quartz four-component dynamometer with transfer of data by telemetry. [Specifications: maximum speed, 10,000 rpm; range for F_z, −4.5 to +4.5 klb (−20 to 20 kN); range for M_z, −147.5 to +147.5 lb-ft (−200 to 200 Nm); threshold for <0.9 lb (<4 N); nat. freq. 1 kHz; weight 11 lb (5 kg).] (after Kistler Instrument Corp.)

Fig. 7.33 Measuring chain for a drilling operation. (after Kistler Instrument Corp.)

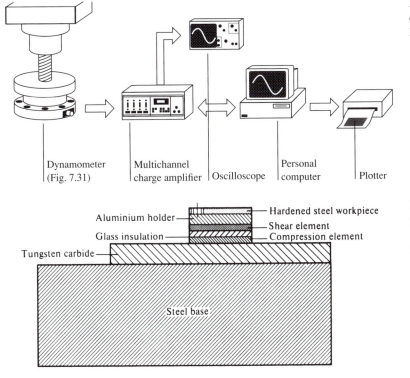

| Dynamometer (Fig. 7.31) | Multichannel charge amplifier | Oscilloscope | Personal computer | Plotter |

Aluminium holder — Hardened steel workpiece
Glass insulation — Shear element
Tungsten carbide — Compression element
Steel base

Fig. 7.34 Piezoelectric dynamometer. Steel base is 3 in (76 mm) long, other dimensions being in proportion except for thickness of piezoelectric elements and glass insulation which have been exaggerated. (after Crisp, Seidel, and Stokey, 1968)

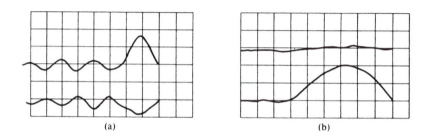

Fig. 7.35 Calibration tests. Upper traces are shear element responses and lower traces are compression. Sweep 10 μs cm^{-1}, right to left. (a) Tangential Force $\frac{1}{4}$ in ball, $\frac{1}{4}$ in drop; peak force, 11.3 lb; sensitivity: normal, 100 mV cm^{-1}; tangential, 500 mV cm^{-1}. (b) Normal Force $\frac{1}{2}$ in ball, $\frac{11}{16}$ in drop peak force, 19.6 lb; sensitivity: normal, 200 mV cm^{-1}; tangential, 500 mV cm^{-1}. (after Crisp, Seidel, and Stokey, 1968)

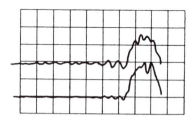

Fig. 7.36 Tangential (upper trace) and normal components of cutting force. Sweep 50 μs cm^{-1}, right to left. Calibration: tangential, 1.43 lb cm^{-1} (6.36 N cm^{-1}); normal, 4.94 lb cm^{-1} (21.98 N cm^{-1}). (after Crisp, Seidel, and Stokey, 1968)

using a polaroid camera. The oscilloscope sweep was triggered by the signal itself.

Calibration of the dynamometer was obtained by dropping small steel balls from different known heights and calculating the height of rebound (needed to obtain the energy absorbed) from the time elapsed between the first and second impacts. Figure 7.35 shows calibration results for tangential and normal crystals respectively. Cross sensitivity is seen to be very low. Figure 7.36 shows force traces (cutting right to left) for a single diamond particle. The normal component of force is seen to be about three times the tangential component.

CONCLUDING REMARK

Many additional dynamometer designs are to be found in the literature. However, it is felt that the units described are sufficient to demonstrate the problems involved and a few representative ways of solving them.

REFERENCES

Bridgman, P. W. (1931). *Dimensional Analysis*, 2nd ed. Yale University Press, New Haven.
Cook, N. H., Loewen, E. G., and Shaw, M. C. (1954). *Am. Mach., N.Y.* **98**, 125.
Crisp, J., Seidel, J. R., and Stokey, W. F. (1968). *Int. J. Prod. Res.* **7**, 159.
Eiss, N. S. (1967). *J. Engng. Ind.* **89**, 463.
Iketa, T. (1990). *Fundamentals of Piezoelectricity*. Oxford University Press, Oxford.
Ito, S., Sakai, S., and Ishikawa, M. (1980). *Bull. Japan Soc. Precision Engrs.* **14**, 23.
Shaw, M. C. (1996). *Principles of Abrasive Processing*. Clarendon Press, Oxford.
Taylor, E. S. (1974). *Dimensional Analysis for Engineers*. Clarendon Press, Oxford.
Van Driest, E. R. (1946). *J. Appl. Mech.* **68**, A34.

8 SHEAR STRAIN IN STEADY STATE CUTTING

In Chapter 3 the mechanics of orthogonal metal cutting with a continuous chip and a sharp tool was discussed from the point of view of the interrelationships that exist between important variables. However, no attempt was made to *predict* cutting performance. In Chapter 3 it was revealed that the bulk of the energy in metal cutting is consumed on the shear plane and that the specific shear energy (u_S) accounts for about $\frac{3}{4}$ of the total. It was further shown in Eq. (3.27) that

$$u_S = \tau \gamma$$

A first step in predicting cutting performance would therefore involve estimating the shear stress on the shear plane (τ) and the shear strain associated with chip formation (γ). Both of these quantities are strongly influenced by the friction on the face of the tool which has been expressed in terms of the friction angle β (Fig. 3.5), where

$$\beta = \tan^{-1}\frac{F_C}{N_C} \tag{8.1}$$

In this and the next two chapters, problems associated with the estimation of γ, τ, and β will be discussed in that order.

In Chapter 3, it was shown in Eq. (3.17) that a relatively simple relationship exists between shear strain (γ), rake angle (α), and shear angle (ϕ):

$$\gamma = \frac{\cos \alpha}{\sin \phi \cos (\phi - \alpha)}$$

Since the rake angle will be known, the problem of predicting γ reduces to the problem of predicting (ϕ).

DIMENSIONAL ANALYSIS

To begin, we might perform a dimensional analysis in an effort to determine the nature of the relationship that is to be expected for ϕ. In addition to ϕ, two quantities are needed to define completely the geometry of a two-dimensional cutting process, namely the rake angle (α) and the depth of cut (t). Two distinct operations are involved in the cutting process: a shearing action along the shear zone, and sliding with friction along the tool face. The mean coefficient of friction ($\mu = \tan \beta$) is usually sufficient to characterize a friction process, and the shear stress τ will usually characterize

TABLE 8.1 Dimensional Analysis for Shear Angle

Quantity	Symbol	Dimensions
Shear angle	ϕ	. . .
Rake angle	α	. . .
Depth of cut	t	L
Friction angle	β	. . .
Shear stress	τ	FL^{-2}
Cutting velocity	V	LT^{-2}

the dynamic aspects of a shear process. The cutting velocity V, together with the geometry of the process, will fix the kinematics of the system. It would then appear that the several quantities that might figure in a determination of the angle ϕ are those listed in Table 8.1, where F, L, and T refer to the fundamental dimensions of force, length, and time. When a dimensional analysis is performed in the usual manner we obtain

$$\phi = \psi(\alpha, \beta) \tag{8.2}$$

which states that the shear angle ϕ is a function of α and β only.

EARLY COMPLETE SOLUTIONS

There have been many notable attempts in the past to derive a complete physical equation corresponding to Eq. (8.2). The reasoning followed in several early analyses will next be briefly outlined followed by a discussion of this work.

Piispanen (1937) presented a graphical analysis for the shear angle, which was equivalent to the second analysis of Merchant given later in this chapter. However, inasmuch as this early work did not lead to an analytical expression, it will not be considered further.

Ernst and Merchant (1941) presented the first quantitative analysis. In Fig. 8.1 the forces at the point of a tool are shown. Assuming the shear stress on the shear plane τ to be uniformly distributed, it is evident that

$$\tau = \frac{F_S}{A_S} = \frac{R' \cos(\phi + \beta - \alpha) \sin \phi}{A} \tag{8.3}$$

where A_S and A are the areas of the shear plane and that corresponding to the width, b, times the depth of cut, t, respectively. Ernst and Merchant (1941) reasoned that ϕ should be an angle such that τ would be a maximum, and a relationship for ϕ was obtained by differentiating Eq. (8.3) with respect to ϕ and equating the resulting expression to zero. This led to the result

$$\phi = 45 - \frac{\beta}{2} + \frac{\alpha}{2} \tag{8.4}$$

However, it is to be noted that in differentiating, both R' and β were considered independent of ϕ.

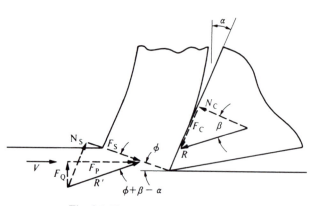

Fig. 8.1 Force relations at tool point.

Merchant (1945) presented a different derivation that also led to Eq. (8.4). This time, an expression for the total power consumed in the cutting process was first written

$$P = F_p V = \tau A V \frac{\cos (\beta - \alpha)}{\sin \phi \cos (\phi + \beta - \alpha)} \tag{8.5}$$

It was then reasoned that ϕ would be such that the total power would be minimum. An expression identical to Eq. (8.4) was obtained when P was differentiated with respect to ϕ, this time considering τ and β to be independent of ϕ.

This is what Piispanen (1937) had done previously in a graphical way. However Piispanen immediately carried this line of reasoning one step further and assumed that the shear stress τ would be influenced directly by normal stress on the shear plane σ as follows:

$$\tau = \tau_0 + K\sigma \tag{8.6}$$

where K is a material constant. Piispanen then incorporated this observation into his graphical solution for the shear angle.

Upon finding Eq. (8.4) to be in poor agreement with experimental data, Merchant also independently (without knowledge of Piispanen's work) assumed a relationship as given by Eq. (8.6) and proceeded to work this into his second analysis as follows:

From Fig. 8.1 it may be seen that

$$\sigma = \tau \tan (\phi + \beta - \alpha) \tag{8.7}$$

or from Eq. (8.6)

$$\tau = \tau_0 + K\tau \tan (\phi + \beta - \alpha) \tag{8.8}$$

hence

$$\tau = \frac{\tau_0}{1 - K \tan (\phi + \beta - \alpha)} \tag{8.9}$$

When this is substituted into Eq. (8.5), we have

$$P = \frac{\tau_0 A V \cos (\beta - \alpha)}{[1 - K \tan (\phi + \beta - \alpha)] \sin \phi \cos (\phi + \beta - \alpha)} \tag{8.10}$$

Now, when P is differentiated with respect to ϕ and equated to zero (with τ_0 and β considered independent of ϕ), we obtain

$$\phi = \frac{\cot^{-1}(K)}{2} - \frac{\beta}{2} + \frac{\alpha}{2} = \frac{C - \beta + \alpha}{2} \tag{8.11}$$

Merchant called the quantity $\cot^{-1} K$ the "machining constant" C. In Fig. 8.2 the quantity C is seen to be the angle the assumed line relating τ and σ makes with the τ-axis.

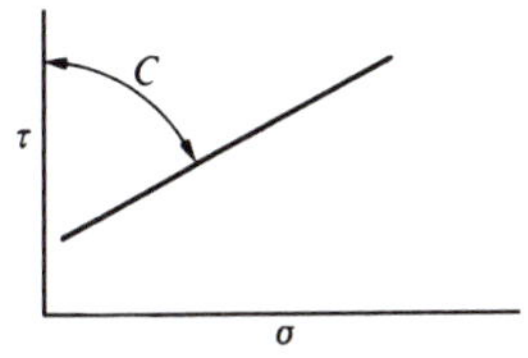

Stabler (1951) has presented an analysis of three-dimensional cutting operations in which an expression for shear angle is also presented. When it was assumed that the maximum shear stress on the shear plane and the resultant shear strain were collinear, then it followed from purely geometrical considerations that

$$\phi = 45 - \beta + \frac{\alpha}{2} \tag{8.12}$$

Fig. 8.2 Relation between flow stress and normal stress assumed by Piispanen (1937) and Merchant (1945).

SLIPLINE FIELD SOLUTIONS

Lee and Shaffer (1951) have derived still other relationships for ϕ. These authors assumed (a) that the material cut behaves as an ideal plastic which does not strain harden and (b) that the shear plane represents a direction of maximum shear stress.

The slip-line field A–B–C in Fig. 8.3a was assumed. Here region A–B–C will be plastically rigid and subjected to a uniform state of stress. This means that the stress at any point in A–B–C may be represented by a single Mohr's stress circle diagram. The diagram in Fig. 8.3b, in fact, satisfies all SLF requirements. Here point a corresponds to the stress on face a in Fig. 8.3a, i.e., on the shear plane. This is seen to follow the assumption that the shear plane is a direction of maximum shear stress. Line B–C in Fig. 8.3a is a line along which the stress is zero. Due to the double-angle feature of Mohr's circle, the zero point of stress b in Fig. 8.3b which is 90° from point a, will be but 4.5° from point a in Fig. 8.3a. Thus B–C makes an angle of 45° with A–B in Fig. 8.3a. In a similar way point c in Fig. 8.3b represents the stress on face c in Fig. 8.3a, and the stress on the tool face is represented as point d in Fig. 8.3b.

From Fig. 8.3b it is seen that angle dbc = η, since angle dOc = 2η. Therefore,

$$\eta = 45 - \beta \tag{8.13}$$

From Fig. 8.3a

$$\eta = \phi - \alpha \tag{8.14}$$

and hence from Eqs. (8.13) and (8.14)

$$\phi = 45 - \beta + \alpha \tag{8.15}$$

It is evident from Fig. 8.3 that the foregoing ideal plastic solution indicates that the shear stress on the shear plane τ, and the normal stress on the shear plane σ, should be equal. This, however, is generally not the case and hence Eq. (8.15) cannot be regarded as a general solution.

Lee and Shaffer presented a second SLF solution based on the presence of a small BUE represented as a circular arc as shown in Fig. 8.4a. This time the slip line field consisted of two parts, a rigid region A–B–C and a region of varying stress A–B–D having lines of maximum shear stress (slip lines) consisting of radial lines and circular arcs. The rigid region A–B–C is in a state of uniform stress and hence as before a single Mohr's circle diagram is applicable for all points within this area. This diagram is given in Fig. 8.4b where τ_1 is the shear-stress axis. The stresses for points along line D–B are given by the circle in Fig. 8.4b if the τ-axis is at τ_4, the stresses

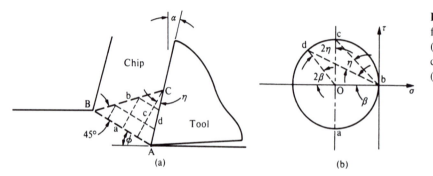

Fig. 8.3 Ideal plastic solution for stress field at tool point. (a) Slip line field. (b) Mohr's circle diagram for field ACB. (after Lee and Shaffer, 1951)

for points along line D–B are given by the circle if the τ-axis is shifted to τ_4 while the stresses along line A–B (as well as all other points in region A–B–C) are given by the circle when the τ-axis is at τ_1.

From Fig. 8.4a it is evident that

$$\phi = \alpha + \eta + \theta \tag{8.16}$$

But from Fig. 8.4b

$$\eta = 45 - \beta \tag{8.17}$$

and hence

$$\phi = 45 + \theta - \beta + \alpha \tag{8.18}$$

This differs from Eq. (8.15) only by the angle θ, which is a measure of the size of the built-up edge. When there is no built-up edge, θ is zero and Eqs. (8.15) and (8.18) give the same result. From Fig. 8.4b it is evident that

$$\sigma = \tau_m(1 + 2\theta) \tag{8.19}$$

$$\tau = \tau_m \tag{8.20}$$

where σ and τ are the normal and shear stresses on the shear plane, respectively. Combining Eqs. (8.19) and (8.20)

$$\theta = \frac{\left(\dfrac{\sigma}{\tau} - 1\right)}{2} \tag{8.21}$$

From this it is evident that σ must no longer be equal to τ as required by the SLF solution without built-up edge. However, it is evident from Eq. (8.21) that the solution with built-up edge requires that σ be greater than τ. Otherwise, θ would be negative and this is physically impossible.

Hücks (1951) presented a derivation which in many respects is similar to that of Lee and Shaffer, but was obviously arrived at without knowledge of the existence of their solution. Hücks reasoned that the stress between the chip and tool along the tool face could be considered uniform in accordance with the principle of Saint Venant. Then as usual, he assumed the shear plane to be in the direction of maximum shear stress. Hücks further reasoned that the resistance to flow in

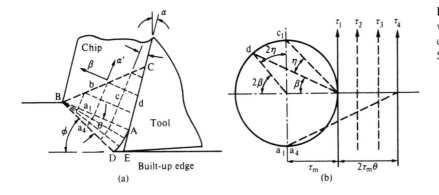

Fig. 8.4 Slip line field solution with BUE. (a) SLF. (b) Mohr's circle diagrams. (after Lee and Shaffer, 1951)

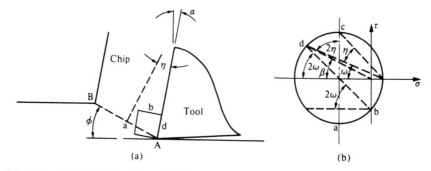

Fig. 8.5 Hücks (1951) solution for tool point stress field. (a) Stress field. (b) Mohr's circle diagram.

the direction of chip motion would be relatively small and thus considered the stress on face b in Fig. 8.5a to be zero. The Mohr's circle diagram corresponding to this set of assumptions is shown in Fig. 8.5b.

Here it is evident that

$$\tan 2\omega = 2\mu \tag{8.22}$$

where μ is the coefficient of friction on the tool face.

From Fig. 8.5a

$$\phi = \eta + \alpha \tag{8.23}$$

and from Fig. 8.5b

$$\eta + \omega = 45° \tag{8.24}$$

Combining Eqs. (8.22) to (8.24) we obtain

$$\phi = 45 - \omega + \alpha \tag{8.25}$$

where

$$\omega = \tfrac{1}{2}\tan^{-1}2\mu \tag{8.26}$$

It may be seen from Fig. 8.5b that $\beta > \omega$ and thus Eq. (8.25) will yield values of ϕ that are larger than those obtained from Eq. (8.15).

In order that the stress on face b in Fig. 8.5a be zero under all conditions, it is necessary that the τ-axis intersect the stress circle at some point b, as in Fig. 8.5b. This requires that the shear stress on the shear plane τ always be greater than σ the normal stress on the shear plane. Inasmuch as σ is frequently observed to be greater than τ_1, Hücks' solution cannot hold, in general.

Hücks (1951) also extended the foregoing treatment to include the influence of normal stress on the shear plane, as Piispanen and Merchant had done. This led to the following equation

$$\phi = \frac{\cot^{-1}K}{2} - \frac{\tan^{-1}2\mu}{2} + \alpha \tag{8.27}$$

where K is the material constant postulated in Eq. (8.6). The several equations that have been reviewed thus far are collected for convenience in Table 8.2.

TABLE 8.2 Collection of Early Equations for Shear Angle

Source	Equation No.	Equation
Dimensional analysis	8.2	$\phi = \psi(\alpha, \beta)$
Ernst and Merchant	8.4	$\phi = 45 - \beta/2 + \alpha/2$
Merchant	8.11	$\phi = \dfrac{\cot^{-1}K}{2} - \beta/2 + \alpha/2$
Stabler	8.12	$\phi = 45 - \beta + \alpha/2$
Lee and Shaffer	8.15	$\phi = 45 - \beta + \alpha$
Lee and Shaffer	8.18	$\phi = 45 + \theta - \beta + \alpha^{\dagger}$
Hücks	8.25	$\phi = 45 - \dfrac{\tan^{-1}2\mu}{2} + \alpha$
Hücks	8.27	$\phi = \dfrac{\cot^{-1}K}{2} - \dfrac{\tan^{-1}2\mu}{2} + \alpha$

† With built-up edge of magnitude θ.

LATER SHEAR ANGLE RELATIONSHIPS

Since the mid-1950s there have been a number of additional attempts to come up with a relatively simple model of the steady state cutting process that will enable the shear angle ϕ to be predicted for a wide range of cutting conditions. While lack of space does not permit a detailed discussion of all of these, a few representative treatments follow:

1. Oxley (1961) presented a SLF solution including strain hardening.
2. Sata (1963) derived a shear angle relation in terms of the angle $\theta = \phi + \beta - \alpha$ (Fig. 8.1) between resultant force R' and the shear plane, the ratio of shear stress on shear plane to that on rake face (K), and the ratio of tool chip contact length (a') to undeformed chip thickness (t) in addition to rake angle (α). The angle θ was used as a boundary condition since it was empirically found to remain constant over a wide range of cutting conditions including work material, rake angle, cutting speed, and controlled contact length. The value of K was also found to be essentially constant (~ 0.8 to 1).
3. Oxley (1963) presented a SLF solution including strain hardening and variation of shear stress on the tool face.
4. Usui and Hoshi (1963) presented a SLF solution for tools of controlled contact length. Figure 8.6a shows the interpretation of flow lines from quick stop photomicrographs and Fig. 8.6b gives the corresponding SLF. This solution involves a centered fan of extent θ which increases in magnitude as the tool face contact length (a') decreases. When $\theta = 0$ this solution reduces to the Lee–Shaffer solution that pertains when cutting without chip curl and with a natural contact length.
5. Rowe and Spick (1967) presented a simple upper bound model of orthogonal cutting with two discontinuities (shear plane and tool face). Instead of using a coefficient of friction, friction on the tool face was expressed in terms of the relative shear stress (m) where

$$m = \frac{\text{mean shear stress on tool face}}{\text{uniaxial flow stress of material in shear}} \tag{8.28}$$

A relatively awkward trigonometric shear angle relationship was then obtained by applying the minimum energy principle.

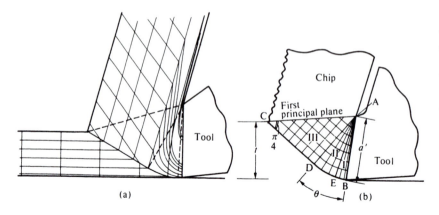

Fig. 8.6 Slip line field solution with controlled tool–chip contact length. (a) Slip line field solution for cut-away tool. (b) Flow lines corresponding to (a). (after Usui and Hoshi, 1963)

(a)

(b)

6. De Chiffre and Wanheim (1981) have also used an upper bound approach but instead of using a friction angle (β), they employed a nondimensional contact length $\eta = a'/t$ (where a' = tool chip contact length) and in place of ϕ, the main dependent variable is taken to be what is called the chip compression factor (λ) = reciprocal of the cutting ratio = t_C/t. Applying the minimum energy principle and assuming rigid plastic work material and sticking friction on the tool face, the following relation was obtained:

$$\lambda = \sqrt{(1 + \eta \cos \alpha)} \tag{8.29}$$

When turning tests were run on an aluminum alloy using tools of HSS with $\alpha = 12°$, $V = 50$ m min^{-1}, no fluid, and various reduced contact lengths (a'_0), results shown in Fig. 8.7 were obtained which are in excellent agreement with Eq. (8.29). In obtaining Eq. (8.29), it was assumed that η does not depend on λ. While this is the case for restricted contact (RC) length tool (Fig. 8.7), it is not the case for natural contact lengths (Fig. 8.8) and for such tools, the agreement between theory and experiment is not as good as that of Fig. 8.7. Apparently the inconsistent points A and B in Fig. 8.7 correspond to a restricted length (a'_0) that is too small for the cutting conditions involved. Further details of the nondimensional contact length approach may be found in De Chiffre (1990).

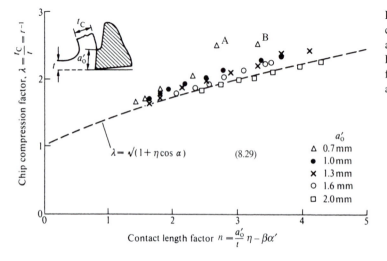

Fig. 8.7 Experimental data for restricted-contact tools. [Work material: 515 WP aluminum alloy. Tool material; HSS. Rake angle: 12°. Cutting speed: 155 f.p.m. (50 m min^{-1}).] (after De Chiffre and Wanheim, 1981)

$\lambda = \sqrt{(1 + \eta \cos \alpha)}$ (8.29)

a'_0
Δ 0.7 mm
● 1.0 mm
✕ 1.3 mm
○ 1.6 mm
□ 2.0 mm

Contact length factor $n = \dfrac{a'_0}{t} \eta - \beta\alpha'$

De Chiffre (1982) has presented a study of restricted contact (RC) tools in which cutting forces and cutting temperatures were measured for different work materials and cutting conditions. When an RC tool is used, the following results are obtained:

- a reduction in the chip compression factor λ
- an increase in shear angle
- a reduction of cutting forces
- a reduction of strain in the chip
- an increase in stress on the tool face by reduction of the sliding region
- a more stable BUE
- a decrease in chip curl
- an increase in tool life
- an improved surface finish
- a reduction of vibration of the tool

It was suggested that RC tools be considered relative to increasing productivity, improved chip control, and when machining difficult materials. It was also suggested that use of RC tools may reduce need for extended processing time (EP) additives in cutting fluids.

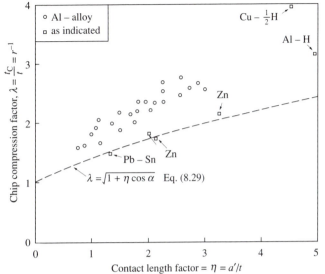

Fig. 8.8 Experimental data for natural contact length tools for different work materials and cutting condition. Contact lengths were estimated visually. (after De Chiffre and Wanheim, 1981)

DISCUSSION

All of the equations in Table 8.2 are in agreement with the dimensional analysis with the exception of Eqs. (8.11), (8.18), and (8.27). These equations include the effect of BUE or normal stress on the shear plane, which was not included in the dimensional analysis.

Before proceeding it may be good to consider briefly the assumptions on which the several early derivations were based.

In Ernst and Merchant's analysis leading to Eq. (8.4), the following assumptions were made:

1. The shear stress is a maximum in the direction of the shear plane.
2. The coefficient of friction is independent of the shear angle.
3. The resultant force R is independent of the shear angle.

All three of these are questionable. Items 1 and 2 will be discussed in the next chapter; item 3 is obviously incorrect since an examination of cutting data reveals a strong dependence of R on ϕ.

In the second derivation of Eq. (8.4), a different set of assumptions were made as follows:

4. The angle ϕ is such that the total power is a minimum.
5. The coefficient of friction is independent of the shear angle.
6. The shear stress on the shear plane is independent of the shear angle.

There is no clear physical reason why item 4 should be true. Many examples may be found in nature in which nonconservative processes, such as that of metal cutting, clearly occur in such a manner that the energy consumed is not a minimum.

Speaking of the derivation of Eq. (8.4), Hill (1950) stated: "The comparative failure of this theory is almost certainly due to the inadequacy of the minimum-work hypothesis." As already mentioned, item 5 will be discussed later. Assumption 6 is only roughly approximate.

In deriving Eq. (8.11), assumption 7, which follows, was introduced in addition to items 4, 5, and 6:

7. The flow stress of the metal on the shear plane is increased by the presence of a normal compressive stress on the shear plane.

Figure 8.9a shows the variation of shear stress on the shear plane versus compressive stress on the shear plane for a range of cutting conditions when NE 9445 steel is turned with a carbide tool.

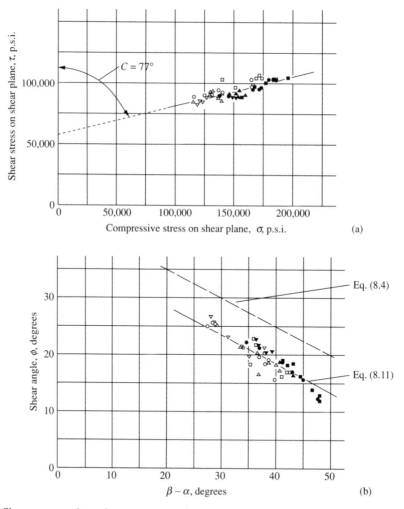

Fig. 8.9 (a) Shear stress on shear plane vs. compressive stress on shear plane for NE 9445 steel machined with sintered carbide tool. (b) Observed shear angle, ϕ, versus $\beta - \alpha$, degrees. [Legend: Open points, $\alpha = +10°$. Solid points, $\alpha = -10°$. ○ ● $V = 542$ f.p.m. (165 mpm), variable t. □ ■ $t = 0.0018$ in (45.7 μm), variable V. △ ▲ $t = 0.0037$ in (94 μm), variable V. ▽ ▼ $t = 0.0062$ in (157 μm), variable V.] (after Merchant, 1945b)

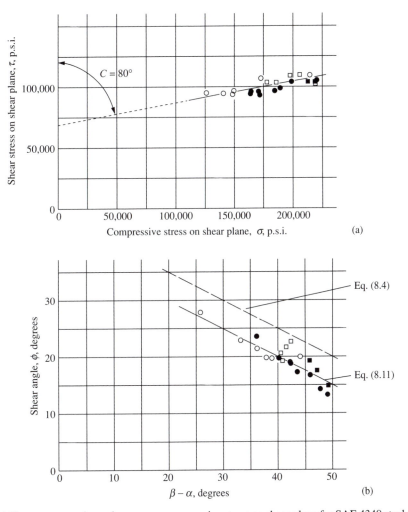

Fig. 8.10 (a) Shear stress on shear plane versus compressive stress on shear plane for SAE 4340 steel machined with sintered carbide tool. (b) Observed shear angle, ϕ, versus $\beta - \alpha$, degrees. [Legend: Open points, $\alpha = +10°$. Solid points, $\alpha = -10°$. ○ ● $V = 542$ f.p.m. (165 mpm), variable t. □ ■ $t = 0.0011$ in (27.9 μm), variable V.] (after Merchant, 1945b)

The value of C (the machining constant) in Eq. (8.11) is seen to be 77° to a good approximation. Figure 8.9b shows a comparison of the experimental data with Eqs. (8.11) and (8.4). Figure 8.10 shows similar results for SAE 4340 steel, where the value of C is found to be 80° to a good approximation.

While it is well established that the rupture stress of both brittle and ductile materials is increased significantly by the presence of a compressive stress, as suggested by Mohr, it is generally believed that a similar relationship for flow stress does not hold. However, in the next chapter a possible explanation is given for this paradox.

As already mentioned, Stabler's equation [Eq. (8.12)] is based on item 7 and the following assumption:

8. The maximum shear stress and shear strain are colinear on the shear plane in three-dimensional cutting operations. Shear must occur on both the shear plane and tool face which are in general not at right angles to each other. This means that both cannot be planes of maximum shear stress. It is therefore to be expected that neither the shear plane nor the tool face will correspond to a direction of maximum shear stress.

In deriving Eq. (8.15), the following were assumed:

9. The material cut is an ideal plastic.

10. The shear plane is along the direction of maximum shear stress.

The strain hardening that occurs when metals are cut is usually important. However, despite this fact, assumption 9 is applied only to a region in which the material is essentially rigid, i.e., A–B–C in Fig. 8.3. All of the strain hardening that takes place occurs as the metal crosses line A–B, and this is before the ideal plastic assumption is applied to the material within area A–B–C.

Equation (8.18) differs from (8.15) only by the introduction of a built-up edge. In the treatment of the built-up edge the ideal plastic assumption must be considered as an approximation, since the material within zone A–B–D of Fig. 8.4 is actually highly deformed. However, if the built-up edge is small, the degree of approximation associated with assumption 9 is probably slight since area A–B–D will then be small compared with A–B–C.

In deriving Eq. (8.25) the following assumptions were made:

11. The stresses along the tool face and shear plane are uniformly distributed.

12. The shear plane is along the direction of maximum shear stress.

13. The normal stress on a plane perpendicular to the tool face is zero.

Item 11 will be discussed in the next chapter and item 12 is the same as item 8.

Assumption 13 appears to be purely arbitrary. While we should expect the normal stress on face b of Fig. 8.5 to be low, there is no reason why it has to be zero. If assumption 13 is adopted, there is no line corresponding to B–C in Fig. 8.3a beyond which the stress in the free chip is zero.

Equations (8.25) and (8.27) differ only by the introduction of the previously discussed concept of influence of normal stress on flow stress.

A general steady state theory of metal cutting must incorporate or at least be consistent with the following phenomena, which are listed in approximate order of importance:

1. Chips are produced primarily by a shear process.

2. There is a strong interaction between the shear deformation occurring in the shear zone and that occurring on the tool face.

3. Most materials strain harden.

4. A built-up edge is usually present at low cutting speeds.

5. At high cutting speeds a secondary shear zone is present in the chip along the tool face.

6. Chips frequently curl and this in turn changes the chip–tool contact length.

7. The chip–tool contact length may be controlled by tool design.

8. Dull tools or tools cutting with a large BUE will have a round cutting edge.

9. Forces exist on the relief face of a tool, especially when an appreciable wear land is present.

All steady state theories of chip formation that have been proposed to date have emphasized one or at most two of these items.

Most workers tend to believe that the problem is to identify the one or two key effects and that the resulting model will hold for all problems. Actually the situation is far more complex than this. All but the first of the nine items listed may be either of first-rank importance or completely insignificant depending upon cutting conditions. Practically all of the limited solutions that have been previously published represent good models over a limited range of operating conditions. The problem now appears to be to recognize the true breadth and complexity of the problem of predicting the shear angle.

The Oxley theory for predicting the shear angle in steady state orthogonal cutting has been elaborated in detail in Oxley (1989) and Arsecularatue and Mathew (2000). In this approach the work material is assumed to behave in a perfectly plastic manner with a flow stress obtained by extrapolating high-speed compression test results from a shear strain of less than one to much higher values that pertain in chip formation. The shear angle is assumed to be in a direction for which the shear stress and strain-rate on the shear plane and along the tool face are maximum. When predicted and experimental values of shear angle are compared for steady state orthogonal cutting, the results are found to be in relatively good agreement. However, it should be noted that the Oxley procedure involves the following questionable items that could cancel each other leading to the degree of agreement between predicted and observed results that have been observed:

(a) the large extrapolation of strain from that of the materials test to that in chip formation

(b) the difference in the strain volumes in the materials test (relatively large) and in chip formation (relatively small)

While item a should be expected to give values of shear stress that are too high, the reverse should be the case for item b (the size effect).

This could explain the difference in the results of Merchant (which involve a machining constant, C) and the results of Oxley and associates (no machining constant effect).

An important consideration regarding prediction of the shear angle is the possibility that the result may not be unique. This was first suggested by Hill (1950, 1954) and subsequently elaborated by Roth (1975), Dewhurst (1978), and several others. It has been suggested that the equilibrium cutting condition for a given set of operating variables depends on how equilibrium is approached. This could involve such items as build-up at the beginning of a cut that could differ from one test to the next. The result of this "lack of uniqueness" of the equilibrium shear angle is that any predicted value would, at best, fall between a range of values due to the inability to completely control the test approach to the equilibrium value of the shear angle.

Stevenson and Stephenson (1996) have submitted a study designed to determine whether initial conditions actually have a significant influence on the uniqueness of results in orthogonal cutting of metals. Tests were performed on a computer numerical control (CNC) machine that enabled test conditions to be repeated to within less than 0.4%. Two test materials were used: 99.8% pure zinc and 99.0% pure aluminum. Cutting forces were accurately measured under a wide range of conditions including values of different constant chip thickness as well as tests with gradually increasing and gradually decreasing undeformed chip thickness. It was found that initial conditions or prior history influenced the steady state cutting forces less than 3% even when there was an initial transient effect of up to 10%. While these test results infer that initial conditions have an essentially negligible effect on orthogonal equilibrium cutting values, it should be kept in mind they were performed at very low cutting speed (< 1 f.p.m.) and apparently without a cutting fluid.

TABLE 8.3 Values of C in Eq. (8.11) for a Variety of Work and Tool Materials in Finish Turning Without a Cutting Fluid*

Work Material	Tool Material	C (deg)
SAE 1035 Steel	HSS[†]	70
SAE 1035 Steel	Carbide	73
SAE 1035 Steel	Diamond	86
AISI 1022 (leaded)	HSS[†]	77
AISI 1022 (leaded)	Carbide	75
AISI 1113 (sul.)	HSS[†]	76
AISI 1113 (sul.)	Carbide	75
AISI 1019 (plain)	HSS[†]	75
AISI 1019 (plain)	Carbide	79
Aluminum	HSS[†]	83
Aluminum	Carbide	84
Aluminum	Diamond	90
Copper	HSS[†]	49
Copper	Carbide	47
Copper	Diamond	64
Brass (60–40)	Diamond	74

* After Merchant, 1950.
[†] HSS = high speed steel.

Concluding Remarks

Of the several attempts at predicting shear angle (ϕ), the one that appears most acceptable is Eq. (8.11). From Figs. 8.9 and 8.10, it is seen that C is not a constant. Merchant has determined the values of C given in Table 8.3 for materials of different chemistry and structure being turned under finishing conditions with different tool materials. From this table it is further evident that C is not a constant. This indicates that in addition to the items listed in Table 8.1 in the dimensional analysis for shear angle, additional items should have been included that reflect the influence of different tool and work properties.

The fact that an increase of compressive stress on the shear plane gives rise to an increase in shear stress on the shear plane suggests that some measure of brittleness of the work material should have been included in the dimensional analysis for shear angle ϕ in Table 8.1. The quantity C is such a factor. The fact that this chapter is limited to steady state chip formation rules out the possibility of periodic gross cracks. However the role of microcracks is a possibility consistent with steady state chip formation and the influence of compressive stress on the flow stress in shear. A discussion of the role microcracks can play in steady state chip formation is presented in the next chapter.

References

Arsecularatne, J. A., and Mathew, P. (2000). *Int. Jour. Science and Technology* **4/3**, 363–397.
De Chiffre, L. (1977). *Int. J. of Mach. Tool Design and Research* **17**, 225–234.
De Chiffre, L. (1982). *Int. J. of Mach. Tool Design and Research* **22/4**, 321–332.

De Chiffre, L. (1990). *Metal Cutting Mechanics and Applications*, Dr. Technices Dissertation, Technical University of Denmark.

De Chiffre, L., and Wanheim, T. (1981). *Proc. 9th North American Manuf. Res. Conf.* Soc. Man. Eng. (Dearborn), p. 231.

Dewhurst, P. (1978). *Proc. of the Royal Soc. of London* **A-360**, 587.

Ernst, H. J., and Merchant, M. E. (1941). *Trans. Am. Soc. Metals* **29**, 299.

Fung, Y. C., and Sechler, E. E. (1960). *Proc. First Symp. on Struct. Mech.* Pergamon Press, New York, p. 115.

Hill, R. (1950). *The Mathematical Theory of Plasticity.* Clarendon Press, Oxford. (Reissued 1983.)

Hill, R. (1954). *J. Mech. Phys. Solids* **3**, 47–53.

Hücks, H. (1951). Dissertation, Aachen T. H.

Kobayashi, S., and Thomsen, E. G. (1962). *J. Engng. Ind.* **84**, 63.

Lee, E. H., and Shaffer, B. W. (1951). *J. Appl. Mech.* **73**, 405.

Merchant, M. E. (1945). *J. Appl. Phys.* **16**, (a) 267 and (b) 318.

Merchant, M. E. (1950). *Machining Theory and Practice.* Am. Soc. for Metals, p. 5.

Oxley, P. L. B. (1961). *Int. J. Mech. Sci.* **3**, 68.

Oxley, P. L. B. (1963). *International Research in Production Engineering.* ASME, New York, p. 50.

Oxley, P. L. B. (1989). *Mechanics of Machining: An Analytical Approach to Assessing Machinability.* Ellis Horwood, Chichester, England.

Piispanen, V. (1937). *Teknillinen Aikakauslehti* (Finland) **27**, 315.

Piispanen, V. (1948). *J. Appl. Phys.* **19**, 876.

Roth, R. N. (1975). *Int. J. of Machine Tool Design and Research* **15**, 161.

Rowe, G. W., and Spick, P. J. (1967). *J. Engng. Ind.* **89**, 530.

Sata, T. (1963). In *International Research in Production Engineering.* ASME, New York, p. 18.

Stabler, G. V. (1951). *Proc. Inst. Mech. Eng.* **165**, 14.

Stevenson, R., and Stephenson, D. A. (1996). Paper submitted for publication in *J. of Eng. for Industry (Trans. ASME).*

Usui, E., and Hoshi, K. (1963). In *International Research in Production Engineering.* ASME, New York, p. 61.

9 SHEAR STRESS IN CUTTING

In this chapter the shear stress that occurs on the shear plane in metal cutting is considered and compared with results from other materials tests.

When shear-stress versus shear-strain orthogonal cutting data obtained by use of Eqs. (3.9) and (3.16) are compared with conventional torsional tests data for the same material, substantial differences are observed (Fig. 9.1). The cutting data do not lie on a single curve (appreciable scatter) and stresses for a given strain are considerably higher than for torsion. Likewise, when cutting data are compared with extrapolated uniaxial tensile data on the basis of equivalent stress and strain, the correlation is relatively poor.

Plastic flow on the shear plane in orthogonal cutting with continuous chip formation may be characterized as follows:

1. large values of strain rate and temperature
2. moderate normal stress on the shear plane giving rise to large values of homogeneous strain without gross fracture in the chip
3. progressive deformation with very small volume deformed at one time
4. close proximity of shear zone to rigid cutting tool which constitutes a flow constraint

Some of these characteristics give rise to entirely different flow conditions than those met in conventional materials tests which results in the conventional tests being of limited value in predicting the flow stress to be expected in cutting. Each of the above characteristics will be discussed in order.

Strain Rate and Temperature

The influences of strain rate and temperature may be considered at the atomic level in terms of dislocations or at the macrolevel in terms of homogeneous material. Becker (1926) has suggested that the thermal energy of a material is available to assist the applied stress τ, in forming dislocations while Orowan (1934, 1935) has indicated that although the thermal energy is insufficient to close the gap between observed τ and calculated τ_0 values of flow stress, a stress concentration factor C' may exist at the tip of a microcrack and thus augment thermal energy to bridge the gap between C'_τ and τ_0 (τ_0 = theoretical strength).

In Fig. 9.2 the stress–strain curve for a perfectly homogeneous material is shown. The energy required to raise a unit volume of material to the stress level τ_0 is represented by the area under the curve up to a stress τ_0.

The portion of this energy of thermal origin is represented by the dark area and is

$$u_\theta = \tfrac{1}{2}\Delta\tau\Delta\gamma \qquad (9.1)$$

but as

$$\Delta\tau = G\Delta\gamma \qquad (9.2)$$

then

$$u_\theta = \frac{1}{2}\frac{(\Delta\tau)^2}{G} = \frac{(\tau_0 - C\tau)^2}{2G} \qquad (9.3)$$

Hence a dislocation will be formed whenever the thermal energy per unit volume, u_θ, in the vicinity of a stress concentration reaches the value required by Eq. (9.3).

Boltzmann (see Sproull, 1956) has shown that the probability, P, of finding a thermal energy $u_\theta V'$ in a volume V' is

$$P = e^{-u_\theta V'/(kT)} \qquad (9.4)$$

where T is the absolute temperature and k is the constant known as Boltzmann's constant.

Since a dislocation will travel across a crystal at a velocity approaching that of the speed of sound, the rate of strain, $\dot\gamma$, will be proportional to the rate of dislocation formation. This in turn will be proportional to the probability of there being sufficient thermal energy to form a dislocation. Hence we may write

$$\dot\gamma = Ae^{-u_\theta V'/(kT)} \qquad (9.5)$$

where A is a proportionality constant. With the value of u_θ, given by Eq. (9.3), Eq. (9.5) becomes

$$\dot\gamma = Ae^{-V'(\tau_0 - C\tau)^2/2GkT} \qquad (9.6)$$

or solving for τ:

$$\tau = \left[\tau_0 - \left(\frac{2GkT}{V'}\ln\frac{A}{\dot\gamma}\right)^{1/2}\right]\Big/ C \qquad (9.7)$$

For a given material τ_0, C, G, k, V', and A will be constants and we may write

$$\tau = C_1 - C_2[T(C_3 - \ln\dot\gamma)]^{1/2} \qquad (9.8)$$

where C_1, C_2, and C_3 are constants.

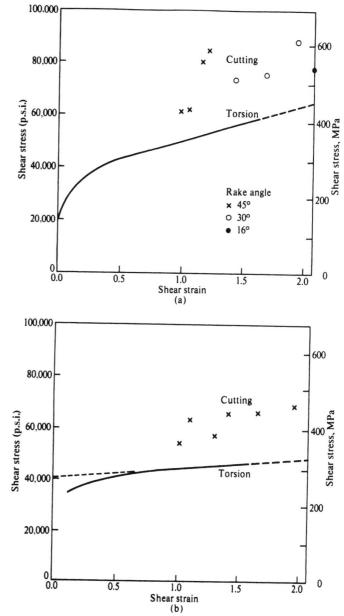

Fig. 9.1 Comparison of shear-stress–shear-strain data for cutting and torsion tests. (a) Material, AISI B1112 steel cut at low speed using several rake angles and fluids. (b) Material, leaded free machining steel cut at low speed using several rake angles and fluids. (after Shaw and Finnie, 1955)

TABLE 9.1   Tensile Data for Annealed AISI B1112 Steel

Absolute Temperature (T), deg R	Strain Rate ($\dot{\gamma}$), s⁻¹	True Ultimate Stress (σ), p.s.i.	Equivalent[a] Shear Stress (τ), p.s.i.
530	0.0009	77,100	38,550
530	482	92,800[b]	46,400
160	0.0009	119,400	59,700

[a] Calculated according to maximum shear theory.
[b] Value corrected for difference in hardness of high- and low-speed test specimens.

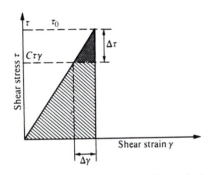

Fig. 9.2 Stress–strain curve for perfectly homogeneous material.

Table 9.1 gives some tensile data obtained by MacGregor (1944) for the ultimate stress of annealed AISI B1112 steel at different temperatures and strain rates.

With the values of Table 9.1, the constants in Eq. (9.8) may be evaluated and thus

$$\tau = 85,500 - 311[T(36.2 - \ln \dot{\gamma})]^{1/2} \qquad (9.9)$$

It should be noted that the effect of temperature is greatly reduced at high strain rates as is the effect of strain rate at low temperatures.

At the macro level, MacGregor and Fisher (1946) combined the effects of temperature and strain rate in the tension-testing of metals by use of a velocity modified temperature:

$$T_m = T\left(1 - k \ln \frac{\dot{\varepsilon}}{\dot{\varepsilon}_0}\right) \qquad (9.10)$$

where T is the absolute test temperature, $\dot{\varepsilon}$ is the constant true strain rate of the test and $\dot{\varepsilon}_0$ and k are constants. While velocity modified temperature had been used previously in studies of secondary creep, MacGregor and Fisher found this concept quite satisfactory for correlation of tensile tests run at different temperatures and strain rates.

It is to be expected that since metal cutting normally involves *both* high strain rates and high temperatures on the shear plane, the two effects will tend to cancel as Drucker (1949) has suggested.

PLASTIC BEHAVIOR AT LARGE STRAIN

There has been remarkably little work done in the region of large plastic strains. Bridgman (1952), using the hollow tubular notched specimen shown in Fig. 9.3, performed tests under combined axial compression and torsion. The specimen was loaded axially in compression as the center section was rotated relative to the ends. Strain was concentrated in the reduced sections, and it was possible to crudely estimate and plot shear stress versus shear strain with different amounts of compressive stress on the shear plane. From these tests, Bridgman concluded that the flow curve for a given material was the same for all values of compressive stress on the shear plane, a result consistent with other materials tests involving much lower plastic strains. However, the strain at gross fracture was found to be strongly influenced by compressive stress, increasing markedly with increase in compressive stress.

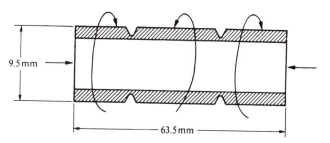

Fig. 9.3 Bridgman (1952) specimen for combined axial load and torsion.

Lankford and Cohen (1969) were interested in the behavior of dislocations at very large plastic strains and whether there was a saturation relative to the strain hardening effect with strain or whether strain hardening continued to occur with strain to the point of fracture. Their experimental approach was an interesting and fortunate one. They performed wire drawing on iron specimens using a large number of progressively smaller dies with remarkably low semi die angle (1.5°) and a relatively low (~ 10%) reduction in area per die pass. After each die pass, a specimen was tested in uniaxial tension and a true stress–strain curve obtained. The drawing and tensile tests were performed at room temperature and low speeds to avoid heating, and specimens were stored in liquid nitrogen between tests to avoid strain aging effects. All tensile results were then plotted in a single diagram, the strain used being that introduced in drawing (0.13 per die pass) plus the plastic strain in the tensile test. The result is shown in Fig. 9.4a. The general overlap of the tensile stress–strain curves gives an overall strain-hardening envelope which indicates that the wire drawing and tensile deformations are approximately equivalent relative to strain hardening. Figure 9.4b shows similar results on the same iron wire tested in uniaxial compression following drawing.

Figure 9.4c shows somewhat similar results obtained earlier by Blazynski and Cole (1960) for AISI 1012 steel but carried to much lower values of total strain. Blazynski et al. were interested in strain hardening in tube drawing and sinking. Drawn tubes were sectioned as shown in Fig. 9.4d and tested in plane strain compression as shown in Fig. 9.4e. Like Fig. 9.4b, Fig. 9.4c shows the flow stress in compression plotted against the total strain. The curves with bent tops in Fig. 9.4c were obtained using graphite grease as a lubricant in the plane strain tests while the data points were obtained in similar tests using a more effective molybdenum disulfide lubricant. The smooth curve drawn through the molybdenum disulfide data points constitutes the flow curve essentially in the absence of friction.

Up to a strain of about 1 (Figs. 9.4a, 9.4b, and 9.4c) the usual strain-hardening curve was obtained that is in good agreement with the generally accepted equation given in Eq. (5.8):

$$\sigma = \sigma_1 \varepsilon^n$$

However, beyond a strain of 1, the curve was linear corresponding to the equation

$$\sigma = A + B\varepsilon \qquad (\varepsilon > 1) \tag{9.11}$$

where A and B are constants. It may be shown that

$$A = (1 - n)\sigma_1 \tag{9.12}$$

$$B = n\sigma_1 \tag{9.13}$$

in order that the curves of Eqs. (5.8) and (9.11) have the same slope and ordinate at $\varepsilon = 1$.

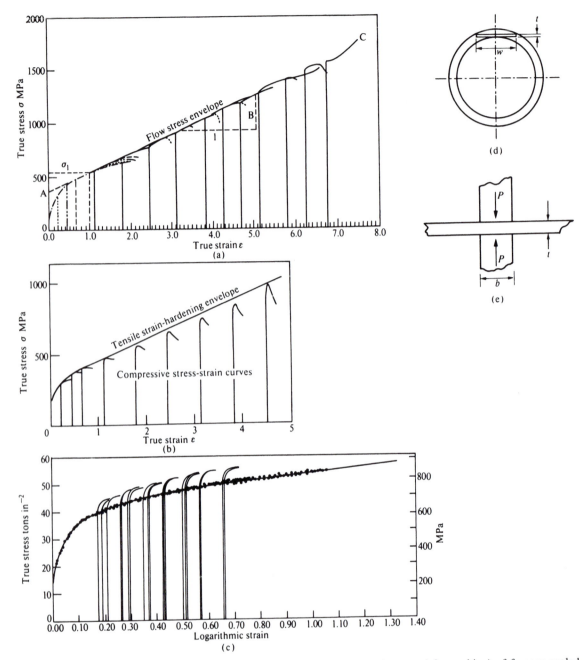

Fig. 9.4 (a) Effect of wire drawing strain on uniaxial stress–strain curves (corrected for necking) of furnace-cooled iron (0.007 $^w/_o$ C) drawn and tested at room temperature. (after Lankford and Cohen, 1969) (b) Uniaxial compression test data following wire drawing of same materials (tested in tension following drawing) shown in (a). (c) Plane strain compression tests following tube drawing of AISI 1012 steel. Curves with hooks at top are with graphite lubricant while data points are for a more effective molybdenum sulfide lubricant. (after Blazynski and Cole, 1960) (d) How plane strain specimen was removed from drawn tube by Blazynski et al.; tube diameter approximately 55 mm, wall thickness 4.8 mm. (e) Plane strain compression test used by Blazynski et al. following tube drawing: $t = 0.045$ in (1.14 mm), $b = 0.115$ in (2.92 mm), w (perpendicular to paper) = 0.690 in (17.53 mm).

While Eq. (5.8) is well known and widely applied, there is relatively little data in the literature for plastic strains greater than 1 and hence Eq. (9.11) is relatively unknown.

From transmission electron micrographs of deformed specimens Lankford et al. found that cell walls representing concentrations of dislocations began to form at strains below 0.2 and became ribbon shaped with decreasing mean linear intercept cell size as the strain progressed. Dynamic recovery and cell wall migration resulted in only about 7% of the original cells remaining after a strain of 6. The flow stress of the cold-worked wires was found to vary linearly with the reciprocal of the mean transverse cell size.

Data that makes one question the wisdom of extrapolating ordinary materials test data into the larger strain regime of metal cutting is given in Fig. 9.5. These are data for the same material cut under the same conditions except for rake angle (α). The shear stress on the shear plane is obviously not constant but appears to decrease with increase in shear strain (negative strain hardening).

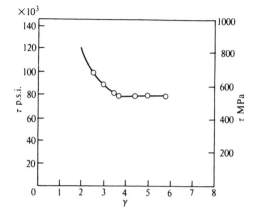

Fig. 9.5 Values of shear stress on the shear plane (τ) versus shear strain in the chip (γ) when cutting the same material with tools of different rake angle. (after Shaw, 1954)

Toward the end of the sixties, it was decided to conduct an acoustical emission study of concentrated shear at Carnegie Mellon University. The initial acoustical studies were on specimens of the Bridgman type but fortunately, lower levels of axial compressive stress than Bridgman had used were employed in order to more closely simulate the concentrated shear process of metal cutting. The apparatus used which was capable of measuring stresses and strains as well as acoustical signals arising from plastic flow is described in the dissertation of Walker (1967). Two important results were obtained:

1. A region of rather intense acoustical activity occurred at the yield point followed by a quieter regime until a shear strain of about 1.5 was reached. At this point, there was a rather abrupt increase in acoustical activity that continued to the strain at fracture which was appreciably greater than 1.5.

2. The shear stress appeared to reach a maximum at the shear strain corresponding to the beginning of the second burst of acoustical activity ($\gamma \cong 1.5$).

The presence of the notches in the Bridgman specimen (Fig. 9.3) made interpretation of stress–strain results somewhat uncertain. Therefore, a new specimen was designed (Fig. 9.6) which substitutes simple shear for torsion with normal stress on the shear plane. By empirically adjusting distance Δx (Fig. 9.6) to a value of 0.25 mm, it was possible to confine all the plastic shear strain to the reduced area, thus making it possible to determine readily the shear strain ($\gamma = \Delta y / \Delta x$). When the width of minimum section was greater or less than 0.25 mm, the extent of the plastic strain observed in a transverse micrograph at the minimum section either did not extend completely across the 0.25 mm dimension or it extended beyond this width.

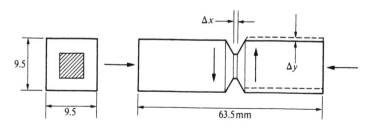

Fig. 9.6 Plane strain simple shear-compression specimen of Walker and Shaw (1969).

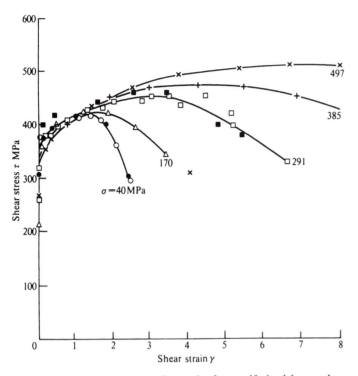

Fig. 9.7 Shear-stress–shear-strain results for resulfurized low carbon steel for specimen of Fig. 9.6. [σ = normal stress on shear plane.] (after Walker and Shaw, 1969)

A representative set of curves is shown in Fig. 9.7 for resulfurized low carbon steel. Similar results were obtained for nonresulfurized steels and other ductile metals. There is little difference in the curves for different values of normal stress on the shear plane (σ) to a shear strain of about 1.5. This is in agreement with Bridgman. However, beyond this strain, the curves differ substantially with compressive stress on the shear plane (σ). At large strains τ was found to decrease with increase in γ, a result that does not agree with Bridgman (1952).

When the results of Figs. 9.4a and 9.7 are compared, they are seen to be very different. In the case of Fig. 9.4a, strain hardening is positive to normal strains as high as 7. In the case of Fig. 9.7, strain hardening appears to become negative above a particular shear strain that increases with normal stress on the shear plane.

From Fig. 9.7, it is seen that for a low value of normal stress on the shear plane of 40 MPa, strain hardening appears to go negative at a shear strain (γ) of about 1.5; that is, when the normal stress on the shear plane is about 10% of the maximum shear stress reached, negative strain hardening sets in at a shear strain of about 1.5. On the other hand, strain hardening remains positive to a normal strain of about 8 when the normal stress on the shear plane is about equal to the maximum shear stress (note curve for σ = 497 MPa in Fig. 9.7).

NEW MECHANISM OF LARGE STRAIN PLASTIC FLOW

It is thus seen that whereas the Bridgman and Lankford and Cohen results are in agreement, these results are completely different from those of Walker and Shaw. The proposed mechanism of large strain plastic flow (Shaw, 1980) suggests that at moderate values of normal stress on the shear plane, discontinuous microcracks begin to appear in a plane of concentrated shear at a shear strain of about 1.5. As strain proceeds beyond this point, the first microcracks are sheared shut as new ones take their place. The sound area on the shear plane gradually decreases until it becomes insufficient to resist the shear load without gross fracture. What seems to be negative strain hardening in Fig. 9.7 is due to what might be described as "internal necking" (i.e., a gradual decrease in sound internal area with load, just as the area in the neck of a tensile specimen decreases with load to give the appearance of negative strain hardening in an engineering stress–strain curve).

The reason such "negative strain hardening" was not observed by Bridgman or Lankford and Cohen appears to be due to the normal compressive stress on the shear plane in their experiments being high enough to prevent the formation of microcracks. The choice of die angle (only 1.50° half-angle) and reduction per pass (0.22) in the Lankford and Cohen drawing experiments provides essentially homogeneous compressive strain in the deformation zone and, under such conditions, one would not expect microcracks to develop.

There is considerable indirect evidence to support the formation of microcracks in metal cutting. While it has been reported (Komanduri and Brown, 1967) that microcracks have been observed on the shear plane of quick-stop chip roots, one should not expect to find many. Such cracks will be of the thin, hairline variety, and most of them should be expected to reweld as the specimen is suddenly unloaded.

The new theory of plastic flow discussed here is an add-on to dislocation theory. As long as microcracks do not occur in appreciable number, a material may be deformed to very large strains with a continuous increase in dislocation density and strain hardening. This is consistent with the experimental results of Bridgman and Lankford and Cohen. However, at a particular value of shear strain, depending upon the ductility of the material and the normal stress on the shear plane, microcracks will begin to form. If the normal stress on the shear plane is tensile, these cracks will spread rapidly over the shear surface leading to gross fracture. If, however, a moderate compressive stress ($\sim \frac{1}{2}$ the shear flow stress for a ductile metal) is present on the shear plane, the new mechanism will pertain. When this occurs there will be an extended stress–strain region exhibiting a decrease in flow stress with strain as the ratio of real to apparent area on the shear plane (A_R/A) decreases from one to the critical value at which gross fracture occurs.

It is suggested that in metal cutting the shear stress on the shear plane is not in general independent of normal stress on the shear plane. The part of the shear plane that involves microcracks should show an increase in shear stress with normal stress as Piispanen and Merchant suggested. However, the part that does not involve microcracks should have a shear stress that is independent of normal stress in keeping with the experimental results of Bridgman and Lankford and Cohen.

It should not be inferred that the foregoing discussion has brought us any nearer to a solution of the shear angle problem. It should, however, serve to further explain why it is unlikely that a simple solution to this persistent problem is apt to be found.

INHOMOGENEOUS STRAIN

When metallic single crystals are plastically deformed, it is found that slip does not occur uniformly on every atomic plane but that the active slip planes are relatively far apart (Fig. 9.8). It is further found that poly-crystalline metals also strain blockwise rather than uniformly. Crystal imperfections are responsible for this inhomogeneous behavior.

When the volume of material deformed at one time is relatively large, there will be a uniform density of imperfections and for all practical purposes, strain (and strain hardening) may be considered to be uniform. However, as the volume deformed approaches the small volume associated with an imperfection, the material will show obvious signs of the basic inhomogeneous character of strain. The mean flow stress will rise and the ends of the active shear planes will be evident in a free surface. This is called the size effect.

In metal cutting, the undeformed chip thickness (t) is small, the width of shear zone (Δy) is very small, but the width of cut (b) is relatively large. It would thus appear that the volume deformed at one time would be

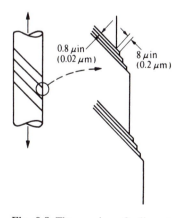

Fig. 9.8 The spacing of adjacent slip planes. Pure aluminum single crystal. (after Heidenreich and Shockley, 1948)

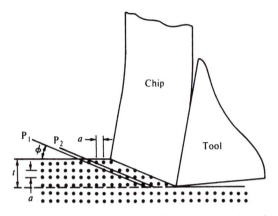

Fig. 9.9 Specimen with uniform distribution of points of stress concentration (weak points). (after Shaw, 1950)

$(bt\Delta y/\sin \phi)$. However, when the back of a very continuous chip is observed under the microscope, it is found (Fig. 6.28) that the edges of the slip bands that are observed are not continuous across the width of the chip but have an extent characteristic of the imperfection spacing (a). Thus, the volume deformed at any one time should be taken to be $[a(t)\Delta y/\sin \phi]$ where $a \ll b$. In metal cutting this volume will aproach a^3 (mean imperfection volume) and there will be a size effect. This is the main reason specific energy (u) increases with decrease in undeformed chip thickness (t) as suggested by Eq. (3.39).

Under ordinary conditions, the shear planes will be very closely spaced corresponding to the closeness of spacing of the weak spots in the metal. It may be assumed that slip planes are so spaced that a single weak spot is present on each plane. Drucker (1949) has employed a random array of weak points to qualitatively demonstrate the increase in unit cutting energy with decrease in depth of cut. However, inasmuch as the spacing of weak points is very small compared with usual depths of cut, an orderly array of weak spots seems justified. The dots shown in Fig. 9.9 represent such an orderly arrangement of weak points to an exaggerated scale. These points have a uniform spacing of a units in each direction.

Let P_1 and P_2 be two shear planes making an angle ϕ with the direction of cut and passing through adjacent points in the first row below the surface. If the depth of cut is t, then t/a planes may be placed between those at P_1 and P_2 so that a single plane passes through each weak spot in the layer in the process of being cut. The number of planes per unit distance in a direction perpendicular to the shear plane will be

$$n = \frac{t}{a(a \sin \phi)} \tag{9.14}$$

or the spacing of successive planes is

$$\Delta y = \frac{a^2 \sin \phi}{t} \tag{9.15}$$

The total slip on a given shear plane will be

$$x = \frac{a^2 \sin \phi}{t} \gamma \tag{9.16}$$

where γ is the unit uniform strain.

Assuming a normal stress on the shear plane that is sufficient to suppress microcrack formation as in the Lankford and Cohen experiments, the flow stress should vary with strain as given by Eq. (9.11). In view of the origin of strain hardening according to dislocation theory and the Lankford and Cohen strain-hardening results for large strains, it appears reasonable to assume that strain hardening will increase linearly with the extent of the slip that occurs on a given shear plane (x given by Eq. (9.16)). Fig. 9.10 shows the assumed variation of shear stress (τ') with x where B' is the linear rate of strain hardening relative to x. The *mean* shear stress on the shear plane during slip may thus be expressed as follows (substituting for γ from Eq. (3.16)):

$$\tau' = A' + B'x = A' + \frac{B'a^2 \sin \phi}{2t}[\cot \phi + \tan (\phi - \alpha)] \tag{9.17}$$

This equation suggests that the shear stress on the shear plane is not constant as for an ideal plastic but depends upon the rate of strain hardening, the strain in the chip, the density of imperfections (as measured by a), and the undeformed chip thickness relative to the imperfection spacing (t/a).

The only convenient method of evaluating the constants in Eq. (9.17) (A' and B') is from metal cutting. To illustrate the application of Eq. (9.17), the orthogonal data of Table 3.1 may be considered. When the shear stress on the shear plane (τ) is plotted against the quantity ($\gamma \sin \phi)/2t$ as in Fig. 9.11 and the best straight line drawn through the points, values for A' and a^2B' may be readily determined:

$$A' = 87,000 \text{ p.s.i. } (601 \text{ MN m}^{-2})$$

$$a^2B' = 34.55 \text{ lb in}^{-1} (60.59 \text{ N cm}^{-1})$$

The maximum difference in shear stress on the shear plane due to the difference in "specimen size" and its influence on strain hardening is 14,000 p.s.i. (96.5 MPa) or 15% of the minimum observed value in this particular case.

When the orthogonal cutting data of Table 3.2 are similarly plotted, Fig. 9.12 is obtained. For this case

$$A' = 52,000 \text{ p.s.i. } (359 \text{ MN m}^{-2})$$

$$a^2B' = 153 \text{ lb in}^{-1} (268 \text{ N cm}^{-1})$$

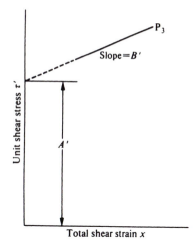

Fig. 9.10 Assumed variation of unit shear stress (τ') versus total shear strain (x) diagram.

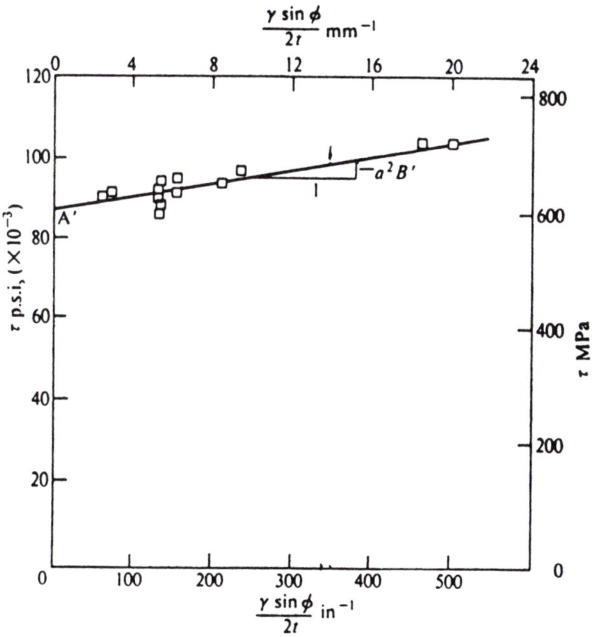

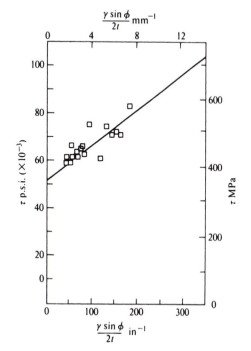

Fig. 9.11 Variation of shear stress on shear plane τ with ($\gamma \sin \phi)/2t$ for data of Table 3.1.

Fig. 9.12 Variation of shear stress on shear plane τ with ($\gamma \sin \phi)/2t$ for data of Table 3.2.

The size-effect strain-hardening influence in this case is larger corresponding to a difference between maximum and minimum shear stress on the shear plane of about 21,000 p.s.i. (144.8 MPa) or 36% of the minimum value.

While the foregoing treatment of orthogonal cutting, which involves a size effect that is a consequence of inhomogeneous strain and strain hardening cannot be used to predict the shear stress on the shear plane since there is no way of estimating the quantities A' and a^2B', it does serve to explain why specific cutting energy (u) varies inversely with the undeformed chip thickness (t). It also explains why the backs of thin chips show evidence of inhomogeneous strain and how this strain inhomogeneity can act to intensify the strain hardening effect provided the normal stress on the shear plane is sufficiently high to prevent extensive formation of microcracks.

INTERACTION BETWEEN SHEAR PLANE AND TOOL FACE

In most materials tests things get very complicated when friction on a die face in close proximity to the plastic flow zone plays a significant role. An example is the barreling that occurs in the uniaxial compression test due to friction on the die faces. While this complication may be handled by an extension of the Bridgman correction to situations with negative radius of curvature at the free surface, the best solution is to reduce the friction to zero or to periodically remove the barrel shape by machining.

In metal cutting, action on the shear plane and on the tool face are closely associated and connected by a single stress field that must be compatible with both processes. Flow will not occur at maximum shear stress on the shear plane unless it can simultaneously occur on a plane of maximum shear stress on the tool face. This is possible only when the shear angle (ϕ) equals the rake angle (α) which is not normally the case.

In the foregoing discussion, it has been suggested that normal stress on the shear plane can play an important role in influencing the resulting flow stress on the shear plane in cutting due to the formation of microcracks. Ordinary materials tests either do not involve microcrack formation or have a ratio of normal stress to shear stress on the shear plane that is quite different from that in cutting. Therefore, ordinary normal materials test results cannot be used to predict metal cutting performance.

PREDICTION VERSUS UNDERSTANDING OF PERFORMANCE

In Chapter 8, it was explained why a simple model of the cutting process is not capable of predicting the shear angle and hence the strain pertaining in a cutting process. In this chapter, it has been similarly argued that the flow stress in metal cutting cannot be predicted in terms of properties derived from ordinary materials tests. In the next chapter, the complexities associated with the friction process on the tool face will be discussed. All this suggests that it is next to impossible to *predict* metal cutting performance. However, it should not be inferred that detailed study of the cutting process is without value. Each fundamental study that is made and properly interpreted adds to our *understanding* of the process, and understanding is the next best thing to the ability to predict.

The practical approach to optimization of a metal cutting process involves an incremental approach. This consists of beginning the operation under conditions representing the best estimate from past experience (a handbook or data stored in a computer) followed by an observation of resulting performance relative to the optimization variable of interest. A small change in operating variables is then made and the effect on the optimization variable noted. Optimum performance may be gradually approached as useful parts are produced. The role of understanding of course lies in helping decide which variables to alter and by how much so that optimum performance may be reached in the shortest time and recognized when achieved. Optimization of a metal cutting process from this point of view is considered in Chapter 19.

REFERENCES

Becker, H. (1926). *Zeit. Tech. Phys.* **7**, 547.

Blazynski, T. Z., and Cole, I. M. (1960). *Proc. Instn. Mech. Engrs.* **174**, 797.

Bridgman, P. W. (1952). *Studies in Large Plastic Flow and Fracture.* McGraw-Hill, New York.

Drucker, D. C. (1949). *J. Appl. Phys.* **20**, 1013.

Heidenreich, R. O., and Shockley, W. (1948). *Report on Strength of Solids.* Physics Society, London, p. 57.

Komanduri, R., and Brown, R. H. (1967). *Metals Mater.* **95**, 308.

Lankford, G., and Cohen, M. (1969). *Trans. Am. Soc. Metals* **62**, 623.

MacGregor, C. W. (1944). *J. Franklin Inst.* **238**, 111, 159.

MacGregor, C. W., and Fisher, J. C. (1946). *J. Appl. Mech.* **68**, A–11.

Orowan, E. (1934). *Zeit. Phys.* **89**, 605, 614, 634.

Orowan, E. (1935). *Zeit. Phys.* **98**, 382.

Shaw, M. C. (1950). *J. Appl. Phys.* **21**, 599.

Shaw, M. C. (1954). *Metal Cutting Principles*, 3rd ed. MIT Press, Cambridge, Mass.

Shaw, M. C. (1980). *Int. J. Mech. Sci.* **22**, 673.

Shaw, M. C. (1982). *ASME PED-3*, 215–226.

Shaw, M. C., and Finnie, I. (1955). *Trans. Am. Soc. Mech. Engrs.* **77**, 115.

Shaw, M. C., Janakiram, M., and Vyas, A. (1991). *Proc. NSF Grantees Conf.*, Austin, Tex. Pub. by SME, Dearborn, Mich., 359–366.

Shaw, M. C., Ramaraj, T. C., and Santhanam, S. (1988). *Internat. Conf. on Behavior of Materials in Machining*, 9.1. The Institute of Metals, England.

Sproull, R. L. (1956). *Modern Physics.* Wiley, New York.

Walker, T. J. (1967). Dissertation, Carnegie Mellon University, Pittsburgh, Pa.

Walker, T. J., and Shaw, M. C. (1969). *Proc. 10th Int. Mach. Tool Des. and Res. Conf.* Pergamon Press, Oxford, p. 291.

10 FRICTION

Friction which is the resisting force one surface experiences when it slides over another is always oppositely directed to the relative velocity vector for the two surfaces. The entire field of friction may be divided into two general regimes—lightly loaded sliders and heavily loaded sliders. Each of these will be briefly discussed followed by a consideration of the special friction conditions that pertain in metal cutting. What constitutes a lightly loaded slider will be defined in a later section. Tribology is the term used to cover the performance of surfaces relative to lubrication, friction, and wear. The complete tribological characteristics of lightly loaded sliding surfaces is treated in detail in the two-volume classical work by Bowden and Tabor (1950, 1964). Suh and Sin (1981) and Suh (1986) have discussed the relation between friction and wear.

LIGHTLY LOADED SLIDERS

Types of Sliding Contact

There are five types of sliding contact:

1. **Fluid film lubrication.** This is the ideal type of lubrication in which there is no metal-to-metal contact. The load is supported by a pressurized fluid film. Viscosity is the physical property of the system of major interest, and the principles of fluid mechanics enable the load capacity and friction characteristics of surfaces that are so lubricated to be computed. Well-designed journal and slider bearings are of this type.

2. **Elastohydrodynamic lubrication.** This is the type of lubrication in which the localized stresses in the bearing surfaces are sufficiently high to cause a significant change in surface geometry and the fluid pressures are sufficiently high to alter the effective viscosity of the fluid. The lubrication of gear teeth, cam surfaces, and rolling contact bearings is of this type. This is a special version of hydrodynamic lubrication.

3. **Boundary lubrication.** Here the film is not complete and the surfaces come together close enough so that the resulting frictional resistance is not entirely due to viscous drag. The surface finish and the physical and chemical properties of the surfaces and lubricant are of major interest. Effective boundary lubricants are usually additives consisting of long-chain polar molecules that are physically adsorbed on the high points of mating surfaces and tend to prevent metal-to-metal contact.

4. Extreme boundary lubrication. This is the type of lubrication that obtains under the most severe sliding conditions (highly loaded bearing surfaces operating at relatively low speed). The additive in this case reacts chemically with one of the bearing metals to form a low-shear-strength solid layer that prevents metal-to-metal contact, welding, and metal transfer (wear). The solid films that are formed are usually inorganic, have a high melting point, and hence are suitable for use at elevated temperatures without melting. Hypoid gear lubricants and heavy-duty type cutting fluids are examples of extreme boundary lubricants.

5. Clean metal surfaces. This represents the extreme in adverse conditions of sliding contact. Hardness, shear strength, and surface finish are the main characteristics of interest in this type of action.

Real Area of Contact

All finished surfaces are found to have irregularities that are very large in comparison with atomic dimensions (i.e., large compared with 10^{-8} in or 3×10^{-10} m).

As a consequence of these micro irregularities, the real area of contact (A_R) is much less than the apparent area of contact (A). For example, if two carefully ground blocks, each of 1 in^2 cross-sectional area, are placed in contact and loaded with 10 lb, the real area of contact will be less than 10^{-4} in^2 (< 0.1 mm^2), while of course the apparent area (A) is actually 1 in^2 (645 mm^2). When such surfaces first make contact, A_R is zero. As load is applied, the relatively few high points that make contact are plastically deformed and A_R increases. Thus, the real area of contact will be independent of the apparent area of contact or the surface finish but will be determined solely by the applied load and the flow stress or hardness of the protuberances. The area A_R developed must be just sufficient to equal the applied load (P) divided by the hardness (H) of the metal asperities, or

$$P = A_R H \qquad (10.1)$$

The characteristic roughness found on most surfaces consists of irregularities with a pitch that is at least an order of magnitude greater than the peak-to-valley distance.

Freshly generated surfaces will be clean but will quickly oxidize or be covered by an adsorbed layer of water vapor or other material depending on the environment and properties of the metal. A freshly generated surface will have a density of electrons and an atom spacing corresponding to the metal in bulk. However, electrons will leave the surface (these are called exoelectrons) as the atom spacing increases to the equilibrium value. The energy associated with the formation of a new surface after equilibrium has been established is called surface energy (T), $[FL/L^2]$. Freshly generated surfaces are extremely active chemically before equilibrium is established.

Beilby (1921) showed that the upper layer of a polished surface is different from the under-lying material. His microscopic observations led him to believe that, in polishing, surface metal is caused to move from point to point as though it were molten. The net effect was to smooth the surface by transferring metal from peaks to valleys. Beilby further observed that the resulting surface layer had a glasslike or amorphous structure. The existence of such an amorphous Beilby layer on polished metals has been disputed for many years, electron diffraction data being used successively to prove and disprove the existence of such a layer. More recently, a high-energy laser beam has been used to traverse surfaces following a rastor pattern to produce a surface texture that resembles the Beilby layer. In this case (laser glazing), a very thin layer of metal is melted and then solidifies so rapidly by heat flowing into the cold subsurface that either an amorphous surface is produced or one that has a crystal size so small as to be unresolvable.

Bowden (1945) has presented the most logical arguments in favor of the existence of the Beilby layer. He points out that, in the polishing process, the polishing agent is generally embedded in a relatively soft lap, which is rubbed against the specimen in the presence of some vehicle such as water. Hot spots will develop at the points of contact between the abrasive and the specimen, and local surface temperatures up to the melting point of the metal may be encountered if the polishing speed is sufficiently high. The molten or softened metal will be smeared over the surface and will quickly solidify to form the layer which Beilby observed. Probably the molten surface layer solidifies so rapidly that the crystals produced are extremely small and the material appears amorphous by present methods of study. Bowden states that this hypothesis may be tested by the following procedure: if polishing is due primarily to mechanical abrasion, then the relative hardness of specimen and polisher is the property of importance; but if the action is caused by surface melting, then the relative melting points are of major concern. In a series of simple experiments, it was observed that camphor, which melts at 350 °F (177 °C), readily polishes metals that melt at a lower temperature, such as Woods metal, but has no effect on tin, lead, white metal, or zinc, all of which melt at higher temperatures. That camphor is softer than Woods metal indicates that relative hardness is unimportant. Similarly, zinc oxide, which is comparatively soft, readily polishes quartz, which is harder, because the melting point of zinc oxide is higher than that of quartz. Many other cases support the view that the relative melting point is the important property, rather than hardness, when choosing a polishing agent.

Origins of Friction

The friction of a lightly loaded slider is due to one or a combination of the following causes (ignoring viscous effects):

1. adhesion (fracture of microwelds formed at the tips of contacting asperities)
2. plowing or burnishing (displacement of metal without generation of wear particles)
3. abrasion (microcutting action producing fine chips, i.e., wear particles)
4. transverse displacement (movement of one surface transversely to the other against the applied load due to interference of hard strong asperities)

Rules of Dry Sliding

When two dry clean surfaces slide together under relatively light load, the following rules pertain over a fairly wide range:

1. The coefficient of friction is independent of the applied load.
2. The coefficient of friction is independent of the sliding speed.
3. The coefficient of friction is independent of surface finish.
4. The coefficient of friction is independent of the apparent area of contact (A).
5. The coefficient of friction is independent of the temperature of the sliding surfaces.

The first of these rules is known as Amontons' law (Amontons, 1699). All of these rules fail to hold at extreme values. For example, the coefficient of friction at zero speed (static friction) is normally greater than the dynamic value. Also, the coefficient of friction will normally decrease with speed above that required to cause thermal softening of the less refractory surface. However, for speeds between these two extremes, the coefficient of friction will be approximately independent of speed.

A lightly loaded slider is one for which Amontons' law represents a good approximation (μ independent of P).

Adhesive Friction

For two surfaces in adhesive sliding contact, the friction force (F) will be

$$F = \alpha \tau A_{\mathrm{R}} \tag{10.2}$$

where τ is the shear strength of the weaker of the two sliding materials and α is the fraction of the real area that involves clean adhering asperities. For an unlubricated case, α will approach unity.

For dry surfaces sliding in air, the coefficient of friction will be from Eqs. (10.1) and (10.2):

$$\mu = \frac{F}{P} = \frac{\tau A_{\mathrm{R}}}{H A_{\mathrm{R}}} = \frac{\tau}{H} \tag{10.3}$$

For low friction, one of the surfaces should have low shear strength and high hardness. Since this is a combination not found in nature, most bearing metals are designed to consist of a composite in which a low-shear-strength material is smeared over a harder substrate. The copper–lead bearing (small particles of insoluble lead in copper) is one of many examples of bearing materials based on this principle.

When two metal surfaces are forced together, the high points that make contact will weld by establishing metallic bonds, provided the surfaces are clean. The surfaces need not be heated, for a *pressure weld* may be established at room temperature if sufficient pressure is applied. However, when such a weld is produced between materials of high stiffness (i.e., materials of large Young's modulus, E), the junctions at the tops of the asperities frequently rupture when the load is released. Elastic recovery in such cases causes the junctions to rupture one by one as the load is removed. The larger the A_{R}/A ratio and the smaller the stiffness of the metals united, the smaller will be the tendency for a pressure weld to rupture when load is released.

The blacksmith normally heats two pieces of steel to be united in order to decrease E and increase A_{R}/A for the limited force that can be applied. Similarly, it is possible to produce a pressure weld between clean indium and silver surfaces at room temperature with very light pressure applied, due to the softness and low E of indium. However, such a weld cannot be formed between steel and silver due to the relatively low A_{R}/A ratio and high values of E involved.

The importance of a large value of A_{R}/A is further illustrated by the relative ease with which a fine, freshly drawn glass filament may be made to adhere to a similar filament, compared with larger more rigid glass rods. The greater conformability of thin specimens is also demonstrated by the relative ease of welding freshly cut sheets of mica of thick and thin sections. A large real area of contact may be obtained with little applied pressure by previously depositing films in the liquid state on the surfaces to be joined. When the adhesive is transformed into a solid film by cooling, by evaporation of a solvent, or by polymerization or other chemical change, a strong joint may be obtained.

In addition to the magnitude of the load, the cleanness of the surfaces is also important in pressure welding. If an oxide or other film is present, it may be broken through only with difficulty when a purely normal load is applied between the surfaces. If, however, there is some tangential motion between two contaminated mating surfaces, it is easier to penetrate the film and establish a weld. A brittle oxide or other surface film is more easily penetrated and the blacksmith frequently uses a flux to react with oxide films on surfaces to be joined to render them more brittle and hence more easily pierced.

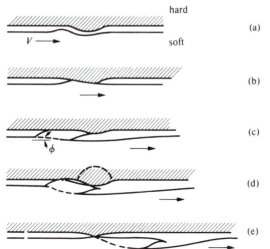

Fig. 10.1 Mechanism of prow formation.

Other Sources of Friction

Cocks (1962, 1964, 1965) has studied the detailed mechanism involved when one dry surface slides over another. Asperities do not simply slide over each other, but what Cocks calls "prows" are produced. These are wedges of plastic metal that tend to force the surfaces apart. By using a flexible system that enabled normal separation of the surfaces by as much as 0.010 in (0.25 mm), large wedges [0.012 in by 0.032 in (0.3 mm by 0.8 mm) in direction of motion] were produced with copper sliding on copper. Figure 10.1 shows the mechanism of prow formation observed when a moving soft asperity encounters a harder one. At (a) the two asperities are about to make contact while (b) shows the asperities shortly after contact. The soft (lower) metal is deformed causing the prow to rise up thus tending to force the surfaces apart. The surfaces deform locally allowing the prow to flow past the hard asperity. With a stiff system, the shear angle (ϕ) will be small and the horizontal prow generating force will be high. The prow generating force will contribute to the friction force and occasionally the strength of the softer material will be exceeded during prow formation giving rise to a wear particle coming from the moving surface. Alternatively, the strength of the harder asperity may be exceeded giving rise to a wear particle coming from the harder (upper) region.

Feng (1952) has suggested that the tips of asperities will interpenetrate (Fig. 10.2) due to the inhomogeneous nature of strain associated with the weak points present in all real materials. Subsequent motion will give rise to a friction component of force associated with

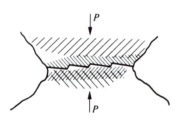

Fig. 10.2 Interface of pair of contacting asperities roughened by plastic deformation. (after Feng, 1952)

1. the transverse displacement of one roughened surface over the other (in the case of hard strong shear plane ends)
2. the shearing-off of the interfering shear plane ends
3. plastic flow of the shear plane ends (burnishing action)

The friction component of force associated with Fig. 10.2 may be due to any combination of the above causes depending upon ductility.

SURFACE TEMPERATURE

When surfaces slide one relative to the other under load, the energy dissipated per unit time is the product of the friction force (F) and sliding speed (V) ($U = FV = \mu PV$). All of this energy is converted into heat. The following approximate analysis is useful in indicating the variables of importance relative to the mean or maximum temperature at the interface for a high speed slider.

Figure 10.3 shows a moving heat source traversing a semi-infinite body subject to the following assumptions:

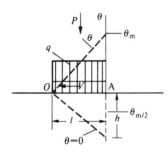

Fig. 10.3 Moving heat source corresponding to a friction slider.

1. Dispersion of thermal energy over contact area is uniform (q = thermal flux density = thermal units per unit area per unit time).

2. Slider is a perfect insulator—all thermal energy goes to extensive member which has thermal conductivity (k) and volume specific heat (ρC).

3. Slider is extensive perpendicular to paper in Fig. 10.3 (two-dimensional heat flow).

4. Temperature varies linearly from leading edge (O) to trailing edge (θ_m) of slider.

5. Depth of penetration of thermal energy varies linearly from leading edge (O) to trailing edge (h).

Equating the two expressions that may be written for total heat flux, Q,

$$Q = \rho C \left(\frac{\theta_m}{2}\right)(bhV) = k(bl)\frac{\theta_m}{h} \tag{10.4}$$

and hence

$$h = \left[\frac{2kl}{\rho CV}\right]^{1/2} = \left[\frac{2lK}{V}\right]^{1/2} \tag{10.5}$$

where $K = k/\rho C$ = diffusivity, b is slider width, and l is slider length. Substituting into Eq. (10.4),

$$Q = \rho C \frac{\theta_m}{2}(bV)\left[\frac{2lK}{V}\right]^{1/2}$$

or

$$\frac{\theta_m}{2} = 0.707 \frac{Q}{b[lV(k\rho C)]^{1/2}} \tag{10.6}$$

The more exact Jaeger (1942) solution leads to the same result except that the coefficient (0.707) is 0.754. However, this agreement is purely fortuitous. What is significant, however, is that the somewhat overly simplified analysis leads to the same functional relationships for ($\theta_m/2$) as the more exact Jaeger approach. Therefore, the coefficient in Eq. (10.6) should be omitted and the mean surface temperature ($\theta_m/2$) merely considered to be proportional to the quantity

$$\left(\frac{Q}{b[lV(k\rho C)]^{1/2}}\right)$$

For a friction slider, q will be constant along the slider if μ is constant and

$$Q = \frac{\mu PV}{J} \tag{10.7}$$

where μ is the mean coefficient of friction, P is the normal load, and J is the mechanical equivalent of heat.

From Eqs. (10.6) and (10.7)

$$\bar{\theta} = \frac{\theta_m}{2} \sim \frac{\mu}{J}\left(\frac{P}{bl}\right)\sqrt{\left(\frac{Vl}{k\rho C}\right)} \tag{10.8}$$

For a high-speed slider, practically all of the thermal energy will be convected away by the extensive member (the slider is a perfect insulator in effect), and Eq. (10.8) will represent a good approximation. This equation states that the mean interface temperature ($\bar{\theta}$) varies

1. directly with the specific friction energy ($\mu P/bl$)
2. directly as $V^{1/2}$
3. directly as $l^{1/2}$
4. inversely as $(k\rho C)^{1/2}$

It is of interest to note that the product ($k\rho C$) of the extensive member is the only thermal property of importance and that the mean and maximum temperatures increase with the length of the slider provided the specific friction energy is constant along the slider.

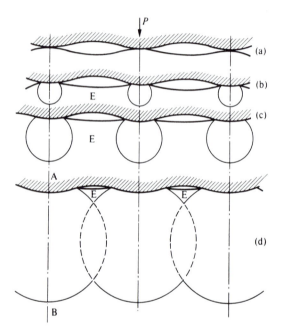

Fig. 10.4 Plane slider under progressively increasing load (P). Upper cross-hatched surface is hard; lower surface is soft. Region E is elastic. (after Shaw, 1963)

HEAVILY LOADED SLIDERS

Moore Effect

As the load on a slider increases, the size of the plastic zone associated with each asperity will increase as indicated in Fig. 10.4. At a critical value of load the plastic zones will join up, and then for further increase in load there will be no further flattening of asperities. Instead, the subsurface will flow plastically. Moore (1948) drew attention to this phenomenon by pressing a hard cylinder into a soft copper surface having small parallel grooves [0.025 mm (0.001 in) apart and having a 100° included angle] machined in the surface. He found that even though the subsurface flowed extensively, the grooves were still substantially present (Fig. 10.5).

For an array of flattened asperities, as shown in Fig. 10.6, the mean normal stress on each of the flat areas will be $3Y$, assuming the areas to be far enough apart so there is no interaction between asperities. The asperities should flatten until the total load on all asperities will be just sufficient to cause the subsurface to flow plastically (i.e., until the mean stress in the subsurface = Y).

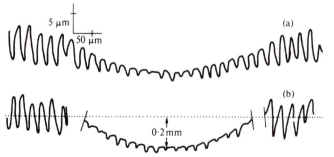

Fig. 10.5 Tallysurf traces of work-hardened copper (H_V 84) having grooves 0.025 mm (0.001 in) apart machined in surface before indentation by 6 mm diameter × 0.18 mm (0.25 × 0.007 in) cylinder. (a) Applied load = 200 kg. (b) Applied load = 2500 kg. (after Moore, 1948)

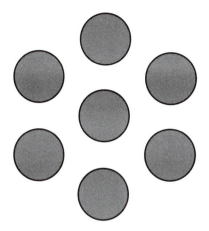

Fig. 10.6 Plan view of asperities flattened by normal stress applied by hard smooth surface.

If asperity interaction is ignored, the real area (A_R) should increase in accordance with Eq. (10.1) until $A_R/A = Y/3Y = \frac{1}{3}$. This represents a relatively crude first approximation since there will be interaction between asperities, and this will cause the limiting A_R/A to be $> \frac{1}{3}$.

Williamson (1968) has performed tests to more exactly identify the transition from a lightly loaded situation to a heavily loaded one. An aluminum specimen with bead-blasted surface was encased in a steel container and loaded by a polished flat piston that closely fitted the steel container. As load was applied, the surface asperities were flattened but did not disappear. Figure 10.7a shows the change in the number of plateaus with load. At first the number increases linearly with load and then remains constant with further increase in load. Figure 10.7b shows the variation in area of contact with load, and this curve is seen to have a knee at about the same load as that where the previous one ceased to rise. Figure 10.7c shows the variation in separation of two surfaces versus load. Again, there is an abrupt change in the slope of this curve at a load of about 900 lb (4009 N) which corresponds to the force required for bulk plastic flow of the specimen. It therefore appears as though the abrupt

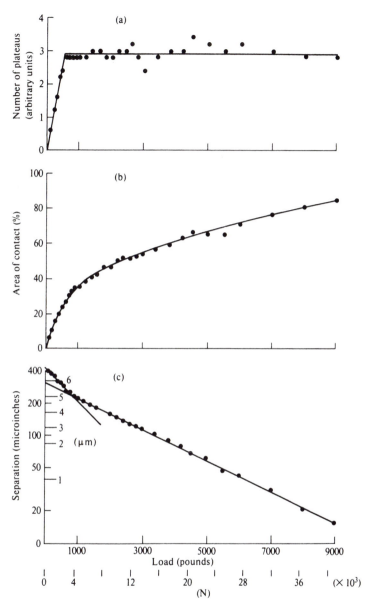

Fig. 10.7 Behavior of contacting surfaces under high normal stress. Bead-blasted aluminum surface encased in a steel container and loaded by a hard-polished steel piston closely fitting the steel container. (after Williamson, 1968)

change in behavior in the curves of Fig. 10.4c to Fig. 10.4d corresponds to a transition from a lightly loaded situation to a heavily loaded one.

From Fig. 10.7b departure from Amontons' law occurs at A_R/A close to the previously predicted value of $\frac{1}{3}$. This suggests that the asperities may in fact act independently of each other as assumed. When the applied load was increased to four times that corresponding to the transition in Fig. 10.7 (4×900 lb $\cong 16,000$ N), the real area of contact was only 50% of the apparent area.

Junction Growth

Moore (1948) found that when a hard smooth indentor slides across a grooved copper surface, the degree of ridge flattening is greatly increased, the extent of the increase depending upon the coefficient of sliding friction and the direction of the ridges relative to the sliding direction. The friction force will of course help satisfy the flow criterion (either Tresca or von Mises) and the normal stress required for any degree of asperity flattening will be less when a friction force is present. The tendency for the ridges to disappear was greatest when the ridges were oriented normal to the sliding direction since this gave the shortest transport distance associated with the movement of material from peaks to valleys.

Burnishing is a surface-refining operation that requires a high friction force. In wire drawing, surface roughness will be excessive if the friction force is not sufficient to iron out the roughness due to strain inhomogeneity (Fig. 10.2). The challenge is to have sufficient frictional force to provide burnishing without galling. The difference in texture of the front and back of a continuous metal cutting chip is due to removal of shear plane roughness due to burnishing on the tool face but not on the free surface of the chip.

The increase in real area of contact that accompanies sliding (friction) is called junction growth in tribology, and this is what Moore found when a hard smooth indentor was slid across a grooved surface.

Heavily Loaded Slider Friction

Shaw et al. (1960) have presented Fig. 10.8 to illustrate the variation in coefficient of friction with change in normal stress (σ) and hence with (A_R/A). Three regimes are identified. Regime I is that where Amontons' law holds ($\mu = \tau/\sigma = $ constant). Regime III is for an internal shear surface of a material that has not yet developed microcracks ($A_R/A = 1$, and τ is independent of σ). Regime II is the transition region between Regimes I and III. In Regime II, the coefficient of friction ($\mu = \tau/\sigma$) decreases with increase in load. Regime II corresponds to the situation on the tool faces of metal-cutting and -forming tools. Surface sliding is accompanied by subsurface plastic flow in Regime II.

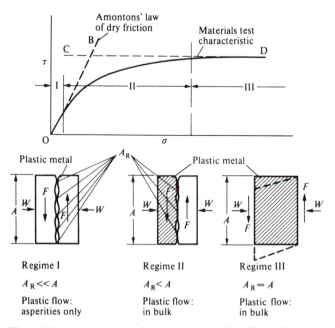

Fig. 10.8 Three regimes of solid friction. (after Shaw, Ber, and Mamin, 1960)

Earlier, Finnie and Shaw (1956) suggested that the ratio of real to apparent area of contact might be approximated as follows:

$$\frac{A_R}{A} = 1 - e^{-BP} \qquad (10.9)$$

where B is a constant for a given material combination, and P is the applied load.

Shaw et al. (1960) have demonstrated the reduction of coefficient of friction with applied load in Regime II using the apparatus shown in Fig. 10.9. A hard steel ball is pressed into a soft steel surface, Brinell fashion, until plastic flow occurs. The torque required to slide the specimen relative to the ball at low velocity is then measured. A small hole is provided at the center to eliminate the singularity that would otherwise exist there. The mean shear stress (τ) and normal stress (σ) may then be estimated for different values of applied load. Figure 10.10 shows representative results where deviation from Amontons' law is clearly indicated as well as the reduction in coefficient of sliding friction with increase in the applied load.

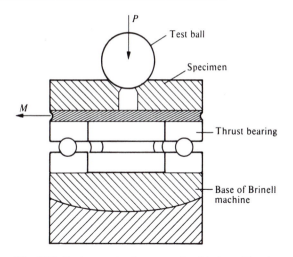

Fig. 10.9 Test apparatus for measuring friction with subsurface plastic flow. (after Shaw, Ber, and Mamin, 1960)

Friction in Metal Cutting

Friction plays an important role on the tool face of a sharp cutting tool and also on the clearance face of a worn tool. However, conditions on the tool face are far different from those for a lightly loaded slider. Amontons' law does not hold over the entire contact area nor do the other rules of dry friction previously discussed for lightly loaded sliders. The complexities of tool-face friction are discussed in a monograph by Bailey (1975).

Stress Distribution on Tool Face

Usui and Takeyama (1960) studied the distribution of shear (τ_C) and normal (σ_C) stress along the tool face of a cutting tool by cutting lead at low speed with a photoelastic tool. Results of this study are shown in Fig. 10.11. The shear stress (τ_C) was found to remain constant over the half of tool–chip contact nearest the cutting edge but to decrease to zero over the other half, reaching

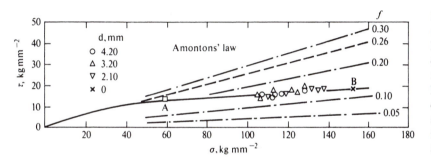

Fig. 10.10 Variation of τ and σ for unlubricated hard steel Brinell ball of $\frac{1}{2}$ in (12.7 mm) diameter tested against mild steel in apparatus of Fig. 10.9. Diameter d is the size of the central hole drilled in the specimen before test. Point A is for subplastic load. (after Shaw et al., 1960)

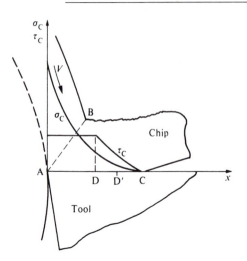

Fig. 10.11 Variation of shear and normal stresses on tool face of tool. (after Usui and Takeyama, 1960)

zero of course at point (C) where the chip leaves contact with the tool. The normal stress (σ_C) was found to increase monotonically from (C) to the cutting edge (A). Zorev (1963) reached similar conclusions by observing that grinding scratches parallel to the cutting edge were transferred to the fully plastic surface of the chip from A to D, but that these markings were replaced by an orthogonal set of scratches running in the direction of chip motion over the second region of contact extending from D to C where sliding actually occurred. Zorev concluded that chip flow was completely within the chip (sub-surface) from A to D but occurred at the interface from D to C. Kato and Yamaguchi (1972) used a special tool dynamometer with divided rake face to verify these results. They found stress distributions similar to those of Fig. 10.11 and that the region of constant shear stress decreased relative to total contact length as the rake angle increased.

Figure 10.11 is consistent with the microcrack theory of plastic flow presented in Chapter 9. From A to D, the normal stress (σ_C) is sufficiently high to suppress microcrack formation, the dislocation mechanism pertains, and conditions are similar to those in the Lankford–Cohen experiments. The shear stress (τ_C) will be independent of normal stress (σ_C) over this no-microcrack region (from A to D) and will remain constant at a value consistent with the secondary shear strain pertaining in the chip adjacent to the tool face. Microcracks will be involved in the region of decreasing shear stress (τ_C) extending from D to C. Initially (from D to D$'$) relatively few microcracks will be involved since the normal stress (σ_C) will be sufficiently high to cause appreciable rewelding. However, as (σ_C) decreases, there will be less and less rewelding relative to fracture, and conditions will approach those for a lightly loaded slider where $A_R/A \ll 1$ and the "microcracks" communicate. The picture of tool-face stresses shown in Fig. 10.11 has been further verified and elaborated recently by other workers (Amini, 1968; Trent, 1977).

In contrast to Fig. 10.11, several studies have revealed sliding at the chip–tool interface near the tool tip but not in the region where the chip leaves contact with the tool (Horne et al., 1977; Doyle et al., 1979; Madhavan et al., 1996). These tests involved transparent single crystal Al_2O_3 (sapphire) tools to cut ductile materials (lead, aluminum, and copper) at very low speed. Under these conditions the chip–tool interface could be directly observed at magnifications as high as 450x, and slip at the tool tip but not at the separation point was clearly evident. The explanation for this paradox appears to be material and speed related.

In the case of lead, the shear strength is so low relative to the bond strength between lead and Al_2O_3 that slip is not prevented even under the relatively high normal stress at the tool tip. At the exit region between chip and tool, the contact stress falls rapidly and A_R/A will be less than one. At low cutting speed, air can penetrate the exit region and react with the chemically active new chip surface to produce lead oxide or lead nitride that has a greater shear strength than the bond strength with Al_2O_3. This apparently accounts for the absence of slip at the exit region when lead is cut at a very low speed. A similar explanation holds for aluminum and copper when cut at very low speed.

Bagchi and Mittal (1988) have performed experiments on steel aluminum and brass using sapphire tools to enable direct observation of the chip–tool interface, but at higher more realistic cutting speeds using a movie camera instead of direct observation. Under steady state conditions two zones were observed corresponding to those of Fig. 10.11. It thus appears that when metals are cut at practical cutting speeds (above that where a BUE disappears), conditions at the tool–chip interface are well approximated by Fig. 10.11. However, lead which does not produce a BUE is a special case where the sticking and sliding zones are reversed for reasons given above.

It has been suggested by De Chiffre (1977) that only region DC (Fig. 10.11) is influenced by a cutting fluid and that an effective fluid causes a decrease in length DC and hence a decrease in length AC which in turn causes an increase in the shear angle ϕ and a decrease in cutting force. The only change the microcrack theory of Chapter 9 would suggest relative to this concept of cutting fluid action is that the primary action of the cutting fluid extends from D' to C (Fig. 10.11) where the microcracks communicate like the holes in a sponge. The results of Fig. 9.7 suggest that point D corresponds to the point where $\sigma_C \cong \tau_C/2$.

As Zorev (1963) pointed out, Amontons' law of sliding friction (coefficient of friction is independent of normal stress) should not be expected to hold in the region extending from A to D but only from D to C in Fig. 10.11. There will obviously be a transition region (D to D'), and it would now appear to be more accurate to suggest that Amontons' law will hold only from D' to C.

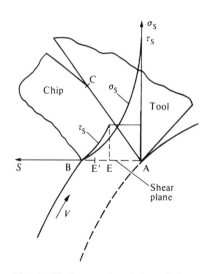

Fig. 10.12 Proposed variation of shear and normal stress across shear plane. (after Shaw, 1980)

It has been suggested by Shaw (1980) that the situation on the shear plane is similar to that on the tool face. Figure 10.12 shows the proposed variation in shear (τ_S) and normal (σ_S) stress on the shear plane with distance across the shear plane. In region AE (nearest the cutting edge), τ_S will be independent of normal stress on the shear plane (σ_S), while in region E'B (nearest the free surface) τ_S/σ_S will be approximately constant corresponding to Amontons' law for a friction slider. These two regions should be connected by a transition region (EE' in Fig. 10.12) where the density of microcracks is increasing to the point of intercommunication. Figure 10.12 explains why the microcracks mechanism of plastic flow of Chapter 9 involving the rewelding of microcracks should be expected to pertain over part of the shear plane in metal cutting even though the *mean* normal stress on the shear plane will generally be $\geq \bar{\tau}_S$.

Variation of Coefficient of Tool–Face Friction with Rake Angle

One of the paradoxes associated with metal cutting involves the variation of tool–face friction coefficient (μ) with rake angle (α). Table 10.1 presents typical orthogonal cutting data for copper and steel cut at low speed with high speed steel (HSS) tools in air. The stresses were calculated from measured forces and contact areas on the tool face. These stresses (τ_C and σ_C) are based on the apparent area and are assumed to be uniformly distributed. It is clearly evident that the coefficient of cutting friction increases markedly with increase in rake angle. This is just the opposite behavior one would expect from experience with lightly loaded sliders that perform in accordance with Amontons' rule. However, it is what should be expected from heavily loaded slider experience where the coefficient of friction ($\mu = \tau_C/\sigma_C$) is found to decrease with increase in normal stress as a consequence of an increase in A_R/A. It may therefore be concluded that the paradox is in part explained by the fact that conditions at the tool face are largely those of a heavily loaded slider.

An additional reason for the tool–face friction paradox appears to lie in the fact that the friction process in cutting interacts strongly with the shear process. As a first crude approximation, we may characterize the shear process in terms of the Lee–Shaffer (1951) model which appears to be one of

TABLE 10.1 Representative Low-Speed Cutting Data

Work Material	Rake Angle α (deg)	Average Stresses at Tool–Chip Interface (p.s.i.[†])		
		Shear Stress (p.s.i.) τ_C	Normal Stress (p.s.i.) σ_C	Mean Friction Coefficient $\mu = \tau_C / \sigma_C$
Electrolytic tough pitch copper; cutting speed (V), 0.3 f.p.m. (0.0016 m s^{-1}); depth of cut (t), 0.002 in (0.05 mm)	30 45 60	35,600 34,700 36,700	43,300 30,000 16,000	0.82 1.15 2.30
SAE B1112 steel; cutting speed (V), 0.04 f.p.m. (0.0002 m s^{-1}) depth of cut (t), 0.003 in (0.076 mm)	16 30 45	46,000 49,300 51,700	70,000 57,000 43,200	0.66 0.86 1.20

[†] 1 p.s.i. = 6905 N m^{-2}.

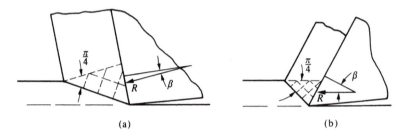

(a)   (b)

Fig. 10.13 Variation in magnitude and direction of resultant cutting force with rake angle.

the least objectionable simple models yet proposed. Figure 10.13 shows the slip line fields for sharp high- and low-rake-angle tools cutting the same materials. Assuming the shear plane to be a plane of maximum shear stress and the resultant cutting force (R) to be displaced 45° from the shear plane in accordance with the Lee–Shaffer solution, angle β ($\beta = \tan^{-1} \mu$) will be smaller for a negative rake angle.

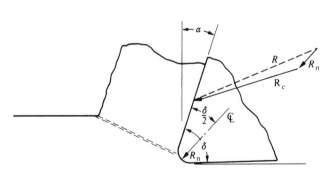

Fig. 10.14 Influence of tool tip radius on cutting forces. [R_n = Indenting force due to tool radius. R_c = Resultant cutting force on remainder of tool face.] (after Masuko, 1953)

Indentation Force at Tool Tip

Masuko (1953) suggested that a sharp cutting tool will deform elastically to give a radius at the tip of even a sharp cutting tool. Figure 10.14 shows a cutting tool with the assumed radius at the tip. Masuko assumed that in addition to the chip forming force (R_c) on the tool face, there would be an indentation force at the tool tip (R_n in Fig. 10.14). He assumed that the indentation force R_n bisects the included angle of the tool (δ). The components of indentation force R_n were substracted from the measured force components F_P and F_Q and the resulting values used to compute stresses and strains as outlined in Chapter 3.

The concept of an indenting component of force was probably suggested to Masuko by the fact that when measured values of F_P and F_Q are plotted against undeformed chip thickness (t), straight lines with an intercept at $t = 0$ are obtained as shown diagrammatically in Fig. 10.15. The cutting (R_{nP}) and feed (R_{nQ}) components of indentation force R_n were assumed to correspond to the intercepts of the F_P and F_Q versus t curves in Fig. 10.15.

The actual values of R_{nP} and R_{nQ} were obtained by an iterative approximate elastic analysis. When the above procedure was applied to data for a duralumin specimen cut dry using HSS tools having rake angles ranging from 0 to 44°, the coefficient of friction for cutting on the tool face was found to be constant relative to α and t. At the same time when the shear stress on the shear plane (τ) was compared with extrapolated values of τ versus γ for torsion tests on the same material, no size effect was found. However, in extrapolating the torsion data to the larger strains pertaining in cutting, Eq. (5.8) was used instead of Eq. (9.11). From these results, Masuko concluded that there is no size effect in cutting and that the unusually high values of shear stress on the shear plane in cutting are due to strain hardening alone.

Results practically identical to those of Masuko have been reported by Albrecht (1960, 1961, 1963).

While the proposal of Masuko is logical, it does not appear to hold up when subjected to critical experiment. Finnie (1963) has performed orthogonal cutting tests on a perfectly sharp tool and one of very small bluntness. Figure 10.16 gives his results. These curves show that whereas the force versus feed curves pass through the origin for the sharp tool, the blunt tool has intercepts. From these results, extrapolation of force versus undeformed chip thickness (t) data to $t = 0$ is seen to be a questionable procedure for a sharp tool. The nonlinearity of the force versus undeformed chip thickness curves for the sharp tool is direct evidence of a size effect (increase in flow stress of material cut with decrease of t). Finnie's results indicate that the height of the rounded nose of a blunt tool must exceed about 20% of the undeformed chip thickness (t) before its influence becomes noticeable.

Finnie (1963) also suggests that the general absence of a deformed layer on surfaces produced by sharp tools argues against the general importance of an indentation or plowing in metal cutting. However, the indentation or ploughing effect should be expected to be important in grinding (very small t and large negative rake angle) or for tools of appreciable dullness.

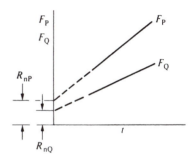

Fig. 10.15 Variation of power (F_P) and feed (F_Q) components of cutting force with undeformed chip thickness (t). R_{Pn} and R_{Qn} are the horizontal (power and vertical components of R_n in Fig. 10.14.

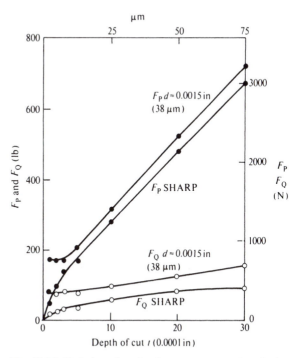

Fig. 10.16 Variation of cutting force components in velocity (F_P) and feed (F_Q) directions with undeformed chip thickness (t) for sharp and dull tools. Dimension d is extent of tool bluntness measured in feed direction. (after Finnie, 1963)

Stevenson and Stephenson (1995) have performed cutting tests on pure zinc to further explore the possibility of a plowing effect existing at the tip of a sharp tool to account for the extrapolated intercepts at $t = 0$ shown in Fig. 10.15. Compression values of stress and strain for cylindrical specimens of the same material (zinc) were extrapolated to obtain corresponding metal cutting results. The von Mises criterion was used in making adjustments for differences in temperature, strain, and strain rate. Using the same analytical approach it was found that when the undeformed chip thickness values (t) for cutting with a sharp tool were extrapolated to zero, any plowing effect was found to be negligable. It was inferred that the following results would hold for other metals as well as for zinc:

(a) negligable plowing for a sharp tool

(b) ability to extrapolate compression test data to cutting conditions using the von Mises criterion

However, while item a is in agreement with the results of Finnie (Fig. 10.16), item b which holds for pure zinc does not agree with results for steel as the experiments of Merchant in Chapter 8 clearly indicate. The probable reason for this is discussed in Chapter 9. This involves the appearance and transport of microcracks as well as dislocations when metals are subjected to very high strains (1.5 or higher) as in the steady state chip formation of steel.

Multiplicity of Mechanisms

Metal cutting is an extremely complex process that cannot be described by a single simple mechanism. While a single mechanism may be predominant over a limited range of operating conditions, experience teaches that no single mechanism holds in general. This accounts for the extremely wide range of views that have appeared in the literature to explain cutting results, most of which are supported by sound experimental data. In most cases, the relatively simple mechanism suggested by an author to explain his or her results fails to hold when applied to substantially different operating conditions. This condition has led Hill (1954) to suggest that no unique solution exists as already discussed in Chapter 8.

REFERENCES

Albrecht, P. (1960). *J. Engng. Ind.* **82**, 348.

Albrecht, P. (1961). *J. Engng. Ind.* **83**, 557.

Albrecht, P. (1963). *International Research in Production Engineering.* ASME, New York, p. 32.

Amini, E. (1968). *J. Strain Anal.* **3**, 206.

Amontons, G. (1699). *Historie de L'Academie Royale des Sciences avec les Memories de Mathematique et de Physique*, p. 206.

Bagchi, A., and Mittal, R. O. (1988). *Proc. of Manufracturing International, Atlanta, Ga.*

Bailey, S. A. (1975). *Wear* **31**, 243.

Beilby, G. (1921). *Aggregation and Flow of Solids.* Macmillan, London.

Bowden, F. P. (1945). *Proc. R. Soc. N.S.W.* **78**, 187.

Bowden, F. P., and Tabor, D. (1950). *The Friction and Lubrication of Solids*, Vol. I. Clarendon Press, Oxford.

Bowden, F. P., and Tabor, D. (1964). *The Friction and Lubrication of Solids*, Vol. II. Clarendon Press, Oxford.

Cocks, M. (1962). *J. Appl. Phy.* **33**, 2152.

Cocks, M. (1964). *J. Appl. Phys.* **35**, 1807.

Cocks, M. (1965). *Int. J. Wear* **8**, 85.

De Chiffre, L. (1977). *Int. J. Mach. Tool Des. Res.*, 225.

Doyle, E. D., Horne, J. G., and Tabor, D. (1979). *Proc. Roy. Soc. of London* **A-366**, 173–187.

Feng, I. M. (1952). *J. Appl. Phys.* **22**, 1011.

Finnie, I. (1963). In *International Research in Production Engineering.* ASME, New York, p. 76.

Finnie, I., and Shaw, M. C. (1956). *J. Engng. Ind.* **78**, 1649.

Hill, R. (1954). *The Mathematical Theory of Plasticity.* Clarendon Press, Oxford. (Reissued 1983.)

Horne, J. G., Doyle, E. D., and Tabor, D. (1977). Proc. of 5th N.A. Metal Working Research Conf. of 5ME, pp. 237–241.

Jaeger, J. C. (1942). *Proc. R. Soc. N.S.W.* **76**, 203.

Kato, H., and Yamaguchi, K. (1972). *Proc. ASME* **94**, 603.

Madhavan, V., Chandrasekar, S., and Farris, T. N. (2002). *ASME J. of Tribology* **124**, 617–626.

Masuko, M. (1953). *Trans. Jap. Soc. Mech. Engrs.* **19**, 32.

Moore, A. J. W. (1948). *Proc. R. Soc.* **A195**, 231.

Shaw, M. C. (1963). *Int. J. Wear* **6**, 140.

Shaw, M. C. (1980). *Int. J. Mech. Sci.* **22**, 673.

Shaw, M. C., Ber, A., and Mamin, P. A. (1960). *J. Engng. Ind.* **82**, 341.

Stevenson, R., and Stephenson, D. A. (1995). *Trans ASME-J. of Eng. Materials and Technology* **117**, 172–178.

Suh, N. P. (1986). *Tribophysics.* Prentice-Hall, Englewood Cliffs, N.J.

Suh, N. P., and Sin, H. C. (1981). *Wear* **49**, 9.

Trent, E. M. (1977). *Metal Cutting.* Butterworths, London.

Usui, E., and Takayama, H. (1960). *J. Engng. Ind.* **B82**, 303.

Williamson, J. B. P. (1968). In *Interdisciplinary Approach to Friction and Wear.* Office of Technology Utilization, Washington, D.C., p. 85.

Zorev, N. N. (1963). In *International Research in Production Engineering.* ASME, New York, p. 42.

11 WEAR AND TOOL LIFE

Wear is usually undesirable and to be minimized. This is certainly the case with tool wear or when machine surfaces rub together and a loss of material from one or both surfaces results in a change in the desired geometry of the system. In some cases wear is useful:

1. writing with a pencil on paper
2. erasing unwanted pencil lines
3. metal removal by honing (fixed abrasive), lapping (free abrasive), or ultrasonic machining
4. liquid honing of surfaces
5. run-in of a bearing surface to improve its load capacity and conformity
6. comminution
7. the self-sharpening action of grinding wheels

Wear may be classified into several types as follows:

1. attritious (small particle) wear associated with adhesion, prow formation and shear plane ends
2. abrasive wear (due to the cutting action of hard particles)
3. erosive wear (cutting action of particles in a fluid)
4. diffusion wear at high surface temperatures
5. corrosive wear (due to chemical attack of a surface)
6. fracture wear (chipping of brittle surfaces)
7. delamination wear (subsurface microcracks join up to produce laminar wear particles)

Practical wear situations rarely involve only one of these types of wear and there are important interactions. For example, adhesive wear is often accompanied by oxidation of the wear debris which may in turn act as a solid lubricant to reduce further adhesive wear or as a hard particle which causes abrasive wear. Fretting corrosion is a well-known composite of several actions. In this case two surfaces have a relative motion of small amplitude so that wear particles generated cannot escape the system. The wear particles first formed oxidize and produce abrasive particles that generate further wear particles by a cutting action. These larger wear particles in turn oxidize and produce further abrasive wear particles, etc. The result is an appreciable number of loose oxidized wear particles (Fe_2O_3 or rouge) between steel surfaces which are evident as a red powder when the surfaces are separated. Possible solutions for fretting corrosion are

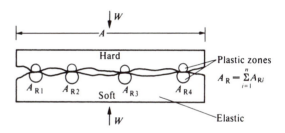

Fig. 11.1 Asperities of lightly loaded surfaces. A_R is the real area of contact and A is the apparent area of contact.

1. prevent motion

2. exclude oxygen (with grease)

3. use metals that either do not oxidize (gold) or form soft oxides (silver)

Wear is always accompanied by a great deal of friction and in some cases, subsurface plastic flow. In general, the milder the sliding conditions, the smaller the wear particles and the lower the wear rate but the greater the specific energy required to generate the wear particles that are formed.

Diffusion wear is a special type of considerable importance in metal cutting. In this case, the tool surface is decomposed, and the decomposition products diffuse into the surface of the chip. This type of wear is most important at high surface temperatures and with systems of high solubility.

In addition to these types of wear, tools may fail due to gross fracture or due to plastic flow of the tool tip in the case of high tool tip temperature. The ratio A_R/A discussed in Chapter 10 plays a very important role relative to all types of tool wear.

In the case of adhesive wear the mating surfaces come close enough together to form strong bonds. If the bonds so established are stronger than the local strength of the material, a particle may transfer from one surface to the other. After this has occurred several times, a loose fragment may be formed and leave the system as a wear particle.

If the particles so removed are very small (submicroscopic), we refer to the process as attritious wear. If the particles are visible under the microscope, the process is referred to as galling. In all cases the mechanism is the same except for the size of particle generated. Figure 11.1 shows a lightly loaded situation where Amontons' law holds.

If a weak point of stress concentration is found close to a point of contact, a particle will transfer if the strength of the asperity at this weak point is less than the strength of the bond between asperities. As a real area of contact increases, the size of the particle transferred increases and so does the volume rate of wear (B). It is empirically found that

$$\frac{B}{L} = KA_R \tag{11.1}$$

where

$\qquad B$ = volume worn away

$\qquad L$ = sliding length

$\qquad A_R$ = real area of contact

$\qquad K$ = probability that a real contact will result in a wear particle

Combining Eqs. (10.1) and (11.1) we have

$$\frac{B}{L} = K\frac{P}{H} \tag{11.2}$$

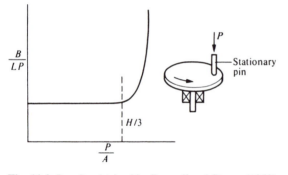

Fig. 11.2 Results obtained by Burwell and Strang (1952) with soft pin sliding on hard disc.

or

$$\frac{B}{LP} = \frac{K}{H}$$ (11.3)

The quantity K/H should be constant for a given pair of lightly loaded surfaces for which $A_R \ll A$. The quantity K will be constant when the hardness is changed as long as the chemistry of the system is not changed.

When the applied load over the apparent area of contact (P/A) becomes equal to the uniaxial flow stress of the material, which is approximately equal to ($H/3$), bulk subsurface flow occurs instead of localized flow at the asperities. The model of Fig. 11.1 then no longer holds. In such a case there is a sudden increase in the real area of contact (A_R) and hence in the volume rate of wear [Eq. (11.2)].

Burwell and Strang (1952) have performed pin-on-disc experiments in which the wear volume B was measured for different sliding lengths (L) and for the different loads (P). Figure 11.2 shows their results plotted in accordance with Eq. (11.3). The quantity K/H should be a constant as long as Eq. (10.1) is operative (i.e., until gross subsurface flow occurs). The quantity B/LP should therefore be a constant until $P/A = H/3$ at which time subsurface flow occurs and A_R rises abruptly as does the wear rate.

It is thus evident that low rates of adhesive wear will result as long as there is no bulk subsurface flow but that the wear rate will increase abruptly when

$$\frac{P}{A} > \frac{H}{3} = Y$$ (11.4)

In some of the experiments of Burwell and Strang, the pins were cylindrical in shape and softer than the moving plate. For this situation it is evident that the subsurface of the pin will flow when $P/A = Y$ (where Y is the uniaxial flow stress) if friction is negligible. However, if the coefficient of friction is 0.3, then the value of P/A (where P is the normal load component) required for bulk subsurface flow of the pin will be less than Y since the friction component of force helps make the pin go plastic. By the maximum shear theory this is found to be $0.86Y$ for $\mu = 0.3$.

When the softer member of the sliding pair is the disc and the coefficient of friction is 0.3, Shaw (1971) has estimated the critical value of P/A for rapid increase in A_R/A to be Y, the uniaxial flow stress of the disc material. To a first approximation, therefore, we may assume that subsurface flow commences when $P/A = Y$ for the softer member in all cases.

Archard (1953) has presented a derivation leading to Eq. (11.3) based upon a relatively simple idealized model. Equation (11.3) may also be obtained by dimensional analysis for a given pair of metals operating with a given lubricant (Shaw, 1977). For a lightly loaded slider, the volume (B) worn away is assumed to depend upon sliding distance (L), applied load (P), the sliding velocity (V), and the hardness of the softer member of the sliding pair (H):

$$B = \psi_1(L, P. H. V)$$

After performing the dimensional analysis in the usual way,

$$\frac{BH}{LP} = N_W \text{ (a constant)}$$

The velocity (V) is seen to drop out and there is finally only one nondimensional quantity which will be called the wear number (N_W)—a constant for a given sliding system. It has been suggested by Shaw (1977) that the wear number (N_W) is proportional to the ratio of volume of metal worn away to the volume plastically deformed. The American Society of Mechanical Engineers (ASME, 1980) has published a *Wear Control Handbook* in which values of N_W are to be found for a wide range of tribological systems.

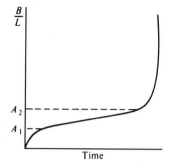

Fig. 11.3 Variation of B/L with time for lightly loaded conical pin in pin-on-disc test (Fig. 11.2).

When a loaded conical slider is subjected to a wear test, the rate of wear (B/L) will vary with time as shown in Fig. 11.3. The wear rate is high at the beginning of the test since $P/A > H/3$ because P is finite but A is zero. After the apparent area reaches a critical value A_1, the wear rate will drop to that corresponding to $P/A < H/3$. Then, at a second critical value of apparent area A_2 the temperature will have increased to the point where thermal softening occurs and P/A is again $> H/3$, this time because of a decrease in H with temperature. The more refractory the softer member of the pair, the larger the value A_2 may be before the wear rate increases abruptly. During the period of low wear rate, there is no subsurface flow, and the load is supported on a relatively small real area of contact as in Fig. 11.1.

In the foregoing discussion clean sliding surfaces are visualized. These surfaces may, however, be of different hardnesses or strengths. The stronger of the two surfaces will then wear at a lower rate than the weaker one, and this is reflected in a lower value of K or N_W for the stronger member of the pair. The wear rate of the stronger surface will not be zero, however, since occasionally a weak point in the generally stronger surface may occur opposite a strong point in the generally weaker surface. Transfer will then involve a particle leaving the stronger surface. This explains why the wear rate of the stronger member of a metal pair is lower than that for the weaker element *but not zero*.

In the presence of a contaminating environment, the value of K will generally be reduced to (αK) where α is a factor less than one. The contaminating action may involve

1. physical adsorption of a low-shear-strength film (example is oleic acid)
2. formation of an inorganic film by reaction (example is oxidation in air)

The first of these actions is often called boundary lubrication while the second is referred to as extreme boundary lubrication.

When Eq. (11.4) pertains, and there is subsurface flow, then a third type of environmental action known as the Rehbinder effect (1947) may occur. In this case a contaminating vapor is adsorbed into microcracks in the deforming subsurface which in turn alters the equilibrium between the generation and rewelding of microcracks. The layer of metal directly beneath the surface then behaves as though it were weaker, and in Russia this is sometimes referred to as "hardness reduction". An example of this type of action is when carbon tetrachloride is applied to a heavily loaded surface subjected to very large plastic strains (> 2). If the subsurface strains are not large enough, microcracks will not be formed and the Rehbinder effect will not be operative.

It is relatively difficult to achieve pure adhesive wear, for the material transferred during adhesive wear will often cause some abrasive wear. Burwell and Strang were very careful to adjust conditions to provide adhesive wear. They used a hardened (H_B 550) steel disc having a finish of 1 μin (R_a). To prevent corrosion and to remove loose wear particles, the surfaces were flooded with an inert liquid (pure hexadecane) and tests were run at relative humidities of less than 20%. The wear track was changed frequently to avoid grooving or abrasive wear due to build up of wear particles on the surface of the disc. Unless such precautions were taken, consistent results were not obtained and pure adhesive wear was not assured.

Under these carefully controlled conditions, very low wear rates were obtained. For example, the value of (B/PL) for the horizontal region of Fig. 11.2 was found to be approximately 10^{-12} in^2 lb^{-1} (1.45×10^{-12} cm^2 N^{-1}) when the Brinell hardness of the $120°$ cone was either 223 or 430 kg mm^{-2}. The value of $K = N_W$ may be estimated from Eq. (11.2). Thus, for the softer pin

$$N_W = \frac{B}{PL}H = 10^{-12}(223)(1420) = 0.317 \times 10^{-6}$$

This suggests that only about one in a million encounters between asperities resulted in a wear particle.

The quantity (B/PL) has a special physical significance. The quantity P may be replaced by F/μ, where F is the friction force and μ is the coefficient of friction. Hence,

$$\frac{B}{PL} = \frac{\mu B}{FL} \tag{11.5}$$

The quantity (FL) is the energy dissipated in sliding a distance L and hence the energy per unit volume for material removal or specific energy will be

$$u = \frac{FL}{B} \tag{11.6}$$

From this it follows that

$$\frac{B}{PL} = \frac{\mu}{u} \tag{11.7}$$

If we assume μ to be 0.5 in the experiments of Burwell and Strang for which $B/PL = 10^{-12}$ in^2 lb^{-1}, 1.45×10^{-12} cm^2 N^{-1}), then from Eq. (11.7)

$$u = \frac{\mu}{B/PL} = 0.5 \times 10^{12} \text{ in lb in}^{-3} (3450 \times 10^6 \text{ MPa})$$

This extremely high value of specific energy is due to the fact that very many bonds between asperities had to be broken before a single wear particle was generated. If we assume that, when each such bond is broken, the material is stressed to the point of generating a wear particle, then the energy per unit volume required to produce a wear particle in the absence of all the unsuccessful attempts may be estimated to be

$$Ku = (0.317 \times 10^{-6})(0.5)(10^{12})$$
$$= 0.158 \times 10^6 \text{ in lb in}^{-3} (1.09 \times 10^6 \text{ MPa})$$

The actual value of specific energy required to produce the one wear particle in the absence of the approximately one million unsuccessful attempts will be somewhat greater than this since the unsuccessful attempts will involve a lower specific energy than the one actually producing a wear particle. If we assume that on the average the bonds formed that do not result in a wear particle are only 1% as strong as those that do, due to the contaminating influence of the lubricant used, then the specific energy associated with the formation of a wear particle would be 15.8×10^6 in lb in^{-3} (109×10^6 MPa).

Abrasive wear involves the loss of material by the formation of chips as in abrasive machining. To have such wear it is necessary that one material be harder (or have harder constituents) than the other member of the sliding pair or that hard particles be formed by chemical reaction of the wear debris, by oxidation for example.

A convenient way of studying abrasive machining or abrasive wear is in terms of specific energy (u in lb in^{-3} or MPa) which is the energy required to remove a unit volume of material. Specific energy is experimentally found to be a strong function of the undeformed chip size $(t)^{\dagger}$ as shown in Fig. 11.4.

The dotted curve represents the transition from abrasive action to adhesive action. For each abrasive particle there is a load below which it will not cut, which depends upon its fine geometry (sharpness). As we move upward on the dotted curve there is more rubbing and adhesive action and less abrasive action.

We should expect the performance of an abrasive particle giving rise to abrasive wear to resemble that of abrasive grains in a grinding wheel. Part of the action in each case will be associated with material removal and part will be associated with rubbing and friction.

Figure 11.5 shows an abrasive particle with radius ρ at its tip cutting a groove of depth t.

The cross-section involved is A' and the tangential force F'_P will be

$$F'_P = uA' \qquad (11.8)$$

where u is the specific energy to produce a "chip."

In grinding, the radial force is approximately twice the tangential force and hence

$$F'_Q = 2uA' \qquad (11.9)$$

This corresponds to a "coefficient of friction" $\mu = F'_P/F'_Q$ of 0.5 for the grinding process.

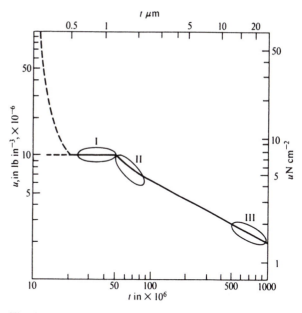

Fig. 11.4 Variation of specific energy (u) with undeformed chip thickness (t) when grinding mild steel. (after Backer, Marshall, and Shaw, 1952)

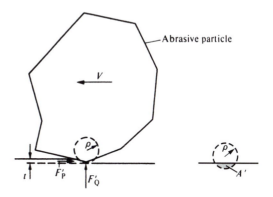

Fig. 11.5 Abrasive particle cutting groove of depth t.

When the particle has traveled a distance (L), the volume of metal removed (B') will be

$$B' = A'L \qquad (11.10)$$

---

† Values of t in Fig. 11.4 are estimates based on mean grinding wheel properties and may be in error by a factor of 2 or more. For our purposes here the relative values are of the greatest concern and these are undoubtedly more accurate than the absolute values.

The quantity (B/PL) which appeared in the discussion of adhesive wear is a convenient parameter for wear studies and may be evaluated for abrasive wear as follows (with the aid of Eqs. (11.9) and (11.10)):

$$\frac{B}{PL} = \frac{\sum B'}{\sum F'_Q L} = \frac{B'}{F'_Q L} = \frac{A'L}{2uA'L} = \frac{1}{2u} = \frac{\mu}{u} \tag{11.11}$$

Substituting values of u from Fig. 11.4 into Eq. (11.11) gives Fig. 11.6. The dotted curve represents the transition from abrasive to adhesive wear.

When the wear rate (B/PL) for fine abrasion (5×10^{-8} in^2 lb^{-1} or 7.25×10^{-8} cm^2 N^{-1}) is compared with that for adhesive wear (10^{-12} in^2 lb^{-1} or 1.45×10^{-12} cm^2 N^{-1}) (Burwell and Strang, 1952), it is evident that abrasive wear rates are very much larger. This is because there is much less rubbing and non-removal work done in the abrasive case. While the total specific energy for adhesive wear was found to be 0.5×10^{12} in lb in^{-3} (0.34×10^{12} N cm^{-2}), it is seen to be about 10^7 in lb in^{-3} (0.69×10^7 N cm^{-2}) for fine abrasive action (region I, Fig. 11.4). However, when the estimated non-removal action in adhesive wear is extracted (15.8×10^6 in lb in^{-3} or 11×10^6 N cm^{-2}), it is found to correspond closely with the specific energy for fine abrasive action (10×10^6 in lb in^{-3} or 6.9×10^6 N cm^{-2}).

It is evident that many wear mechanisms exist, and frequently several play an important role in a given test situation. This makes the study of wear phenomena one of the most challenging engineering activities.

The specific wear parameter (B/PL) is proportional to the ratio of coefficient of friction to specific energy (μ/u) and will generally increase when the size of the wear particles increase. Figure 11.7 shows a composite specific wear diagram, where t may be considered to be either the depth of scratches produced or the size of wear particles. The relationship between the polishing or run-in, adhesive wear, abrasive wear, and abrasive machining regimes is clearly evident in this diagram. The specific energy varies over about eight orders of magnitude in going from one extreme (polishing) to the other (abrasive machining), decreasing of course as the size of particle removed increases.

To minimize wear the hardness of the sliding surfaces should be as high as possible, and the materials should be chosen to be compatible and hence to give a low K. The surfaces should be well lubricated to further reduce the effective value of K and provide a

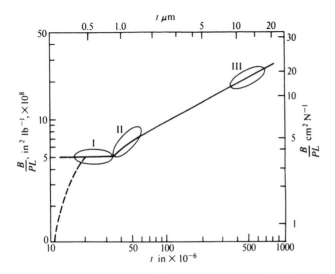

Fig. 11.6 Variation of specific wear parameter (B/PL) with undeformed chip size (t). (after Shaw, 1977)

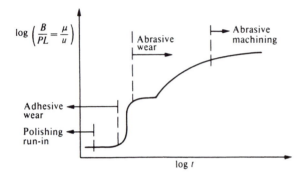

Fig. 11.7 Schematic diagram showing variation of specific wear rate (B/PL) with undeformed size of material removed (t).

low coefficient of friction. The surface roughness should be as low as possible, and the asperities that remain should be large in number and be as blunt as possible (large ρ). Polishing or run-in will tend to provide the desired surface texture. The atmosphere should be controlled to decrease wear (sometimes oxygen is beneficial and sometimes detrimental to wear), and means should be provided to remove abrasive wear particles from the system when they tend to form (grooving and copious oil supply with filtering). Finally, the load on the sliding surfaces should be as low as possible.

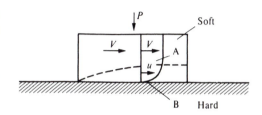

Fig. 11.8 Heavily loaded slider showing boundary layer of subsurface plastic flow.

When surface temperatures become very high and surface velocities are low, it is possible for solid state diffusion to play a role in the wear process. This unusual combination (high surface temperature and low surface velocity) is found only with heavily loaded sliders involving appreciable subsurface flow. In such a case there is a boundary layer of material across which there is a large velocity gradient (Fig. 11.8). High temperatures are then possible due to the high velocity pertaining at the upper surface of the boundary layer (A) while long contact times are available for diffusion along the lower surface of the boundary layer (B).

When diffusion wear is important, the relative solubilities of the two sliding materials become an important consideration as discussed by Roach et al. (1956).

At very high speeds the boundary layer may actually become molten. Under such conditions the thickness of the boundary layer will increase in the direction of sliding motion (Fig. 11.8), and this is capable of developing hydrodynamic pressure within the molten layer in the same way that load supporting pressure is developed in a journal bearing (DeSalvo and Shaw, 1968). When the boundary layer is liquid, the real area of contact will equal the apparent area ($A_R = A$) and diffusion rates will be very large. These conditions will give rise to very high wear rates if the system tends to wear by diffusion. Hence, a liquid boundary layer is to be avoided in such cases even though it gives rise to a load-supporting hydrodynamic film.

Diffusion wear is often accompanied by the decomposition of a component of one of the sliding surfaces. For example, in cutting a ferrous alloy with a tungsten carbide tool at high speed (temperature), Opitz (1963) presents evidence in support of a transformation from α-iron to γ-iron on the surface of the chip. The γ-iron has a strong affinity for carbon—the tungsten carbide (WC) crystals in the surface decompose and the carbon released diffuses into the surface of the chip. According to Opitz, the increased carbon concentration strengthens the surface of the chip which in turn increases the wear rate.

The fact that the wear rate of cutting tool constituents increases with increased solubility in the chip (Kramer and Suh, 1980) suggests that formation of a solid solution precedes decomposition. Transport by diffusion depends on the chemistry and structure of the materials, temperature, time, and the concentration gradient of the diffusing species. Secondary shear flow on the tool face plays an important role relative to all aspects of diffusion transport. Research on this type of tool wear is very difficult to carry out since it is not possible to duplicate all conditions except by use of a cutting test and this will involve other types of wear in addition to that due to diffusion.

Static diffusion studies of pieces of tungsten carbide in contact with various tool materials held at constant temperature in a furnace for a long time (1 or 2 hours) have been somewhat useful in screening the compatability (solubility) of different pairs of materials (Dawihl, 1940a, 1940b, 1941). However, such tests involve conditions far different from those in cutting (no freshly generated surface, no high local deformation strains, and temperatures that are too low, times that are too long, and the absence of realistic temperature and concentration gradients).

Suh (1973) has published a valuable monograph that considers in detail the physics of friction and wear, particularly for heavily loaded sliders. In his discussion of different types of wear, subsurface microcracks that initiate at second phase particles subjected to high shear stresses play a major role. Delamination is a special type of wear where subsurface microcracks join up in the direction of maximum shear stress to produce wear particles in the form of thin sheets. This type of wear was first described by Suh (1973) and elaborated by Johamir (1977) and by Fleming and Suh (1977). This type of wear is of particular interest for heavily loaded sliders whereas adhesive wear, leading to small spherical wear particles, has greater application to lightly loaded sliders. Important monographs concerning lightly loaded sliders are Bowden and Tabor (1950, 1964) and for very lightly loaded surfaces (electrical contacts), Holm (1950).

TYPES OF CUTTING TOOL WEAR

Tool wear may be classified as follows:

   1. adhesive wear

   2. abrasive wear

   3. diffusion wear

   4. fatigue

   5. delamination wear

All of these are generally present in combination, the predominant wear mechanism depending upon cutting conditions. In addition to the above sources of tool failure, the following also pertain:

   6. microchipping

   7. gross fracture

   8. plastic deformation

These, however, are usually readily identified and the solution is apparent. The fracture modes result from subjecting the tool to too high a cutting force, operating with too large a BUE or too much stored elastic energy, or using a tool material that is too brittle. Plastic flow of the tool tip arises when the temperature is too high relative to the softening point of the tool material. When plastic flow occurs at the tool tip, tool clearance is lost, the temperature rises abruptly, and total tool failure occurs rather rapidly. The obvious solution to the latter difficulty is to use a lower cutting speed or a tool material that is more refractory. While cemented tungsten carbide is a very refractory material, even this material will suffer from plastic flow at the tool tip if the temperature is too high (Trent, 1967).

Cutting tools wear predominantly in different ways, depending on cutting conditions (principally cutting speed V and undeformed chip thickness t). Opitz (1956) has identified the principal types of tool wear, together with the approximate ranges of values of V and t where each type of wear is predominant (Fig. 11.9). At low values of the product ($Vt^{0.6}$), tool wear consists predominantly of a

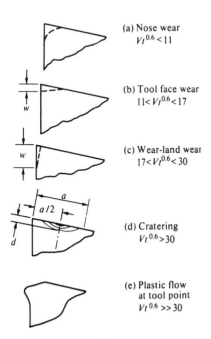

(a) Nose wear
$Vt^{0.6} < 11$

(b) Tool face wear
$11 < Vt^{0.6} < 17$

(c) Wear-land wear
$17 < Vt^{0.6} < 30$

(d) Cratering
$Vt^{0.6} > 30$

(e) Plastic flow at tool point
$Vt^{0.6} \gg 30$

Fig. 11.9 Types of predominant tool wear depending on product ($Vt^{0.6}$). (after Opitz, 1956)

rounding of the cutting point and a loss of tool sharpness. This is the main type of tool wear in a broaching operation where both V and t are small. As the product ($Vt^{0.6}$) increases, the predominant type of tool wear shifts as shown in Fig. 11.9. The most important types of tool wear for a single point turning tool is wear-land formation (Fig. 11.9c) and crater formation (Fig. 11.9d). Crater formation is apt to be more important than wear-land formation in situations where cutting temperatures are very high.

WEAR-LAND WEAR

Wear-land wear results in a loss of relief angle on the clearance face of the tool. This gives rise to increased frictional resistance. The wear land is normally a lightly loaded slider with maximum temperature at the trailing edge which increases with increase in wear-land length (w). The mean tool face temperature varies in accordance with Eq. (10.8). Wear-land wear is usually followed by observing the change in w with time, w being determined by a measuring microscope.

Figure 11.10 shows a typical set of curves of w versus cutting time for different values of cutting speed (V). These curves are generally similar to Fig. 11.3, the extent of the low wear region decreasing with increased cutting speed. The wear rate rises abruptly when the temperature at the trailing edge of the wear land reaches the thermal softening point of the work material. The A_R/A ratio then increases rapidly with time along with the wear rate.

The tool life (T min) for a given cutting speed is either the cutting time at the limit of the low wear rate region (time corresponding to points A in Fig. 11.10) typically for a high speed steel (HSS) tool or the time to reach a given wear-land value (time corresponding to T_1, T_2, T_3 in Fig. 11.10) typically for a tungsten carbide tool. A common limiting wear-land value for tungsten carbide cutting mild steel is 0.030 in (0.75 mm). However, as the work material becomes more difficult to machine, as the tool material used is more refractory and hence brittle, or as the undeformed chip thickness decreases, it is generally advantageous to employ a smaller limiting wear-land value.

When the tool life (T) based on either total destruction (points A in Fig. 11.10) or a limiting wear land (points T) is plotted using log–log coordinates, a result as shown in Fig. 11.11 will be

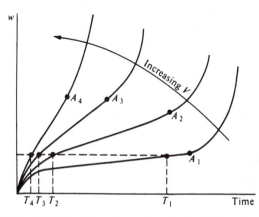

Fig. 11.10 Variation of wear land (w) with time for different cutting speeds (V).

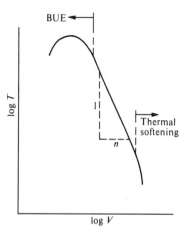

Fig. 11.11 Typical tool life curves based on wear-land wear.

obtained. For the practical range of cutting speeds, this curve is generally linear corresponding to the equation

$$VT^n = C \tag{11.12}$$

where n and C are constants for a given work and tool material and machining conditions other than cutting speed (feed, depth of cut, fluid, tool geometry, etc.). This is called the Taylor equation after F. W. Taylor (1907) who first employed it to characterize his HSS tools. Since T is usually measured in minutes, C is the cutting speed that gives a one-minute tool life. The value n increases as the tool material becomes more refractory approximately as follows:

Tool Material	*n*
High speed steel	0.1
Tungsten carbide	0.2
Al_2O_3 (ceramic)	0.4

The Taylor equation may be rearranged and generalized as follows:

$$TV^{1/n}t^{1/m}b^{1/l} = C' \tag{11.13}$$

where t is feed and b is depth of cut. It is found experimentally that tool life is most sensitive to changes in speed and least sensitive to changes in depth of cut. Hence

$$n < m < l$$

The volume worn away on the relief face (B_r) for a given wear land (w), depth of cut (b), and relief angle (θ) will be ($\alpha = 0°$):

$$B_t = \frac{bw^2 \tan \theta}{2} \tag{11.14}$$

For a given wear land (tool life), the wear volume is seen to increase with $\tan \theta$. Thus, from the point of view of wear volume, a large clearance angle is advantageous. However, the strength of the cutting edge, capacity to absorb thermal energy, and ability to maintain dimensional accuracy decreases with an increase in relief angle. Therefore, an optimum relief angle exists in the range of 5° to 10°.

CRATER WEAR

Conditions on the tool face are much more severe than those on the clearance face. Temperatures, pressures, and A_R/A are much higher and the tool face constitutes a very heavily loaded slider. Diffusion wear is predominant at high cutting speeds and the characteristic wear pattern is a crater (Fig. 11.9d). The point of greatest crater depth usually occurs near the midpoint of contact length (a) since this is where the tool face temperature is normally maximum.

Crater wear is followed by plotting crater depth versus time, and results similar to those of Fig. 11.12 are obtained when this is done.

When the crater approaches the cutting edge, the tool temperature increases as does the rate of increase of crater depth (d). Crater depth versus time curves for different cutting speeds (Fig. 11.12) are very similar to wear-land curves (Fig. 11.10). Crater tool life may be taken just short of total destruction when the crater approaches the cutting edge and edge chipping occurs (points A in Fig. 11.12) or when d reaches a critical value (points T_1, T_2, T_3). The tool-life curve for crater wear will have the same appearance as Fig. 11.11. A Taylor type equation with different values of n and C also holds for tool life based on crater wear.

Fortunately, the volume available to be worn away before total destruction is much greater for crater wear than for wear-land wear to compensate for the much greater rate of wear on the tool face. As a consequence, tool life due to crater wear and due to wear-land development are approximately the same despite the very large difference in volume rates of wear.

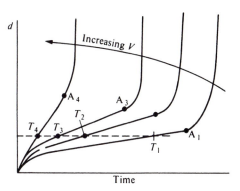

Fig. 11.12 Variation of crater depth (d) with time for different cutting speeds (V).

TOOL-LIFE TESTING

Unfortunately, due to the complexity of the cutting process, it is not possible to use noncutting tests to evaluate cutting tools, work material structures, cutting fluids, etc., or to appreciably accelerate or truncate tool-life tests. This is because cutting temperatures and pressures play such an important and complex role.

Soon after radioactive tracers were first used in wear studies (Burwell, 1947), Shaw and Strang (1950) cut AISI 4027 steel that had been made radioactive by neutron irradiation at the Oak Ridge Laboratory. It was found by the autoradiograph technique that even under the most ideal cutting fluid conditions, a small amount of work material was transferred to the cutting edges of an end-milling cutter.

Merchant et al. (1953) first used radioactive tools to study the rate of tool wear. Cutting tests of very short duration (~ 10 s) were made and the γ-radiation of tool material transferred to the chips was measured to obtain *relative* wear rates under different cutting conditions.

Since that time, there have been a number of major radioactive tool-life studies, but the technique has failed to provide a satisfactory substitute for conventional cutting tests for the following reasons:

1. Cost and inconvenience are associated with safety precautions.

2. Tool wear is not uniform with time but frequently "cascades" making it unreliable to extrapolate a very short time radioactive determination to predict the life of a tool.

3. Wear is not uniform across a cutting edge or with regard to the relative amounts of different wear patterns. Grooving (presently to be discussed) and the nonuniform development of the wear land along the cutting edge are examples of why it is difficult to interpret radioactive data in terms of tool life.

GROOVE FORMATION

Deep groove wear (Fig. 11.13) frequently occurs when machining high temperature alloys, very soft steels, or other materials having a strong tendency to strain harden. Hovenkamp and van Emden (1952) in an early study of groove wear discussed several possible causes for this phenomenon.

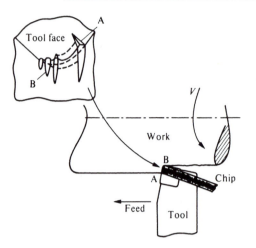

Fig. 11.13 Turning operation showing wear grooves formed at free edges of chip (A and B).

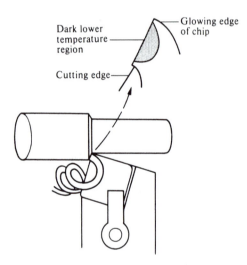

Fig. 11.14 Turning operation on high-temperature alloy showing glowing edges of chip.

Turning tests on a low carbon steel were performed without cutting fluid using carbide inserts. It was found that when a second cut follows, a first cut groove formation is greater for the second cut than for the first cut. This was believed due to strain hardening that occurred during the first cut and was confirmed by annealing the bar between the first and second cuts which then gave the same degree of groove formation for both cuts.

A large groove will generally form at the free edge of the chip on the main cutting edge while smaller grooves form at the free surface on the secondary cutting edge. When there is more than one groove on the end-cutting edge of the tool, these have a spacing corresponding to the feed per revolution. The work material that is not removed by the first groove on the end-cutting edge rubs without clearance on the tool, generating the second groove, etc. The presence of grooves on the end-cutting edge is a source of surface roughness of the machined surface and frequently dictates tool life in finish machining.

Moltrecht (1964) performed a number of experiments to identify the main source of groove wear and concluded that the flow of chip material at the cutting edges was the main cause.

Shaw et al. (1966) made an unexpected discovery when a high temperature alloy (Waspalloy) was being turned under normal conditions. The edges of the chip glowed while the central region did not, suggesting that the edges had a much higher temperature than the central region. This was confirmed by repeating the experiment in total darkness using radiation from the chip to make a photograph on high-speed infrared film.

The photograph revealed at the point of chip generation two red-hot chip edges with a dark cooler central region (Fig. 11.14). Since one should expect a greater rate of heat transfer from the edges of the chip than from the central region, the observed high-temperature edges can only mean that more energy was expended in chip formation in the vicinity of the edges of the chip than elsewhere.

Shaw et al. (1966) have explained the high temperatures observed at the free edges of the chip by noting that whereas the edges of the chip will deform in plane stress, the central region will deform in plane strain. The Lee–Shaffer approach (Chapter 8) is employed in what follows as a first approximation. Figure 11.15 shows the corresponding Mohr's stress circles for region ABC for plane stress at (d) pertaining at the edges of the chip (Fig. 11.15a) and for plane strain at (c) pertaining in the central region of the chip. According to the Lee–Shaffer solution σ_1 is the principal stress in the direction of resultant force R in both cases. Evaluating the von Mises flow criterion, we have for plane stress

$$\sigma_e^* = \frac{1}{\sqrt{2}} \left[(\sigma_1^* - 0)^2 + (0)^2 + (0 - \sigma_1^*)^2 \right]^{1/2} = \sigma_1^* \qquad (11.15)$$

where σ_1^* is plus when compressive and the * identifies the effective stress for plane stress. For plane strain (no *)

$$\sigma_e = \frac{1}{\sqrt{2}}\left[\left(\sigma_1 - \frac{\sigma_1}{2}\right)^2 + \left(\frac{\sigma_1}{2} - 0\right)^2 + (0 - \sigma_1)^2\right]^{1/2} = \frac{\sqrt{3}}{2}\sigma_1 \qquad (11.16)$$

For the same material, flow should occur when $\sigma_e^* = \sigma_e$ or when

$$\sigma_1^* = \frac{\sqrt{3}}{2}\sigma_1 = 0.87\sigma_1 \qquad (11.17)$$

This means the material at the edges of the chip will be sheared twice—once parallel to the cutting edge before the main shear plane is reached and again on crossing the shear plane when the main body of the chip is also sheared in direction AB.

The additional energy expended in double shear at the edges of the chip will be only slightly greater for a material that does not strain harden significantly (such as ordinary steel) but will be significantly greater for a material having a strong tendency to strain harden at high strain rate (such as Waspaloy). This explains why hot chip edges are not observed with relatively nonstrain-hardening steel but are observed with Waspaloy.

Uehara studied groove formation when a titanium alloy was turned with carbide tools. Video pictures of chips leaving the cutting edge clearly showed glowing edges but a nonglowing central region.

A number of additional reasons for groove formation have been proposed (Albrecht, 1956; Solaja, 1958; Leyensetter, 1956; Lambert, 1961):

1. presence of a work-hardened layer on the previously cut surface

2. stress concentration due to the stress gradient at the free surface

3. formation of thermal cracks due to the steep temperature gradient at the free surface

4. presence of a burr at the edge of the freshly machined surface

5. a higher velocity at the outside diameter [This obviously cannot be the answer since a groove also forms at the point of lowest speed, B, in Fig. 11.13.]

6. presence of an abrasive oxide layer on the previously cut surface

7. flow of built-up edge (BUE) material parallel to the cutting edge [This cannot be the major source since the most troublesome grooves are those which form at such high speeds and temperatures that a BUE will not be present.]

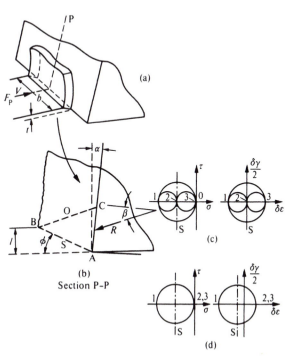

Fig. 11.15 (a) Orthogonal cutting operation. (b) Corresponding Lee–Shaffer slip line field. (c) Mohr's circle for plane strain. (d) Mohr's circle for plane stress.

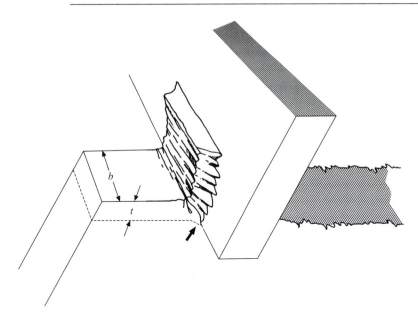

Fig. 11.16 Sketch of high-temperature superalloy chip being formed in orthogonal cutting process with groove generating side flow. (after Lee et al., 1979)

8. fatigue of the tool material due to the fluctuation of force at the free surface which accompanies the small lateral motions of the edge of the chip

9. particles of tool material left behind on the previously cut surface which act as small cutting tools to induce wear [This obviously cannot be the primary cause of groove formation.]

Lee et al. (1979) studied groove initiation at the beginning of a cut. This was done using commercial polycrystalline boron nitride (PCBN) tools to cut a nickel base superalloy (Inconel 718) in a two-dimensional orthogonal cutting mode. Figure 11.16 is a sketch showing one of their 0° rake angle tools producing a chip with substantial side flow at both free edges. The cutting speed was 197 f.p.m. (1.0 m/s) and the depth and width of cut were $t = 0.050$ in (1.25 mm) and $b = 0.125$ in (3.13 mm), respectively. The length cut was 4 in (16 mm), and a new cutting edge was used for each cut. Chip and used tool surfaces were examined using a scanning electron microscope.

The conclusion was that a groove at the depth of cut line when machining a superalloy is initiated by seizure and pullout of tool material by the chip as it flows outward at the edge of the chip in the direction of the cutting edge, and not an excessive edge temperature. The Lee et al. (1979) tests were performed under an unusually low b/t ratio of $0.125/0.050 = 2.5$ as represented in Fig. 11.16. This suggests that chip formation was under plane stress conditions at all points along the cutting edge. In order to have plane strain, chip formation at essentially all points along the cutting edge requires a b/t ratio of 10 or more. The tests of Lee et al. (1979) are therefore not representative of most turning and milling conditions. It appears that under normal cutting conditions ($b/t > 10$), the unusually high temperatures pertaining at the cutting edges with plane stress flow followed by plane strain flow will constitute a major cause of edge grooving. Under plane stress conditions and a large undeformed chip thickness (t), the normal stress on the tool face will be high and then the primary source of grooving might be seizure and pullout and not excessively high temperature at the cutting edges.

Shaw et al. (1966) suggested that an effective way of reducing groove formation is to use the largest permissible side cutting edge and negative rake angles. When cutting a nickel base superalloy with a carbide tool, Muetze (1967) found the optimum rake angle to be $-15°$.

GROSS TOOL FRACTURE

Gross fracture and edge chipping are two of the principal modes of failure of metal cutting tools. This is particularly true for unusually brittle tool materials such as polycrystalline boron nitride (PCBN), sintered aluminum oxide, titanium carbide, and finishing grades of tungsten carbide having a low (~ 3 $^w/_o$) cobalt content. Fracture is also a more severe problem in interrupted cutting or when the work material is unusually hard and strong. As machining speeds are increased to increase productivity, there is need to use more refractory (greater hot hardness) tool materials. More refractory tool materials are more brittle and their use generally involves a greater increase in failure due to fracture. Tool failure due to fracture inherently gives a wide dispersion of tool life. With increased use of numerical control, it is important that tool fracture be reduced or eliminated since use of more capital-intensive equipment calls for an ability to predict tool life more precisely.

The entire field of tool fracture may be subdivided into several categories:

1. fracture due to an excessive steady load

2. fracture due to shock loading on entering a cut or in interrupted cutting

3. fracture due to suddenly unloading a cutting system with substantial stored elastic energy

4. fracture due to thermal stresses or thermal shock

5. fracture due to repeated stress cycles (fatigue) of mechanical or thermal origin

6. fracture due to lack of surface integrity of tool surface (of geometrical, structural, or chemical origin or due to an adverse residual surface stress)

Linear elastic fracture mechanics (LEFM), discussed in Chapter 6, is now widely used in design with brittle materials. The critical plane strain stress intensity factor (K_{IC}), a material property, enables the safe level of nominal tensile stress to be determined for a given crack length.

The stress intensity factor (K_I) is found to increase with crack length (a) (Fig. 11.17) and to approach a constant value (K_{IC}) for values of crack length (a) greater than a critical value (a_c). Thus, K_{IC} is a material constant independent of crack length, provided $a > a_c$.

The American Society for Testing Materials (ASTM, 1981) has empirically established that K_{IC} is a valid concept only when the critical crack length is

$$a_c \gtrless 2.5 \left(\frac{K_{IC}}{\sigma_y} \right)^2 \qquad (11.18)$$

where σ_y' is the 0.2% offset yield stress in tension.

For values of sharp crack length smaller than a_c in Eq. (11.18), the apparent fracture toughness (K_1) decreases with decrease in crack length. This is because any sharp cracks present will not be sufficiently large compared with the size of the plastic zone at the tip of the advancing sharp crack for (K_{1c}) to be a valid property.

Sampath and Shaw (1986) have determined the value of $K_{1c} = 8500$ p.s.i.(in)$^{0.5}$ and $\sigma_y = 190,000$ p.s.i. for a typical carbide tool material (6 $^w/_o$) Co tungsten carbide). When these values are substituted into Eq. (11.18), the limiting value a_c is found to be 0.005 in (125 μm). However, when a diamond ground specimen of this material was examined at 100× using the apparatus of Fig. 11.18, no cracks were observed before gross fracture. An indirect method of estimating the crack length based on the work

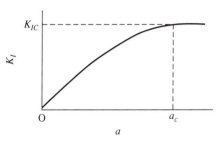

Fig. 11.17 Variation of stress intensity factor K_1 with crack length a.

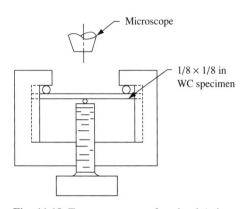

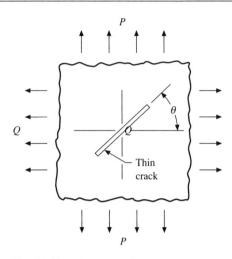

Fig. 11.18 Test arrangement for visual (micro-scopic) observation of ground and polished speci-ment surface during loading. (after Sampath and Shaw, 1986)

Fig. 11.19 Orientation of thin two-dimensional crack in biaxial stress field in Griffith analysis.

of Pickens (1977) estimated the mean crack length to be 5.5 μin (220 μm) in a diamond ground carbide tool specimen. Since this is far smaller than the value of a_c of 5000 μin from Eq. (11.18), it follows that K_{1c} is not a valid value for use with diamond ground carbide tools, and this makes the LEFM approach to tool fracture very questionable.

Nevertheless, application of LEFM to the study of tungsten carbide has received considerable attention. Lueth (1972), Pickens (1977), and Chermant and Osterstock (1977) have studied the fracture toughness of tungsten carbides. Kalish (1982) has recommended fracture toughness as a quality control test for tungsten carbide alloys. Shibasaka et al. (1983) made an attempt to apply LEFM principles to the analysis of brittle fracture of cutting tools.

It is more reasonable to begin modeling the brittle fracture of tools in terms of the Griffith (1924) approach discussed in Chapter 6 and in the review and the extension given below based on Shaw and Avery (1986).

Griffith (1924) postulated the presence of many sharp, randomly oriented, essentially closed hairline cracks in real materials. He employed a modification of the classical maximum tensile stress criterion in which the *intensified* principal tensile stress at a crack tip was substituted for the *nominal* (unintensified) principal tensile stress. Griffith assumed a two-dimensional (plane stress) model in a biaxial stress field, the crack being oriented at an angle θ to the minor principal stress direction Q (Fig. 11.19). By use of an elastic solution published much earlier by Inglis (1918), Griffith calculated the maximum intensified tensile stress at or near the crack tip and took the orientation θ that gave the largest value for this tensile stress. The maximum intensified tensile stress was then assumed to have a constant value for extension of the initial crack. In this way, the fracture locus shown in Fig. 11.20 was obtained where T corresponds to uniaxial tension and C to uniaxial compression. The classical nominal maximum tensile stress criterion was found to hold until the compressive stress Q reached a value of three times the tensile

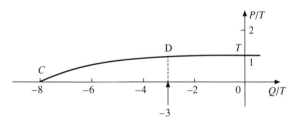

Fig. 11.20 Plane stress fracture locus due to Griffith, in second quadrant (P plus and Q minus).

stress at fracture (point D in Fig. 11.20). The curve then fell gradually to point C which shows that the same material should be eight times as strong in uniaxial compression as in uniaxial tension. For negative values of $(3P + Q)$, the value of critical crack direction (θ) was

$$\cos 2\theta = -\frac{1}{2}\frac{P - Q}{P + Q} \quad (11.19)$$

This means that when $P = 0$ (uniaxial compression), $\theta = 30°$.

When experiments are performed on sintered tungsten carbide with a cobalt binder, the results are found to be in good qualitative but poor quantitative agreement with Griffith's predictions (Takagi and Shaw, 1983). For example, Fig. 11.21a shows Weibull plots for 6 and 12 $^w/_o$ cobalt sintered tungsten carbide while Fig. 11.21b gives the resulting fracture loci for quadrant II loading (σ_1 plus, σ_2 minus). Points D in Fig. 11.21b are for the disk test. In this test, brittle fracture is found to initiate at the center of the disk (Sampath et al., 1986) where the nominal stresses are σ_1 (tensile) in a horizontal direction and $\sigma_2 = -3\sigma_1$ (compressive) in the vertical direction (Fig. 11.22). T and C in Fig. 11.21b correspond to uniaxial tension and uniaxial compression, respectively.

The tungsten carbide test results are seen to be very different from the predictions of the Griffith theory in the following respects:

- The ratio of C/T is closer to 3 than the value 8 predicted by the Griffith theory.
- The value of σ_1 for the disk test is closer to $T/2$ than to T as predicted by the Griffith theory.
- The fracture surface was found to be parallel to the compressive stress direction ($\theta = 0$) in the uniaxial compression test than the value of 30° predicted by the Griffith theory.

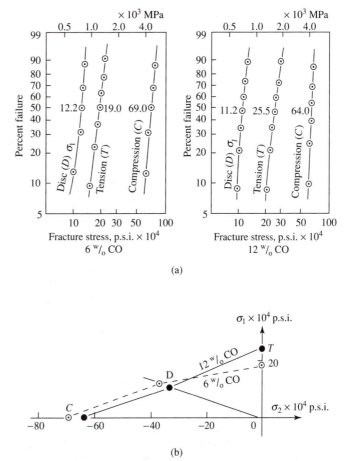

(a)

(b)

Fig. 11.21 Weibull plots and second quadrant fracture loci based on median values for 6 and 12 $^w/_o$ cobalt sintered tungsten carbide. (after Takagi and Shaw, 1983)

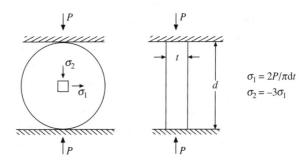

Fig. 11.22 Plane stress disk test for brittle fracture.

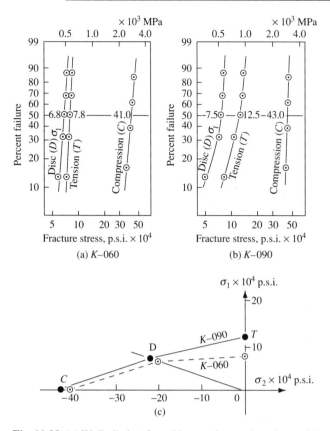

Fig. 11.23 (a) Weibull plots for cold pressed ceramic tool material. (b) Weibull plots for hot pressed ceramic tool material. (c) Fracture loci for cold and hot pressed ceramic tool material—experimental median F_{50} values from (a) and (b). (after Takagi and Shaw, 1983)

Results similar to those of Fig. 11.21 for tungsten carbide are given in Fig. 11.23 for cold (K–060) and hot (K–090) pressed Al_2O_3 cutting tool materials.

In consulting the literature one finds values of uniaxial compressive strength divided by uniaxial tensile strength (C/T) to range from about 3 to 18. For example, Brace (1964) found C/T to be about 18 for granite while Babel and Sines (1968) found C/T to be about 7 for 31% porous zirconia. It appears that the reason for this wide range of values is due to variation in the size and shape of the imperfections causing stress concentrations. In the case of sintered carbides and porous zirconia, these are probably open voids in the binder (Co phase) of the carbides and in the pores of the sintered zirconia bodies. Since the cobalt binder phase in carbides will be at or near the melting point during sintering, these voids should be nearly circular in shape due to surface tension while for sintered ZrO_2, the voids are probably elliptical in shape. Granite and other igneous rocks are multiphase materials and the critical defects are probably relatively long, straight hairline regions of stress concentration in the phase boundaries separating material of significantly different values of Young's modulus. These may actually not be cracks but may be thought of as essentially closed sharp cracks similar to those employed by Griffith in his analysis.

Support for the idea that the defects in the case of sintered tungsten carbide are circular voids is the fact that there is an ASTM quality scale for rating the size and number of spherical voids observed at high magnification (ASTM, 1981). Sharp hairline cracks are usually not observed in a sound carbide and thus there is no comparable quality rating for sharp cracks. In the case of glass, two types of defects must be distinguished—surface defects (scratches and nicks) and internal defects. The internal defects are gas-filled spherical voids called "seeds" in the glass industry. A seed-free ordinary glass is usually considered to be one with circular voids less than 0.005 in (1/8 mm) in diameter, while a seed-free optical glass will have voids smaller than 0.001 in (1/40 mm). Since surface defects are usually predominant relative to values of T for glass, values of C/T vary over a wide range depending on the geometry of the surface defects. However, values of C/T as low as three may be obtained for glass if the surface of specimens used to determine T are not diamond ground before testing but are carefully handled in the fire polished state. The surface condition for specimens used to determine C are not nearly as critical.

Other examples of relatively low C/T values for specimens expected to have open, relatively circular defects are

- $C/T \cong 6$ by Salmassey et al. (1955) for thick walled cylinders of hydrostone plaster loaded axially and with internal pressure
- $C/T \cong 7$ by Cornet and Grassi (1961) for tubes of high silicon cast iron loaded axially and with internal pressure

It thus appears important to distinguish two types of defects as Adams and Sines (1978) have suggested:

- open, relatively circular defects
- thin, relatively closed hairline defects

Sintered tungsten carbide typifies the first of these while granite rock typifies the second.

SIMPLIFIED GRIFFITH THEORY FOR CIRCULAR VOIDS

When cylindrical defects are assumed instead of sharp ones, the fracture locus obtained is much simpler to derive and in much better agreement with experiments for sintered tungsten carbides. Since two-dimensional solutions such as that originally used by Griffith represent good approximations to three-dimensional ones, the defect may be assumed to be a cylindrical void. It is then easy to find the resultant intensified stress on the surface of a defect subject to a biaxial state of stress.

Figure 11.24a shows a circular defect in a body subjected to a nominal elastic compressive stress (σ_2). An intensified stress equal to but of opposite sign to σ_2 will occur at A on the surface of the defect. Similarly, Fig. 11.24b shows a circular defect in a body subjected to a nominal elastic tensile stress (σ_1). An intensified stress equal to three times σ_1 will occur at A on the surface of the defect. If two nominal elastic principal stresses act together (Fig. 11.25), the intensified stress (σ_1) at A will be the sum of the two individual intensified stresses. The intensified stresses (σ_i) will thus be as follows for the T, D, and C test conditions:

$$T \text{ test: } \sigma_1 = T, \ \sigma_2 = 0 : \sigma_i = 3T$$
$$D \text{ test: } \sigma_1 = D, \ \sigma_2 = \text{minus } 3D : \sigma_i = 3D - (-3D) = 6D \qquad (11.20)$$
$$C \text{ test: } \sigma_1 = 0, \ \sigma_2 = C : \sigma_i = C$$

When these values of σ_i are equated in pairs at fracture,

$$D = T/2 \qquad (11.21)$$

$$C = 3T \qquad (11.22)$$

In addition, the direction of the fracture surface should be perpendicular to σ_i in all cases according to the maximum intensified tensile stress criterion described above. All of these results relative to both magnitude and direction are in good agreement with a wide range of experiments (Takagi and Shaw, 1983).

It is well established experimentally that in the first quadrant (both σ_1

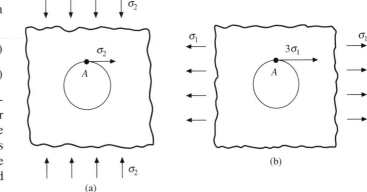

Fig. 11.24 Intensified tensile stress in a body subjected to a circular defect. (a) In a direction normal to the nominal uniaxial compressive stress (σ_2). (b) In a direction parallel to the nominal uniaxial tensile stress (σ_1). (after Takagi and Shaw, 1983)

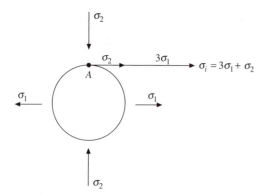

Fig. 11.25 Two-dimensional circular void with nominal elastic principal tresses σ_1, and σ_2. The intensified stress at A is $\sigma_i = 3\sigma_1 + \sigma_2$. (after Takagi and Shaw, 1983)

and σ_2 tensile) only the maximum nominal stress (σ_1) is responsible for fracture. The reasons for this is readily apparent in terms of the circular defect model. For a circular defect oriented as shown in Fig. 11.26a with its axis in the $\sigma_3 = 0$ (plane stress) direction, the intensified stress at A will be

$$\sigma_i = 3\sigma_1 - \sigma_2 \qquad (11.23)$$

However, for a circular void oriented with its axis in the σ_2 direction (Fig. 11.26b), the intensified stress at B will be

$$\sigma_i = 3\sigma_1 \qquad (11.24)$$

Since this is a larger value than that given by Eq. (11.23) and corresponds to the intensified stress for uniaxial tension, the fracture locus for all states of biaxial tension (quadrant 1) is a horizontal line.

A similar argument holds for the third quadrant (σ_3 and σ_2 both compressive). The critical point for plane stress fracture will be at C in Fig. 11.27b and fracture will occur when σ_3 reaches a critical value regardless of the magnitude of stress σ_2.

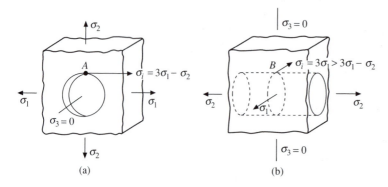

(a)   (b)

Fig. 11.26 Circular defects subjected to first quadrant loading (σ_1 and σ_2 plus) with axis in (a) σ_3 direction and (b) σ_2 direction.

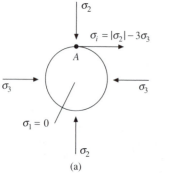

(a)

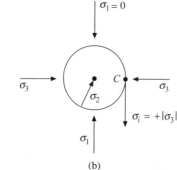

(b)

Fig. 11.27 Circular defects subjected to third quadrant loading (σ_2 and σ_3 minus) with axis in (a) σ_1 direction and (b) σ_2 direction.

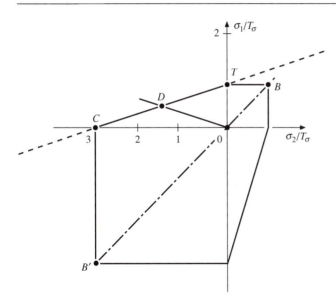

Fig. 11.28 Complete plane stress fracture locus for two-dimensional circular void model. [B = balanced biaxial tension. B′ = balanced biaxial compression. T = uniaxial tension. D = tensile stress in disk test. C = uniaxial compression.] (after Takagi and Shaw, 1983)

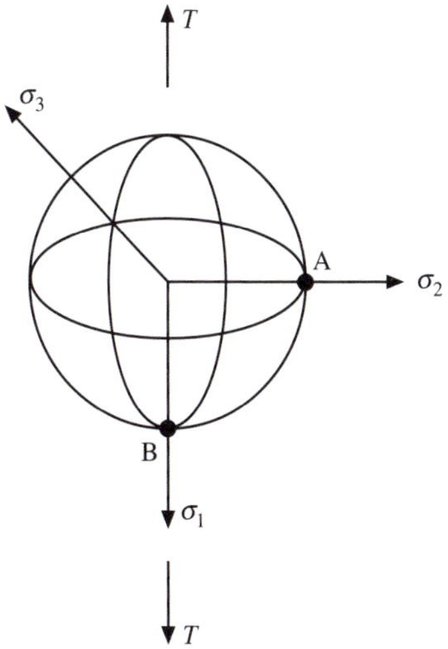

Fig. 11.29 Spherical cavity in infinite body subjected to uniaxial tensile stress T in σ_1 direction. (after Shaw and Avery, 1986)

The complete two-dimensional plane stress fracture locus for tungsten carbide is shown in Fig. 11.28 where $B - B'$ is an axis of symmetry. In this plot σ_3 is taken to be zero even though in quadrant III σ_1 would be zero according to the convention ($\sigma_1 > \sigma_2 > \sigma_3$). The coordinates are made nondimensional by dividing the nominal stresses (σ_1 and σ_2) by the uniaxial plane stress fracture stress (τ_σ).

The maximum intensified tensile stress approach to brittle fracture having two-dimensional circular defects in next extended to bodies in triaxial states of stress having spherical defects.

STRESS INTENSIFICATION FOR A SPHERICAL VOID

Goodier (1933) and Timoshenko and Goodier (1951) have presented equations for elastic stresses around a spherical cavity as shown in Fig. 11.29. For present purposes, stresses at points A and B on the surface of the void are of interest for a large body subjected to a nominal tensile stress (T) in the σ_1 direction. At point A, the principle stresses are

$$\sigma_1 = \frac{27 - 15v}{2(7 - 5v)} T = k_1 T$$

$$\sigma_2 = 0 \qquad\qquad\qquad\qquad (11.25)$$

$$\sigma_3 = \frac{15v - 2}{2(7 - 5v)} T = k_2 T$$

TABLE 11.1 Values of k_1, k_2, and k_3 for Different Values of Poisson's Ratio v

v	k_1	k_2	k_3
0.2	2.000	0	−0.500
0.3	2.050	0.136	−0.682
0.4	2.100	0.300	−0.900
0.5	2.170	0.500	−1.167

and at point B, the principal stresses are

$$\sigma_1 = 0$$

$$\sigma_2 = \sigma_3 = \frac{3 + 15v}{2(7 - 5v)} T = k_3 T \qquad (11.26)$$

where v = Poisson's ratio. Values of k_1, k_2, k_3 for different values of v are given in Table 11.1.

When three principal stresses (σ_1, σ_2, and σ_3) are present simultaneously as in Fig. 11.30, points A, B, and C are potentially sites of

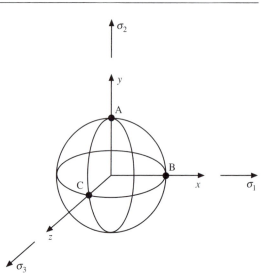

Fig. 11.30 Critical points (A, B, C) on surface of spherical void subjected to principal stresses σ_1, σ_2, σ_3. (after Shaw and Avery, 1986)

maximum intensified tensile stress (σ_i). It turns out, however, that the resultant intensified stress at point C in the x-direction has the maximum value, provided the principal stresses are designated in the usual way ($\sigma_1 > \sigma_2 > \sigma_3$) in each quadrant. At point C

$$\sigma_x = k_1\sigma_1 + k_2\sigma_2 + k_3\sigma_3 = \sigma_i \qquad (11.27)$$

PLANE STRESS FRACTURE LOCI

Substituting appropriate values of σ_1, σ_2, σ_3, and k from Table 11.1 into Eq. (11.27) for quadrants I (σ_1, and σ_2 both plus and $\sigma_3 = 0$ for plane stress); II (σ_1 plus, $\sigma_2 = 0$, and σ_3 minus); and III ($\sigma_1 = 0$, σ_2 and σ_3 minus), values of C (uniaxial compressive stress), D (nominal tensile stress at center of disk), B (equal biaxial tensile stress), and B' (equal biaxial compressive stress) at fracture may be found. These values of C, D, B, and B' may be made nondimensional by dividing each by the nominal uniaxial plane stress fracture stress (τ_σ). The intensified tensile stress corresponding to a nominal uniaxial plane stress of τ_σ is $k_1\tau_\sigma$. The values given in Table 11.2 are obtained for different values of Poisson's ratio (v), and these values yield the family of fracture loci shown in Fig. 11.31. In this figure, σ_3 is considered zero for all quadrants even though this does not correspond to the usual convention ($\sigma_1 > \sigma_2 > \sigma_3$) which was used in the individual derivations for each quadrant.

Poisson's ratios for brittle materials such as glass, ceramics, rock, and sintered tungsten carbide will be close to 0.2; and when the fracture locus for $v = 0.2$ (Fig. 11.31) is compared with the approximate one given in Fig. 11.28 the two

TABLE 11.2 Values of Fracture Stress Ratios for Brittle Material in Plane Stress for Different Values of v

v	C/T_σ	D/T_σ	B/T_σ	B/T_σ
0.2	4.00	0.57	1.00	4.00
0.3	3.01	0.50	0.94	3.75
0.4	2.33	0.44	0.88	3.50
0.5	1.86	0.38	0.81	3.25

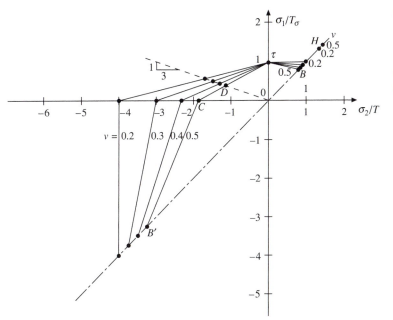

Fig. 11.31 Fracture loci for plane stress fracture based on spherical voids and for materials of different Poisson's ratio (v). H = hydrostatic tension T_σ = uniaxial tensile stress in plane stress. (after Shaw and Avery, 1986)

are seen to be remarkably similar. It may therefore be concluded that the plane stress fracture locus of Fig. 11.28 which involves two-dimensional defects is sufficiently close to the more exact one involving three-dimensional defects provided the Poisson's ratio is between 0.2 and 0.3.

PLANE STRAIN FRACTURE LOCI

Substituting appropriate values of σ_1, σ_2, σ_3, and k from Table 11.1 into Eq. (11.27) for quadrants I (σ_1 and σ_3 plus and $\sigma_2 = v(\sigma_1 + \sigma_3)$); II ($\sigma_1$ plus, σ_3 minus and $\sigma_2 = v(\sigma_1 + \sigma_3)$); and III ($\sigma_1$ and σ_3 minus and $\sigma_2 = v(\sigma_1 + \sigma_3)$), values of C, D, B, and F at fracture may be found. These values are made nondimensional by dividing each by the nominal uniaxial plane strain fracture stress (T_e). The intensified tensile stress corresponding to nominal stress T_e is $T_e(k_1 + vk_2)$. The plane strain values in Table 11.3 are for different values of v, and these yield the family of fracture loci shown in Fig. 11.32. In this plot the intermediate plane strain stress ($v[\sigma_1 + \sigma_2]$) is taken to be σ_3 even though this does not correspond to the usual convention ($\sigma_1 > \sigma_2 > \sigma_3$) for all quadrants. However, in deriving the values for each quadrant, the usual convention was employed.

The main difference between the plane stress and plane strain fracture loci (Figs. 11.31 and 11.32) for brittle material (v close to 0.2) is in the third quadrant where a lateral constraining stress (σ_1) is seen to cause a much greater increase in compressive strength (σ_2) in plane strain than in plane stress. Except for this, the approximate locus of Fig. 11.28 is a close approximation to the plane strain locus of Fig. 11.32 for $v = 0.2$.

TABLE 11.3 Values of Fracture Stress Ratios for Brittle Material in Plane Strain for Different Values of v

v	C/T_σ	D/T_e	B/T_e	E/T_e^*
0.2	4.00	0.57	1.11	0.50
0.3	3.20	0.51	1.18	0.84
0.4	2.69	0.46	1.32	1.11
0.5	2.37	0.40	1.61	1.27

* E is the value of σ_1 in quadrant III when $\sigma_2 = -6T_e$ (Fig. 11.31).

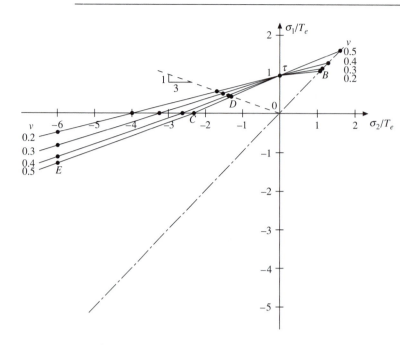

Fig. 11.32 Fracture loci for plane strain fracture based on spherical voids and for materials of different Poisson's ratio (v), E = points for which $\sigma_2/T_e = -6$, T_e = uniaxial tensile stress in plane strain. (after Shaw and Avery, 1986)

TABLE 11.4 Values of H/T_σ and H/T_e for Different Values of v

v	H/T_σ	H/T_e
0.2	1.33	1.33
0.3	1.37	1.39
0.4	1.40	1.48
0.5	1.45	1.61

HYDROSTATIC LOADING

For a hydrostatic tensile loading ($\sigma_1 = \sigma_2 = \sigma_3 = H$), the resultant intensified stress will be $H(k_1 + k_2 + k_3)$. Values of H/τ_σ and H/τ_e at fracture for different values of v are given in Table 11.4 and values of H/τ_σ are plotted in Fig. 11.31. The hydrostatic tensile stress at fracture for a brittle material (v near 0.2) is seen to be about $\frac{1}{3}$ greater than the maximum nominal tensile stress at fracture for either a plane stress or plane strain loading.

Experimental verification that a more balanced triaxial state of tensile stress gives a greater tensile stress at fracture than the corresponding uniaxial case is discussed in Shaw and Avery (1986).

For hydrostatic compression ($\sigma_1 = \sigma_2 = \sigma_3 = H$, all minus), the value of σ_1 will be negative, and hence a brittle material should not fail regardless of the magnitude of the hydrostatic compressive stress. This is why glass may be used for deep submersion vessels in the ocean.

For tool materials such as sintered tungsten carbide and ceramics that contain defects characterized as spherical voids, the two-dimensional void model leading to Fig. 11.24 appears adequate for planning purposes.

TOOL FRACTURE AT ENTRANCE AND EXIT

Tool fracture probability for 6 $^w/_o$ cobalt tungsten carbide tools turning AISI 4340 steel [H_B = 235 Kg/mm^2 at 200 f.p.m. (61 m/min)] was studied by Sampath, Lee, and Shaw (1984) for entrance and steady state conditions and by Lee, Sampath, and Shaw (1984) for exiting conditions using the brittle fracture locus of Figure 11.28. Cutting forces were measured under a wide range of values of

feed (0.011 i.p.r. = 0.28 mm/rev to 0.100 i.p.r. = 2.54 mm/rev) at
a depth of cut of 0.05 in (1.27 mm). The stress distribution on
the tool face shown in Fig. 11.33 was assumed which is close to
that suggested by experiments of Usui and Takeyama (1960) and
Zorev (1963). Values of a and a' were obtained by observing the
tool face after cutting, and values of σ_m and τ_m were obtained from
a, a', and the components of cutting force. Values of principal
stresses σ_1 and σ_3 (where σ_1 is the algebraically greatest stress
and σ_3 is the algebraically smallest stress) were obtained using a
finite element code assuming plane strain conditions and elastic
behavior of the tool. From these principal stresses, values of
intensified tensile stress (σ_i) were determined from

$$\sigma_i = 3\sigma_1 - \sigma_3 \tag{11.28}$$

There was no sign of failure at entrance regardless of entrance
angle (ξ in Fig. 11.34) or feed rate. There was no steady state
fracture at feeds up to 0.069 i.p.r. (1.75 mm/rev) with or without
chatter. Only with a feed rate of 0.100 i.p.r. (2.54 mm/rev) and
heavy chatter did fracture occur during steady state cutting. There
was a greater tendency for fracture at exit, and the exit angle (ψ)
had to be greater than 75° if fracture at exit were to be avoided.

Figure 11.35 is a map showing types of fracture that occurred
with different combinations of exit angles (ψ) and rates of feed
(t). Fracture of types 1 or 2 were found to occur in all areas except
for those corresponding to the largest exit angle. Fracture of type 1

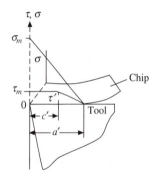

Fig. 11.33 Assumed shear (τ) and normal
(σ) stress distribution on tool face for section
parallel to resultant chip flow direction.
(Sampath, Lee, and Shaw, 1984)

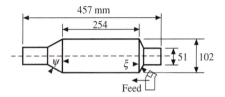

Fig. 11.34 Workpiece geometry for cutting
tests on AISI 4340 steel showing entrance
(ξ) and exit ψ) angles.

Fig. 11.35 Map showing type of
fracture occurring at end of the cut
for different combinations of workpiece exit angle (ψ) and feed rate (t):
(1) due to sudden unloading; (2) due
to strongly adhering (BUL); (2') due
to strongly adhering BUE; and (3)
no chipping or gross fracture. (after
Y. M. Lee et al., 1984)

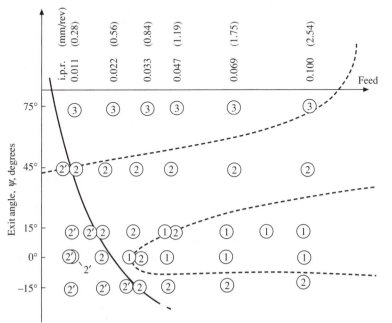

due to a rapid rate of unloading was found to occur when the exit angle was close to zero and for all but the lowest feeds. The formation of a roll-over burr and subsequent cutting with this burr were found to reduce this type of fracture.

When cutting with a low exit angle (0°), the chip flow direction was observed to change abruptly through about 90° as the end of the cut was approached (Fig. 11.35). This was due to the decrease in resistance of the work material in the feed direction as the end of the cut was approached. While the significance of this change in chip flow direction is not understood, it is undoubtedly related to the tendency toward type 1 fracture.

Foot Formation

Pekelharing (1978) has emphasized the strong tendency for fracture to occur as a tool exits a cut. He attributed this to a rotation of the shear plane as the end of the workpiece is approached. A finite

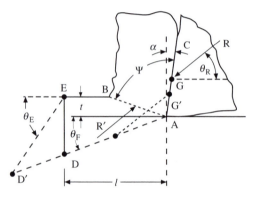

Fig. 11.36 Force diagram for tool approaching end of workpiece. (after Ramaraj et al., 1988)

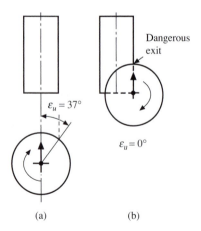

Fig. 11.37 (a) Orientation of plane-milling cutter relative to work giving rise to safe exit with a burr. (b) Unsafe exit from work relative to crack formation. (after Pekelharing, 1984)

element analysis was employed to show an increase in stresses in the tool face as the shear angle rotated. Ramaraj et al. (1988) have reported that the shear plane does not rotate from AB to AD (Fig. 11.36) but suddenly *jumps* to gross fracture along AD with no intervening deformation as the end of the workpiece ED is approached. This occurs so rapidly that resultant forces R and R′ do not change in magnitude or direction. Since AD is more nearly parallel to resultant force R than AB, the mean shear force on AD will be greater than on AB, and the mean normal stress on AB will be greater than on AD. These two effects enable fracture to occur along AD instead of shear along AB even though the area corresponding to AD is greater than that corresponding to AB. The sudden shift in R′ from the midpoint of AB to the midpoint of AD causes R to suddenly move closer to the tool tip (from G to G′) and hence to decrease the area of tool–chip contact. This in turn causes a sudden increase in the stresses at the tool point that can give rise to chipping or gross fracture at the tool tip. The chip root formed at the end of a cut (CADEBC in Fig 11.36) resembles a foot; therefore, Pekelharing (1978) refers to this action at the end of a cut as foot formation.

The paper by Pekelharing (1978) that revealed that chipping of tools in turning occurs primarily when a tool exits a cut due to "foot formation" was extended to milling in a two-part publication. Part 1 (Pekelharing, 1984) was concerned with the fundamentals, while part 2 (van Luttervelt and Willemse, 1984) discussed experimental results for commercial carbide milling cutters.

Figure 11.37a shows the relation between cutter and work that is usually recomended for face milling, if the cutter diameter (d) is 1.6 times the width of the work (w). This gives an exit angle (ε_u) of 37° and the "foot" produced does not give rise to chipping of the tool as it exits the work. A helical burr, which is not troublesome, will usually form as the tool leaves contact with the work. Figure 11.37b shows the same cutter and work situated with exit angle $\varepsilon_u = 0°$. This is a situation that is very likely to give a "foot" that causes chipping of the cutting edge as the tool exits the work. It was found that as the exit angle decreases, the tendency for crack formation increases. Figure 11.38 shows two situations: the one at (a) being dangerous relative to crack formation at exit (ε_u small), while that at (b) has a safe exit angle (37°). Most face-milling situations are much more complex than those in Figs. 11.37 and 11.38. Pekelharing gives a number of valuable suggestions for dealing with more complex cases.

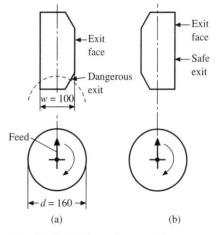

It was found that the probability of cracking at exit decreased for tools that had been first used under safe exit conditions ($\varepsilon_u = 37°$ or greater) and then used with the exit angle closer to 0°.

It was suggested that if exit cracks cannot be avoided, a reduction in feed and an increase in cutting speed would be a good choice. This is based on the fact that an increase in wear rate is not as serious as an increase in the probability of fracture.

Fig. 11.38 (a) Case where tool fracture at exit is much more likely than with the orientation at (b). (after Pekelharing, 1984)

In the second paper, von Luttervelt and Willemse (1984) discuss face-milling tests using different carbide grades to cut AISI 1045 steel to determine the dangerous exit region. A wide range of cutting conditions were employed, and the results were presented in the form of bar charts showing the effect of the following items on probable exit crack formation:

- cutting speed
- cutting feed
- exit angle
- cutter diameter
- carbide grade
- insert shape

It was found that

- The cutter diameter has a relatively small influence on the extent of the dangerous region.
- The extent of the dangerous region generally increases with increase in thickness of cut.
- The dangerous region becomes smaller with increase in cutting speed.
- At low cutting speeds where a stable BUE is formed, this protects the tool tip and there is no chipping of the cutting edge regardless of the size of the exit angle.
- At high operating temperatures the toughness of the carbide increases, and there is no cracking at exit regardless of the size of the exit angle.
- A reduction in the length of cut gives a smaller danger zone of crack formation.
- Results of the same grade of carbide from different manufacturers gave different tendencies toward cracking at exit.
- Edge preparation (chamfering and rounding of the tool tip) have a significant effect on fracture at exit.
- Entrance conditions in milling are far less important than exit conditions.

In addition to the increased concentration of stress at the tool tip due to gross fracture of the work along a downward sloping plane AD as the end of the workpiece is approached, there are two additional possible sources of tool fracture at exit:

- differential contraction as a tool having a patchy built-up layer on its rake face suddenly cools as it leaves a cut
- a sudden reversal of stress in a tool as the cutting forces are suddenly relaxed and the tool is carried past the unloaded equilibrium position when the load on the tool is suddenly relaxed

THERMAL CONTRACTION

Tool fracture due to thermal contraction of a thin built-up layer attached to the tool face of a carbide tool has been described by Y. M. Lee et al. (1984). If a thin patchy layer of work material strongly bonded to the tool face has a high temperature as the tool is rapidly separated from the work, both the adhering layer of metal and the tool will shrink on cooling. The work material ($\alpha = 11 \times 10^{-6}/°C$) will shrink more than the tool material ($\alpha = 7 \times 10^{-6}/°C$). Two adjacent patches of work material will tend to shrink more toward their centers than the tool material, giving rise to a tensile stress in the tool face across the gap which will act as a notch to intensify the nominal tensile stress to the point where a tensile crack is initiated.

This mechanism of fracture involving differential contraction of a strongly adhering metal film was demonstrated by cementing thin aluminum blocks to the surface of a previously scribed glass microscope slide using silicone cement (Fig. 11.39) and then reducing the temperature of both metal and glass to a low value by immersion in liquid ntrogen. The differential contraction of the aluminum ($\alpha = 24 \times 10^{-6}/°C$) and of the glass ($\alpha = 8 \times 10^{-6}/°C$) induced a tensile stress in the glass high enough to initiate a sharp crack at the tip of the notch which spontaneously propagated across the thickness of the glass. Development of a substantial pattern of surface stress in tension in a material of low α in contact with a thin layer of higher α was also shown by finite element (FEM) analysis. Uehara (1977) has found progressive tool chipping to accompany adherence of metal on the tool face in milling.

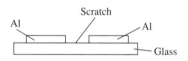

Fig. 11.39 Arrangement used in simulated tool fracture test involving aluminum blocks cemented to glass surface on either side of a scratch. (Y. M. Lee et al., 1984)

STRESS REVERSAL

Shaw (1979) has suggested that a reversal of the normally compressive stress at the tool tip may occur on unloading to cause tool tip fracture provided the unloading rate is sufficiently high. Kamarruddin (1984) has verified the finding of Pekelharing regarding foot formation and suggests that the most dangerous situation relative to tool-tip fracture involves both stress reversal and foot formation.

To verify the possibility of tool reversal upon rapid unloading requires a dynamometer of very high frequency response. An ordinary strain gage dynamometer or a conventional piezoelectric dynamometer is completely inadequate mainly because of the large mass between the transducers and the point of action of the tool. Since an ordinary tool holder has a natural frequency in the direction of the major cutting force that is much lower than that of the carbide or ceramic insert itself, it becomes necessary to mount the transducer directly on the insert or on the carbide shim normally located directly below an insert. Lindberg and Lindstrom (1983) have used a piezoelectric element inserted between a carbide insert and shim to study chip formation in continuous cutting. Budynas (1977) has shown that a wire or foil resistance strain gage which is essentially massless can respond to time-varying signals within 0.1 μs. Therefore, Sampath et al. (1986) have mounted a strain gage on the shim beneath a carbide insert to monitor the loading/unloading of tools in interrupted cutting (Fig. 11.40).

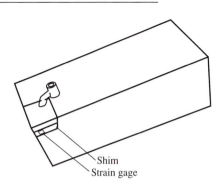

By use of the arrangement of Fig. 11.40, it was found that at high cutting speeds (> 500 f.p.m. = 150 m/min), overshoot on loading and unloading at the beginning and end of a cut was observed. The overshoot on entering is not of importance since it leads to an increase in compressive stress which does not lead to fracture. However, the overshoot at exit gives a tensile stress that can lead to fracture. In contrast, Pekelharing (1978) reported that

Fig. 11.40 Position of strain gage used to record stress reversal when a tool is suddenly unloaded. (after Sampath et al., 1986)

in none of his tests was an overshoot noted either at the beginning or at the end of a cut. This was because the dynamometer he used was a commercial piezoelectric unit that has far too low a natural frequency to resolve such detail due to the relatively large mass between the transducers and the point of cutting action. It was observed by Sampath et al. (1986) that the frequency of fracture at exit increased with the extent of overshoot.

Tungsten carbide tools tended to have a greater tendency toward stress reversal on sudden unloading than ceramic tools due primarily to a greater cutting force under the same machining conditions. The greater density of tungsten carbide would also contribute to the greater tendency for tungsten carbide to show a stress reversal on unloading than ceramic tools. This greater tendency of tungsten carbide toward stress reversal on unloading tends to offset the somewhat greater ductility of tungsten carbide relative to ceramic tools.

A review of work done on tool fracture at Arizona State University is given in Shaw and Ramaraj (1989).

CONCLUDING REMARKS

From the foregoing discussion, it is evident that brittle (tensile) fracture depends upon the principal stresses, Poisson's ratio, the shape and distribution of defects (voids) involved, and the uniaxial tensile strength of the material. A plane stress fracture locus based on cylindrical voids (Fig. 11.28) represents a conservative criterion for use with complex states of stress. For triaxial stress states, it is advisable to use the maximum and minimum algebraic stresses present when approximating plane stress and using Fig. 11.28. The frequently employed maximum nominal tensile stress criterion should be used only when all three principal stresses are tensile which is not the case for metal cutting tools. The nominal principal stresses interact with the defects (voids) present in all real materials to give rise to an intensified tensile stress which is the item of importance relative to brittle fracture. The fracture locus of Fig. 11.28 is based on a maximum intensified tensile stress criterion first proposed by Griffith (1924).

If one or more of the principal stresses is compressive, there is less tendency for fracture to occur than if all stresses are tensile. The beneficial role of compression in postponing fracture increases as the defects present become less circular and as the principal stresses become more balanced (hydrostatic).

Metal cutting tools have a much greater tendency to fracture on rapidly exiting a cut than at the beginning of a cut or when cutting under steady state conditions. Three reasons for this have been identified:

- a sudden decrease in the tool–chip contact area as the direction of shear jumps from that corresponding to the ordinary shear plane to a direction that is more parallel to the direction of the resultant force on the tool (foot formation) [This sudden change in the direction of shear fracture is associated with the fact that the shear strength of a material increases with normal stress on the shear plane (Ramaraj et al., 1988).]

- differential contraction on sudden cooling of thin patchy flakes of metal strongly bonded to the tool face and the tool itself which will normally have a lower coefficient of expansion than the metal cut

- the reversal of stress in the tool from compression to tension due to inertia-associated over-shoot upon rapid unloading of the tool

DIMENSIONAL ANALYSIS

When the input to a brittle body is a static force, the appropriate fracture criterion is the maximum tensile stress criterion provided a uniaxial tensile stress is involved at the critical point of fracture. However, when the input is an energy, the fracture criterion is the maximum tensile specific strain energy criterion (Shaw, 1977). This means that for an energy input, fracture will occur when (σ_A^2/E) reaches a critical value where σ_A is the maximum uniaxial tensile stress in the loaded body. For a tool of given geometry and material, operating at a given cutting speed, we should expect

$$\sigma_A = \psi_1(E, F_P, \delta, b, t) \tag{11.29}$$

or

$$\frac{\sigma_A^2}{E} = \psi_2\left(\frac{F_P\delta}{b}, t\right) \tag{11.30}$$

where,

F_P = cutting force component in direction of cutting speed

δ = total deflection of tool relative to work in the same direction

t = undeformed chip thickness

b = width of cut

ψ_2 = some function of $F_P\delta/b$ and t

and after performing a dimensional analysis, we have

$$\frac{\sigma_A^2 b t^2}{E F_P \delta} = \text{constant} = N_F \tag{11.31}$$

where N_F is the nondimensional fracture number. But, we may write

$$\delta = \frac{F_P}{K} \tag{11.32}$$

and

$$F_P = ubt \tag{11.33}$$

where

$$K = \text{spring constant of system in the } F_P \text{ direction } [FL^{-1}]$$

$$u = \text{specific cutting energy.}$$

Substituting into Eq. (11.31)

$$N_F = \frac{\sigma_A^2(K)}{Eu^2(b)} \tag{11.34}$$

The nondimensional fracture number (N_F) will be a constant at fracture for a given tool material and geometry when the tool is operating at a given speed and with a given cutting fluid. Just as the Reynolds number in fluid mechanics is proportional to the ratio of the inertia to viscous forces acting on a fluid particle, N_F is proportional to the ratio of the specific elastic energy at the critical point at fracture (σ_A^2/E) to the specific elastic energy stored in the cutting system $F_P \delta/2 = u^2 b/2K$. For a given system stiffness (K) and width of cut (b), the maximum allowable specific cutting energy (u) will vary as (σ_A^2/E)$^{1/2}$ where σ_A is the critical value of uniaxial tensile stress at fracture as obtained from a "static" tensile test. As harder materials are cut (higher u), a stronger work material (> σ_A) and/or a stiffer system (greater K) is called for if fracture at exit is to be avoided.

EXOELECTRONS AND COUNTER CURRENTS

Electrons (exoelectrons) are emitted from a freshly scratched cut or fractured surface as Kramer demonstrated at the PTB in Braunschweig about 1950. This emission appears to be due to the slightly greater equilibrium atom spacing at a surface than internally. Associated with the denser internal atom spacing is a greater density of free electrons. When a new surface is generated there will be an excess of electrons in the surface initially which must leave as exoelectrons in order that the atom spacing in the surface expand to its equilibrium value. The energy involved in this adjustment is surface energy.

A freshly generated surface will have an excess of free electrons and hence will be in a highly active chemical state relative to oxidation reactions (reactions involving a loss of electrons by the surface as for example: $Fe \rightarrow Fe^{++} + 2e^-$).

It would appear that the flow of electrons coming from a freshly scratched asperity could have an important influence on bond formation between asperities making contact as well as on surface chemical reactions. This subject is therefore of importance to those interested in wear and is particularly applicable to the wear of heavily loaded sliders where more new surface will be generated as a result of an increased wear rate.

The enormous influence that the generation of fresh surfaces has on surface reactions is dramatically illustrated by the mechanical activation process (Shaw, 1948), discussed in detail in Chapter 24.

Over the years there have been many claims and counter-claims concerning the role of the heavy thermoelectric currents that are generated when a tungsten carbide tool cuts steel (Opitz and Axer, 1956). The current flow in these cases is from chip to tool and it is claimed that wear may be reduced if this current flow may be prevented by insulating the tool from the system (Fig. 11.41a) or by introducing a counter EMF that reduces the current flow to zero (Fig. 11.41b).

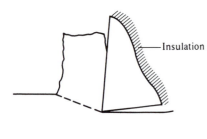

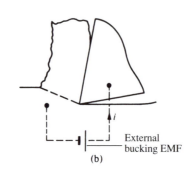

Fig. 11.41 Prevention of current flow between chip and tool. (a) By insulating tool. (b) By use of a bucking potential.

(a)

(b)

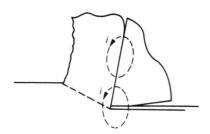

Fig. 11.42 Eddy current flow from tool to chip.

Great claims have been made for the method of preventing tool wear in Fig. 11.41b, but these have generally not been substantiated. It would appear the explanation for this anomoly lies in the fact that in general the heavy thermoelectric currents are eddy currents as shown in Fig. 11.42. Insulating the tool or applying a counter EMF will have little influence on these eddy currents. Occasionally the path of resistance may be less through the machine and work than for the path of an eddy current, which would be the case if a semi-conducting oxide were developed between chip and tool. In such instances it would be possible to alter the current flow by tool insulation or by use of a bucking circuit. It is probable that in such cases good results have been obtained and reported in the literature using these techniques. However, it is believed that in general the situation is as shown in Fig. 11.42. The only thing that will reduce the extremely high current flow here is the formation of an insulating oxide film that greatly increases the electrical resistance between chip and tool, or use of a nonconducting (ceramic) tool.

There have been a number of reports of greatly improved tool life (from 2 to 10 times) due to the formation of stationary oxide films on the surface of titanium-carbide-containing tools. This oxide has been shown to be due to the build-up of inclusions in the steel resulting from special deoxidation techniques (ferrosilicon and calcium oxide) (Opitz et al., 1962). While these oxide films have generally been thought to act as a diffusion barrier to prevent the loss of carbon from the WC crystals in the tool face to the austenitic layer on the chip face, it is also possible that the oxide is preventing the large eddy currents that will be present in the absence of an insulating film.

The complete absence of crater wear for ceramic tools could be due to the absence of eddy currents with a tool material that is electrically nonconducting.

Tool Condition Monitoring

There have been many studies in the recent past to monitor in-process tool performance by measurement of a variety of quantities. A few of these include the following with the item of primary interest in brackets:

- Koenig et al. (1972)—[cutting forces]
- Cook and Subramaniam (1978)—[tool wear sensing]
- Lan and Dornfeld (1984)—[acoustic emission]

- Jiaa and Dornfeld (1990)—[acoustic emission]
- Hanchi and Klamecki (1991)—[tool wear]
- Uehara et al. (1992)—[thermoelectric current]
- Stephenson (1993)—[tool life]
- Tarng and Hwang (1994)—[forces]
- Donovan and Scott (1995)—tribological EMF
- Kim and In-Hyu (1996)—[multi-sensors]
- Dimla et al. (1997)—[neural networks]

A comprehensive review of tool condition monitoring as of 1995 was presented by Byrne et al. (1995).

In the 1950s and 1960s there was considerable interest in use of nuclearly radiated tools for monitoring tool wear. However, this required special shielding and cumbersome safety precautions so that interest was gradually lost in this technique. One of the last papers concerning nuclear tool monitoring was by Cook and Subramanian (1978).

When tool wear is continuously monitored on line, it is difficult to determine when tool face wear reaches a given value. Murugan (2000) has developed a clever way of doing this that employs special modified tools.

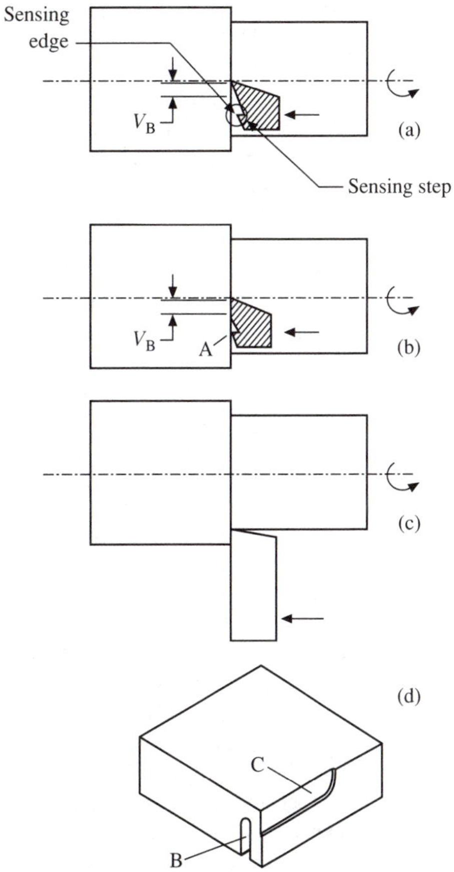

Figure 11.43a is a side view of a modified tool in an orthogonal machining operation. When the wear land corresponds to a particular value (V_B), point A contacts the work as shown in Fig. 11.43b. There is then a sudden increase in the tool-work contact length. Initially an attempt was made to monitor this by detecting a sudden increase in vibration. This was not found satisfactory; therefore, measurement of the acoustical signal emitted was tried and found to be more useful. Figure 11.43c is a top view for either Fig. 11.43a or Fig. 11.43b.

Fig. 11.43 Modified tool for monitoring when wear land reaches the predetermined value V_B. (after Murugan, 2000)

In order to increase the sensitivity to when contact at A occurs, the tool was further modified as shown in Fig. 11.43d. In this case fracture occurred with contact at A (Fig. 11.43b), giving rise to a strong acoustical signal. This weakening of the tool involved diamond grinding a slot at B and one at C in Fig. 11.43d by electrical discharge machining. Finite element modeling was employed to determine the dimensions and locations of the slots that would ensure fracture when V_B was reached for any machining condition.

Considerable additional work was performed on error analysis to assertain the effects of clearance angle, work diameter, step dimensions, and cutting conditions on the design requirements to assure fracture when a specific V_B was reached.

REFERENCES

Adams, M., and Sines, G. (1978). *Tectonophysics* **49**, 97.
Albrecht, P. (1956). *Microtecnic* **10**, 45.
Archard, J. F. (1953). *J. Appl. Phys.* **24**, 981.

ASME (1980). *Wear Control Handbook.*

ASME/ANSI (1981). *Standard B276–279*, p. 93.

ASTM (1981). *Standard B276–279*, p. 93.

Babel, W. F., and Sines, G. (1968). *ASME J. Basic Eng.* **90**, 285.

Backer, W. R., Marshall, E. R., and Shaw, M. C. (1952). *J. Engng. Ind.* **74**, 61.

Bowden, F. P., and Tabor, D. *The Friction and Lubrication of Solids* (Vol. 1, 1950; Vol. 2, 1964). Clarendon Press, Oxford.

Brace, W. F. (1964). In *State of Stress in Earth's Crusts*, ed. W. Judd. Elsevier, New York, pp. 111–174.

Budynas, S. (1977). *Advanced Strength and Applied Stress Analysis.* McGraw-Hill, New York.

Burwell, J. T. (1947). *Nucleonics* (Dec), 38.

Burwell, J. T., and Strang, C. D. (1952). *J. Appl. Phys.* **23**, 18.

Byrne, G., Dornfeld, D., Inasaki, I., Kettler, W., Koenig, W., and Teti, R. (1995). *Annals of CIRP* **44/2**, 541–567.

Chermant, J. L., and Osterstock, F. (1977). *Proc. 4th Int. Conf. on Fracture.* Waterloo, Canada, pp. 229–235.

Cook, N. H., and Subramaniam, K. (1978). *Annals of CIRP* **27/1**, 73–78.

Cornet, I., and Grassi, R. C. (1961). *ASME J. Basic Eng.* **83**, 39.

Dawihl, W. (1940a). *Z. Metallkunde* **32**, 320.

Dawihl, W. (1940b). *Z. Tech. Phys.* **21**, 44.

Dawihl, W. (1941). *Stahl U. Eisen* **61**, 210.

DeSalvo, G. J., and Shaw, M. C. (1968). *Proc. of Int. Conf. on Mach. Des. and Res.,* Pergamon Press, Oxford, p. 961.

Dimla, D. E., Lister, P. M., and Leighton, N. J. (1997). *Int. J. Mach. Tools and Manufacture* **37/9**, 1219–1241.

Donovan, A., and Scott, W. (1995). *Int. J. Mach. Tools and Mfg.* **35/11**, 1523–1535.

Fleming, J. R., and Suh, N. P. (1977). *Wear* **44**, 39.

Goodier, J. N. (1933). *ASME Trans.* **55**, 39.

Griffith, A. A. (1924). *Proc. 1st Int. Conf. on Applied Mech.*, Delft, Netherlands, p. 55.

Hanchi, J., and Klamecki, B. E. (1991). *Wear* **145**, 1–27.

Holm, R. (1950). *Electrical Contacts.* Springer, Berlin.

Hovenkamp, L. H., and Van Emden, E. (1952). *Annals of CIRP* **1/1**, 7–12.

Inglis, C. E. (1918). *Trans. Instn. Naval Arch.* **55**, 219–230.

Jiaa, C. L., and Dornfeld, D. A. (1990). *Wear* **139**, 403–424.

Johanmir, S. (1977). "A Fundamental Study of the Delamination Theory of Wear," Ph.D. thesis, MIT.

Kalish, H. S. (1982). *Mfg. Engng.* **89**, 97–99.

Kamarruddin, A. G. (1984). "Tool Life in Interupped Cutting," Ph.D. dissertation, UMIST, Manchester, England.

Kim, J., and In-Hyu, T. S. (1996). *Int. J. Mach. Tools and Mfg.* **36/8**, 861–870.

Koenig, W., Langhanner, K., and Schemmel, H. U. (1972). *Annals of CIRP* **21/1**, 19–20.

Kramer, B. M., and Suh, N. P. (1980). *J. Engng. Ind.* **102**, 303.

Lambert, H. J. (1961). *Annals of CIRP* **10**, 246.

Lan, M. S., and Dornfeld, D. A. (1984). *ASME J. of Eng. Mat. and Tech.* **106**, 111–118.

Lee, M., Horne, J. G., and Tabor, D. (1979). *General Electric Technical Information Series*, No. 78CRD246.

Lee, Y. M., Sampath, W. S., and Shaw, M. C. (1984). *J. Eng. for Industry (Trans. ASME)* **106**, 168.

Leyensetter, W. (1956). *Z. Ver Deutscher Ing.* **98**, 957.

Lindberg, B., and Lindstrom, B. (1983). *Annals of CIRP* **32/1**, 17.

Lueth, R. C. (1972). "A Study of the Strength of WC–Co Alloys From a Fracture Mechanics Point of View," Ph.D. dissertion, Michigan State University.

Merchant, M. E., Ernst, H., and Krabacher, E. J. (1953). *J. Engng. Ind.* **75**, 549.

Moltrecht, K. H. (1964). Paper 637. Am. Soc. Tool Manuf. Eng., Dearborn.

Muetze, H. (1967). DIng. dissertation. Aachen T.U., Germany.

Murugan, M. (2000). Ph.D. dissertation, IIT, Madras, India.

Opitz, H. (1956). *Werkstattstch. U. Maschinenbau* **46**, 210.

Opitz, H. (1963). In *International Research in Production Engineering*. ASME, New York, p. 107.

Opitz, H., and Axer, A. (1956). *Forschungs Berichte des Wirtschafts und Verkehrsministeriums Nordrhein-Westfallen*, Nr. 271.

Opitz, H., Gappisch, M., Koenig, W., and Wicker, A. (1962). *Archiv. fur Eisenhuttenwesen* **33**, 841.

Pekelharing, A. J. (1978). *Annals of CIRP* **27/1**, 5.

Pekelharing, A. J. (1984). *Annals of CIRP* **33/1**, 47–50.

Pickens, J. R. (1977). "The Fracture Toughness of WC–Co Alloys as a Function of Microstructural Parameters," Ph.D. dissertation, Brown University.

Ramaraj, T. C., Santhanem, S., and Shaw, M. C. (1988). *J. Eng. for Ind. (Trans. ASME)* **108**, 222.

Rehbinder, P. A. (1947). *Nature* **159**, 866.

Roach, A. E., Goodzeit, C. L., and Hunnicutt, R. P. (1956). *Trans. Am. Soc. Mech. Engrs.* **78**, 1639, 1669.

Salmassy, O. K., Duckworth, W. H., and Schwoke, A. D. (1955). *Tech. Report* WADC-TR543-30, Parts I and II.

Sampath, W. S., Lee, Y. M., and Shaw, M. C. (1984). *J. Eng. for Ind. (Trans. ASME)* **106**, 161.

Sampath, W. S., Ramaraj, T. C., and Shaw, M. C. (1986). *J. Eng. for Ind. (Trans. ASME)* **108**, 232.

Sampath, W. S., and Shaw, M. C. (1986). *J. Eng. for Ind. (Trans. ASME)* **108**, 11–14.

Shaw, M. C. (1948). *J. Appl. Mech.* **70**, A137.

Shaw, M. C. (1971). *Annals of CIRP* **19**, 533.

Shaw, M. C. (1977). *Int. J. Wear* **43**, 263.

Shaw, M. C. (1979). *Annals of CIRP* **28/1**, 19.

Shaw, M. C., and Avery, J. P. (1986). *J. Eng. Materials and Tech. (Trans. ASME)* **108**, 222.

Shaw, M. C., and Ramaraj, T. C. (1989). *Annals of CIRP* **38/1**, 59–63.

Shaw, M. C., and Strang, C. D. (1950). *J. Appl. Phys.* **21**, 349.

Shaw, M. C., Thurman, A. L., and Ahlgren, H. J. (1966). *J. Engng. Ind.* **88**, 142.

Shibasaka, T., Hashimoto, H., Veda, K., and Iwata, K. (1983). *Annals of CIRP* **32/1**, 37.

Solaja, V. (1958). *Int. J. Wear* **2**, 40.

Stephenson, D. A. (1993). *ASME J. of Eng. for Ind.* **115**, 432–437.

Suh, N. P. (1973). *Wear* **25**, 111.

Suh, N. P. (1973a). *Tribophysics*. Prentice-Hall, New York.

Suh, N. P. (1977). *Wear* **44**, 1–16.

Takagi, J., and Shaw, M. C. (1983). *J. Eng. for Ind. (Trans. ASME)* **105**, 143.

Tarng, Y. S., and Hwang, T. S. (1994). *Computers and Structures* **53/4**, 937–945.

Taylor, F. W. (1907). *Trans. Am. Soc. Mech. Engrs.* **28**, 31.

Timoshenko, S., and Goodier, J. N. (1951). *Theory of Elasticity*. McGraw-Hill, New York, p. 359.

Trent, E. M. (1963). *Proc. Iron and Steel Inst.* 1001.

Trent, E. M. (1967). *Proc. Eighth Mach. Tool Des. and Res. Conf.* Pergamon Press, Oxford, p. 629.

Uehara, K. (1977). *Annals of CIRP* **25/1**, 11.

Uehara, K., Sakurai, M., and Ikeda, T. (1992). *Annals of CIRP* **41/1**, 75–78.

Usui, E., and Takeyama, H. (1960). *J. Eng. for Ind. (Trans. ASME)* **82**, 303.

van Luttervelt, C. A., and Willemse, H. R. (1984). *Annals of CIRP* **33/1**, 51–54.

Zorev, N. N. (1963). In *International Research in Production Engineering*. ASME, New York, pp. 42–49.

12 CUTTING TEMPERATURES

There are several temperatures of importance in metal cutting. The shear plane temperature (θ_S) is of importance for its influence on flow stress and since it has a major influence on temperatures on the tool face (θ_T) and on the relief surface (θ_R). The latter two temperatures (θ_T and θ_R) are very important to crater wear rate and rate of wear-land development, respectively. The temperature on the tool face also plays a major role relative to the size and stability of the BUE. The ambient temperature of the work approaching the cutting zone (θ_o) is also important since it directly affects θ_S, θ_T, and θ_R.

THERMAL ENERGY IN CUTTING

Practically all of the mechanical energy associated with chip formation ends up as thermal energy. One of the first measurements of the mechanical equivalent of heat (J) was made by Benjamin Thomson (better known as Count Rumford). Rumford (1799) measured the heat evolved during the boring of brass cannon in Bavaria. He immersed the work, tool, and chips in a known quantity of water and measured the temperature rise corresponding to a measured input of mechanical energy. These experiments not only provided a good approximation to the mechanical equivalent of heat that stood as the accepted value for several decades but also provided new insights into the nature of thermal energy at a time when most people believed that heat was a special form of fluid called "caloric." It is well known that some of the energy associated with plastic deformation remains in the deformed material. Taylor and Quinney (1934, 1937) using a very accurate calorimetric technique measured the residual energy involved when metal bars were deformed in torsion. It was found that the percentage of deformation energy retained by the bars decreased with increase in strain energy involved. When these results are extrapolated to strain energy levels in chip formation, it is estimated that the energy that is not converted to thermal energy is only between 1% and 3% of the total cutting energy. Bever et al. (1953) have directly measured the residual energy stored in metal cutting chips, and Bever et al. (1974) have discussed the stored energy in plastically deformed bodies from a broad point of view.

All of these results suggest it is safe to assume as a first approximation that all of the energy associated with chip formation is converted to thermal energy. The energy retained in the chips and that associated with the generation of new surface area is negligible relative to the total energy expended in chip formation.

EXPERIMENTAL TECHNIQUES

Tool temperatures could not be treated quantitatively until the 1920s when means were devised by Shore (1924) in the United States and almost simultaneously by Gottwein (1925) in Germany and Herbert (1926) in Great Britain for measuring the mean temperature along the face of a cutting tool by the tool–work thermocouple technique. This method is based on the fact that an electromotive force (EMF) is generated at the interface of two dissimilar metals when the temperature of the interface changes.

The laws of thermoelectric circuits that are applicable here may be summarized as follows:

1. The EMF in a thermoelectric circuit depends only on the difference in temperature between the hot and cold junctions and is independent of the gradients in the parts making up the system.

2. The EMF generated is independent of the size and resistance of the conductors.

3. If the junction of two metals is at uniform temperature, the EMF generated is not affected if a third metal, which is at the same temperature, is used to make the junction between the first two.

Just how these principles are applied in practice is illustrated in Fig. 12.1, which shows a lathe equipped for measuring tool-face temperatures. The chip and tool junction H constitutes the hot junction, while A and B, the cold junctions, remain at room temperature. Should these points be heated appreciably above room temperature during tests, special procedures are available to compensate for this. The mercury contact is used to make electrical contact with the rotating member without introducing extraneous voltages that frequently accompany the use of slip rings. The wires C and D may be ordinary copper wires since both ends are at the same temperature and thus have no influence on the EMF generated at H (rule 3), as measured with the potentiometer P.

The simplest and possibly the most accurate method for calibrating the tool–work thermocouple is shown in Fig. 12.2. The most convenient form of work-piece is a long chip. The end of the tool is best ground to a $\frac{1}{8}$ in (3 mm) diameter, both to ensure uniform temperature and to limit the amount of heat transferred to the "cold" end of the tool. To keep the "cold" end at room temperature, it is advisable to use a long tool, or possibly several similar carbide inserts brazed together; otherwise corrections may be necessary. If the workpiece does not form a long chip, it must be prepared similar to the tool. As long as the lead bath is at uniform temperature, its presence will have no effect on the EMF measured, according to the third law of thermoelectricity. Its temperature

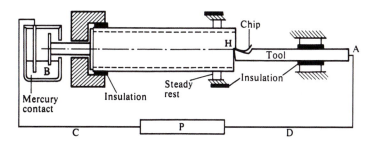

Fig. 12.1 Schematic diagram showing the manner in which tool–chip interface temperatures are measured by the thermoelectric technique.

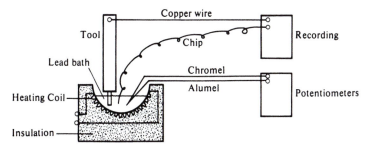

Fig. 12.2 Arrangement for calibration of tool–work thermocouple.

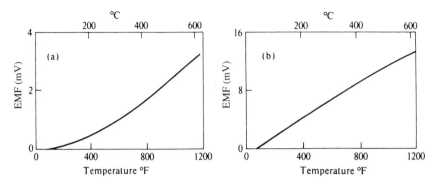

Fig. 12.3 Temperature calibration curves of AISI 1113 steel against (a) 18–4–1 HSS and (b) K2S cemented carbide.

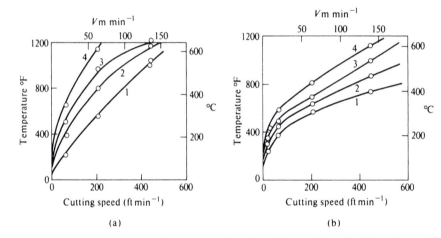

Fig. 12.4 Representative temperature data for two-dimensional dry cutting of AISI 1113 steel. (a) Tool: 18–4–1 HSS, 15° rake angle, 5° clearance. (b) Tool: K2S cemented carbide, 20° rake angle, 5° clearance. Feed: (1) 0.0023 i.p.r. (0.05 mm/rev); (2) 0.0044 i.p.r. (0.11 mm/rev); (3) 0.0065 i.p.r. (0.17 mm/rev); (4) 0.0116 i.p.r. (0.29 mm/rev).

is measured with a standard chromel–alumel couple. In case a large number of calibrations are to be made with various tool and work materials, it is often more convenient to measure the EMFs of the tool and work materials separately, both against alumel wire and then obtain the tool–work calibration by difference. That this will yield the desired calibration is also evident from the foregoing third law of thermoelectricity. Two typical calibration curves obtained in the manner described are shown in Fig. 12.3. They are for the same low-carbon, free machining steel (0.32% sulphur), but one is for 18–4–1 high speed steel while the other is for K2S cemented carbide.

Representative cutting temperatures measured with the tools and work material of Fig. 12.3, over a range of cutting speeds and at several depths of cut, are given in Fig. 12.4. The cutting temperatures are seen to be noticeably higher with the HSS tool, than with the carbide in Fig. 12.4 at the same values of speed and feed. The carbide tool had a rake angle of 20°, while the HSS tool had a rake angle of 15°. Both tools had clearance angles of 5°, and all cuts were made under orthogonal conditions.

While the tool–work thermocouple technique is relatively simple to apply, it is not without its limitations. The method estimates the temperature over the entire contact area between chip and tool including the wear land. Also, misleading values are apt to be obtained if a BUE is present because then dissimilar materials do not exist over the entire contact area. There is a further question about whether a static calibration is satisfactory for the dynamic cutting situation involving exoelectrons. Also, oxide layers tend to form on carbide tools when low-melting nonmetallic inclusions resulting from certain deoxidation practices are employed (discussed in Chapter 15). Such oxide layers will of course change the static calibration of a tool–work thermocouple if present.

Stephenson (1993) considered several issues associated with implementation of the chip–tool thermocouple technique. He points out it is necessary to isolate only the tool, since a potential differ-ence is measured and the machine is usually well grounded. Removing the necessity of insulating the work eliminates the possibility of reducing the stiffness of the system. He also suggests use of a lead wire whose thermoelectric output is low (alumel) in contact with the tool in order to introduce less extraneous EMF due to secondary junctions.

Thermocouples embedded in cutting tools have also been used to measure cutting temperatures. Figure 12.5 is an example of such a study. The hole in which the thermocouple is placed represents a disturbance and may appreciably change the temperature field being measured unless it is very small. Mapping the temperature field by use of a thermocouple embedded in the tool is an extremely

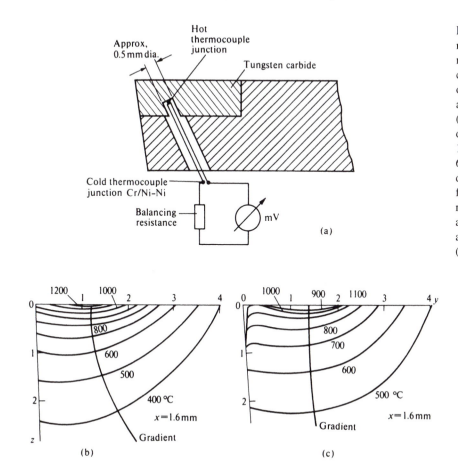

Fig. 12.5 Embedded ther-mocouple technique for mapping temperatures in cutting tool used in turning operation. (a) Experimental arrangement. (b) Sharp tool. (c) Worn tool. Cutting con-ditions: work, 0.3 $^w/_o$ C, 1 $^w/_o$ Mn steel; cutting speed, 600 f.p.m. (180 m min^{-1}); depth of cut, 0.12 in (3 mm); feed, 0.03 i.p.r. (0.74 mm/ rev); rake angle, 0°; relief angle, 4°; side cutting edge angle, 45°; fluid, none. (after Küster, 1956)

tedious procedure involving many tools with thermocouples mounted at different points. In the study leading to Fig. 12.5, the very small holes were made by electroerosion and temperatures within 0.2 mm of the surface could be measured. Figure 12.5 shows the maximum temperature in the tool surface to occur an appreciable distance from the cutting edge. Figure 12.5b is for a new tool and Fig. 12.5c is for a worn tool. The center of the crater that develops is seen to occur very near the point of maximum surface temperature.

Schwerd (1937) measured the infrared radiation from the tool, work, and chip in orthogonal cutting to establish the temperature field on the outside surfaces of these regions. While this method did not reveal internal temperatures, it did provide a valuable early insight into the temperature distribution on shear plane, tool, and chip. The field covered by each measurement was 0.2 mm.

Boothroyd (1963) used a somewhat similar technique to check analytical studies of cutting temperatures. He heated workpieces to about 600 °C and then photographed the radiation from work, chip, and tool in orthogonal cutting. A microdensitometer was used to obtain the point-to-point intensity of radiation with the aid of a suitable calibration negative. The fact that the specimen had to be preheated limited the cutting speed that could be used without overheating the HSS tools and made interpretation difficult. In the discussion of the Boothroyd paper, Professor Opitz suggested that the experiments be repeated using tungsten carbide tools at higher speeds so that the workpiece need not be preheated.

The main difficulty with radiation techniques is that they are generally limited to accessible surfaces although Bickel (1963) has described techniques in which a hole in work or tool uncovers the surface of interest periodically so that a photo diode may produce a calibrated signal on an oscilloscope.

Trigger (1963) has similarly used a lead sulfide radiation sensor to map temperatures on the flank of a cutting tool (Fig. 12.6). The peak flank temperature was found to occur about 2.5 feed distances from the cutting edge and was about 15% greater for a worn tool [0.020 in (0.5 mm) wear land] than for a sharp tool. Using a geometric electronic analog the tool flank temperature distribution was used to compute the tool face temperature curve shown in Fig. 12.6b (see Chao et al., 1961, for method). These curves show peak temperature about 2000 °F (1100 °C) located near the midpoint of the contact length for sharp tools. These results are in good agreement with those of Fig. 12.5 for tool-face temperatures but not for tool-flank temperature distribution. Figure 12.5 shows the peak temperature on the tool flank to be at the cutting edge while Fig. 12.6a shows the peak temperature to lie well beyond the cutting edge for both a sharp and dull tool. The peak temperature on the clearance face should be expected to occur at the tool point for a sharp tool but at the end of the wear land for a worn tool.

Ueda and his coworkers have written a number of papers concerning the use of optical fibers to conduct infrared radiation to a pyrometer to measure cutting and grinding temperatures. In earlier work indium antimonide (InSb) cells were used. Calibration was performed by comparing the radiation from a heated specimen with that from a black body at the same temperature to determine the emissivity of the specimen (Ueda et al., 1985). However, emissivity compensation may be achieved by making radiation measurements at two different wave lengths and fusing filters before amplification. A two-color pyrometer with a fused fiber coupler developed by Ueda et al. (1995) is used in present day radiation temperature measurements.

Al Huda et al. (2000) have used a two-color pyrometer to measure tool–chip interface temperatures in a turning operation. A translucent hot pressed Al_2O_3 tool was used to turn annealed ($H_V = 210$) AISI 1045 steel both dry and when using a water-soluble cutting fluid. The experimental arrangement and positioning of the optical fiber is shown in Fig. 12.7a and Fig. 12.7b, respectively. The optical fiber located at the tool is divided into two fibers by a fused fiber coupler located about 79 in (2 m) from the tool. Each of these sub-fibers is led to a different detector—one being germanium

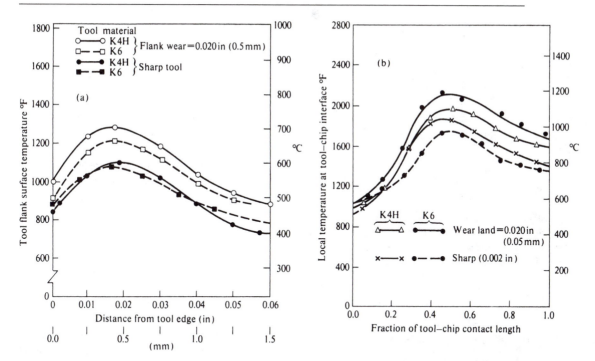

Fig. 12.6 Temperature distribution on (a) tool flank and (b) tool face as determined by radiation measurement (flank) and computation (face) when turning AISI 52100 steel (380 Brinell hardness) with sharp and dull carbide tools. Machining conditions: Tool geometry, 0, 5, 5, 5, 5, 5, 0.045 in (1.1 mm); cutting speed, (a) 150 f.p.m. (47 m min^{-1}); (b) 200 f.p.m. (63 m min^{-1}); feed, 0.0074 i.p.r. (0.19 mm/rev). (after Trigger, 1963)

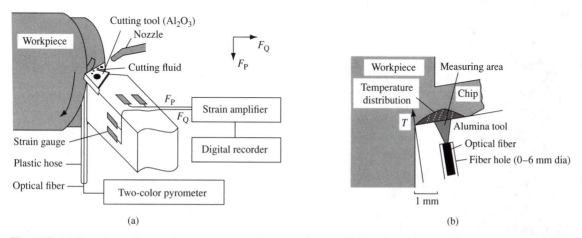

Fig. 12.7 (a) Experimental setup for measuring cutting temperatures. (b) Position of optical fiber to measure tool-face temperature. (after Al Huda et al., 2000)

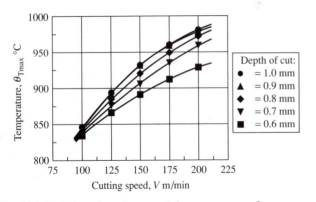

Fig. 12.8 Variation of maximum tool-face temperature θ_{Tmax} versus cutting speed for different values of depth of cut. Work material, AISI 1045 steel of hardness H_V 250; feed, 0.008 i.p.r. (0.2 mm/rev); rake angle, −5°; clearance angle, 5°; fluid, none. (after Al Huda et al., 2000)

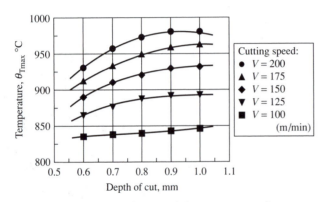

Fig. 12.9 Variation of maximum tool-face temperature θ_{Tmax} versus depth of cut for different cutting speeds. Cutting conditions same as Fig. 12.8. (after Al Huda et al., 2000)

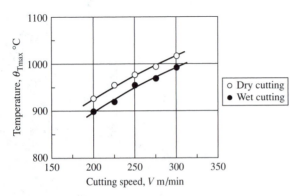

Fig. 12.10 Variation of maximum experimental tool face temperature θ_{Tmax} versus cutting speed for dry and wet cutting. The cutting conditions are the same as for Fig. 12.8 except the workpiece hardness was slightly lower (H_V 210). (after Al Huda et al., 2000)

(Ge) and the other indium antimonide (InSb). Calibration was obtained by sighting on an AISI 1045 radiating surface of known uniform temperature. Figure 12.8 shows the variation in maximum tool-face temperature (θ_{Tmax}) with cutting speed for different values of depth of cut. Figure 12.9 shows the variation of maximum tool-face temperature (θ_{Tmax}) with depth of cut for different values of cutting speed (V). Figure 12.10 shows values of maximum tool-face temperature (θ_{Tmax}) versus cutting speed when cutting dry and with a water-base cutting fluid discharge from above at 5.4 l/min. At a cutting speed of 330 f.p.m. (100 m/min) and above, there is insufficient time for cooling and the value of (θ_{Tmax}) is essentially the same for dry and wet cutting.

A finite element method (FEM) analysis was performed assuming the energy on the shear plane to be uniformly distributed and based on experimental cutting forces, shear angle (28.5°), cutting ratio ($r = 0.57$) and tool chip contact length (0.032 in = 0.8 mm). The FEM analysis followed the procedure of Tay et al. (1974, 1976), Childs et al. (1988), and Li et al. (1995) using a triangular mesh element.

Figure 12.11 shows isothermal lines (°C) for three cutting speeds when cutting dry. Figure 12.12 shows the calculated chip–tool contact temperature along the tool face for several cutting speeds. The maximum tool-face temperature at all speeds is seen to occur at about 75% of the contact length of 0.032 in (0.8 mm) from the tool tip. Figure 12.13 shows a comparison of measured and FEM calculated results for dry and wet cutting for the cutting conditions of Figure 12.8.

The FEM results were found to be in excellent agreement with similar results by Tay et al. (1974, 1976), Childs et al. (1988), and Li et al. (1995).

Wright and Trent (1973) have mapped the softening of a HSS tool when operated at high speed for a short time (30 s). When an M-34 HSS tool (0.87 $^w/_o$ C, 3.75 $^w/_o$ Cr, 9.5 $^w/_o$ Mo, 1.65 $^w/_o$ W, 1.15 $^w/_o$ V, 8.25 $^w/_o$ Co) is operated above the maximum

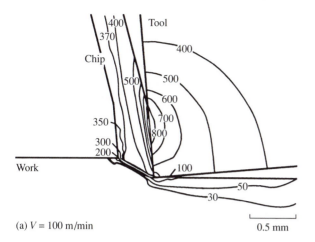

(a) $V = 100$ m/min   0.5 mm

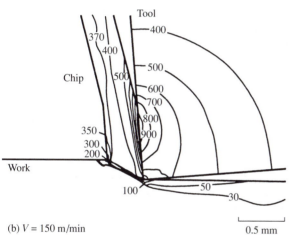

(b) $V = 150$ m/min   0.5 mm

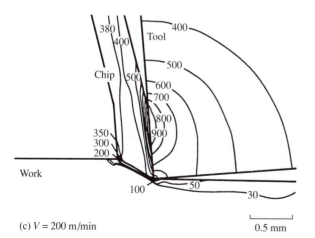

(c) $V = 200$ m/min   0.5 mm

Fig. 12.11 FEM calculated isothermal lines (°C) for three cutting speeds (V) when cutting dry under the conditions of Fig. 12.8 (after Al Huda et al., 2000)

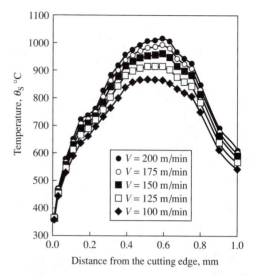

Fig. 12.12 Temperature variation along the shear plane (θ_S) for different cutting speeds when machining dry under conditions otherwise those of Fig. 12.8. (after Al Huda et al., 2000)

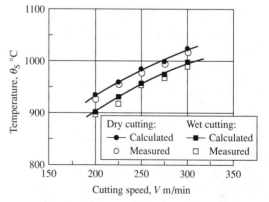

Fig. 12.13 Comparison of experimental and FEM calculated maximum shear plane temperatures from FEM analysis for dry and wet cutting for cutting conditions of Fig. 12.8. (after Al Huda et al., 2000)

tempering temperature (550 °C), a change in structure and hardness results depending upon the time at temperature. Structure is revealed by sectioning, polishing, and etching (2 $^w/_o$ nital for 30 s) a quick-stop chip root with tool attached. The section is taken at the midpoint of the depth of cut. If the normal stress on the tool face is not too high (as when machining pure iron) and the tool is operated above the maximum temperature for a time long enough for temperature equilibrium to be established but not long enough for plastic flow and wear to change tool geometry, useful changes in structure and hardness may be observed in the tool in the vicinity of the cutting edge.

Interpretation of change in structure or microhardness in terms of temperature is accomplished with the aid of calibration samples. These are thin (1.5 mm) platelets of the tool material that are plunged into a salt bath and held at a known temperature for 30 s (same as cutting time) and then water quenched. It is assumed that when the change in structure and hardness in calibration specimen and used tool are the same, the times at temperatures correspond. Three assumptions are involved in this procedure:

1. Equilibrium is reached in 30 s.

2. The presence of a steep temperature gradient in the tool and the essential absence of one in the calibration specimen is not important.

3. Iron cut at an excessive speed for HSS for a very short time will yield a temperature pattern that is characteristic of other more practical tool–material combinations and cutting speeds.

While assumption 1 is perfectably reasonable, there are questions concerning 2 and 3. Nevertheless the method appears to work. Figure 12.14 is an example for iron (0.04 $^w/_o$ C steel) having a Vickers hardness of 83 machined at 600 f.p.m. (183 m min^{-1}).

This temperature pattern is in good agreement with those of Fig. 12.5 and 12.6b with regard to the magnitude and location of the peak temperature. Wright claims an accuracy of ±25 °C for the method.

Wright (1978) has extended the method for use with less refractory metals (aluminum and copper) by substituting tools of less refractory die steel (01) for HSS. This enables temperatures as low as 150 °C to be mapped. The most important application of the technique is to explain the unique cutting performance of nickel (Smart and Trent, 1975) as discussed in Chapter 15 and the behavior of cutting fluids (Smart and Trent, 1974) as discussed in Chapter 13.

Thermosensitive paints have been used by Schallbrock and Lang (1943), Pahlitzsch and Helmerdig (1943), Bickel and Widmer (1951), and others. This technique is limited to accessible surfaces under steady state conditions. This method is of limited value to estimate relative tool temperatures and is not capable of giving accurate temperatures at actual wear surfaces.

A somewhat related method of assessing cutting temperatures is to note the temper color of a chip. The temper color of a chip is due to optical interference that depends on the thickness of oxide produced on the surfaces of the chip. This thickness depends on time at temperature and oxygen concentration as well as temperature which makes interpretation difficult. For example, a silver white chip produced with a flood of cutting fluid may actually have involved a higher cutting temperature than a blue chip produced in air. The fluid would exclude oxygen from the surface of the chip in this case and hence prevent the temper color from developing.

There are no simple reliable methods of measuring the temperature field in the chip–tool and workpiece for even steady state orthogonal

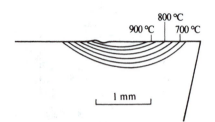

Fig. 12.14 Isotherms for machining of iron (0.04 $^w/_o$ C) at 600 f.p.m. (183 m min^{-1}), feed of 0.010 i.p.r. (0.25 mm/rev), depth of cut of 0.050 in (1.25 mm) and cutting time of 30 s. (after Wright and Trent, 1973)

metal cutting. Therefore, analytical approaches, even though often more complex than we would like, must be relied on to obtain a broad insight into the all important thermal aspects of metal cutting.

ANALYTICAL APPROACH

There are two major sources of thermal energy in metal cutting with a sharp tool—in the shear zone (1) and along the tool face (2) as shown in Fig. 12.15a. As a first approximation, the following are assumed:

1. All of the energy expended at (1) and (2) is converted to thermal energy.

2. The energy at (1) and (2) is concentrated on a plane surface (Fig. 12.15b).

3. The energy at (1) and (2) is uniformly distributed (i.e., the heat flux along (1) and (2) is constant).

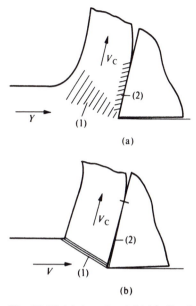

Fig. 12.15 (a) Actual and (b) idealized heat sources in orthogonal metal cutting.

Even with these assumptions, the problem of estimating the mean temperatures on the shear plane ($\bar{\theta}_S$) and tool face ($\bar{\theta}_T$) is complex. This is because part of the energy at (1) will be convected away by the chip and part will flow into the work. Also, part of the energy at (2) will usually go to the chip and part to the tool. There are thus two partition coefficients to be evaluated. R_1 is the fraction of energy (1) going to the chip; R_2 is the fraction of energy at (2) going to the chip. A fourth assumption is that none of the thermal energy goes to the environment during the chip forming process. This means that the energy per unit volume going to the chip on the shear plane (1) is $u_{C1} = R_1 u_S$ while the energy per unit volume going to the chip at (2) is $u_{C2} = R_2 u_F$ where u_S and u_F are the specific energies involved in shear and friction, respectively.

The method of estimating the partition coefficients is due to Blok (1938). This method evaluates the mean interface temperature from two points of view:

1. assuming all of the heat to flow into a stationary member

2. assuming all of the heat to flow into the extensive member [This latter calculation involves the moving heat source solution of Jaeger (1942) for an insulated slider moving across a semi-infinite solid.]

When the mean temperature from calculation (1) using a heat flux of $(1 - R)q$ (q = total thermal units per unit area per unit time) is equated to that of calculation (2) using a heat flux of Rq, an equation with R as the only unknown is obtained which may be solved for R.

FRICTION SLIDER

Before treating shear-plane and tool-face temperatures, the aforementioned method of obtaining the heat flow partition coefficient (R) will be illustrated by a simpler application—that of a friction slider corresponding to a moving rectangular heat source of dimensions $2l \times 2m$ of uniform strength $q[FL/L^2T = FL^{-1}T^{-1}]$ moving over a semi-infinite body insulated everywhere except across the interface (Fig. 12.16a with $V = 0$).

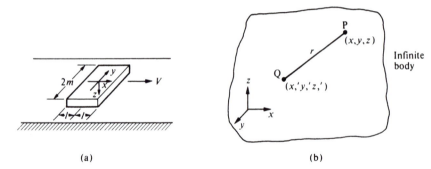

Fig. 12.16 (a) Insulated slider on conducting surface. (b) Infinite body with sudden energy release at Q at time zero.

The partial differential equation that relates the temperature and energy input is

$$\frac{k}{\rho c}\left(\frac{\partial^2\theta}{\partial x^2}+\frac{\partial^2\theta}{\partial y^2}+\frac{\partial^2\theta}{\partial z^2}\right)+\frac{q}{\rho c}=\frac{d\theta}{dt} \tag{12.1}$$

When a quantity of heat Q is liberated instantaneously at a point x', y', z' in an infinite body (Fig. 12.16b), the temperature at any point $P(x, y, z)$ in the body after a time t will be (Carslaw, 1921).

$$\theta(x, y, z, t) = \left[\frac{QK}{8k(\pi Kt)^{3/2}}\right]e^{-r^2/4Kt} \tag{12.2}$$

where

$$r^2 = (x - x')^2 + (y - y')^2 + (z - z')^2 \tag{12.3}$$

For the case of a continuous heat source extending over a finite area, both time and area integration are required. If the time integration is performed first, for the semi-infinite body, and if we concentrate our interest on the steady state solution (i.e., let $t \to \infty$), then the equation for the steady state temperature rise anywhere in the semi-infinite body with a uniform heat source, of strength $q\,[FL^{-1}T^{-1}]$ extending over $-l < x' < l$ and $-m < y' < m$ is given by

$$\Delta\theta = \frac{q}{2\pi k}\int_{-l}^{l}dx'\int_{-m}^{m}\frac{dy'}{[(x-x')^2+(y-y')^2+z^2]^{1/2}} \tag{12.4}$$

The main interest lies in the temperature rise in the plane containing the source. Then, upon integrating Eq. (12.4) and letting $z = 0$, we have

$$\begin{aligned}
\Delta\theta = \frac{q}{2\pi k}\Bigg[&|x+l|\left\{\sinh^{-1}\left(\frac{y+m}{x+l}\right)-\sinh^{-1}\left(\frac{y-m}{x+l}\right)\right\} \\
&+|x-l|\left\{\sinh^{-1}\left(\frac{y-m}{x-l}\right)-\sinh^{-1}\left(\frac{y+m}{x-l}\right)\right\} \\
&+|y+m|\left\{\sinh^{-1}\left(\frac{x+l}{y+m}\right)-\sinh^{-1}\left(\frac{x-l}{y+m}\right)\right\} \\
&+|y-m|\left\{\sinh^{-1}\left(\frac{x-l}{y-m}\right)-\sinh^{-1}\left(\frac{x+l}{y-m}\right)\right\}\Bigg]
\end{aligned} \tag{12.5}$$

The mean surface-temperature rise over the area of the source may be obtained by integrating Eq. (12.5) over the area of the source and dividing the result by the area, thus

$$\Delta\bar{\theta} = \frac{\int_{-l}^{l}\int_{-m}^{m}(\Delta\theta)\,dx\,dy}{4lm} \qquad (12.6)$$

$$\Delta\bar{\theta} = \frac{ql}{k}\frac{2}{\pi}\left[\sinh^{-1}\left(\frac{m}{l}\right) + \left(\frac{m}{l}\right)\sinh^{-1}\left(\frac{l}{m}\right) - \frac{1}{3}\left(\frac{m}{l}\right)^2\right.$$

$$\left. + \frac{1}{3}\left(\frac{l}{m}\right) - \frac{1}{3}\left\{\left(\frac{l}{m}\right) + \left(\frac{m}{l}\right)\right\}\left\{1 + \left(\frac{m}{l}\right)^2\right\}^{1/2}\right] \qquad (12.7)$$

This equation may be written

$$\Delta\bar{\theta} = \frac{ql}{k}\bar{A} \qquad (12.8)$$

where $\bar{A}$ is the area factor (a function of the aspect ratio of the surface area m/l only), plotted in Fig. 12.17. The function A_m, also shown in Fig. 12.17, has a similar significance for the maximum temperature rise in the surface ($\Delta\theta_m$) and is defined as

$$\Delta\theta_m = \frac{ql}{k}A_m \qquad (12.9)$$

For values of (m/l) greater than 20, the quantities $\bar{A}$ and A_m may be found to a good approximation from the equations

$$\bar{A} = \frac{2}{\pi}\left(\ln\frac{2m}{l} + \frac{1}{3}\frac{l}{m} + \frac{1}{2}\right) \qquad (12.10)$$

$$A_m = \frac{2}{\pi}\left(\ln\frac{2m}{l} + 1\right) \qquad (12.11)$$

The nature of the temperature distribution over the surface as well as below it is shown in Fig. 12.18 for a stationary heat source.

Jaeger (1942) has presented a more exact solution to that given in Chapter 10 for the problem where a perfect insulator slides over the surface of a semi-infinite body with velocity V (Fig. 12.16a) and heat flux q at the interface. For $m/l > 2$, the influence of the changes in m will be negligible over the velocity range of interest here and results will involve a single dimensionless velocity parameter L defined as follows:

$$L = \frac{Vl}{2K} \qquad (12.12)$$

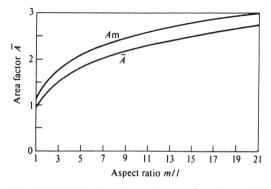

Fig. 12.17 Variation of area factors $\bar{A}$ and A_m with aspect ratio m/l.

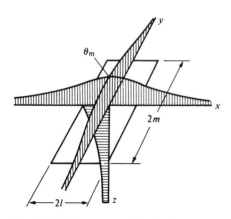

Fig. 12.18 Temperature distribution across and below a stationary heat source on a semi-infinite body.

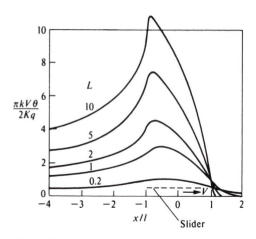

Fig. 12.19 Temperature distribution curves for a nonconducting friction slider of infinite width sliding on a plane conducting surface.

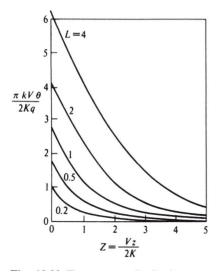

Fig. 12.20 Temperature distribution perpendicular to the plane of sliding at the trailing edge of slider.

Provided $L > 5.0$ the mean $(\Delta\bar{\theta})$ and maximum $(\Delta\theta_m)$ surface temperature rises will be

$$\Delta\bar{\theta} = 0.754\frac{ql}{k\sqrt{L}} \tag{12.13}$$

$$\Delta\theta_m = 1.130\frac{ql}{k\sqrt{L}} \tag{12.14}$$

where $q[FL^{-1}T^{-1}]$ is the frictional heat developed at the interface in the case of a slider which is assumed to be uniformly distributed. If, however, q is not uniform, neither of the above equations is seriously affected as long as the heat input remains the same.

The manner in which the temperature is distributed along the surface of the slider and below the surface at the trailing edge is shown in Figs. 12.19 and 12.20, respectively.

An actual slider will not be a perfect insulator; and part of q (Rq) will flow into the extensive member, and $(1 - R)q$ will flow into the member that is stationary relative to the heat source. The quantity R may be found by equating Eqs. (12.8) and (12.13) after making the appropriate substitutions for q in each case.

$$R = \frac{1}{1 + \dfrac{0.754(k_2/k_1)}{\bar{A}\sqrt{L}}} \tag{12.15}$$

where k_1 and k_2 are the thermal conductivities for the extensive and nonextensive members of the sliding pair respectively.

R is seen to be a function of three nondimensional quantities:

1. a geometric factor ($\bar{A}$)
2. a ratio of conductivities (k_2/k_1)
3. a velocity-diffusivity factor (L)

Once R is known, $\Delta\bar{\theta}$ may be found by substituting into either of the following equations:

$$\Delta\bar{\theta} = \frac{0.754Rql}{k_1\sqrt{L}} \tag{12.16}$$

$$\Delta\bar{\theta} = \frac{(1-R)ql}{k_2}\bar{A} \tag{12.17}$$

where $\Delta\bar{\theta}$ is the temperature rise above ambient (θ_0).

It should be noted that equating Eqs. (12.16) and (12.17) is not exact. From Figs. 12.18 and 12.19, it is evident that the temperature distributions for the stationary and moving heat source situations are quite different. The procedure of equating average temperatures is an approximation to a very complicated situation. However, experimental results are found to be in reasonably good agreement with calculated mean temperature values and hence the procedure is useful.

EXAMPLE: FRICTION SLIDER

The following example illustrates the application of the foregoing analysis to estimate the mean temperature rise for a stainless steel slider moving across a plane carbon steel infinite surface:

$$l = 0.05 \text{ in } (1.27 \text{ mm})$$

$$m = 0.25 \text{ in } (6.35 \text{ mm})$$

$$V = 20 \text{ in s}^{-1} \text{ (508 mm s}^{-1})$$

$$P = \text{nominal load} = 100 \text{ lb } (445 \text{ N})$$

$$\mu = \text{coefficient of friction} = 0.5$$

Conductivity:

$$k_1 = 2.7 \times 10^{-4} \text{ BTU/in}^2/\text{s}/(°\text{F}/\text{in})$$
$$(0.2018 \text{ J/cm}^2/\text{s}/°\text{C}/\text{cm})$$

$$k_2 = 7.1 \times 10^{-4} \text{ BTU/in}^2/\text{s}/(°\text{F}/\text{in})$$
$$(0.5307 \text{ J/cm}^2/\text{s}/°\text{C}/\text{cm})$$

Diffusivity:

$$K_1 = 0.02 \text{ in}^2 \text{ s}^{-1} \text{ (0.13 cm}^2 \text{ s}^{-1})$$

$$q = \frac{\mu PV}{J(4lm)} = \frac{(0.5)(100)(20)}{(9340)(4)(.05)(.25)} = 2.14 \text{ BTU/in}^2/\text{s} \text{ (350 J/cm}^2/\text{s)}$$

From

$$\text{Fig. 12.17: } \bar{A} = 1.80$$

$$\text{Eq. (12.12): } L = 25$$

$$\text{Eq. (12.15): } R = 0.82$$

$$\text{Eq. (12.16): } \Delta\bar{\theta} = 49 \text{ °F (27.2 °C)}$$

If the materials are interchanged (extensive member, stainless steel), the temperature rise would then be only 13 °F (7.2 °C). As sliding speed increases, R increases and for the above friction slider

$$\Delta\bar{\theta} = \frac{0.754\,\mu P}{J(4ml)} \left[\frac{V(2l)}{(k\rho C)_1} \right]^{1/2} \tag{12.18}$$

when $R = 1$.

SHEAR-PLANE TEMPERATURE

The first analytical treatment of cutting tool temperatures was that of Trigger and Chao (1951). The treatment that follows was originally presented by Loewen and Shaw (1954). Somewhat different treatments of shear-plane temperature than that presented here were presented by Hahn (1951) and Leone (1954).

The rate at which shear energy is expended along the shear plane will be

$$U_S = F_S V_S \tag{12.19}$$

where F_S is the component of force directed along the shear plane and V_S is the velocity of the chip relative to the workpiece, which also is directed along the shear plane. The rate at which energy is expended per unit area on the shear plane for orthogonal cutting conditions will be

$$U_{S'} = \frac{F_S V_S}{tb \, \csc \, \phi} \tag{12.20}$$

where t is the undeformed chip thickness, b is the width of the workpiece, and ϕ is the shear angle.

To a good approximation, it may be assumed that all of the mechanical energy associated with the shearing process is converted into thermal energy, and the heat which flows from the shear zone per unit time per unit area will be

$$q_1 = \frac{F_S V_S}{Jtb \, \csc \, \phi} = \frac{u_S V \sin \, \phi}{J} \tag{12.21}$$

where J is the mechanical equivalent of heat, u_S is the shear energy per unit volume of metal cut, and V the cutting speed.

If $R_1 q_1$ is the heat per unit time per unit area which leaves the shear zone with the chip, then $(1 - R_1)q_1$ is the heat per unit time per unit area that flows into the workpiece. The mean temperature of the metal in the chip, in the vicinity of the shear plane, will be

$$\bar{\theta}_S = \frac{R_1 q_1 (bt \, \csc \, \phi)}{C_1 \rho_1 (Vbt)} + \theta_0 = \frac{R_1 u_S}{J C_1 \rho_1} + \theta_0 \tag{12.22}$$

where θ_0 is the ambient workpiece temperature and $C_1\rho_1$ is the volume specific heat at the mean temperature between $\bar{\theta}_S$ and θ_0.

Equation 12.13 may be used to compute the shear-plane temperature rise as shown in Fig. 12.21. Here the chip may be considered to be a perfect insulator if the total heat flowing from the interface is $(1 - R_1)q_1$. The velocity of sliding is taken as V_S and not V in accordance with the generally adopted Piispanen picture of metal cutting shown in Fig. 12.22. By this picture, metal when cut behaves as a deck of cards, one card at a time sliding a finite distance across its neighbor. Heat is generated only when sliding occurs in direction V_S, and we might consider that one plate at a time is always in motion. An approximation involved in the use of Fig. 12.21 is that the metal which is assumed present in wedge A is not there, while that in wedge B is ignored. Since the heat flows associated with wedges A and B will tend to cancel each other, it is considered that the idealized picture of Fig. 12.21 should provide a good approximation.

When Eq. (12.13) is used to compute the shear-plane temperature, it is found that

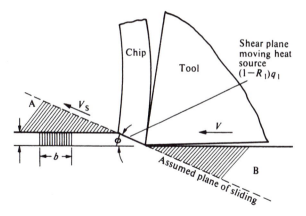

Fig. 12.21 Idealized diagram of shear plane moving heat source.

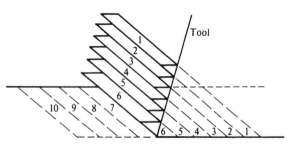

Fig. 12.22 Piispanen's mechanism of chip formation.

$$\bar{\theta}_S = 0.754 \frac{(1 - R_1)q_1 \left(\dfrac{t \csc \phi}{2} \right)}{k_2 \sqrt{L_1}} + \theta_0 \quad (12.23)$$

and

$$L_1 = \frac{V_S \left(\dfrac{t \csc \phi}{2} \right)}{2K_1} = \frac{V\gamma t}{4K_1} \quad (12.24)$$

where k_1 is the conductivity, K_1 the diffusivity of the workpiece material at temperature $\bar{\theta}_S$, and γ is the strain in the chip. Then, equating Eqs. (12.22) and (12.23), and solving for R_1, it is found that

$$R_1 = \frac{1}{1 + 1.328 \left[\dfrac{K_1\gamma}{Vt} \right]^{1/2}} \quad (12.25)$$

Once R_1 is known, $\bar{\theta}_S$ may be calculated from Eq. (12.22). Figure 12.23 is convenient for us in determining R_1. It is evident that the percentage of energy going to the chip (R_1) does not increase with increased cutting speed (V) alone, but rather with the nondimensional quantity (Vt)/($K_1\gamma$). From Eq. (12.22) it may be seen that the temperature rise at the shear plane varies directly with the shear energy per unit volume going into the chip (R_1u_S), and inversely with the volume specific heat of the workpiece ($C_1\rho_1$).

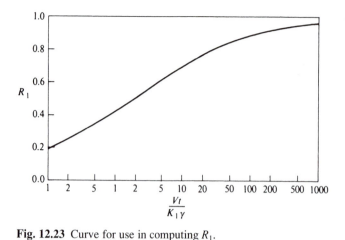

Fig. 12.23 Curve for use in computing R_1.

TOOL-FACE TEMPERATURE

From an analytical viewpoint, the friction between chip and tool can be regarded as a heat source that is moving in relation to the tool. As before, it is a double treatment that allows calculation of the partition of friction energy between chip and tool, and hence the resulting temperature.

The total friction energy $q_2[FL^{-1}T^{-1}]$ that will be dissipated at the chip–tool interface per unit time per unit area will be

$$q_2 = \frac{F_C V_C}{Jab} = \frac{u_F Vt}{Ja} \tag{12.26}$$

where F_C is the friction force along the tool face, V_C is the velocity of the chip relative to the tool, a is the length of contact between chip and tool in the direction of motion, b is the width of the chip, and u_F is the energy per unit volume of metal cut that is consumed by friction on the tool face.

If R_2 is the fraction of q_2 that flows into the chip, then the average temperature rise in the surface of the chip due to friction $\Delta\bar{\theta}_F$ is from Eq. (12.13):

$$\Delta\bar{\theta}_F = \frac{0.754(R_2 q_2)a/2}{k_2\sqrt{L_2}} \tag{12.27}$$

where k_2 is the thermal conductivity of the chip at its final temperature, the parameter L_2 is defined as

$$L_2 = \frac{V_C a/2}{2K_2} \tag{12.28}$$

and K_2 is the thermal diffusivity of the chip at its final temperature.

The mean temperature of the chip surface along the tool face ($\bar{\theta}_T$) will be the sum of the mean shear-plane temperature ($\bar{\theta}_S$) and the mean temperature rise due to friction ($\Delta\bar{\theta}_F$), and

$$\bar{\theta}_T = \bar{\theta}_S + \Delta\bar{\theta}_F = \bar{\theta}_S + \frac{0.377(R_2 q_2)a}{k_2\sqrt{L_2}} \tag{12.29}$$

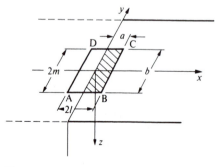

Fig. 12.24 Relation between two-dimensional tool and semi-infinite body.

To find R_2, the stationary heat-source solution must be applied to the tool. The shaded area in Fig. 12.24 represents the area of contact between a chip and a two-dimensional cutting tool, the cutting edge being along the y-axis and the tool face in the xy-plane. The tool represented by the solid lines may be considered a quarter-infinite body relative to the shaded area of contact. This is seen to represent a good approximation of an orthogonal cutting tool, provided the rake and clearance angles are not too large. By symmetry it is evident that the temperature at any point in the surface of the quarter-infinite body subjected to the uniform heat source represented by the shaded area would be the same as the temperature of the corresponding point in the semi-infinite body when subjected to a uniform heat source extending over A–B–C–D, provided that the yz-plane is a perfect insulator, just as the xy-plane, which is a valid assumption. Thus, the aspect ratio to be used in obtaining the mean tool-face temperature rise is

$$\frac{m}{l} = \frac{b}{2a} \qquad \text{(orthogonal tool)} \qquad (12.30)$$

By similar reasoning the aspect ratio for a lathe tool which cuts at the corner of the tool would be

$$\frac{m}{l} = \frac{b}{a} \qquad \text{(lathe tool)} \qquad (12.31)$$

With the aspect ratio thus defined for any tool, the shape factor $\bar{A}$ may be determined from Fig. 12.16 and the temperature $\bar{\theta}_T$ solved for as a point in the tool from Eq. (12.8)

$$\bar{\theta}_T = \frac{(1 - R_2)q_2 a}{k_3} \bar{A} + \theta'_0 \qquad (12.32)$$

where k_3 is the thermal conductivity of the tool material at the final tool temperature $\bar{\theta}_T$, and θ'_0 is the ambient temperature of the tool.

By equating the two values of $\bar{\theta}_T$, given in Eqs. (12.29) and (12.32), it is possible to solve for R_2:

$$R_2 = \frac{q_2(a\bar{A}/k_3) - \bar{\theta}_S + \theta'_0}{q_2(a\bar{A}/k_3) + q_2(0.377a/k_2\sqrt{L_2})} \qquad (12.33)$$

or, substituting from Eqs. (12.26) and (12.28), and noting that $V_C = rV$, where r is the chip–thickness ratio (i.e., the ratio of the depth of cut t to chip thickness when taking an orthogonal cut), we have

$$R_2 = \frac{(u_F V t \bar{A}/J k_3) - \bar{\theta}_S + \theta'_0}{\{(u_F V t \bar{A}/J k_3) + (0.754 u_F/J \rho_2 C_2)\}\{V t^2/a r K_2\}^{1/2}} \qquad (12.34)$$

When R_2 is determined, $\bar{\theta}_T$ may be found readily from either Eq. (12.29) or Eq. (12.32).

As usual, it has been necessary in the foregoing derivations to consider all thermal quantities to be constants. Actually the significant thermal quantities are functions of temperature, so that the question arises about just what sort of mean values should be substituted when evaluating the final equations. In the foregoing equations the following quantities occur: ρ_1, C_1, K_1, ρ_2, C_2, K_2, k_2, and k_3. The thermal quantities ρ_1, C_1, and K_1 refer to the shear plane and, to obtain a good approximation, should be evaluated at a temperature halfway between the shear-plane temperature $\bar{\theta}_S$ and the ambient temperature θ_0. The thermal quantities having the subscript 2 refer to the workpiece material in the vicinity of the tool face. Inasmuch as the heat flow in the chip by conduction and convection will be controlled predominantly by the properties very near the tool face, because of the large temperature gradient that exists in the surface, it would appear that the best temperature to use in connection with ρ_2, C_2, K_2, and k_2 would be the mean temperature at the surface ($\bar{\theta}_T$). By similar reasoning, it appears that the value of the thermal conductivity of the tool k_3 also should be taken as that corresponding to temperature $\bar{\theta}_T$. In view of the uncertainties involved, it is fortunate that the numerical solutions obtained are not sensitive to the particular mean temperatures that are used. This is confirmed in the calculations that follow.

EXAMPLE: SHEAR-PLANE AND TOOL-FACE TEMPERATURES

To show how this analysis can be applied, a sample calculation will be carried through and the result compared with the temperature measured by the tool–work thermocouple method.

Experimentally measured quantities:

- work material: AISI 1113 steel
- tool material: K2S carbide, 20° rake, 5° clearance angle
- type of cut: orthogonal
- cutting speed: $V = 445$ f.p.m. = 139 m min^{-1}
- undeformed chip thickness: $t = 0.0023$ in = 0.06 mm
- width of cut: $b = 0.151$ in = 3.84 mm
- chip contact length[†]: $a = 0.009$ in = 0.23 mm
- cutting force: $F_P = 80$ lb = 356 N
- feed Force: $F_Q = 28$ lb = 125 N
- chip–thickness ratio: $r = 0.51$
- thermal properties of steel: given in Fig. 12.25

Calculated values:

- average shear strain in chip: $\gamma = 1.91$
- shear energy per unit volume: $u_S = 153,000$ in lb in$^{-3} = 1056$ N mm^{-2}
- shear angle: $\phi = 30°$
- friction energy per unit volume: $u_F = 78,000$ in lb in$^{-3} = 544$ N mm^{-2}

Outline of solution:

(a) Estimate the value of $\bar{\theta}_S$, e.g., 500 °F (260 °C) and determine corresponding values of K_1 and $\rho_1 C_1$ at $(500 + 75)/2 = 288$ °F (142 °C). From Fig. 12.25, $K_1 = 0.023$ in^2 s^{-1} (14.84 mm^2 s^{-1}) and $\rho_1 C_1 = 0.034$ BTU in^{-3} °F^{-1} (3.94 J cm^{-3} °C^{-1}).

(b) Calculate quantity $Vt/K_1\gamma = 4.65$.

(c) Obtain R_1 from Fig. 12.23: $R_1 = 0.62$.

(d) Calculate $(\bar{\theta}_S - \theta_0)$ from Eq. (12.22), $\bar{\theta}_S - \theta_0 = 299$ °F (156 °C) and if $\theta_0 = 75$ °F (24 °C) then $\bar{\theta}_S = 374$ °F (190 °C).

(e) Repeat steps (a) to (d) if original estimate of $\bar{\theta}_S$ is in error by more than 25 °F (14 °C). In this case K_1 remains at 0.023 in^2 s^{-1} (14.84 mm^2 s^{-1}) and $\rho_1 C_1$ is 0.033 BTU in^{-3} °F^{-1} (3.82 J cm^{-3} °C^{-1}). The $\bar{\theta}_S$ changes to 383 °F (195 °C).

(f) Estimate value of $\bar{\theta}_T$, e.g., 800 °F (444 °C) and determine K_2 and $\rho_2 C_2$ from Fig. 12.25 at this temperature: $K_2 = 0.016$ in^2 s^{-1} (0.103 cm^2 s^{-1}) and $\rho_2 C_2 = 0.042$ BTU in^{-3} °F^{-1} (4.87 J cm^{-3} °C^{-1}).

(g) Calculate aspect ratio $m/l = b/2a = 8.4$ and determine area factor $\bar{A}$ from Fig. 12.17: $\bar{A} = 2.15$.

(h) Calculate quantity $(u_F VtA)(Jk_3) = C'$, where k_3 is conductivity of carbide tool, taken as $k_3 = 7.63 \times 10^{-4}$ BTU s^{-1} in^{-2} °F^{-1} in (0.570 J s^{-1} cm^{-2} °C^{-1}): $C' = 4860$ °F (2700 °C).

[†] Obtained by measuring a thin layer of metal deposited on tool face with a microscope.

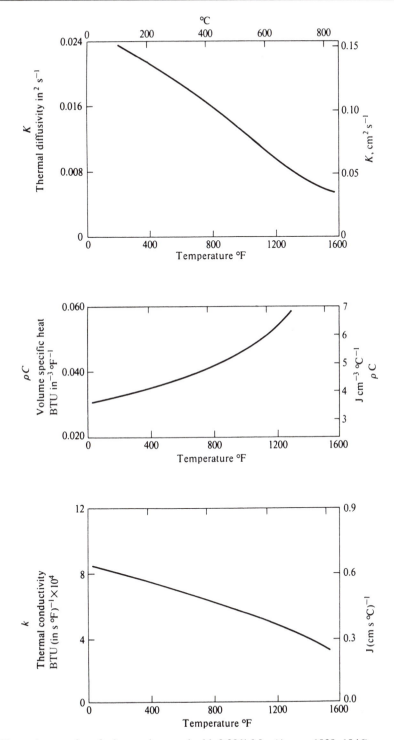

Fig. 12.25 Thermal properties of a low-carbon steel with 0.39% Mn. (Anon., 1939, 1946)

(i) Calculate quantity

$$\frac{0.754\,u_F}{J\rho_2 C_2}\left(\frac{Vt}{arK_2/t}\right)^{1/2} = B' = 383 \text{ °F (213 °C)}$$

(j) Determine R_2 from Eq. (12.34)

$$R_2 = \frac{C' - \theta_S + \theta_0'}{C' + B'} = 0.87$$

If θ_0' is taken as 75 °F (24 °C)

(k) From Eq. (12.27), $\Delta\bar{\theta}_F = R_2 B'$, hence $\Delta\bar{\theta}_F = 333$ °F (185 °C).

(l) From Eq. (12.29), $\bar{\theta}_T = \bar{\theta}_S + \Delta\bar{\theta}_F = 716$ °F (380 °C).

(m) Compare this value with value assumed in (f) and repeat steps (f) to (l) until agreement is observed between the assumed and calculated values of $\bar{\theta}_T$. When this is done in this case, $\theta_T = 723$ °F (384 °C). The experimentally determined value of $\bar{\theta}_T$ for the foregoing conditions was found to be 735 °F (391 °C). This represents a far better agreement than is significant in view of experimental accuracy, approximate data on thermal properties, the many assumptions made in the analysis, and the question of what mean value is measured by the chip–tool–thermocouple method. In the foregoing example it is evident that the calculated values of both $\bar{\theta}_S$ and $\bar{\theta}_T$ are not very sensitive to the temperature variation of the thermal properties and hence this variation does not need to be known with great accuracy, but should be taken into account.

Cutting Speed

The manner in which the shear and friction components as well as the total tool temperature vary with cutting speed in a representative case is shown in Fig. 12.26, for conditions identical with those of the sample calculations. The calculated temperature curve is seen to follow the measured values closely. The corresponding variation of factors R_1 and R_2 is shown in Fig. 12.27. At very low speeds the friction component ($\Delta\bar{\theta}_F$) may become negative. This merely means that not only all the frictional energy, but also some of the shear energy is flowing into the tool. This apparently happens at normal turning speeds for nickel as

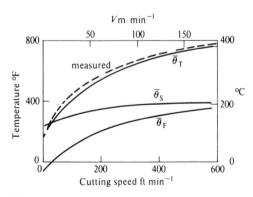

Fig. **12.26** Comparison between observed and computed cutting temperatures for AISI B1113 steel when cut with a K2S cemented carbide tool having 20° rake angle, $t = 0.0023$ in (0.06 mm), $b = 0.151$ in (3.84 mm).

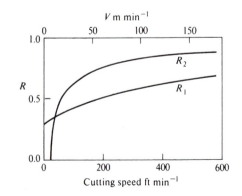

Fig. **12.27** Variation of R_1 and R_2 for data of Fig. 12.26.

discussed in Chapter 15. Both the observed and calculated resultant shear-plane temperatures $\bar{\theta}_T$ are seen to vary approximately parabolically with cutting speed (i.e., as $V^{1/2}$).

CUTTING GEOMETRY

Experimental results of tests in which the undeformed chip thickness t and width of cut b were varied are shown in Figs. 12.28 and 12.29, respectively. The cutting temperature is very nearly independent of the width of cut b.

ENERGY BALANCE

The energies going to the chip (u_C), tool (u_T), and workpiece (u_W) per unit volume of metal removed are functions of R_1 and R_2, and the specific energies of the cutting process (u_S = shear energy per unit volume, u_F = friction energy per unit unit volume). Thus

$$u_C = R_1 u_S + R_2 u_F \qquad (12.35)$$

$$u_T = (1 - R_2)u_F \qquad (12.36)$$

$$u_W = (1 - R_1)u_S \qquad (12.37)$$

When the approximate values are substituted into these equations for the tests in Figs. 12.26 and 12.27, the distribution of energy shown in Fig. 12.30 is obtained. The percentage of the total energy going to the chip increases with increased speed although the percentages going to the tool and workpiece decrease. At very high cutting speeds, practically all of the energy is carried away in the chip, a small amount going into the workpiece and a still smaller amount to the tool.

Calorimetric measurements of the three energies would be an ideal method for checking the validity of the analysis were it not for the experimental difficulties involved.

Schmidt and Roubik (1949) have drilled metal in a calorimeter mounted in a drill press as shown in Fig. 12.31. Total energy was measured by noting the temperature rise of the water as $\frac{3}{8}$ in (9.53 mm) diameter rods of magnesium alloy with 0.110 in (2.79 mm) bore were drilled using a 0.440 in (11.18 mm)

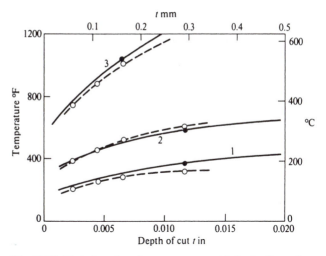

Fig. 12.28 Variation of cutting temperature with depth of cut when cutting AISI 1113 steel orthogonally with a K2S tool having 20° rake angle. Width of cut 0.151 in (3.84 mm). Cutting speeds: (1) 10 f.p.m. (3.13 m min⁻¹), (2) 60 f.p.m. (18.75 m min⁻¹), (3) 450 f.p.m. (141 m min⁻¹). Dotted curves are drawn through experimentally measured temperatures. Points marked • were determined analytically and solid curves passing through these points were obtained by allowing t to vary in Eq. (12.32).

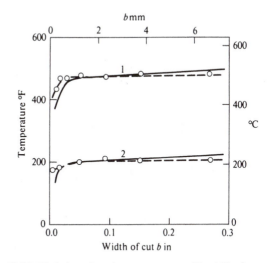

Fig. 12.29 Variation of cutting temperature with width of cut when cutting AISI 1113 steel orthogonally with K2S carbide tool having 20° rake angle. Cutting speeds: (1) 226 f.p.m. (70.6 m min⁻¹) and (2) 15 f.p.m. (4.7 m min⁻¹). Depth of cut: 0.0023 in (0.06 mm). Dotted curves are drawn through experimentally measured temperatures. Temperatures for $b = 0.151$ in (3.84 mm) were determined analytically in each case, and solid curves passing through these points were obtained by allowing only b to vary in Eq. (12.32).

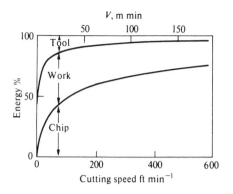

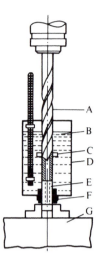

Fig. 12.30 Calculated variation of energy distribution with cutting speed for cutting conditions of Fig. 12.28.

Fig. 12.31 Calorimetric apparatus for determining energy distribution in drilling. [A = Drill. B = Distilled water. C = Agitator. D = Container. E = $\frac{3}{8}$ in (9.5 mm) specimen. F = Rubber grommet. G = Chuck. (after Schmidt and Roubik, 1949)

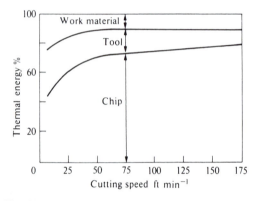

Fig. 12.32 Variation of energy distribution with drilling speed. (after Schmidt and Roubik, 1949)

drill. The energy to the chips was obtained by drilling in air, letting the chips fall into the water, again noting the temperature rise. Finally, the same small quantity of metal was again cut in air and the drill quickly plunged into water to obtain the energy going into the drill. The results obtained for a feed rate of 0.0091 in s⁻¹ (0.23 mm s⁻¹) are given in Fig. 12.32. Although absolute values differ since different materials are involved, the trends of Fig. 12.32 are similar to those of Fig. 12.30. Schmidt and Roubik (1954) have repeated the calorimetric drilling experiments when milling an aluminum alloy and have obtained similar results to those of Fig. 12.32 for energy partition versus cutting speed.

AMBIENT TEMPERATURE RISE

When a bar is turned in a lathe, a small portion of the cutting energy is conducted into the workpiece, and this energy tends to raise the temperature of the material approaching the shear plane. This increase in "ambient" temperature will depend upon the amount of heat which flows into the workpiece per unit time, the number of revolutions of cutting preceding the instant under question, the thermal properties of the workpiece, and the geometry of the system.

Detailed analysis reveals the following:

1. $\Delta\theta_1$ increases proportionately with the shear energy per unit volume (u_S), which is normally approximately proportional to the Brinell hardness of the work material and will generally increase with tool dulling.

2. $\Delta\theta_1$ varies inversely with the workpiece diameter (D) as has been found experimentally on many occasions.

3. $\Delta\theta_1$ varies linearly with the depth of cut (b).

4. $\Delta\theta_1$ varies inversely with the thermal conductivity of the workpiece material (k).

5. $\Delta\theta_1$ varies approximately as the 0.8-power of the feed (i.e., $\Delta\theta_1 \sim t^{0.8}$).

6. The influence of cutting speed is complex resulting in a relatively small change in $(\Delta\theta_1)$ with (V) which may be either positive or negative.

7. A coolant should decrease $\Delta\theta_1$ and this is observed experimentally.

While the ambient workpiece temperature rise is the lowest of those considered, it will directly influence the temperatures of importance to tool wear on the tool face and tool flank. It is frequently reported, for example, that tool life decreases when all cutting conditions are maintained the same except the workpiece diameter which is decreased.

ANALYTICAL REFINEMENTS

Since the mid 1950s there have been a number of analytical refinements in the approach to cutting-temperature analysis first introduced by Trigger and Chao (1951). However, these refinements have generally been accompanied by increased complexity.

Chao et al. (1952) and Chao and Trigger (1953) improved on the original 1951 analysis placing considerable emphasis upon a nondimensional thermal number (Vt/K, where K is the diffusivity of the work material). Chao and Trigger (1958) and Chao et al. (1961) discussed the temperature distributions along the chip–tool and chip–work interfaces. Rapier (1954), Weiner (1955), and Nakayama (1956) have also considered the temperature distribution in orthogonal cutting.

Boothroyd (1963) assumed the thermal source on the tool face to be distributed in a direction normal to the tool face and concluded that use of a planar heat source on the tool face leads to a significant overestimate of tool-face temperatures. However, based on the discussion of Zorev (1963), it appears that the depth of secondary shear zone (normal to the tool face) considered by Boothroyd is unusually large probably due to the high ambient temperature (1100 °F; 600 °C) required by Boothroyd's radiation measuring technique. Also, as Zorev pointed out, the rate of energy generation in the secondary shear zone will be far from the uniform distribution assumed by Boothroyd. Both of these discrepancies tend to magnify the difference between a planar source on the tool face and one that is distributed normal to the tool face. Arndt and Brown (1967) have further considered the temperature distribution in orthogonal machining.

Stephenson (1991a and 1991b) has compared analytical results from several steady state models for shear-plane temperature with experimental results from chip–tool thermocouple and infrared measurements.

INFRARED EXTRAPOLATION METHOD

The infrared method devised by Stephenson measures temperatures on the side surface of a tube being machined that are essentially constant in the circumferential direction but decrease exponentially in the axial direction. Figure 12.33a shows the machining arrangement used in orthogonal tube turning. Figure 12.33b shows a representative infrared video image of the side of a tube being machined at constant feed rate (f). Temperatures at different axial distances from the source were determined by use of a thermal video system, and video tapes were analyzed to yield temperatures along the tube axis that may be extrapolated back to obtain the mean temperature rise at the source (shear plane).

In the heat transfer analysis the following assumptions were made:

- Thermal radiation may be neglected.
- The work material is homogeneous and isotropic.
- Thermal properties are temperature independent.
- Temperatures are uniform across the tube thickness.
- Convection from the exposed surface may be modeled as a heat sink.

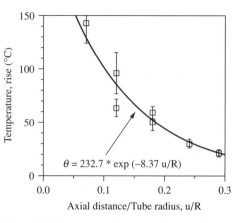

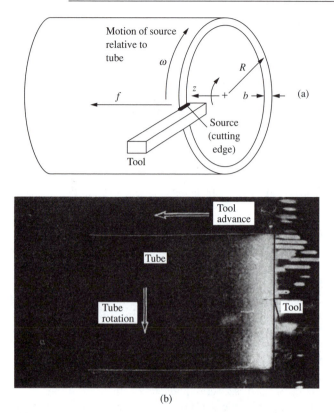

Fig. 12.34 Exponential temperature profile fit to measured temperatures for an AISI 1018 steel tube being machined with a tungsten carbide tool having a rake angle of $-5°$, a cutting speed of 610 f.p.m. (3.1 m/s) and feed of 0.0043 i.p.r. (0.11 mm/rev). (after Stephenson, 1991a)

The infrared values of remote temperatures were measured using a full field infrared thermal imaging system (Guerrieri, 1987). Tubes were painted with an acrylic lacquer having an emissivity of 0.93. Figure 12.34 shows the exponential temperature profile for a typical infrared temperature test.

This back extrapolation method of estimating shear plane temperatures is limited to cutting speeds below about 984 f.p.m. (5 m/s). At higher speeds the tool moves down the tube more rapidly than heat diffuses along the

Fig. 12.33 (a) Thin walled tube heated by helically moving heat source under orthogonal cutting conditions. (b) Infrared video image of a dry end turning test on a 202A aluminum tube using a tungsten carbide tool having a rake angle of $-5°$, a cutting speed of 3.1 m/s (610 f.p.m.) and a feed of 0.005 i.p.r. (0.13 mm/rev). (after Stephenson, 1991a)

tube. This results in too much of the material heated in one revolution being removed in the next revolution. Below the upper cutting speed limit, the method yields reasonable results for a wide range of tube diameters and for most work materials. It should also be noted that the upper speed limit will be less than the above value for high feed rates.

COMPARISON OF ANALYTICAL AND EXPERIMENTAL SHEAR-PLANE TEMPERATURES

Tubes of the following four materials were cut with tungsten carbide tools as shown in Fig. 12.33a:

- AISI 1018 steel
- 2024 aluminum
- free machining brass
- gray cast iron (3.5% C, 2.5% Si)

Tests were performed as follows to provide a fractional factorial set of experiments:

- cutting speeds: 173, 348, 606, 770 f.p.m. (0.88, 1.77, 3.08, 3.91 m/s)
- feeds for steel: 0.004, 0.006, 0.0087 i.p.r. (0.11, 0.15, 0.2 mm/rev)
- feeds for aluminum: 0.005, 0.008, 0.013 i.p.r. (0.13, 0.2, 0.3 mm/rev)
- rake angles: 0 and −5°

Figure 12.35 shows plots of 3 types of calculated mean shear-plane temperatures plotted against experimental values from extrapolated infrared imaging tests for each of the four materials tested. The analytical models used were as follows:

A: Loewen and Shaw (1954)   ●
B: Boothroyd (1975)   □
C: Wright et al. (1980)   △

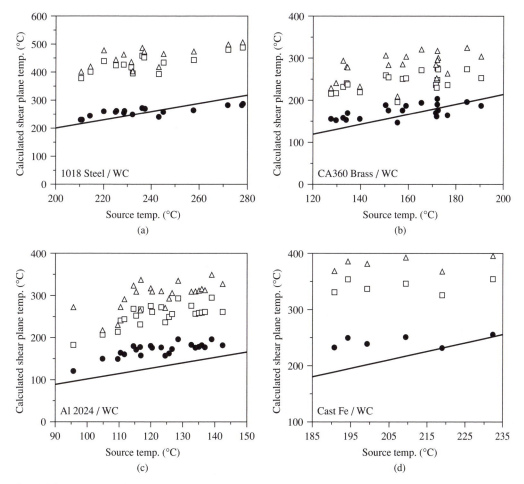

Fig. 12.35 Comparison of calculated average shear-plane temperatures with experimental source (shear plane) temperatures on tungsten carbide tools based on infrared imaging for (a) AISI 1018 steel, (b) CA 360 brass, (c) Al 2024, and (d) cast iron. The solid line represents equal values for calculated and measured results. (after Stephenson, 1991a)

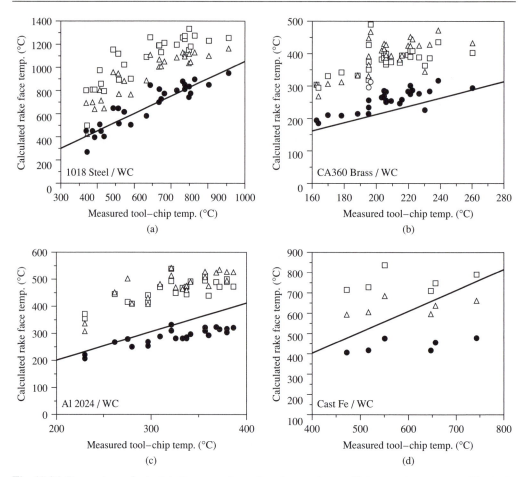

Fig. 12.36 Comparison of calculated average shear-plane temperatures with temperatures measured by tool–chip thermocouple method (Fig. 12.1) for (a) AISI 1018 steel, (b) CA 360 brass, (c) Al 2024, and (d) cast iron. The solid line represents equal values for calculated and measured results. (after Stephenson, 1991b)

Figure 12.36 shows plots of values from the above three analytical models plotted against experimental tool–chip thermocouple values. (Figure 12.1 shows the tool–chip thermocouple test employed.)

All of the models give reasonable qualitative results, but only model A gives acceptable quantitative results. This is probably due to the fact that only A takes changes in thermal property values with cutting conditions into account. The other models use thermal properties that are the same for all cutting conditions. None of the analytical models give results for discontinuous chip formation (i.e., for the cast-iron tests) that are satisfactory.

The infrared back extrapolation experimental method gives results that are reasonably close to those of the tool–chip thermocouple method both of which are in reasonable agreement with analytical method A. The most important result of this study is that it enhances the legitimacy of the widely used chip–tool thermocouple method of estimating the mean tool-face temperature in metal cutting.

Model A has been extended to three-dimensional cutting conditions by Venuvinod and Lau (1986). However, this extended model gives practically the same results as model A for the two-dimensional cutting situation involved here.

FINITE ELEMENT ANALYSIS

Tay et al. (1974) have applied the finite element method (FEM) to obtain temperature fields for three specific cases. While this technique is a very powerful one enabling variable material properties to be included in the analysis, the results will be only as good as the properties fed in. For example, it was assumed throughout the Tay et al. analysis that the flow stress of the material corresponds to Eq. (5.8), $\sigma = \sigma_0 \varepsilon^n$, and that uniaxial stress and strain are related to shear stress and shear strain in cutting through the concept of equivalent stress and equivalent strain. As discussed in Chapter 9 these are very questionable assumptions for the large plastic strains involved in cutting. Nevertheless, this analysis is very comprehensive and provides a valuable insight into the thermal fields associated with metal cutting.

However, the theory presented is not predictive since it requires inputs that cannot be arrived at from material properties but which must be obtained from metal cutting tests for the specific combination of tool and work materials, tool geometry, and cutting conditions pertaining. For example, the extent of secondary shear was estimated from the distortion of lines on a gridded specimen, and the magnitude of the internal secondary shear stress was estimated from measured values of tool-face friction and chip–tool contact length.

The secondary shear zone that occurs on the tool face plays a very important role relative to the magnitude and distribution of thermal energy and hence with regard to tool-face temperatures as Boothroyd (1963) has indicated. This shear zone may be of three types:

1. No subsurface shear flow, in which case there is sliding and friction along the entire chip tool contact plane.

2. Subsurface plastic flow, in which case the surface velocity of the chip will be less than V_C at the cutting edge but will increase as the chip moves up the tool reaching a value of V_C at about half the contact length.

3. Subsurface fluid flow in situations where a molten layer actually forms on the tool face.

An important example of the latter type will be discussed in Chapter 15 in connection with the performance of specially deoxidized steels.

Tay et al. (1974) studied the thermal fields produced when a free-machining (0.13 $^w/_o$ C, 1.45 $^w/_o$ Mn, 0.25 $^w/_o$ S and 0.19 $^w/_o$ P) steel was machined with a brazed tungsten carbide tool having a rake angle of 20° and a relief angle of 6°. Orthogonal cuts of 9.50 mm width (b) were made at a feed rate (t) of 0.274 mm/rev. Three cutting speeds (V) were employed: 29.6, 78.0, and 155.4 m min^{-1}. The specimens were provided with a grid that could be analyzed in the chip root after suddenly interrupting the cut using an explosive device. This enabled strain rates in the primary and secondary shear zones to be determined from which the rate of energy dissipation and hence the heat flux could be found.

Interpretation of the deformed grid pattern on the shear plane was relatively easy but much more difficult for the secondary shear zone. This was due to the thinness of the secondary shear zone (~ 0.10 mm) and the fact that the grid on the free chip surface had been sheared three times in different directions (first in plane stress, then in plane strain in the primary shear zone, and finally in the secondary shear zone). When cutting at the lowest speed (29.6 m min^{-1}), there was no subsurface flow in the chip, but there was at the two higher speeds.

Figure 12.37 shows deformed grid lines in the chip root obtained at the highest cutting speed (155.4 m min^{-1}). The velocity in the main body of the chip ($V_C = 68.9$ m min^{-1}) is seen to be uniform except in the secondary shear zone of width δ. The velocity at the cutting edge in the direction of V_C was assumed to be $V_C/3$ which gradually accelerated to V_C as the chip moved

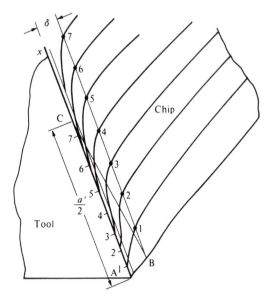

Fig. 12.37 Secondary shear zone for dry orthogonal cutting of free machining steel with a carbide tool (α = 20°) at 497 f.p.m. (155.5 m min^{-1}) and a feed of 0.01 i.p.r. (0.274 mm/rev). (after Tay et al., 1974)

over about half of the chip–tool contact length (a = chip tool contact length = 0.91 mm, and δ = width of the secondary shear zone = 0.10 mm). The region over which secondary internal shear occurs is triangle ABC. As the chip passes up the tool face, the extent of the internal shear decreases to zero as the velocity of the surface of the chip accelerates. Beyond point C the velocity of the surface of the chip equals V_C and the heat source then lies in the tool face.

This interpretation follows that of Nakayama (1959) and is in agreement with the stress distribution on the tool face discussed in Chapter 10 where the shear stress was found to be constant over the first half of the chip tool contact and to fall to zero over the second half of the contact area. The value of the internal shear stress (τ_i) was obtained by Tay et al. as follows by assuming τ_i to decrease linearly over the second half of the contact area:

$$F_C = \tau_i(a/2b) + \tau_i/2(a/2b) \qquad (12.38)$$

or

$$\tau_i = 4F_C/3ab \qquad (12.39)$$

where F_C is the measured amount of force along the tool face, b is the width of cut, and a is the total chip–tool contact length obtained by noting the extent of the scar on a polished tool face.

It should be noted that the foregoing interpretation of secondary flow of Tay et al. is not consistent with the observation of Zorev (1963) that grinding scratches parallel to the cutting edge were not removed over the first half of the contact zone but were replaced by scratches in the cutting direction over the second half of contact. It is also not in agreement with the experimentally based views of Trent (1976). The foregoing interpretation of flow lines in a chip root assumes there is some sliding of the chip over the tool face all the way to the cutting edge (merely lower sliding velocities over the first half of the contact area). Zorev's observation would require a thin layer of completely stationary chip material between the main body of the chip and the tool face. This thin layer of metal would be strongly bonded to the tool face, would remain on the tool in a quick-stop test, and would not be evident if only the chip and not the tool were subsequently examined at high power. Such built-up layers (BUL) have been observed on the face of a carbide tool after cutting (Schaller, 1964). If a stationary BUL is present between the main body of the chip and the tool face, this could appreciably alter calculated temperature distributions that do not take the BUL into account.

Figure 12.38 shows isothermal lines for V = 155.4 m min^{-1} and t = 0.274 mm/rev due to Tay et al. The maximum temperature was found to occur just beyond the midpoint of the contact length (at 0.68a from the cutting edge) but was closer to the midpoint at lower cutting speeds. The following measured values were used in obtaining Fig. 12.38:

- cutting force: F_P = 842 lb (3750 N)
- feed force: F_Q = 267 lb (1190 N)
- bulk chip speed: V_C = 220 f.p.m. (68.9 m min^{-1})

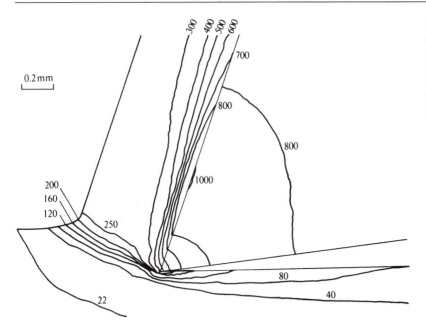

Fig. 12.38 Isothermal lines for dry orthogonal cutting of free machining steel with a carbide tool ($\alpha = 20°$) at 497 f.p.m. (155.4 m min^{-1}) and a feed of 0.01 i.p.r. (0.274 mm/rev). (after Tay et al., 1974)

- shear angle: $\phi = 27.8°$
- chip–tool contact length: $a = 0.0358$ in (0.91 mm)
- maximum width of secondary shear zone: $\delta = 0.004$ in (0.10 mm)

The following calculated values are obtained:

- average shear-plane temperature: 378 °F (192 °C)
- maximum tool-face temperature: 1861 °F (1016 °C)

The temperature of the shear plane in the vicinity of the tool tip is seen to be considerably higher than the mean temperature on the shear plane.

Childs et al. (1988) have applied FEM to extend experimental values of cutting force, shear angle, and chip–tool contact length to a number of metals to obtain a wide variety of valuable distributions of temperature along the shear plane and cutting tool face and flank.

Hybrid machining processes (HMPs) combine two or more processes of material removal in order to enhance the efficiency of individual processes when they stand alone. An example of an HMP is hot machining discussed later in this chapter. Kozak and Rajurkar (2000) have presented a general discussion of HMP with an emphasis on electrochemical discharge machining (ECDM). HMPs represent an area that will probably receive considerable attention in the future.

TEMPERATURE ON THE TOOL FLANK

Hirao (1987) has measured temperatures on the flank face of a P10 carbide tool when cutting AISI 1045 steel. A small constantan wire was mounted in a hole drilled perpendicular to the surface to be machined and the maximum temperature on the wear land determined in terms of the thermoelectric EMF generated as the wire traverses the wear land. Figure 12.39 shows the arrangement before and after

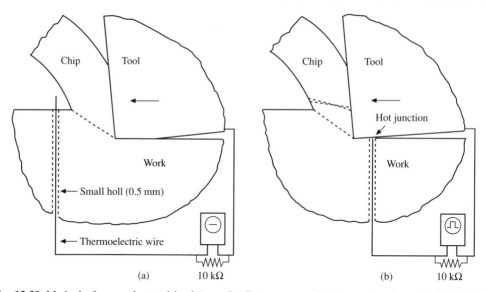

Fig. 12.39 Method of measuring tool land (wear land) temperature by thermoelectric method. (a) Before cutting wire. (b) After cutting wire. (after Hirao, 1987)

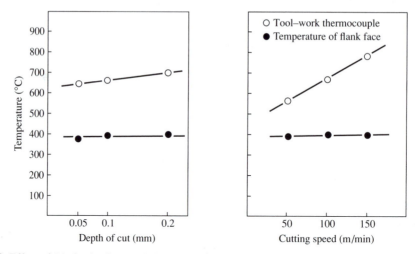

Fig. 12.40 Effect of (a) depth of cut and (b) cutting speed on tool-face and wear-land temperatures. (after Hirao, 1987)

a wire is cut. Wires from 0.004 to 0.020 in (0.1 to 0.5 mm) in diameter gave the same tool-flank temperature. The temperature on the tool face (tool–work thermocouple temperature) has essentially no influence on the temperature on the tool flank (wear land). Figure 12.40 shows essentially no change in tool-flank temperature with changes in depth of cut (feed) or cutting speed but the usual increase in tool-face temperature with depth of cut and cutting speed. Figure 12.41 shows the specific wear rate to be essentially independent of cutting speed on the tool flank but to decrease exponentially with sliding distance. All of these results suggest that tool-flank temperatures and wear characteristics are those of a friction slider and not those of the shear plane and tool face of a cutting tool.

DIMENSIONAL ANALYSIS

It is obvious that a simpler approach to cutting temperatures, although more approximate, is needed for practical application in the workshop. Dimensional analysis satisfies this need.

Kronenberg (1949) was apparently the first to apply dimensional analysis to the tool-face temperature problem. Based on his broad experience and intuitive skill, he reasoned that a first approximation to the magnitude of mean tool-face temperature ($\bar{\theta}_T$) could be obtained by considering the variables listed in Table 12.1. There are four dimensionally independent quantities in this list which may be chosen to be V, u, k, and ρC. When these four quantities are combined with each of the remaining two ($\bar{\theta}_T$ and A for area), two nondimensional quantities are obtained and hence by the principle of dimensional homogeneity

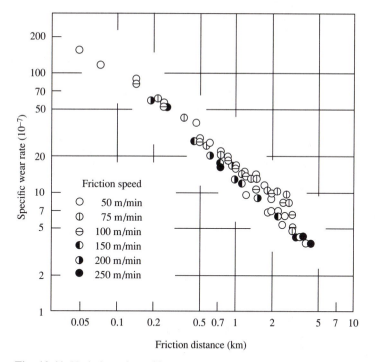

Fig. 12.41 Variation of specific wear rate with sliding distance on wear land at different cutting speeds. (after Hirao, 1987)

$$\frac{\bar{\theta}_T(\rho C)}{u} = \psi_1\left(\frac{AV^2(\rho C)^2}{k^2}\right) \tag{12.40}$$

Data for all of these quantities were available from Gottwein (1925), and Fig. 12.42 shows the first nondimensional group plotted against the second (Kronenberg, 1954). If the dimensional analysis were correct, Fig. 12.42 should show a smooth curve. Actually considerable scatter is observed (note plot is log–log) which suggests that the choice of quantities might be improved.

TABLE 12.1 Kronenberg List of Quantities for Mean Tool-Face Temperature Dimensional Analysis

Quantity	Symbol	Dimensions[†]
Mean tool-face temperature	$\bar{\theta}_T$	θ
Cutting speed	V	LT^{-1}
Undeformed chip area	A	L^2
Specific cutting energy	u	$ML^{-1}T^{-2}$
Thermal conductivity of work	k	$MLT^{-3}\theta^{-1}$
Volume specific heat of work	ρC	$ML^{-1}T^{-2}\theta^{-1}$

[†] In terms of fundamental set: temperature (θ), length (L), time (T), and mass (M).

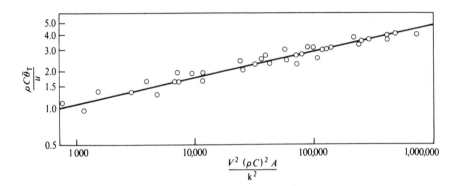

Fig. 12.42 Nondimensional plot of Gottwein data. (after Kronenberg, 1954)

TABLE 12.2 Improved List of Quantities for Mean Tool-Face Temperature Dimensional Analysis

Quantity	Symbol	Dimensions[†]
Mean tool-face temperature	$\bar{\theta}_T$	θ
Cutting speed	V	LT^{-1}
Undeformed chip thickness	t	L
Specific cutting energy	u	FL^{-2}
$k\rho C$ of work	$k\rho C$	$F^2 L^{-2} T^{-1} \theta^{-2}$

[†] In terms of fundamental set: temperature (θ), length (L), time (T) and force (F).

An obvious improvement is to substitute undeformed chip thickness (t) for undeformed chip area (bt) since b has a negligible influence on $\bar{\theta}_T$ while t has an appreciable influence. Another improvement is to consider the product ($k\rho C$) as a single variable instead of including both k and ρC. All of Kronenberg's other variables appear to be good choices.

Table 12.2 gives an improved list of quantities which leads to a single nondimensional quantity and hence

$$\frac{\bar{\theta}_T}{u}\left[\frac{k\rho C}{Vt}\right]^{1/2} = \text{nondimensional constant} \tag{12.41}$$

or

$$\bar{\theta}_T \sim u\left[\frac{Vt}{k\rho C}\right]^{1/2} \tag{12.42}$$

where the constant is for a given tool–environment (fluid) combination (Shaw, 1958).

At first glance Eq. (12.42) suggests that V and t are of equal importance relative to $\bar{\theta}_T$. However, as discussed in Chapter 3, u is approximately independent of V but varies inversely as $t^{0.2}$ [Eq. (3.36)]. Thus, for a given tool–workpiece–environment combination

$$\bar{\theta}_T \sim V^{0.5} t^{0.3} \tag{12.43}$$

This suggests that the classification of dominant wear types due to Opitz in terms of $Vt^{0.6}$ (Fig. 11.9) is of thermal origin.

In the detailed analysis of shear-plane temperature rise presented above, the model adopted was based on the Piispanen card model (Fig. 12.22), as shown in Fig. 12.21. Here the work material below the shear plane was considered to be the extensive member while the shear plane itself was considered to be the moving heat source. When q_1 from Eq. (12.21) and L_1 from Eq. (12.24) are substituted into Eq. (12.23), the following expression is obtained:

$$\bar{\theta}_S = 0.754 \frac{R_1 u_S}{J} \sqrt{\left[\frac{Vt/\gamma}{(k\rho C)_1}\right]} + \theta_0 \qquad (12.44)$$

An equally valid interpretation of Fig. 12.21 is one in which the *chip* is the extensive member, and then the following equation for θ_S is obtained directly from Eq. (12.13):

$$\bar{\theta}_S = 0.754 \frac{R_1' u_S}{J} \sqrt{\left[\frac{V_S l_S}{(k\rho C)_1}\right]} + \theta_0 \qquad (12.45)$$

where l_S is the length of the shear plane ($t/\sin\phi$) and R_1' is not the same as R_1. Due to the convective heat transfer associated with chip velocity (V_C) (not considered in [Eq. (12.45)]), R_1' in this equation will be essentially 1 and therefore

$$\Delta\bar{\theta}_S \sim u_S \sqrt{\left[\frac{V_S l_S}{(k\rho C)_1}\right]} \qquad (12.46)$$

From Eq. (12.27), it may be similarly shown that

$$\Delta\bar{\theta}_F \sim u_F \sqrt{\left[\frac{V_C a}{(k\rho C)_1}\right]} \qquad (12.47)$$

where a is the chip–tool contact length.

The mean temperature rise on the tool face will be

$$\Delta\bar{\theta}_T \sim \bar{\theta}_S + \Delta\bar{\theta}_F \qquad (12.48)$$

if it is assumed $\Delta\bar{\theta}_S$ and $\Delta\bar{\theta}_F$ are about equal. Noting that $u_S + u_F = u$ (the total specific cutting energy) and that $V_S l_S \sim V_C a \cong Vt$, it follows from Eq. (12.48) that

$$\Delta\bar{\theta}_T \sim u \sqrt{\left[\frac{Vt}{(k\rho C)_1}\right]} \qquad (12.49)$$

which is the result obtained directly by dimensional analysis [Eq. (12.42)].

Equation (12.46) for the mean shear-plane temperature appears to be more realistic than Eq. (12.22) since the former indicates a substantial rise in $\Delta\bar{\theta}_S$ with speed (i.e., as $\sqrt{V_S}$ since u_S is essentially independent of speed) in agreement with experience while the latter indicates a small increase in $\Delta\bar{\theta}_S$ with speed (due only to a small increase in R_1 with speed).

TIME EFFECTS

While equations in the foregoing discussions were derived on the basis of a steady state, most practical applications involve a nonsteady state. When a slider starts from rest, a certain time must elapse before a steady equilibrium temperature distribution obtains. The steady state is approached asymptotically, and theoretically the absolute steady state is obtained at time $T = \infty$. However, the temperature distribution is indistinguishable from the absolute distribution after an elapsed time of but

$$T = \frac{6LK}{(V)^2} \qquad (12.50)$$

When values are substituted into this equation for a typical machining case, T is found to be about 10^{-3} s. This corresponds to a tool movement of but 0.040 in (1 mm) at a cutting speed of 200 f.p.m. (63 m min^{-1}). Therefore the degree of approximation involved in applying steady state temperature theory to nonsteady state machining problems should be relatively insignificant.

The foregoing conclusions have been experimentally verified by Opitz who found, by use of oscillograms of thermoelectric voltages generated with a milling cutter, that equilibrium temperatures are reached in the tool surface in about 10^{-4} s.

Even though tool temperatures obtained in intermittent and continuous cutting will be the same, the effects of these temperatures upon tool life will not be the same. The rest period that occurs between cuts in an intermittent operation such as milling is very significant for the temperature will drop very rapidly when cutting ceases by heat flow from the surface of the tool into its interior. For this reason, milling cutter teeth can operate at significantly higher peak temperatures than can a lathe tool. This point is well illustrated by some thermoelectric measurements made by Herbert (1926).

He found that there was no softening of the teeth of a saw when subjected to a temperature of 1830 °F (1000 °C) for $\frac{1}{3}$ second during an actual sawing operation whereas the teeth were completely softened when held in a lead pot at 1650 °F (900 °C) for 3 s.

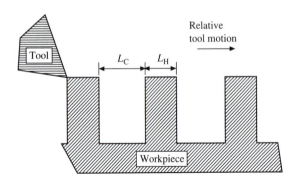

Fig. 12.43 Interrupted cutting process analyzed. L_c = cooling length, L_h = heating (cutting) length. (after Stephenson and Ali, 1992)

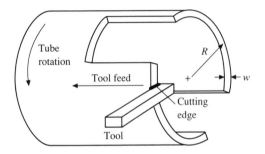

Fig. 12.44 End turning of tube in interrupted cutting tests. (after Stephenson and Ali, 1992)

TOOL-FACE TEMPERATURE IN INTERRUPTED CUTTING

Stephenson and Ali (1992) performed theoretical and experimental studies of tool-face temperatures for cyclic interrupted cutting. Figure 12.43 shows the cutting process analyzed. The results of a theoretical heat transfer study were compared with tool–chip thermocouple measurements on 2024 aluminum and gray cast iron at cutting speeds up to 3542 f.p.m. (18 m/s). The tube-cutting test used experimentally is shown in Fig. 12.44.

Figure 12.45 shows time-temperature values for a typical case for different depths (z) below the surface for a case where the heat input is uniform over the surface and heating and cooling times are equal. Figure 12.46 shows a comparison of calculated and chip–tool thermocouple values for gray cast iron machined with carbide tools for the different heating/cooling cycles. Figure 12.47 gives similar results for 2024 aluminum machined with carbide tools.

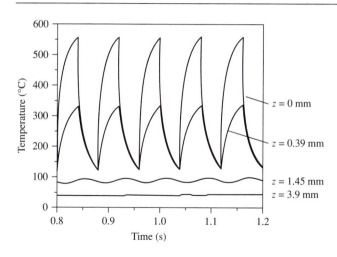

Fig. 12.45 Typical analytical time-temperature results for four depths (z) below the surface. (after Stephenson and Ali, 1992)

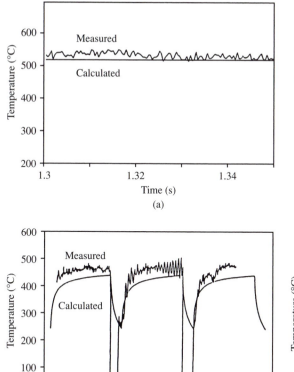

(a)

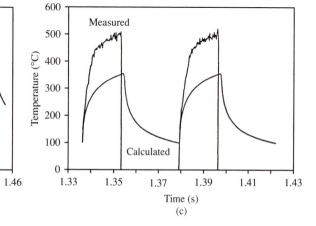

(b)   (c)

Fig. 12.46 Comparison of calculated temperatures and temperatures measured by the tool–chip thermocouple method when cutting gray cast iron with carbide tools. (a) Continuous cutting. (b) Interrupted cutting with relatively short cooling time. (c) Interrupted cutting with relatively long cooling time. (after Stephenson and Ali, 1992)

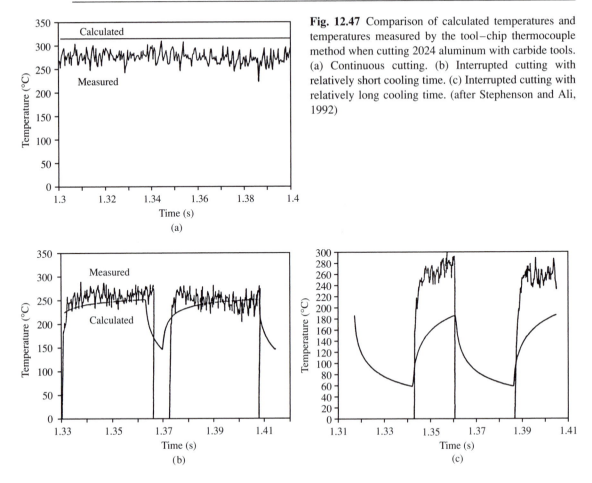

Fig. 12.47 Comparison of calculated temperatures and temperatures measured by the tool–chip thermocouple method when cutting 2024 aluminum with carbide tools. (a) Continuous cutting. (b) Interrupted cutting with relatively short cooling time. (c) Interrupted cutting with relatively long cooling time. (after Stephenson and Ali, 1992)

The following observations may be made:

1. For continuous cutting of cast iron, the measured values are slightly higher than the calculated values, but for aluminum the measured values are lower than the calculated values.

2. For both cast iron and aluminum, the calculated temperature values underestimate the measured temperatures, particularly when the heating time is less than the cooling time.

3. Temperatures for continuous cutting are higher than for intermittent cutting.

4. Measured temperatures are higher than calculated values particularly for short cooling times.

HIGH-SPEED MACHINING (HSM)

It is well established that mean tool-face temperature has a very important influence on the rate of tool wear and tool life even for relatively refractory tool materials such as sintered tungsten carbide and ceramics. In a continuous cutting operation such as turning, it has been previously explained why the mean tool-face temperature ($\bar{\theta}_T$) varies approximately as follows:

$$(\bar{\theta}_T) \sim u(Vt/k\rho C)^{0.5} \tag{12.51}$$

where

u = specific cutting energy of work material

V = cutting speed

t = undeformed chip thickness (feed per revolution in turning)

k = thermal conductivity of work

ρC = volume specific heat of work

Since u is approximately independent of V but varies inversely with t to the 0.2 power in turning, it follows from Eq. (12.51) that

$$(\bar{\theta}_T) \sim V^{0.5} t^{0.3} \qquad (12.52)$$

Thus, it should be expected that the mean tool-face temperature in turning will increase monotonically with cutting speed, and this is found to be the case in practice.

Solomon (1931) suggested that a critical cutting speed existed beyond which mean tool-face temperature would drop and hence tool life would rise with increase in cutting speed. This intriguing idea caught the imagination of many engineers who tried in vain to implement it. While it is possible to successfully turn materials at unusually high speeds if u is sufficiently low and $(k\rho C)^{0.5}$ for the work is sufficiently high, this is generally not the case. It was demonstrated many years ago at ALCOA (Templin, 1948) that there is essentially no limit to the machining speed at which aluminum may be machined because aluminum is relatively weak and has a low value of $u/(k\rho C)^{0.5}$ and hence gives a low temperature rise. However, all this is not the case for ferrous alloys.

Attempts to find a region where an increase in tool life results from an increase in turning speed (sometimes referred to as traversing the "valley of death") have always been unsuccessful. While it may be reasoned that the decrease in u due to thermal softening could more than offset the effect of an increase in V in Eq. (12.52), this does not turn out to be the case, even when machining aluminum alloys.

High-speed machining is a topic that has periodically attracted considerable interest usually fueled by generous government funding. High-speed machining is attractive when making long cuts (cutting time is an appreciable percentage of cycle time) on relatively easily machined material having favorable thermal properties (aluminum alloys, for example). This is often the case in the aerospace industry. The decision to employ high-speed machining should, of course, be based on whether the savings in reduced machining time (labor and machine cost per part) more than offset the increased tool, capital, and maintenance costs.

During the late 1950s the U.S. Air Force sponsored a project supervised by R. L. Vaughn to explore high-speed machining (20,000 to 240,000 f.p.m.) of high-strength materials. The method used to reach such high cutting speeds was a 20 mm smooth bore canon (called ballistic cutting) and rocket sleds. Results of this study were presented in Vaughn (1960). This triggered a number of high-speed experiments around the world:

- Colwell and Quackenbush (1962)
- Recht (1964)
- Okushima (1965)
- Tanaka et al. (1967)

In 1970 the U.S. Navy sponsored a study of ultra-high-speed machining (UHSM) at the Lockheed Missiles and Space Company. This project emphasized the high-speed machining of aluminum alloys for the aircraft industry.

Arndt (1971) made a very thorough literature survey of UHSM (cutting speed > 500 f.p.s.). He found that this involved primarily empirical studies with relatively little attention to scientific implications of deformation at extremely high strain rates. Arndt examined momentum effects in UHSM and found momentum forces to be significant. Arndt also showed that the resultant cutting force in UHSM depends upon $(u/\rho)^{0.5}/V_a$ where

$$u = \text{energy per unit volume of metal cut}$$

$$\rho = \text{density of work material}$$

$$V_a = \text{deformation velocity at onset of adiabatic shear}$$

This is based on the concept that with increasing cutting speed the action on the shear plane shifts from deformation of continuous solid material to deformation involving localized molten areas and voids generated by microcracking (referred to as adiabatic shear). The term adiabatic shear arises from the concept that as the deformation rate in the shear zone increases, a point is reached where the rate of energy dissipated exceeds the rate of energy diffusion from the shear zone.

Recht (1960) discussed the fact that above a certain cutting speed (V_a = onset of adiabatic shear), the temperature of the shear plane will begin to rise above the asymptotic value reached in ordinary machining. Arndt (1973) further suggested that the cutting force (F_P) will in general be as shown in Fig. 12.48 where the solid curve (F_P) is the resultant of deformation (Curve F_s) and inertia (curve F_m) components. Arndt suggests that what he calls the "adiabatic trough" shown in Fig. 12.48 should only be significant with low melting point, low density materials such as aluminum alloys.

The General Electric Research and Development Division conducted an extensive investigation of HSM sponsored by the Defence Advance Research Projects Agency (DARPA). The main results from this project were presented at a conference on high-speed machining (Komanduri et al., 1984). The objective of this project was to lower weapon systems cost by use of science-based metal removal operations. The following materials of interest to defense applications were investigated:

- aluminum alloys
- nickel base alloys
- a titanium alloy (Ti–6Al–4V)
- an alloy steel (AISI 4340)

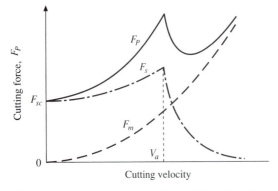

Fig. 12.48 Variation of cutting force (F_P) with cutting velocity in high-speed machining where F_P has two components, one due to plastic deformation (F_s) and the other due to inertia (F_m). (after Arndt, 1973)

Tests were performed on modified conventional machine tools at cutting speeds up to 25,000 f.p.m. and on ballistic equipment at speeds up to 240,000 f.p.m. Using conventional tool materials it was confirmed that aluminum alloys could be machined at all speeds available on high-power stable machine tools and that Ti–6Al–4V had to be machined at a relatively low removal rate in order to achieve a reasonable tool life. Nickel base alloys and alloy steels, such as AISI 4340, of course, fell between the two extremes. Titanium alloys are particularly difficult to machine because their poor thermal properties give unusually high cutting temperatures and their tendency to react with essentially all tool materials gives high rates of wear particularly at high temperatures (high cutting speeds).

The Ledge Tool

Lee (1980) invented a special carbide tool design to enable a titanium alloy to be machined at 3 to 5 times the cutting speed normally used. This was achieved by controlling flank wear. Figure 12.49a shows a carbide insert with the tool flank altered to provide a thin (0.015 to 0.50 in, 0.38 to 1.25 mm) over-hanging ledge extending 0.15 to 0.060 in (0.38 to 1.50 mm) beyond the main flank surface of the insert. This is called a ledge tool where the depth of the overhang equals the depth of cut employed. Figure 12.49b is an image of a ledge insert, while Fig. 12.49c shows just the ledge portion of a ledge insert in a turning operation. The concept behind this design is that the ledge will wear back by microchipping and without gross fracture. The ledge tool is an extension of what has been done on the rake face since Klopstock (1925) introduced a tool design that restricted contact between chip and tool to increase the shear angle, decrease cutting forces, and increase chip curl to improve chip control. However, in the ledge tool case the alteration is on the tool flank to limit the area of the flank wear land to enable cutting to occur without gross frac-ture for a longer period of time. This is useful for enabling an increase in removal rate for difficult-to-machine materials such as titanium alloys and nickel base superalloys.

    The choice of carbide for ledge tool use is important. A relatively brittle carbide gives best results. The optimum thickness of the wedge

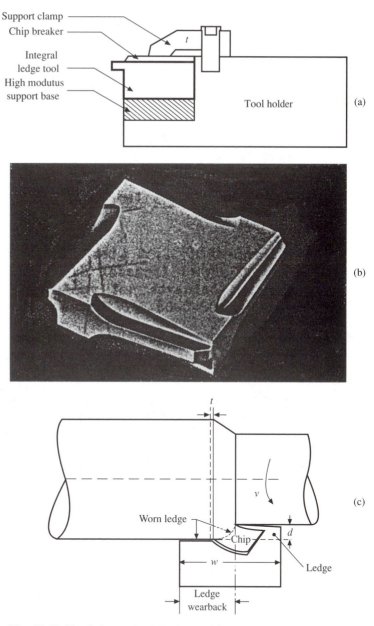

Fig. 12.49 The ledge tool. (a) Ledge tool insert mounted in conventional tool holder. (b) Image of ledge tool insert. (c) Ledge tool shown in turning operation with ledge partially worn back. (after Komanduri and Lee, 1984)

depends upon the material machined, the cutting conditions, and the flank wear that is tolerable. The following example of improved performance when using a ledge insert to turn a 6 in (150 mm) diameter bar of Ti–6Al–4V is given in Komanduri and Lee (1984). The dimensions of the ledge

were 0.03 in (0.75 mm) overhang × 0.04 in (1 mm) thick and 0.50 in (1.27 mm) wide and the
cutting conditions were as follows:

- side and back rake angle = −5°
- SCEA = 0°
- ECEA = 1°
- clearance angle = 5°
- cutting speed = 600 f.p.m. (183 m/min)
- depth of cut = 0.03 in (0.75 mm)
- feed Rate = 0.009 i.p.r. (0.23 mm/rev)
- water-base fluid

The tool life (for wear back of the ledge by microchipping) was 30 min. The cutting speed to
obtain the same tool life with a conventional tool with all conditions except for cutting speed the
same was 150 f.p.m. (55 mm/min). This represents a fourfold increase in tool life.

Ledge inserts have also been successfully used in face milling a titanium alloy at a speed of
515 f.p.m. (157 m/min). In general the cutting speed could be increased by a factor of 3 to 5 times
that for conventional inserts for the same life.

Two limitations of the ledge tool are

- may be used only for straight cuts
- in long cuts the small ECEA (~1°) employed will give rise to slight taper unless compensated
  for on an numerical controlled (NC) machine

The ledge tool produces finishes of R_A = 40 to 120 μin (1 to 3 μm) for both turning and face
milling as long as wear back involves microchipping.

HIGH-SPEED TEMPERATURES IN INTERRUPTED CUTTING

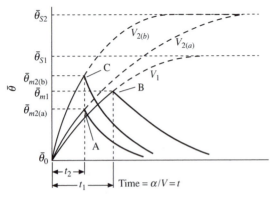

Fig. 12.50 Approach to steady state temperature at the
beginning of a cut for different cutting speeds (V_1 and
V_2) and cooling curves for a face-milling situation.
(after Lezanski and Shaw, 1990)

It has been pointed out by Palmai (1987) that
Solomon's experiments were for milling and not with
continuous cutting (as in turning). Palmai suggests
that the tool temperature in milling builds up gradually
as the cut proceeds and that the peak temperature is
a function of the rate of increase of temperature with
time and the total time of tool–work contact per cut.
Figure 12.50 shows schematically the situation for
two cutting speeds V_1 and V_2 where $V_2 = 2V_1$. For a
face-milling cutter and a constant arc of tool–work
contact, the contact times will be t_1 and t_2 for cutting
speeds V_1 and V_2 and since $t \sim 1/V$, $t_1 = 2t_2$. Temperatures $\bar{\theta}_{s1}$ and $\bar{\theta}_{s2}$ are the steady state temperatures
reached asymptotically at speeds V_1 and V_2. Since
$V_2 = 2V_1$, $\bar{\theta}_{s2} = \sqrt{2}\bar{\theta}_{s1}$ in accordance with Eq. (12.42).
When cutting stops, the mean tool temperature will
decrease rapidly to an ambient value ($\bar{\theta}_0$).

Figure 12.50 shows two heating curves for speed V_2. Curve (a) assumes that equilibrium is approached less rapidly than in the case of curve (b). The maximum mean tool-face temperatures are designated $\bar{\theta}_{m1}$ for speed V_1 and $\bar{\theta}_{m2(a)}$ and $\bar{\theta}_{m2(b)}$ for speed V_2 for cases (a) and (b), respectively. Comparing values of $\bar{\theta}_m$ it is seen that for case (a) the maximum temperature will be less at the high speed V_2 than at the low speed V_1 while the opposite holds for case (b). That is, $\bar{\theta}$ at A $<\bar{\theta}$ at B but $\bar{\theta}$ at C $>\bar{\theta}$ at A or B. The abscissa of Figure 12.50 is cutting time which is equal to the arc of tool–work contact (α) divided by the cutting speed (V).

Based on numerous assumptions concerning the manner in which mean tool-face temperature approaches an equilibrium value as well as the rate of cooling after cutting ceases, Palmai concluded that in face milling, the maximum tool-face temperature $\bar{\theta}_m$ could decrease with increase in cutting speed and a constant arc of tool–work contact. The purpose of the present study is to determine experimentally whether the Solomon concept (decrease in maximum cutting temperature with increased cutting speed) actually holds in intermittent cutting.

Figure 12.51 shows diagramatically the arrangement actually employed on a lathe in the study being considered. The solid black regions represent nonconducting elements. A mercury rotating contact provided means for connecting a storage oscilloscope to the rotating workpiece with minimum contact resistance. Output from the oscilloscope was fed to an x–y recorder which plotted EMF versus time.

Initial tests were performed on plain carbon steels (AISI 1018 and AISI 1045) using M–2 HSS and K45 carbide tool materials. Cuts were made at relatively light feed, usually ~ 0.005 i.p.r. (0.125 mm/rev) or ~ 0.010 i.p.r. (0.25 mm/rev) to simulate the undeformed chip thickness in a milling operation. All cuts were made without use of a cutting fluid. The HSS tool had a rake angle of plus 15° and a clearance angle of 5°. All carbide tools had a rake angle of −5° and a clearance angle of 5°.

Figure 12.52 shows representative data from an x–y plotter for three duplicate tests at

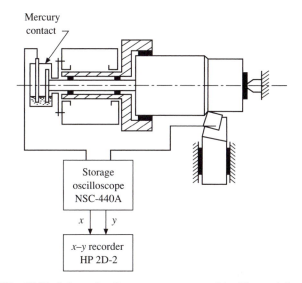

Fig. 12.51 Schematic of test arrangement used in chip–tool thermocouple study. (after Lezanski and Shaw, 1990)

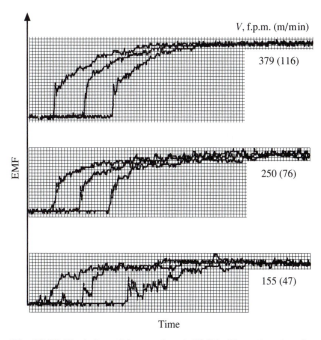

Fig. 12.52 Variation of thermoelectric EMF with cutting time for three different cutting speeds (V). Three duplicate tests for each speed. Work material, AISI 1018 steel; tool material, K45 carbide; feed, 0.0048 i.p.r. (0.122 mm); depth of cut, 0.100 in (2.54 mm). (after Lezanski and Shaw, 1990)

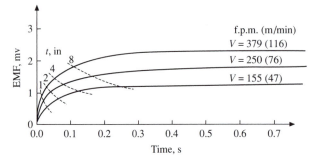

Fig. 12.53 Variation of EMF (~temperature) with cutting time (solid curves) and lines of constant cutting arc length *l* (dash lines). (after Lezanski and Shaw, 1990)

each of three cutting speeds. These results were used to obtain the mean solid curves of Fig. 12.53. The dashed curves are for different constant tool–work contact lengths (*l*). It is seen that at the higher cutting speed (379 f.p.m. = 116 m/min) it takes only about $\frac{3}{8}$ s for the tool temperature to reach the steady state value. In no case was there evidence of the tool temperature at the end of a cut decreasing with increased cutting speed. The calibration curves used are given in Fig. 12.3.

The measured EMF in mv was not converted to temperature since to a good approximation, the EMF is proportional to temperature over the temperature range of interest, and we are here interested only in relative values.

Figure 12.54 shows plots similar to that of Fig. 12.53 for different combinations of tool and ferrous work materials. In these plots, the following are evident:

- In all cases, the temperature at the end of a cut is found to increase with cutting speed regardless of the magnitude of the cutting arc.

- The time to reach a steady state temperature decreases with increase in cutting speed.

Figure 12.55 shows plots for aluminum and bronze alloys cut using HSS and carbide tools. These results are very similar to those for steel. Again, in no case did the temperature at the end of a cut decrease with increase in cutting speed.

The thesis of Palmai (1987) that in milling the tool temperature at the end of an intermittent cut will decrease with increase in cutting speed has been experimentally tested and found to be false. Using a variety of tool and work materials, the approach of tool temperature to equilibrium in turning has been recorded using the tool–chip thermocouple technique. Such temperature-time plots may be used to simulate maximum temperatures in milling with different arc length of tool–work contact. In no case investigated, did the temperature on exiting the cut decrease with increased speed.

It would thus appear that the Solomon "valley of death" concept not only does not hold for continuous chip formation but also does not hold for milling.

In Europe high-speed machining (HSM) research was conducted at the University of Damstadt and at a number of companies from 1979 to 1988. The results of these studies were published in a book (Schultz, 1989), and a general overview of HSM studies is available in Schulz and Moriwaki (1992), where items emphasized for HSM machining included

- cutting mechanics
- cutting tools
- machining tools

Figure 12.56 shows the relation suggested to exist between HMS technology and the principle cutting components.

Cutting mechanics involved in HSM depends primarily upon the work material. For soft materials (Al and Cu alloys and polymers) continuous chip mechanics are involved, while for cast irons discontinuous chips pertain (Chapter 15) and for hard difficult materials (titanium and nickel based alloys) wavy and sawtooth chips (Chapters 21 and 22) are involved. While early HSM research

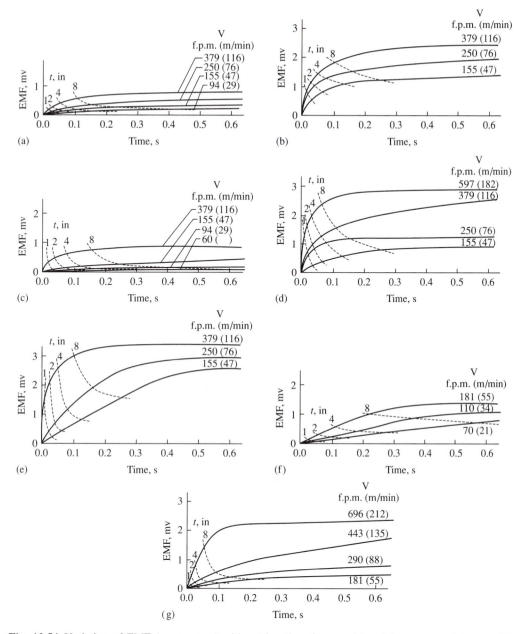

Fig. 12.54 Variation of EMF (temperature) with cutting time for a variety of ferrous materials cut with HSS and carbide tools at different speeds (*V*). (a) AISI 1018 steel cut with M–2 HSS tool; feed, 0.0048 i.p.r. (0.122 mm/rev); depth of cut, 0.05 in (1.27 mm). (b) AISI 1018 steel cut with K45 carbide tool; feed, 0.0048 i.p.r. (0.122 mm/rev); depth of cut, 0.100 in (2.54 mm). (c) AISI 1045 steel (soft) cut with M–2HSS tool; feed, 0.0048 i.p.r. (0.122 mm/rev); depth of cut, 0.05 in (1.27 mm). (d) AISI 1045 steel (soft) cut with K45 carbide tool; feed, 0.0153 i.p.r. (0.389 mm/rev); depth of cut, 0.06 in (1.52 mm). (e) AISI 1045 steel (H_{RC} 42) cut with K68 carbide tool; feed, 0.0048 i.p.r. (0.122 mm/rev); depth of cut, 0.05 in (1.27 mm). (f) Gray cast iron cut with M–2 HSS tool; feed, 0.0048 i.p.r. (0.122 mm/rev); depth of cut, 0.05 in (1.27 mm). (g) Gray cast iron cut with K–68 carbide tool; feed, 0.00448 i.p.r. (0.122 mm/rev); depth of cut, 0.05 in (1.27 mm). (after Lezanski and Shaw, 1990)

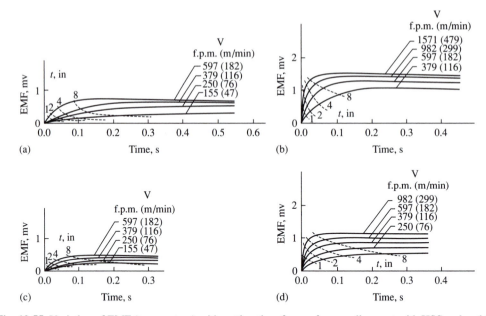

Fig. 12.55 Variation of EMF (temperature) with cutting time for nonferrous alloys cut with HSS and carbide tools. (a) AISI 2040 aluminum cut with M–2 HSS tools; feed, 0.0096 i.p.r. (0.244 mm/rev); depth of cut, 0.10 in (2.54 mm). (b) AISI 2024 aluminum cut with K68 carbide tool; feed, 0.0096 i.p.r. (0.244 mm/rev); depth of cut, 0.10 in (2.54 mm). (c) AISI 31600 bronze cut with M2 HSS tool; feed, 0.0192 i.p.r. (0.488 mm/rev); depth of cut, 0.05 in (1.27 mm). (d) AISI 31600 bronze cut with K68 carbide tool; feed, 0.0192 i.p.r. (0.488 mm/rev); depth of cut, 0.05 in (1.27 mm). (after Lezanski and Shaw, 1990)

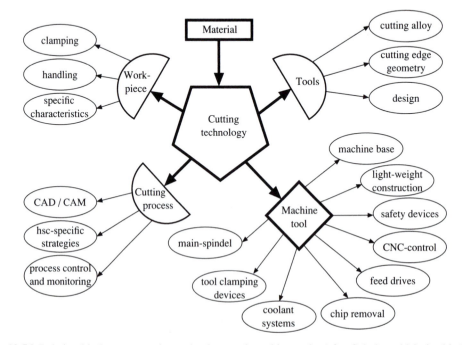

Fig. 12.56 Relationship between cutting technology and machine tools. (after Schulz and Moriwaki, 1992)

was directed toward aluminum and other relatively soft materials, later work involved HSM nickel-based alloys and other difficult-to-machine materials. The range of speeds involved in HSM depends primarily on the material as follows:

- soft materials and plastics: 3300 to 30,000 f.p.m. (1000 to 9000 m/min)
- steel and cast iron: 2500 to 16,000 f.p.m. (700 to 5000 m/min)
- titanium alloys: 300 to 3000 f.p.m. (160 to 1600 m/min)
- nickel base alloys: 160 to 1600 f.p.m. (50 to 500 m/min)

Since a major problem in HSM is wear and tool life, this usually calls for application of the most advanced tool materials and tool geometry. Modern cutting tool materials such as microcarbide structures, reenforced ceramics, SiALONs, coatings, and superhard materials (D and PCBN) play an important role in the success of HSM. All of these tool materials are discussed in Chapter 14.

HSM also involves special machine tool requirements such as

- high spindle speeds: up to 8000 to 12,000 r.p.m.
- high feed rates: up to 60 to 200 f.p.m. (20 to 25 m/min)
- low friction and stable guideways
- high chip removal capability
- protection against hot high-speed chips, coolant spray, tool breakage, rigid and strong work-holding apparatus
- CNC process control since there is usually insufficient time for manual control

Because of the cost of all of these special HSM machine tool requirements it is too costly to have universal machine tools for all HSM applications, and the added cost of need for special machines for HSM production represents an added problem related to high-speed machining.

CRACK FORMATION IN INTERRUPTED CUTTING

Wang et al. (1996) have investigated the reason for the development of cracks in tools that undergo cyclic temperature changes as in milling. This often involves an engagement (heating) period of about 0.05 s followed by a period of disengagement (cooling period) of about 0.1 s. It was first shown that cooling by convection to the air is negligible compared to conduction to the tool and hence may be ignored. An FEM analysis for a typical milling situation was performed leading to the temperature distribution in the tool shown at exit in Fig. 12.57a and at the end of the cooling period in Fig. 12.57b. Experiments were then conducted to test whether the results of Fig. 12.57 were reasonable. This involved placing a NiCr–NiAl thermocouple different distances below the flank surface of the tool. A chip–tool contact area corresponding to that in the milling

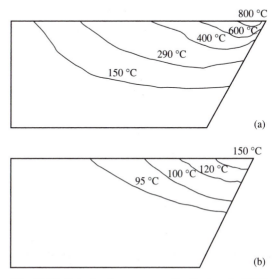

Fig. 12.57 Calculated temperature distributions by FEM analysis. (a) At end of cutting period. (b) At end of noncutting period. (after Wang et al., 1996)

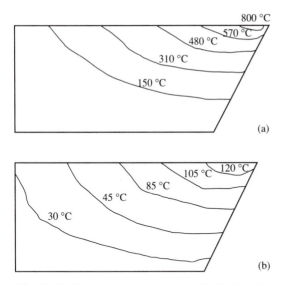

Fig. 12.58 Experimental temperature distributions by thermoelectric measurements. (a) At end of cutting period. (b) At end of noncutting period. (after Wang et al., 1996)

operation being considered was then cyclically heated by a 500 W pulsed CO_2 laser that could be adjusted to cycle at up to 382 Hz. The experimental temperature distributions corresponding to those of Fig. 12.57 are shown in Fig. 12.58 for the same periods of heating and cooling. The agreement between Figs. 12.57 and 12.58 is more than sufficient for the purpose intended (i.e., to show the nature of the temperature distributions during the heating and cooling phases of a cycle).

Since the temperature at the end of the cooling period does not go to zero, it was reasoned that stress at the end of the cooling period would be compressive and hence would not induce tensile cracks. It was further suggested that any cracks that form would occur at the end of the heating period due to thermal expansion of the heated zone giving rise to tensile stresses.

Finally, cyclic heating was continued until gross fracture occurred so that the sequence of actions associated with cracking due to cyclic heating could be followed. It was found that there were four cyclic heating stages leading to gross fracture:

1. development of cracks perpendicular to the cutting edge (comb cracks) after a number of thermal cycles
2. development of transverse cracks parallel to the cutting edge and running between comb cracks or originating at points of stress concentration (grain boundaries, impurity atoms, or second phase particles)
3. crossing cracks that tend to outline a finite area
4. gross fracture with a sizeable particle popping out

The conclusions derived from this useful study include the following:

- Rapid cooling of a tooth exiting a cut is a conduction-dominant process and cooling in air is negligible.
- Cracks do not form during tool exit.
- Cyclic heating on entry is the source of the tensile stress leading to fatigue fracture.
- Pulsed laser cyclic heating appears to be a useful tool to evaluate tool materials relative to resistance to thermally induced fatigue.

WHITE LAYER (WL) FORMATION ON HARD MACHINED SURFACES

Akcan (1998) has conducted a thorough study of the white layer (WL) formed on hard steel surfaces that were machined at high speed. Three steels were included in the study: AISI 52100, AISI 4340, and AISI M2 tool steel. These were machined using polycrystalline cubic boron nitride (PCBN) tools. The nonetching white layers were of nanometer thickness (100 to 500×10^{-9} m). X-ray diffraction showed them to consist of approximately 90% martensite and 10% austenite, and electron micrographs

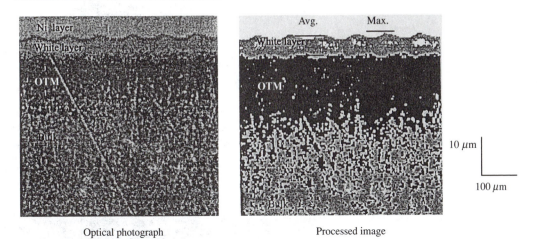

Optical photograph   Processed image

Fig. 12.59 Optical micrograph and processed version of an AISI 52100 finished surface showing the white layer, overtempered mamtensite (OTM) layer, and the base material. Hardness of work, H_{RC} 61; extent of wear land, 0.012 in (300 μm); cutting speed, 492 f.p.m. (150 m/min); depth of cut, 0.008 in (0.2 mm); feed, 0.004 i.p.r. (0.1 mm/rev). (after Akcan, 1998)

revealed an extremely fine grain size (100 to 300 nm). The nature of the WL and the overtempered martensite layer below was obtained from optical photomicrographs of chip samples. The outer surfaces of these samples were protected by depositing a layer of electroless nickel before polishing. The optical photographs were digitized and subjected to contrast enhancement. Figure 12.59 shows a typical optical photograph and the corresponding processed image.

Since the study involves a WL left on the finished surface, it is included in this chapter because temperatures on the tool flank (wear land) are of importance. White layers that are formed in hard turned sawtooth chips are also important and these are discussed in Chapter 22.

In his analysis of temperature rise on the tool flank, Akcan made the following assumptions:

- Only heat developed on the flank face of the tool is involved, i.e., all heat developed on the shear plane and tool face is carried off by the chip.

- The magnitude of the tool flank energy equals the increase in total energy as the size of the wear land increases.

- Heat on the wear land is uniformly distributed.

- The residual temperature rise from a previous revolution is negligible (i.e., the cooling rate of the workpiece surface is so great that by the time a given point on the bar is again cut, the surface temperature has dropped to essentialy ambient).

- The transformation rate from α iron to γ iron is so great that any point below the surface reaching the transformation temperature will transform.

- The mean coefficient of friction on the wear land for AISI 52100 steel was between 0.1 and 0.24 which is in good agreement with cubic boron nitride (CBN) sliding on hard steel in a sliding contact experiment. In contrast, the coefficient of friction on the tool face is much higher (0.25 to 0.7) since this involves some sticking contact.

The author presents feasible arguments, based on experimental data for AISI 52100 steel, for why all of these assumptions are reasonable.

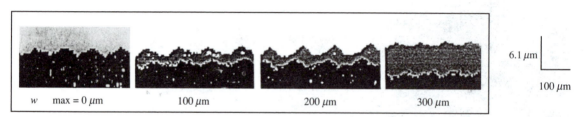

Fig. 12.60 Processed micrographs showing variation in WL thickness for AISI 52100 steel with extent of wear land (*w*). Hardness of work, H_{RC} 61; cutting speed, 656 f.p.m. (200 m/min); depth of cut, 0.004 in (0.1 mm); feed, 0.004 i.p.r. (0.1 mm/rev). (after Akcan, 1998)

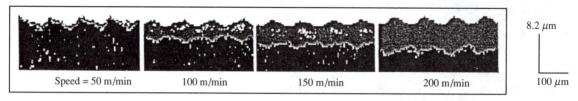

Fig. 12.61 Processed micrographs showing variation of WL thickness for AISI 52100 steel with cutting speed. Hardness of work, H_{RC} 61; depth of cut, 0.008 in (0.2 mm); feed, 0.004 i.p.r. (0.1 mm/rev); extent of wear land, 0.012 in (300 μm). (after Akcan, 1998)

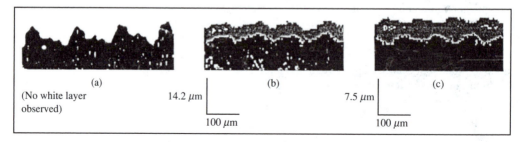

Fig. 12.62 Processed micrographs showing variation in WL on finished surface for different work materials. (a) AISI: M2. (b) AISI 4340. (c) AISI 52100. Machining conditions: cutting speed, 492 f.p.m. (150 m/min); depth of cut, 0.008 in (0.2 mm); feed, 0.004 i.p.r. (0.1 mm/rev). (after Akcan, 1998)

Calculated values based upon a sliding heat source model developed by Ju (1997) were found to be in excellent agreement with the experimental values. The thickness of the WL on the finished surface was found to increase with the size of the wear land (Fig. 12.60) and with increase in cutting speed (Fig. 12.61). A WL was found for AISI 52100 and AISI 4340 steels but not for AISI M2 steel (Fig. 12.62). This is believed to be due to the difference in transformation temperature between AISI 4340 and AISI 52100 steels (775–845 C) and AISI M2 (~ 1200 C). Microhardness values were found to be higher for a WL than for untempered martensite for the same materials. This was probably due to the very high quenching rate, very small crystal size, and the large deformation involved. The residual stress in the WL on a finished surface cut at high speed was tensile and increased with WL thickness and cutting speed.

HOT MACHINING

Hot machining was first investigated in Germany and Japan during the later part of World War II. Although beneficial effects were noted when metal was cut at elevated temperatures, the major drawback was the lack of heat-resistant tooling. With the introduction of tungsten-carbide tooling, however, this is no longer a problem. There are still practical difficulties that have kept the process of hot machining in the laboratories and out of the machine shop and production operations.

Tests have been made in several laboratories to determine the practicability of artifically heating the surface of the workpiece just before a cut is taken. This procedure will raise the temperature of the workpiece closer to or slightly above the recrystallization temperature and hence reduce the tendency to strain harden. While the method does not appear to be of general interest, it may prove to be of particular value in the machining of high temperature alloys where strain hardening is serious. The power of this method is strikingly illustrated when the machining characteristics of vitallium are compared at room temperature and at 2000 °F (1903 °C). At room temperature, this material is almost completely unmachinable while at 2000 °F (1903 °C) a good continuous chip is obtained. Although less spectacular results are obtained when machining other high temperature alloys, the method deserves careful study.

Several methods of heating the workpiece have been employed including: furnace, gas torch, induction coil, carbon arc, plasma arc, laser, and electrical resistance. When a furnace is used, the part is heated throughout. This is only feasible for small parts as in the drilling of turbine buckets. In the other methods, one object is to supply enough heat to the surface to raise the temperature of a layer about equal to the depth of cut to the desired value. This heat must be supplied just in front of the cutting tool so that there is not sufficient time for a large portion of it to flow into the inner regions of the bar.

It will be recalled that the tool temperature is the sum of the shear, friction, and ambient temperatures. The friction temperature ($\bar{\theta}_F$) is probably not greatly influenced by heating the workpiece. Obviously the base temperature θ_0 is raised in hot machining, since it represents the temperature of the workpiece before cutting. Increased tool life can be expected on this basis only if the shear temperature increment ($\bar{\theta}_S - \theta_0$) decreases faster with temperature than θ_0 increases. This should be expected to be the case for materials having a strong tendency to strain harden during cutting. For materials that do fall in this category, there should be an optimum hot-machining ambient workpiece temperature. This was actually found by Krabacher and Merchant (1951) in milling tests on hot Timken alloy (16 $^w/_o$ Cr, 25 $^w/_o$ Ni, 6 $^w/_o$ Mo). Milling allowed the tests to be carried on at higher temperatures than would be possible in a lathe.

It is interesting that the temperature for maximum tool life varies with the cutting speed as shown in Fig. 12.63. The higher the cutting speed the higher the optimum ambient temperature.

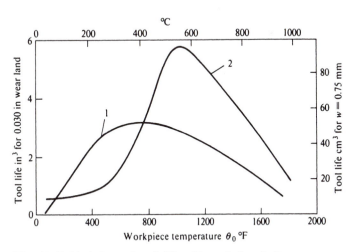

Fig. 12.63 Variations of tool life with workpiece bulk temperature when milling Cr-Ni-Mo steel at speeds of (1) 150 f.p.m. (47 m min^{-1}) and (2) 200 f.p.m. (63 m min^{-1}). Tool characteristics: fly-milling end-mill of KM carbide; rake angle, $-14°$; inclination, $0°$; relief angle, $7°$; corner angle, $30°$; feed per tooth, 0.01 in (0.25 mm); depth of cut, 0.05 in (1.27 mm). (after Krabacher and Merchant, 1951)

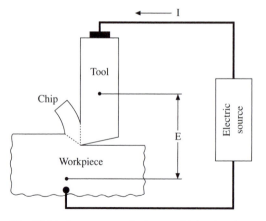

Fig. 12.64 Arrangement for hot machining by electric current. (after Okoshi and Uehara, 1963)

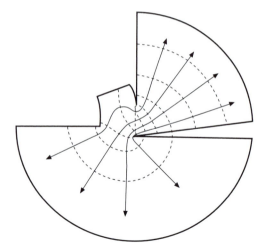

Fig. 12.65 Equipotential lines (dotted) and paths of electric current flow (solid lines). Interval of equipotential lines is 5 mV. (after Okoshi and Uehara, 1963)

The data of Fig. 12.63 illustrate the nature of the improvement in tool life that may be achieved by hot machining. The best cutting conditions were found to be $\theta_0 = 1000$ °F (538 °C) and a cutting speed of about 200 f.p.m. (69 m min^{-1}) for this particular material (16–25–6). Under these conditions the tool life was found to be over ten times the corresponding value when θ_0 was at room temperature. All of the tool life values in Fig. 12.63 are considerably lower than would be obtained under similar conditions in turning. These low values are probably due in part to the discontinuous nature of the cutting operation studied.

Considerable work was performed by Okoshi and his coworkers on hot machining by electric current (HMEC) during the 1960s. A summary of these studies was presented by Okoshi and Uehara (1963). The principle of this method is shown in Fig. 12.64. The area through which the current passes is the deformation zone. A tin foil model of the tool–chip work area was placed on a nonconducting surface and connected to a low voltage source. Equipotential lines were established using a vacuum tube voltmeter. Figure 12.65 shows equipotential lines and current paths for a zero rake angle tool. Different models were used to determine the relative amounts of heat generated and its distribution along the shear plane and chip–tool interface for different values of rake angle (α) and chip–tool contact length (l). Using models as shown in Fig. 12.65, the following observations were made:

- Under ordinary conditions more heat is generated at the chip–tool interface than along the shear plane.
- For very small shear angles (small rake angles) and very large values of l more heat is generated along the shear plane than at the chip–tool interface.
- The percentage of heat generated in the cutting tool increases as the rake angle (and shear angle) increases.
- The depth of cut has no effect on the distribution of heat.
- More of the heat is generated in the tool when its conductivity is less than that of the work.

Thermal analysis resulted in the following for the temperature rise in the chip (θ_c):

$$(\theta_c) \sim EI/Vbt\rho C \tag{12.53}$$

where

E and I are voltage and current in Fig. 12.63

v, b, and t are cutting speed, depth of cut, and feed, respectively

ρ and C are mass density and specific heat of the chip

Experiments performed when turning with and without electric current heating were found to be in good agreement with the above results. Figure 12.66 shows the effect of the rate of current flow on forces in the F_P and F_Q directions when turning a stainless steel. The point at which the forces are a maximum (250 A) was found to correspond to the current where the BUE no longer exists and the surface finish improves. Beyond this current the forces fall off rapidly with increase in current flow. Barrow (1966) has confirmed results of Okoshi and Uehara (1963) in an independent study.

Hot machining by electric current was also found to be effective in drilling. Figure 12.67 shows chips produced when drilling hard (H_{RC} 50) steel with a carbide-tipped twist drill with and without electric current heating. Hot machining by electric flow was also found useful when milling, provided at least one tooth remained in contact with the work.

Uehara (1968) has studied the effect of electric current heating on finish, forces, and tool life. He found that electric current heating was sometimes beneficial to surface finish (Fig. 12.68), cutting forces (Fig. 12.69) and rate of tool flank wear (Fig. 12.70). Figure 12.70a shows the three composite inserts tested, while Fig. 12.70b shows the results. This figure shows that the influence of electric heating depends on the composition of the carbide insert. In general, the rate of flank wear first decreases and then increases as the electric current is increased. The current flow for minimum rate of flank wear changes with the structure and composition of the carbide insert as shown in Fig. 12.70b.

It should be noted that the examples discussed in connection with Figs. 12.66, 12.68, and 12.69 are for relatively low cutting speeds, feed rates, and depths of cut (finishing type cuts). It appears that for high cutting speeds, electric current hot machining is often disadvantageous since the increase in tool

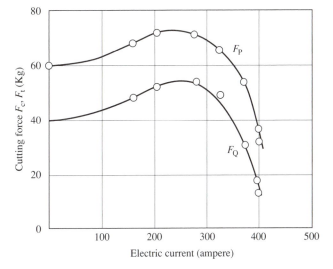

Fig. 12.66 Variation of cutting forces F_P and F_Q with electric current. Work material, stainless steel (13% Cr); cutting speed, 49 f.p.m. (15 m/min); feed, 0.005 i.p.r. (0.13 mm/rev); depth of cut, 0.040 in (1.0 mm). (after Okoshi and Uehara, 1963)

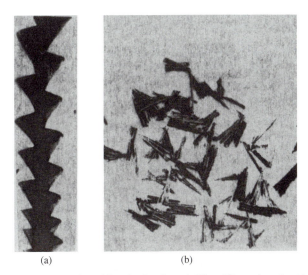

(a)   (b)

Fig. 12.67 Drilling chips for hard steel (H_{RC} 50) produced with 0.034 in (8.5 mm) diameter carbide tipped drill at 155 r.p.m.; feed, 0.0032 i.p.r. (0.08 mm/rev). (a) Heated by 600A electric current. (b) Unheated. (after Okoshi and Uehara, 1963)

temperature then has a greater adverse effect on tool life than it does on reduction in cutting forces by reducing the strain hardening effect in chip formation. It would appear that the most promising application of electric current hot machining is to applications involving low cutting speeds (as in broaching) and light chip loads (low feeds and depths of cut as in finishing operations).

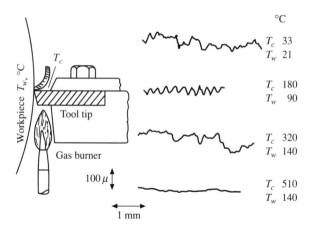

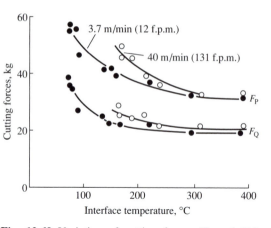

Fig. 12.68 Variation of surface finish traces for carbon steel machined with different temperatures of workpiece (T_w) and tool face (T_c) obtained by use of gas flame positioned as shown. Cutting conditions: cutting speed, 7.6 f.p.m. (2 m/min); feed, 0.0096 i.p.r. (0.24 m/rev); depth of cut, 0.004 in (0.1 mm); tool material, K20 carbide. (after Uehara, 1968)

Fig. 12.69 Variation of cutting forces (F_P and F_Q) versus chip–tool interface temperature. Cutting conditions: work material, copper; cutting speed, as shown; feed, 0.0052 i.p.r. (0.13 mm/rev); depth of cut, 0.0032 in (0.8 mm); tool material, K20 carbide; orthogonal turning with 0° rake angle. (after Uehara, 1968)

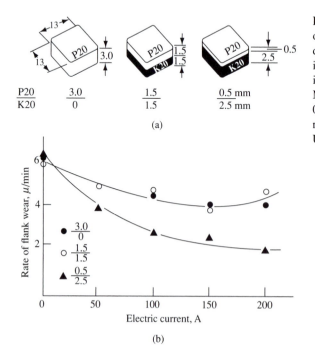

Fig. 12.70 (a) Composite inserts consisting of layers of P20 and K20 carbide of different thickness (all dimensions in mm). (b) Variation of rate of flank wear in turning with electric current for the three composite inserts in (a). Cutting conditions: work material, high Mn steel; cutting speed, 46 f.p.m. (14 m/min); feed, 0.004 i.p.r. (0.1 mm/rev); depth of cut, 0.060 in (1.5 mm); electric source, dc with tool negative. (after Uehara, 1968)

Laser-assisted machining employs a laser beam focused on the work material just in front of the cutting tool. The laser beam extends over the entire depth of cut and is positioned and adjusted to soften the material on the shear plane without allowing appreciable laser energy to flow into the tool. This constitutes a difficult control problem and requires a relatively expensive (and potentially dangerous) high-power (15 kW) continuous wave or pulsed laser. The high power required is due in part to the fact that the absorptivity of a high-temperature alloy is only about 10%. A sprayed-on coating of material such as potassium silicate may be used to increase the absorptivity somewhat and thus decrease the power of the laser required (Rajagopal, 1982). However, this represents an added complication. All things considered, it does not appear likely that the complications (adjustment control and safety) and high cost of laser-assisted machining will allow this approach to hot machining to be applied except in very unusual situations.

While hot machining can be valuable when cutting very difficult materials such as high-temperature alloys, it has often not proven to be worth the additional complication when machining conventional materials. Even with high-temperature alloys such as nickel and cobalt base materials, great care must be exercised in applying the technique to be sure the work-softening advantage is not more than offset by a tendency for the peak tool-face temperature to move closer to the tool tip as when machining nickel under conventional conditions.

HOT MACHINING OF VERY BRITTLE MATERIALS

Plasma arc hot machining has been used to improve the machinability of glasses and ceramics. Figure 12.71a shows the location of the plasma jet when turning a glass such as Pyrex at temperatures of 1800 °F (1000 °C). Figure 12.71b shows a water-cooled plasma torch used to direct a small (0.14 in; 3.5 mm diameter) plasma arc against the workpiece at the heating position shown in Fig. 12.71a. The working gas is a mixture of argon and helium. Temperature is increased by increasing the hydrogen content and/or increasing the flow of current from the tungsten electrode

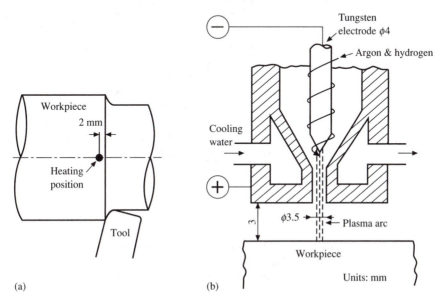

Fig. 12.71 Arrangement for plasma arc heating. (a) Heating location. (b) Plasma torch. (after Kitagawa and Maekawa, 1990)

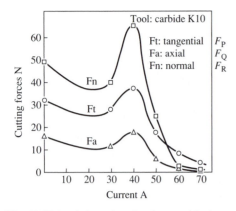

Fig. 12.72 Variation of cutting forces with plasma arc current when turning Pyrex glass with K10 carbide insert. [F_P = tangential force component. F_Q = axial force component. F_R = radial force component.] (after Kitagawa and Maekawa, 1990)

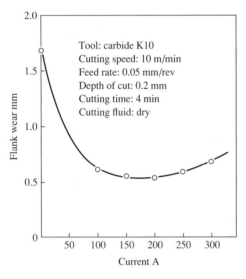

Fig. 12.73 Flank wear of carbide tool versus plasma arc current when turning Pyrex glass. (after Kitagawa and Maekawa, 1990)

to the torch. Kitagawa and Maekawa (1990) have presented turning results for cutting forces, chip morphology, surface roughness, and tool wear for a number of glasses and ceramics employing plasma arc heating. Figure 12.72 shows changes in cutting force components with flow of current when turning a Pyrex specimen 0.79 in (20 mm) in diameter under the following conditions:

- insert geometry: $-6°$, $5°$, $6°$, $15°$, $15°$, nose rad = 0.032 in (0.8 mm)
- insert material: K–10 carbide
- cutting speed: 33 f.p.m. (10 m/min)
- feed rate: 0.002 i.p.r. (0.05 mm/rev)
- depth of cut: 0.008 in (0.2 mm)
- cutting fluid: dry

Beyond the point of maximum cutting forces (about 40 A), the chips began to be continuous and the surface finish improved. Similar results were obtained for mullite, alumina, zirconia, and silicon nitride.

An important contribution in the Kitagawa and Maehawa (1990) article is the concept of producing a deteriorated layer of microcracks extending just short of the depth to be removed by heating and cooling the surface before machining. Figure 12.73 shows the change in flank wear for a cutting time of 4 min when a Pyrex specimen having a deteriorated surface of just short of 0.008 in (0.2 mm) was turned with a variable plasma arc current.

Catastrophic failure often resulted when alumina, silicon nitride, and zirconia were hot machined at temperatures of 2700 °F (1500 °C) and higher using plasma arc heating. This occurred not while cutting when the thermal surface stress was compressive but upon rapid cooling when the surface stress is tensile. The material beneath the heated surface layer resists contraction of the surface due to heat transfer to the air giving rise to a troublesome tensile stress. The tendency for gross fracture upon cooling was greatest with silicon nitride (Si_3N_4), least for alumina (Al_2O_3), and intermediate for zirconia (ZrO_2). If thermal shock is avoided by slowly decreasing the surface temperature by continued thermal heating during cooling, then plasma arc hot machining may be useful when machining ceramics.

When machining ceramics at room temperature, the radial stress F_R is greater than the tangential stress F_P and the chips produced are very discontinuous. However, when hot machining these materials above the temperature where continuous chips are formed, $F_P > F_R$. The change from discontinuous to continuous chips occurs when $F_P = F_R$.

Miyasaka et al. (1991) have published a report concerning the machining of Si_3N_4 at temperature over 1800 °F (1000 °C) with the application of plasma arc heating using an arrangement similar to Fig. 12.71. The water-cooled tool shown in Fig. 12.74 was used to turn Si_3N_4 at temperatures of 2820 °F (1550 °C) and above (where continuous chips are first formed) using a CBN insert. The tool life of a water-cooled tool was an order of magnitude greater than that of an uncooled tool under the same plasma arc heating conditions.

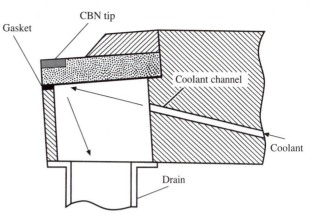

HARD TURNING

Kishawy (2002) has discussed an experimental study of dry turning temperatures obtained when machining hardened type D2 tool steel (H_{RC} 62) with PCBN inserts having a clearance angle of 0° and a side cutting angle of 0° [orthogonal turning of 3.94 in (100 mm)

Fig. 12.74 Water-cooled tool for use in hot machining with plasma arc. (after Miyasaka et al., 1991)

diameter bars]. Thermocouples were mounted in a groove at the bottom of the chip breaker at B and C as shown in Fig. 12.75. The thermocouple at A was used for calibrating the thermocouples at B and C. The cutting temperature was found to reach the same temperature of 1470 °F (800 °C) at B and C after cutting for 30 s at a cutting speed of 1148 f.p.m. (350 m/min), a feed of 0.002 i.p.r. (0.05 mm/rev), and a depth of cut of 0.004 in (0.1 mm). The following items were explored and the effects on cutting edge temperature, θ_c (shown in Fig. 12.76) were obtained:

- cutting speed (V): θ_c increased from 840 °F (450 °C) at 460 f.p.m. (140 m/min) to 1800 °F (1000 °C) at 1640 f.p.m. (500 m/min) (Fig. 12.76a)

- feed (f): 212 °F (100 °C) or less increase in θ_c for change in feed from 0.002 i.p.r. (0.05 mm/rev) to 0.004 i.p.r. (0.1 mm/rev) (Fig. 12.76a)

- a slight increase in θ_c for an increase in tool nose radius (Fig. 12.76b) (due to increase in cut length with increase in nose radius)

- a slight increase in θ_c with an increase in effective rake angle from −35° to 0° (Fig. 12.76c)

- a sizeable increase in θ_c with increase in cutting speed for a honed edge but not for a. sharp edge (Fig. 12.76d)

- an increase in θ_c and for increase in wear land (Fig. 12.76e)

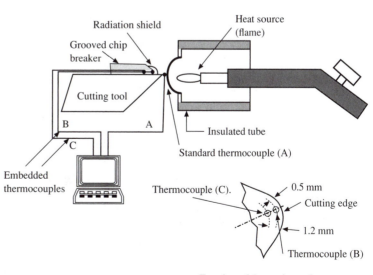

Fig. 12.75 Experimental setup for measuring temperatures at cutting edge of tool. (after Kishawy, 2002)

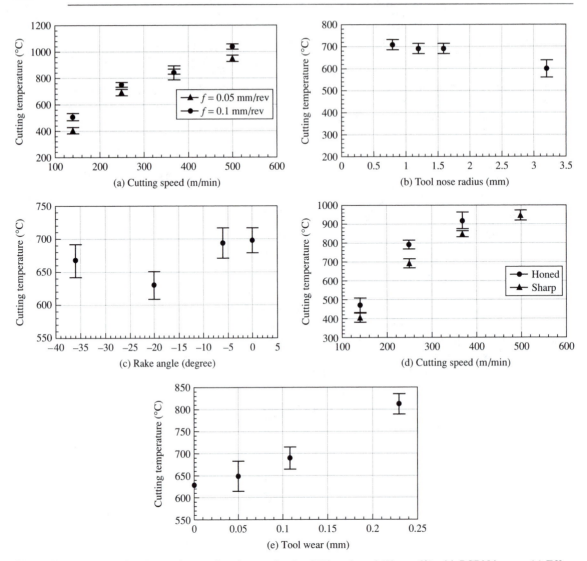

Fig. 12.76 Representative dry turning results when cutting hard D2 tool steel ($H_{RC} = 62$) with PCBN inserts. (a) Effect of cutting speed (V) and feed (f) on temperature at cutting edge (θ_c): depth of cut, 0.004 in ($b = 0.1$ mm); nose radius, $NR = 0.064$ in (1.6 mm); and sharp tool. (b) Effect of NR: $V = 820$ f.p.m. (250 m/min), $f = 0.002$ i.p.r. (0.05 mm/rev), $b = 0.004$ in (0.1 mm), sharp tool. (c) Influence of effective rake angle: $V = 820$ f.p.m. (250 m/min), $f = 0.004$ i.p.r. (0.1 mm/rev), $b = 0.004$ in (0.1 mm). (d) Effect of edge preparation: $f = 0.002$ i.p.r. (0.05 mm/rev), $b = 0.004$ in (0.1 mm), $NR = 0.064$ in (1.6 mm). (e) Effect of wear land (w): $V = 850$ f.p.m. (250 m/min), $f = 0.004$ i.p.r. (0.1 mm/rev), $b = 0.008$ in (0.2 mm), $NR = 0.128$ in (3.2 mm) and with chamfered edge. (after Kishawy, 2002)

REFERENCES

Akcan, N. S. (1998). M.Sc. thesis. Purdue University, West Lafayette, Ind.

Al Huda, M., Yamada, K., Hosokana, A., and Ueda, T. (2000). *ASME Transactions J. Mfg. Sc. and Eng.*

Anon. (1939). *The Iron and Steel Institute, Sp. Rep.* **24**, 215.

Anon. (1946). *Proc. Iron and Steel Inst.* **154**, 83.

Arndt, G. (1971). *Proc. H. Armstrong Conf. on Prod. Science in Ind.* at Monash University, Australia.

Arndt, G. (1973). *Proc. Inst. of Mech. Engrs., UK* **187 44/73**, 625–633.

Arndt, G., and Brown, R. H. (1967). *Int. J. Mach. Tool Des. Res.* **7**, 39.

Barrow, G. (1966). *Annals of CIRP* **14/1**, 145.

Bever, M. B., Holt, R., and Tichener, L. B. (1974). *Prog. Mater. Sci.* **17**.

Bever, M. B., Marshall, E. R., and Tichener, L. B. (1953). *J. Appl. Phys.* **24**, 117.

Bickel, E. (1963). In *International Research in Production Engineering*. ASME, New York, p. 89.

Bickel, E., and Widmer, W. (1951). *Industrial Organization* **8**.

Blok, H. (1938). In *Proc. General Discussion in Lubrication and Lubricants*. Instn. Mech. Eng., London, p. 222.

Boothroyd, G. (1963). *Proc. Instn. Mech. Engrs.* **177**, 789.

Boothroyd, G. (1975). In *Fundamentals of Metal Cutting and Machine Tools* (Chapter 3). Hemisphere Pub., Washington.

Carslaw, H. S. (1921). *Introduction to the Mathematical Theory of the Conduction of Heat in Solids.* Macmillan, London.

Chao, B. T., Li, H. L., and Trigger, K. J. (1961). *J. Eng. Ind. Trans. Am. Soc. Mech. Engrs.* **83**, 496.

Chao, B. T., and Trigger, K. J. (1953). *Trans. Am. Soc. Mech. Engrs.* **75**, 109.

Chao, B. T., and Trigger, K. J. (1955). *Trans. Am. Soc. Mech. Engrs.* **77**, 1107.

Chao, B. T., and Trigger, K. J. (1958). *Trans. Am. Soc. Mech. Engrs.* **80**, 311.

Chao, B. T., Trigger, K. J., and Zylstra, L. B. (1952). *Trans. Am. Soc. Mech. Engrs.* **74**, 1039.

Childs, T. H. C., Maekawa, K., and Maulik, P. (1988). *Materials Science and Technology* **4**, 1006–1019.

Colwell, L. V., and Quackenbush, L. J. (1962). *The University of Michigan Report 05038—1 and 2*.

Gottwein, K. (1925). *Maschinenbau* **4**, 1129.

Guerrieri, M. (1987). *Infrared Temperature Sensing, SME Technical Paper* MS, 87–408.

Hahn, R. S. (1951). *Proc. First U.S. Nat. Cong. Appl. Mech.*, p. 661.

Herbert, E. G. (1926). *Proc. Instn. Mech. Engrs.* **1**, 289.

Hirao, M. (1987). *J. Mech. Eng. Lab., Japan* **41**, 22–29.

Jaeger, J. C. (1942). *Proc. R. Soc. N.S.W.* **78**, 203.

Ju, Y. (1997). Ph.D. thesis. Purdue University, West Lafayette, Ind.

Kishawy, H. A. (2002). *Int. J. Mach. Sc. Tech.* **6/1**.

Kitagawa, T., and Maekawa, K. (1990). *Wear* **139**, 251–267.

Klopstock, H. (1925). *Trans. ASME* **47**, 345–377.

Komanduri, R., Flom, D. G., and Lee, M. (1984b). In *High Speed Machining, ASME PED Vol. 12*, pp. 16–36.

Komanduri, R., and Lee, M. (1984). In *High Speed Machining, ASME PED Vol. 12*, pp. 217–229.

Komanduri, R., Subramanian, K., and von Turkovich, B. F. (1984a). *High Speed Machining, ASME PED Vol. 12*.

Kozak, J., and Rajurkar, K. P. (2000). *Advances in Manufacturing and Technology, Polish Academy of Sciences* **24/2**, 25–50 and **24/3**, 5–24.

Krabacher, E. J., and Merchant, M. E. (1951). *Trans. Am. Soc. Mech. Engrs.* **73**, 761.

Kronenberg, M. (1949). *Am. Mach.* **93**, 104.

Kronenberg, M. (1954). In discussion to Loewen and Shaw (1954).

Kusters, K. J. (1956). Dissertation, T. H. Aachen.

Lee, I. M. (1980). *Adv. Machining Res. Prog. Ann. Tech Report SRD* **80–118**, 8.1–8.3.

Leone, W. C. (1954). *Trans. Am. Soc. Mech. Engrs.* **76**, 121.

Lezanski, P., and Shaw, M. C. (1990). *Trans. ASME, J. Eng. for Ind.* **112**, 132.

Li, X., Kopalinsky, E. M., and Oxley, P. L. B. (1995). *Proc. Inst. of Mech. Engrs.* **B209**, 33–43.

Loewen, E. G., and Shaw, M. C. (1954). *Trans. ASME* **76**, 217.

Miyasaka, K., Kasuya, U., Inkuai, K., Hikao, M., Ono, T., and Sakata, O. (1991). *J. Mech. Eng. Lab., Japan* **45–6**, 239–248.

Nakayama, K. (1956). *Bull Fac. Engng. Yokohama Nat. Univ.* **5**, 1.

Nakayama, K. (1959). *Bull Fac. Engng. Yokohama Nat. Univ.* **8**, 1.

Okoshi, M., and Uehara, K. (1963). *International Research in Production Engineering*. ASME, New York, pp. 264–271.

Okushima, K. (1965). *Bull. Japan Soc. Mech. Engrs.* **8**, 702.

Oxley, P. L. B. (1976). *Int. J. of Mach. Tool Design and Research* **16**, 335.

Pahlitzsch, G., and Helmerdig, H. (1943). *A des Ver. Deutscher Ing.* **81**, 564 and 691.

Palmai, Z. (1987). *Int. J. of Mach. Tools and Manufacture* **27**, 261–274.

Rajogopal, S. (1982). *Laser Focus* **18**, 49.

Rapier, A. C. (1954). *Br. J. Appl. Phys.* **5**, 400.

Recht, R. F. (1960). M.Sc. thesis. Denver University, Colo.

Recht, R. F. (1964). *Trans. ASME, J. Appl. Mech.* **31**, 189–193.

Rumford, Count. (1799). In *Essays Political, Economical and Philosophical*. David West, Boston.

Schallbrock, H., and Lang, M. (1943). *Z. Ver. Deutscher Ing.* **87**, 15.

Schaller, E. (1964). Dissertation, T. H. Aachen.

Schmidt, A. O., and Roubik, J. R. (1949). *Trans. Am. Soc. Mech. Engrs.* **71**, 245.

Schmidt, A. O., and Roubik, J. R. (1954). In discussion to Loewen and Shaw (1954), p. 229.

Schulz, H. (1989). *Hochgeschwindigkeitsfrassen Metalischer und Nichtmetalischer Werkstoffe*. C. Hanser Verlag.

Schulz, H., and Moriwaki, T. (1992). *Annals of CIRP* **41/2**, 637–643.

Schwerd, F. (1937). *Z. des Ver. Deutscher Ing.* **77**, 211.

Shaw, M. C. (1958). *Technische Mitteilungen* (Essen) **5**, 211.

Shore, H. (1924). M.S. dissertation, MIT. Also *J. Wash. Acad. Sci.* **15**, 85 (1925).

Smart, E. F., and Trent, E. M. (1974). In *Proc. Int. Conf. Machine Tool Des. Res.* Macmillan, London, p. 187.

Smart, E. F., and Trent, E. M. (1975). *Int. J. Prod. Res.* **13**, 265.

Solomon, C. (1931). Deutsche Patentschrift No. 523594K149b.

Stephenson, D. A. (1991a). *Trans. ASME, J. Eng. for Ind.* **113**, 121–128.

Stephenson, D. A. (1991b). *Trans. ASME, J. Eng. for Ind.* **113**, 129–136.

Stephenson, D. A. (1992). *Trans. ASME, J. Eng. for Ind.* **114**, 432–437.

Stephenson, D. A. (1993). *Trans. ASME, J. Eng. for Ind.* **115**, 432–437.

Stephenson, D. A., and Ali, A. (1992). *Trans. ASME, J. Eng. for Ind.* **114**, 127.

Tanaka, Y., Tsuwa, H., and Kitano, M. (1967). *ASME paper No. 67, Prod. 14.*

Tay, A. O., Stevenson, M. G., and de Vahl Davis, G. (1974). *Proc. Inst. Mech. Engrs.* **188**, 627–628.

Tay, A. O., Stevenson, M. G., de Vahl Davis, G., and Oxley, P. L. B. (1976). *Int. J. of Mach. Tool Design and Research* **16**, 335–349.

Taylor, G. I., and Quinney, H. (1934). *Proc. R. Soc.* **A143**, 307.

Taylor, G. I., and Quinney, H. (1937). *Proc. R. Soc.* **A163**, 157.

Templin, R. I. (1948). *Trans. ASME* **70**, 887.

Trent, E. M. (1976). *Metal Cutting*. Butterworths, London.

Trigger, K. J. (1963). In *International Research in Production Engineering*. ASME, New York, p. 95.

Trigger, K. J., and Chao, B. T. (1951). *Trans. Am. Soc. Mech. Engrs.* **73**, 57.

Ueda, T., Hosokawa, A., and Yamoto, A. (1985). *ASME Trans., J. Eng. Ind.* **107**, 127.

Ueda, T., Iriyama, T., Sugita, T. (1995). *J. of JSPE* **61**, 278–282.

Ueda, T., Sato, M., and Nakayama, K. (1998). *Annals of CIRP* **47/1**, 41–44.

Uehara, K. (1968). *Annals of CIRP* **16/1**, 85–91.

Vaughn, R. L. (1960). *Ultra High Speed Machining*. Final Report, Lockheed Aircraft Co., Burbank, Calif.

Venuvinod, P. K., and Lau, W. S. (1986). *Int. J. Mach. Tool Design and Research* **26**, 1–14.

Wang, Z. Y., Sahay, C., and Rajurkar, K. P. (1996). *Int. J. Mach. Tools and Mfg.* **36**, 129–140.

Weiner, J. (1955). *Trans. Am. Soc. Mech. Engrs.* **77**, 1331.

Wright, P. K. (1978). *J. Engng. Ind.* **100**, 131.

Wright, P. K., McCormick, S. P., and Miller, T. R. (1980). *Trans. ASME, J. Eng. for Ind.* **102**, 123–128.

Wright, P. K., and Trent, E. M. (1973). *J. Iron Steel Inst.* **211**, 361.

Zorev, N. N. (1963). In *International Research in Production Engineering*. ASME, New York, p. 42.

13 CUTTING FLUIDS

The two main functions of cutting fluids are lubrication at relatively low cutting speeds and cooling at relatively high cutting speeds. At high cutting speeds there is not time for fluid to penetrate the chip–tool interface, the wear land, or microcracks on the back of the chip to provide lubrication; and at low cutting speeds cooling is relatively unimportant. Typical low-speed cutting operations where lubrication is important are broaching and tapping while typical high-speed machining operations where cooling is important are turning and milling. Water-base cutting fluids are more likely to be used in the high-speed cooling regime while oil-base fluids are more apt to be used in the low-speed lubrication regime.

There are many other secondary considerations associated with cutting fluid selection (Shaw, 1970) such as

- chip disposal
- corrosion
- health, safety, and aesthetic considerations
- cost

The main objective in use of a cutting fluid is either reduction in total cost per part or an increase in the rate of production.

LOW-SPEED CUTTING FLUID ACTION

The data presented in Table 13.1 illustrates the very large difference a cutting fluid can make when used in an orthogonal cutting operation performed at a very low cutting speed. In these tests the workpiece was moved past the stationary tool (planer fashion) at very low speed ($V = 1$ i.p.m. = 0.025 m min^{-1}). The undeformed chip thickness was 0.003 in (0.08 mm) and the width of cut 1.000 in (25.4 mm). Forces were measured using a two-component dynamometer, and fluid was applied dropwise at the chip–tool interface. Only those values are recorded that correspond to a continuous chip, and no BUE was evident in any of the tests recorded in Table 13.1.

Figure 13.1 shows values of shear stress (τ) and normal stress (σ) on the tool face plotted with shear stress positive (clockwise) and with shear stress negative (counterclockwise) for points on the shear plane. The numbers refer to the test numbers in Table 13.1. Single Mohr's circles passing through the origin and corresponding points (same test number) for tool face and shear plane are shown in

TABLE 13.1 Cutting Data for Annealed AISI 1112 in Orthogonal Machining Tests at Low Cutting Speed

Test No.	Fluid	α deg	F_P lb	F_Q lb	r	ϕ deg	γ	$\tau(\times 10^{-3})$ p.s.i.	$\sigma(\times 10^{-3})$ p.s.i.	μ	$\beta = \tan^{-1}\mu$ deg
1	Benzene	45	699	84	0.602	36.5	1.20	85.6	84.0	1.33	53.0
2	Air	45	504	46	0.629	38.6	1.14	70.1	72.7	1.20	50.2
3	Ethanol	45	374	−6	0.680	42.7	1.04	63.1	56.5	0.97	44.1
4	CCl$_4$	45	329	−41	0.739	47.4	0.96	62.1	52.5	0.78	37.9
5	Benzene	30	912	251	0.380	22.2	2.31	94.4	72.7	1.01	45.4
6	Air	30	740	142	0.454	27.0	1.91	89.9	69.8	0.86	40.9
7	Ethanol	30	541	40	0.521	31.4	1.66	76.4	54.8	0.68	34.2
8	CCl$_4$	30	437	−30	0.625	38.2	1.41	74.9	50.7	0.49	26.1
9	Air	16	1115	347	0.301	17.5	3.20	96.5	66.9	0.66	33.3
10	Ethanol	16	850	168	0.362	21.1	2.68	87.9	55.6	0.51	27.2
11	CCl$_4$	16	576	19	0.485	28.1	2.07	78.9	46.2	0.32	17.9
12	CCl$_4$	0	816	108	0.340	18.8	2.28	79.2	39.2	0.13	7.6

Note: 1 lb = 4.45 N; 1 p.s.i. = 0.69 N cm^{-2}; V = 1 i.p.m. (0.025 m min^{-1}); t = 0.003 i.p.r. (0.08 mm/rev); b = 1 in (25.4 mm). (after Shaw, Cook, and Finnie, 1953)

each case, only the smallest and largest circles being shown completely. From Fig. 13.1, it is evident that for the same Mohr's circle to hold for shear plane and tool face (no strain hardening) neither the shear plane nor the tool face can be a plane of maximum shear stress. Figure 13.2a shows a slip line field corresponding schematically to any of the points in Fig. 13.1 while Fig. 13.2b is the corresponding Mohr's circle. Here the maximum shear direction deviates by an angle η' from the shear plane and an angle η from the tool face.

Several observations may be made concerning the data of Table 13.1:

1. Benzene which is a very stable compound does not contaminate a clean metal surface by reacting with it to prevent adhesion but merely excludes air which is a useful (friction-reducing) contaminant. Thus benzene is a negative "lubricant" relative to dry cutting in air.

2. Ethanol and carbon tetrachloride are positive lubricants relative to air since they lower cutting forces and increase shear angle.

3. The coefficient of tool face friction (μ) decreases as the effectiveness of the cutting fluid increases and also as the rake angle decreases.

4. The force component F_Q tends to be negative with increased effectiveness of cutting fluid and increased rake angle.

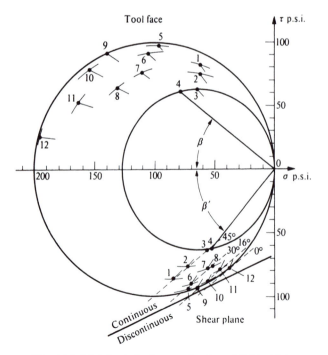

Fig. 13.1 Mohr's circle diagram for data of Table 13.1.

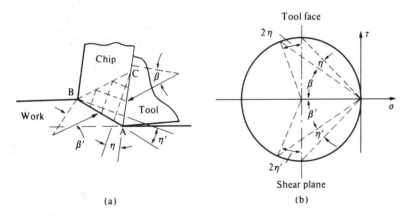

A negative F_Q represents an unstable condition in that tool deflection will tend to cause an increase in $-F_Q$, a further increase in t, etc. Thus, when F_Q is negative, the tool will tend to cut deeper with time until its stiffness becomes nonlinear or the tool fractures.

5. The tendency to form a discontinuous chip increases with decrease in rake angle but decreases with effectiveness of the cutting fluid. The only fluid that was capable of producing a continuous chip with a rake angle of zero degrees under the conditions of Table 3.1 was CCl_4 (carbon tetrachloride).

6. A BUE will not form at very low cutting speeds such as that for Table 13.1. The tendency to form a BUE at first increases and then decreases with cutting speed reaching a maximum at a speed of about 10 f.p.m. (3 m min^{-1}), when cutting in air. In the speed range where a BUE forms (say 5 f.p.m. (1.5 m min^{-1})) it will be suppressed by effective cutting fluids such as CCl_4. A BUE will not form if η is negative. For test nos. 7–12, η is positive and a BUE could form were it not for the very low cutting speed.

7. The shear and normal stresses on the tool face are far from constant.

8. The shear stress on the shear plane increases with normal stress on the shear plane.

9. All values of η are negative; that is, the material approaching the shear plane appears to shear before reaching the maximum value of shear stress which is within the chip.

Carbon Tetrachloride

Carbon tetrachloride (CCl_4) is one of the most effective cutting fluids known for a wide range of work materials at low cutting speeds. Other effective low-speed fluids perform in a similar way to CCl_4 but less decisively. It is therefore useful to study the performance of CCl_4 as a model low-speed cutting fluid. Unfortunately, CCl_4 is very toxic and must not be used in the workshop since small amounts entering the blood stream through the lungs or skin can lead to serious consequences and even death under certain conditions.

The fact that the coefficient of tool face friction (μ) is greatly reduced with CCl_4 (see Table 13.1) suggests that the reaction products are merely acting as a lubricant. However, it is also observed that the shear stress (τ) and shear strain (γ) on the shear plane are also reduced with CCl_4. The question then arises as to which is cause and which, effect. Is the main action of CCl_4 on the tool face (which effects a decrease in shear strain and shear stress on the shear plane) or is the main action on the shear plane with the reduction in μ on the tool-face a consequence due to the close coupling between the shear and friction processes?

TABLE 13.2   Friction Experiments on Freshly Cut Surfaces

Surface Treatment	Fluid	Mean Coefficient of Friction μ
Cut with 0.010 in (0.25 mm)	CCl_4	0.23
feed using CCl_4;	Air	0.18
allowed to dry	Oleic Acid	0.14
Cut with 0.010 in (0.25 mm)	CCl_4	0.32
feed using oleic acid;	Air	0.22
washed with CCl_4 and allowed to dry	Oleic Acid	0.19
Cut with 0.005 in (0.125 mm)	CCl_4	0.29
feed in air	Air	0.65
	Oleic Acid	0.17
Cut with 0.002 in (0.05 mm)	Water	0.39
feed using water;	CCl_4	0.38
allowed to dry	Air	0.38
	Graphite	0.15
	Molybdenum disulfide	0.10

Carbon Tetrachloride as a Lubricant

There appears to be no evidence that either carbon tetrachloride or its reaction product with steel ($FeCl_3$) exhibits good lubrication action in the sense that these materials give low sliding friction. For example, when CCl_4 is used as a wire-drawing lubricant for steel wire, very high friction results whereas materials such as oleic acid give low die friction.

A number of friction tests have shown that both CCl_4 and $FeCl_3$ may give much higher friction than oleic acid between sliding steel surfaces and in fact even higher friction than that pertaining for dry surfaces sliding in air. These tests were run by first making a planing type cut of $\frac{1}{4}$ in (6.35 mm) width on a specimen 3 in (76.2 mm) long. The clearance face of the high speed steel tool used to make the cut was then immediately loaded against the freshly cut surface and drawn back after first applying the appropriate lubricant. The coefficient of friction recorded using a cutting dynamometer was found to be independent of the normal load. Values given in Table 13.2 are averages of five or more tests each and are representative of a large number of such sliding tests.

As long as the cut surface is smooth (i.e., when cut using CCl_4 or oleic acid), the coefficient of sliding friction in air is small and much below that pertaining for carbon tetrachloride. When, however, the surface is rough (as when cut in air), the oxide breaks down and the friction is very high. In some tests the surface cut with CCl_4 was first washed with water before testing with air, or oleic acid, but this made no difference in the results obtained. In no case was the friction coefficient obtained with CCl_4 in the same class with such good lubricants as oleic acid, graphite, or molybdenum disulfide.

In Table 13.3 friction values are given for a steel ball rubbing on a steel turntable. Before each test a new spot on the ball was selected and both sliding surfaces were washed with benzene. The steel plate was then abraded with 4/O aluminum oxide paper immediately before the test lubricant was applied. The plate was somewhat rougher than the surfaces mentioned in Table 13.3 and hence the friction results in air were high. In this test the CCl_4 values were of intermediate magnitude

TABLE 13.3 Friction Tests with $\frac{1}{4}$ in (6.35 mm) Diameter AISI 51100 Hardened Steel Ball Sliding on an AISI 1020 Steel Turntable at 6 f.p.m. (1.88 m min^{-1})

Fluid	Mean Coefficient of Friction μ
Air	0.52
FeCl$_3$	0.49
CCl$_4$	0.29
Mineral oil	0.18
Oleic acid	0.13
Graphite	0.11
Molybdenum disulfide	0.08

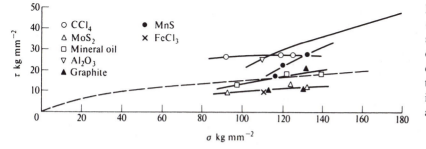

Fig. 13.3 Variation of τ and σ for friction tests involving subsurface plastic flow with different lubricants present. The dotted curve is the curve for tests in dry air given previously in Fig. 10.10. (after Shaw, Ber, and Mamin, 1960)

but distinctly higher than those for mineral oil or the better lubricants: oleic acid, graphite, or molybdenum disulfide.

From these considerations it would appear that CCl$_4$ is not a lubricant in the sense that oleic acid or graphite is or in fact even in the sense that a thin oxide film present on a smooth metal surface is.

The apparatus of Fig. 10.9 in which a smooth steel ball is rotated in an AISI 1020 steel socket was used with different lubricants applied to the socket before the test (Fig. 13.3). CCl$_4$ was again found to be poor lubricant relative to graphite, MoS$_2$, or even air. The results shown in Fig. 13.3 were initial values corresponding to a fraction of 1 revolution of the ball in the socket.

A new subsurface plastic flow friction apparatus was constructed in which the ball could be rotated continuously at low speed. When running a test in air (AISI 52100 ball against AISI 1020 steel), it was found that the low initial friction torque rose gradually to a steady equilibrium value that was about six times as high as the initial value (Fig. 13.4). During the period of rising torque, the soft steel surface was severely galled. Figure 13.5 shows mean frictional stress (τ) versus mean normal stress (σ) when sliding in air under initial (smooth) and steady state (galled) conditions.

Figure 13.6a shows results similar to those of Fig. 13.4 while Fig. 13.6b shows results similar

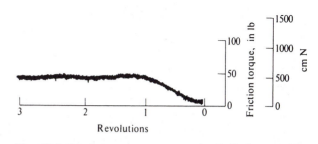

Fig. 13.4 Friction torque versus rotational displacement for sliding in air. Load on sphere = 2200 lb (1000 kg); mean sliding speed = 0.16 f.p.m. (0.05 m min^{-1}); sphere diam. = 0.500 in (12.7 mm); hole diam. = 0.125 in (3.8 mm). (after Shirakashi, Komanduri, and Shaw, 1978)

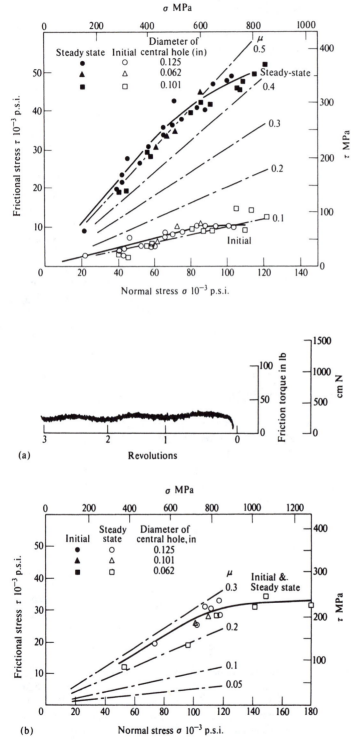

Fig. 13.5 Variation of mean shear stress (τ) with mean normal stress σ for sliding in air under initial (open points) and steady state (solid points). Conditions same as Fig. 13.4. (after Shirakashi, Komanduri, and Shaw, 1978)

Fig. 13.6 Tests with CCl_4. (a) Friction torque versus displacement. (b) Variation of mean shear stress (τ) with mean normal stress (σ) for sliding with CCl_4. There is no difference between initial and equilibrium values in this case. Conditions same as Fig. 13.4. (after Shirakashi, Komanduri, and Shaw, 1978)

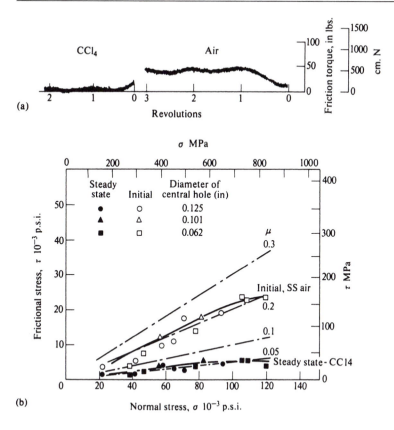

Fig. 13.7 Test to equilibrium surface condition in air followed by CCl$_4$. (a) Friction torque versus displacement. (b) Variation of mean shear stress (τ) with mean normal stress (σ). Conditions same as Fig. 13.4. (after Shirakashi, Komanduri, and Shaw, 1978)

to Fig. 13.5 for CCl$_4$. It is evident that with CCl$_4$ the steady state value is reached immediately. It should also be noted that whereas the steady state value of coefficient of friction (μ) for air is about 0.5, that for CCl$_4$ is now only about 0.28. CCl$_4$ now looks like a lubricant relative to air.

Additional tests were run in which sliding in air was continued until equilibrium (galling) was obtained in air followed by sliding with CCl$_4$. Figure 13.7a shows the torque versus displacement curves for a typical test while Fig. 13.7b shows that the coefficient of friction (μ) for CCl$_4$ is very low (~ 0.05) when sliding on a very rough surface. The CCl$_4$ might be said to be a super lubricant under these conditions.

Figure 13.8 shows values of τ and σ for surfaces of different initial roughness with CCl$_4$ as lubricant.

The foregoing results are consistent with the following explanation. Carbon tetrachloride is ineffective as a boundary lubricant and is, in fact, a negative boundary lubricant if it does not react chemically with the surface being lubricated to form a brittle compound (FeCl$_3$ in the case of steel). The negative action comes from the fact that it excludes oxygen in air. Since metal oxides are somewhat beneficial in lowering frictional resistance, the presence of air is important.

Surface asperities are merely flattened by an indentor but not removed completely unless the indentor is slid across the surface (Chapter 10). This is because the plastic zones beneath each asperity grow as they are flattened under the increased load, but only until adjacent plastic zones interact. After that, there is gross plastic flow beneath the indentor instead of localized flow beneath individual asperities. The tendency for CCl$_4$ to react with iron will increase with pressure. For a smooth surface, the pressure will be insufficient to cause CCl$_4$ to react to form FeCl$_3$ and the role of

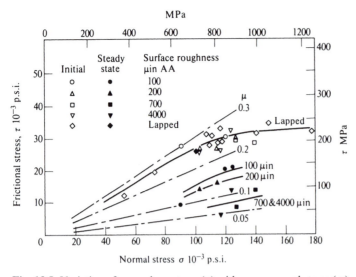

Fig. 13.8 Variation of mean shear stress (τ) with mean normal stress (σ) for sliding with CCl_4 when the initial surface roughness has different values in μ in AA. Conditions same as Fig. 13.4. (after Shirakashi, Komanduri, and Shaw, 1978)

CCl_4 will be to exclude oxygen which in turn results in high friction. This explains why the friction was so high when a tool used to cut steel or aluminum using CCl_4 was slid back over a freshly cut surface. The surface produced was so smooth that the CCl_4 did not react to form a metal chloride on the backward pass. The foregoing explanation is also consistent with the results of Fig. 13.8. The pressure is too low on asperities of a smooth surface to cause CCl_4 to react, and hence its role is the negative one of excluding oxygen.

When local pressure is sufficiently high, CCl_4 will react with the metal and the reaction product will prevent galling and junction growth. This then causes friction with CCl_4 to be less than in air. It appears probable, as suggested by De Chiffre (1977), that an effective low-speed cutting fluid such as CCl_4 can function as a lubricant only in the tool face region away from the cutting edge where $A_R/A < 1$ and the chip slides over the face of the tool. Even in this region, it is not clear whether the action will be positive (low μ) or negative (high μ) because of the decreasing normal stress on the tool face with distance form the cutting edge.

It is necessary for CCl_4 to penetrate between chip and tool before it can react with the chip surface to prevent junction growth and galling. Vapor would have a greater chance of penetrating the intercommunicating voids in the region of contact further from the cutting edge than liquid. It has been demonstrated that CCl_4 vapor is just as effective as the liquid in slow-speed cutting (Shaw, 1951). This was shown by cutting metal at low speed in a heated box into which superheated CCl_4 vapor was introduced. By maintaining the temperature of the box and tool above the boiling point of CCl_4 (80 °C), it was certain that only vapor reached the cutting zone. When pure aluminum was cut at low speed in the presence of CCl_4 vapor, the same dramatic improvement of Table 13.1 was obtained, and the following reaction was found to occur on the freshly cut and highly active surfaces contacting the vapor:

$$2\,Al + 6\,CCl_4 \rightarrow 2\,AlCl_3 + 3\,C_2Cl_6 \tag{13.1}$$

The reaction products on the right-hand side of this equation were identified.

Carbon Tetrachloride on Lead

Lead is about the only metal for which CCl_4 does not give reduced cutting forces relative to cutting in air. The reason normally given to explain this is that $PbCl_3$ has a higher shear strength than lead itself whereas all other metal chlorides are weaker than the corresponding metal. Tests were run on lead using the improved apparatus of Fig. 10.9 to better understand the performance of CCl_4 on lead. Figure 13.9 gives results of tests on lead performed in air and Fig. 13.10, results for tests on lead with CCl_4. While the initial value of μ is slightly greater for CCl_4 than for air, the steady state

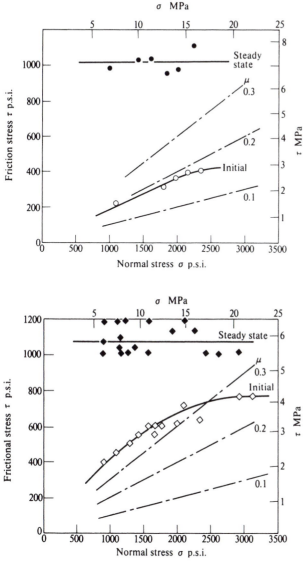

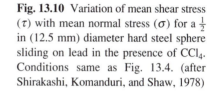

Fig. 13.9 Variation of mean shear stress (τ) with mean normal stress (σ) for a $\frac{1}{2}$ in (12.5 mm) diameter hard steel sphere sliding on lead in air. Conditions same as Fig. 13.4. (after Shirakashi, Komanduri, and Shaw, 1978)

Fig. 13.10 Variation of mean shear stress (τ) with mean normal stress (σ) for a $\frac{1}{2}$ in (12.5 mm) diameter hard steel sphere sliding on lead in the presence of CCl_4. Conditions same as Fig. 13.4. (after Shirakashi, Komanduri, and Shaw, 1978)

values are seen to be about the same. The explanation for this is that the air-lubricated specimen gives low friction until the oxide film initially present on the asperities is worn away. The friction in air is then that of clean metal resulting in high friction and galling. In the case of CCl_4, the $PbCl_3$ formed is too brittle on a soft substrate (lead) and fractures exposing clean lead as in the equilibrium air case. To check this, tests were performed using a hydrogen peroxide solution (H_2O_2) as a source of oxygen to replace the oxide worn away in the tests of Fig. 13.9. Figure 13.11 shows that in this case the steady state value is the same as the initial value.

It would thus appear that CCl_4 is a negative lubricant for lead due primarily to the fact that the $PbCl_3$ layer is too brittle to support the normal load without fracture when the substrate is as soft as lead.

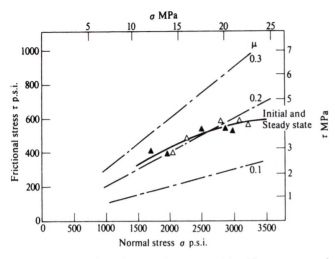

Fig. 13.11 Variation of mean shear stress (τ) with mean normal stress (σ) for a $\frac{1}{2}$ in (12.5 mm) diameter hard steel sphere sliding on lead in the presence of hydrogen peroxide. Conditions same as Fig. 13.4. (after Shirakashi, Komanduri, and Shaw, 1978)

Other Fluids at Low Cutting Speed

A wide variety of organic chemicals covering a range of chain length and several homologous series have been found to give low coefficient of tool-face friction and a high cutting ratio relative to air as shown in Table 13.4. All fluids were put through a small laboratory still (nine theoretical plates, 5 ml holdup). Low-speed (5.5 in m^{-1} = 0.14 m min^{-1}) planer type cuts of 0.005 in (0.125 mm) undeformed chip thickness and 0.250 in (6.35 mm) width were made on commerically pure aluminum, the keenness of the cutting edge being checked periodically using CCl$_4$ as a standard. The first three cuts with each new fluid were ignored being considered as part of the tool cleaning process. The values in Table 13.4 are average values for five cuts.

The disulfides, mercaptans, and long chain length esters were found to be outstanding in their ability to reduce tool-face friction. The lowest coefficient of friction obtained was 0.237. A general decrease in μ with increased chain length was observed for all homologous series. The cutting ratio (r) and corresponding shear angle (ϕ) are shown plotted against the coefficient of tool-face friction (μ) in Fig. 13.12. All points except those for water and pentachlorethane fall on a smooth curve which suggests that in general a relation between r and μ exists.

It is evident from the data in Table 13.4 for the several chlorinated hydrocarbons that there is no correlation between the chlorine content of the molecule and the observed coefficient of friction;

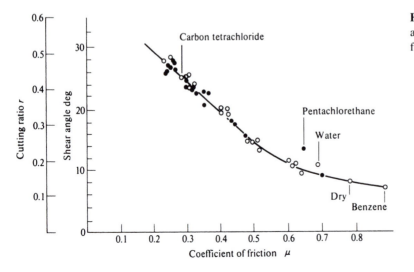

Fig. 13.12 Variation of cutting ratio and shear angle with coefficient of friction for data of Table 13.4.

TABLE 13.4 Friction Data for a Variety of Organic Compounds[†]

Fluid	Formula	Cutting Ratio r	Coefficient of Friction μ	Fluid	Formula	Cutting Ratio r	Coefficient of Friction μ
Chlorinated				Mercaptans:			
Hydrocarbons:				Ethyl mercaptan	H_5C_2-SH	0.437	0.308
Carbon tetrachloride	CCl_4	0.427	0.288	Propyl mercaptan	H_7C_3-SH	0.440	0.276
Chloroform	$CHCl_3$	0.401	0.312	Butyl mercaptan	H_9C_4-SH	0.473	0.267
Dichloromethane	CH_2Cl_2	0.265	0.477	Amyl mercaptan	$H_{11}C_5-SH$	0.473	0.265
Pentachlorethane	$Cl_2HC-CCl_3$	0.230	0.648	Hexyl mercaptan	$H_{13}C_6-SH$	0.477	0.262
S. tetrachlorethane	$Cl_2HC-CHCl_2$	0.320	0.419	Heptyl mercaptan	$H_{15}C_7-SH$	0.483	0.255
β Trichlorethane	$ClH_2C-CHCl_2$	0.344	0.407	Lauryl mercaptan	$H_{25}C_{12}-SH$	0.473	0.237
Methyl chloroform	Cl_3C-CH_3	0.340	0.422				
Ethylene dichloride	ClH_2C-CH_2Cl	0.250	0.481	Disulfides:			
Ethylidene dichloride	Cl_2HC-CH_3	0.330	0.406	Methyl disulfide	$H_3C-S-S-CH_3$	0.417	0.302
Ethyl chloride	H_3C-CH_2Cl	0.255	0.511	Ethyl disulfide	$H_5C_2-S-S-C_2H_5$	0.448	0.271
Tetrachloroethylene	$Cl_2C=CCl_2$	0.195	0.603	Propyl disulfide	$H_7C_3-S-S-C_3H_7$	0.457	0.256
Trichloroethylene	$Cl_2C=CHCl$	0.188	0.623	Butyl disulfide	$H_9C_4-S-S-C_4H_9$	0.457	0.257
T. dichloroethylene	$ClHC=CHCl$	0.185	0.612	Amyl disulfide	$H_{11}C_5-S-S-C_5H_{11}$	0.462	0.250
Hexyl choride	$C_6H_{13}Cl$	0.308	0.440				
Lauryl chloride	$C_{12}H_{25}Cl$	0.385	0.369	Alcohols			
				Ethyl alcohol	H_5C_2-OH	0.327	0.425
				Hexyl alcohol	$H_{13}C_6-OH$	0.396	0.320
Esters:				Decyl alcohol	$H_{21}C_{10}-OH$	0.400	0.325
				Undecyl alcohol	$H_{23}C_{11}-OH$	0.412	0.325
Ethyl acetate	$H_3C-C{\underset{O-C_2H_5}{\overset{O}{}}}$	0.227	0.518	Lauryl alcohol	$H_{25}C_{12}-OH$	0.385	0.331
				Hydrocarbons:			
				Hexane	C_6H_{14}	0.163	0.639
Hexyl acetate	$H_{11}C_5-C{\underset{O-C_2H_5}{\overset{O}{}}}$	0.403	0.299	Dodecane	$C_{12}H_{26}$	0.156	0.703
				Miscellaneous:			
Lauryl acetate	$H_{23}C_{11}-C{\underset{O-C_2H_5}{\overset{O}{}}}$	0.440	0.240	Benzene	C_6H_6	0.123	0.892
Ethyl caproate	$H_3C-C{\underset{O-C_6H_{13}}{\overset{O}{}}}$	0.391	0.353	Acetic acid	$H_3C-C{\underset{OH}{\overset{O}{}}}$	0.247	0.496
				Hexanoic acid	$H_{11}C_5-C{\underset{OH}{\overset{O}{}}}$	0.349	0.353
Ethyl laurate	$H_3C-C{\underset{O-C_{12}H_{25}}{\overset{O}{}}}$	0.447	0.244	Heptaldehyde	$H_{13}C_6-C{\underset{OH}{\overset{O}{}}}$	0.430	0.298
				Water	H_2O	0.187	0.690
				Air	–	0.140	0.785

[†] When aluminum is cut at 5.5 in min⁻¹ (14 cm min⁻¹) using a high-speed steel tool of rake angle, 15°, relief angle, 5°. (after Shaw, 1951)

TABLE 13.5 Comparisons of the Relative Reactivities of Several Chlorinated Hydrocarbons Toward Aluminum with Their Performance as Cutting Fluids

Fluid	Formula	Boiling Point, °C	Time to React, min	Coefficient of Friction μ	Chip Shape
Carbon tetrachloride	CCl_4	76.7	8.5	0.288	⌒
Chloroform	$CHCl_3$	61.2	135	0.312	⌒
Dichloromethane	CH_2Cl_2	39.8	360	0.477	⌒
Pentachlorethane	Cl_3C-CCl_2H	161.9 (180 mm)	6 (125 mm)	0.648	—
S. tetrachlorethane	HCl_2C-CCl_2H	146.5 (360 mm)	2 (200 mm)	0.419	⌒
β. Trichlorethane	$H_2ClC-CCl_2H$	113.5 (700 mm)	1.5 (500 mm)	0.407	⌒
Methyl chloroform	H_2C-CCl_3	74.1	2	0.422	⌒
Ethylene dichloride	$H_2ClC-CClH_2$	83.7	not in 24 hr.	0.481	⌒
Ethylidene dichloride	H_3C-CCl_2H	57.3	not in 24 hr.	0.406	⌒
Ethyl chloride	$H_3C-CClH_2$	12.5	not in 24 hr.	0.511	—
Tetrachloroethylene	$Cl_2C=CCl_2$	120.8	not in 24 hr.	0.603	—
Trichloroethylene	$HClC=CCl_2$	86.7	not in 24 hr.	0.623	—
T. dichloroethylene	$HClC=CClH$	48.4	not in 24 hr.	0.612	—

nor is there any indication that a long chain length or highly polar molecule is essential for low friction, since carbon tetrachloride gave the lowest coefficient of friction of all of the chlorinated hydrocarbons tested. The unsaturated chlorinated hydrocarbons gave particularly high values for the coefficient of friction.

A further series of tests was performed upon the chlorinated hydrocarbons listed in Table 13.4 in order to determine the relative reactivities of these compounds toward aluminum, at their boiling points under a pressure near 1 atm. Each fluid (20 mls) was refluxed in a glass apparatus with six discs of aluminum, measuring 0.25 in (6.4 mm) diameter by 0.06 in (1.5 mm) thick, in contact with the boiling liquid. The condenser was attached to the flask with a ground glass joint and moisture was kept from entering the top of the condenser by means of a drying tube. When the fluid was boiling, the aluminum discs were introduced and the time noted. The fluid was refluxed until visible signs of a reaction were noted or for 24 hours in those cases where no reaction occurred. The reactions were all autocatalytic (some violently so) and the reaction time, therefore, could be easily recognized and reproduced. Since three of the compounds decomposed on boiling at atmospheric pressure, it was necessary to test these compounds at reduced pressure. The pressures used in these cases are stated with the corresponding times in Table 13.5 in which the results of the reflux tests are summarized.

It is evident that those fluids which did not react are the poorer cutting fluids and those which reacted readily are the better cutting fluids. It is not possible, however, to correlate the reaction time directly with the coefficient of friction. This is not surprising, inasmuch as the conditions of temperature, pressure, residual stress, and cleanness of surface are quite different along the tool face from those on the surface of the aluminum discs in the reflux experiments. The observation that the unsaturated chlorinated hydrocarbons are relatively stable toward aluminum in both the cutting and reflux experiments is consistent with the known reactivity of these compounds. In general, it is found that a halogen atom attached to a carbon atom holding a double bond is very stable, relative to one in a saturated compound, whereas a halogen atom attached to a carbon atom one removed from a carbon atom holding a double bond is relatively active.

Further evidence of the physico-chemical nature of cutting fluid action at low cutting speeds is presented in Fig. 13.13. Here values of cutting force (F_P) are shown plotted against the chain length of a series of pure normal primary monohydric alcohols when machining pure aluminum at a speed of 5.5 i.p.m. (0.14 m min^{-1}), rake angle of 15°, and undeformed chip thickness of 0.005 in (0.125 mm). A systematic variation of cutting properties is noted with chain length, and a significant difference in performance occurs for compounds having even and odd numbers of carbon atoms.

An interesting phenomen was observed when aluminum was cut at 0.5 i.p.m. (0.01 m min^{-1}) using a 15° rake tool and a 3% mixture of n-decanol in benzene as the fluid. The chip thickness and the forces were observed to vary periodically as shown in Fig. 13.14. The finished surface was found to be alternately rough and smooth with a period corresponding to the variation in thickness of the chip. From the time scale that is shown it is evident that the fluctuations observed have too high a period (6 s) for the variation to be due to chatter or any other periodic variation of the mechanical system.

This periodic cutting action can be explained in terms of the chemico-physical action of cutting fluids as follows. It is reasonable to expect that a certain minimum quantity of reaction product will be required to produce an appreciable reduction in the coefficient of friction between chip and tool and consequently a noticeable improvement in finish and a decrease in cutting force. At the beginning of a cut the surface temperature and the force increase gradually. At 3% decanol, the concentration of the active ingredient is just sufficient at the higher surface temperature and force to produce enough reaction product for effective cutting fluid action, thus decreasing the frictional resistance and the cutting force. At this reduced cutting force the surface temperature and area of contact are decreased, and thus the quantity of the chemical product formed is insufficient for effective cutting fluid action; therefore, the force and surface temperature once more increase. This cycle is repeated periodically.

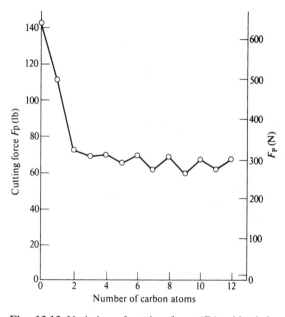

Fig. 13.13 Variation of cutting force (F_P) with chain length when machining aluminum with pure normal primary monohydric alcohols. (after Shaw, 1944)

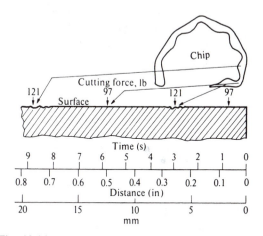

Fig. 13.14 Drawing of chip and periodic rough surface produced when tool cutting pure aluminum was lubricated with 3 $^w/_o$ decanol in benzene. (after Shaw, 1942)

Specificity

The fact that a cutting fluid in the vapor phase may be as effective as the fluid in the liquid phase supports the view that cutting fluid action is at least in part chemical rather than totally physical in nature. The fact that the same cutting fluid does not give equally good results on all metals offers

TABLE 13.6 Values of Cutting Force (F_P) in Pounds for Several Metal–Fluid Combinations at Slow Cutting Speeds[†]

Fluid	Pure Iron		Pure Aluminum		Copper		AISI 1020 Steel		18–8 Stainless Steel	
	F_P	% Dry	F_P	% Dry	F_P	% Dry	F_P	% Dry	F_P	% Dry
Dry	370	100	35	100	292	100	200	100	430	100
n-Diamyl Sulphide	153	41	12	34	31	11	110	55	215	50
Carbon tetrachloride	119	32	18	51	28	10	108	54	192	45
Oleic Acid	122	33	16	46	111	38	115	58	245	57
n-Propyl Alcohol	192	52	18	51	200	68	110	55	369	86
Carbon Disulphide	253	68	29	83	80	27	122	61	339	79
Acetic Acid	160	43	32	91	130	45	108	54	377	88
Turpentine	245	66	14	40	46	16	130	65	384	89
Paraffin Oil	208	56	23	66	153	52	130	65	291	68
Kerosene	340	92	31	89	139	48	165	83	384	89
Water	200	54	18	51	123	42	139	70	284	66
Benzene	340	92	37	106	122	42	240	120	407	95

[†] Cutting conditions: undeformed chip thickness, 0.002 in (0.05 mm); width of cut, 0.25 in (6.35 mm); cutting speed, 5.5 in min⁻¹ (0.14 m min⁻¹); rake angle, 15°.

further evidence. The data shown in Table 13.6 illustrates the variety of results that may be obtained when different fluids are used with different metals. Benzene is seen to be uniformly poor with all metals except copper. Turpentine is very good with copper but poor with stainless steel. Diamyl sulphide and carbon tetrachloride are seen to be good in all cases (lead not included here), and acetic acid is unusually poor with aluminum.

Shear-Plane Action

In addition to lubrication action on the tool face, there is evidence that cutting fluids may also play an important role on the shear plane at low cutting speeds by preventing microcracks from rewelding as the material passes along the tool face. Russian workers appear to favor such an action to explain low-speed cutting behavior (Rehbinder, 1948a, 1948b, 1949, 1951, 1954a, 1954b). A large number of microcracks is assumed to be produced in the chip along the shear plane when a new surface is generated. The absorbed fluid is believed to keep the microcracks from welding under compressive load. The increased number of points of stress concentration corresponding to the unhealed cracks is believed to yield a closer spacing of active slip planes. This action has been termed "hardness reduction" in Russia.

Epifanov et al. (1955) also suggest that an active cutting fluid makes the metal more brittle in a manner similar to hydrogen embrittlement.

A shear test (shown in Fig. 13.15) was devised to study the influence of a fluid on the shearing action in low-speed cutting in the absence of sliding. The protruding element of dimensions b and d is made $\frac{1}{8}$ in (3.18 mm) high and resembles a partly formed chip. However, this is not a chip but a piece of undeformed work material rising out of the surface. The specimen is clamped horizontally in a dynamometer, and the tool is fed downward, shaving the front of the specimen to ensure a perfectly flat surface against which to push. The tool is then set exactly at the level of the rear surface with the aid of a microscope. When all is ready, the horizontal feed is engaged and the protruding member is pushed to the right as the horizontal (F_P) and vertical (F_Q) components of force are recorded.

The shearing test resembles a two-dimensional cutting operation performed by a tool with zero rake angle making a cut with zero shear angle and with negligible motion of the chip along the tool face. The dimension b is comparable to a typical width of cut while dimension d is made to correspond to a representative shear plane length (about 0.05 in, or 1.27 mm).

Before these tests were performed it was expected that

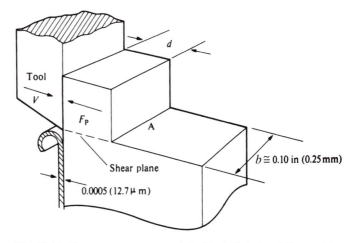

Fig. 13.15 Shear test arrangement. (after Usui, Gujral, and Shaw, 1961)

1. the "chip" could be broken free from the specimen after a very short displacement (of the order of about 0.01d)

2. CCl_4 applied at A would be without influence on the force displacement curves

Figure 13.16, representative of a large number of similar tests, reveals that both of these expectations are false. The force does not reach zero until the "chip" has been displaced a distance considerably in excess of d.

The curves obtained with CCl_4 are seen to fall to zero more quickly than those obtained in dry air. While the presence of the CCl_4 does not influence the maximum force or the displacement corresponding to this force, it has a distinct influence on the shape of the falling portion of the curve. The area under the CCl_4-curve is about 10% less than that for the corresponding test in air.

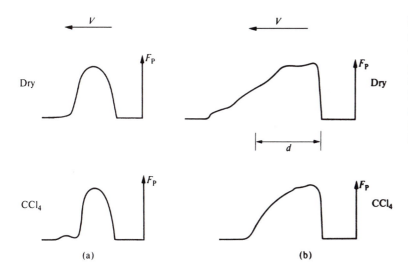

Fig. 13.16 Variation of shearing force (F_P) with tool displacement with and without drop of CCl_4 placed at A in Fig. 13.15. (a) Low carbon steel. (b) Soft copper. Speed = 0.4 i.p.m. (10.2 mm min^{-1}); d = 0.05 in (1.25 mm); b = 0.1 in (2.5 mm); relief angle = 20°. (after Usui et al., 1961)

The explanation of these two unexpected results appears to lie in the tendency for initial rupture to be a local action rather than taking place across the entire width b. After each local rupture, rewelding appears to occur with further motion. The role of CCl_4 appears to lie in its tendency to reduce the rewelding portion of the fracture-welding cycle which naturally leads to earlier total fracture.

Photomicrographs of partially completed shear tests (Fig. 13.17) reveal that

1. the specimens are still firmly welded after a displacement as large as length d

2. the first visible crack appears initially at the rear of the specimen where the hydrostatic compressive stress is least and spreads towards the tool tip as successive fracture regions fail to reweld

3. the extent of the region failing to reweld is greater when CCl_4 is present than in air

4. the volume of material that is plastically deformed is far greater for the specimen sheared in air than for the CCl_4 specimen

All the foregoing results appear to support the idea that CCl_4 can prevent the rewelding of local cracks which form during a shearing process involving large strains. They further suggest that the formation of "continuous" metal cutting chips may also be accompanied by the formations of many small local cracks which tend to weld and refracture in cyclic fashion in a manner analagous to that apparently pertaining in the shear test (Fig. 13.15).

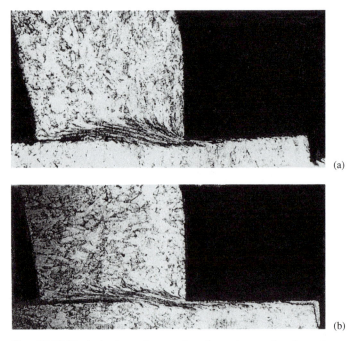

(a)

(b)

Fig. 13.17 Typical photomicrographs of specimens that have been sheared a distance that is approximately equal to the shear plane length (d). (a) In air. (b) With drop of CCl_4 applied at point A in Fig. 13.15 before beginning shearing action. Speed, 0.5 i.p.m. (13 mm/min); width (b), 0.003 in (1.6 mm); rake angle and shear angle, 0°; clearence angle, 20°; material, soft copper. (after Usui et al., 1961)

Tensile and Torsion Tests

True-stress–true-strain tensile tests were performed upon very small (diameter $d = 0.010$ in = 0.25 mm) specimens of mild steel and copper with and without CCl_4 applied to the test surfaces. The presence of CCl_4 was found to have negligible influence on the shapes of the flow curves. The only observed influence of CCl_4 was to increase the strain at fracture a small amount. A mineral oil gave tensile results that were identical to the tests in air.

Miniature torsion tests were next performed upon cylindrical specimens to provide further materials test data on specimens of large area to volume ratio. These tests were unsuccessful since the small specimens strained in erratic ways due to the large influence of grain boundaries and other crystal imperfections. It was not until a specimen of the shape shown in Fig. 13.18a was adopted that reproducible results were obtained.

There was no difference in the torsional curves with and without a thin film of CCl_4 on the copper surface as

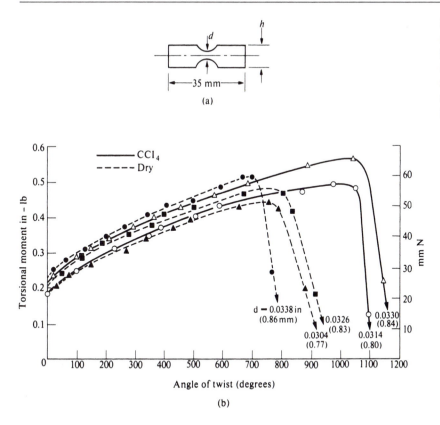

Fig. 13.18 Torsion test for copper with axial compressive load of 10 lb (44.5 N). (a) Specimen. (b) Torsional moment vs. angle of twist with and without CCl_4 on surface. (after Usui et al., 1961)

long as the axial load was zero or tensile. However, when an axial compressive load was present, there was a considerable difference in the strain at fracture as shown in Fig. 13.18b. The fracture strain with CCl_4 corresponded to 800 degrees of twist while that without CCl_4 corresponded to about 1100 degrees of twist.

These tests seem to indicate that although CCl_4 has no important influence in conventional tensile or torsion tests, it does have a significant influence upon the combined torsion-compression characteristics at large strains. Apparently a torsion specimen develops small surface cracks which subsequently weld and refracture before total failure if a compressive stress is also present on the shear plane to induce rewelding. When CCl_4 is present, rewelding is largely prevented and total failure comes earlier.

Microcracks in Machined Surface

A number of orthogonal cutting tests were performed at low speed (1 i.p.m. = 0.025 m min⁻¹) with a 20° rake angle HSS tool. In these tests several cuts ($t = 0.005$ in = 0.127 mm) were first made in air. Then a small amount of CCl_4 was placed on the second half of the last surface produced. A final cut was then taken and the forces recorded. The first half of this cut was produced with a dry tool face while the second half of the cut still had a dry tool face, but what became the back surface of the chip had been coated with CCl_4. Figure 13.19 shows the surprising result representative of many others. At the first point where CCl_4 had been placed, the chip curled abruptly and force (F_P) dropped about 10%. This occurred whether the CCl_4 was allowed to evaporate or not. The obvious

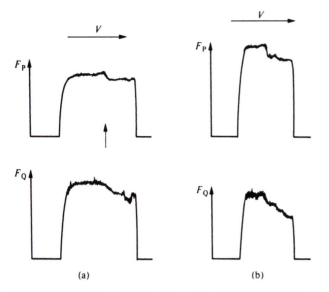

Fig. 13.19 Variation of cutting forces F_P and F_Q with tool displacement. First half of cut is for untreated surface previously cut in air while second half was coated with fluid. (a) CCl_4. (b) Trichlorethane. (after Usui et al., 1961)

explanation of this unexpected result is that the machined surface contained microcracks which absorb CCl_4 which in turn keeps the microcracks from rewelding as the material passes through the shear zone.

When the surface was heated to the boiling point of the fluid after evaporation (80 °C in the case of CCl_4) the same result was obtained. Only when the surface was heated about 20 °C above the boiling point of CCl_4 did the sudden change in cutting performance at the midpoint of the cut fail to develop. Apparently, the boiling point of the fluid in small metal capillaries is 20° higher than in bulk at atmospheric pressure. Another observation in agreement with the microcrack contamination explanation for the unexpected change in cutting performance seen in Fig. 13.19 was the fact that a surface that was carefully ground before coating with CCl_4 failed to exhibit the effect. Apparently, the grinding operation removed microcracks that had developed during previous cutting.

It appears from all of the foregoing tests that at low cutting speeds important cutting-fluid action occurs on both the tool face and shear plane. On the tool face extreme boundary lubrication involving reaction between chip and tool is important over the part of the chip–tool contact area where sliding occurs. The function of the fluid here is apparently to prevent galling and to shorten the total chip–tool contact length which in turn causes an increase in the shear angle. The cutting fluid appears also to play an important role in preventing rewelding of microcracks that are formed during initial chip formation. Since the tendency for microcrack formation increases with decrease in normal stress on the shear plane, a greater density of microcracks should be expected on the free surface of the chip away from the tool face. This should promote chip curl which in turn will further decrease chip–tool contact length.

It would thus appear that the spectacular benefit observed with CCl_4 at low cutting speed is due to the combined effect of a decreased chip–tool contact area, a reduction in the flow stress on the shear plane and tool face due to microcrack contamination (hardness reduction), and a reduction in the coefficient of friction on the sliding portion of the chip–tool contact due to the prevention of junction growth and galling.

De Chiffre et al. (2001) have used hole reaming tests to evaluate the performance of cutting fluids at low speeds. The following items may be readily measured in the evaluation of a fluid:

- reaming torque and thrust
- surface finish (R_a and R_t)
- oversize of hole relative to reamer diameter

Tests were made at different depths of cut when holes in AISI 316L austenitic stainless steel were reamed using water-base cutting fluids. All tests employed 6 flute reamers, having a 7° left helix angle at a cutting speed of 20 f.p.m. (6.1 m/min) and a feed of 0.016 i.p.r. (0.4 mm/min).

Mechanical Activation

The unusual conditions at the tip of a cutting tool (high pressure and temperature, and clean active surface conditions) may be used to encourage chemical action as in the production of organometallic reagents (Shaw, 1948) and the formation of small seed diamonds (Komanduri and Shaw, 1974). These and other unusual applications of metal cutting are discussed in Chapter 24.

HIGH-SPEED CUTTING FLUID ACTION

If a series of tests is performed using CCl_4 as the fluid over a range of cutting speeds, it is found that this fluid loses its effectiveness at speeds above about 40 f.p.m. (12.5 m min^{-1}). The complex action of CCl_4 including penetration and reaction with the chip takes time and at high cutting speeds there just is not time for the fluid to be effective. This is also true to other less effective low-speed cutting fluids and additives. CCl_4 will actually give a poorer tool life than air at cutting speeds of 100 f.p.m. (30 m min^{-1}) or higher due to the fact that at the elevated temperature pertaining, chemical action is excessive and the cutting edge of the tool is weakened by chemical attack.

At cutting speeds from about 10 f.p.m. (3.0 m min^{-1}) to 100 f.p.m. (30 m min^{-1}) for steels, a BUE is apt to be troublesome from the standpoint of surface finish. Oil-base cutting fluids containing fatty acids for boundary lubrication and more active chlorine, sulfur, or phosphorus compounds for extreme boundary lubrication (chemical reaction with the surface of the chip) are effective in decreasing surface roughness due to a BUE. An increased rake angle (α) and a reduction in the undeformed chip thickness (t) are also effective ways of controlling BUE roughness at relatively low cutting speeds.

As the cutting speed increases, the tool-face temperature increases rapidly. In the practical speed range for turning, the Taylor equation ($VT^n = C$) represents a good approximation. For a HSS tool, the exponent n is about 0.1 and from Eq. (12.43), the tool-face temperature varies as $V^{0.5}$. It thus follows from these two results that for a HSS tool the tool life (T) should vary with tool temperature as follows:

$$T \cong \theta_t^{-20} \tag{13.2}$$

This result was originally obtained empirically by Shallbrock et al. (1938). The corresponding exponent for a carbide tool will be −10.

At speeds between about 100 f.p.m. (30 m min^{-1}) and 200 f.p.m. (60 m min^{-1}) BUE ceases to be a problem, but another problem takes its place—temperature-accelerated tool wear. Above this speed range there is not time for the low-speed cutting-fluid action on the tool face or the back of a chip to occur, and the sole function of the fluid gradually becomes one of cooling. At very high speeds (> 500 f.p.m., 150 m min^{-1}) there is not even time for a fluid to be effective in cooling. When turning at 500 f.p.m. (150 m min^{-1}) and a feed rate of 0.010 i.p.r. (0.25 mm/rev) the total time available for cooling is about 10^{-3} s.

Just as there is the possibility of negative cutting-fluid action at low cutting speeds (benzene, for example), there is also the possibility of negative fluid performance at high cutting speeds. This results from the fluid sometimes causing excessive chip curl decreasing the chip–tool contact area to the point where the maximum tool temperature is brought too close to the cutting edge. This aspect of high-speed cutting-fluid action will be presently discussed in detail.

Another negative aspect of high-speed cutting-fluid action is thermal shock and thermal fatigue. Certain cutting operations are periodically intermittent (milling, for example). In such a case, use of a coolant may give too great a temperature range from the cutting (sudden approach to high tool-tip temperatures) to noncutting (strong cooling) parts of the cycle with a high range of fluctuating tensile stress due to differential thermal expansion. In such cases, it is best to cut in air. In general, dry cutting in air is recommended for carbide milling cutters.

De Chiffre (1978) has discussed the laboratory evaluation of cutting fluids. It was suggested that four operations be employed to evaluate tool life, cutting forces, and product quality: drilling, boring, reaming, and tapping. It was found that the performance ranking of commercial fluids varied with the type of operation as well the performance criteria employed. It was assumed that results in boring reflect results to be expected in single-point turning where cutting speeds are high and cooling is more important than reduction of friction. In a subsequent article by De Chiffre, an outline of a general testing procedure to be followed in evaluating a given fluid is presented.

The lubrication action of cutting fluids is considered by De Chiffre (1981) from the point of view of

- the establishment of a soft film between chip and tool
- a reduction in the shear strength of the work material (Rehbinder effect)
- a reduction in the chip–tool contact length

It was concluded that force and temperature measurements on tools of reduced contact length provide a means for separating the friction reduction and cooling action performed on a cutting fluid.

De Chiffre (1988a) has described a method of using frequency analysis of machined surfaces as a measure of cutting-fluid performance. A high-resolution signal analyzer is used to obtain the power spectrum characteristics of surfaces measured in the direction of cutting to evaluate lubricating action.

De Chiffre (1988b) has performed a series of tests on aluminum, copper, and low- and high-carbon steels to ascertain the principal influence a cutting fluid has on cutting performance. The conclusion reached was that the main effect on cutting performance is due to a reduction in the tool–chip contact length. This reduction in contact length results in an improvement in surface finish and a reduction in cutting forces.

Chip–Tool Thermocouple Tests

The chip–tool thermocouple technique may be used in conjunction with a cutting fluid provided the fluid is a poor electrical conductor. Most water-base cutting fluids provide this requirement. In a series of tests on AISI 1015 steel tubing using an orthogonal tool with a 15° rake angle, the following fluids were investigated:

1. dry tool
2. commercial emulsifiable water-base cutting fluid in concentration of 40:1
3. commercial soluble water-base cutting fluid in concentration of 40:1
4. water plus 0.1% sodium nitrite (rust inhibitor)
5. water plus 0.1% sodium nitrite plus wetting agent

The results of these tests are shown in Fig. 13.20.

The fluids are seen to be most effective in lowering the chip–tool interface temperature at the smaller values of feed. In all cases the effectiveness of the fluid in lowering the tool temperature on a percentage basis is seen to decrease with cutting speed. At the larger feeds and cutting speeds, the fluids are seen to be completely ineffective. The cutting speed at which the fluid becomes ineffective decreases as the feed is increased. At a feed of 0.0023 i.p.r. (0.06 mm/rev), the fluid was effective at all speeds tested; at a feed of 0.0052 i.p.r. (0.13 mm/rev), the fluid was effective to 400 f.p.m.

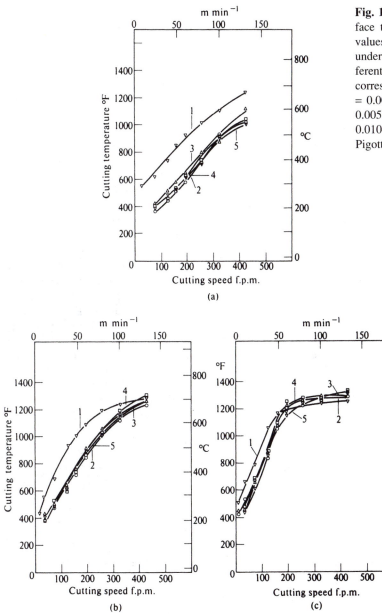

Fig. 13.20 Cutting speed versus mean tool face temperature (tool–chip thermocouple values) curves for AISI 1015 steel turned under orthogonal cutting conditions at different speeds and feeds. Numbers on curves correspond to fluids defined in text. (a) Feed = 0.0023 i.p.r. (0.06 mm/rev). (b) Feed = 0.0052 i.p.r. (0.13 mm/rev). (c) Feed = 0.0104 i.p.r. (0.26 mm/rev). (after Shaw, Pigott, and Richardson, 1951)

(122 m min^{-1}); and at a feed of 0.0104 i.p.r. (0.26 mm/rev), the fluid became ineffective at about 200 f.p.m. (60 m min^{-1}).

In these tests the fluids were applied to the chip, tool, and workpiece from above (i.e., toward the back of the chip). The increase in heat transfer path with undeformed chip thickness (t) is evidently responsible for the decreased effectiveness of the fluid with increase in t. The decreasing effectiveness of the fluids with cutting speed is undoubtedly due to the shorter time available for meaningful cooling of the chip with increase in cutting speed.

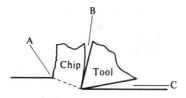

Fig. 13.21 Directions of cutting-fluid application in high-speed orthogonal machining.

Direction of Fluid Application

There are three main directions of fluid application as shown in Fig. 13.21. In turning with a tool having a nose radius, there is a fourth direction C′ which is along the end clearance, direction C then being along the side clearance. Over the years much has been written about the most effective direction of cutting fluid application. Taylor (1907) was one of the first engineers to recognize the importance of a cutting fluid. Using soda water as a coolant, he demonstrated that tool life could be increased up to 40% over cutting in air. He found that better results were obtained when the fluid was directed on the back of the chip (direction A in Fig. 13.21) than from directions B or C. Niebusch and Strieder (1951) advocated a heavy flow from direction A but also a heavy flow from direction C when heavy cuts are taken. A total flow of one gallon per horsepower consumed at the tool tip was advocated. Lauterbach (1952) demonstrated an increase in tool life when the lubricant was introduced in direction C. Pigott and Colwell (1952) used very high velocity jets of fluid and found these were most effective when aimed in direction C.

Smart and Trent (1974) have used the tool-softening technique described in Chapter 12 to study the influence of cutting fluids and the direction of fluid application on the isotherms found in a cutting tool when machining iron and nickel. Figure 13.22 shows the isotherms when machining iron in air and with the fluid from directions A and C. There was little change in the peak temperature with use of the fluid or with the fluid application. However, the peak temperature area is seen to be substantially reduced when using the fluid and in particular when the fluid was from direction C. Since the peak temperature area is at about the same distance from the cutting edge for all three cases, it would appear there was little change in chip–tool contact length or chip curl.

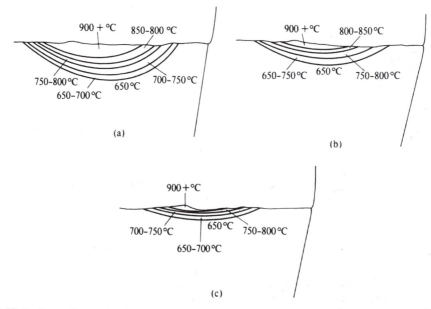

Fig. 13.22 Isotherms for cutting iron (a) Dry. (b) With 30:1 soluble oil from direction A. (c) With 30:1 soluble oil from direction C. Cutting conditions: cutting speed, 600 f.p.m. (183 m min⁻¹); feed, 0.010 i.p.r. (0.25 mm/rev); tool material, equivalent of M34 HSS; cutting time 30s. (after Smart and Trent, 1974)

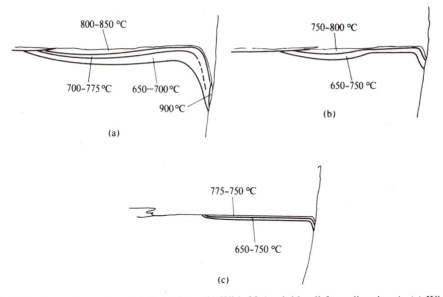

Fig. 13.23 Isotherms for cutting nickel. (a) Dry. (b) With 30:1 soluble oil from direction A. (c) With 30:1 soluble oil from direction C. Cutting conditions: cutting speed, 150 f.p.m. (46 m min^{-1}); feed, 0.010 i.p.r. (0.25 mm/rev); tool material, equivalent of M34 HSS; cutting time, 30 s. (after Smart and Trent, 1974)

Figure 13.23 shows similar results for pure nickel. When machining nickel, the maximum temperature is at the tool tip (as discussed in Chapter 15) resulting in rounding of the cutting edge. With a cutting fluid, wear at the cutting edge is much less particularly when the fluid is directed along C. The peak cutting temperature is seen to be less when a fluid is used and particularly so when applied from direction C.

Many other papers concerning the optimum direction of cutting fluid application may be found in the literature—some advocating flow from direction A as in Taylor's case (1907), some from direction C as in the case of Smart and Trent (1974), and some advocating flow from both directions A and C as with Niebusch and Strieder (1951). There is little support for direction B.

There appears to be no single answer for the best method of fluid application. A strong flow from direction A can have an important influence on chip curl, tool–chip contact length, and the location of the point of maximum temperature relative to the cutting edge. As will be shown in the next section, cooling the back of a chip (flow from direction A) reduces the chip–tool contact length for cutting in air, and this may have a beneficial or a detrimental influence on tool life. If application of the fluid from direction A causes overcurling of the chip bringing the maximum temperature too close to the cutting edge, then application of the fluid from direction C or C′ or both will be preferable. If, however, the maximum temperature is already at the tool point in dry cutting (as when cutting nickel as discussed in Chapter 15), then direction C will give best results as Smart and Trent (1974) suggest.

Chip Curl

While the cooling function of a fluid may be beneficial relative to tool life, this is not always the case. Figure 13.24 is a case in point. Here the wear land size (w) is plotted against cutting time (T_c) when turning with a HSS tool. The rate of wear with water is considerably higher than that for a dry tool.

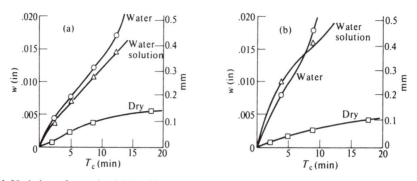

Fig. 13.24 Variation of wear land (w) with cutting time T_c for different cutting fluids applied from direction A. (a) AISI 1020 steel machined with M–2 HSS tool at V = 350 f.p.m. (109 m min⁻¹); t = 0.0052 i.p.r. (0.13 mm/rev); depth of cut, 0.100 in (254 mm); tool geometry, 0, 15, 10, 10, 10, 20, 0.030 (N.R. = 0.75 mm). (b) AISI 4340 steel machined with M–2 HSS tool at V = 250 f.p.m. (76 m min⁻¹); t = 0.0026 (0.07 mm); depth of cut = 0.100 in (2.54 mm); tool geometry, same as (a). (after Feng, Gujral, and Shaw, 1961)

Figure 13.25 shows the wear land and crater development when machining AISI 1020 steel with and without a coolant (water). Each time the wear land was measured, a diamond stylus was drawn across the tool face, and the shape of the build-up and crater on this face was recorded. The vertical magnification of motion is made five times the horizontal value in order to allow craters to be more conveniently represented.

The numbers next to each test-point correspond to the numbers of the crater traces. At the points indicated, the clearance face of the tool was carefully reground to decrease the extent of the wear land without disturbing the crater on the tool face. The crater produced with water is seen to form more rapidly than when cutting dry and the center of the crater is found to be closer to the tool point. This latter observation is an important one. Even though the water carries some heat away from the wear land, it simultaneously shifts the maximum temperature closer to the tool tip which results in more heat flowing into the wear land. The latter of these two effects is the predominant one in Fig. 13.25.

In order to check this thesis, the temperature of the newly developed surface was measured when cutting with and without a flood of water-base cutting fluid from direction A in Fig. 13.21. In Fig. 13.26, the change in temperature of the workpiece surface in a turning operation is shown diagrammatically in developed form, from a point one revolution before the cutting edge to a point one revolution after the cut is completed. The temperature at (1) is ($\Delta\theta_1$) above the ambient temperature of the bar. The temperature at (2) is the workpiece temperature at the cutting edge. The workpiece surface temperature will rise to a maximum value at (3) due to the heat which flows across the wear land from the tool. The temperatures at (4) and (5) correspond to surface temperatures at points 180° and 360° from the cutting tool.

Nakayama (1956) measured the temperature at point (4) for dry, two-dimensional cutting on the end of a tube and noted a significant decrease in this temperature with increase in workpiece diameter. In making these measurements Nakayama allowed several strands of constantan wire to rub upon the freshly cut surface at a point 180° from the cutting tool. In repeating these tests for bar-turning, it was found that this method worked quite well for dry cutting, but when a fluid was applied, unsatisfactory results were obtained which were traced to a change in the rubbing friction of the wires on the workpiece surface with the lubricating properties of the fluid. It was found that there was an EMF generated when the wires were rubbing without cutting. When cutting dry, this initial EMF was very constant and could be subtracted from the EMF when cutting to obtain the temperature of the workpiece at point (4). However, when a fluid was present, the amount to be subtracted was variable and it was not clear how to make the correction.

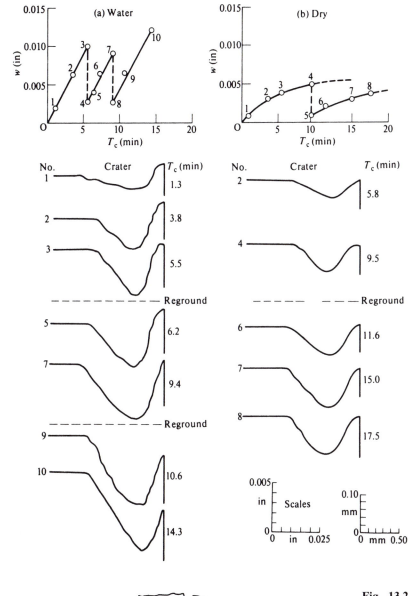

Fig. 13.25 Variation of wear land and crater with cutting time (T_c). Cutting speed = 450 f.p.m. (141 m min^{-1}); feed, 0.0052 i.p.r. (0.13 mm/rev); depth of cut = 0.100 in (2.54 mm); tool, M–2 HSS; tool geometry, 0, 15, 10, 10, 10, 20, 0.03 (N.R. = 0.75 mm). (a) Water applied from A at $2\frac{1}{2}$ g.p.m. (9.5 1 min^{-1}). (b) Cutting in air (dry).

Fig. 13.26 Developed view of workpiece showing temperatures of interest in region extending one revolution on either side of cutting tool.

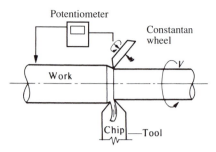

Fig. 13.27 Arrangement of apparatus for measuring cutting workpiece surface temperature one half-revolution from point of cutting.

After investigating rubbing wires of several different sizes and shapes a wheel was designed of solid constantan which ran upon the shoulder left by the cutting tool as shown in Fig. 13.27. In this way rolling friction was substituted for rubbing friction and there was no longer a troublesome variable correction to be made.

Some representative data are given in Table 13.7 for relatively light cuts ($t = 0.0046$ i.p.r. or 0.12 mm/rev) made with and without a cutting fluid (water in this case). The temperature ($\bar{\theta}_t$) is the chip–tool thermocouple temperature, while (θ_4) is that measured by the constantan wheel at a point 180° from the cutting tool. It is evident that while there is no significant difference in (F_P), (F_Q), or ($\bar{\theta}_t$) when the fluid is applied, there is a reproducible difference in (θ_4). Furthermore, the change in (θ_4) that was observed was in the opposite direction to that expected, i.e., when the coolant was applied, the temperature of the workpiece was found to increase rather than to decrease.

After considering several possible explanations for this paradox, the answer was found to lie in chip curl. It was found that the radius of curvature of the chip decreased appreciably when the fluid was applied, and returned to its original value when dry cutting was resumed. While there was a negligible difference in tool–chip thermocouple temperature ($\bar{\theta}_t$) with and without the fluid, the maximum temperature on the tool face was closer to the cutting edge with a tightly curled chip and hence more heat was caused to flow from the tool and into the workpiece.

All of the data of Table 13.7 are for a sharp tool. The data for Table 13.8 are for the same tool with a wear land present. As cutting progressed, the fluid was shut off every few minutes. The values of radius of chip curvature (ρ) were the same for the worn and sharp tools. The only difference between the tests of Tables 13.7 and 13.8 lies in the wear land being either 0.008 in (0.2 mm) or 0.12 in (0.3 mm) for Table 13.8 and essentially zero for Table 13.7.

Presence of a wear land caused a reversal in the influence of the fluid upon θ_4. It is seen in Table 13.8 that θ_4 is less when the fluid is present whereas the reverse was true in Table 13.7. This result was also found in other instances with small feed, and the explanation would appear to be as follows. With a sharp tool, the increased curl that accompanies chip cooling shifts the maximum tool-face temperature closer to the cutting edge, and this gives rise to more heat flow into the work as shown in Fig. 13.28. For the dry case the maximum temperature is too far from the sharp edge for much heat to flow into the workpiece. The net effect of the direct cooling action of the fluid and of the increased

TABLE 13.7 Values of θ_4 for Dry and Wet Cutting with a Sharp Tool[†]

Fluid	F_P		F_Q		θ_t		θ_4		Average
	lb	N	lb	N	°F	°C	°F	°C	
Dry	220	975	170	757	975	524	117	47	
Dry	220	975	170	757	975	524	123	51	120 °F 49 °C
Water	225	1001	175	779	975	524	139	59	
Water	230	1024	180	801	975	524	136	58	
Water	240	1068	185	823	975	524	136	58	137 °F 58 °C

[†] Cutting conditions: work material, AISI 4340 steel; cutting speed, 400 f.p.m. (125 m min^{-1}); feed, 0.0046 i.p.r. (0.12 mm/rev); depth of cut, 0.100 in (2.54 mm); side rake angle, −7°; tool material, C–5 carbide. (after Shaw, Cook, and Smith, 1958)

TABLE 13.8 Values of θ_4 for Dry and Wet Cutting with Tool Having a Wear Land (w)[†]

Fluid	w			F_P		F_Q		θ_4	
	in	mm		lb	N	lb	N	°F	°C
Dry	0.008	0.20		240	1068	180	801	121	49
Water				235	1046	180	801	107	42
Dry				245	1090	180	801	125	52
Water				245	1090	180	801	115	46
Dry				255	1135	185	823	128	53
Water	0.012	0.30		255	1135	190	846	118	48

[†] Cutting conditions; same as Table 13.7. (after Shaw, Cook, and Smith, 1958)

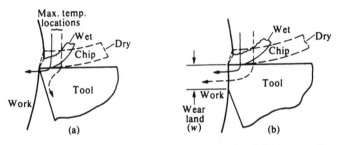

Fig. 13.28 Paths of heat flow for (a) sharp tool and (b) worn tool.

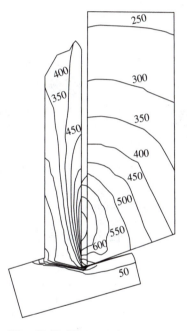

Fig. 13.29 Temperature contours (50 K intervals) near cutting edge at cutting speed of 150 f.p.m. (46 m/min) and a heat transfer coefficient (h) of 10^3 Wm^{-2}K^{-1}. (after Childs et al., 1988)

flow of heat into the workpiece gives rise to a higher wear-land temperature than for dry cutting. However, after an appreciable wear land has developed, heat can flow into the workpiece from the tool face in dry cutting as well (see Fig. 13.28) and then the direct cooling action of the fluid causes the resultant wear-land temperature to be less when the fluid is used.

Sharma et al. (1971) injected cutting fluids directly into the chip–tool interface through a small hole in the tool. While this gave an initial increase in chip curl due to cooling, the hole soon became blocked by chip material.

FEM Analysis

Childs et al. (1988) used FEM analysis coupled with experimental results to study the influence of a water-base cutting fluid on tool-face and tool-flank temperatures. The material cut was AISI 1045 steel using a high-speed steel tool having a rake angle of 14° and a clearance angle of 6°. A range of speeds was used from 108 to 200 f.p.m. (33 to 60 m/min) and a feed of 0.010 i.p.r. (0.254 mm/rev). Cutting forces were measured with a dynamometer and chip shapes obtained from quick-stop chip roots. The heat transfer from tool to fluid was assumed to be 10^3 Wm^{-2}K^{-1} for a flow rate directed on the tool of 0.25 l/min) and 5×10^3 Wm^{-2}K^{-1} for a flow rate of 2.51 l/min. The cutting fluid gave a greater reduction on the tool flank (60 to 70 °C) than on the tool face, as shown by temperature contours in Fig. 13.29 for a representative case. Figure 13.30 shows the change in maximum temperature on

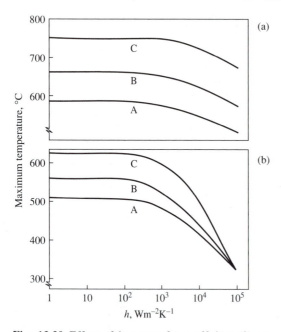

Fig. 13.30 Effect of heat transfer coefficient (h) on (a) maximum rake-face temperature and (b) maximum flank-face temperature for three cutting speeds: A, 108 f.p.m. (33 m/min); B, 151 f.p.m. (46 m/min); C, 200 f.p.m. (61 m/min). (after Childs et al., 1988)

(a) the tool face and on (b) the tool flank for three different cutting speeds. Childs et al. (2000) have presented many valuable temperature contour plots for metal cutting based on similar FEM analyses.

Vapors and Mist

Gaseous lubricants appear very attractive when the cutting fluid penetration problem is considered. Operations which are performed dry are actually carried out using air as a cutting fluid; and although the presence of the air is usually taken for granted, it is a very effective boundary lubricant. When air is considered as a coolant, however, it is not very attractive since all gases have relatively poor cooling capacity compared with liquids.

Attempts to improve the cooling capacity of air by refrigerating it have yielded different results. Olson (1948), using cooled air in a milling operation, reported a 400% increase in tool life over air at room temperature; while Pahlitzsch (1951) found a small difference in tool life in turning with air introduced from position B (Fig. 13.21) at 50 °F and −19 °F (10 °C and −28 °C).

Pahlitzsch has also presented test results using carbon dioxide gas and nitrogen gas as cutting fluids. He found an improvement in tool life over air by 150% using carbon dioxide and 240% using nitrogen. Tests using carbon dioxide gas at different temperatures gave a negligible difference in tool life, and Pahlitzsch concluded from these results that the favorable action of carbon dioxide in these tests is not a cooling effect but is due to the exclusion of air. He believed that oxidation promotes friction; therefore, the exclusion of oxygen by carbon dioxide is beneficial.

At the same time, if this is assumed to be the explanation, then it follows that nitrogen must have some sort of positive boundary lubrication effect. It has been suggested (Wister, 1936) that nitrides could form on a sliding metal surface in the presence of nitrogen which would provide lower friction as a result of the increased hardness. Further German work on this problem (Axer, 1954) confirmed Pahlitzsch's observations regarding nitrogen as a boundary lubricant. Axer used a carbide tool equipped with a small hole in the clearance surface of the tool to enable gas to be injected in direction C (Fig. 13.21) at high pressure. It was found that a jet of oxygen gave lower tool life than dry cutting while a jet of nitrogen gave increased tool life. Furthermore, the tool life was observed to increase with the nitrogen pressure while it tended to show lack of improvement with increase in the oxygen pressure. Argon gave slightly better results than dry cutting, while carbon dioxide gas gave slightly inferior results to dry cutting. It should be noted that this latter observation is not in agreement with Pahlitzsch's results.

While carbon dioxide gas was used as a cutting fluid a long time ago (Reitz, 1919), it appears that it is only relatively recently that very rapid expansion of liquid carbon dioxide through a nozzle has been used. When this is done, the workpiece is coated with carbon dioxide snow, which upon subsequently subliming into the atmosphere extracts the latent heat of sublimation of carbon dioxide (250 BTU lb^{-1}) from the workpiece. In such instances, the cooling action of the carbon dioxide is thought to play a major role.

Many articles have appeared in the technical press (Bingham, 1954, for example) on the use of high-velocity jets of carbon dioxide. It is reported that better results are obtained when the jet is directed on the work and along the clearance face of the tool (i.e., from direction C (Fig. 13.21)) than when the stream is applied to the chip (direction A or B). The relatively poor performance obtained when the chip rather than the tool face is cooled is thought to be due to the chip becoming stronger relative to the tool face which, in turn, results in an increased tendency for a crater to develop on the tool face. The widely different results that are reported with a carbon dioxide jet are undoubtedly due in part to differences in the application of the fluid.

Mist lubrication, in which very small droplets of a liquid are distributed in a large volume of air (aerosol), has been used to lubricate high-speed ball bearings for a long time (Faust, 1952). A similar scheme involving a larger concentration of liquid and coarser particle size has been adopted in workshops to combine the attractive aspects of gas and liquid cutting fluids. The very large surface-to-volume ratio for each particle of liquid provides the possibility of rapid vaporization. While the convective cooling capacity of an aerosol will be about the same as that for the gas, due to the very small amount of liquid present, rapid vaporization makes it possible to have a larger measure of evaporative cooling. For this purpose, a material of low boiling point and high latent heat of vaporization should give best results. Water is an ideal substance in this regard and oils are much less attractive.

A number of commerical mist generators are available. These units usually have filters for both air and liquid, a pressure regulator, a mixing valve, discharge nozzle, and a reservoir for the fluid. Most units can be used with either cutting oils or water-base fluids, but fluids containing solid additives in suspension are not recommended. Some manufacturers strongly recommend water-base materials over oil-base materials. The air pressure used with these devices ranges from about 1 to 5 atmospheres.

Mist lubrication has been used on essentially all machining operations, and a number of articles have been written that describe industrial experience with this method of fluid application (e.g., Brosheer, 1953; Thuma, 1954; Chamberland, 1950). While the cost of the air used is significant, the liquid cutting fluid is conserved by this method. There is no drag-out problem on chips and the method may be used on operations where fluids are normally not employed, for example, on high-speed aluminum routing operations in the aircraft industry and on large planers. The contamination of the air can be a problem, depending on the nature and amount of the liquid that is employed. While there appears to be no tangible evidence that mist lubrication represents any health hazard, this problem has been raised and debated. In this connection it would seem wise to use as little liquid entrained in the air as is necessary and to install an automatic solenoid shut-off valve which closes whenever the machine is not in operation.

A water mist can be more effective than a flood of coolant with carbide tools operated at speeds in the range where the shear-plane temperature exceeds the boiling point of water and an insulating steam blanket tends to form between fluid and chip. With the mist, each water droplet striking the surface of the chip will evaporate and absorb the corresponding latent heat of vaporization. If the concentration of water droplets is not too great, the steam generated has an opportunity to escape and the insulating steam blanket obtained with a flood of fluid does not develop. At cutting speeds below the temperature where a steam blanket develops, a flood of liquid provides better cooling than a mist. At very high speeds there is not time for the mist to evaporate and extract heat while the chip is still in contact with the tool.

Cryogenic Cooling

Dhar (2001) has presented a valuable study concerning cryogenic cooling with liquid nitrogen. Fig. 13.31 is a schematic representation of the apparatus used. Two nozzles were positioned to

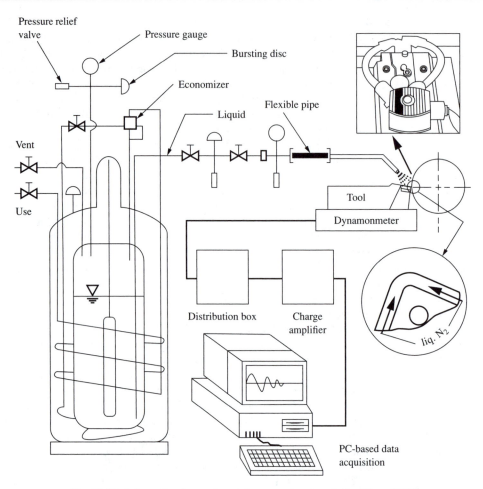

Fig. 13.31 Schematic of experimental cryogenic setup. (after Dhar, 2001)

direct liquid nitrogen along grooves as shown in the insert. Five plain carbon and alloy steels were turned dry and with nitrogen cooling on a 15 h.p. (11 kW) lathe under the following conditions:

- work materials: AISI 1040, 1060, E4340C, 4320 and 4140
- carbide insert geometry: −6, −6, 6, 6, 15, 75, 0.8 mm
- cutting speeds: 197 to 492 f.p.m. (60 to 150 m/min)
- feed rates: 0.0048, 0.0067, 0.0080, 0.0096 i.p.r. (0.12, 0.16, 0.20, 0.24 mm/rev)
- depths of cut: 0.060 and 0.080 i.p.r. (1.5 and 2.0 mm)

Cutting forces, types of chip, chip length ratios, cutting temperatures, surface finish, tool wear, and dimensional deviation were measured. Cutting temperatures were obtained by the chip–tool thermocouple method. Cutting was interrupted periodically to measure tool-face and tool-flank wear patterns. The shape and color of the chips obtained with two grades of carbide when turning AISI 1040 steel at lower feeds are shown in Fig. 13.32. Variation of the chip reduction coefficient,

V m/min	env.	Feed, *f*, mm/rev							
		SNMG 120408-26 TTS				SNMM 120408 TTS			
		0.12		0.16		0.12		0.16	
		shape & color		shape & color		shape & color		shape & color	
66	Dry	◖	light blue	◖	bluish gray	●	golden	●	bluish gray
	Cryo	□	metallic	◖	metallic	●	metallic	□	metallic
85	Dry	□	bluish	◖	blue	●	light blue	●	bluish gray
	Cryo	□	metallic	◖	metallic	●	metallic	●	metallic
110	Dry	○	deep blue	◖	deep blue	●	blue	●	gray
	Cryo	◖	metallic	◖	metallic	●	metallic	●	metallic
144	Dry	□	burnt blue	◖	burnt gray	●	deep blue	○	burnt blue
	Cryo	●	metallic	◖	metallic	●	metallic	●	metallic
Chip shape									
Group		◖	half turn arc	●	tubular/ helical	□	spiral	○	ribbon

Fig. 13.32 Shape and color of AISI 1040 chips at low feeds. (after Dhar, 2001)

ζ (reciprocal of the cutting ratio, *r*) versus cutting speed (*V*) are given in Fig. 13.33 when turning AISI 1040 steel with two different grades of carbide.

The variation of mean flank wear (V_B) with turning time is shown in Fig. 13.34 for AISI 1040 steel when turning dry and under cryogenic cooling conditions. The variation of surface roughness with cutting speed (*V*) when turning AISI 1040 steel with two different grades of carbides under dry and cryogenic cooling conditions is given in Fig. 13.35. Similar results to those of Figs. 13.32 to 13.35 are also given (Dhar, 2001) for the four other materials tested.

An FEM model incorporating the many improvements that have been made over the years was used to estimate the mean chip–tool interface temperature. Figure 13.36 shows the relation between the measured and predicted values for four of the five steels tested under dry turning conditions for two different carbide inserts. In general, the predicted values (solid curves) are higher than the measured values, but the agreement is generally satisfactory. Figure 13.37 shows the computed temperature distribution in the chip and tool when turning AISI 1040 steel under dry and with cryogenic cooling for SNMG carbide, while Fig. 13.38 is for SNMM carbide inserts. Figure 13.39 shows the variation in the average interface temperature with cutting speed for AISI 1040 steel turned at different feeds with and without cryogenic cooling for two different carbides.

As in the case of measured values, FEM-predicted results similar to those for AISI 1040 steel in Figs. 13.37, 13.38, and 13.39 are also presented in Dhar (2001) for the other steels tested. The

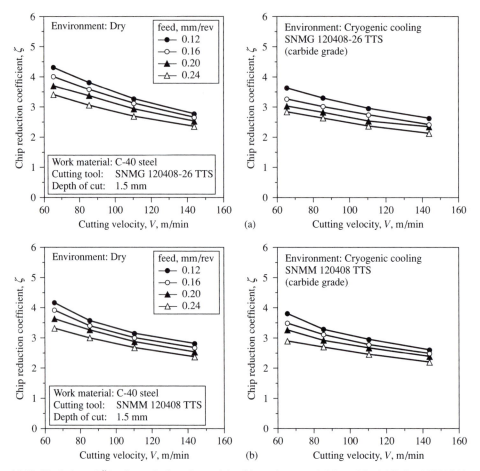

Fig. 13.33 Variation of ζ (reciprocal of cutting ratio) with cutting speed (V) and feed (f) for AISI 1040 steel turned dry with cryogenic cooling and with two different grades of carbide. (after Dhar, 2001)

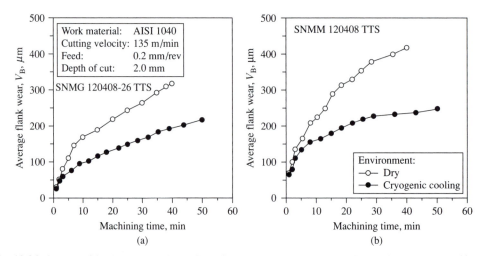

Fig. 13.34 Growth of flank wear, V_B, in (a) SNMG and (b) SNMM inserts while machining AISI 1040 steel at $V = 135$ m/min under dry and cryogenic conditions. (after Dhar, 2001)

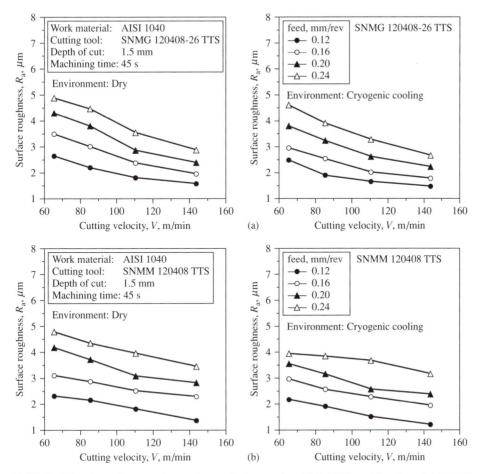

Fig. 13.35 Variation in surface roughness observed after turning AISI 1040 steel by sharp (a) SNMG and (b) SNMM inserts at different cutting speeds and feeds under dry and cryogenic conditions. (after Dhar, 2001)

patterns of estimated temperature distributions for dry and cryogenic-cooled values were found to be quite similar in all cases with the cryogenic values being substantially lower. The mean shear-plane temperatures were found to be considerably lower than the tool-face temperatures.

The cryogenic cooling system shown in Fig. 13.31 enables the average chip–tool interface temperature to be reduced up to 34% depending on the

- work material
- tool geometry and composition
- cutting conditions

Similarly, cryogenic cooling reduced cutting forces up to 30%.

The results given in Dhar (2001) should serve a useful purpose when considering whether use of cryogenic cooling is attractive based on economic or productivity considerations. In a time when environmental concerns are becoming more important, the environmentally friendly aspect of cryogenic cooling is an important consideration.

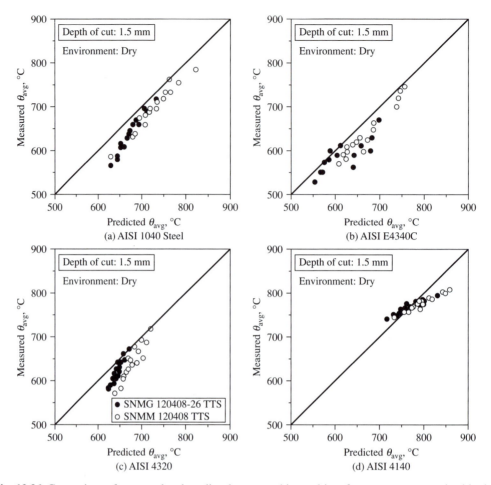

Fig. 13.36 Comparison of measured and predicted average chip–tool interface temperature attained in dry machining of different steels by SNMG and SNMM inserts. (after Dhar, 2001)

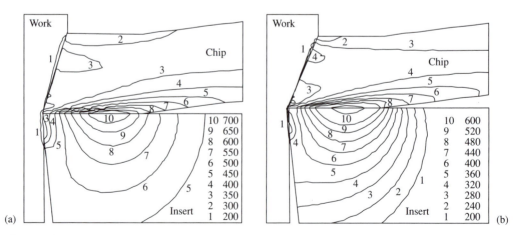

Fig. 13.37 Computed temperature distribution in chip and tool for turning AISI 1040 steel by SNMG insert under (a) dry and (b) cryogenic conditions. Cutting speed, 144 m/min; feed, 0.24 mm/rev; depth of cut, 1.5 mm. (after Dhar, 2001)

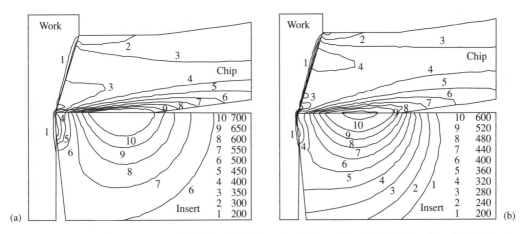

(a) (b)

Fig. 13.38 Computed temperature distribution in chip and tool for turning AISI 1040 steel by SNMM insert under (a) dry and (b) cryogenic cooling conditions. Cutting speed, 144 m/min; feed, 0.24 mm/rev; depth of cut, 1.5 mm. (after Dhar, 2001)

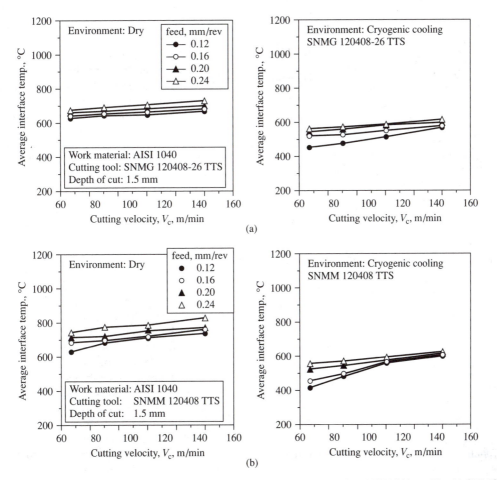

Fig. 13.39 Variation in θ_{avg} for different cutting speeds and feeds when turning AISI 1040 steel by (a) SNMG and (b) SNMM inserts under dry and cryogenic conditions. (after Dhar, 2001)

BENCH TESTS

It would be convenient if cutting fluids could be evaluated in bench tests designed to more clearly characterize a fluid or fluid additive relative to a single item such as extreme boundary lubrication or cooling ability. This has been tried many times in the past but with only limited success. For example, the four-ball wear test has been used to evaluate lubrication characteristics of a fluid (Feng et al., 1961). Also Hain (1952) has measured the temperature rise of coolants when flowing through a heated capillary tube under standard conditions as a measure of their cooling capability. However, the performance of a fluid in cutting is often related to subtle secondary changes in cutting geometry such as that due to chip curl more than to direct lubrication or cooling action. These secondary actions presently cannot be predicted. While bench tests often serve a useful role as screening tests, there is unfortunately no satisfactory substitute for direct machining trials under conditions that are close to those to be used in practice. The selection and application of metal cutting fluids is very much an art rather than a science as are so many other aspects of metal cutting. The main role of theory in metal cutting is often not to predict what will happen but to understand what is observed in order to reduce the number of trials required to reach the desired objective.

Wertheim et al. (1992) investigated high-pressure (up to 25 bars) application of fluid to the chip control groove of a tool in a grooving operation. It was found that the flow rate has a significant influence on tool life and on the shape and metallurgical structure of the chip. The BUE was also found to be smaller when grooving stainless steel and high alloy steels.

Figure 13.40 shows three views of a chip control groove in a parting operation. Figure 13.41 shows the tool life (T) versus the flow rate $\dot{Q}$ for three fluid supply conditions. It is evident that tool life is greatest when fluid is introduced below the chip and increases with the rate of fluid flow. Figure 13.41 also shows that tool life is better with fluid flow beneath the

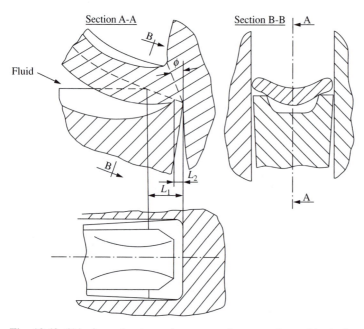

Fig. 13.40 Chip formation in parting or grooving operations with single groove for chip control and fluid supply. (after Wertheim et al., 1992)

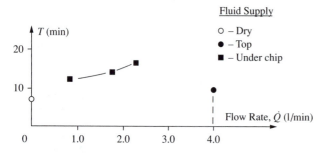

Fig. 13.41 Tool life (T) as a function of flow rate $\dot{Q}$ and different types of fluid supply when grooving AISI 4140 steel with a P–40 carbide tool. (after Wertheim et al., 1992)

tool than when applied to the top of the chip and requires a much lower flow rate. It was also found that with fluid supplied beneath the chip, the chip was more brittle giving rise to better chip control.

PRACTICAL CONSIDERATIONS

The foregoing discussion has been concerned with two important aspects of cutting fluid action–energy reduction and cooling. Where the major problem is to reduce friction and to prevent adhesion between tool and work as in broaching, tapping, and gear cutting, oil-base fluids are generally used. However, for cutting performed at higher speeds, where cooling is of greater importance, water-base fluids are most effective. The reason for this division into the two types lies chiefly in the different chemical characteristics of the additives used with each type.

An effective cutting fluid must get to the tool face and quickly react with the chip to form a low shear strength solid or act as a contaminent to prevent rewelding of microcracks. To get to the tool point rapidly, the important properties of the fluid are interfacial tension between the fluid and the metal being cut and the viscosity of the fluid. Organic additives containing sulphur and chlorine form strongly adhering low-shear-strength films, but are relatively slow to react. Inorganic reactions are generally more rapidly completed than organic ones, and the reaction products have higher melting points. The reactivity of a useful cutting fluid must be selective with regard to temperature and pressure. In other words, the fluid should be highly reactive under conditions obtaining at the tool face, but be perfectly stable when in contact with metals under atmospheric conditions to prevent corrosion.

Since water has the highest specific heat of all liquids, it is natural to use this material as the carrier for fluids designed for high-speed operations where cooling is most important. Many of the additions that are made to a water-base fluid are insoluble in water, and hence an emulsifying agent is frequently necessary. Inasmuch as water in different localities contains different amounts of solids (have different degrees of hardness), it is usually necessary to employ a water conditioner to prevent the solids from causing other ingredients from separating out. In addition to friction-reducing additives, emulsifying agents, and water conditioners, other additives employed include bactericidal agents, rust inhibitors, wetting agents, antifoaming agents, and antimisting additives.

Cutting Fluid Literature

There are many articles in the technical press dealing with all aspects of cutting-fluid selection, application, and troubleshooting. A useful collection of such articles is available in a book edited by Byers (1994).

Cutting Fluid Types

There are several types of cutting fluids:

- *straight oils*, which are of petroleum, animal, marine, vegetable or synthetic origin (with or without additives)
- *soluble oils*, which are straight oils containing an emulsifying agent with or without performance enhancing additives
- *semisynthetic fluids*, which are emulsions that contain little mineral oil, are chlorine free, and are designed to be diluted with water
- *synthetics*, which are clear water-base fluids that contain no mineral oil or emulsifying agents

Mixing

When a concentrate is to be mixed with water, water should be added to the concentrate while stirring. Otherwise, an inverted emulsion (water in oil) may be formed. A water-base cutting fluid must be maintained on the alkaline side of the hydrogen ion concentration scale (pH < 7). Otherwise, corrosion, microbial infestation, and a very unpleasant odor may result. An easy way of remembering what should be added to what is provided by the fact that dilution of a cutting fluid follows the same rule as that employed in freshman chemistry where it was stressed that the safe way of diluting an acid is to slowly add the concentrate to a large volume of water. Otherwise, the heat developed may cause the concentrate to be ejected from the container.

Cutting Fluid Selection

Sluhan (1986) has discussed problems in selecting a cutting fluid for use in a flexible machining cell (FMC), where a variety of operations may be performed including

- turning
- boring
- milling
- drilling
- reaming
- threading

The first three of these primarily require cooling while the latter three require mostly lubrication. The problem of cutting-fluid selection is even more difficult when a variety of materials are involved such as

- a variety of steels
- cast and nodular irons
- a variety of nonferrous materials
- titanium and high-temperature alloys

The rule generally followed is to select the fluid that best meets the *most important* situation. Special considerations associated with cutting management include

- bacteria control
- use of deionized water particularly in areas where the mineral content of city water is high
- periodic cleaning of coolant sumps
- control of concentration of additives

Sluhan discusses these important considerations and the compromises that are required when selecting a fluid for an FMC. A rule of thumb for the flow rate to be used is 2.5 to 3 gal/min/h.p.

Evaluation of Cutting-Fluid Performance

This is considered from the point of view of repeatability, resolution, and cost in De Chiffre and Belluco (2000). It was concluded that tool-life and surface-finish tests are of limited repeatability while tests based on cutting forces have better repeatability.

Health and Safety Considerations

One of the concerns associated with use of cutting fluids in the workshop is possible deleterious effects on those coming in contact with their chemical ingredients. Lucke (1992) has presented a convincing view that toxicology experiments based on the reaction of rats to high doses to estimate safe levels of exposure for humans are often overly pessimistic. This is important since it is very costly to prevent fluids that are effective in increasing productivity from being used if their use is prevented based on a false assumption they may be hazardous to an operator's health.

Three commonly used ingredients are discussed from a chemical perspective:

- ethanolamine
- formaldehyde
- N-nitrosodiethanolamine

The case is made that when properly used and when realistic animal tests are employed and reasonably extrapolated, these materials should not represent a health concern.

In very large facilities, as in the automotive industry, the presence of mist consisting of very small droplets of cutting fluid associated with high-speed machining operations has been a problem. Extremely fine particles are difficult to control by vented enclosures and filtering systems. Possible mechanisms for the generation of unwanted cutting-fluid particles by spinoff from a workpiece at high cutting speeds or by splash due to the impact of a cutting fluid as it contacts the workpiece are considered from a theoretical point of view by Chen et al. (2000).

A useful approach has been to employ mist suppression technology that prevents extremely fine particles from being generated. It has been found that small concentrations of high molecular weight polymers added to cutting oils prevent formation of extremely small particles of oil under mist-forming conditions.

Small additions of polyisobutylene (PIB) have been very effective in preventing formation of very small particles of mist. A molecular weight of about 10^6 is employed in concentrations below 1000 ppm, which has an negligible influence on the surface tension and viscosity of a cutting oil (Marano et al., 1997).

The way in which PIB acts to suppress mist formation is described in Smolinski (1994) and Smolinski et al. (1996). The PIB molecules in solution in the cutting oil are initially coiled, act as springs as they are extended, and inhibit elongation of the fluid before breakup. The result is to produce particles of larger size. Energy in the fluid that normally causes misting is absorbed by the PIB molecules as they are elongated and is released to the fluid as heat as they return to their coiled condition.

A similar approach to suppress misting of water-base cutting fluids using high molecular weight polyethyleneoxide (PEO) is described in Kalhan et al. (2000). Addition of 150 ppm of PEO to a water-base cutting fluid has been found to reduce misting by an order of magnitude. This is of special importance since the quantity of water-base fluid in use is about four times the amount of cutting oil in use.

However, PEO tends to degrade under the very high shear stresses encountered by a water-base fluid in high-speed cutting operations. This requires fresh PEO to be added periodically. Alternatively a new polymer with greater shear resistance has been found. This is derived from the monomer 2-acrylamine-2 methylpropane sulfonic acid. Large-scale plant tests with this polymer added to a soluble oil emulsion were found to reduce misting by an average of 50%. In the first large-scale plant test, 1000 ppm of the polymer (designated polymer B) was employed at an automotive transmission plant machining cast-iron parts at high speed on several transfer lines. A second large-scale plant test was at an engine plant involving six transfer lines machining aluminum cylinder heads. This second test involved two central fluid systems (27,000 gallons total capacity) to which 750 ppm of polymer B was added. Again misting was reduced by 50%, and there were no adverse effects on the product or the system and no reduction in the decrease in misting for the duration of the test (6 weeks). In both cases the operators were favorably impressed by the improvement of air quality in the shop and the reduction of oil deposits on the floor and on the machines.

The problem of mist suppression is soon apt to become more important, since the Occupational Safety and Health Administration (OSHA) is expected to lower the permissible exposure limits of oil mist from 5 mg/m^3 to 0.5 mg/m^3 in the near future.

DRY MACHINING

With increased interest in environmental concerns, there is interest in the possibility of machining without cutting fluids. Klocke and Eisenblaetter (1997) have discussed the possibilities of dry machining, which is not an option for all materials. Dry machining is possible for some aluminum alloys and cast irons. New coating materials can play a role in making dry machining a possibility. When considering the dry machining of aluminum alloys, an important consideration is the increased recycling value of swarf that is uncontaminated by residual cutting fluid. In drilling operations the removal of swarf from the hole without the help of a cutting fluid can be a problem.

Alternatives to the use of a free-flowing cutting fluid in high-speed cutting operations, where cooling is important, include the following:

- use of internal passages in the tool and/or the tool holder to confine the coolant
- use of a free-flowing cryogenic gas such as CO_2 or N_2
- thermoelectric cooling of the tool by use of an electric current

In general, solutions of this type that have been tried are too costly to be practical.

The machining of titanium and superalloys involves such a high temperature rise that dry cutting is not an option.

Graham (2000) has discussed problems and solutions involved in changing from operation with a cutting fluid to cutting dry. When making such a change, there will be need to alter operating conditions (speed, feed, depth of cut, rake angle, side cutting edge angle, and chip control grooving) to compensate for the higher temperatures involved when changing from wet to dry machining. The changes to be made are complex and operation specific (e.g., turning and milling). It is best to arrive at a new set of optimum operating conditions by systematic change, keeping in mind that chip breaking is more difficult without coolant. Maintenance of tolerance when changing from wet to dry cutting may be achieved by adjusting the depth of cut to offset the greater contraction on cooling.

AN UNEXPECTED RESULT

While the cooling action of a cutting fluid is usually advantageous with respect to tool life, this is not always the case. In a turning operation at relatively light feed (0.005 i.p.r.), it was found that the rate of tool wear increased over cutting in air even though the specific cutting energy decreased when a strong flow of coolant was applied to the back of the chip. Subsequent study revealed that this unexpected result was due to an increase in chip curl (decrease in tool–chip contact length) when the coolant was applied to the relatively thin chip. Even though the resultant cutting force was less with increased chip curl and the maximum tool face temperature was less, the fact that the point of maximum temperature moved closer to the cutting edge with increased chip curl produced a net increase in wear rate.

REFERENCES

Axer, H. (1954). In *Fortschrittliche Fertigune und Moderne Werkzeugmaschinen*. Giradet (Essen).

Bingham, H. L. (1954). *Scient. Lubric.* **6**, 22.

Brosheer, B. C. (1953). *Am. Mach.* **97**, 137.

Byers, J. P. (1994). *Metal Working Fluids*. Marcel Dekker, New York.

Chamberland, H. J. (1950). *Lubric. Engng.* **6**, 21.

Chen, Z., Atmadi, A., Stephenson, D. A., and Liang, S. Y. (2000). *Annals of CIRP* **49/1**, 53–56.

Childs, T. H. C., Maekawa, K., and Maulik, P. (1988). *Material Science and Technology* **4**, 1006.

Childs, T. H. C., Maekawa, K., Obikawa, T., and Yamane, Y. (2000). *Metal Machining Theory and Applications*. Wiley, New York.

De Chiffre, L. (1977). *Int. J. Mach. Tool Des. Res.* **17**, 225.

De Chiffre, L. (1978). *Lub. Eng.* **34**, 344–351.

De Chiffre, L. (1981). *ASLE Trans.* **24**, 340–344.

De Chiffre, L. (1988a). *ASLE Trans.* **27/3**, 220–226.

De Chiffre, L. (1988b). *Lub. Eng.* **44**, 514–518.

De Chiffre, L., and Belluco, W. (2000). *Annals of CIRP* **49/1**, 57–60.

De Chiffre, L., Belluco, W., and Zeng, Z. (2001). *Lub. Eng.* **58**, 24–28.

Dhar, N. R. (2001). Ph.D. thesis, IIT Kharagpur, West Bengal, India.

Doyle, E. D. (1977). Private communication.

Epifanov, G. I., Soloshko, E. P., and Rehbinder, P. A. (1955). *Dokl. Akad. Nauk.* **104**, 68.

Faust, D. G. (1952). *Lubric. Engng.* **8**, 183.

Feng, I. M., Gujral, A., and Shaw, M. C. (1961). *Lubric. Engng.* **17**, 324.

Graham, D. (2000). *Mfg. Eng. (SME)* **1**, 72–78.

Hain, G. M. (1952). *Trans. Am. Soc. Mech. Engrs.* **74**, 1077.

Kalhan, S., Twining, S., Denis, S., Marino, R., Messick, R., and Johnson, R. (2000). *Lub. Eng.* (Sept.).

Klocke, E., and Eisenblaetter, G. (1977). *Annals of CIRP* **46/2**, 519–526.

Komanduri, R., and Shaw, M. C. (1974). *Nature* **248**, 582.

Lauterbach, W. E. (1952). *Lub. Engng.* **8**, 135.

Lucke, W. E. (1992). *Lub. Eng.* (May), 425–428.

Marino, R. S., Smolinski, J. M., Manke, C. W., Gulari, E., and Messick, R. L. (1997). *Lub. Eng.* (Oct.), 25–34.

Nakayama, K. (1956). *Bull. Fac. Engng. Yokohama Nat. Univ.* **5**, 1.

Niebusch, R. B., and Strieder, E. H. (1951). *Mech. Eng.* **73**, 203.

Olson, G. B. (1948). *Machinery*, **43**.

Pahlitzsch, G. (1951). *Industrie Anzeiger* **69/70**, 796.

Pigott, R. J. S., and Colwell, A. T. (1952). *Trans. Soc. Automot. Engrs.* **6**, 547.

Rehbinder, P. A. (1947). *Nature* **159**, 866.

Rehbinder, P. A. (1948a). *Dokl. Akad. Nauk.* **62**, 053.

Rehbinder, P. A. (1948b). *Dokl. Akad. Nauk.* **63**, 159.

Rehbinder, P. A. (1949). *Dokl. Akad. Nauk.* **64**, 874.

Rehbinder, P. A. (1951). *Dokl. Akad. Nauk.* **80**, 781.

Rehbinder, P. A. (1954a). *Dokl. Akad. Nauk.* **97**, 270.

Rehbinder, P. A. (1954b). *Dokl. Akad. Nauk.* **99**, 801.

Reitz, P. (1919). DRP (German Patent) Nr. 336960.

Schallbroch, H., Schaumann, H., and Wallichs, R. (1938). *Vortrage der Haptversammlung der Deutsche Gesellschaft fur Metallkunde*. VDI, p. 34.

Sharma, C. S., Rice, W. B., and Salmon, R. (1971). *J. Engng. Ind.* **93**, 441.

Shaw, M. C. (1942). *Metal Prog.* **42**, 85.

Shaw, M. C. (1944). *J. Am. Chem. Soc.* **66**, 2057.

Shaw, M. C. (1948). *J. Appl. Mech.* **15**, 37.

Shaw, M. C. (1951). *Ann. N.Y. Acad. Sci.* **53**, 962.

Shaw, M. C. (1959). *Wear* **2/3**, 217.

Shaw, M. C. (1970). *Proc. Soc. Manuf. Engrs.* **MR70**, 277.

Shaw, M. C., Ber, A., and Mamin, P. A. (1960). *J. Basic Engng.* **82**, 342.

Shaw, M. C., Cook, N. H., and Finnie, I. (1953). *Trans. Am. Soc. Mech. Engrs.* **75**, 273.

Shaw, M. C., Cook, N. H., and Smith, P. A. (1958). Research Report No. 19. Soc. Man. Eng.

Shaw, M. C., Pigott, J. D., and Richardson, L. P. (1951). *Trans. Am. Soc. Mech. Engrs.* **75**, 45.

Shirakashi, T., Komanduri, R., and Shaw, M. C. (1978). *J. Engng. Ind. Trans. Am. Soc. Mech. Engrs.* **100**, 244.

Sluhan, C. A. (1986). SME Technical Paper MS86-124.

Smart, E. F., and Trent, E. M. (1974). In *Proc. Int. Conf. Machine Tool Des. Res.* Macmillan, London, p. 187.

Smolinski, J. M. (1994). Ph.D. dissertation. Wayne State University Detroit, Mich.

Smolinski, J. M., Galuri, E., and Manke, C. (1996). *J. Am. Inst. of Chem. Engrs.* **42**, 1201.

Taylor, F. W. (1907). *Trans. Am. Soc. Mech. Engrs.* **28**, 31.

Thuma, R. F. (1954). *Am. Mach.* **98**, 117.

Usui, E., Gujral, A., and Shaw, M. C. (1961). *Int. J. Mach. Tool Des. Res.* **1**, 187.

Wertheim, R., Rotberg, J., and Ber, A. (1992). *Annals of CIRP* **41/1**, 101–106.

Wister, H. J. (1936). *Archiv. fur des Eisenhutten Wesen Gruppe E*–Nr. **513**, 523.

14. TOOL MATERIALS

THE MACHINING SYSTEM

Table 14.1 shows the variety of inputs and internal machine considerations involved when metals are cut. The cutting tool material is one of the important elements of the machining system. Tool material and geometry must be carefully chosen in relation to the workpiece material to be machined, the kinematics and stability of the machine tool to be employed, the amount of material to be removed, and the required accuracy and finish.

The most satisfactory tool will usually be the one corresponding to the minimum total cost of performing a required operation to the specified accuracy. This total cost includes

1. initial tool cost
2. tool grinding cost
3. tool life
4. labor cost as influenced by cycle time, machine, operator, and labor cost
5. the proportion of tool changing cost assignable to one part

Thus, the best tool material will not necessarily be the one that gives the longest life; such factors as grindability, tool material cost, and the practical levels of cutting speeds and feeds for a given tool material play important roles in the selection of the best tool material for a particular operation.

TABLE 14.1   Inputs, Outputs, and Internal Items in the Machining System When Metals Are Cut[†]

Inputs ⟶	Machine Internal Items ⟶	Output
Workpiece	Forces	Parts/cost
Tools	Energy	
Motions	Temperatures	Parts/time
Control	Wear	
Fluids	Stationary zones	Required geometry
	Vibration	Surface integrity

[†] After Shaw, 1968.

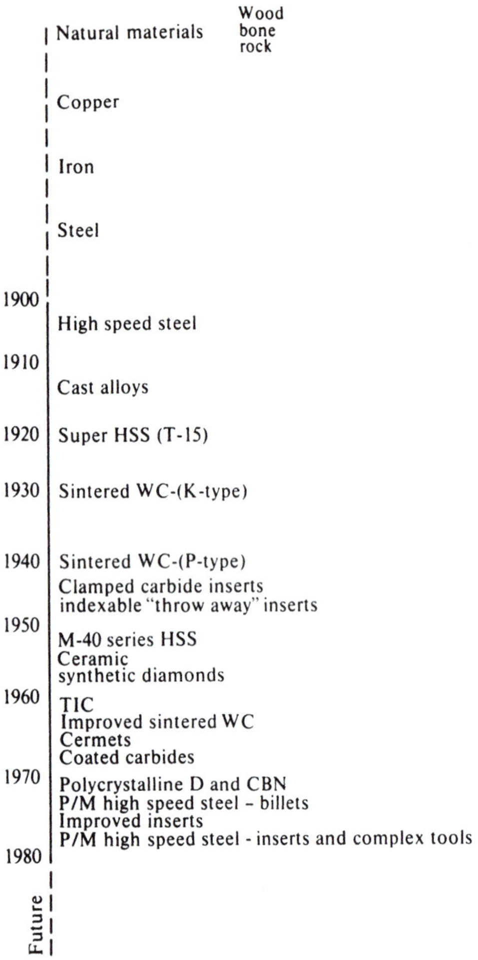

```
| Natural materials    Wood
|                      bone
|                      rock
|
| Copper
|
|
| Iron
|
|
| Steel
|
1900
|     High speed steel
1910
|     Cast alloys
1920 | Super HSS (T-15)
1930 | Sintered WC-(K-type)
1940 | Sintered WC-(P-type)
|      Clamped carbide inserts
|      indexable "throw away" inserts
1950
|      M-40 series HSS
|      Ceramic
|      synthetic diamonds
1960 | TIC
|      Improved sintered WC
|      Cermets
|      Coated carbides
1970 | Polycrystalline D and CBN
|      P/M high speed steel – billets
|      Improved inserts
|      P/M high speed steel - inserts and complex tools
1980 |
     |
Future
     |
```

Fig. 14.1 Approximate dates of introduction of different cutting-tool materials and concepts. Important, more recent cutting-tool developments (after 1980) and *possible* new ones beyond the present are covered in the rest of this chapter. (after Shaw, 1980a)

With the increased use of numerically controlled machine tools, reliability and predictability of performance are of greater significance than before, and these items must be given greater weight in selecting tool materials for such applications.

HISTORICAL BACKGROUND

History is a useful liberal art for explaining the present and anticipating the future relative to human events. It is similarly useful to consider the progression of developments in a given area of technology.

Figure 14.1 is a chronological list of major developments relative to tool materials from 1900 to 1980. While one might question the time when each of these major developments entered our technology or whether other landmark events should have been included, it is believed that this presentation is reasonably complete and accurate. Only those items that appear to have a long (100 years) future have been included in Fig. 14.1. After making a few general observations concerning this figure, each item will be discussed in detail.

The most striking aspect of Fig. 14.1 is the increasing rate of introduction of new concepts from 1900 onward. This was due in part to an accelerating improvement in materials technology and in part to the need for special machining characteristics.

Some of the events responsible for the acceleration of change evident in Fig. 14.1 are listed in Fig. 14.2. The new areas of technology that have appeared since 1945 have each introduced new materials to be machined to which the cutting-tool industry has had to respond. The decade of the 1970s was one of rapidly changing manufacturing conditions that has been reflected in major changes in the cutting-tool industry in the 1980s and beyond.

TOOL MATERIAL REQUIREMENTS

The second general observation that may be made is that there is need for a wide spectrum of cutting-tool materials.

Three important considerations are high-temperature physical and chemical stability, abrasive wear resistance, and resistance to brittle fracture. A given tool material is generally not outstanding relative to all three of these attributes. In general, as a material is made more refractory, it becomes more brittle; or if it is made more abrasion resistant, it also becomes more brittle. This is true not only when comparing different classes of tool material such as high speed steel (HSS), tungsten carbide (WC), and titanium carbide (TiC) but also when comparing different compositions within a given class such as different grades of sintered tungsten carbide. Tungsten carbide tool manufacturers frequently indicate the area of usefulness of a given product relative to other carbides on a triangular plot similar to that shown in Fig. 14.3.

A similar plot may be used to show the relative areas of usefulness of HSS, WC, and TiC tools. Such a diagram can give only a rough indication of the area of usefulness of a given material since the exact location of the lines will depend upon many specific items such as

1. relative hardness of tool and work material (tool should be four times as hard as work)
2. abrasive particles in work or on work surface (scale)
3. chemical compatibility of tool and work material
4. tool-tip temperature (speed, feed, tool geometry, coolant, etc.)
5. condition of machine (rigidity)
6. type of operation (continuous or interrupted cut)

In general, there will be regions of overlap (A, Fig. 14.3b) and regions that are not covered by the three types of tools considered (B, Fig. 14.3b). This explains the need for a wide range of tool materials and suggests that as a new tool material is introduced, and has found its place, an old material will not completely disappear but will simply yield some of the area it once dominated.

As the rate of metal removal has increased over the years, there has been need for more refractory tool materials which accounts for the progression from HSS to tungsten carbide to titanium carbide to ceramic tool materials. At the same time such a progression calls for an increase in machine-tool horsepower and rigidity. Since an appreciable number of tools must operate under conditions that are not suitable for the newer more refractory but more brittle tools, HSS is still very much in demand. The conditions under which more ductile tool materials such as HSS are called for include

1. older, less rigid machine tools
2. under-powered low-speed machines (in such cases, temperatures do not warrant more expensive, more refractory tooling)

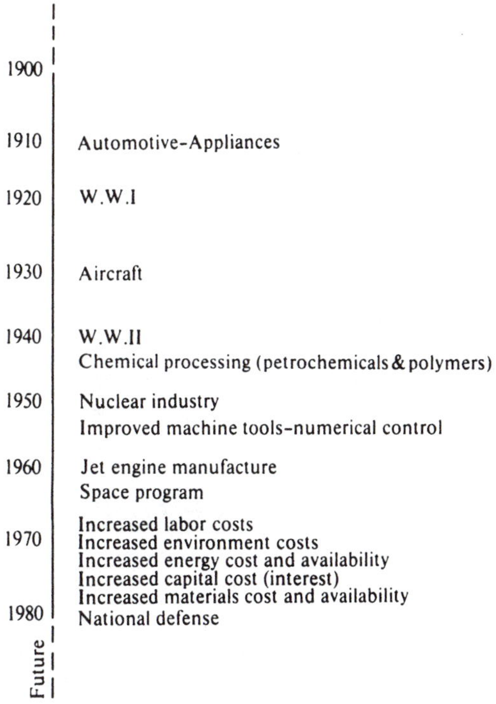

1900
1910
Automotive-Appliances
1920
W.W.I
1930
Aircraft
1940
W.W.II
Chemical processing (petrochemicals & polymers)
1950
Nuclear industry
Improved machine tools-numerical control
1960
Jet engine manufacture
Space program
1970
Increased labor costs
Increased environment costs
Increased energy cost and availability
Increased capital cost (interest)
Increased materials cost and availability
1980
National defense
Future

Fig. 14.2 Approximate dates of introduction of principal product classes and conditions influencing the need for new and improved cutting-tool materials. (after Shaw, 1980a)

High temp stabilty
Composition A
Composition B
Resistance to brittle fracture
Abrasive wear resistance
(a)
High temp stability
TiC
WC
A
Resistance to brittle fracture
Abrasive wear resistance
HSS
B
(b)

Fig. 14.3 Diagram showing (a) the relative high-temperature stability, resistance to brittle fracture, and abrasive wear resistance of a cutting tool material and (b) approximate application of (a) to high speed steel, tungsten carbide, and titanium carbide tool materials. (after Shaw, 1980a)

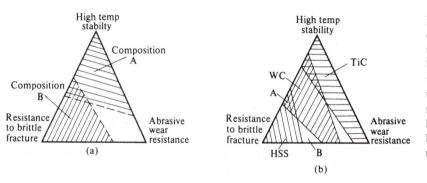

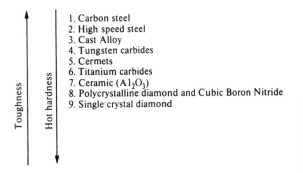

1. Carbon steel
2. High speed steel
3. Cast Alloy
4. Tungsten carbides
5. Cermets
6. Titanium carbides
7. Ceramic (A1$_2$O$_3$)
8. Polycrystalline diamond and Cubic Boron Nitride
9. Single crystal diamond

Fig. 14.4 Principal classes of cutting-tool materials and their relative toughness and hardness at elevated temperature.

3. small-diameter work that cannot be operated in the speed range to completely avoid a built-up edge

4. workpieces involving interrupted cuts or heavy surface scale

In general, as a tool material becomes more brittle, it becomes more important to avoid the low-speed range where an unstable built-up edge (BUE) tends to form. If a ductile metal that forms a BUE, such as mild steel, must be machined in the built-up edge speed range, then it may be necessary to use HSS instead of a WC or ceramic tool material to avoid chipping of the cutting edge.

MAJOR TOOL MATERIAL TYPES

The major classes of cutting tool materials may be listed in order of increasing hot hardness as shown in Fig. 14.4. As this list is descended, the strain at fracture decreases. All of these materials except the last one have a two-phase structure—a softer continuous phase separating very hard particles. As the list is descended, the spacing between hard particles decreases. Initially (first three items) the spacing of the hard particles is sufficiently great so that Young's modulus of elasticity corresponds to that of the continuous phase (a substitutional solid solution of iron in the case of HSS). However, beginning with cemented tungsten carbide (item 4), the hard carbides are so close together that Young's modulus approaches that of the tungsten carbide particles instead of the cobalt binder-phase [i.e., 90×10^6 p.s.i. instead of about 30×10^6 p.s.i. (6.21×10^5 MPa instead of 2.07×10^5 MPa)]. In the case of items 7 and 8 of Fig. 14.4, the space between hard particles is essentially that corresponding to crystal boundaries. In the case of ceramic tools, no binder material or sintering aid is normally employed, the only difference in chemistry in the crystal boundaries being that due to the very small impurity content. In the case of item 8, a small amount of cobalt is present in the crystal boundaries that evidently diffuses there from the cemented tungsten carbide substrate on which the diamond or cubic boron nitride (CBN) crystals are supported. Item 9 is the only truly single-phase material. This is single crystal (natural) diamond that is extremely brittle due to the complete lack of crack-arresting mechanism as in the polycrystalline counterpart (item 8) and the presence of many structural defects.

High Speed Steel

Before 1870 all lathe tools were forged from plain carbon steel consisting of about 1% carbon, 0.2% manganese, and the remainder iron. These tools had a low hardness at elevated temperature (low hot-hardness) and could only be used to machine steel at relatively low speeds (up to about 15 f.p.m. = 0.08 m s^{-1}). This material also required a water quench in hardening which frequently resulted in quenching cracks. In 1870 Robert Mushet introduced a steel in England having approximately the composition 2% carbon, 1.6% manganese, 5.5% tungsten, and 0.4% chromium. This steel was air-hardening and retained its hardness to higher cutting temperatures. As a result it could be used to speeds of about 25 f.p.m. (0.13 m s^{-1}). Mushet steel with modification (mainly replacement of manganese by chromium) was used for lathe tools until the turn of the twentieth century.

Taylor and White (1901) produced tools of greatly increased stability which allowed cutting speeds of about 60 f.p.m. (0.32 m s^{-1}) and consequently this material became known as high speed steel (HSS). Contrary to common belief this was not the development of a new steel but rather a new heat treatment for the existing material. Taylor and White found that if tools were heated quickly through the brittle temperature range of 845 to 930 °C to a temperature just short of the melting point of the steel before quenching, a steel of improved hot-hardness resulted. It was also found that such tools were improved by a higher tempering temperature. The composition of the steel used by Taylor and White in 1901 was approximately 1.9% carbon, 0.3% manganese, 8% tungsten, and 3.8% chromium. This is seen to differ from the original Mushet steel mainly in the increased amount of tungsten and the replacement of manganese by chromium.

The period from 1900 to 1906 saw the rapid development of HSS. Taylor found that improved tools were produced by using less carbon and more tungsten (the carbon decrease was necessary to make the steel forgeable when tungsten was increased to improve hot-hardness). By 1903 the carbon content of HSS had fallen to 0.7% while the tungsten content had risen to 14%. In 1904, Dr. J. A. Matthews found that the abrasive resistance of HSS could be increased by additions of vanadium, and by 1906 Taylor was using about 0.3% vanadium in his tools. By 1910 the tungsten content had increased to 18%, the chromium content to 4%, and the vanadium content to 1%, thus providing the well known 18–4–1 HSS which was the standard HSS for the next 40 years. This steel is designated AISI T–1.[†]

In the short span from 1900 to 1910 the allowable cutting speed had been increased from 25 f.p.m. (0.13 m s^{-1}) to over 100 f.p.m. (0.52 m s^{-1}), and as might be expected this had an immediate and far-reaching influence on machine tool design.

The production of higher quality HSS was made possible by the introduction of electric furnace melting in 1907 but it was not until about 10 years later that such furnaces came into wide use. In 1912 it was found that the red-hardness of HSS could be improved by additions of 3–5% cobalt. However, such materials were not extensively used because the speed and power available in lathes still had not been increased to values called for by 18–4–1 HSS.

By 1920 there were but three high speed steels in common use:

AISI Symbol	W	Cr	V	Co
T–1	18%	4%	1%	0
T–7	14%	4%	2%	0
T–4	18%	4%	1%	5%

The first of these was the standard, the second was used where greater abrasive resistance was called for, while the third was used where tool temperatures were high and greater hot-hardness was desired.

About 1923 the first of the so-called super high speed steels appeared. This steel (T–6) had the composition

$$0.7\% \text{ C}, \quad 4\% \text{ Cr}, \quad 2\% \text{ V}, \quad 20\% \text{ W}, \quad 12\% \text{ Co}$$

and was followed by steels with even greater amounts of carbon and vanadium. However, these steels could not be made commercially due to the difficulty of forging, rolling, and heat-treating them. It was not until 1939 that Gill showed that steels of high carbon and vanadium content could

---

[†] The American Iron and Steel Institute (AISI) has introduced symbols for the two major classes of HSS: the symbol T designates a tungsten-base steel while the symbol M designates a molybdenum-HSS.

be readily hot-worked without cracking if the tungsten content were decreased. This led to such super high speed steels as T–15:

$$1.5\% \text{ C}, \qquad 4\% \text{ Cr}, \qquad 5\% \text{ V}, \qquad 12\% \text{ W}, \qquad 5\% \text{ Co}$$

While it was known as early as 1900 that molybdenum could be substituted for tungsten in HSS (half the weight of molybdenum need be used since the atomic weights of molybdenum and tungsten stand approximately in the ratio of 1:2), this was not done commercially until about 1923 due to difficulties associated with a narrower heat-treating range and a greater tendency for molybdenum bearing steels to decarburize. The molybdenum steels are essentially equivalent to their tungsten counterparts but were not widely used until World War II when there was a shortage of tungsten. Since this time, the molybdenum high speed steels have largely replaced the tungsten steels due to a significant cost advantage and the greater strategic availability of molybdenum. Today, M–2 HSS (composition below) is the standard in the United States (having displaced T–1 about 1950):

$$0.8\% \text{ C}, \qquad 4\% \text{ Cr}, \qquad 2\% \text{ V}, \qquad 6\% \text{ W}, \qquad 5\% \text{ Mo}$$

Current HSS tools enable steel to be machined at speeds that are often in excess of 300 f.p.m. (1.45 m s^{-1}).

It is well known that the hardness and wear resistance of HSS depends on the composition, size, and distribution of the carbides in the steel and upon the stability of the matrix at high temperatures. The latter is increased by addition of cobalt. Harder mixed carbides are provided by increased vanadium and carbon content. A major development in the HSS area in the 1950s was the discovery that steels of increased hardness (R_c 70 instead of the R_c 65 for the more conventional HSS) and reasonable toughness may be produced by use of compositions such as

$$1.4\% \text{ C}, \qquad 4\% \text{ Cr}, \qquad 4\% \text{ V}, \qquad 9\% \text{ W}, \qquad 4\% \text{ Mo}, \qquad 12\% \text{ Co}$$

These steels have a large concentration of finely and uniformly divided carbides in a rather refractory matrix. They are rendered forgeable by substitution of Mo for W. However, like wrought T–15 HSS, they are difficult to grind.

During the 1970s there were two additional significant developments in the HSS area. The first involved the atomization of a prealloyed HSS such as T–15, followed by consolidation into billets by hot isostatic compaction.

The main characteristics of this material are a smaller carbide size and a more uniform carbide distribution than that for the more conventionally produced wrought billets. These structural changes made it much easier to grind normally difficult-to-grind HSS such as T–15. The atomized material has carbides that are typically less than 3 μm in size. The importance of this is that carbides of this size are displaced to the side by the abrasive particles in a grinding wheel instead of resisting the motion of the abrasive through the material being ground resulting in less loss of abrasive from the grinding wheel when materials with small carbides are ground. The atomized material has less carbide segregation and produces tougher tools.

The second important development consists of a cold impact fracture process as the final comminution step before sintering. This results in individual particles having many sharp points instead of the rounded surfaces produced in ball milling. The advantage of this, of course, is that less pressure is required in preform production. This has made it possible to produce preforms in dies by cold pressing instead of by the more expensive hot isostatic pressing required for directly atomized material. The so-called cold-stream impact comminution process (Friedman et al., 1965) has the added advantage of making it possible to use scrap HSS chips as feed material instead of 100% atomized material. However, the chief advantage of this latest HSS-P/M[†] development is

---

[†] P/M = powder metallurgy.

that HSS *inserts* for use in standard carbide tool holders and having pressed-in chip curlers may be economically produced. Complex tools such as taps, hobs, and milling cutters may also be produced by the new HSS-P/M process in near net shape, thus greatly reducing the energy and productivity loss associated with the slitting of such a tool from a solid billet. As the cost and scarcity of energy and strategic materials such as chromium, cobalt, and tungsten increases, the HSS-P/M approach to inserts and tools of complex shape is bound to play an increasing role.

A development in the HSS area not included in Fig. 14.1 is the production of precision-cast HSS tools. This technique like the HSS-P/M approach makes it possible to use compositions of higher carbon and alloy content than would be possible if forging and rolling were employed. Krekeler (1957) has shown that a heat treatment of 3 hours at 1100 °C is beneficial in removing the embrittling microcracks that are normally closed in forging. However, cast HSS tools have in general not produced tools as shock-resistant as wrought HSS tools.

Table 14.2 gives compositions for a number of commercially available high speed steels.

TABLE 14.2 Composition of Some Commerically Available High Speed Steels

AISI Designation	Weight Percent					
	C	W	Cr	V	Mo	Co
M1	0.80	1.75	3.75	1.15	8.75	
M2, Class 1	0.85	6.25	4.00	2.00	5.00	
M3, Class 1	1.05	6.25	4.00	2.50	5.75	
M4	1.30	5.50	4.00	4.00	4.75	
M6	0.80	4.25	4.00	1.50	5.00	12.00
M7	1.02	1.75	3.75	2.00	8.75	
M8	0.80	5.00	4.00	1.50	5.00	(1.25 Cb)
M10, Class 1	0.89	0.70	4.00	2.00	8.00	
M15	1.50	6.50	4.00	5.00	8.50	5.00
M30	0.80	1.80	4.00	1.20	8.25	5.00
M34	0.90	1.75	3.75	2.10	8.75	8.25
M36	0.85	6.00	4.00	2.00	5.00	8.25
M41	1.10	6.75	4.25	2.00	3.75	5.25
M42	1.08	1.60	3.75	1.15	9.60	8.25
M43	1.20	2.70	3.75	1.60	8.00	8.20
M44	1.15	5.25	4.25	2.00	6.50	11.75
M45	1.27	8.25	4.20	1.60	5.20	5.50
M46	1.24	2.10	4.00	3.20	8.25	8.25
T1	0.73	18.00	4.00	1.00		
T2	0.85	18.00	4.00	2.00		
T3	1.05	18.00	4.00	3.00	0.60	
T4	0.75	18.00	4.00	1.00	0.60	5.00
T5	0.80	18.00	4.25	2.00	0.90	8.00
T6	0.80	20.50	4.25	1.60	0.90	12.25
T7	0.75	14.00	4.00	2.00		
T8	0.80	14.00	4.00	2.00	0.90	5.00
T9	1.20	18.00	4.00	4.00		
T15	1.55	12.50	4.50	5.00	0.60	5.00

TABLE 14.3 Compositions of Some Commercially Available Cast Alloy Tool Materials

Trade Designation	Weight Percent			
	Co	Cr	W	C
Blackalloy 525	44	24	20	2
Blackalloy T × 90	42	24	22	2
Crobalt 1	48	30	14	2
Crobalt 2	40	33	18	2.5
Crobalt 3	40	33	20	3
Haynes Stellite R Star J	43	32.5	17.5	2.5
			(1.2 B, 4 V,	4 Ni)
Haynes Stellite R3	50	31	12.5	2.4
Haynes Stellite R19	53	31	10.5	1.8
Haynes Stellite R98M2	38	30	18.5	2
Tantung G	47	30	15	3
Tantung 144	45	28	18	3

Cast Alloy Tools

About 1915 nonferrous high-temperature alloys containing significant amounts of cobalt, chromium, and tungsten were devised by Elwood Haynes. These materials were not heat-treatable but were used as cast and are, therefore, sometimes known as cast alloy tools. However, it was not until the period 1920 to 1925 that cast alloy tools were commonly used in cutting. While the cast alloy tools are not quite as hard as the high speed steels at room temperatures, they do retain their hardness to high temperatures and hence are occasionally used for such special applications as form tools.

Cast alloy tools in the past were used in the form of relatively large bars just as HSS turning tools and brazed carbides were used before clamped sintered carbide inserts were introduced about 1950. The appearance of HSS-P/M inserts suggests that cast alloy inserts that do not exhibit the same casting segregation problems as HSS and have greater hot-hardness should be reexamined. The fact that such materials contain appreciable amounts of cobalt, chromium, and tungsten would be partially offset by the fact that an insert requires a much smaller volume of material than a massive tool bar. In addition, some of the work done in substituting nickel for cobalt in high-temperature turbine alloys might be applicable to cutting-tool technology.

Table 14.3 gives compositions of a number of cast alloy tool materials. The room temprature hardness values of these materials is generally in the range R_c 60 to R_c 65.

Cemented Tungsten Carbide

Tungsten carbide tools were first developed in Germany in 1927 and produced by the Widia Corporation. [The word *Widia* was derived from the idea that, relative to HSS, carbide is like diamond in hardness (*wie diamente* in German).] The first successful tungsten carbide tools were brought to the United States from Germany in 1928. These tools consisted of finely ground tungsten carbide particles sintered together with cobalt binder. Early work with these new tool materials was hampered by

1. their brittleness and tendency to chip

2. brazing difficulties

3. difficulty of grinding

4. lack of lathe rigidity, power, and speed to use them adequately

5. tendency for a crater to form on the tool face at high cutting speeds, especially when machining steels

Gradually during the 1930s, tools of greater shock resistance were produced, but carbide tools were used mainly to turn cast iron and nonferrous metals due to the greater tendency for steel to cause tool-face cratering. Diamond wheels for grinding carbide tools were introduced in the thirties which greatly aided tool preparation. About 1938 it was found that the tendency for a carbide tool to crater when cutting steel could be reduced by additions of titanium and tantalum carbides. During World War II the use of carbide tools expanded rapidly.

Titanium and tantalum carbides are more stable than tungsten carbide and have a greater resistance to decomposition in the presence of γ (FCC) iron. The temperature of the surface of the chip in contact with the tool in high-speed machining will generally lie above the allotropic transformation temperature for steel, and the great affinity of austenite (γ-iron) for carbon will cause a loss of carbon from WC crystals in the tool surface. This results in the development of a crater having its greatest depth at the point of maximum tool-face temperature. Titanium carbide having a greater resistance to decomposition has a dramatic effect on crater resistance of carbide tools when a steel is machined at a high cutting speed (high tool-face temperature). The addition of titanium carbide decreases the strength and abrasive wear resistance of a sintered tungsten carbide tool. Tantalum carbide also provides crater resistance with less loss of impact strength than titanium carbide since it gives rise to less grain growth during sintering but costs more than an equivalent amount of titanium carbide.

The net result of all of this is that there are two types of sintered tungsten carbide—one for machining gray cast iron, nonferrous metals, and abrasive nonmetals such as fiber glass and graphite (ISO K-type) and one for machining ferrous metals (ISO P-type). The cast-iron type consists of WC crystals with normally 3–12% cobalt as a binder while the steel-cutting grades of carbide have some of the WC substituted by TiC, TaC, or NbC, all of which render greater crater resistance with different degrees of loss of strength, abrasive wear, impact, and corrosion (oxidation) resistance. The strength (strain at fracture) of a carbide tool increases with increase in cobalt content, and tools of higher cobalt content are required in roughing operations where the rate of material removal is higher than in finishing operations.

During the 1960s there was a gradual increase in the quality of carbide tools. Carbide tool materials were made less brittle by reduced particle size (0.1 to 0.5 μm) and improved binding and sintering techniques. By milling the tungsten carbide powder with the sintering material, it was possible to coat each particle with the binder phase before pressing and thus provide a more homogeneous material. Tungsten carbide tools having a smaller void content and improved shock resistance were produced by subjecting the material to high hydrostatic pressures before sintering.

In recent years the U.S. Cemented Carbide Producers Association (CCPA), which was organized in 1955 to promote the effective use of cemented carbides, has been instrumental in standardizing test procedures for cemented carbides in cooperation with the American Society for Testing Materials (ASTM) and other standardizing organizations. However, there is much that remains to be done in the area of standardization. One of the most pressing needs is a system of carbide classification that could serve as a guide to the user in the selection of a carbide grade for a particular application. There are four systems of classification of cemented carbides in use in the world today:

1. the U.S. unofficial "C" classification system which is based on performance

2. the British Hard Metal Association (BHMA) system which is based on properties

3. the International Standards Organization (ISO) system (R513—*Application of carbides for machining by chip removal*)

4. the Russian system which is based on composition

The "C" classification system now used in the United States was introduced by automotive engineers to obtain some indication of equivalence of products made by different manufacturers. This system employs eight numbers C–1 to C–8 which cover the grades of carbide manufactured for cutting-tool applications. The C–1 to C–4 grades are for machining cast iron and materials that yield short chips such as the nonferrous alloys. In going from C–1 (roughing cuts) to C–4 (finishing cuts), the cobalt content and hence the shock resistance decreases. The C–5 to C–8 grades are steel-cutting grades which contain TiC, TaC, and NbC. In going from C–5 (roughing cuts) to C–8 (finishing cuts), the cobalt content decreases and the shock resistance decreases.

The BHMA designation consists of three numbers. The first number (1 to 9) gives the hardness category of the carbide (1 = low hardness, 9 = high hardness) and is meant to reflect wear resistance. The second number (1 to 9) gives the transverse rupture strength (TRS) of the carbide (1 = low TRS = 150,000 p.s.i. = 1035 MPa, and 9 = high TRS = 500,000 p.s.i. = 3449 MPa) and is meant to reflect shock resistance. The third number (0 to 9) gives the TiC/TaC content (0 = low TiC/TaC content, 9 = high TiC/TaC content) and is meant to reflect crater resistance. The difficulty with this system is that hardness is not a good measure of edge-wear resistance and TRS is not a good measure of shock resistance. It also fails to accommodate coated carbide tools to be discussed later.

The ISO system (Table 14.4) is somewhat like the "C" system and classifies all cutting grade carbides into three categories: P (steel-cutting grades), K (cast-iron-cutting grades), and M (an intermediate category for ductile irons, hard steels, and high-temperature alloys).

TABLE 14.4 ISO Carbide Classification System

Symbol	Category	Color Code	Designation	Cutting Conditions
P	Ferrous metals with long chips	Blue	P01	
			P10	
			P20	
			P30	Inc. Speed → Inc. Speed → Inc. Speed ↑
			P40	
			P50	
M	Intermediate	Yellow	M10	
			M20	
			M30	
			M40	Inc. Feed ← Inc. Feed ← Inc. Feed ↓
K	Nonferrous metals and ferrous metals with short chips (gray cast iron)	Red	K01	
			K10	
			K20	
			K30	
			K40	

The Russian system is based on composition of the carbide. This does not help the relatively uninformed user choose the proper carbide and is unacceptable in the West for proprietary reasons.

It appears as though the ISO system is gaining ground and that it may someday be a world standard.

Inserts

When tungsten carbide was introduced as a tool material, it was so expensive that it became necessary to braze a thin wafer of the material to a steel shank. Brazed tools of this type were used throughout World War II even though they did not generally

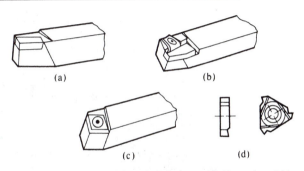

Fig. 14.5 Various turning-tool designs. (a) Brazed carbide tool. (b) Single edge clamped tool. (c) Multi-edge clamped tool. (d) Special screw thread cutting insert.

allow the full potential of tungsten carbide to be realized. Frequently carbide tips would fail through the braze due to the abrupt change in Young's modulus and coefficient of thermal expansion in this region (sintered carbide has about one half the coefficient of expansion of steel). This difficulty was minimized by making carbide wafers thicker and using a copper shim at the brazed surface to decrease concentrations of stress. A representative brazed carbide tool is shown in Fig. 14.5a.

Following World War II when improved lathe power and rigidity enabled larger cuts to be taken, clamped carbide holders came into use. The first of these (Fig. 14.5b) resembled the brazed type of carbide tool having but one cutting edge and several surfaces to be ground and honed. The next development to lower tool preparation cost was a multi-edge clamped tool. The tool shown in Fig. 14.5c has four cutting edges on each end of the blank, and a new cutting edge may be rotated into position without disturbing the tool setting. The decrease of grinding and tool-changing downtime costs associated with tools of this type is evident. Another feature associated with clamped carbide tools is the use of disposable inserts. In this instance an insert having several cutting edges (eight in the case of Fig. 14.5c) is ground at the factory; and when all cutting edges have been used, the insert is discarded rather than reconditioned. Today special inserts are even available for cutting a variety of screw thread forms (Fig. 14.5d).

Regrinding of "throw away" carbide inserts is done to further reduce carbide tool cost, and scrap carbide inserts are being recycled by mixing some finely divided used tool material with virgin powder in new tool manufacture.

Indexable carbide inserts are now available in a wide variety of shapes including triangles, squares, circles, rectangles, hexagons, pentagons, and diamonds having a variety of angles (80°, 55°, 42°, and 35°). The American National Standards Institute (ANSI) has introduced a special numbering system for inserts and tool holders. For example, the following designation has the meaning indicated below:

$$\begin{array}{ccccccccc} \text{Designation:} & T & N & M & G & 3 & 2 & 3 & E \\ & (1 & 2 & 3 & 4 & 5 & 6 & 7 & 8) \end{array}$$

1 = shape of insert (T = triangle)

2 = relief angle (N = 0°)

3 = accuracy of point location, relative to pin and thickness

M = point location accuracy of ±0.002 in to ±0.004 in (±0.05 to 0.10 mm) and thickness accuracy of ±0.005 in (±0.13 mm)

4 = type of insert (G = insert with hole and molded chip breaker)

5 = size of inscribed circle (IC) in eighths-of-an-inch (3 = IC = 0.375 in = 9.53 mm)

6 = thickness in sixteenths-of-an-inch (2 = $\frac{1}{8}$ in thickness = 3.18 mm)

7 = cutting point radius in 64ths of an inch (3 = $\frac{3}{64}$ in = 1.19 mm rad)

8 = other (E = unground insert with honed edge)

This insert is shown in Fig. 14.6. A similar ANSI numbering system is used to specify tool holders, and details on both of these systems are to be found in most tool manufacturers' catalogs.

Fewer brazed tools are used today since clamped inserts eliminate brazing stresses, eliminate the need for tool grinding by the user, enable hardened tool holders of long life to be used, and decrease tool-setting time when changing tools. A carbide shim is usually employed beneath the indexable carbide for improved support and to protect the tool holder against damage. Since cemented carbides have a stiffness (Young's modulus) that is about three times that of the steel used in the tool holder, it is important to use carbide of considerable thickness to prevent a stress concentration that could cause the carbide to fracture. The carbide thickness is usually divided into two parts—that in the shim and that in the indexable cutting tool.

Chip breakers (chip curlers) may be of two types—a separate carbide plate clamped on top of the indexable carbide or a molded groove in the face of the cutting tool. Figure 14.6 shows a triangular insert with molded top surface. The geometry of the top surface of the tool into which the chip curler is molded may be as shown in Fig. 14.7a (negative rake) or Fig. 14.7b (positive rake). Both of these tools provide a limited contact on the tool face which tends to reduce cutting forces and cutting temperatures and will have improved tool life when machining low strength steels. Chip curlers of the type shown in Fig. 14.6 give good chip control over a wide range of feeds and depths of cut.

With the passage of time, molded-in chip control geometry in the surfaces of inserts has become more and more complex. Figure 14.8 shows a variety of insert geometries. So-called wiper inserts used in finishing operations have little or no clearance angle to improve finish by burnishing action by the flank face of the insert.

Titanium Carbides

In the 1960s efforts were made to find hard refractory materials that are more effective as tools than tungsten carbide. Titanium carbide is a material that has attracted considerable interest in this connection. This material is more oxidation resistant and more readily available at low cost. Best results have been obtained with TiC when it is bonded with nickel. However, such materials have been nonuniform and extremely brittle. The reason for this was not known until Humanik and Parikh (1954) demonstrated the importance of *wetting* in the liquid-phase sintering of TiC. While liquid cobalt was found to have a wetting angle

Fig. 14.6 Triangular carbide insert with center hole and molded chip curler in top surface.

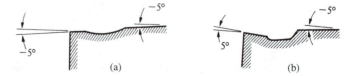

Fig. 14.7 Side elevations of inserts with molded top surface. (a) Negative rake angle. (b) Positive rake angle.

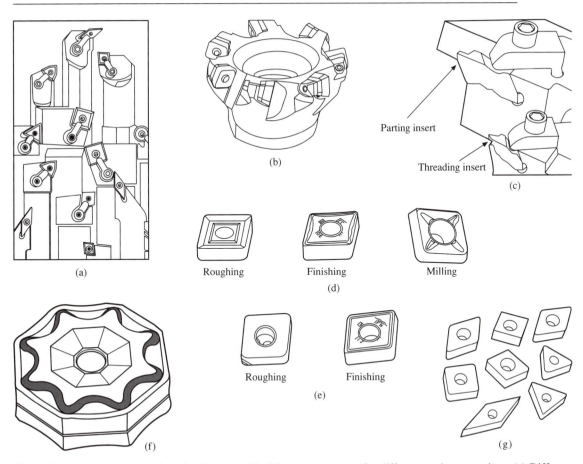

Fig. 14.8 Representative metal-cutting inserts and holding arrangements for different cutting operations. (a) Different insert shapes mounted in tool holders used in turning operations. (b) Square inserts mounted in a milling cutter. (c) Inserts mounted in tool holders for parting and threading. (d) Inserts for machining steel. (e) Inserts for machining cast iron. (f) Insert with sixteen useable cutting edges. (g) Polycrystalline cubic boron nitride (PCBN) for machining hard steel.

of zero degrees against WC (indicating perfect wetting), nickel was found to have an angle of 30° against TiC. Additives to nickel that would reduce the wetting angle against TiC were systematically studied, and it was found that metals that were weak carbide formers (i.e., metals having low negative free-energy of carbide formation) had the greatest influence. When 10% molybdenum powder is mixed with nickel before sintering TiC, it is found that complete wetting takes place and the bonding material is thoroughly dispersed. Details of this interesting development are discussed by Humanik and Parikh (1954).

The iron-group materials in the periodic table (Co, Ni, Fe) have been found useful as binder materials for tungsten carbide and other refractory carbides such as TiC. The solubility of WC in Co is low (~ 1%) while the room-temperature solubility of WC in Ni and Fe are much higher (25% and 5%, respectively). The higher the carbide solubility in the binder phase at room temperature, the more brittle the cemented carbide. Low solubility of WC in Co is one of the main reasons why cobalt is so widely used as the binder for tungsten carbide. The greater solubility of titanium carbide in nickel than WC in Co is one of the reasons TiC–Ni is more brittle than WC–Co.

Ceramic and Cermet Tools

Ceramic cutting tools were put into use in the United States in the mid-1950s following earlier use in Russia, England, and Germany. Most of these tools were made from finely divided α-alumina that was sintered without binders or other additives. Ceramic tools are more refractory and harder than carbide tools but also much more brittle. In order to provide reasonable toughness it is important that aluminum oxide tools have very small grain size and be sintered to essentially maximum density. Hot pressing in graphite molds yields superior tools, but the dies do not last long and substantially increase the cost of the product. While the cost of the aluminum oxide used to make ceramic tools is low, manufacturing costs are high, primarily due in part to the necessity of slicing the large blocks first produced into small pieces, using diamond saws.

Ceramic tools have proved to be of greatest usefulness in machining cast iron and hard steel. Cast iron gives good results for the same reason it does not cause straight tungsten carbide tools to crater—i.e., due to the formation of an oriented graphite layer of low shear strength on the surface of the chip. Hard steel will also cause less cratering of tungsten carbide since the weld areas formed between chip and tool will be small. The brittleness of ceramic tools makes them susceptible to chipping when machining soft steels under conditions that produce a large unstable built-up edge (BUE) or when making discontinuous cuts. Ceramic tools are particularly disappointing when used to machine aluminum or titanium alloys, due to the unusually high affinity of these metals for oxygen or an oxide surface. Strong bonds are then established between chip and tool and the wear rate of the tool is accelerated.

Cermets are ceramic–metal composites. They consist of two phases—a ceramic phase and a metal phase. Composite tools consisting of Al_2O_3 and TiC but no metallic binder have two ceramic phases but no free metallic phase and hence are not cermets. Tools consisting of 15–30 $^w/_o$ TiC and the remainder Al_2O_3 are less brittle than Al_2O_3 tools but less refractory than Al_2O_3 tools. These are made by hot pressing and sintering the mixture. Such tools are of greatest usefulness in high-speed machining of hard cast irons ($H_B = 350$ to $600 \ kg/mm^2$) and relatively hard steels under conditions of limited mechanical shock.

While Al_2O_3 is more chemically stable than WC in contact with hot iron, both WC and Al_2O_3 tools will crater. The frequency of transport from tool to work is greater for WC, but the amount of material per transfer is less. Al_2O_3 being a very brittle material will fracture deeper beneath the tool surface even though the stress there is lower than nearer the surface and hence larger chunks of Al_2O_3 are transferred to the chip than in the case of WC. However, a WC tool with a thin Al_2O_3 coating is a composite. In this case the Al_2O_3 coating gives a lower rate of thermal transport and the thinness of the coating limits the amount of material transferred per event. Thus, Al_2O_3 is an effective anti-crater coating material for WC–Co.

As a cutting tool becomes more refractory, it becomes more brittle and as a consequence brittle (tensile) fracture plays an important role. Ceramic tools are generally more inclined to chipping and fracture than WC or HSS tools, and as a consequence there is a greater dispersion in tool life. Ceramic tools are very sensitive to adhesion particularly at low cutting speeds where the strength of the adhering chip material is frequently stronger than the tool material beneath the surface. This results in tool life frequently being greater at low cutting speeds for a tungsten carbide tool than for a ceramic tool even though the reverse is true at higher cutting speeds. Similarly, there is a critical temperature below which HSS gives a longer tool life than WC. Brittle tool materials are not only sensitive to a built-up edge or a thin built-up layer attached to the tool but also to a suddenly applied load or an abrupt unloading of the tool at the end of a cut. For this reason, the probability of a tool failing at the end of a cut is far greater than at the beginning of a cut or under steady state cutting conditions (Sampath et al., 1985).

Once a sharp crack is initiated in a relatively homogeneous brittle material, such as glass or even a fine-grained aluminum oxide, it will penetrate deep into the specimen without a change of direction as the energy stored in the system is released. Since the maximum tensile stress is generally near the surface and parallel to the surface, the crack will usually run perpendicular to the surface leading to gross facture.

A solution to this is to introduce second-phase particles that will deflect a newly initiated sharp crack back to the surface before it has penetrated very far. The result is the formation of a relatively small wear particle instead of gross fracture. One method of doing this is to incorporate into a ceramic matrix from 5 to 20 volume percent of a stronger, stiffer material having a higher coefficient of expansion, for example, by mixing SiC whiskers (SiC_w) (approximately 0.5 μm diameter by 30 μm length) into fine-grained ($\sim$ 5 μm) Al_2O_3. The whiskers should be randomly distributed with their long dimension parallel to the surface. When a sharp crack initiates at a defect in the matrix, it will not go very far before it encounters a stronger SiC cylinder and be deflected away from its original path. Thermally decomposed rice hulls (outer shells of rice grain) are usually the source of SiC whiskers. This involves heating the hulls to 700–900 °C and then to 1300–1500 °C in a flowing inert gas. This ceramic whisker strengthening method was introduced in the early 1980s and has proven to be very successful.

Another way of redirecting sharp cracks is to use a less refractory second-phase particle that extracts energy from the crack tip, thus making it easier to redirect it. An example of this approach is to incorporate fine particles of partially stabilized zirconium oxide (PSZ) into a fine-grained Al_2O_3 matrix. Zirconia undergoes a phase transformation on cooling that can be marginally prevented by adding yttrium oxide. When such a particle is shocked, it will transform instantaneously into its equilibrium state, absorbing substantial energy from the surroundings as it does. The PSZ concentration is usually in the range of 20–45%.

The topic of crack deflection in brittle materials is an extremely complex one that is largely empirical. A number of tribology studies on materials containing crack-deflecting second-phase particles have been presented. Yust et al. (1987) is typical of these studies where it was found that the rate of wear (wear coefficient) was several orders of magnitude less for Al_2O_3 sliding dry on Al_2O_3 in a pin-on-disk test when silicon carbide whiskers were present than in their absence. Ceramic tools are available that incorporate either the whisker or the PSZ technique or both and are subjected to a pressure of 6000 p.s.i. before sintering.

Ceramic tools based on Al_2O_3 were introduced in the 1950s for machining steel and cast iron in the automotive industry. These tools are more refractory and enable higher cutting speeds to be used. However, to fully utilize this possibility machines having higher cutting speeds must be available. A material called SiAlON is a composite ceramic, developed in England in the late 1970s, and widely used in the United States and throughout Europe. It is a silicon nitride, aluminum, oxygen solid solution. A typical composition is 87% Si_3N_4 + 13% Al_2O_3 + 10% Y_2O_3 which is sintered at 1800 °C. The yttrium oxide acts as a sintering aid. SiAlONs are often used for machining nickel base super alloys. Depending on the application and the problems encountered (face wear, flank wear, depth of cut notching, chipping, and fracture) Al_2O_3 + TiC, Al_2O_3 + titanium nitride composites, Al_2O_3 + $(SiC)_w$ or SiAlONs are materials to be considered.

Wick (1987) has discussed use of cermets for gear cutting tools, end mills, drills, and taps.

Superhard Tool Materials

The composition tetrahedron (C, B, N, Si) in Fig. 14.9 shows unusually hard materials and compounds used as cutting tools beginning in the late 1950s. Diamond (D) is of special interest because its properties are of special importance for cutting-tool materials (extremely high hardness, high

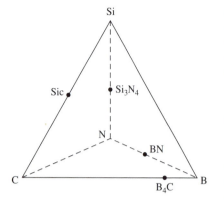

Fig. 14.9 Hard materials in the composition tetrahedron C–B–N–Si. Other hard materials not shown include amorphous hydrocarbons and compounds with other B–C ratios, B–O compounds, and C–N compounds. (after NMAB, 1990)

thermal conductivity, and low sliding friction with metals). A valuable review of the properties of diamond and its application has been published by the National Materials Advisory Board of the National Research Council (NMAB, 1990). Use as a cutting-tool material is just one of the many applications of diamond discussed in NMAB (1990).

Natural diamond has been in use for grinding very hard non-ferrous materials, notably glass and ceramic, since about 1890 for saws and since about 1940 for cutting tools.

Harris (1982) has presented a discussion of the geology of natural diamond, which is found in deposits in the Earth's upper mantle. The age of these diamond deposits is estimated to be greater than 20,000 years, and methane is believed to play an important role in the formation of natural diamond. It is estimated that natural diamond was developed slowly at temperatures from 900 to 1300 °C and pressures from 40 to 60 atm. Kimberlite is the errupted rock which transports diamond particles to the Earth's surface in the form of nodules that are as large as four feet across. Natural diamond is extracted from these nodules of kimberlite. Kimberlite is a variety of mica peridodite, low in silica and high in magnesium, in which natural diamonds are formed and grow. The word *kimberlite* is derived from Kimberley, the city in South Africa where large deposits of this mineral are mined.

Synthetic Diamond

Even before the end of the nineteenth century, many attempts were made to synthesize diamond without success. It gradually became evident in the United States that extremely high temperatures and pressures would be required to convert graphite into diamond (Rossini and Jessops, 1938).

Much of the pioneering experimental work that led to the first successful synthesis of diamond was performed by Bridgman (1947). While Bridgman was able to maintain temperatures near 3000 °K at about 29 kbar pressure for short intervals of time, this was not sufficient to convert graphite directly to diamond (1 kbar = 1000 atm). It remained for a team of innovative researchers at the General Electric Company to discover that certain catalysts made it possible to increase the rate of conversion to a practical level at relatively low temperatures and pressures (Hall, 1960a; Suits, 1964). Subsequently, Bundy (1963) discovered that the direct conversion of graphite to diamond without a catalyst is possible, but only at pressures above about 125 kbar and temperatures in the vicinity of 3000 °K. Under these conditions graphite spontaneously collapses into polycrystalline diamond.

The elements that are effective as catalysts include chromium, manganese, and tantalum, plus all elements of Group VIII of the periodic table. Carbides and compounds of these elements that decompose at or below these diamond synthesis conditions may also be used. These elements are believed to play a dual role: (a) as a catalyst and (b) as a good solvent for graphite, but a poor solvent for diamond. It appears that graphite first dissolves into the catalyst and is then converted into diamond at the appropiate conditions of pressure and temperature within the diamond stable region. Being relatively insoluble in the molten catalyst, diamond precipitates and thus allows more of the nondiamond form of carbon to go into solution.

Each catalyst has a different pressure–temperature region of effectiveness. Figure 14.10 is the pressure–temperature equilibrium diagram, according to Bovenkerk et al. (1959), when nickel is

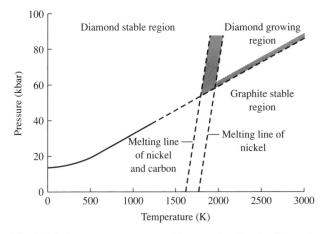

Fig. 14.10 A pressure-temperature diagram showing the diamond-growing region with nickel as catalyst. (after Bovenkirk et al., 1959)

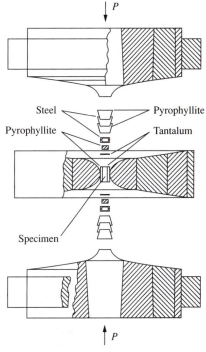

Fig. 14.11 The "belt." (after Hall, 1960a)

the catalytic solvent. This shows the diamond–graphite equilibrium line, as well as the melting lines of nickel and nickel–carbon eutectic. While graphite is a preferred starting material, since it is available in a very pure form, other types of carbonaceous material may be used.

The higher the pressure above the equilibrium line at a given temperature, the greater is the rate of diamond nucleation and growth. Diamonds formed at pressures substantially above the equilibrium line develop from many nuclei and have a skeletal structure. Such diamonds are very friable. On the other hand, by subjecting the reaction mixture to pressures and temperatures closer to the equilibrium line for a longer time, fewer nucleation sites develop, and larger and more perfect single crystals of diamond are formed. Cubo-octahedra are predominant at intermediate temperatures, while octahedra form at the highest temperatures.

The conversion of graphite to diamond is carried out in a high-temperature, high-pressure apparatus called a "belt" (Fig. 14.11) by its designer, Hall (1960b). Nickel appears to be the most common metal used with graphite, although other catalytic solvents can be used. The common operating range, with nickel as catalytic solvent, appears to be 75–95 kbar and up to about 2000 °C. Since diamond size increases with time, the reaction conditions are normally maintained for several minutes when larger grits are desired. The crystal structures of hexagonal graphite and cubic diamond are shown in Fig. 14.12. The synthesis of diamond was announced by General Electric on February 16, 1955, with more details following in articles prepared by the research team (Bundy et al., 1955a, 1955b).

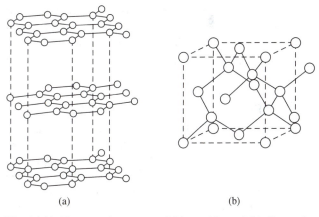

(a)   (b)

Fig. 14.12 The crystal structure of (a) graphite and (b) diamond.

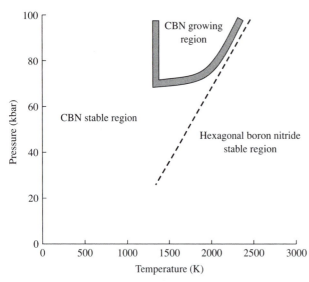

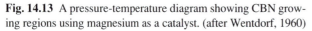

Fig. 14.13 A pressure-temperature diagram showing CBN growing regions using magnesium as a catalyst. (after Wentdorf, 1960)

Cubic Boron Nitride (CBN)

Similarities in the layer-type crystal structure of hexagonal boron nitride and graphite prompted Wentdorf to investigate the possibility of a high-temperature, high-pressure stable *cubic* form of boron nitride similar to diamond, but not found in nature. Early work using hexagonal boron nitride powder with the then already familiar diamond-forming catalysts did not result in a cubic form, even at pressures as high as 100 kbar and temperatures up to 2000 °C (Wentdorf, 1960). The catalyst solvents useful for CBN production were found to be the alkali metals, the alkaline earth metals, their nitrides, antimony, tin, and lead, or combinations of all of these.

Larger size CBN crystallites (300 μm or larger) of commercial importance were found to form in the presence of a catalyst. Figure 14.13 is a pressure–temperature equilibrium diagram due to Wentdorf (1960), indicating the regions in which hexagonal boron nitride and CBN are stable in the presence of magnesium as catalyst. Larger crystals were produced when pressures and temperatures were close to the equilibrium line. The particular catalyst used was found to influence the pressure-temperature necessary for conversion. The higher the atomic weight of the catalyst, the higher was the pressure necessary to effect the transformation. The most effective catalysts were found to be nitrides of magnesium, calcium, or lithium. A mixture of magnesium nitride and sodium metal, when used as a catalyst at 80 kbar and 1700–1900 °C, was found to result in large crystals. The size of cubic boron nitride particles increases with time, and hence it is advantageous to maintain reaction conditions from 3 to 5 min, even though the reaction time is only about 0.5 min. CBN grits were available commercially in 1967.

In the high-pressure–high-temperature (HPHT) synthesis of D and CBN, it is important that the temperature be decreased before the pressure to avoid conversion of the D back to the hexagonal form of carbon (graphite).

While diamond, which is the hardest known substance, has a hardness of about 6000 kg mm^{-2}, the hardness of CBN (the second hardest known substance) is only 4500 kg mm^{-2}. The main reason why CBN is of interest is that it is much more chemically stable than diamond in the presence of hot iron. CBN is also more refractory than diamond, as the hardness-temperature plots in Fig. 14.14 indicate. CBN is stable in air to about 1300 °C, but diamond is stable in air to only about 800 °C.

Polycrystalline Diamond (PD)

For metal-cutting applications the active tips of carbide inserts are often provided with a layer of sintered D or CBN. This is accomplished by pressing particles of synthetic diamond or CBN into a WC–Co surface and subjecting this to high temperature and pressure in the Hall "belt" apparatus (Fig. 14.11). No binder is used. However, a small amount of cobalt diffuses to the surface and acts as a binder phase for the superhard coating. Tools of this sort with aggregates of D or CBN are called polycrystalline diamond (PD) and polycrystalline cubic boron nitride (PCBN) inserts, respectively; and they have the same properties in all directions.

The size of a WC–Co substrate coated with D or CBN is limited by the size of the Hall "belt" apparatus (about 2 in $\cong$ 50 mm in diameter). These aggregates are cut into different sizes and shapes by wire electro discharge machining (EDM) and brazed to the corners of steel inserts. Brazing must be done below 700 °C to avoid damage to the diamonds. The coated layer of an aggregate is about 0.020 in (0.5 mm) thick and the carbide substrate is 0.1 to 0.17 in (2.5 to 4.3 mm) thick. These superabrasive tipped inserts that became commercially available in the late 1960s may be reground and may be used with or without cutting fluids. Use without a cutting fluid is of special importance when there are health and environmental concerns.

Polycrystalline diamond (PD) inserts give poor results on soft steel, titanium alloys, and high-temperature superalloys due to the tendency of these materials to react with diamond at high temperature (high speeds and feeds). However, PD inserts are very satisfactory for machining nonferrous alloys, fiber glass, silicon-aluminum alloys, graphite, and other abrasive nonferrous materials.

PCBN tools are used primarily to machine high-speed steels, stainless steels, and high-temperature alloys. PD or PCBN tools are used for drills, end mills, and other milling cutters as well as for turning tools, since they give less burn, enabling cutting speeds and feeds to be increased. They are particularly useful on transfer lines since their longer life reduces downtime for tool changing.

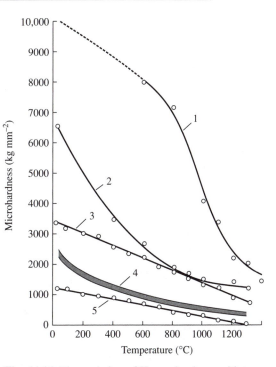

Fig. 14.14 The variation of Knoop hardness with temperature for several hard materials: 1, diamond; 2, CBN; 3, SiC; 4, varieties of Al_2O_3; 5, tungsten carbide (92 $^w/_o$ WC, 8 $^w/_o$ Co). (after Loladze and Bockuchava, 1972)

PCBN tools for cutting steel are provided with an edge preparation such as that shown in Fig. 14.15. These tools can normally be operated at cutting speeds about 50% higher than would be used with carbide. High speeds and light feeds are usually best. Figure 14.16 gives suggested starting conditions for PD-tipped tools depending on the work material. These values will usually have to be adjusted for optimum performance by an amount depending upon a number of details.

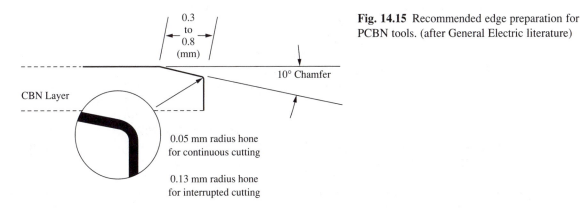

Fig. 14.15 Recommended edge preparation for PCBN tools. (after General Electric literature)

Material Machined	Cutting Speed f.p.m. (m/min)	Feed Rate i.p.r. (mm/rev)	Depth of Cut in (mm)
Nonferrous Metals:			
Aluminum alloys, low to medium silicone	1000–6000 (305–1829)	0.002–0.020 (0.05–0.51)	0.002–0.100 (0.05–2.54)
Aluminum alloys, high silicone	500–3000 (152–914)	0.002–0.010 (0.05–0.25)	0.002–0.100 (0.05–2.54)
Copper alloys	700–4000 (213–1219)	0.001–0.015 (0.03–0.38)	0.005–0.100 (0.13–2.54)
Tungsten carbide, sintered	50–350 (15–107)	0.004–0.020 (0.10–0.51)	0.002–0.040 (0.05–1.00)
Nonmetallic Materials:			
Plastics	1000–3000 (305–914)	0.004–0.020 (0.10–0.51)	0.002–0.100 (0.05–2.54)
Fiberglass	750–2000 (229–610)	0.004–0.015 (0.10–0.38)	0.001–0.100 (0.03–2.54)
Carbon	500–3000 (152–914)	0.003–0.015 (0.08–0.38)	0.005–0.100 (0.13–2.54)
Composite Materials:			
Plastic–fiberglass	300–2500 (91–762)	0.001–0.010 (0.03–0.25)	0.002–0.050 (0.05–1.27)
Plastic–carbon	500–2000 (152–610)	0.005–0.015 (0.13–0.38)	0.005–0.100 (0.13–2.54)
Epoxy–graphite	500–2000 (152–610)	0.002–0.010 (0.05–0.25)	0.002–0.050 (0.05–1.27)

Fig. 14.16 Recommended starting parameters for machining with PCD tool. (after Wick, 1987)

Vapor Growth of Diamond

Amborite, which is DeBeers Industrial Diamond Division's version of PCBN, was introduced in 1980 for machining hard ferrous materials. It is available as round and square inserts with sintered CBN particles employing a ceramic binder and no carbide backing material. This product, called Syndite, is described in Roberts (1979), and its properties are given in Pipkin et al. (1982).

Vapor growth of diamond was initiated around 1949 by Eversole at Union Carbide Labs (Eversole, 1962). Angus et al. (1968) had proved that diamond may be grown from hydrocarbons such as methane + hydrogen. However, growth rates were very low. The first practical growth rates on *nondiamond* substrates was achieved in Russia probably in the early 1960s (Deryagin et al., 1977), but the technique was not revealed.

In June of 1971 the author visited the Research Institute for Superhard Materials (ISM) in Kiev, as a guest of the Soviet Academy of Science. The institute was under the Ministry of Planning and at that time was about 10 years old and under the direction of Dr. V. H. Bakul. The first synthetic diamonds were produced there in December 1960. At the time of this visit, a new superhard alloy had just been developed called Slavotich (the ancient name of the Dnieper River). It was said this would be used primarily to cut rock and that a single tool was capable of a linear advance through 600 m of rock. Visitors were conducted through a very elaborate set of exhibit rooms that illustrated all aspects of diamond and tungsten carbide applications, then on a tour of the *outside* of the diamond development laboratory buildings which were located in a beautiful park-like setting. It is undoubtedly in these laboratories where a practical low-pressure method of growing diamonds was developed. This important work in Kiev was followed by work in Japan (Matsumoto et al., 1982). Further development in Japan led to growth rates of 10 μm/h.

Vapor diamond growth processes involve an energetically assisted aspect. The important detail in the chemical vapor deposition (CVD) formation of diamond involves the formation of atomic hydrogen that converts the carbon in a hydrocarbon (such as methane), with plasma activation, into cubic carbon (diamond). The overall reaction according to Hinterman and Chattopadhyay (1993) is

$$CH_4(s) + H_2 \xrightarrow[\text{Plasma Activation}]{T} C_D(s) + 2H(g)$$

The plasma activation may take several forms:

- hot filament (Fig. 14.17a)
- microwave plasma (Fig. 14.17b)
- oxygen-acetylene torch (Fig. 14.17c)
- high-frequency plasma torch (Fig. 14.17d)

(A plasma is a high-temperature ionized gas composed of an electrically neutral mixture of electrons and positive ions.)

In all of these cases the input is a 99:1 H_2:CH_4 mixture or O_2/C_2H_2, in the case of Fig. 14.17c. C_D is activated carbon that deposits on the substrate as diamond. The exact mechanism may be more complex than this, but this seems to be the best explanation presently available.

Diamond-Like Carbon (DLC)

DLC is an amorphous (noncrystalline) form of carbon that was discovered in 1973. It is produced only at low pressure and at a temperature that is lower than that required to produce crystalline carbon. These dense carbon deposits are produced by an energy-assisted means such as a radio frequency plasma or a low-energy ion beam. It is believed these amorphous diamond-like structures are related to quenching associated with a cold substrate. This is of special interest for coating HSS tools where the surface temperature must be lower than the transformation temperature of the tool.

DLC films have several diamond-like properties: high hardness, low friction, high thermal conductivity, and undistorted image transmission. DLC films provide a useful scratch-resistant coating for eyeglass and other optical lenses.

Carbonado

This is a polycrystalline aggregate of very small diamond particles, graphite, and other impurities. This black material is found in natural form in Brazil with diamond particles sintered together to form large polycrystalline lumps. It is as hard as single crystal diamond but much less brittle. A crack formed in one particle propagates only until it reaches the next particle having a different crystal orientation.

Carbonado may be produced from graphite at a pressure of about 70,000 atm when the temperature is raised to about 2700 °C for a very short time. This combination of pressure and temperature is at the boundary between diamond and graphite in the equilibrium diagram.

Hill (a member of the General Electric diamond research team) proposed producing synthetic carbonado, and in 1966 founded the Megadiamond Corp. with B. J. Pope and M. D. Horton to produce synthetic carbonado. This product was used for wire drawing dies since it gives a more uniform wear pattern than is obtained with single crystal diamond dies.

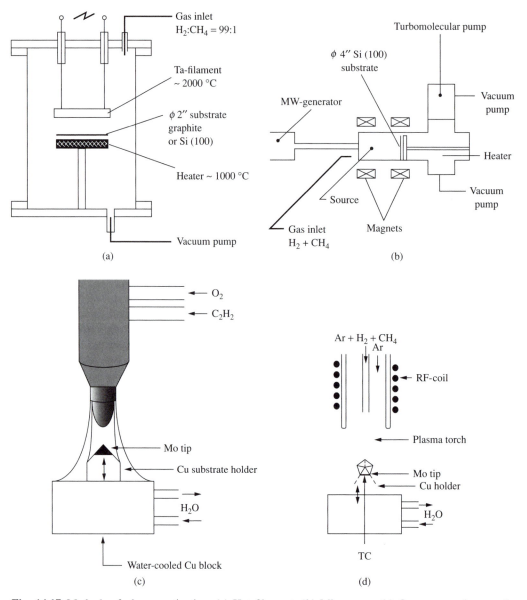

Fig. 14.17 Methods of plasma activation. (a) Hot filament. (b) Microwave. (c) Oxygen-acetylene torch. (d) High frequency plasma torch. (after Hintermann and Chattopadhyay, 1993)

Large Synthetic Diamond

Since 1985 Sumitomo Electric Industries, Inc., has grown single crystals of diamond up to 1.2 carats in size by what is called the thermal gradient method using a "belt" apparatus as shown in Fig. 14.18. Doping with nitrogen produces a yellow single crystal, while aluminum eliminates N and produces a single crystal that is colorless or light blue. A seed is placed in the HPHT apparatus with a carbon source and a solvent. It is then heated to 1450 °C at 50,000 atm. Electrical

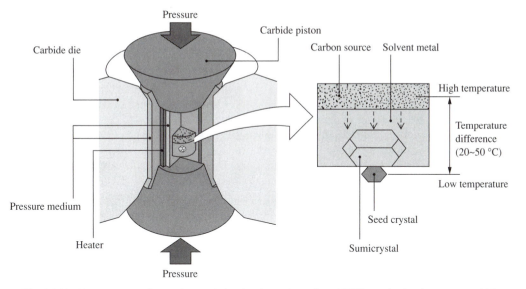

Fig. 14.18 Single crystal diamond growth by the thermal gradient HPHT method. (after Anon., 1989)

conductivity of the single crystal may be adjusted by additions of boron. Grown crystals as large as 0.03 in (0.75 mm) in diameter (limited only by the size of the "belt" apparatus) are used primarily for optical and electronic applications.

Coated Tools

The most important tool development of the 1960s was the introduction of coated carbide tools about 1969. The initial efforts in this direction involved ordinary steel-cutting grade carbide tools that were provided with a thin film of TiC, applied by chemical vapor-phase deposition (CVD). The thickness of TiC coating was about 0.0002 in (0.005 mm). Such coatings reduce the tendency of a cemented carbide tool to crater when machining mild steel and thus enable low-alloy steels to be machined at higher values of temperature (i.e., at higher speeds and feeds) without excessive cratering. The effect of the very thin coating exists long after a crater depth is reached corresponding to the thickness of the coating. For example, the wear-reducing effect of a 0.0002 in (0.005 mm) coating will still be evident at crater depths of several thousandths of an inch. This suggests that the coating material is carried into the crater by the moving chip.

Chemical Vapor Deposition (CVD)

CVD involves the application of solid material from a vapor by chemical reaction occurring on or near a heated substrate to form a thin film. Such coatings can be applied to surfaces having complicated shapes that are not planer. CVD is applicable only when the required substrate temperature is acceptable. An alternative is to produce an electrical discharge in the reactive gas which generates free radicals that require a lower surface temperature. This is called plasma enhanced chemical vapor deposition (PECVD).

Although other carbides and nitrides (for example, hafnium carbide and hafnium nitride) have been used as coatings to a limited extent only three coating materials are presently in wide commercial use (TiC, TiN, Al_2O_3). While TiC and Al_2O_3 appear to provide the most chemically stable screening layer between chip and tool, TiN appears to offer the lowest tool friction. This has been attributed to a lower chip surface strengthening action of nitrogen than carbon when the coating material is decomposed in the presence of the austenitic surface of the chip (Rao et al., 1978). Although Al_2O_3 should be the most effective diffusion barrier, it is much more difficult to bond to the substrate than either TiN or TiC. Bond strength is probably the overriding attribute of a coating and, in this regard, TiC excels.

Coated tools are inherently more brittle than uncoated tools. The increased brittleness associated with a coating is in part due to the difference in properties between the substrate and the coating, and it is important to provide a graded transition from one material to the other if undue brittleness is to be avoided. Some manufacturers today use a multiple coating technique. The layer next to the carbide substrate is TiC to provide a stable diffusion barrier to prevent carbon leaving the substrate or iron from entering the cobalt phase. The outer layer is a 0.0002 in (5 μm) layer of TiN for low tool–face friction with transition layers of titanium carbonitride between the TiC and TiN coatings.

In general, coated carbides are not useful for machining high-temperature alloys (either nickel- or cobalt-base), titanium alloys, or nonmetals such as graphite, fiber glass, plastics, or nonferrous metals. Coated carbides are most useful for finishing cuts at high speed on ferrous alloys of all types including stainless steel. For such applications, only a few varieties of coated carbide are required to cover the range of materials normally encountered. In other words, coated carbides are generally more versatile than noncoated carbides. When first introduced in the late 1960s, coatings were applied only to steel-cutting grades of WC–Co. Most manufacturers now include at least two types of coated steel-cutting grades in their catalogs and one cast-iron grade of coated carbide.

While coatings may be applied by sputtering or other physical deposition techniques (PVD) better bonding is generally obtained by chemical vapor deposition (CVD) where vapors such as $TiCl_4$ and methane are reacted on the hot tool surface (1000 °C or higher) to produce the coating.

Venkatesh (1984) has presented comparative wear studies on (a) uncoated WC–Co, (b) TiC-coated WC–Co, (c) solid TiC, and (d) TiC-coated solid TiC tools. Turning performance was found to improve in going from (a) to (b) to (c) to (d). The smaller grain size of TiC in CVD coatings was believed responsible for the unusually good performance whenever a coating was present. These comparative results were obtained when turning AISI 1020 steel with a feed of 0.004 i.p.r. (0.1 mm/rev) and a depth of cut of 0.080 in (2 mm). The solid TiC and TiC-coated tools were found to be more likely to fracture in interrupted cutting than the uncoated WC–Co tools. It was also found that coatings of TiN gave better overall performance than coatings of TiC. Although not demonstrated, it was assumed the best results would be obtained with a TiN coating on a solid TiC tool.

Chemical vapor deposition (CVD) coatings were first applied to WC–Co substrates in the 1970s. This normally involves a substrate temperature of 1000 °C or higher which is too high for HSS tools and hence coatings were not applied to HSS inserts until about 10 years later when physical vapor depostion (PVD) processes became available.

CVD is the principal method employed on WC–Co and ceramic substrates. Tools that are provide with some sort of coating are approaching 100%. The principal coating materials and reasons for their use are

- TiN, to lower friction and build-up
- TiC, to increase hardness
- Al_2O_3, to provide a thermal barrier

CVD coatings are applied singly, in combination, or as multiple layers to increase toughness. CVD coatings have a total thickness of 5–20 μm. About 1980 use of a cobalt-enriched substrate surface 10–40 μm thick was introduced to increase fracture resistance in roughing and interrupted cutting operations. This enables a more refractory insert to be used at high cutting speeds with a lower loss of toughness and crater resistance.

Salik (1984) has discussed application of TiC, TiN, and Al_2O_3 deposits. These coatings involve surface reactions such as the following:

$$TiCl_4 + CH_4 + H_2 \rightarrow TiC + 4\ HCl \tag{14.1}$$

$$TiCl_4 + N_2 + H_2 \rightarrow 2\ TiN + 4\ HCl \tag{14.2}$$

$$2\ AlCl_3 + 3\ H_2O + H_2 \rightarrow Al_2O_3 + 6\ HCl \tag{14.3}$$

The H_2 shown has a much higher concentration than CH_4, N_2, or H_2O and is highly activated to produce atomic hydrogen. Since all of these processes involve a substrate temperature of over 800 °C, they are limited to carbides. When applied to HSS tools, softening of the substrate that occurs is unacceptable.

Most coated tools have several layers, some of which are only a nanometer ($0.04\ \mu in \cong 10^{-9}$ m) in thickness. Kubel (1998) has discussed the role coatings play in providing higher production rates and longer tool life.

Pfouts (2000) has presented a valuable review of coating technology and applications. Most inserts have some form of coating: CVD, PVD, or a combination. The most popular CVD coatings are titanium nitride (TiN), titanium carbide (TiC), aluminum oxide (Al_2O_3), and diamond (D). The most popular PVD coatings are titanium nitride (TiN), titanium carbonitride (TiCN), and diamond (D). As previously mentioned, CVD involves a chemical reaction between a material such as $TiCl_4$, a hydrocarbon such as CH_4, and hydrogen [Eq. (14.1)]. The temperature of the substrate with CVD is normally relatively high. This can give rise to a thin (2 μm) carbon-deficient (η) phase at the surface of a WC–Co substrate that is very brittle. Since the tendency of obtaining a brittle η phase increases with increase in substrate temperature, one method of solving this problem is to lower the substrate temperature. This has led to moderate-temperature CVD techniques where the maximum substrate temperature is 850 °C or lower.

Ezugwu et al. (2001) have conducted a comprehensive study of the wear of carbide turning inserts. Three types of tests designated T1, T2, and T3 were performed under conditions given in Table 14.5 to tool life values in Table 14.6. The work material was quenched and tempered martensitic stainless steel having the following composition ($^w/_o$):

$$C = 0.105, \qquad Si = 0.35, \qquad Cr = 11.75$$

$$Mo = 1.75, \qquad Ni = 2.5, \qquad Fe = bal$$

The hardness of the work material was H_{RA} 91. Roughing tests were performed until one of the conditions in Table 14.6 was obtained. The following Taylor equations were statistically established using multiple linear regression:

$$T1: W_1 = 1.83 \times 10^{-8}\ V^{3.60} f^{2.36} \tag{14.4}$$

$$T2: W_2 = 2.18 \times 10^{-8}\ V^{4.83} f^{4.70} \tag{14.5}$$

$$T3: W_3 = 4.02 \times 10^{-8}\ V^{2.64} f^{1.32} \tag{14.6}$$

where W is the flank wear plus nose wear per minute of cutting time (mm/min), and V and f are also expressed in SI units.

The following material removal index (Q) is a measure of the relative performance of the three cases investigated, where

$$Q = (Vfd)/(1000W),\ m^2 \tag{14.7}$$

is the volume of metal removed per unit of wear rate on both tool face and tool nose.

TABLE 14.5 Cutting Parameters for Tests T1, T2, and T3[†]

Tool	Substrate	Other
T1	WC grain size: 1–7 μm	Coating: Outer layer, Al_2O_3 – 1.5 μm
	Hardness: 91 Rockwell A	Intermediate layer, TiC – 4 μm
	Grade: P20–P40	Bottom layer, Ti(C,N) – 2 μm
	Binder content (wt%): 6.3%	Coating technique: CVD
	Cubic carbides (wt%): 7.0%	Chip geometry: Roughing
	Shape: Trigon	Insert geometry: Back rake angle, −6°
		Side rake angle, −6°
		Nose radius: 1.2 mm
T2	WC Grain size: 1–8 μm	Coating: Outer layer, TiN – 10 μm
	Hardness: 90 Rockwell A	Intermediate layer, Al_2O_3 – 10 μm
	Grade: P20–P40	Bottom layer, Ti(C,N) – 10 μm
	Binder content (wt%): 8.0%	Coating technique: CVD
	Cubic carbides (wt%): 8.4%	Chip geometry: Roughing
	Cobalt-enriched substrate	Insert geometry: Back rake angle, −6°
	Shape: Trigon	Side rake angle, −6°
		Nose radius: 1.2 mm
T3	WC grain size: 1–6 μm	Coatings: Outer layer, TiN – 4 μm
	Hardness: 92.1 Rockwell A	Intermediate layer, Al_2O_3 – 2.5 μm
	Grade: P05–P25	Bottom layer, TiC – ~ 4 μm
	Binder content (wt%): 6.0%	Coating technique: CVD
	Cubic carbides (wt%): 9.1%	Chip geometry: Roughing
	Cobalt-enriched substrate	Insert geometry: Back rake angle, −5°
	Shape: 80°-Rhomboid	Side rake angle, −5°
		Nose radius: 1.2 mm

[†] After Ezugwu et al., 2001.

Figure 14.19 gives results for Q under the best machining conditions when rough turning under the three situations investigated (T1, T2, and T3). The T3 case is seen to be superior to the other cases relative to removal index, Q. This is probably due to the slightly higher substrate hardness, the greater strength of the 80° rhomboid insert over the trigon, and the low friction of the relatively thin TiN outer layer of coating.

TABLE 14.6 Tool Life Corresponding to When One of the Following Conditions First Occurs[†]

Average flank wear ≥ 0.4 mm

Maximum flank wear ≥ 0.7 mm

Notching ≥ 1.0 mm

Nose wear ≥ 0.5 mm

Surface roughness (R_a) ≥ 6.0 μm

[†] After Ezugwu et al., 2001.

At high cutting speeds (656–820 f.p.m. = 200–250 m/min), wear for all three cases was predominantly nose wear and chipping/fracture of the cutting edge. At lower cutting speeds (328–492 f.p.m. = 100–150 m/min) and with an outer layer of Al_2O_3 (T1 case), plucking of material from the tool face was predominant. Such removal of small chunks of material from the tool face at low speed due to tool–chip adhesion was not observed for cases T2 and T3 where the outer coating was TiN which gave lower friction.

Physical Vapor Deposition (PVD)

PVD is performed by evaporation, sputtering, or ion plating. Evaporation is performed in a vacuum. Atoms or molecules from a heated source vaporize and reach a substrate without collision with residual molecules in the vacuum chamber. To avoid such collision and to prevent residual molecules from contaminating the substrate, a high vacuum is required. Evaporative deposition is the most energy efficient of the PVD processes and is often the PVD process of choice.

Sputtering involves physical vaporization of a target by bombardment by argon ions produced in a glow discharge. This is distinctly different from evaporation, since it involves momentum transfer between the bombarding particles and those in the target, which if sufficient, results in the atoms released going to the substrate. An AC field between electrodes on either side of a tube will produce the glow discharge. Arc vapor deposition is a PVD process that employs vaporization of an electrode (target) under arcing conditions.

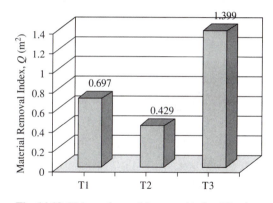

Fig. 14.19 Values of material removal index (Q) when machining hard stainless steel with carbide inserts of different shape and multilayer coatings. $V = 492$ f.p.m. (150 m/min); feed = 0.002 in (0.2 mm); and depth of cut = 0.080 in (2 mm). (after Ezugwu et al., 2001)

In ion plating, evaporated material passes through a gaseous glow discharge that ionizes some of the vaporized atoms before reaching the substrate. This PVD process is sometimes referred to as ion-assisted deposition. The bombardment of the substrate by energetic particles modifies the film produced by enhancing the rate of film growth and improving adhesion. Such bombardment of the substrate not only improves adhesion of the deposited film by cleaning the surface, but also in densifying the deposit by atomic peening and in promoting chemical reaction in the case of reactive deposition. Ion bombardment can be too intensive resulting in the developmen of voids or the generation of adverse residual stress in the deposited film.

A number of factors that influence the quality of a PVD film include

* initial cleanliness of the substrate surface
* finish and soundness (absence of flaws) of the initial substrate surface
* chemical properties of the substrate surface
* geometry of the substrate surface
* post-processing procedures (for example, burnishing, heat treatment, and peening)

PVD is such a complex process with so many variables controlling the result that it is in general not possible to predict the outcome of a projected solution despite the large collection of literature available pertaining to PVD film growth and properties. Some exploration is usually required.

Physical vapor deposition of TiC coatings on WC–Co tools was introduced in the early 1980s and TiN coatings about 1985. Both of these enabled higher cutting speeds. These coatings have been applied to milling, drilling, threading, and cut-off of soft steels, titanium- and nickel-based alloys, and nonferrous metals in order to lower cutting forces. PVD soft coatings of MoS_2 are also in use to lower friction between chip and tool.

Combinations of CVD and PVD coatings are also used—CVD for improved coating adherence and PVD to enable higher cutting speeds and lower cutting forces.

PVD coatings are usually applied by DC magnetron vacuum sputtering using an inert gas that accelerates the deposited material to the substrate. Adhesion of the coating and subsequent fracture and peel-off at the interface are the principal problems with PVD coatings. Adherence of a PVD coating becomes a greater problem the higher the temperature of the substrate. It has been reported (Kubel, 1998) that improved technology has enabled PVD to be achieved at a substrate temperature as low as 250 °C. This probably includes pretreatment of the surface of the substrate (e.g., etching, preplating).

Bruno et al. (1988) have discussed the performance of turning, drilling, and tapping HSS tools coated with thin PVD deposited layers of titanium nitride. The TiN coatings were found to be most beneficial when turning at high speed but of little value on taps which cannot be operated at speeds high enough to take advantage of this coating. TiN coatings were less useful on drills than in turning.

In the application of PVD coatings to tools involving loading and unloading (discontinuous cutting), fatigue of a coating is often a problem. Bouzakis et al. (1996) and Bouzakis and Vibakis (1997) have discussed this problem.

Owen (1994) discusses use of TiN and Al_2O_3 PVD coatings on (WC–Co)-based tool materials to provide a wider range of applications than uncoated tool materials do. It is interesting to note that a tool with a TiN coating has a gold appearance. By use of such coatings, cutting speed may be increased, without reducing the Co content to maintain resistance to thermal softening. As use of near-net-shape technology increases, thinner chips are involved which increases the difficulty of chip control. This calls for sharper chip control grooves in the rake face particularly when machining soft, gummy, low-carbon steels. The increased toughness of a coated WC–Co tool usually enables an increase in rake angle that results in lower cutting forces. However, this also results in thinner chips increasing need for good chip control (discussed in Chapter 18).

Coatings require a substrate temperature below the transformation temperature for HSS (500 °C or lower). The most commonly employed coating for HSS is TiN. The main advantage of a TiN coating is to decrease adhesion and friction. TiN-coated HSS tools for gear cutting were introduced about 1980. TiN-coated HSS tools are not effective for use on titanium or high-nickel alloys because of their tendency to react with these work materials.

Caselle (2001) has discussed the possibility of recoating tools after microblasting, regrinding, and cleaning, on the users shop floor. This is particularly attractive for small and medium-size users where it is more convenient and cost effective to eliminate need for sending tools to a reconditioning company. The three basic coatings that represent about 92% of those in use in 2001 are

- TiAlN, for dry high-speed machining
- TiCN, for interrupted cutting
- TiN, for general-purpose use

In addition, friction-reducing glide coatings such as MoS_2 may also be applied in the users' plant. Valuable references concerning coatings are Bunshaw (1994), Pierson (1999), and Mattox (1998).

A technique used to increase HSS tool life and to improve finish and productivity before PVD became available for HSS was electrospark alloying (ESA). This consists of moving an electrode that remains in continuous contact under a light load (10 lb) as a pulsed current is caused to flow from the electrode to the tool surface being coated. Small globules of the electrode transfer to the tool and alloy with it to increase its wear resistance. One of the best electrode materials was found to be 69% WC, 20% TiC, 6% Ni.

An electron beam physical vapor deposition unit (EB-PVD) has been developed at Pennsylvania State University for coating engine components. This unit is described in *Mfg. Eng.*

(2001). It employs six electronbeam guns in a vacuum chamber. Four of the beams evaporate the coating material while the other two are used to preheat the substrate. The value of this unit would be greatly enhanced if it could be used to apply PVD diamond coatings to HSS tools.

Komanduri (1997) has presented an extensive review of tool materials and their application in the *Encyclopedia of Chemical Technology*.

Diamond Coatings

Diamond-coated tools are of two types:

- a substrate coated with a thin CVD or PVD polycrystalline diamond film produced by a low-pressure, low-temperature (LPLT) process
- a platelet of polycrystalline diamond produced in the high-pressure, high-temperature (HPHT) "belt" apparatus and brazed on a substrate (General Electric approach)

The first of these readily conforms to tools of complex shape. The second is applicable only to cutting edges in the form of a plane surface.

Craig (1992) has discussed the use of thin film diamond coatings. The main problems of LPLT diamond film formation is poor adhesion of the coating to the substrate and a low rate of deposition. Both of these are being solved by proprietary research, and use of CVD diamond coatings appears to have a bright future in the United States, Japan, and Europe.

Sprow (1995) has discussed the use of PCD diamond cutting tools in the automotive industry for machining high-silicon aluminum alloys. Here, diamond-coated tools have been replacing tungsten carbide with greatly improved tool life, improved surface finish, and decreased need for a cutting fluid.

One of the advantages of CVD diamond coatings is that the coating ends up with a beneficial compressive stress on cooling because of its low coefficient of expansion relative to the substrate. Another advantage is that the batch size with CVD is close to an order of magnitude greater than for a HPHT case. HPHT coatings of PCBN can be used for machining titanium and ferrous alloys. At present there is no method of producing thin CBN films. Therefore, only thick brazed-on layers of PCBN are available.

In the CVD of diamond, *activation* of the carbon released from the hydrocarbon is required to cause the deposit to be diamond rather than graphite. Molecular hydrogen is a nonreactive diluent which dissociates to atomic hydrogen at elevated temperature. The presence of small amounts of a halogen decreases the substrate temperature for diamond formation. Atomic hydrogen stabilizes the diamond formed and removes unwanted graphite without attacking (etching) the diamond.

Nucleation and growth are important steps in the formation of diamond coatings discussed by Davis (1993). Nucleation involves the establishment of diamond particles on the substrate. This may be accomplished by seeding (embedding submicron diamond crystals in the surface) or scratching the surface preferably with diamond, in which case diamond wear particles can act as seeds. When scratching is performed by a nondiamond, then it is believed the minute grooves produced provide a crevice with sharp sides constituting energy sites that promote formation and bonding of solid diamond. It is also possible that if scratching occurs in the presence of the activated carbon evolved from the hydrocarbon being decomposed in the atomic hydrogen atmosphere, exoelectrons (discussed in Chapters 11 and 24) emitted from the freshly produced scratched surface may play a role. In any case, a diamond surface in an atomic hydrogen-rich atmosphere containing activated atomic carbon will grow to produce a polycrystalline film. Liu and Dandy (1995) have presented a useful discussion of the thermodynamics, kinetics, and chemistry of the nucleation of diamond by chemical vapor deposition.

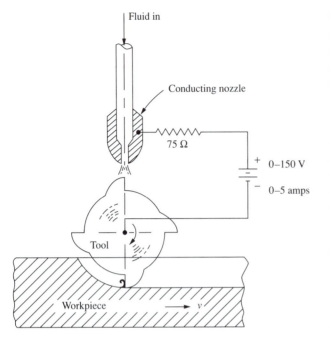

Fig. 14.20 Schematic plan view of end milling setup. (after Cook et al., 1966)

Continuously Renewable Coatings

Cutting operations that involve interrupted chip formation offer an opportunity for continuously depositing a material that lowers the adhesion of the chip to the tool, thus lowering friction. Such a film need not be hard and abrasive resistant since it is being continuously replaced. An example of this is described in Cook et al. (1966) where a soft metal film is deposited to faces of the tool electrolytically when the tool is out of contact with the work. An arrangement used to test this idea is shown in Fig. 14.20. A 1 in (25 mm) diameter M–2 HSS end mill having 4 flutes and a 30° helix angle was mounted in a vertical milling machine. A titanium alloy workpiece was held in a special transparent fixture that enabled the cutting zone to be completely immersed. Fluid was pumped from a small sump through an electrically conducting nozzle directed toward the noncutting side of the tool. A variable DC circuit enabled current to flow between the tool (cathode, −) and the nozzle (anode, +). Sparking was prevented by the 75 ohm resistance. Encouraging results were obtained when a titanium alloy was machined using a layer of Zn or Pb. Figure 14.21 shows a comparison of chips produced with and without an electro-deposited film. Accelerated tool life tests were also performed showing improvement of from 2 to 8 times.

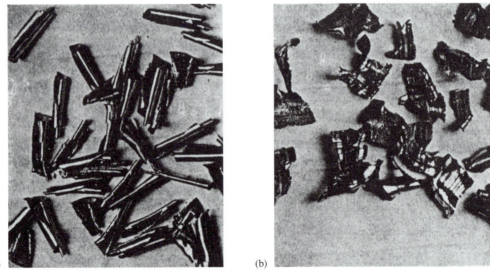

Fig. 14.21 Milling chips produced (a) with electoplated film and (b) with water. (after Cook et al., 1966)

Another investigation to explore the possibilities of using a soft renewable coating was carried out at Carnegie Mellon University. This involved coating a ceramic tool with a soft metal coating having a tendency to form a strong bond with an oxide surface (aluminum is such a material). The strong tendency for aluminum to bond to an oxide surface may be demonstrated by using an aluminum stylus to write on glass as one would write on paper with a lead pencil. When this is done, a strongly bonded line of small aluminum wear particles are left behind on the glass which can be "erased" only by abrading off a thin layer of glass. A thin layer of aluminum tends to form a particularly strong bond to the surface of an Al_2O_3 insert containing partially stabilized zirconia (PSZ) when machining aluminum at a relatively high speed (temperature). Such a layer is very thin and continuous without cracks. Since aluminum tends to contract much more (~ 2.5 times more) than an Al_2O_3 coating, the thin layer of Al will end up in tension on cooling while the surface of the Al_2O_3 will be in compression. This should increase the resistance of the tool to chipping beyond the already high toughness due to the PSZ in the tool.

The soft aluminum layer also has a low shear strength that should reduce the tendency for chipping due to adhesion, particularly when machining at low speed with a normally brittle material (Al_2O_3). Of course such a soft coating has a short life when machining difficult materials such as titanium alloys and, therefore, must be frequently renewed. A challenging consideration is how the deposition of the aluminum layer is best incorporated into a production process. This will be different for each machining process. For example, when broaching gas turbine firtree grooves, a backup material (cast iron) is sometimes used to reduce the fracture-inducing, high rate of unloading of the tool at exit. By substituting aluminum for the backup material, both gradual unloading and tool recoating may be achieved.

It has been demonstrated that when 6061 aluminum is milled at high speed using an Al_2O_3 tool, a thin hard layer of the aluminum alloy adheres to the tool and remains a surprisingly long time when subsequently used to mill hard (H_{RC} 60) steel. Similar results were not obtained for turning (continuous cutting). This is undoubtedly due to the opportunity for oxidation of the aluminum coating when the tool is out of contact in intermittent chip formation.

An alternative method of renewing an aluminum coating is to wire brush the tool periodically with a brush having 6061 aluminum wires to remove any buildup especially when broaching, drilling, or milling titanium or other difficult-to-machine work materials that are machined at relatively low speeds.

While the possibility of improved productivity in processes involving discontinuous chip formation is offered by the two preceeding examples, no reports of their use in an industrial setting have been found.

Miscellaneous Tool Materials

Some interesting tool materials have been developed that are technically sound but either cost too much, have too limited an area of application, or have appeared at a time when another more advantageous material is also available.

One of these materials is a columbian (niobium) based material called UCON (tradename of Union Carbide Co.). These tools which consist of 50 $^w/_o$ columbium, 30 $^w/_o$ titanium, and 20 $^w/_o$ tungsten are cast, rolled into sheets, and slit into blanks. This material is relatively soft (H_B 200) but difficult to machine. Nitrogen is finally diffused into the surface at a temperature of 3000 °F (1650 °C) to produce a hard surface layer (about 3000 kg mm^{-2}) but a relatively ductile core having a hardness of about 500 kg mm^{-2}. These blanks thus produced are not reground or brazed but are used as clamped tools. This material is used primarily to machine steel (not useful for cast iron, titanium alloys, stainless steels, or high-temperature alloys) at large depths of cut but light and medium feeds, and high cutting speeds (2500 f.p.m. = 900 m min^{-1}). The final tools are relatively expensive (> carbide).

Castable carbide tools (Leverenz, 1973) represent another type that have had difficulty competing with WC–Co tools. These materials consist of a hard carbide alloy in a high-strength refractory-metal matrix. The resulting tool material is best suited for use in the rough cutting [feeds up to 0.061 i.p.r. (1.55 mm/rev) and depths-of-cut of 1 in (25 mm), and over] of low-alloy and stainless steels at moderate and high cutting speeds. These tools are particularly resistant to crater formation and to plastic flow at high operating temperatures. Cast carbide is fabricated by consumable and nonconsumable electrode arc melting and spincasting of the melt into graphite molds. A typical composition is 20 $^w/_o$ titanium, 58 $^w/_o$ tungsten, and 22 $^w/_o$ carbon. The resulting material contains about equimolar amounts of titanium and tungsten carbides in a tungsten-rich fine-grained (0.5 μm) metal matrix. The metal matrix is dispersion-strengthened by finely distributed interlocking carbides. The superior hot-hardness of this material at the high temperatures encountered under severe cutting conditions is responsible for its unusual performance. These cast tungsten-titanium carbide tools are not particularly hard at room temperature but are unusually resistant to thermal softening under severe machining conditions. Compared with coated sintered tungsten carbide, these materials are found to crater more rapidly at first; but after equilibrium temperature is reached, the cast tungsten-titanium carbide material exhibits a slower rate of crater development. However, it does not appear that the marginal benefits over conventional WC–Co tools under an extremely limited range of conditions will warrant their adoption. For this reason, they have not been included in Fig. 14.1.

These are but two examples of tool materials that offer potential benefits over too limited a range of application to be widely adopted. Still another example is Baxtron, developed by the DuPont Co. primarily to provide a material superior to HSS for multispindle screw-machining applications. In this case, a limited range of optimum field performance was compounded by no tool marketing expertise of the developing company.

Owen (1994) has discussed several important details concerning the successful application of metal cutting inserts. It is indicated that with the movement of the industry toward having one operator tend several machines, avoiding tool breakage becomes very important. It is suggested that in this regard decreasing feed while increasing speed to maintain production is advisable. This is based on the observation that increased wear rate with increased cutting speed is less costly than cost associated with loss of production associated with tool breakage.

In addition, several useful suggestions related to cutting edge problems are given:

• If flank wear is too rapid, decrease the cutting speed.

• If troublesome edge grooving occurs, use a more wear-resistant tool material and/or increase the side cutting edge angle.

• To increase crater resistance, decrease cutting speed and/or use a more refractory tool material.

• If built-up material on the cutting edge causes poor finish and/or edge chipping, increase the speed and the side cutting edge angle and improve coolant selection and application.

• If interrupted cutting causes thermally induced comb cracks along the cutting edge, use a tougher tool material and/or improve coolant effectiveness.

• If gross fracture occurs, change insert size and grade, and change cutting parameters to reduce cutting forces.

Owen (1994) has also discussed a number of items concerning the design and application of inserts. The size of an insert is very important relative to cost, and the recent trend has been toward inserts having a smaller inscribed circle (IC) [$\frac{3}{8}$ in (9.5 mm)] instead of larger values. Also, the shape of inserts have been changed in order to reduce cost without sacrificing performance. Trigons shown in Fig. 14.22 are an example. In this case the tool tip has an included angle of 80° instead of 60° as in the case of a triangular insert, with very little increase in the volume of the insert (~ 3%).

Several useful suggestion concerning chip control are also presented in Owen (1994), and these are presented in Chapter 18.

Fig. 14.22 Trigon inserts with molded chip control faces. (after Owen, 1994)

DESIGN AND PERFORMANCE

Tool Material Properties

Strength, hardness, and ductility are extremely important properties of a tool material. In general, as the hardness and strength of a material increases, it becomes less ductile. The tensile strength of a tool material is usually measured in a three- or four-point bending test (Fig. 14.23), and the maximum stress at fracture is referred to as the transverse rupture strength (TRS). The TRS values for tungsten carbide tools are considerably lower than those for HSS, and the TRS of a tungsten carbide tool increases with percentage of cobalt binder used.

The hardness and Young's modulus values for different tool materials are approximately as given in Table 14.7. In general, as the material becomes more refractory (has a higher softening temperature), it has a higher value of Young's modulus (stiffness) and a higher hardness.

The hardness and strength of HSS tools falls off rapidly above the tempering temperature (~ 1000 °F or 535 °C) as shown in Fig. 14.24. It is of interest to note that the cast-cobalt alloys while softer than HSS at room temperature are harder at temperatures above 1000 °F (535 °C). The wide

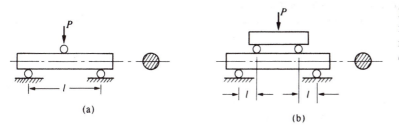

(a)   (b)

Fig. 14.23 (a) Three-point and (b) four-point bending test. Transverse rupture strength = $8\,Pl/\pi d^3$ in each case.

TABLE 14.7   Relation of Young's Modulus (*E*) and Vickers Hardness (*H*$_V$) of Tool Materials

Material	*E* p.s.i. (× 10⁶)	*E* MPa (× 10⁶)	*H*$_V$ (kg mm⁻²)
Steel	30	0.207	850
Cast alloy	30	0.207	800
Cemented carbide	60–90	0.414–0.621	1400–1800
Ceramic	60	0.414	2200
Polycrystalline diamond	>90	>0.621	7800
Single crystal diamond	>90	>0.621	8000

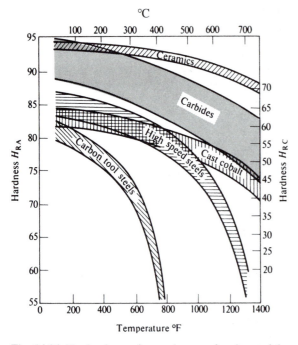

Fig. 14.24 Hot-hardness of several types of tool materials. (after Shaw, 1968)

range of values indicated for the carbides is primarily due to the range of cobalt content (~ 3 to 12 $^w/_o$). As the cobalt content is increased, the carbide becomes more ductile and shock resistant but less refractory.

When heat-treatable tool materials (carbon steel and HSS) are cooled to room temperature following exposure to a temperature above their tempering temperature, there is a loss of hardness. Other tool materials such as tungsten carbide, titanium carbide, cast alloys, ceramics, and cermets recover their hardness when returned to room temperature following exposure to a high temperature.

Table 14.8 summarizes some of the important characteristics of different classes of tool materials. As shown in this table, the typical rake angle increases with impact strength (or ductility) while the useful range of cutting speeds increases as the tool material becomes more refractory.

Several other properties not given in Table 14.8 are of importance. These include the coefficient of expansion (α), thermal conductivity (k), and volume specific heat (ρC). The coefficient of thermal expansion influences the thermal stresses and thermal shock on a material. Carbides have a coefficient of expansion about one-half that of steel, but this value increases with cobalt content. However, the carbides are so brittle that they are much more sensitive to thermal shock than high speed steels. As a result, cutting fluids are often used when milling with high speed steel cutters, but are rarely used with carbide cutters.

TABLE 14.8 Characteristics of Several Types of Tool Materials[†]

Cutting Tool Material	Hardness			Typical Transverse Rupture Strength ($\times 10^3$) p.s.i. (MPa)	Impact Strength	Wear Resistance	Typical Conditions for Single-Point Turning of 1045 Steel	
	Room Temperature	1000 °F (535 °C)	1400 °F (760 °C)				Rake Angles	Cutting Speed f.p.m. (m min^{-1})
High speed steel	63–70 R_C (85–87 R_A)	50–58 R_C (77–82 R_A)	Very low	600 (4140)	Increasing ↑	Increasing ↓	+5° to +30°	75–180 (23–56)
Cast alloy	60–65 R_C (82–85 R_A)	48–58 R_C (75–82 R_A)	40–48 R_C (70–75 R_A)	300 (2070)			0° to +20°	75–260 (23–81)
Carbide	89–94 R_A	80–87 R_A	70–82 R_A	250 (1724)			−6° to +10°	150–1800 (47–560)
Ceramic	94 R_A	90 R_A	87 R_A	85 (586)			−15° to −5°	500–2500 (156–781)
Diamonds	7000 Knoop	7000 Knoop	7000 Knoop	40 (276)				

[†] After Shaw, 1968.

Unfortunately as the wear resistance of most cutting tool materials improves, their grindability generally decreases. For example, one of the major problems in using a highly wear-resistant high-hardness high speed steel is the high cost of grinding. Very frequently one high speed steel will be chosen over another that gives somewhat better tool life, due to the differential cost of grinding finished tools from the two steels.

Special Tool-Tip Geometry

Klopstock (1926) was the first to show that tool life and cutting forces could be favorably altered by restricting the contact length between chip and tool. This was done by use of a composite rake face —small positive primary rake angle of +15° followed by secondary rake surface having a greater positive rake angle of 30°. This was found to give a more stable BUE and hence better surface finish with HSS tools. Tool life was also improved. When applying this concept to carbide tools, the primary rake surface is usually made slightly negative (< 10°) in order to provide a larger included angle at the tool tip for added strength. However, when this is done, the negative primary rake angle gives high cutting forces, tending to neutralize the positive effect of the controlled contact.

A compromise solution for carbide tools is shown in Fig. 14.25a. Here the primary rake angle is positive (~ +30°), but the cutting edge is given a large included angle for strength and thermal capacity by employing a land extending from a quarter to a half of the undeformed chip thickness in the feed direction. The rake angle of the land is usually between −5 and −10°. The contact length is controlled by providing a third negative rake surface (Fig. 14.25a).

When the rake angle of the land is −30° or more, it is called a chamfer (Fig. 14.25b). A chamfer has the added advantage of providing a very stable BUE and better surface finish.

When the cutting tool is relatively brittle and subjected to an unusually hard work material (such as surface scale) or to shock-loading, the very tip of the tool is frequently rounded to prevent chipping (Fig. 14.25c). This is referred to as a honed tool. Edge honing may be provided mechanically by hand stoning or by tumbling cutting inserts with shaped abrasives in a rotating basket. A lightly honed edge will have a radius of about 0.001 in (0.03 mm) while a heavily honed edge will have a radius of about 0.005 in (0.125 mm).

The purpose of the limited-contact tool is to reduce cutting forces by as much as 30% and hence the mean tool-face temperature. In doing this, it is important to recognize there is an optimum contact length. When the contact length becomes too small, the maximum tool-face temperature will be too close to the tool tip and even though the mean cutting temperature is reduced, the tool life will then be decreased. Limited-contact cutting has been studied by Takeyama and Usui (1958), Chao and Trigger (1959), Usui and Shaw (1962), Hoshi and Usui (1962), and Usui et al. (1964).

Mason (1995) has discussed the move toward more positive rake angle inserts as a result of increased toughness

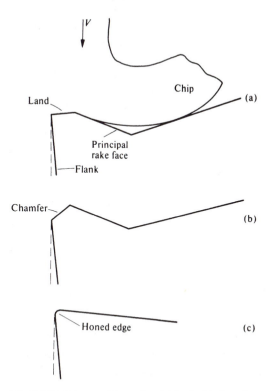

Fig. 14.25 Diagram illustrating (a) land, (b) chamfer, and (c) honed edge at the tip of a cutting tool.

of the substrate. Higher rake angles (20° or higher) result in lower force and power requirements, quieter milling operations, and better finish. Also, this is consistent with the move toward near net shape production.

Using the Built-Up Edge

Sorin (1955) demonstrated that unhardened steel in the shape of a tool was capable of producing chips in a turning operation. While tool life was short, the soft tool was definitely capable of producing chips. This suggests that a built-up edge (BUE) could be utilized in cutting with reduced tool wear if the BUE is continuously replaced.

K. Hoshi (1939) extensively studied the characteristics of the BUE and, in the mid-fifties, invented what he called the silver white chip (SWC) cutting technique as reported by Hitomi (1961) and T. Hoshi (1980). The SWC technique involves tool geometry that produces a BUE that is caused to flow away continuously in the form of a separate secondary chip (Fig. 14.26a). The BUE is stabilized by chamfering the cutting edge (Fig. 14.26b) and the contact length is controlled by a chip curler. The metal trapped by the chamfer acts as the cutting edge and is continuously replaced as the secondary chip flows in a direction approximately parallel to the principal cutting edge. Figure 14.27 is a photomicrograph of a partially formed chip that shows the BUE trapped by the chamfer. Figure 14.28 shows the geometry of SWC tools designed to cut low-carbon steel and Table 14.9 gives recommended tool dimensions for different feed rates in mm/rev. The SWC tool is reported to give 15% lower specific energy and about 20% longer tool life (Hitomi, 1961) than conventional tools. The lower cutting forces result in lower temperatures in the primary chip that often do not give a temper color but a shiny silver surface, hence the name "silver white chip."

The SWC technique has been found to be most useful in Japan in rough-turning and planing operations at high feed rates and low to moderate cutting speeds ($V < 100$ m min^{-1}). A relatively high feed rate is important since the extent of the chamfer should be a fraction of the undeformed chip thickness.

A relatively large percentage of the cutting energy is convected away by a secondary chip and, at high cutting speeds, this may be red hot. This then causes a softening of the cutting edge and loss of tool geometry in the case of tungsten carbide. More refractory ceramic or cermet tools should be better in this regard. The cutting speed of SWC cutting is limited primarily by the temperature of the secondary chip and its influence on the temperature of the tool tip.

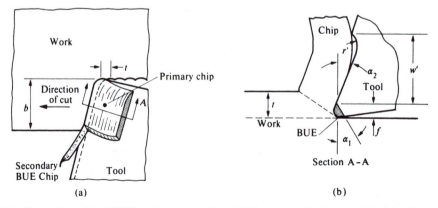

Fig. 14.26 Silver white chip (SWC) cutting mechanism. (a) Plan view. (b) Section A-A normal to principal cutting edge.

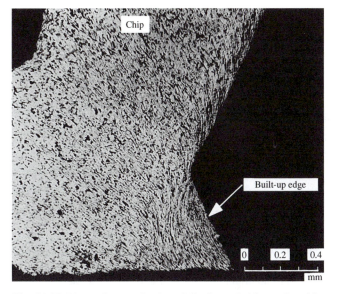

Fig. 14.27 Photomicrograph of partially formed chip produced by SWC tool. (after T. Hoshi, 1980)

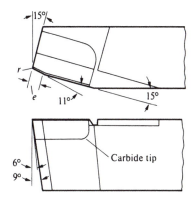

Fig. 14.28 Geometry of SWC cutting tool. (after Hitomi, 1961)

TABLE 14.9 SWC Cutting Tool Dimensions†

Depth of Cut mm	Feed mm/rev	Flat Width f (mm)	Chip Former (mm)		Sec. Cut Edge Length (e), mm
			Width (w′)	Radius (r′)	
1–3	0.2–0.4	0.1	2.0	0.5	0.0
2–5	0.3–0.5	0.2	2.5	1.0	1.0
3–6	0.4–0.7	0.3	4.0	1.5	2.0
5–8	0.5–0.9	0.4	6.0	2.0	3.0
5.5–9	0.7–1.2	0.5	8.0	4.5	4.0
6.5–10	0.8–1.4	0.6	9.0	5.0	5.5
7–15	2.0–15.0	0.7	10.0	5.5	6.0

† End relief angle = 6°; end clearance angle = 9°; side relief angle = 6°; side clearance angle = 9°; first rake angle (α_1) = −30°; second rake angle (α_2) = 15° to 30°; nose radius (r) = 1 to 1.5 mm; cutting speed = 70 to 120 f.p.m. (21 to 37 m min⁻¹); fluid = none. (after Hitomi, 1961)

Brittle Fracture of Tools

Brittle fracture is frequently the cause of cutting tool failure, particularly for the more refractory materials. This is when the input to the body is a force that is gradually increased until fracture occurs when the stress at the critical point reaches a certain value.

In the case of cutting-tool engagement, the magnitude of the contact force that develops depends on the stiffness of the material. The higher the Young's modulus, the higher the contact force, and hence the higher the stress at the critical point. Therefore, for a dynamic loading situation, a more appropriate brittle fracture criterion appears to be a maximum elastic strain energy criterion which

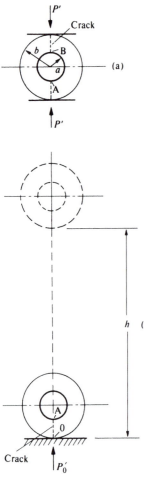

Fig. 14.29 (a) Static fracture test on disc with central hole. (b) Dynamic drop test of disc with central hole.

states that fracture will occur when $\frac{1}{2}\sigma_c e = \sigma_c^2/2E$ reaches a critical value where σ_c is the stress at fracture in a static test. This is equivalent to considering the input to be a specific energy for the dynamic case just as the input is a force in the case of static loading.

Such a maximum tensile strain energy criterion has been suggested by Shaw (1980b) based on dimensional reasoning.

Consider a thin brittle disc with a central hole (Fig. 14.29a). If this is loaded statically as shown, fracture will occur at either A or B and in accordance with the maximum tensile stress criterion, the load at fracture (P_f') per unit axial length before and after dimensional analysis will be

$$P_f' = \psi_1(a, b, \sigma_c) \tag{14.8}$$

$$\frac{P_f'}{\sigma_c b} = \psi_2\left(\frac{a}{b}\right) \tag{14.9}$$

If the disc is dropped from successively higher heights (h), it will fracture at A (Fig. 14.29b) at a critical drop height (h_c). The maximum unit force at O (P_O') will depend upon the maximum deformation at O on impact and

$$P_O' \sim \gamma' h_c E \tag{14.10}$$

(where γ' and E are the specific weight and Young's modulus of the disc material) assuming completely elastic deformation, a uniform rate of deceleration, and that all of the kinetic energy is absorbed on impact. Now, before and after dimensional analysis:

$$\gamma' h_c E = \psi_3(\sigma_c, a, b) \tag{14.11}$$

$$\frac{\gamma' h_c E}{\sigma_c^2} = \psi_4\left(\frac{a}{b}\right) \tag{14.12}$$

Comparing this with the static equivalent Eq. (14.9), it is seen that

1. the input is $\gamma' h_c$ instead of P_f'/b
2. the equivalent fracture criterion for the dynamic case will be $\sigma_c^2/E =$ constant, instead of $\sigma_c =$ constant for the static case

Table 14.10 gives representative values of σ_c (transverse rupture stress), Young's modulus (E), and $u_c = \sigma_c^2/2E$ for different classes of cutting tool materials. From these values, it is evident why the more refractory tool materials are so much more brittle than HSS under dynamic conditions of loading. (u_c is two orders of magnitude greater for HSS than ceramic.)

THE FUTURE

Beginning about 1970 a new basic concept was introduced—the use of composite tool materials in the form of coatings or a layer of very hard particles on a less brittle substrate (polycrystalline diamond and CBN tools). These developments should have been anticipated and introduced much earlier because a similar concept had been in use for many years in tribology. It is well known that a good

TABLE 14.10 Representative Values of Critical Specific Strain Energy at Fracture (u_c) for Different Classes of Tool Materials

Tool Material	Transverse Rupture Strength σ_c		Young's Modulus E		$u_c = \sigma_c^2/2E$	
	p.s.i. $\times 10^{-3}$	MPa	p.s.i. $\times 10^{-6}$	MPA $\times 10^{-4}$	p.s.i.	MPa
HSS	600	4138	30	20.7	6000	41.4
Cast alloy	300	2069	30	20.7	1500	10.3
Tungsten carbide	250	1724	75	51.8	420	2.90
Ceramic	85	586	60	41.4	60	0.41
Polycrystalline diamond	60	414	90	62.1	20	0.14
Single crystal diamond	40	276	90	62.1	9	0.06

bearing surface is one that is hard (to give a small ratio of real to apparent area of contact to prevent galling and high friction) and which has low shear strength at the surface (to give low friction) Unfortunately shear strength increases with hardness for all materials. The solution, of course, is to provide a low-shear-strength film, that is preferably renewable from within, on a hard substrate. This is the reason cast iron, copper–lead, babbit, silver–lead–indium, etc., are such good bearing materials. High hardness is a bulk requirement while low shear strength is a surface requirement. By providing a thin film of graphite or lead of low shear strength on a harder substrate (iron, copper, or silver), the opposing requirements of high hardness and low shear strength may be met.

Similarly, there are opposing material requirements in the case of cutting tools. For machining steel at high speed, a more chemically stable carbide (TiC) is required than tungsten carbide. However, when TiC is substituted for WC, a more brittle tool results even when great care is taken in selecting the binder phase. A useful solution is to use a relatively ductile substrate on which a thin layer of the more stable carbide (TiC) is deposited. The polycrystalline diamond and CBN tools introduced about 1973 represent a second application of the same principle (composite). Single-crystal (natural diamond) tools have been used for years to machine nonferrous materials, but their use was limited by the extreme brittleness of a single-crystal superhard material. Cubic boron nitride does not appear in nature and only small particles previously used in grinding wheels can be synthesized. The polycrystalline tool development represents a second example of use of composites. In this case small, ultrahard particles are sintered together on a relatively more ductile tungsten carbide substrate. The composite concept is relatively new. Just as the idea of using harder and more refractory hard particles has led to a progression of new tool materials, we should expect the composite concept to lead to a similar progression of new tool-material concepts over at least the next half century.

One form of technology transfer is to apply concepts that have proven useful in one engineering field to another. As already mentioned, the use of the composite concept should have appeared much earlier relative to tool materials in view of its earlier use in tribology (Bowden and Tabor, 1950).

In recent years, there have been important developments in the area of structural materials, and it would appear useful to consider some of these developments in seeking clues to some of the next generation of cutting-tool materials. Argon (1980) has reviewed new structural materials with emphasis on microstructural concepts. While many of the concepts of importance in structural materials development cannot be transferred to tool material development (such as those pertaining to plastics and other nonrefractory materials), the following ideas appear to have some relevance to possible new tool material concepts:

1. thermomechanical treatments to produce textural or orientation hardening

2. internal oxidation to produce very small (100 Å) hard particles that strengthen by precipitation or dispersion hardening

3. multiphase materials with each phase interconnected, produced by spinoidal decomposition with a phase wavelength of 100 Å or less

4. improved fracture-toughness of ceramics by precipitation of a crack-arresting phase

5. directional solidification of cast tool materials and casting techniques that eliminate the embrittling action of microshrinkage cracks and centers of tensile stress due to the dimensional change associated with the last material to solidify between dendrites

6. application of SPLAT cooling techniques to produce material of very fine grain size approaching an amorphous structure

7. continued and extended application of the composite concept

Some of these concepts are already being applied to the development of new tool materials. For example, tool materials are now being evaluated that involve hardening by spinoidal decomposition. These materials consist of TiC + MoC hard particles in a rather brittle matrix and have cutting characteristics similar to those for sintered TiC tools. High-speed steels are being produced from powder that is rapidly cooled to yield an extremely fine carbide spacing. This is one example of item 6. Another possibility in this direction is laser-glazing of tool surfaces. In this case, the surface of a tool is scanned by a high-powered continuous wave (CW) CO_2-laser beam of 10,600 Å wavelength at a speed sufficient to melt a thin layer of material on the surface. This requires about 10^6 W cm^{-2}. The rate of heating is so high relative to the thermal diffusivity of the material that only the surface is heated and the molten layer is "self quenched" by heat flowing into the bulk metal at a very rapid rate. The result is to produce a surface layer of very fine grain size or one that is truly amorphous. This may be considered a further extension of the composite concept.

Still another possible extension of the composite concept involves the formation of a layer of metal on the surface of a cutting tool with many diamond or CBN particles in the coating formed *in situ*. It has been demonstrated that small diamonds can be produced when a sandwich of nickel and graphite is subjected to the high temperatures and pressures associated with cutting. It is conceivable that this technique might be used to produce a thin coating of nickel with dispersed diamonds on a tungsten carbide substrate by simply coating the tool with layers of nickel and graphite and perhaps a "protective" coating of a material such as TiC before the tool is put into use. Alternatively, thin layers of nickel and graphite on a tungsten carbide substrate might be subjected to an explosive shock-wave (another way of producing small diamond particles) before the tool is put into use.

Any one of the above-mentioned developments could lead to the next breakthrough which leads to the new idea of the decade. However, it is not possible to predict which of the many possible new developments will produce a tool material that outperforms existing tools in any corner of the huge field of cutting-tool applications. It is not sufficient that a new tool material concept perform successfully to be included in Fig. 14.1 when it is updated fifty to a hundred years from now. It must also meet a particular need better and more economically than a tool material already in existence. When a new idea is introduced, such as the coated carbide, it takes a relatively long time to find out where and how it should be used, an even longer time to disseminate the information to the public, and a still longer time to overcome the inertia and resistance to change that unfortunately is a deep seated characteristic of human behavior.

In conclusion, it may be said that there are many exciting new possibilities for further improvement of cutting tools. It is unlikely that a universal tool material that meets all needs and displaces all present materials will be found. One approach to the problem of predicting the future is to ask a large group of tool-material experts for their views. This should be expected to yield a result of limited or negative value since it automatically reflects the collected bias and self-interest of the group of experts and will only extrapolate the present without sufficient regard for the fact that the future will be dominated by breakthroughs very difficult to anticipate.

REFERENCES

Angus, J. C., Will, H. A., and Stanko, W. S. (1968). *J. Appl. Phys.* **39**, 2915.

Anon. (1989). *Sumitomo General Catalog*. Tokyo, Japan.

Argon, A. S. (1980). In *Fundamentals of Tribology*. MIT Press, Cambridge, Mass.

Bouzakis, K. D., and Vidakis, N. (1997). *Wear* **206**, 197–203.

Bouzakis, K. D., Vidakis, N., Leyendecker, T., Lemmer, O., Fuss, H. G., and Erkens, G. (1996). *Surface Coatings and Technology* **86/87**, 549–556.

Bovenkerk, H. P., Bundy, F. P., Hall, H. T., Strong, H. M., and Wentdorf, R. H. (1959). *Nature* **184**, 1094–1098.

Bowden, F. P., and Tabor, D. (1950). *The Friction and Lubrication of Solids*. Clarendon Press, Oxford.

Bridgman, P. W. (1947). *J. Chem. Phys.* **15**, 92.

Bruno, M., Bugliosi, S., and Chiara, R. (1988). *J. of Materials and Tech. (Trans. ASME)* **110**, 274–277.

Bundy, F. P. (1963). *J. Chem. Phys.* **38**, 631.

Bundy, F. P., Hall, H. M., Strong, H. M., and Wentdorf, R. H. (1955a). *Chem. Engng. News* **33**, 718.

Bundy, F. P., Hall, H. M., Strong, H. M., and Wentdorf, R. H. (1955b). *Nature* **176**, 55.

Bunshah, R. F. (1994). *Handbook of Deposition Technologies for Films and Coatings*, 2nd ed. Noyes Publications, Norwich, N.Y.

Caselle, T. (2001). *Mfg. Eng.* (March), 138.

Chao, B. T., and Trigger, K. J. (1959). *Trans. Am. Soc. Mech. Engrs.* **81**, 139.

Cook, N. H., Rabinowicz, E., and Vaughn, R. L. (1966). *Lub. Eng.* (Nov.), 447–452.

Craig, P. (1992). *Cutting Tool Eng.*, **44/1**.

Davis, R. F., ed. (1993). *Diamond Films and Coatings*. Noyes Publications, Norwich, N.Y.

Deryagin, B. V., Spitsyn, B. V., Builov, L. L., Klochkov, A. A., Gorodetskil, A. E., and Smolyaninov, A. V. (1977). *Sov. Phys., Dokl* **21**, 676.

Eversole, W. G. (1962). U.S. Patents 3,030,137 and 3,030,188.

Ezuggwu, E. O., Olajire, K. A., and Jawaid, A. (2001). *Mach. Science and Technology* **5/1**, 115–121.

Friedman, L. S., Friedman, I. L., and Zagielski, K. C. (1965). U.S. Patent 3,184,169.

Griffith, A. A. (1924). *Proc. 1st Int. Cong. of Appl. Mech.*, Delft, Netherlands, pp. 55–62.

Hall, H. T. (1960a). U.S. Patent 2,947,608.

Hall, H. T. (1960b). U.S. Patent 2,947,610.

Harris, J. W. (1982). In *Ultrahard Materials Applications Technology*. DeBeers Industrial Diamond Division, Charters, England.

Hintermann, H. E., and Chattopadhyay, A. K. (1993). *Annals of CIRP* **42/2**, 769–783.

Hitomi, K. (1961). *J. Engng. Ind.* **83**, 509.

Hoshi, K. (1939). *Trans. Jap. Soc. Mech. Engrs.* **5**, 137.

Hoshi, K., and Usui, E. (1962). *Proc. Int. Machine Tool Des. Res. Conf.*, p. 121.

Hoshi, T. (1980). In *Proc. Int. Conf. Cutting Tool Mater.*, Am. Soc. for Metals Cincinnati, Ohio, p. 199.

Humanik, M., and Parikh, N. H. (1954). *J. Am. Ceram. Soc.* **37**, 18.

Kalish, H. S. (1982). *Mfg. Eng.* **89**, 97–99.

Klopstock, H. (1926). In *Berichte des Versuchsfeldes fur Workzeugmaschinen an der Tech. Hochschule Berlin*, Vol. 8.

Komanduri, R. (1997). "Tool Materials" in *Kirk-Othmer Encyclopedia of Chemical Technology*, 4th ed. Wiley, New York.

Krekeler, K. A. (1957). Dissertation, T. H. Aachen.

Kubel, E. (1998). *Mfg. Eng.* (Jan.), 40–46.

Lee, Y. M., Sampath, W. S., and Shaw, M. C. (1984). *Trans. ASME, J. Eng. for Industry* **106**, 161.

Leverenz, R. V. (1973). *Tech Paper* MR73-924. Society of Manufacturing Engineers, Dearborn, Mich.

Liu, H., and Dandy, D. S. (1995). *Diamond Chemical Vapor Deposition*. Noyes Publications, Norwich, N.Y.

Loladze, T. M., and Bokuchava, G. V. (1972). *Proc. Internat. Grinding Conf., Carnegie Mellon University*, Carnegie Press, Pittsburgh, Pa., p. 432.

Mason, F. (1995). *Mfg. Eng.* (Jan.), 27.

Matsumoto, S. Y., Sato, M., Tsutsumi, M., and Setaka, N. (1982). *J. Mat. Sci.* **17**, 3106.

Mattox, D. M. (1998). *Handbook of Physical Vapor Deposition*. Noyes Publications, Norwich, N.Y.

Mfg. Eng. (2001). (July), 326.

NMAB. (1990). *Status and Application of Diamond and Diamond-Like Materials*, NMAB-445. National Academy Press, Washington, D.C.

Owen, J. V. (1994). *Mfg. Eng.* (July), 57–62.

Pekelharing, J. (1978). *Annals of CIRP* **27/1**, 8.

Pfouts, W. R. (2000). *Mfg. Eng. (SME)* **7**, 98–107.

Pierson, H. O. (1999). *Handbook of Chemical Vapor Deposition-Principles, Technology, and Applications*, 2nd ed. Noyes Publications, Norwich, N.Y.

Pipkin, N. J., Roberts, D. C., and Wilson, W. L. (1982). "Amborite" in *Ultrahard Materials Applications Technology*. DeBeers Industrial Diamond Division, Charters, England.

Rao, S. B., Kumar, K. V., and Shaw, M. C. (1978). *Wear* **49**, 353.

Roberts, D. C. (1979). *Ind. Diamond Review* **39** (July), 237–241.

Rossini, F. P., and Jessofs, R. S. (1938). *J. Res. Nat. Bur. Stds.* **21**, 491.

Salik, J. (1984). NASA Tech. Memo. TM-83512.

Sampath, W. S., Lee, Y. M., and Shaw, M. C. (1981). *Trans. ASME, J. Eng. for Industry* **106**, 161–167.

Sampath, W. S., Ramaraj, T. C., and Shaw, M. C. (1985). *Proc. 12th NSF MFG Sys. Research Conf.* SME, Dearborn, Mich., pp. 371–373.

Shaw, M. C. (1968). In *Cutting Tool Material Selection*. SME, Dearborn, Mich., p. 1.

Shaw, M. C. (1980a). *Bull. Jap. Soc. Mech. Engrs.* **23**, 324.

Shaw, M. C. (1980b). In *Proc. of Third Int. Prod. Eng. Conf.* (Tokyo), p. 492.

Shaw, M. C. (1989). Chapter on tool life in *Ceramic Cutting Tools*, ed. D. Whitney. Noyes Publications, Park Ridge, N.J.

Sorin, P. (1955). *Microtechnic* **9**, 125.

Sprow, E. E. (1995). *Man. Engineering* (Feb.), 41–46.

Suits, C. G. (1964). *Amer. Sci.* **52**, 395.

Swinehart, H. J. (1968). *Cutting Tool Material Selection*. ASTME, Dearborn, Mich., pp. 89–97.

Takagi, J., and Shaw, M. C. (1983). *Trans. ASME, J. Eng. for Industry* **105**, 143–149.

Takeyama, H., and Usui, E. (1958). *Trans. Am. Soc. Mech. Engrs.* **80**, 1089.

Taylor, F. W., and White, M. (1901). U.S. Patent 668,270.

Topinka, W. (1988). M.Sc. thesis, Washington State University.

Usui, E., Kikuchi, K., and Hoshi, K. (1964). *J. Engng. Ind.* **86**, 95.

Usui, E., and Shaw, M. C. (1962). *Trans. Am. Soc. Mech. Engrs.* **84**, 87.

Venkatesh, V. C. (1984). *Trans. ASME* **106**, 84–87.

Wentdorf, R. H. (1960). U.S. Patent 2,947,617.

Wick, C. (1987). *Mfg. Eng.* (May), 38–42.

Wilson, F. G., Alworth, R., and Ramalingam, S. (1980). *Proc. Int. Conf. Manu. Eng.* (Melbourne), p. 3.

Yust, C. S., Leitnaker, J. M., and DeVere, C. E. (1987). *Proc. Int. Conf. on Wear of Materials*, Houston, Tex.

15 WORK MATERIAL CONSIDERATIONS

Machinability is a general term used to rate ease of machining a material relative to tool life, surface finish produced, or specific power consumed. Due to the wide range of meaning involved, it is best to state the basis of comparison when two materials are compared relative to their machinability.

The basic machinability of a material is a function of its

1. chemistry
2. structure
3. compatibility with tool material

The basic machining characteristics of iron, aluminum, titanium, nickel, cobalt, and copper, and their alloys are quite different due to the difference in chemical and physical properties of the base metal. Such a wide variety of materials will also have different machinability ratings depending upon the tool material used. Steels of the same chemical composition but different metallographic structure may have widely different machining characteristics. Some work-material–tool-material combinations are well suited to each other while others are relatively incompatible.

Machined materials may be classified as

1. easy-machining materials (aluminum and copper alloys)
2. ordinary wrought steels and cast irons
3. difficult-to-machine materials

Since the wrought steels and cast irons are of major importance, these will be considered first.

STEELS

Steel is basically an alloy of iron and up to about 1.5 $^w/_o$ carbon. Alloy steels contain additions of one or more elements introduced to provide a variety of special properties. Low-carbon steels contain up to about 0.3 $^w/_o$ carbon, high-carbon steels contain 0.9 $^w/_o$ carbon or higher, and the medium-carbon steels are in between.

The hardness, flow, and fracture characteristics of steel can be altered appreciably by heat treatment provided there is a sufficient carbon content (greater that about 0.3 $^w/_o$ C). These characteristics may also be altered by cold working (strain hardening) and by recrystallization to alter grain size. All of these changes are due generally to a change in the structure of the material.

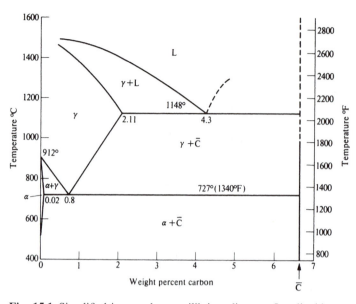

Fig. 15.1 Simplified iron–carbon equilibrium diagram. L = liquid, γ = austenite, α = ferrite, M = martensite, C = Fe₃C.

Pure iron is too soft for good machining due to the large strain associated with chip formation (low shear angle and low cutting ratio) and relatively high tool-face friction. Its machinability can be improved by cold working (large strain plastic deformation at a temperature below an homologous temperature of 0.5 = 1200 °F or 650 °C for iron) or by refinement of crystal size prior to machining. All practical metals are polycrystalline with a mean crystal diameter in the range of 0.001 to 0.010 in (0.025 to 0.250 mm). A decrease in crystal size may be accomplished by generating a large number of dislocations by plastically working the material followed by the removal of dislocations at a temperature above T_H = 0.5. This may be done by hot-working the material or by a recrystallization heat treatment following cold work.

The properties of a steel capable of being heat treated (C > 0.3 $^w/_o$) may be appreciably altered by changing the shape and spacing of insoluble carbides distributed in an iron matrix. Iron undergoes an allotropic transformation from a body-centered cubic (BCC) atomic arrangement to a face-centered cubic (FCC) arrangement at a temperature of 1670 °F (910 °C). Carbon is essentially insoluble in BCC-iron but is quite soluble in FCC-iron. All of the carbon may be put into solid solution by heating and holding above the transformation temperature. Then, by altering the cooling path and rate to room temperature, a wide variety of carbide shapes and spacings may be obtained. Carbide particles (Fe₃C for a plain carbon steel) grow on nuclei, the number of nuclei increasing with an increase in cooling rate. Thus, the spacing of carbides generally decreases with an increase in cooling rate and the lower the carbide spacing, the greater the resistance to dislocation formation and motion and hence the greater the hardness of the material.

A phase diagram is a chart showing the number and nature of phases (physically identifiable constituents) present in a system at any temperature and composition under equilibrium conditions. The iron–carbon diagram (Fig. 15.1) is such a diagram for steel. This chart merely indicates which of the following phases are present at a given temperature and carbon content:

L = liquid

γ = austenite = a solid solution of up to 2 $^w/_o$ C in FCC-iron

α = ferrite = essentially pure (< 0.02 $^w/_o$ C) BCC-iron

$\bar{C}$ = iron carbide (Fe₃C), also called cementite.

This diagram shows that transformation from α to γ generally occurs on heating over a range of temperatures the magnitude of which depends on the carbon content. At the particular composition of about 0.8 $^w/_o$ transformation begins and ends at the same temperature (727 °C). This is called the eutectoid composition.

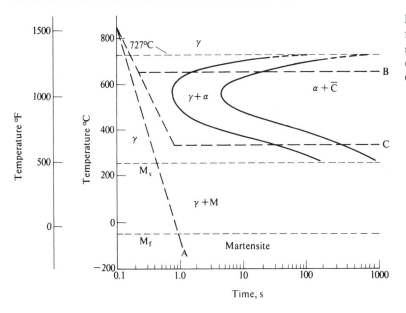

Fig. 15.2 Time–temperature–transformation diagram for plain carbon steel of eutectoid composition (0.8 $^w/_o$ C). γ = austenite, α = ferrite, $\bar{C}$ = Fe$_3$C, M = martensite.

An equilibrium phase diagram does not indicate the shape and sizes of the phases present. This depends on the rate of cooling. Another diagram that shows temperature plotted against time [the time-temperature-transformation (TTT) diagram] for a given composition provides information on the shape and size of the precipitated carbides present. Figure 15.2 is the TTT diagram for a eutectoid steel (AISI 1080 steel). The three dotted curves represent three cooling paths. Path A represents a quench in brine to produce a phase called martensite (M). Path B (quenched into a high-temperature bath and held until transformation is complete) leads to a soft structure called pearlite (P), while path C leads to a hard structure called bainite (B). Paths B and C are called isothermal transformations.

The various structures encountered in a heat-treated steel may be briefly summarized as follows:

Martensite (M) is a metastable body-centered tetragonal iron atomic arrangement with carbon in solid solution in the elongated tetrahedron. The M$_s$ and M$_f$ lines in Fig. 15.2 correspond to the temperatures at which the martensitic transformation starts and finishes, respectively. Cooling path A (Fig. 15.2) will yield 100% M. Martensite is an extremely hard structure that is too brittle for use. The practical structure is tempered martensite (T) which is produced by heating M to an elevated temperature below the critical temperature where transformation of γ begins and holding for a given time such as one hour. This treatment is called tempering. During tempering, the body-centered tetragonal M collapses to a BCC structure as the carbon comes out of solution in the form of spherical iron carbide particles (Fe$_3$C = $\bar{C}$). The size and spacing of carbide particles in a T-structure can be adjusted over a wide range depending on the tempering temperature and time to produce a wide spectrum of hardness and ductility.

Pearlite (P) is a structure consisting of alternate plates of $\bar{C}$ and α. It is produced by slow cooling or by an isothermal transformation such as curve B in Fig. 15.2. It is a relatively soft structure— the coarser the spacing of platelets, the softer the material. Fine P is obtained by transforming isothermally at a low temperature near the knee of the TTT curve, and it has a higher hardness and strength than coarse P. The entire structure will consist of P for a eutectoid steel (0.8 $^w/_o$ C), but it will consist of islands of P in a matrix of α for a hypo-eutectoid steel (C < 0.8 $^w/_o$) and of islands of P in a matrix of $\bar{C}$ for a hyper-eutectoid steel (C > 0.8 $^w/_o$).

Bainite (B) is a relatively hard structure consisting of fine jagged needles of $\bar{C}$ in an α matrix. It can be produced by an isothermal transformation at a temperature below the knee of the TTT curve (as for cooling path C in Fig. 15.2).

Spheroidite (S) is a structure consisting of large, widely spaced, spherical $\bar{C}$ particles in an α matrix. An S structure may be produced by heating P for a long time just below the initial transformation temperature or by tempering M for a long time at high temperature.

Figure 15.3 shows photomicrographs of a few representative structures for wrought steels.

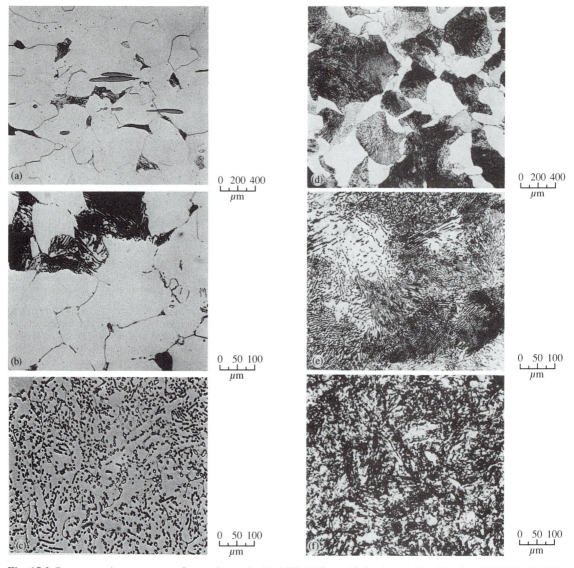

Fig. 15.3 Representative structures of wrought steels. (a) AISI 1112 resulfurized steel, H_B 135. (b) AISI 8620 (30% P, 70% α), H_B 135. (c) AISI 8640 (100% S), H_B 180. (d) AISI 8640 (50% P, 50% α), H_B 170. (e) AISI 4340 (100% P), H_B 221. (f) AISI 8640 (100% T), H_B 300. (after Field, 1963)

The several structural components found in steel may be arranged as follows in order of increasing hardness:

1. austenite (γ)
2. ferrite (α)
3. spheroidite (S)
4. coarse pearlite (P_c)
5. fine pearlite (P_f)
6. bainite (B)
7. tempered martensite (T), wide hardness range
8. martensite (M)
9. iron carbide ($\bar{C}$)

The important consideration relative to hardness is the mean ferrite path (mean spacing of adjacent carbide particles along a straight line on a photomicrograph). Figure 15.4 shows the variation in ultimate stress (and hence hardness) with mean ferrite path for spheroidite and pearlite for steels of a given carbon content. From this, it is evident that only the spacing and not the shape of the particles is important and that the ultimate stress (hardness) varies approximately linearly with the log of the mean ferrite path.

For a given carbon content, there is an optimum hardness relative to tool life for plane carbon steels. In general, a low-carbon steel (C < 0.3 $^w/_o$) should be machined in as hard a state as possible. Since little increase in hardness is possible by heat treatment, this means that the material should be cold-drawn if possible and have a fine grain size.

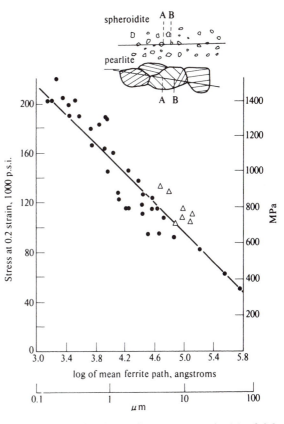

Fig. 15.4 Variation in tensile stress at strain (ε) of 0.2 (~ ultimate stress) with log of mean ferrite path (mean of AB in insert) for steels of pearlitic (●) and spheroidal (△) structure. (after Gensamer, Pearsall, Pellini, and Low, 1942)

For a high-carbon steel (C = 0.8 $^w/_o$ or higher), the best structure is the softest (spheroidite); otherwise, the many carbides are too firmly anchored leading to excessive, abrasive tool wear.

While hardness is an important variable relative to tool life, it is not the only item of importance. Figure 15.5 shows values of V_{30} (cutting speed in f.p.m. for 30 min tool life) plotted against Brinell hardness number for a variety of wrought alloy steels (AISI 1020, 1112, 8630, 3140, 4140, 4340, 8640) for HSS and C–6 grade tungsten carbide turning tools. It is seen that a very large range is to be expected for hardnesses below H_B 300, depending on the structure. The range of values above H_B 300 is relatively narrow since in all cases, the structure is the same (spherical $\bar{C}$ particles in an α matrix, corresponding to a quenched and tempered condition).

Steel has a relatively strong tendency to form a BUE, and it is difficult to obtain a satisfactory finish at cutting speeds below about 100 f.p.m. (31 m min^{-1}). Carbide tools have a tendency to form strong bonds with the chip in the speed range where a BUE forms and this frequently leads to chipping. According to Trent (1963), this is particularly the case for steel-cutting grades of carbide. In general, low-carbon steels which tend to form a strongly adherent BUE are best machined with HSS tools in the speed range where a large BUE tends to form.

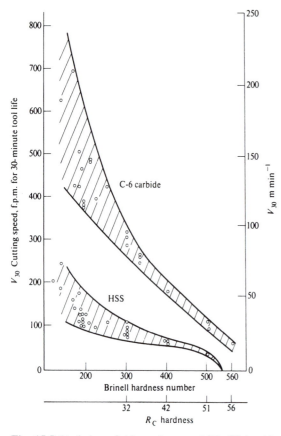

Fig. 15.5 Variation of thirty-minute tool life (V_{30}) with hardness depending on structure of wrought alloy steels. Tool materials: C–6 tungsten carbide, T-1 HSS. Side rake: 6° (WC), 15° (HSS). Back rake: 0°. Feed: 0.010 in. Depth of cut: 0.100 (WC 0.062 (HSS). Fluid: dry (WC), 1:20 sol. oil (HSS). Life: $w = 0.015$ in (WC), 0.030 in (HSS). (after Field, 1963)

At high cutting speeds, the surface of a steel chip will be austenitic, and this tends to reduce WC particles by removing carbon. The carbon absorbed thus strengthens the surface of the chip, and the combined effect is the rapid formation of a crater if a straight WC tool is used to machine a steel. As discussed in Chapter 14, use of mixed carbides (WC + TiC + TaC + NbC) and coatings of TiC, TiN, or Al_2O_3 offer a useful solution to the cratering problem when machining steel.

If a hardened steel (quenched and tempered to above H_B 350) must be machined, a ceramic tool used with a light feed rate is often a good solution.

While it is difficult to predict accurately the best machining conditions for a given workpiece material in a given machining operation, handbook data usually represents a good starting point from which to proceed to the optimum by progressive change. *Machining Data Handbook* (1980) is a very comprehensive source of such data.

ALLOY STEELS

Elements other than carbon are frequently alloyed with iron for one or more of the following reasons, and the elements used for each of these purposes are given in parenthesis:

1. to delay the rate at which austenite transforms to pearlite upon quenching to allow sufficient time for thick sections to be hardened throughout without the use of a quench that is so drastic as to induce damaging cracks in the steel (Mn, Cr, Mo, W, Ni, Si) [This is referred to as increasing hardenability.]

2. to provide additional hard abrasive particles to improve wear resistance (V, Mo, Cr, W)

3. to provide higher hardness at high temperatures (greater hot-hardness) (Mo, Cr, V, W)

4. to provide greater strength at elevated temperatures (Mn, Ni, Si, C, Cr, Mo)

5. to inhibit grain growth in austenite during heat treatment (V, Al)

6. to provide corrosion resistance (Cr, Ni)

7. to combine with oxygen to prevent blowholes (Si, Al, Ti)

8. to combine with sulfur which otherwise causes brittleness at hot-rolling temperatures (Mn)

9. to improve machining properties (S, P, Pb, Te, Bi, graphite)

These alloying elements perform their intended function in one or more of three basic ways:

1. by forming carbides that are insoluble in ferrite (W, Mo, V, Cr)

2. by going into solid solution in the ferrite (Si, Mn, Ni, Co, Cr)

3. by forming inclusions which are insoluble in ferrite (Si, Al, Ti, Mn, S, P, Pb)

Of the materials that go into solid solution in the ferrite, manganese and silicon tend to make the matrix more brittle (decrease the strain to rupture) while nickel makes it less brittle. All of the materials going into solution in the ferrite increase the tendency to strain harden. Of the materials which form inclusions, silicon, aluminum, and titanium will be in the form of oxides MnS or MnP. Lead (insoluble in iron) will be present as small globules in elementary form.

The elements manganese, nickel, and cobalt have a pronounced tendency to lower the initial transformation temperature and if present in sufficient quantity may make it possible to have an austenitic structure at room temperature. Examples of alloy steels that can have an austenitic matrix at room temperature are Hadfield's steel, which contains 12% manganese, and 18–8 stainless steel, which contains 18% chromium and 8% nickel.

The stainless steels contain chromium in amounts from 11 to 27 $^w/_o$ to provide corrosion resistance. All stainless steels contain some carbon, and the greater the carbon content, the greater the amount of chromium required for corrosion resistance, inasmuch as carbon combines with about 17 times its weight of chromium to form carbides. Chromium in the form of carbides is not available for corrosion resistance. Nickel is another important element in stainless steels, and if sufficient nickel is present the steel is austenitic at room temperature.

Stainless steels may be classified into three basic groups:

1. low-chromium stainless steels
2. high-chromium stainless steels
3. chromium-nickel stainless steels

A low-chromium stainless steel contains from 11 to 14 $^w/_o$ chromium, less than 2.5 $^w/_o$ nickel, and less than 0.4 $^w/_o$ carbon. This material is magnetic and can be heat treated to produce a martensitic structure as with ordinary steels to provide a hardness range from about H_B 140 to 600.

High-chromium stainless steels contain from 14 to 27 $^w/_o$ chromium and less than 0.35 $^w/_o$ C. They are also magnetic but cannot be heat treated. The matrix is ferritic, and the hardness of materials in this class is usually from H_B 140 to 187.

The chromium–nickel stainless steels contain 16 to 26 $^w/_o$ chromium, 6 to 22 $^w/_o$ nickel, and up to 0.25 $^w/_o$ C. They are nonmagnetic, have an austenitic structure at room temperature, and can only be hardened by cold work. The hardness of this type of stainless steel usually falls in the range from H_B 140 to 187. Representative alloys from each of the three basic categories are as follows:

Type	AISI No.	C	Cr	Ni	Cb	Fe
1	410	0.15	12			Balance
2	430	0.12	16			Balance
3	347	0.08	18	10	0.8	Balance

From the foregoing considerations it is evident that just as the presence of carbon complicates pure elementary metals, the presence of additional alloying elements further complicates plain carbon steels in the following ways:

1. Additional hard carbides appear which have an even stronger tendency than cementite to precipitate in small size and hence tend to provide a smaller mean ferrite path and greater hardness and strength.
2. The harder inclusions such as SiO_2 and Al_2O_3 will provide additional hard particles to reduce the mean ferrite path.

3. Elements which go into solution in ferrite will cause a hardening of the ferrite and an increased tendency to strain harden. Nickel is particularly pronounced in this respect.

4. Certain elements such as manganese and silicon tend to make the ferrite more brittle which, as we will see later, is beneficial in machining while nickel tends to make the matrix less brittle.

5. The weakening effect of inclusions (such as manganese sulfide and lead) upon the ferrite is apt to be negligible unless the quantity present is excessive or the inclusions are poorly distributed.

6. The presence of manganese, nickel, and cobalt which increase the tendency to obtain an austenitic rather than ferritic matrix will tend to produce a material with greater tendency to strain harden as the percentage of retained austenite is increased.

It is thus evident that the general effect of alloying elements is to produce a harder and stronger matrix having a greater tendency to strain harden. The optimum hardness for an alloy steel will generally be somewhat lower than that for a plain carbon steel of the same carbon content.

Alloy steels have been classified and assigned four-digit numbers by the American Iron and Steel Institute (AISI). The first digit designates the principal alloying element and the last two digits the number of hundreths of one weight-percent of carbon. A few examples of common AISI steel numbers follow:

AISI 1020: plain carbon steel with 0.20 $^w/_o$ C

AISI 1045: plain carbon steel with 0.45 $^w/_o$ C

AISI 1112: resulfurized steel with 0.1 $^w/_o$ S and 0.12 $^w/_o$ C

AISI 8620: Ni-Cr-Mo steel with about 0.5 $^w/_o$ Ni, 0.8 $^w/_o$ Mn, and 0.20 $^w/_o$ C

AISI 3140: Ni-Cr steel containing about 1.25 $^w/_o$ Ni, 0.6 $^w/_o$ Cr, 0.8 $^w/_o$ Mn, and 0.40 $^w/_o$ C

AISI 4140: molybdenum steel containing about 0.80 $^w/_o$ Cr, 0.20 $^w/_o$ Mo, and 0.40 $^w/_o$ C

AISI 4340: Ni-Cr-Mo steel containing about 1.8 $^w/_o$ Ni, 0.65 $^w/_o$ Cr, 0.25 $^w/_o$ Mo, and 0.40 $^w/_o$ C

AISI 8640: Ni-Cr-Mo steel containing about 0.50 $^w/_o$ Ni, 0.60 $^w/_o$ Cr, 0.15 $^w/_o$ Mo, and 0.40 $^w/_o$ C

AISI 52100: ball-bearing steel containing about 1.75 $^w/_o$ Ni, 0.80 $^w/_o$ Cr, and 1.00 $^w/_o$ C

The AISI numbers, compositions, and properties of all metals in common use are to be found in the *Metals Handbook*, 9th ed., published by the American Society for Metals (ASM).

CAST IRONS

Iron alloys which contain from about 2 to 4 $^w/_o$ carbon are referred to as cast irons. Just as in the case of steel there are many characteristic structures that are found in cast iron (Fig. 15.6). If cast iron is cooled rapidly, a product known as white cast iron is obtained which consists of cementite particles in a matrix of pearlite. If the same iron is cooled at a slow rate, a product known as gray cast iron results, which consists primarily of graphite flakes in ferrite or pearlite. If the percentage of silicon or carbon is decreased (or if carbide-forming alloys such as Cr, Mo, or V are present to tie up carbon), a slower cooling rate is required to prevent the formation of cementite ($\bar{C}$) and to produce the graphite flakes that are characteristic of gray iron. Nickel in amounts up to 5 $^w/_o$ will tend to promote the formation of graphite while sulfur has the opposite effect. For this reason when a cast iron is high in sulfur, an excess of manganese is usually added to tie up the sulfur in the form of harmless MnS inclusions.

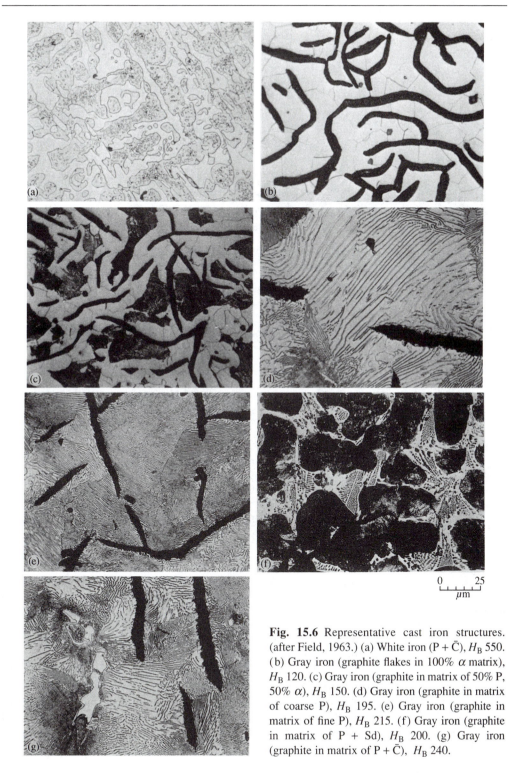

0   25
μm

Fig. 15.6 Representative cast iron structures. (after Field, 1963.) (a) White iron (P + $\bar{\text{C}}$), H_B 550. (b) Gray iron (graphite flakes in 100% α matrix), H_B 120. (c) Gray iron (graphite in matrix of 50% P, 50% α), H_B 150. (d) Gray iron (graphite in matrix of coarse P), H_B 195. (e) Gray iron (graphite in matrix of fine P), H_B 215. (f) Gray iron (graphite in matrix of P + Sd), H_B 200. (g) Gray iron (graphite in matrix of P + $\bar{\text{C}}$), H_B 240.

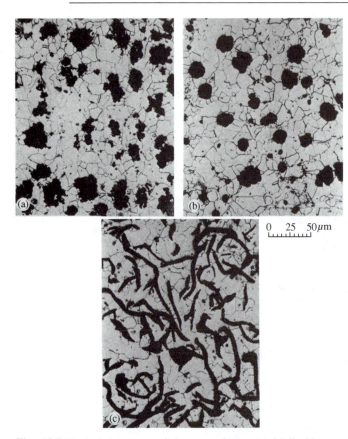

0   25   50μm

Fig. 15.7 Typical structures of three cast irons. (a) Malleable cast iron (3.1 $^w/_o$ C, 2.39 $^w/_o$ Si, 0.37 $^w/_o$ Mn), σ_u = 50,000 p.s.i. (345 MPa), $\sigma_{0.2}$ = 37,000 p.s.i. (255 MPa); E = 21.7 × 10⁶ p.s.i. (150,000 MPa); H_B 150. (b) Innoculated cast iron (3.1 $^w/_o$ C, 2.41 $^w/_o$ Si, 0.43 $^w/_o$ Mn), σ_u = 65,000 p.s.i. (448 MPa), $\sigma_{0.2}$ = 47,500 p.s.i. (328 MPa); E = 23.5 × 10⁶ p.s.i. (162,000 MPa); H_B 150. (c) Gray cast iron (3.2 $^w/_o$ C, 2.43 $^w/_o$ Si, 0.32 $^w/_o$ Mn), σ_u = 20,000 p.s.i. (138 MPa), $\sigma_{0.2}$ = 17,500 p.s.i. (121 MPa); E = 11.6 × 10⁶ p.s.i. (79,980 MPa); H_B 105. (after Cook, Finnie, and Shaw, 1954)

In addition to cementite, pearlite, graphite, carbides of chromium, molybdenum, or vanadium, and inclusions of manganese sulfide, other structures such as free ferrite, steadite, and inclusions of sand and slag are sometimes present in cast iron. Free ferrite will cause a marked weakening of cast iron if present in a concentration greater than about 10 $^w/_o$ and indicates too slow a cooling rate. Steadite is a hard low-melting point (1800 °F or 982 °C) ternary eutectic of iron, cementite, and iron phosphide that forms at the grain boundaries. Its concentration is usually kept below 10 $^w/_o$ by keeping the phosphorus content of the charge less than 1 $^w/_o$. Steadite in amounts less than about 5% has a negligible influence on tool life.

Malleable cast iron is formed by slowly heating white iron to 1600 °F, (870 °C) holding for two or more days and slowly cooling. The long time cycle makes this process expensive. During the soaking period cementite is converted to graphite and austenite; and upon cooling, the austenite transforms to cementite and ferrite, the cementite in turn decomposing to small nodules of graphite in a matrix of ferrite (Fig. 15.7a) and the resulting product is relatively tough.

Cast iron is sometimes innoculated in the ladle to provide a finer dispersion of spheroidal graphite and a less critical cooling rate (Fig. 15.7b). Meehanite is the trade name for cast iron that is innoculated with calcium silicide. Powdered magnesium is another widely used innoculent.

Innoculated irons are generally referred to as ductile irons and appear to provide better properties than malleable iron at decreased cost.

Hardness values of the major phases present in cast irons are given in Table 15.1.

Figure 15.8a shows the variation of tool life with cutting speed when turning different cast irons while Fig. 15.8b gives similar data for several malleable irons.

Gray cast iron does not tend to cause a crater to form on the tool face and consequently the more wear-resistant straight tungsten carbides may be used. Wright (1976) has attributed the lack of crater formation when machining gray cast iron to the formation of a built-up layer of carbonaceous material on the tool face that unlike the usual BUE decreases in thickness as the cutting edge is approached. It is suggested this stationary carbide layer prevents diffusion wear from occurring when gray cast iron is machined. However, it should be noted that Wright's observations were made

TABLE 15.1 Range of Hardness Values
of Major Constituents of Cast Iron[†]

Microconstituent	Knoop Hardness (100 g load) kg mm^{-2}
Graphite	15–40
Ferrite	215–270
Pearlite	300–390
Steadite	600–1200
Carbide	1000–2200

[†] After Field, 1963.

on a HSS tool operated at unusually high
temperature (speed) and it has not been shown
that similar results would be obtained under
more normal operating conditions (a carbide
tool operating at the same speed or a HSS
tool operating at a much lower speed).

Gray cast iron normally gives best results
when machined dry. If a coolant is used, it
is important that it be oil-free. Apparently an
oily film prevents the spreading of graphite
over the rake face, the graphite apparently
acting as a combined solid lubricant and diffusion barrier for carbon moving from tool
to chip. A commonly used coolant for turning cast iron is water plus an inorganic rust
inhibitor (0.2 $^w/_o$ NaNO$_2$). Gray cast iron
gives particularly good results with ceramic
(Al$_2$O$_3$) cutting-tool materials used without
a coolant.

Practical considerations in the machining
of cast irons are discussed in DeBenedictis
(1997).

FRACTURE DURING
CHIP FORMATION

When metal is removed by a single-point
tool, chips are observed to be in the form
of continuous ribbons, a series of individual
segments, or a combination of these two
extremes in which cracks do not penetrate

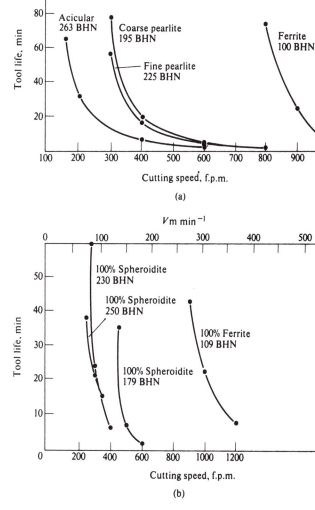

Fig. 15.8 Variation of tool life (0.030 in (0.75 mm) wear land)
with cutting speed for different cast and malleable irons using a
C–2 carbide tool dry. Side rake, 6°; back rake, 0°; feed, 0.010 in
(0.25 mm/rev); depth of cut, 0.100 in (2.5 mm). (a) Cast irons.
(b) Malleable irons. (after Field, 1963)

the chip completely. In all of these cases some fracture of the material cut is involved for even the
perfectly continuous chip that shows no cracks in its surface was formed in conjunction with the
development of new surface and thus fracture.

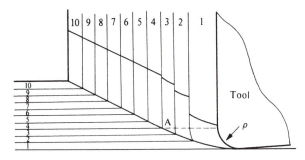

Fig. 15.9 Stack of plates cut with tool having finite radius of curvature at its tip. (after Cook, Finnie, and Shaw, 1954)

The fracture that occurs in all of these cases is either of the ductile shear type which predominates in metal cutting or the brittle tensile type. The very limited fracture that occurs at the point of a tool in continuous cutting is of the tensile type and a possible mechanism for the generation of new surface at the tool tip is discussed below.

No matter how carefully a tool may be ground, there always will be a finite radius of curvature at its point as shown exaggerated in Fig. 15.9. If the metal cut consisted of a number of separate plates (1, 2, 3, etc.), then the shear surface might be expected to be as shown in Fig. 15.9. The rake angle associated with plate 1 is very small and hence the corresponding shear angle also should be expected to be small. The chip from plate 1 would be bent toward the tool by plate 2. The shear angle would increase for subsequent plates inasmuch as the rake angles associated with them increase. This variation in shear angle would occur to point A beyond which the rake angle and hence the shear angle should remain constant. Chips 1, 2, 3, and 4 would be of different lengths as a consequence of the different shear angles obtaining.

In practice, the metal cut does not consist of separate plates. The chip near the tool face cannot be shorter than at other points since the chip is restrained from curling by the tool face. Hence the chip in the vicinity of the tool face must deform plastically in tension. The region subjected to tension usually will extend a very small percentage of the way across the chip (to about point A in Fig. 15.9). The tensile field of stress that is developed as a result of the curvature at the tool point can thus produce the crack necessary in the development of the new surface. The material beyond point A in Fig. 15.9 is subjected to large compressive stresses, and the crack will be quenched upon reaching this region in the case of continuous cutting.

Considerable controversy is to be found in the engineering literature (Sturney, 1925; Schwerd, 1932; Ernst, 1938) concerning the existence of cracks in front of a tool when a continuous chip is produced. It would appear that when a very sharp tool is used, microcracks develop that are so small as to be invisible. An occasional crack may be observed to run a short distance in front of the tool when the cutting of an inhomogeneous ductile material is observed under the microscope. The possibility of observing cracks in front of a tool increases as the radius of curvature at the tool point is increased. The existence of cracks at the point of a sharp cutting tool when a continuous chip is produced is of interest only with regard to the generation of new surface. Their presence is of negligible importance with regard to the analysis of the shear process and the mechanics of cutting.

When a ductile metal that is free of gross inhomogeneities is cut, the microcracks responsible for the new surface will form at the microvoids present in all materials owing to the concentration of stress. The spacing of the individual cracks should then be expected to correspond to the linear spacing of the microvoids, which is indicated to be about 40 μin (1 μm) by several independent studies. The individual plates in the Piispanen model are thus seen to be very thin.

Materials that are cut differ widely in their ability to deform plastically and to sustain tensile stresses. The cutting of β-brass may best be appreciated by studying motion pictures taken through the microscope in conjunction with records from a recording dynamometer. The sketches in Fig. 15.10 are reproduced from pictures taken at 16 frames per second when a 0.007 in (0.18 mm) depth of cut was taken at a cutting speed of 0.5 i.p.m. (12.7 mm min^{-1}), using a 15°-rake-angle tool in a two-dimensional planning operation. The width of cut was 0.161 in (4.09 mm) and no fluid was used. Frame numbers are indicated beneath the sketches, and the distance traversed by the tool per frame was 0.0005 in (0.01 mm). The chips were completely discontinuous and the horizontal and vertical components of force varied periodically as shown in Fig. 15.10, with a period of 39 frames.

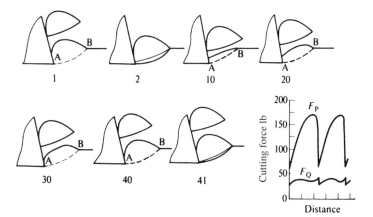

In frame 1 rupture along the dotted line is impending and actually occurs between frames 1 and 2. When failure occurs, the load on the tool is released suddenly, and it is driven rapidly forward by the elastic energy stored in the system as a result of tool deflection. The depth of cut is, in effect, never zero and hence the forces do not drop to zero. Frame 2 shows the tool after its elastic motion has occurred, and metal is seen piled up at the tool point. As the tool moves forward, metal is

Fig. 15.10 Sketches of discontinuous cutting action of β-brass with corresponding cutting force data. (after Cook, Finnie, and Shaw, 1954)

not removed by the normal shearing process represented by the Piispanen model, but is rather extruded between the tool and free surface into a parabolic shape in the manner of Fig. 15.11. During this extrusion the chip does not slide along the tool face but, rather, rolls down upon it. The shear stress on the dotted line, which represents the region of plastic flow, increases as extrusion proceeds, until a value sufficient to cause rupture is obtained. The entire process is then repeated.

The force data in Fig. 15.10 was obtained from a strain-gage dynamometer and, when used to compute the coefficient of friction on the tool face, the results shown in Fig. 15.12 were obtained. Here it is evident that the coefficient of friction actually decreases as the cut proceeds. The values of coefficient of friction given in Fig. 15.12 are unusually low for ordinary metal cutting with a 30°-rake-angle tool. This should not be disturbing, however, when it is realized that the significance of the static friction on the tool face in the extrusion process considered here is entirely different from the dynamic coefficient of friction in continuous cutting.

Sketches and force data are given in Fig. 15.13 for cuts made at different values of rake angle. The views shown are those corresponding to the frame just preceding rupture. The previously described extrusion process is seen to hold in all cases where cutting is discontinuous. Continuous chips are obtained at higher values of rake angle (30° and 45°). Comparison of the mean cutting force F_P for the 15° tool with the values of F_P for the 30° and 45° tools indicates that the mean cutting force in discontinuous cutting is certainly not greater than would be obtained in continuous cutting under comparable conditions. In order to determine the influence of normal stress on fracture shear stress, the stresses were computed on the dotted plane along which fracture would occur, AB, at different stages in the development of the chip. The shear and normal stresses on plane AB for the 15° tool were found to vary as the cut proceeded as shown by the curve marked 15 in Fig. 15.14. Fracture finally occurred at the point marked X.

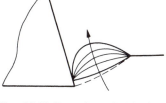

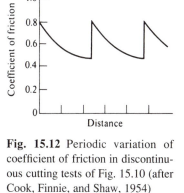

Fig. 15.11 Extrusion of chip into parabolic shape in discontinuous cutting operation. (after Cook, Finnie, and Shaw, 1954)

Fig. 15.12 Periodic variation of coefficient of friction in discontinuous cutting tests of Fig. 15.10 (after Cook, Finnie, and Shaw, 1954)

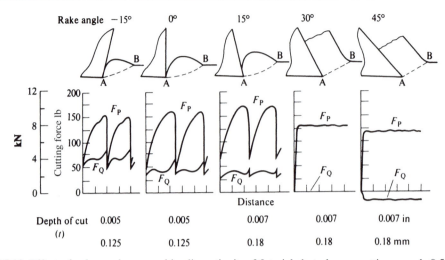

Fig. 15.13 Effect of rake angle upon chip discontinuity. Material, beta brass; cutting speed, 0.5 i.p.m. (12.7 mm min^{-1}); width of cut, 0.161 in (4.09 mm); fluid, none. (after Cook, Finnie, and Shaw, 1954)

Similar curves are shown for other rake angles in Fig. 15.14. Only one point each could be obtained for the 30° and 45° tools inasmuch as cutting was continuous in these cases. The fracture point for each rake angle has an X over it, and it is evident that a line DE can be drawn which separates the conditions where no fracture occurred from those where a crack ran clear across the chip. Line DE is seen to be inclined upward, which means that as the compressive stress on the plane where rupture will occur is increased, there is a corresponding higher value of critical fracture stress. Inasmuch as the normal stress on the shear plane is inherently relatively less for small values of rake angle, the tendency to cut discontinuously increases with decreased rake angle.

When the depth of cut was reduced from 0.007 to 0.005 in (0.18 to 0.13 mm) in cutting with the 15°-rake-angle tool, the chip formation was initially as shown in Fig. 15.15a which is seen to be similar to the action in Fig. 15.10. However, occasionally a crack was found to occur in a direction roughly perpendicular to the potential fracture surface, i.e., along FG in Fig. 15.15b, to suddenly produce a built-up edge on the tool. Since the surface of this built-up edge corresponded to an increased rake angle, the chip formation tended to become more continuous as shown in Fig. 15.15c. Only the back surface of the chip was then irregular. This example illustrates how a built-up edge may quickly develop as a result of plastic fracture within the chip.

Data for magnesium similar to those previously given for β-brass are shown in Fig. 15.16. The action in this case is similar in many respects to that for β-brass. The fracture surface is, however, much higher and consequently the extrusion action is even more pronounced. In the case of magnesium, fracture does not occur at the

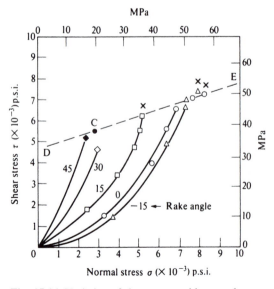

Fig. 15.14 Variation of shear stress with normal stress on fracture surface when cutting under conditions of Fig. 15.13. (after Cook, Finnie, and Shaw, 1954)

boundary of plastic flow but within the region of plastic flow. In Fig. 15.16 the potential fracture surfaces are indicated by dotted lines while the regions of plastic flow are outlined by dashed lines. As in the discontinuous cutting of β-brass, the direction of the fracture surface is little influenced by the rake angle.

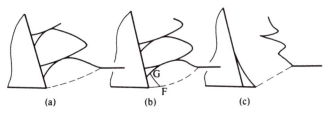

Fig. 15.15 Development of built-up edge when machining β-brass. Rake angle, 15°; depth of cut, 0.005 in (0.13 mm); cutting speed, 0.5 i.p.m. (12.7 mm min⁻¹); width of cut, 0.161 in (4.09 mm); fluid, none; cutting forces under condition (c), F_P = 140 lb (623N), F_Q = 65 lb (289N). (after Cook, Finnie, and Shaw, 1954)

Metal cutting chips frequently contain cracks which go part way across the chip. When β-brass is cut with a 30°-rake-angle tool at a depth of cut of 0.015 in (0.38 mm), the quasi-discontinuous chip shown in Fig. 15.17 is obtained. In this case a crack is found to develop periodically, but the crack goes only part way across the chip. A ripple in the force trace is observed, but the large fluctuations characteristic of the completely discontinuous cut are absent. When the normal and shear stresses for this case are plotted in Fig. 15.14, the point C is found to lie on line DE. This is as it should be, for in this case there is a shear failure, even though the crack does not extend all the way across the chip. In order for a crack to propagate spontaneously across a chip as in the case of completely discontinuous cutting, there must be sufficient elastic energy stored in the tool to supply the energy associated with the propagation of the crack. If the system is very stiff and there is not sufficient stored elastic energy, the tool may lose contact with the chip as the crack grows and hence there may be no source for the energy required for crack propagation.

Several operating variables are found to influence the nature and extent of the fracture involved during chip formation:

1. tool sharpness (fracture tendency decreases with sharpness)
2. rigidity and capacity of tool work system to store elastic strain energy (fracture tendency increases with energy stored)
3. rake angle (fracture tendency increases with decrease in rake angle)

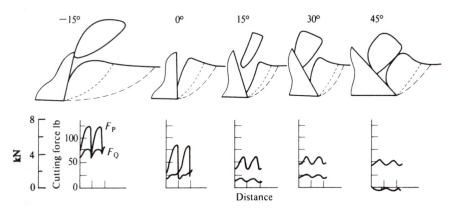

Fig. 15.16 Effect of rake angle upon chip formation when cutting magnesium. Cutting speed, 0.5 i.p.m. (12.7 mm min⁻¹); depth of cut, 0.010 in (0.25 mm); width of cut, 0.159 in (4.04 mm), fluid, none. (after Cook, Finnie, and Shaw, 1954)

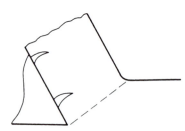

Fig. 15.17 Partially continuous chip produced when β-brass is machined at depth of cut of 0.015 in (0.38 mm) using a 30°-rake-angle tool. Cutting speed, 0.5 i.p.m. (12.7 mm min^{-1}); width of cut, 0.161 in (4.09 mm); cutting fluid, none; cutting forces, F_P = 270 lb (1202N), F_Q = 15 lb (67N). (after Cook, Finnie, and Shaw, 1954)

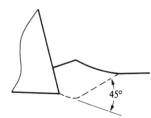

Fig. 15.18 Direction of tensile crack (dotted curve) sometimes observed in brittle materials. (after Cook, Finnie, and Shaw, 1954)

4. tool-face friction (fracture tendency increases with increase in tool-face friction)

5. cutting speed (fracture tendency decreases with increase in cutting speed)

6. depth of cut (fracture tendency increases with depth of cut)

7. material characteristics (number, size, shape and relative hardness of inhomogeneities, hardness produced by heat treatment or by strain hardening)

The importance of inhomogeneities in the metal cut is immediately evident when the cutting characteristics of gray cast iron, nodular iron, and malleable iron are compared. In Fig. 15.7 the structures of specimens of these materials are shown. The carbon content of each of these materials is essentially the same. In each case we observe a ferrite matrix with inclusions of graphite. The graphite in the gray iron is in the form of plates, that in the innoculated iron is in the form of spheres, while that in the malleable iron is in the form of rougher spherical clumps of very fine graphite particles. The tensile strength of the gray iron is relatively low inasmuch as it is not difficult to find plates oriented across the direction of maximum normal stress. A much larger stress concentration is thus associated with the plates than with either of the spherical structures. Similar behavior is observed when hardness readings are taken on these materials or when they are cut.

The gray iron produces a brittle chip as a result of the stress concentration associated with the graphite plates. While most cracks are in the direction of maximum shear stress to the very point of the tool and hence indicate that fracture was of the plastic type, cast iron is sometimes observed to fail in tension, a crack extending downward from the tool point and then swinging into the direction of maximum shear stress as shown in Fig. 15.18.

It is only with very brittle materials that a tensile crack is observed initially to extend downward from the tool point. When this occurs, it is difficult to maintain dimensional accuracy because of the variable depth of cut. By far the majority of chip fracture observed in cutting is of the shear type. While this is true even with such materials as cast iron, tensile-type cracks may be observed when (a) the inhomogeneities are unusually large, (b) the cutting temperature is low, and (c) large depths of cut are taken.

In addition to graphite, inclusions of other soft materials at elevated temperatures such as lead or manganese sulfide frequently lead to discontinuous cutting action.

SECONDARY SHEAR ZONE

The secondary shear zone which forms at high cutting speeds along the tool face has been extensively studied by Trent (1963) and Opitz (1963). The secondary layer is a portion of the chip that has been rendered unusually plastic as a result of high temperatures. The temperatures in this layer will definitely be above the strain recrystallization point and even lie above the $\alpha \rightarrow \gamma$ phase-transformation temperature, in the case of steels. Temperatures in the layer will be sufficiently high to enable carbon and tungsten from the tool material to diffuse into the chip. (Carbon will diffuse particularly rapidly if the zone has transformed to γ–iron.) If the secondary shear zone is moving relative to the tool face, wear is apt to be rapid since the carbon will be rapidly convected away by the chip. A crater will then rapidly form which may weaken the cutting edge as it grows and ultimately cause tool point breakage. It is highly advisable that the secondary shear zone become affixed to the tool face in order to avoid the rapid removal of diffused material. Titanium carbide appears to help form a thin stationary secondary shear zone on the tool face and is used as an additive to tungsten carbide tools to decrease the crater-forming tendency of such tools. At the same time TiC lowers the thermal conductivity of the tool and hence the speed at which the protective (stationary) secondary shear zone will form.

Opitz (1963) has further found that certain oxides, when present in steel as a result of particular deoxidation practices, play an important role in secondary shear zone formation. In such cases the oxides apparently lower the softening (melting) point of the chip material adjacent to the tool face and thus make it possible to provide a protective stationary secondary shear zone at lower cutting speeds. Such additives to steel are found to be effective in increasing tool life only when used in conjunction with tools containing TiC. This is apparently due to the fact that early secondary shear zone formation is important only if the zone can be rendered stationary at the tool face. Since titanium has a greater affinity for oxygen or oxides than for iron or itself, it forms a strong adhesive bond with oxide-bearing steels.

Manganese sulfide appears to influence secondary shear zone formation in a way similar to that for the special oxides (i.e., by lowering the softening point of the metal immediately adjacent to the tool face). Manganese sulfide also appears to have a strong affinity for both WC and TiC, and hence is effective in promoting the formation of a protective stationary shear zone with both types of tools.

From the point of view of crater formation the properties of the work material immediately adjacent to the tool face are more important than the bulk properties of the work. The material in the secondary shear zone is subjected to enormous strains (20–50 in in^{-1} or more), and if oxides or sulfides are present as inclusions, they will be drawn out into extremely thin plates. Thus, the manganese sulfide particles in the secondary shear zone will be transformed into a layered structure of very thin plates. Such a structure may have very different strength, thermal, and chemical properties from those of the usual undeformed material or even the deformed material in the main body of the chip. In addition, the temperature of the secondary shear zone will be very high. It is therefore important that the properties of highly strained materials containing layered structures of sulfides and oxides be studied at a wide variety of temperatures. It is to be expected that the properties of materials in this special structural condition may prove to be more useful in explaining the behavior of work materials containing different additives than the ordinary bulk physical properties of these materials.

It would appear that useful properties of a secondary shear zone include

1. high affinity for the tool material
2. poor thermal conductivity
3. low melting point
4. low solubility for tungsten or carbon

Some of these are the reverse of those to be expected from the more conventional view that the chip should glide freely and with low friction over the tool face. The easy sliding view is correct for low-speed cutting (broaching) but not for high-speed cutting (turning) where crater formation is a major problem. To avoid crater formation a stationary or low-speed protective layer is essential to prevent the major tool-face shear energy from reaching the tool face. The secondary shear zone should be capable of acting not only as a thermal barrier but also as a diffusion barrier. A highly oriented, relatively low-melting point, layered structure of an inorganic inclusion (oxide, sulfide, phosphide, etc.), with the layers running parallel to the tool face, should be useful from all points of view.

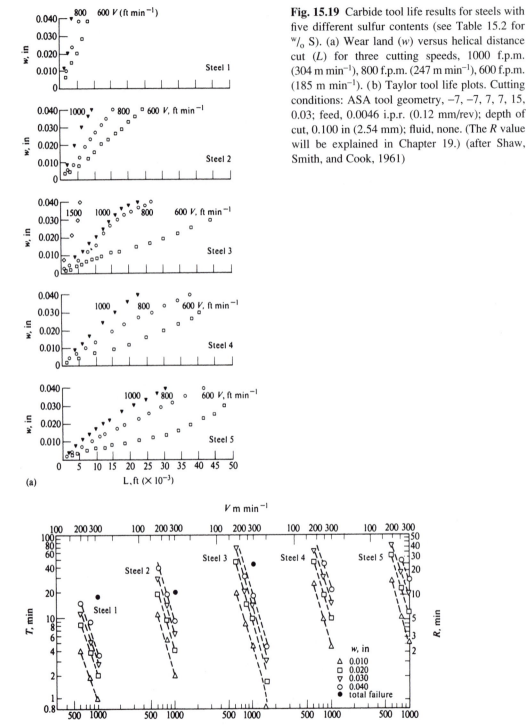

Fig. 15.19 Carbide tool life results for steels with five different sulfur contents (see Table 15.2 for $^w/_o$ S). (a) Wear land (w) versus helical distance cut (L) for three cutting speeds, 1000 f.p.m. (304 m min^{-1}), 800 f.p.m. (247 m min^{-1}), 600 f.p.m. (185 m min^{-1}). (b) Taylor tool life plots. Cutting conditions: ASA tool geometry, −7, −7, 7, 7, 15, 0.03; feed, 0.0046 i.p.r. (0.12 mm/rev); depth of cut, 0.100 in (2.54 mm); fluid, none. (The R value will be explained in Chapter 19.) (after Shaw, Smith, and Cook, 1961)

Properties of the tool must also be considered in relation to the layered structure of the secondary shear zone and not relative to the work material in bulk. A good tool material should have constituents possessing a high affinity for the layered structure material, low conductivity (to promote secondary shear zone formation), and a low tendency for its constituents to diffuse into the layered structure. It is conceivable that the best material for low-speed cutting (no secondary shear zone present) may not be the best for high-speed cutting.

At very high cutting speeds a secondary flow zone will also form on the clearance face, and attention should also be directed to this region of the cutting tool.

BEHAVIOR OF SULFUR AND LEAD

The role of manganese sulfide has been touched upon in the previous section in connection with crater formation, and the results in Fig. 15.19 illustrate this action. Here five steels from the same heat, but having different sulfur contents, were machined using a tungsten carbide tool of 6° rake angle and a feed of 0.0046 i.p.r. (0.12 mm/rev). The cutting performance as measured by either V_{60} (speed corresponding to a 60 min tool life) or V_{240} is seen to improve with sulfur content (Table 15.2). In these tests the cutting speeds are high, no built-up edge is present, and crater wear plays an important role.

The overall role of manganese sulfide is far more complex than this influence on secondary shear zone formation would suggest. This is illustrated by some similar tests on the same steels using high speed steel (HSS) tools (Fig. 15.20). Here it is evident that sulfur does not always lead to improved tool life. Table 15.3 shows that although V_{60} increases with sulfur content, V_{240} does not. The reason for this lies in the fact that manganese sulfide actually promotes BUE formation and when running at speeds where the BUE is apt to be large (speeds in the vicinity of V_{240} for HSS), the presence of sulfur is apt to reduce tool life.

This result, interesting as it is, is not in agreement with general experience, where it is found that the presence of manganese sulfide in steel caused an improvement in the life of automatic screw machine tools. The cause of the discrepancy is to be found in the feed used in the tests of Fig. 15.20 (0.0046 i.p.r. or 0.12 mm/rev). Screw machines actually operate at much smaller feeds. When these tests were repeated at a feed of 0.0023 i.p.r. (0.06 mm/rev), it was found that V_{240} then improved with increase in sulfur content. The size of the BUE for all steels was so small, at the small feed, that it had relatively little influence on tool life even at small speeds (V_{240} or V_{480}). At larger feeds (0.010 i.p.r. or 0.25 mm/rev) V_{240} was found to increase with sulfur content. In this case the BUE had already disappeared at speeds corresponding to V_{240} for HSS.

TABLE 15.2 Tool-Life Results for Free-Machining Steels of Different Sulfur Content Machined with Tungsten Carbide Tools

Steel	Sulphur Content, %	$VT^{0.33} (= C)$	Cutting Speed ft min^{-1} (m min^{-1})	
			V_{60}	V_{240}
1	0.033	1400(427)	330(101)	230(70)
2	0.11	1800(549)	490(149)	295(90)
3	0.18	2300(701)	630(192)	377(115)
4	0.26	2400(732)	640(195)	394(120)
5	0.37	2700(823)	680(207)	442(135)

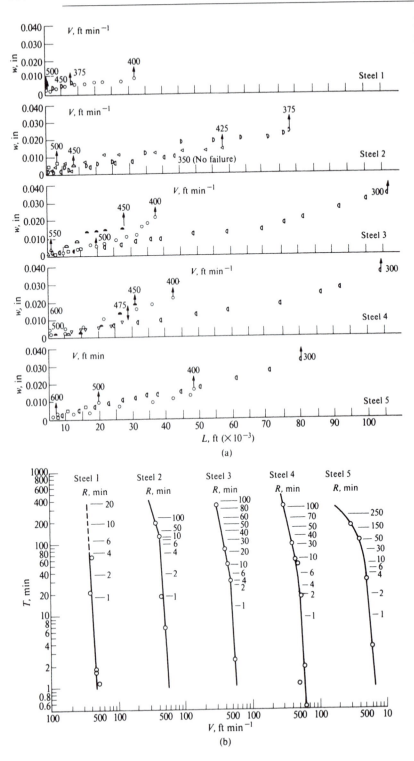

Fig. 15.20 HSS (M2) tool-life results for steels with five different sulfur contents (see Table 15.3 for $^{w}/_{o}$ S). (a) Wear land (w) versus helical distance cut (L) for different cutting speeds. (b) Taylor tool-life plots. Cutting conditions: ASA tool geometry, 0, 15, 10, 10, 10, 20, 0.032; feed, 0.046 i.p.r. (0.12 mm/rev); depth of cut, 0.100 in (2.54 mm); fluid, none. (The R values will be explained in Chapter 19.) (after Shaw, Smith, and Cook, 1961)

TABLE 15.3 Tool-Life Results for Free-Machining Steels of Different Sulfur Content Machined with HSS Tools

Steel	Sulphur Content, %	Equation	Cutting Speed ft min^{-1} (m min^{-1})	
			V_{60}	V_{240}
1	0.033	$VT^{0.05} = 470(143)$	380(116)	360(110)
2	0.11	$VT^{0.05} = 560(171)$	440(134)	350(107)
3	0.18	$VT^{0.07} = 600(183)$	430(131)	320(100)
4	0.26	$VT^{0.10} = 550(168)$	450(137)	320(100)
5	0.37	$VT^{0.10} = 660(201)$	460(140)	220(67)

The foregoing results illustrate how complex things can be in machinability studies again because several mechanisms are generally at work simultaneously.

Specifically, the foregoing example shows that at intermediate feeds (0.005 i.p.r. or 0.125 mm/rev) the tool life for a resulfurized steel is apt to be poorer than that for a non-resulfurized steel, depending on the tool material used, and the speed of interest. The poorer HSS tool life at V_{240} with the 0.0046 i.p.r. (0.12 mm/rev) feed could have been eliminiated by use of a carbide tool in place of HSS. This would cause V_{240} to fall above the speed at which the BUE forms, and the influence of sulfur would then be positive (as is evident in Table 15.2).

Evidence that MnS does not function as a lubricant in the usual sense (i.e., by lowering the coefficient of tool-face friction) is offered by some tests performed using tools of controlled tool-face contact length. These tools were ground as shown in Fig. 15.21a with values of C ranging from one feed distance to five feed distances. Values of tool-face shear stress (τ_c) and tool-face normal stress (σ_c) were determined from dynamometer results and plotted as shown in Fig. 15.21b. Here it is seen that the steel of lowest sulfur content generally gave the lowest ratio $\tau_c/\sigma_c = \mu$ (coefficient of tool-face friction). However, when the value of feed (t) was very small, the curves were found to cross and μ was then slightly less for the steel with lowest sulfur content.

These results may be interpreted by assuming manganese sulfide to have two roles:

1. tendency to reduce strain on the shear plane during chip formation (a bulk effective)
2. tendency to increase tool-face friction (a surface effect)

Under normal cutting conditions (feeds) the first role is predominant. Hence, when the strain in the chip is reduced by added sulfur, the ratio of τ_c to σ_c decreases, just as though the MnS caused low sliding friction. At very low feeds, surface effects become more important than bulk effects and item 2 becomes predominant. It then appears as though MnS is a poor lubricant, which is actually the case in all instances.

Similar results are shown in Fig. 15.21c for leaded and nonleaded steels. In this case the action of lead is seen to increase as the feed decreases. This is because lead acts primarily by reducing tool-face friction. Since this is a surface effect, it is apt to be more pronounced as the chip surface-to-volume ratio increases (i.e., at low feeds).

From the results that have been presented we may conclude that sulfur has its major effect on the shear plane and in conjunction with the size of the BUE (both of these are bulk actions), while lead has its major action on the tool face (surface action). This is also consistent with the

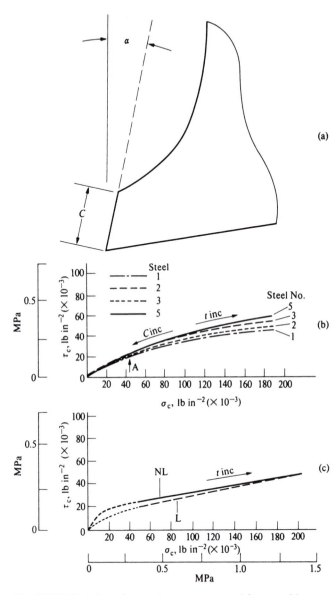

fact that sulfur is most effective at large feeds (small surface-to-volume ratio) while lead is most effective at small feeds.

In practice it proves beneficial to add both sulfur and lead to steels which must be used over a wide range of feeds.

Other free-machining additives to steel include bismuth, which appears to function in the manner of lead, and selenium and tellurium, which appear to function in the manner of sulfur. Tellurium is a particularly effective free-machining agent, but creates a major rolling problem even when used in small quantities.

There is evidence that lead associates with MnS making the latter more effective in its action when both are present.

Free-machining additives normally increase chip curl and hence decrease the chip–tool contact length. While a decrease of chip–tool contact length represents improved machinability (tool life) for a relatively weak material such as low-carbon steel, the reverse is true for a material involving higher specific cutting energy (u) such as AISI 4340 steel. This is because a decrease in chip–tool contact length brings the maximum temperature too close to the cutting edge in the case of the stronger AISI 4340 steel. Thus, while lead is found to improve the machinability (tool life) of a low-carbon steel, it is found to give poorer machinability when added to AISI 4340 steel.

The amount of sulfur and lead added to steel to improve machinability is small. In the case of sulfur from 0.10 to 0.03 $^w/_o$ is normally added while the amount of lead added is usually about 0.035 $^w/_o$. Sulfur has a high affinity for iron and will tend to form low-melting-point FeS in the grain boundaries. This makes the steel difficult to roll without intergranular fracture. The solution is to add sufficient manganese to tie up the sulfur as MnS and prevent the

Fig. 15.21 Variation of mean shear stress on tool face τ_c with mean normal stress on tool face σ_c for various artifical contact lengths C and feeds t. (a) Tool with controlled contact length. (b) τ_c versus σ_c for resulfurized steels (see Table 15.2 for S-contents). (c) τ_c versus σ_c for leaded steel (L) and nonleaded steel (NL). (after Shaw, Usui, and Smith, 1961)

formation of FeS. As an added precaution the amount of manganese added is generally twice that required to convert all of the sulfur added to MnS.

The influences of sulphur and lead on cutting forces are given later in Fig. 17.11.

SPECIAL DEOXIDIZED STEELS

An important development that was pioneered in Germany (Opitz and Koenig, 1965) and later extended in Japan (Japanese Working Group, 1968) and the United States (Tipnis and Joseph, 1971) is the use of special deoxidation methods to produce steel that has a lesser tendency to cause carbide tools to crater in high-speed machining. This technique involves the use of calcium and ferro silicon as deoxidizing materials resulting in a relatively low-melting-point ternary ($SiO_2-Al_2O_3-CaO$) inclusion which spreads over the tool face in high-speed machining and acts as a diffusion barrier.

Calcium-deoxidized steels appear to be most effective in turning, face-milling, and gun-drilling operations but show little improvement in drilling with HSS tools or grinding using white aluminum oxide wheels (Japanese Working Group, 1968, 1969). The protective layer is very thin and the presence of TiC appears to be necessary in the tool for the layer to become firmly attached. Aluminum oxide (ceramic) tools also give good results with calcium-deoxidized steels but HSS tools do not. A similar MnS protective layer is reported to form at lower speeds than required for calcium-deoxidized steels (Opitz and Koenig, 1965).

It has been suggested (DeSalvo and Shaw, 1969) that hydrodynamic action is possible on the tool face. However, this is not a result of penetration of a liquid film of cutting fluid between chip and tool, but rather due to the formation of a wedge-shaped liquid or semi-liquid layer between chip and tool due to thermal softening and spreading of an ingredient in the work material. The shape of such a "fluid" film will be as shown in Fig. 15.22. At first glance it may be concluded that the inclination present is such as to give a negative rather than a positive hydrodynamic pressure. However, the resultant motion of the solid portion of the chip will be parallel to the tool face (AE in Fig. 15.23a) and not to the inclined surface of the molten layer (AF). This gives rise to a squeeze-film action (DeSalvo and Shaw, 1969) that more than offsets the sliding action associated with the negative inclination present to provide the generation of a positive hydrodynamic pressure between chip and tool. An alternative way of looking at this problem is to

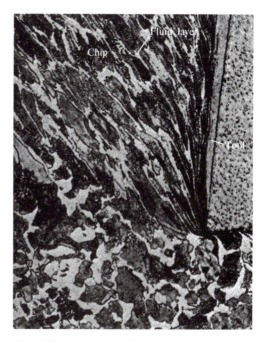

Fig. 15.22 Formation of "fluid" layer immediately adjacent to tool face in high-speed machining. (after Schaller, 1962)

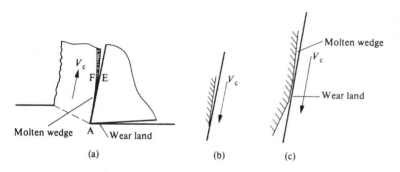

(a)   (b)   (c)

Fig. 15.23 Mechanism of positive-pressure generation on tool face.

draw the kinematically equivalent diagram in Fig. 15.23b where it is immediately clear that a positive pressure will be generated.

Figure 15.23c shows a developed view of the rake and clearance portions of the tool which suggests that the role of the wear land is to provide a composite bearing configuration and a shroud action. Composite bearings are analyzed by Shaw and Macks (1949).

While the role of the inclusion material present in specially deoxidized steels is generally believed to be that of a diffusion barrier for carbon, it could also tend to act as a thermal diffusion barrier just as the glass layer in the Sejournet (1954, 1962) process used in hot extrusion of steel. In this case a hot billet is coated with glass powder which melts to provide a continuous glass coating on the surface of the material to be extruded. An important function of the glass is to decrease heat transfer between the hot billet and the cold die. Similarly, the molten layer of oxide coming from the specially deoxidized steel provides a poor thermal conducting layer that tends to decrease heat transfer from chip to tool.

It has been reported that specially deoxidized steels tend not only to reduce crater formation on the tool face but also to reduce the rate of wear-land formation. However, this is not believed to be due to a diffusion barrier action, since the temperature on the clearance face is very much lower than that on the tool face. A possible alternative explanation for the beneficial effect of oxide on the clearance surface is a hydrodynamic one associated with the liquid oxide that would leak past the cutting edge (Fig. 15.23c).

It is possible that a hydrodynamic action is also responsible for the improved machinability of steels containing such free-machining additives as lead and bismuth. These materials which are insoluble in steel will be squeezed from the chip to provide a thin wedge-shaped liquid layer on the tool face as in the case of the complex low-melting-point oxides in specially deoxidized steels.

Analysis of the hydrodynamic aspects of metal cutting is extremely difficult since the inclination angle of the fluid wedge is not predetermined by the geometry of bearing surfaces but by a melting action that depends on the rate of heat generation, the rate of heat transfer, and the rate of melting. From this one might expect the existence of an optimum cutting speed for each tool–work–inclusion system. If the cutting speed is too low the energy level will be insufficient to cause melting; whereas if it is too high there will not be time for melting during chip–tool contact. This is obviously a problem that is best approached experimentally rather than analytically.

HOT-HARDNESS OF WORKPIECE

It frequently happens that one steel behaves in a superior manner to another in the workshop even though the chemistry and structures of the two steels are essentially the same.

This was found to be the case for two stainless steels which are designated 303 and 303A. The latter steel contained a small amount of very finely divided (unresolvable microscopically) Al_2O_3 which improved tool life in screw machines.

An investigation was made to determine the fundamental reason for the improved performance of the steel containing the fine particles of Al_2O_3.

Form tools are frequently the critical ones which determine cycle time in a screw machine. These tools are usually required to take a relatively wide cut on work of small diameter and tool chatter is frequently a problem. In order to produce the finish required these tools must operate at light feeds and relatively low surface speeds. These conditions and the relative complexity of the tools will normally indicate high-speed steel to be the most economical tool material. Therefore, HSS tools were used in this investigation.

The procedure followed was to look for differences in machining performance of the two steels. The items investigated included chemistry, physical properties, surface finish, cutting forces, chip formation, tool temperatures, and wear rates.

TABLE 15.4 Chemical Analyses of Two Stainless Steels

	C	Mn	Si	S	Cr	Ni	Cu	Mo	Co	Al
303A	0.085	1.63	0.65	0.15	18.14	9.24	0.22	0.56	0.21	0.82
303	0.094	1.67	0.62	0.31	17.36	8.80	0.21	0.32	0.13	0.023

TABLE 15.5 Mechanical Properties of Two Stainless Steels

	303A	303		303A	303
Ultimate tensile strength (lb in^{-2})	100,300	105,000	(MPa	692	724)
Yield strength (lb in^{-2})	71,500	68,000	(MPa	493	469)
Elongation (%)	49.5	55.3			
Reduction of area (%)	67	66			
Charpy impact (ft lb)	79.0	58.5	(Nm	107	79.3)
Hardness (Rockwell B)					
Outside	102.8	103.6			
Mid-radius	98.6	97.0			
Center	95.1	89.1			

The chemistry and physical properties of the two stainless steels studied are given in Tables 15.4 and 15.5.

The analyses show no significant differences except for sulfur and aluminium. Only small differences in mechanical properties are evident in Table 15.5.

The work material was in the form of cold-drawn bars, $1\frac{1}{16}$ in (27 mm) diameter.

The tools investigated were M–2, HSS (5% Mo, 4% Cr, 2% V, 6% W) having a rake angle of 10° and a clearance angle of 8°.

The nature of the cut taken is shown in Fig. 15.24. After each plunge cut was completed the remaining material was removed and a new groove produced by an auxiliary tool.

No significant differences were observed between the two materials for surface finish, cutting forces, specific energy, cutting ratio, shear angle, chip–tool interface temperature, or chip root details obtained from the quick-stop arrangement shown in Fig. 15.25. Only the difference in tool wear shown in Fig. 15.26 was observed. This was found to correlate with a difference in hot-hardness for the two materials (Fig. 15.27). These results suggest that the 303A-material is more refractory than the 303-material and that the tool material at the wear

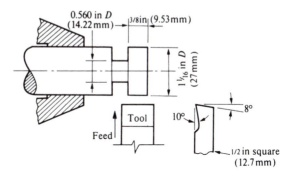

Fig. 15.24 Test arrangement for stainless steel machinability tests. (after Vilenski and Shaw, 1970)

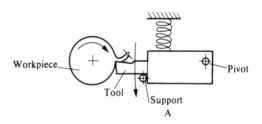

Fig. 15.25 Schematic arrangement for quick-stop tests.

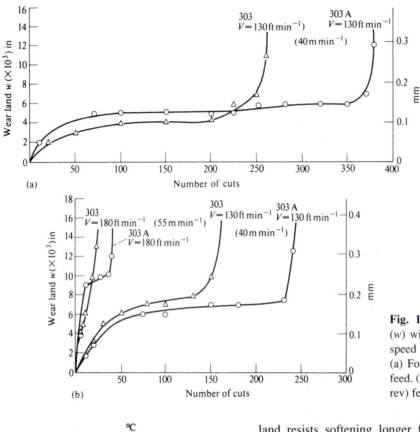

(a)

(b)

Fig. 15.26 Variation of wear land (w) with number of cuts for cutting speed of 130 f.p.m. (39.6 m min^{-1}). (a) For 0.002 in/rev (0.05 mm/rev) feed. (b) For 0.004 in/rev (0.10 mm/rev) feed.

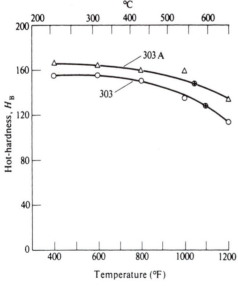

Fig. 15.27 Hot-hardness (Brinell) values versus temperature for 303 and 303A stainless steels.

land resists softening longer for the 303A-material than the 303-material. When thermal softening occurs, the ratio of A_R/A increases abruptly as does the wear rate.

This work suggests that the high wear rates at the beginning and end of a tool life test are due to the fact that the mean pressure on the wear land then exceeds the effective flow stress of the work which is approximately one-third of its indentation hardness. It is thus not the tool that softens when the wear rate increases abruptly at the end of a test but the work.

TITANIUM ALLOYS

Titanium is a member of the tin group of the periodic system. In many ways it is similar to iron. Although sixth in abundance in the Earth's crust, titanium is unusually difficult to convert from the naturally found oxides (TiO_2 and $FeTiO_2$) to the metallic state. Titanium ore is first converted to $TiCl_4$ by treatment with chlorine in conjunction with carbon at high temperature. This in turn is reduced to Ti by treatment with molten magnesium or sodium. The sodium or magnesium chloride is leached from the titanium which is then milled to a fine powder, melted in an arc furnace (m.p. = 1760 °C compared with 1430 °C for structural steel), and cast into ingots. Titanium is a very expensive structural material compared with steel because of the high cost of winning the metal from its ore.

Titanium alloys have a high strength-to-weight ratio and are unusually corrosion resistant. The specific weight of titanium is only $\frac{2}{3}$ that of steel and only 60% greater than aluminum. Titanium has a strong tendency to react with oxygen, nitrogen, carbon, and halogens particularly at high temperatures. Titanium normally has a hexagonal close packed structure and unusually low thermal conductivity and volume specific heat. Titanium alloys have about the same strain hardening tendency as ordinary structural steels.

While the coefficient of friction of two identical metals in dry sliding contact is normally about one, dissimilar metals will have a coefficient of friction of about 0.2. When steel slides over a titanium surface,

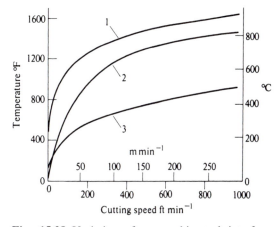

Fig. 15.28 Variation of mean chip–tool interface temperature with cutting speed for (1) titanium alloy, (2) 18–8 stainless steel, (3) AISI 1113 steel. Turning with K2S carbide tool with rake angle of 0° except for item (3), which has rake angle of 20°. Undeformed chip thickness = 0.0023 in (0.06 mm).

the initial coefficient of friction is about 0.2 but this soon rises to a value of about 0.45. This is due to the formation of a built-up layer of titanium oxide on the steel surface resulting in titanium oxide sliding on titanium oxide instead of the dissimilar pair—steel on titanium. Boundary and extreme boundary lubricants that perform well on steel are relatively ineffective on titanium surfaces. This is due to the strong tendency of titanium to form an oxide and the poor tendency for lubricants to adsorb or react with the oxidized surface. For this reason, titanium surfaces are very difficult to lubricate.

Titanium alloys are among the most troublesome materials to machine at practical cutting speeds. The greatest difficulty in machining these alloys stems from the very high cutting temperatures experienced under conditions that are ordinary for most other materials. The extent to which cutting temperatures for titanium alloys are found to exceed those for other metals is shown in Fig. 15.28. The fact that the stainless steel temperatures are greater than those for the AISI 1113 steel may be attributed to the greater specific energy (u) obtained when stainless steel is cut. However, the still-greater titanium temperatures cannot be explained on the basis of a greater specific energy alone.

The very high tool temperatures experienced when machining titanium have been shown to be due to the very low $k\rho C$ value for titanium (Shaw, 1958). When different materials are machined with the same tool, cutting fluid, and feed rate, Eq. (12.42) suggests that

$$\bar{\theta}_t \sim u \left(\frac{Vt}{k\rho C} \right)^{1/2} \tag{15.1}$$

The speed required to give the same tool temperature will be

$$V_\theta \sim \frac{k\rho C}{u^2} \tag{15.2}$$

TABLE 15.6   Relative Cutting Speeds for Equal Tool Temperature for Different Work Materials

Work Material	u in lb in^{-3} (10^{-6})	$k\rho C$ (BTU in^{-2} °F^{-1})^2 × s^{-1} × 10^6	$\dfrac{V}{V_{St}} = \left(\dfrac{k\rho C}{k\rho C_{St}}\right)\left(\dfrac{u_{St}}{u}\right)^2$
1020 steel	0.3	27	1.0
75ST aluminum	0.1	34	11.3
140A titanium	0.3	6.4	0.24

Table 15.6 gives approximate values of u and $k\rho C$ for steel, aluminum, and a titanium alloy as well as values for the ratio of

$$\frac{V}{V_{St}} = \frac{k\rho C}{k\rho C_{St}}\left(\frac{u_{St}}{u}\right)^2 \tag{15.3}$$

From this comparison it is obvious why aluminum may be machined at very high speeds relative to steel and why titanium must be machined at very much lower speeds than steels.

When a titanium alloy is machined at a speed of about one-fourth the satisfactory speed for steel of about the same hardness, comparable tool life results are obtained. However, the rate of removal (Vbt) will then be only one-fourth that for steel. The basic reason for this is the relatively low $k\rho C$ for titanium. As long as this is the case, it is fruitless to try to machine titanium at speeds that are comparable to those used for steel. At the same time the comparison given in Table 15.6 suggests that high-speed machining efforts, where economical should concentrate on aluminum alloys and other materials having high ($k\rho C/u^2$) values.

One of the characteristics of titanium is to give very inhomogeneous chips relative to strain. This is due to two reasons as illustrated in Fig. 15.29. Figure 15.29a shows a titanium chip produced at very low speed. This chip fractures periodically, and block-wise sliding occurs followed by rewelding as a new fracture occurs. The result is a very inhomogeneous chip with a saw toothed

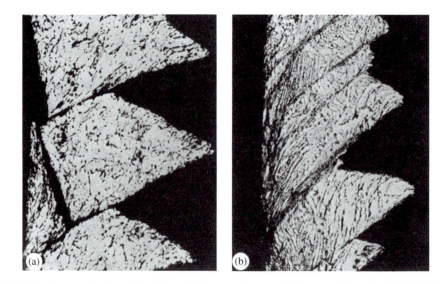

Fig. 15.29 Continuous titanium alloy chips. (a) Low cutting speed (1 i.p.m = 25 mm/min). (b) High cutting speed (175 f.p.m. = 53 m/min).

back. Figure 15.29b shows a titanium chip produced at relatively high cutting speed (175 f.p.m., 53 m min⁻¹). It is seen that this chip does not involve gross block-wise sliding and rewelding but instead bands of very concentrated shear develop. This is called adiabatic shear since the heat associated with initial slip causes thermal softening and hence more shear strain and more deformation, etc., until the shear band jumps to the next point of weakness. The concentrated shear does not extend clear along the chip but is concentrated in a rather narrow band. The result is again a very inhomogeneous sawtoothed chip but for a different reason than in Fig. 15.29a.

Figure 15.30 shows a titanium chip produced at the intermediate speed of 50 f.p.m. (15 m min⁻¹). This chip is a continuous ribbon that is extremely homogeneous relative to shear strain and relatively smooth on the back surface.

It is thus evident that a wide variety of chip types relative to the homogeneity of strain may be obtained with titanium. The adiabatic shear type of chip is more likely to be observed when machining titanium than with other metals due to the unusually low value of $k\rho C$ for titanium.

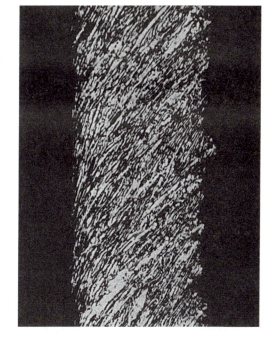

Fig. 15.30 Continuous titanium alloy chip showing homogeneous shear. Cutting speed, 50 f.p.m. (15 m min⁻¹).

NICKEL

Nickel is a member of the platinum group of the periodic system. Nickel-base alloys are refractory materials used in the gas-turbine industry. Nickel is also an important ingredient of stainless steels where it stabilizes austenite, decreases the rate of strain hardening, provides a negative thermal coefficient of friction, and decreases the tendency for the friction force to decrease with rise in temperature. Nickel is about as abundant in the Earth's crust as copper (Ni is 26th in abundance). Nickel is not particularly difficult to reduce from sulfide and silicate ores and is moderate in cost.

Nickel has a somewhat higher specific weight than iron [0.322 lb in⁻³ compared with 0.284 for steel (0.087 N cm⁻³ compared with 0.076 N cm⁻³ for steel)]. Nickel has a face-centered cubic structure and a melting point of 2650 °F (1454 °C) compared with 2793 °F (1534 °C) for iron.

Nickel and nickel-base alloys are relatively difficult to machine because of high tool-tip temperatures. This is not apparently due to unusual strength or hardness (pure nickel has about the same hardness as ingot iron $\cong$ 85 kg mm⁻² Brinell hardness).

Smart and Trent (1975) have found experimentally that when machining nickel, the highest tool-face temperature is at the cutting tool tip. Figure 15.31 shows isotherms obtained when machining steel and nickel. These were obtained by the thermal-softening method of Wright and Trent (1973) described in Chapter 12 and clearly show that whereas the maximum temperature occurs above halfway along the contact length for steel, the peak temperature occurs at the tool point or even on the clearance face when similarly machining nickel. This is apparently due to the fact that instead of the chip heating up as it passes along the tool face, it is cooled. Childs (1978) has pointed out that when this occurs the partition coefficient (R_2) between chip and tool will be negative and has suggested an inequality that determines when this will be so. The following derivation produces the equivalent of the Childs inequality to produce a negative R_2 (condition for cooling of chip as it passes over tool face) in terms of the notation used here.

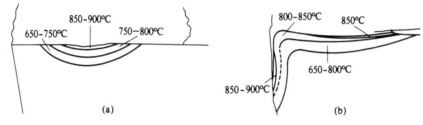

Fig. 15.31 Sections through HSS tools used at ultra-high speed for a short time (30 s) to reveal isotherms by thermal softening. (a) Low-carbon iron cut at 500 f.p.m. (152 m min⁻¹). (b) CP nickel cut at 175 f.p.m. (53 m min⁻¹). Feed, 0.010 i.p.r. (0.25 mm/rev); cutting fluid, none. (after Smart and Trent, 1975)

From Eq. (12.34), for R_2 to be negative

$$\frac{u_F V t \bar{A}}{J k_3} + \theta_o' < \bar{\theta}_S \tag{15.4}$$

Substituting from Eq. (12.22) for $\bar{\theta}_S$:

$$\frac{u_F V t \bar{A}}{J k_3} + \theta_o' < \frac{R_1 u_S}{J (C\rho)_1} + \theta_o \tag{15.5}$$

Ignoring the two ambient temperatures θ_o' (tool) and θ_o (work) and rearranging the inequality we have for negative R_2

$$\frac{u_F}{u_S} \frac{V t \bar{A}}{R_1} \frac{(C\rho)_1}{k_3} < 1 \tag{15.6}$$

For ingot iron cutting with a chip of the same proportion m/l or $\bar{A}$ as one for nickel, (Vt) will be somewhat larger for iron than for nickel but so will the value of R_1 (see Fig. 12.23). Therefore, the quantity $(Vt\bar{A}/R_1)$ will be about the same for iron and nickel. The value of $(C\rho)$ for nickel will be only about 15% greater than for steel and hence when cutting with the same tool material (same k_3), the quantity $(C\rho)_1/k_3$ will be about the same for iron and nickel. Therefore, the unusual tendency for nickel to give a negative R_2 would appear to lie in a tendency for u_S to be large relative to u_F. It may further be shown that to a relatively good approximation

$$\frac{u_F}{u_S} \cong r \tag{15.7}$$

Thus, nickel would appear to be difficult to machine because of a tendency for the maximum tool-face temperature to be at the very tip of the tool which in turn is due to the combination of unusually low value of coefficient of tool-face friction or cutting ratio (r) or both.

If Eq. (12.45) is used for $\bar{\theta}_S$ instead of Eq. (12.22), it still follows that R_2 will tend to become negative as u_F/u_S becomes smaller.

It is also evident from Eq. (15.6) that a low value of tool conductivity (k_2) will tend to offset a low value of u_F/u_S. Table 15.7 gives values of k_2 for a number of tool materials. This explains in part why HSS tools (low k_3) often give unexpectedly good results when machining nickel-base alloys. A more refractory tool material such as polycrystalline SiC should be even more effective than HSS provided it may be prevented from chipping. A high-cobalt-containing tungsten carbide

TABLE 15.7 Thermal Conductivities of Cutting-Tool Materials

Tool Material	k (BTU in^{-2} s^{-1} ($°$F in^{-1})$^{-1}$ × 10^4)	Relative Values
T–1 HSS (18–4–1)	3.7	2.3
Steel grade carbides[†]	3.6 to 7.8	2.3 to 4.9
C 1 grade carbides[‡]	7.3 to 12	4.6 to 7.5
Silicon carbide	1.6	1

[†] The higher the TiC and TaC content, the lower the k.
[‡] The higher the Co content, the lower the k.

containing relatively large amounts of TiC and TaC should be the most effective cemented carbide type for machining refractory nickel-base alloys.

Practical considerations in the machining of nickel-base superalloys are discussed in Koenig and Gerschwiler (1999).

MACHINING HIGH-STRENGTH MATERIALS

The trend in the aircraft industry during the past twenty-five years has been toward materials of construction that

1. are of greater strength
2. are of lighter weight
3. are resistant to oxidation, particularly at high temperatures
4. exhibit small deformations at high temperatures
5. are not brittle at low temperatures

The materials that have been introduced to meet these requirements have generally been more difficult to machine.

While an ever-increasing quantity of ordinary structural materials will be machined in mass-production industries in the future, an increasing percentage of the productive effort will involve the processing of less conventional materials. This section is concerned with some of the basic principles associated with the machining of materials that might be classified as difficult to machine for one reason or another. Such materials are sometimes also referred to as space-age materials; or high-strength, temperature-resistant (HSTR) materials; or corrosion and oxidation resistant materials.

In approaching this problem it is difficult to know which materials to include. It has been decided to consider only materials that have been used in actual engineering structures and whose usefulness has been demonstrated. The materials to be considered may be classified as follows:

1. low-alloy and hot-work die steels
2. stainless steels
3. light-weight metals
4. high-temperature alloys
5. refractory metals

There are many materials and alloys that could be discussed under each of these headings. However, at the risk of over-simplification only a few representative examples from each category will be considered. Table 15.8 gives the chemical compositions of a number of representative materials.

TABLE 15.8 Chemical Compositions of a Variety of Materials[†]

Designation	

Designation	C	Mn	Si	Cr	Ni	Co	Mo	W	Cb	Ti	Al	Fe	Other
Alloy steels													
AISI 4340	0.4	0.7	0.3	0.8	1.8	–	0.2	–	–	–	–	Bal	–
AISI 610 (Mod H–11)	0.4	0.4	1.0	5	–	–	1.4	–	–	–	–	Bal	0.5V
Stainless steels													
Austenitic													
AISI 302 (18–9)	0.15	2	1	18	9	–	–	–	–	–	–	Bal	–
Martensitic													
AISI 420	0.15	1	1	13	–	–	–	–	–	–	–	Bal	–
Ferritic													
AISI 405	0.08	1	1	13	–	–	–	–	–	–	0.2	Bal	–
Precipitation-hardening													
AISI 631 (17–7 PH)	0.07	0.5	0.3	17	7	–	–	–	–	–	1	Bal	–
Light-weight metals													
Ti–8Al–1Mo–1V							1			Bal	8		IV
Zirconium													100Zr
Beryllium													100Be
High-temperature alloys													
Iron-base													
AISI 662 (Discaloy)	0.04	0.9	0.8	13.5	26		2.75			1.75	0.07	Bal	–
Nickel-base													
AISI 685 (Waspaloy)	0.07	0.1	0.1	20	Bal	13.5	4.45	–	–	3	1.4	0.75	–
Cobalt-base													
AISI 671 (S 816)	0.42	1	0.45	20	20	43.5	4	4	4	–	–	–	–
Refractory metals													
Columbium									100				
Molybdenum							100						
Tatalum													100Ta
Tungsten								100					

[†] After Shaw and Nakayama, 1967.

Next in importance to chemical composition is workpiece structure and hardness. The alloy steels may be hardened to strengths in the vicinity of 300,000 lb in^{-2} (210 kg mm^{-2}) and hardness levels of H_B 550. To machine such materials is obviously a very challenging problem which has not been completely solved. Some of the difficult-to-machine materials have low thermal conductivity or low specific heat, both of which give rise to higher cutting temperatures. Other materials are brittle, tend to strain harden badly, have a low Young's modulus of elasticity, or tend to weld to ordinary tool materials. A high coefficient of thermal expansion makes it difficult to maintain dimensional accuracy. A few of the physical properties of the materials of Table 15.8 are given in Table 15.9.

The principal machining characteristics of these difficult-to-machine materials are also encountered when machining ordinary metals, but usually to a lesser degree. Each class of materials has its particular combination of difficulties.

The alloy steels must often be machined after heat treatment; this imposes very large forces on cutting tools, and may result in tool fracture and high cutting temperatures. Feeds and speeds

TABLE 15.9 Mechanical and Thermal Properties of a Variety of Materials[†]

Material	σ_u kg cm^{-2} ($\times 10^4$)	σ_y kg cm^{-2} ($\times 10^4$)	$E \times 10^{-6}$ kg cm^{-2}	k cal cm^{-2} cm^{-1} °C^{-1} s^{-1}	ρ g cm^{-3}	ρC cal cm^{-3} °C^{-1}	$k\rho C$ cal^2 cm^{-4} (°C)$^{-2}$ s^{-1}	α °C^{-1} $\times 10^6$	m.p. °C
Alloy steels									
AISI 4340	2.02	1.80	2.1	0.089	7.8	0.83	0.074	11.3	1500
AISI 610	2.18	1.69	2.15	0.069	7.8	0.90	0.062	11.9	
Stainless steels									
Austenitic									
AISI 302	0.63	0.25	1.95	0.039	7.8	0.94	0.037	17.3	1400
Martensitic									
AISI 420	1.75	1.41	2.0	0.060	7.8	0.86	0.052	10.3	
Ferritic									
AISI 405	0.70	0.62	2.0	0.064	7.8	0.86	0.055	10.8	1480
Precipitation-hardening									
AISI 631	1.86	1.83	2.1	0.040	7.8	0.94	0.038	11.2	1400
Light-weight alloys									
Ti–8A1–1Mo–IV	1.12	1.05	1.1	0.017	4.5	0.59	0.010	8.5	1760
Zr	0.56	0.49	0.96	0.039	6.5	0.44	0.017	5.85	1850
Be	0.35	0.53	3.1	0.35	1.8	0.81	0.28	15.0	1280
High-temperature alloys									
Iron-base									
AISI 662	1.02	0.75	1.99	0.038	8.1	0.97	0.037	15.6	
Nickel-base									
AISI 685	1.34	0.91	2.15	0.022	8.23	1.02	0.023	12.2	
Cobalt-base									
AISI 671	0.98	0.46	2.48	0.029	8.68	0.77	0.022	13.4	
Refractory metals									
Columbium	0.28	0.21	1.05	0.12	8.6	0.60	0.072	7.4	2468
Molybdenum	0.67	0.53	3.30	0.32	10.2	0.67	0.21	5.9	2610
Tantalum	0.37	0.28	1.90	0.13	16.6	0.56	0.073	6.5	2996
Tungsten	0.84	–	3.66	0.28	19.3	0.64	0.18	4.6	3410

Notes: σ_u = ultimate tensile strength; σ_y = yield stress; E = Young's modulus of elasticity; k = thermal conductivity; ρ = specific weight; ρC = volume specific heat; $k\rho C$ = cal^2 cm^{-4} °C^{-1} s^{-1}; α = coefficient of linear expansion; m.p. = melting point. (after Shaw and Nakayama, 1967)

must normally be reduced and tools that are very refractory are called for. Ceramic tools have been particularly successful for this class of work. Hard materials call for rigid machine tools that are stable when operating at high energy levels.

The austenitic stainless steels have somewhat lower conductivities, and this causes temperatures to be higher than for ordinary steels. These materials also tend to strain harden appreciably and may even transform to martensite while being cut. This results in relatively high values of cutting force unless the strain in the chip can be reduced. Tools of relatively large rake angle and low tool–chip friction are called for. Going hand in hand with large strain hardening is built-up edge (BUE) formation and poor finish. The machinability of austenitic stainless steels decreases markedly with carbon content.

Another problem associated with the austenitic stainless steels is the tendency to leave a disturbed subsurface layer and residual stresses in the finished surface. If a second cut must be taken, it is then necessary not to deform previously strain hardened material. In order that the weakest portion of the tool (the tip) will not have to engage strain hardened materials, it is important that the undeformed chip thickness exceed the depth of the disturbed layer. The chips produced when machining austenitic stainless steel are ductile, and chip disposal or chip breaking is apt to be a problem. The large thermal coefficient of expansion and the dimensional change with phase-transformation makes it difficult to machine to close tolerances. An increased rake angle helps both of these problems.

Best results are therefore obtained with austenitic steels when they are machined at moderate values of feed using tools of high rake angle and having a low tendency to weld to stainless steel.

Ceramic tools are not well suited to these materials. Cast-iron grades of carbide (straight tungsten carbide) give best results and the tool hardness should be as high as possible, consistent with edge chipping, in order to decrease the tendency for weld formation.

The martensitic and ferritic stainless steels do not have as great a tendency to strain harden, and their machining characteristics are similar to those of AISI 4340 or a hot-work die steel (AISI 610) at the same hardness level. The precipitation-hardened stainless steels can be austenitic or martensitic. The AISI 631 variety considered here is normally partly austenitic when used. However, by heat treatment it may be machined in either the austenitic or martensitic states. If machined in the austenitic state, it behaves like an austenitic stainless steel and strain hardens greatly. If machined in the martensitic state, it behaves like a martensitic stainless steel, and the problems are associated with high cutting forces as with a hardened alloy steel. AISI 631 is best machined in the semi-austenitic state and under such a condition is relatively easy to machine.

The titanium alloys are difficult to machine for several reasons:

1. low thermal conductivity and specific heat gives high cutting temperatures
2. strong tendency for chips to weld to tool and cause attritious wear
3. low strain in the chip which yields an unusually high shear angle and hence a high chip velocity and small contact length between chip and tool
4. low value of Young's modulus

The strength and strain hardening characteristics of titanium alloys are about the same as those for steel of the same hardness. Hence, high cutting forces are not a major problem. The main problem is associated with an unusually high cutting temperature due to items 1 and 3 above. The best solution is to use a good coolant and to lower the cutting speed. A water-base fluid is preferable due to the tendency for titanium chips to burn.

Despite the strong tendency for welding, a large unstable BUE (such as is formed with steel at low speed) is not produced. Surface finish at low speed is thus not a problem. Clearance angles should be 10° instead of 5° due to the welding tendency when titanium is machined.

Ceramic tools form particularly strong bonds to titanium and hence have a poor life. Cast-iron-cutting grades of carbide generally preform better than grades containing TiC, since they tend to weld less to the work material.

Titanium alloys are relatively expensive and hence trial-and-error experimental work is costly.

The less refractory titanium alloys tend to form chips with inhomogeneous strain. Due to the poor thermal properties, the heat developed on a given shear plane is not dissipated and softens the metal causing further shear deformation at the same point. With the less refractory alloys (such as commercially pure titanium), this procedure can continue to the point where the chip consists of alternate bands of high and low strain. This, of course, gives rise to a periodic fluctuation in cutting force and can lead to a forced vibration.

Pure titanium has a hexagonal close packed (HCP) structure at room temperature. At 890 °C an allotropic transformation occurs where the low-temperature HCP structure (α) changes to a body centred cubic structure (β) similar to that found in iron at room temperature. Titanium alloys may be all α, all β, or a mixture of the two. The all-β-alloys are generally more difficult to machine than the mixed structure, and all alloys are more difficult to machine than the commercially pure material.

Zirconium is next to titanium in the periodic table and has essentially the same machining characteristics as titanium.

Beryllium has an unusually low density (0.065 lb in^{-3} or 1.80 g cm^{-3}) for its Young's modulus (44 × 10^6 lb in^{-2} or 31 × 10^3 kg mm^{-2}) and tensile strength and therefore is sometimes used where a high strength-to-weight ratio is desired. Its main characteristics of importance to machining are its toxicity and brittleness due to its hexagonal close packed structure. Machining dust of beryllium is extremely harmful if ingested into the lungs and special ventilation and clothing procedures must be followed. In addition, the metal is expensive.

The brittleness of beryllium gives rise to discontinuous cutting and hence a tendency for chatter. There is also a tendency for sub-surface cracks to develop in machined beryllium surfaces which may be highly undesirable for certain space applications. Otherwise beryllium is not particularly difficult to machine and can be turned at 200–300 f.p.m. (60–90 m min^{-1}) using ordinary carbide tools.

Austenitic stainless steels are oxidation resistant only to about 1200 °F (650 °C). For higher operating temperatures the so-called high-temperature alloys must be used. The nickel-base alloys, which are presently the most stable against oxidation, may be used to temperatures of 1850 °F (1010 °C). In addition to stability against oxidation at high temperatures the high-temperature alloy must have high structural strength at elevated temperatures. The cobalt alloys exhibit the greatest strength at operating temperatures in the vicinity of 1800 °F (980 °C). The iron-base high-temperature alloys are not as effective as the nickel-base and cobalt-base alloys with respect to oxidation stability and high-temperature strength, but are more easily machined and less expensive.

The iron-base high-temperature alloys are austenitic and hence behave like austenitic stainless steel in machining but to an even greater degree. Therefore, essentially all of the discussion pertaining to austenitic stainless steels applies to these materials as well. The best structure for machining is one produced by annealing, cold-working, and finally stress-relieving. This structure machines better than either an annealed or solution-treated material, since it strain hardens to a far smaller degree than the former and is less hard than the latter.

Work-hardening, welding and BUE formation, and low thermal properties are the chief difficulties when machining the high-temperature alloys.

High rake angles should be used due to the strong tendency for strain hardening. Since larger rake angles may generally be used with HSS tools than with carbide, HSS tools frequently give the best results when machining high-temperature alloys. In addition, low values of speed and feed should be used, and machine tools should be as rigid as possible. Cutting fluids containing active chlorine and sulfur generally aid in the machining of high-temperature alloys. As the maximum operating temperature of a high-temperature alloy rises, the difficulty of machining increases. The nickel-base alloys presently have the highest operating temperatures, and they are among the worst materials to machine. They tend to strain harden very badly and give rise to very large values of shear stress at machining strain rates. Although all materials tend to strain harden in the hot working region when strained at high rates, this is particularly true of the nickel and cobalt high-temperature alloys. This strong tendency to strain harden gives rise to deep grooves at both edges of a chip. The nickel-base alloys machine best in the solution-treated state, since they then strain harden less, but this best structure is far from satisfactory. Also, wrought materials tend to machine better than cast structures. The main difficulties in machining the nickel- and cobalt-base materials are due to

1. tendency for maximum tool-face temperature to be close to tool tip (particularly for Ni)
2. high work-hardening rates at machining strain rates leading to high machining forces
3. strong tendency to weld to the tool and to form a BUE
4. low thermal properties leading to high temperatures

The cobalt high-temperature alloys have machining characteristics that parallel the nickel-base alloys but are not quite so difficult.

The high-temperature alloys and high-strength steels should not be machined with tools having flank wear values that are as high as those normally employed in cutting steel. Tools should be taken from service when flank wear reaches 0.25–0.38 mm instead of the usual 0.75 mm.

The refractory metals (Cb, Ta, Mo, W) generally have melting points in excess of 2200 °C. All four metals have low values of volume specific heat. In addition, columbium and tantalum have low values of thermal conductivity. This results in high machining temperatures for all four materials, and particularly for columbium and tantalum. Columbium and tantalum are quite ductile and give rise to a large and troublesome BUE. They are usually not machined with carbide tools since the speeds employed are usually not high enough to avoid BUE formation.

Molybdenum and tungsten on the other hand are brittle and give rise to discontinuous chips. Like beryllium it is difficult to machine these materials without leaving small cracks in the finished surface. HSS tools are usually not used for machining molybdenum or tungsten.

Tungsten which is the most refractory of the group is used for rocket nozzles. Additions of 10% silver infiltrated into a powder compact of tungsten greatly improves its machinability and at the same time cools the surface of the nozzle in use by ablation of the silver. Tungsten has been successfully machined at low temperatures (−40 °F, −40 °C) which tends to offset its tendency to give high tool-face temperatures due to its high strength and poor thermal properties. It has been discovered that dispersion of 2% thorium oxide in a tungsten matrix triples the machinability of tungsten without changing its refractory character.

Specific recommendations concerning feeds and speeds for use in machining high-strength materials are to be found in many publications, and hence no attempt will be made here to give more than a general impression concerning the relative machinability of the various materials. The results obtained in any particular case depend not only on the chemistry of the workpiece but also on its structure (hardness), the tool materials and geometry used, and the rigidity and condition of the machine tool. Tables of recommended operating conditions should always be looked on as being approximate and as points of departure for further study.

Table 15.10 gives optimistic values of cutting speed for turning and drilling. The turning values are based on a depth of cut of 0.25 mm and a feed of 0.25 mm/rev with the exception of the values for Cb, Ta, and Mo which are for a depth of cut of 1.25 mm and a feed of 0.125 mm/rev. The drilling speeds are for high-speed steel drills (M–2 or M–10) having a feed consistent with

TABLE 15.10 Recommended Machining Speeds for a Variety of Materials

Material	Brinell Hardness	Turning Speed HSS f.p.m.	Turning Speed HSS m min^{-1}	Turning Speed Carbide f.p.m.	Turning Speed Carbide m min^{-1}	Drilling Speed f.p.m.	Drilling Speed m min^{-1}
Alloy steels							
AISI 4340	300	80	25	500	165	70	23
AISI 610							
Stainless steels							
Austenitic							
AISI 302	160	95	30	300	100	30	10
Martensitic and precipitation-hardening							
AISI 420							
AISI 631	200	100	35	450	150	30	25
Light-weight metals							
Ti–8A1–1Mo–1V	320	50	15	150	50	40	13
Zirconium		200	65	450	150	80	25
Beryllium		90	30	300	100	40	13
High-temperature alloys							
Iron-base							
AISI 662	200	40	13	125	40	20	6
Nickel-base							
AISI 685	250	15	5	40	13	15	5
Cobalt-base							
AISI 671	200	25	8	75	25	30	10
Refractory metals							
Columbium	150	50	15	–	–	75	25
Molybdenum	200	–	–	300	100	80	25
Tantalum	150	50	15	–	–	40	13
Tungsten	250	–	–	200	65	150[†]	50

[†] Tungsten carbide.

the drill diameter from the point of view of chip disposal. For this class of materials the feed should be about 1% of the drill diameter (i.e., a 0.2 in drill should have a feed of about 0.002 i.p.r. or a 5 mm drill should have a feed of 0.05 mm/rev). Drills made of T–15 will yield results that are about 30% greater than those for the more usual high speed steels (M–2 or M–10).

The expected tool life in Table 15.10 is about $\frac{1}{2}$ hr in turning, or 50 holes of two-diameters depth in drilling.

The amount of nickel in a nickel-base high-temperature alloy is very important. For example, the alloy of Table 15.10 has a nickel content of about 60% and a recommended carbide tool speed of 13 m min^{-1}. This speed would be increased to about 20 m min^{-1} for work containing about 50% Ni and to 26 m min^{-1} for work of 45% nickel. This shows the extremely important role of work-piece chemistry, which is the most important variable relative to machinability.

The next most important variable is hardness. High-speed steel tools should not be used to machine martensitic low-alloy steels or hot-work die steels at hardness levels above H_B 450. For very hard steels (H_B 550) the feed and depth of cut should be reduced to 0.125 mm and 0.0125 mm/rev, respectively, and the carbide cutting speed dropped to about 50 m min^{-1}. In general as workpiece hardness increases the feed and depth of cut should be reduced to avoid tool chipping or breakage. Ceramic tools are well suited for turning high-hardness alloy steels at light feeds.

In general, carbide tools are not recommended for use in machining columbium or tantalum due to the strong tendency for these materials to form welds at the low cutting speeds that must be used with these materials. On the other hand high-speed steel tools are not recommended for use in machining molybdenum or tungsten except in drilling. Even in drilling it is highly preferable to use carbide in connection with tungsten. Tapping of all of these materials is normally done at 1.5−8 m min^{-1}.

Machining of materials that are difficult to machine in the aerospace industries is discussed in Noaker (1994).

REFERENCES

Childs, T. H. C. (1978). *Int. J. Wear* **50**, 321.

Cook, N. H., Finnie, I., and Shaw, M. C. (1954). *Trans. Am. Soc. Mech. Engrs.* **76**, 153.

DeBenedictis, K. (1997). *Mfg. Eng.* (Sept.), 48−54.

DeSalvo, G. J., and Shaw, M. C. (1969). In *Advances in Machine Tool Design and Research*. Pergamon Press, Oxford, p. 961.

Ernst, H. (1938). In *Machining of Metals*. Am. Soc. for Metals, Metals Park, Ohio.

Field, M. (1963). In *International Research in Production Engineering*. Am. Soc. Mech. Engrs., New York, p. 188.

Gensamer, M., Pearsall, E. S., Pellini, W. S., and Low, J. R. (1942). *Trans. Am. Soc. Metals* **30**, 1003.

Japanese Working Group on Machinability. (1968). *J. Jap. Soc. Precs. Engrs.* **34**, 680 (1st report); **35**, 169 (second report, 1969); **35**, 227 (third report, 1969).

Koenig, W., and Gerschwiler, K. (1999). *Mfg. Eng.* (Mar.), 102−106.

Machining Data Handbook, 3rd ed. (1980). Compiled by U.S. Machinability Data Center. Available from Metcut Research Associates, Inc., Cincinnati, Ohio.

Metals Handbook, 9th ed. "M−1 of *Properties and Selection of Iron and Steel*" (1978), "M−2 *of Nonferrous Metals*" (1979), "M−3 *of Stainless Steel and Special Metals*" (1980). Am. Soc. for Metals, Metals Park, Ohio.

Noaker, P. M. (1994). *Mfg. Eng.* (Oct.), 47−50.

Opitz, H. (1963). In *International Research in Production Engineering*. American Society of Mechanical Engineers, New York, p. 107.

Opitz, H., and Koenig, W. (1965). *Ind. Anz.* **87**, 26 (part I); **87**, 43 (part II); **87**, 51 (part III).

Schaller, E. (1962). Dissertation. T. H. Aachen.

Schwerd, F. (1932). *Z. Ver. Deutscher Ing.* **76**, 1257.

Sejournet, J. (1954). *Engineering* **177**, 463.

Sejournet, J. (1962). *Lub. Engng.* **18**, 324.

Shaw, M. C. (1958). *Technische Mitteilungen*. Essen.

Shaw, M. C. (1967). *Machinability*. Report 94. The Iron and Steel Institute, London, p. 1.

Shaw, M. C., and Macks, E. F. (1949). *Analysis and Lubrication of Bearings*. McGraw-Hill, New York.

Shaw, M. C., and Nakayama, K. (1967). *Annals of CIRP* **15**, 45. International Institution for Production Engineering Research, Paris.

Shaw, M. C., Smith, P. A., and Cook, N. H. (1961). *Trans. Am. Soc. Mech. Engrs.* **83**, 163.

Shaw, M. C., Usui, E., and Smith, P. A. (1961). *Trans. Am. Soc. Mech. Engrs.* **83**, 181.

Smart, E. F., and Trent, E. M. (1975). *Int. J. Prod. Res.* **13**, 265.

Sturney, A. C. (1925). *Proc. Instn. Mech. Engrs.* **1**, 141.

Tipnis, V. A., and Joseph, R. A. (1971). *J. Eng. Ind., Trans. ASME* **93**, 571.

Trent, E. M. (1963). In *International Research in Production Engineering*. ASME, New York, p. 161.

Vilenski, D., and Shaw, M. C. (1970). *Annals of CIRP* **18**, 623.

Wright, P. K. (1976). *J. Aust. Inst. Metals* **21**, 34.

Wright, P. K., and Trent, E. M. (1973). *J. Iron Steel Inst.* **211**, 364.

16 COMPLEX TOOLS

Thus far only two-dimensional cutting operations have been considered in which the cutting edge is perpendicular to the velocity vector. Such cutting operations are referred to as orthogonal. While most commercial tools are three-dimensional, there are some practical orthogonal cases. These include some finishing planer cuts, surface broaching, the lathe cut-off operation, and some plain-milling operations. Representative three-dimensional cutting tools are a conventional lathe tool, a face-milling cutter, and a twist drill. The geometrical aspects of these more complex tools are considered in this chapter.

The simplest three-dimensional cutting tool is a straight cutting edge that is inclined to the velocity vector (see Fig. 16.1b). Inclination i is the distinguishing feature of all three-dimensional cutting operations and represents the point of departure from orthogonal cutting. Inclination significantly alters the chip flow direction and hence the performance of a tool.

Two distinct rake angles are frequently distinguished:

1. The normal rake angle α_n is the angle measured from a normal to the finished surface in a plane perpendicular to the cutting edge (plane OA in Fig. 16.2). This is the angle most frequently specified and most easily measured, but it is not the rake angle of greatest significance. With the clearance angle, α_n determines the amount of material available at the tool point to support the load and absorb the heat generated during cutting. A greater included angle at the tool point (smaller α_n) is generally required for cutting edges subjected to shock (i.e., milling or other interrupted cuts).

2. The velocity rake angle α_v unfortunately sometimes referred to as the true rake angle, is the angle measured from a normal to the finished surface in a plane containing the cutting-velocity vector (plane OB in Fig. 16.2). This particular rake angle is insignificant and is mentioned here only because it has received so much attention from previous investigators.

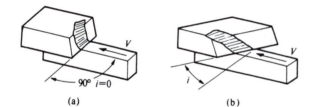

(a)   (b)

Fig. 16.1 Comparison of orthogonal and oblique cutting operations. (a) Orthogonal cut. (b) Oblique cut with inclination angle i.

Angle α_n may be expressed in terms of α_v as follows:

$$\tan \alpha_v = \frac{\tan \alpha_n}{\cos i} \quad (16.1)$$

While the normal and velocity rake angles are easily measured, they are not of fundamental significance in the cutting process.

In Fig. 16.3 the complement to the rake angle in orthogonal cutting $(90 - \alpha)$ is seen to measure the angle through which the metal is deflected by the tool. In general, the greater this angle, the greater

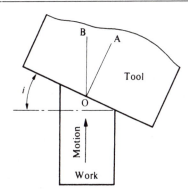

Fig. 16.2 Plan view of cutting tool with inclination i.

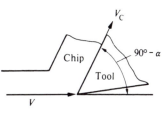

Fig. 16.3 Angle through which metal is deflected in orthogonal cutting $(90 - \alpha)$.

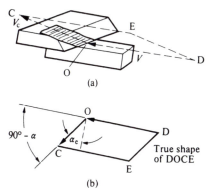

will be the cutting work required. It is unfortunate that the importance of this more natural angle was not originally recognized and the rake angle so defined. In comparing three-dimensional cutting data with orthogonal data, an effective rake angle comparable to that used in the two-dimensional case should be adopted. The effective rake angle for the oblique tool of Fig. 16.4 should be related to the cutting-velocity vector DO and the chip-flow direction OC, and measured in the plane of DO and OC. The effective rake angle is labeled α_e and it is evident in Fig. 16.4 that the metal is deflected through an angle $(90 - \alpha_e)$ as it passes across the oblique tool face. Hence α_e should play the same role in oblique cutting as α does in orthogonal cutting.

Fig. 16.4 Effective rake angle for oblique cutting tool.

CHIP-FLOW DIRECTION

In order to determine the effective rake angle α_e for any tool with inclination, it is necessary to know the direction the chip takes as it crosses the tool face. This is most effectively specified by the angle η_C between the chip-flow direction and the normal to the cutting edge, in the plane of the tool face (Fig. 16.5). Stabler (1951) has shown that when angle η_C is known, the effective rake angle α_e may be determined as follows:

$$\sin \alpha_e = \sin \eta_C \sin i + \cos \eta_C \cos i \sin \alpha_n \quad (16.2)$$

Angle η_C may be directly measured in a number of ways, including

1. directly with a protractor as the chip flows across the tool face

2. from chip-width measurements

The second method has been found the most precise, particularly at higher cutting speeds and hence will be described in detail. In Fig. 16.6 it is evident that for any arbitrary chip-flow direction η_C,

$$\cos \eta_C = \frac{b_C}{b/\cos i} \quad (16.3)$$

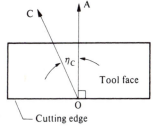

Fig. 16.5 View of tool normal to tool face showing manner of specifying chip flow direction OC.

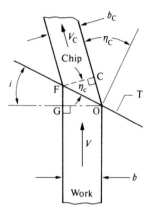

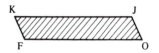

Fig. 16.7 Trapezoidal shape of cross-section of chip in Fig. 16.6.

Fig. 16.6 Plan view of oblique cutting tool.

where b_C is the width of the chip (FC in Fig. 16.6); b is the width of the work (OG in Fig. 16.6). Thus chip flow direction η_C may be determined simply from measurements of chip and work width. In making the chip-width measurement (b_C), it is preferable to use a measuring microscope to determine distance OF on the face of the chip (Fig. 16.7), rather than a micrometer, which will measure horizontal distance OK.

Stabler (1951) reported that angle η_C was approximately equal to the inclination angle i for a variety of tool and work materials, rake angles, and speeds. In the following discussion this relationship will be referred to as Stabler's rule of chip flow. Although Stabler's rule is valid to a first approximation, it has been found that η_C increases relative to inclination angle i when

1. rake angle α_n is decreased

2. a more efficient cutting fluid is used

3. the friction characteristics of the metal cut improve (cold-drawn steel as opposed to soft aluminum for example)

The physical significance of Stabler's rule may be appreciated by reference to Eq. (16.3). Here it is evident that when η_C is equal to i, this is the same as taking chip width b_C the same as work width b. Thus Stabler's rule merely means that the chip will take a direction relative to the cutting edge so that there is no change in width as the metal crosses the cutting edge. Negligible side flow is not only a good approximation in orthogonal cutting, it also represents a first approximation for a tool having inclination.

When Eq. (16.2) is simplified by use of Stabler's rule, Eq. (16.4) is obtained:

$$\sin \alpha_e = \sin^2 i + \cos^2 i \sin \alpha_n \qquad (16.4)$$

This equation is shown graphically in Fig. 16.8.

VELOCITY RELATIONS

Once the chip-flow direction is established, the velocity and strain relations for oblique cutting follow directly from geometry. While most of the equations of this section have been stated correctly by Merchant (1944), they will be briefly reviewed as a matter of convenience. As in orthogonal cutting, there is a shear plane when cutting with an oblique tool. This shear plane will contain the cutting edge

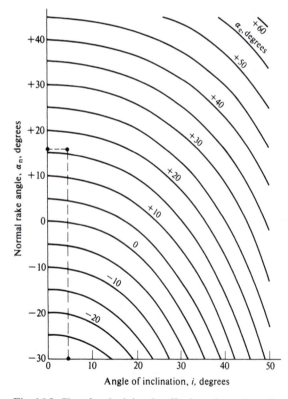

Fig. 16.8 Chart for obtaining the effective rake angle α_e for any combination of inclination i and normal rake α_n angles.

and will rise from the finished surface when proceeding into the workpiece in front of the cutting edge. The direction of the shear plane is most conveniently defined in terms of a normal shear angle (ϕ_n) measured in a plane normal to the cutting edge, i.e., in a vertical plane along OA in Fig. 16.2. In Fig. 16.9 a section view of a partially formed chip is shown in a plane normal to the cutting edge. The angles, ϕ_n and α_n, are as shown and a relationship between these two angles and the chip and workpiece thicknesses may be derived just as in orthogonal cutting (Chapter 3):

$$\tan \phi_n = \frac{(t/t_C) \cos \alpha_n}{1 - (t/t_C) \sin \alpha_n} \tag{16.5}$$

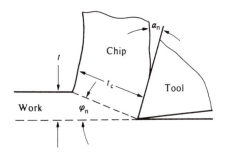

Fig. 16.9 Section of partially formed chip in plane normal to cutting edge.

where t and t_C are the undeformed chip thickness and chip thickness, respectively. If the back of the chip is very rough, it is usually more precise to use a cutting ratio rather than a chip-thickness ratio. This may be done by applying the volume-continuity relationship of plasticity:

$$lbt = l_C b_C t_C \tag{16.6}$$

where l and l_C are corresponding work and chip lengths. From Eq. (16.6),

$$\frac{t}{t_C} = \frac{l_C b_C}{lb} \tag{16.7}$$

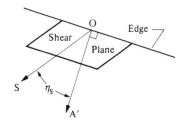

Fig. 16.10 View of shear plane from a direction normal to shear plane showing manner of specifying shear flow direction η_S.

The direction of shear flow is given by a line in the shear plane that makes an angle η_S with a line that lies in the shear plane and is normal to the cutting edge (see Fig. 16.10). This shear-flow angle η_S is

$$\tan \eta_S = \frac{\tan i \cos (\phi_n - \alpha_n) - \tan \eta_C \sin \phi_n}{\cos \alpha_n} \tag{16.8}$$

In orthogonal cutting the shear strain γ is

$$\gamma = \cot \phi + \tan (\phi - \alpha) \tag{16.9}$$

while for an oblique tool this becomes

$$\gamma = \frac{\cot \phi_n + \tan (\phi_n - \alpha_n)}{\cos \eta_S} \tag{16.10}$$

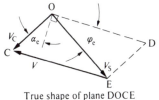

True shape of plane DOCE

Fig. 16.11 Velocity components shown in plane DOCE in Fig. 16.4.

In accordance with the principles of kinematics, the three velocity vectors involved in oblique cutting must form a closed velocity triangle. These three velocities shown in Fig. 16.11 are

1. V, the cutting velocity which is directed along DO in Fig. 16.4a
2. V_C, the chip velocity which is directed along OC in Fig. 16.4a
3. V_S, the shear velocity which must be directed from O toward E in Fig. 16.4a in order that the three vectors form a closed triangle

The relationship between these velocities may be found readily by use of a three-dimensional model of the velocity triangle to give

$$\frac{V_C}{V} = \frac{\cos i \sin \phi_n}{\cos \eta_C \cos (\phi_n - \alpha_n)}$$ (16.11)

$$\frac{V_S}{V} = \frac{\cos i \cos \alpha_n}{\cos \eta_S \cos (\phi_n - \alpha_n)}$$ (16.12)

Just as an effective rake angle α_e was defined previously for an oblique cutting tool, an effective shear angle ϕ_e may be defined similarly (see Fig. 16.11). This effective shear angle lies in the plane DOCE in Fig. 16.4a. From Fig. 16.11,

$$\sin \phi_e = \left(\frac{\cos \eta_S \cos \alpha_e}{\cos \eta_C \cos \alpha_n} \right) \sin \phi_n$$ (16.13)

SHEAR AND FRICTION RELATIONSHIPS

As in orthogonal cutting, the shear and friction processes occur simultaneously and in close proximity. In order that static equilibrium of forces be satisfied it is necessary that

$$R' = R$$ (16.14)

where R' is the resultant force on the shear plane and R is the resultant force on the tool face. In accordance with continuity, Eq. (16.8) must also be satisfied. In order to combine Eqs. (16.8) and (16.14) the relationships between force and velocity for both the shear and friction processes must be known. For this, either of the following may be assumed that

1. force and velocity vectors are collinear for both the friction and shear cases or

2. in either or both cases the vectors are not collinear

If the first assumption (which is intuitively appealing) is adopted and, at the same time Stabler's rule is employed, then Eqs. (16.8) and (16.14) lead directly to the following shear-angle relationship:

$$\phi_n = \Psi(\beta, i, \alpha_n)$$ (16.15)

where Ψ is some function of the friction angle (β), inclination angle (i), and normal rake angle (α_n). However, Eq. (16.15) is not found to be in agreement with experiment. This can only mean that assumption 1 and/or Stabler's rule is invalid.

Since in many cases Stabler's empirical rule is found to represent a good approximation (for example, see Fig. 16.12), the validity of assumption 1 is questioned despite its intuitive appeal.

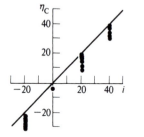

Fig. 16.12 Data illustrating degree of approximation associated with Stabler's rule ($\eta_C = i$). Material, leaded steel; tool, 18–4–1 HSS; cutting speed, 100 f.p.m. (30.5 m min⁻¹); depth of cut, 0.100 in (2.54 mm); back rake, −20, 0, 20, 40; side rake, −20, 0, 20, 40; side-cutting edge, 0; feed, 0.0026, 0.0052, 0.0104 i.p.r. (0.07, 0.18, 0.26 mm/rev).

FORCE RELATIONS

In order to evaluate a three-dimensional cutting operation, it is necessary to determine three mutually perpendicular components of force. Consider the measured force components to be those along the x-, y-, and z-axes of Fig. 16.13. Here y extends along the work surface parallel to cutting velocity V; x is along the work surface normal to V, while z is in the vertical direction. In order to find the component of the resultant force in the plane of the tool face, it is convenient to change to the x', y', z' right-hand orthogonal set in Fig. 16.13 where x' is along the cutting edge, y' is normal to the tool face, and z' is normal to the cutting edge, but in the plane of the tool face. If a_1, b_1, and c_1 are the direction cosines of the x'-axis relative to the x-, y-, z-coordinate system, and a_2, b_2, c_2, and a_3, b_3, c_3, the direction cosines of y' and z', respectively, then the transformation may be readily made as follows:

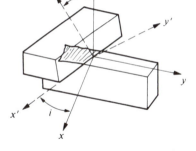

Fig. 16.13 Sketch of inclined cutting tool showing axes of reference.

$$F'_x = a_1 F_x + b_1 F_y + c_1 F_z$$
$$F'_y = a_2 F_x + b_2 F_y + c_2 F_z \qquad (16.16)$$
$$F'_z = a_3 F_x + b_3 F_y + c_3 F_z$$

where

$$
\begin{array}{lll}
a_1 = \cos i & b_1 = -\sin i & c_1 = 0 \\
a_2 = \cos \alpha_n \sin i & b_2 = \cos \alpha_n \cos i & c_2 = \sin \alpha_n \\
a_3 = \sin \alpha_n \sin i & b_3 = -\sin \alpha_n \cos i & c_3 = \cos \alpha_n
\end{array}
$$

If δ_c is the angle that the force in the plane of the tool face F_C makes with the normal to the cutting edge in this plane, i.e., the z'-axis in Fig. 16.13, then

$$\tan \delta_C = \frac{F_{x'}}{F_{z'}} \qquad (16.17)$$

The angle the force in the shear plane F_S makes with the normal to the cutting edge in this plane may be similarly found

$$\tan \delta_S = \frac{-\cos i\, F_x + \sin i\, F_y}{\cos \phi_n \sin i\, F_x + \cos \phi_n \cos i\, F_y + \sin \phi_n\, F_z} \qquad (16.18)$$

where in this case the positive z'-axis is in the direction opposite that used in the derivation of Eq. (16.17).

There are several forces of interest in an oblique cutting process. These include

1. the resultant force acting on the tool face R, or on the shear plane R', ($R = R'$):

$$R^2 = F_x^2 + F_y^2 + F_z^2 \qquad (16.19)$$

2. the component of force along the tool face, F_C:

$$F_C = (F_{x'}^2 + F_{z'}^2)^{1/2} \qquad (16.20)$$

3. the components of force normal to the tool face, F_{Cn}:

$$F_{Cn} = F_{y'} \qquad (16.21)$$

4. the components of force in the tool face that are parallel to the chip-flow direction ($F_C \cos \zeta_C$) and normal to the chip-flow direction ($F_C \sin \zeta_S$), respectively, where $\zeta_C = \eta_C - \delta_C$ and $\zeta_S = \eta_S - \delta_S$

5. the component of force along the shear plane, (F_S):

$$F_S = [(-F_x \cos i + F_y \sin i)^2 + (F_y \cos \phi_n \cos i - F_z \sin \alpha_n)^2]^{1/2} \tag{16.22}$$

6. the component of force normal to the shear plane, F_{Sn}:

$$F_{Sn} = -F_x \sin \phi_n \sin i - F_y \sin \phi_n \cos i + F_z \cos \phi_n \tag{16.23}$$

7. the components of force in the shear plane that are parallel to the shear-flow direction $(F_S \cos \zeta_S)$ and normal to the shear flow direction $(F_S \sin \zeta_S)$, respectively

The coefficient of friction μ is defined as the tangential force divided by the normal force. We must generalize this definition here and refer to the tangential-force component in the direction of relative motion to take care of the fact that the friction force and the velocity vectors are not collinear; thus

$$\mu = \frac{F_C \cos \zeta_C}{F_{Cn}} \tag{16.24}$$

There are two stress components associated with the shear plane that are of interest—a shear component τ and a normal component σ. These may be computed as follows:

$$\tau = \frac{F_S}{bt} \sin \phi_n \cos i \tag{16.25}$$

$$\sigma = \frac{F_{Sn}}{bt} \sin \phi_n \cos i \tag{16.26}$$

Finally, there are three specific energies involved in the oblique cutting process. These are

1. the friction energy per unit volume:

$$u_F = \frac{F_C \cos \zeta_C V_C}{btV} = \frac{F_C}{A_C} \cos \zeta_C \tag{16.27}$$

where A_C is the cross-sectional area of the chip

2. the shear energy per unit volume:

$$u_S = \frac{F_S \cos \zeta_S V_S}{btV} = \tau \gamma \cos \zeta_S \tag{16.28}$$

3. the total energy per unit volume:

$$u = u_F + u_S = \frac{F_y}{by} \tag{16.29}$$

FURTHER ANALYTICAL APPROACH

Usui et al. (1978a, 1978b, 1978c) have presented an analytical cutting model that enables two-dimension orthogonal cutting data to be extended to three-dimensional operations such as oblique turning, plain milling, and groove cutting. This is designed to extend the foregoing approach that involves the concept of effective rake and shear angles by use of plasticity and the finite element method. Possible limitations of this approach to the very high strain deformations involved in metal cutting are discussed in Chapter 20.

EXPERIMENTAL RESULTS

Representative cutting data for oblique cutting tools are given in Table 16.1, and the following observations may be made:

1. The quantity $(\eta_C - i)$ which is a measure of the deviation from Stabler's rule is seen to decrease with increased coefficient of friction or to decrease with decreased shear strain.

2. As would be expected from geometrical consideration, η_S increases whenever η_C increases, but at a lower rate.

TABLE 16.1 Representative Cutting Data for Oblique Cutting Tools

α_n degrees	0	10	30	10	10	10
i degrees	30	30	30	0	15	30
b in (mm)	0.25 (6.35)	0.25 (6.35)	0.25 (6.35)	0.25 (6.35)	0.25 (6.35)	0.25 (6.35)
b_C in (mm)	0.19 (4.83)	0.22 (5.59)	0.24 (6.10)	0.249 (6.32)	0.24 (6.10)	0.22 (5.59)
l in (mm)	3.95 (100.3)	3.95 (100.3)	3.95 (100.3)	3.95 (100.3)	3.95 (100.3)	3.95 (100.3)
l_C in (mm)	1.71 (43.3)	2.30 (58.4)	2.84 (72.1)	1.32 (33.5)	1.96 (49.8)	2.30 (58.4)
F_x lb (N)	−130 (−578.2)	−100 (−444.8)	−90 (−400.3)	0 (0)	−50 (−222.4)	−100 (−444.8)
F_y lb (N)	−330 (−1468)	−270 (−1201)	−200 (−890)	−300 (−1334)	−290 (−1290)	−270 (−1201)
F_z lb (N)	80 (356)	55 (244.6)	20 (89.0)	120 (534)	95 (422.6)	55 (244.6)
V i.p.m. (m min^{-1})	20 (0.51)	20 (0.51)	20 (0.51)	20 (0.51)	20 (0.51)	20 (0.51)
V_C i.p.m. (m min^{-1})	9.56 (0.24)	11.4 (0.29)	13.6 (0.35)	6.74 (0.17)	9.84 (0.25)	11.4 (0.29)
V_S i.p.m. (m min^{-1})	18.6 (0.97)	18.1 (0.46)	15.5 (0.39)	20.0 (0.51)	20.0 (0.51)	18.1 (0.46)
η_C degrees	48.6	40.0	33.3	0	21.2	40.0
δ_C degrees	33.3	25.6	9.9	0	10.9	25.6
ζ_C degrees	15.3	14.4	23.4	0	10.3	14.4
η_S degrees	11.0	8.6	8.2	0	4.7	8.6
δ_S degrees	9.8	12.1	8.6	0	7.1	12.1
ζ_S degrees	1.2	−3.5	−0.4	0	−2.4	−3.5
ϕ_n degrees	18.1	28.5	42.0	19.4	27.0	28.5
ϕ_e degrees	25.3	34.2	44.2	19.4	28.4	34.2
α_e degrees	22.1	25.9	39.6	10	14.5	25.9
F_C lb (N)	95.6 (425)	114 (507)	124 (552)	170 (756)	147 (654)	114 (507)
F_{Cn} lb (N)	−351 (−1561)	−270 (−1201)	−179 (−796)	−274 (−1219)	−271 (−1205)	−270 (−1201)
$F_C \sin \zeta_C$ lb (N)	25.3 (112.5)	28.5 (126.8)	49.4 (219.7)	0 (0)	26.3 (117.0)	28.5 (126.8)
$F_C \cos \zeta_C$ lb (N)	92.2 (410.1)	111 (493.7)	114 (507.1)	170 (756.2)	144 (640.5)	111 (493.7)
F_S lb (N)	313 (1392)	236 (1050)	151 (671.7)	244 (1085)	219 (974.1)	236 (1050)
F_{Sn} lb (N)	185 (822.9)	184 (818.4)	180 (800.6)	213 (947.4)	218 (970.0)	184 (818.4)
μ	0.26	0.41	0.64	0.62	0.53	0.41
σ p.s.i. ($\times 10^{-3}$) (MPa)	39.8 (274)	62.7 (432)	83.5 (576)	56.7 (391)	82.5 (569)	62.7 (432)
τ p.s.i. ($\times 10^{-3}$) (MPa)	67.2 (463)	80.7 (556)	70.0 (483)	65 (448)	77 (531)	80.7 (556)
γ	3.44	2.20	1.33	2.98	2.28	2.20
u_F p.s.i. ($\times 10^{-3}$) (MPa)	35.8 (247)	50.8 (350)	62.2 (429)	46.2 (319)	57.0 (393)	50.8 (371)
u_S p.s.i. ($\times 10^{-3}$) (MPa)	233 (1607)	171 (1179)	94 (579)	196 (1351)	176 (1214)	171 (1179)
u p.s.i. ($\times 10^{-3}$) (MPa)	268 (1848)	222 (1531)	156 (1076)	242 (1669)	233 (1607)	222 (1531)
u_F/u	0.13	0.23	0.40	0.19	0.25	0.23

Work material: AISI 1015 cold-rolled steel; tool material, T–1 HSS; cutting fluid, CCl$_4$; undeformed chip thickness, 0.005 in (0.125 mm).

3. The inclination angle i is seen to have a very significant influence upon the effective rake angle, particularly at high values of the inclination angle. As might be expected, the effective shear angle ϕ_e is seen to increase with α_e.

4. The deviation between the chip-flow vector and the resultant-force vector in the plane of the tool face (as measured by angle ζ_S) is seen to increase with both an increase in the effective rake angle α_e and an increase in the inclination angle i.

5. The deviation between the shear-velocity vector and the resultant-force vector in the shear plane (as measured by angle ζ_S) is seen to be insignificant in all of the tests.

6. The fact that ζ_S is insignificant while ζ_C is substantial would indicate that the shear process predominates over the friction process. That this is actually so may be seen by a comparison of the ratio u_F/u.

7. The coefficient of friction μ on the tool face is seen to decrease with increased inclination angle i. This is particularly significant inasmuch as the coefficient of friction will increase with increased rake angle in orthogonal cutting. However, as already mentioned, when inclination angle is increased, the effective rake angle is seen to increase. Thus the special geometry associated with an inclination angle has a greater effect upon the coefficient of friction than does the effective rake angle.

8. The shear stress on the shear plane is seen to increase with inclination angle while the shear stain decreases. An explanation for this paradox will be considered later.

The total energy per unit volume, u, is seen to decrease with increased inclination. This decrease is due to a decrease in the shear strain which in turn is due to an increase in the effective rake angle with increased inclination. Obviously, the decrease in u is not due to a decrease in the friction energy per unit volume, u_F, with increased inclination.

Views of representative chips obtained using the three angles of inclination of the tests of Table 16.1 are shown in Fig. 16.14. All chips are for a length of cut of approximately 4 in (102 mm). Here the change from the flat spiral chip of orthogonal cutting to the helical chip that is characteristic of a tool with inclination is evident.

One of the functions of inclination is to alter the chip-flow direction, and the manner in which side flow increases with the inclination of the cutting edge is evident in Fig. 16.14.

While there are some production tools that employ simple inclined cutting edges, such as those just discussed, most tools in use are geometrically more complex. However, when carefully studied, these tools may be reduced to equivalent inclined cutting tools. Were this not so, it is unlikely that we should be as interested in the inclined tool to the extent the foregoing discussion would indicate.

The various angles that are used in practice to describe a lathe tool, milling cutter, or drill will now be considered, and the relationships between these angles and

1. normal rake angle (α_n)
2. inclination angle (i)
3. effective rake angle (α_e)

will be presented.

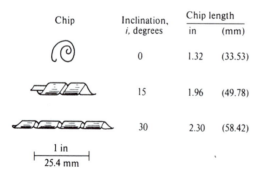

Chip	Inclination, i, degrees	Chip length in	(mm)
	0	1.32	(33.53)
	15	1.96	(49.78)
	30	2.30	(58.42)

1 in
25.4 mm

Fig. 16.14 Representative chips. Rake angle, $\alpha_n = 10°$; length of cut, 4 in (102 mm); depth of cut, 0.005 in (0.13 mm); cutting speed, $V = 20$ i.p.m. (50.8 cm min^{-1}); cutting fluid, carbon tetrachloride.

TURNING

Since each three-dimensional cutting tool evolved during a different period of technological development, the nomenclature that has come to be accepted is far from uniform among the different tools. The same angle on one tool is frequently found to have an entirely different name for a second type of tool. Furthermore the quantities that have come to be recognized in two-dimensional cutting as the most important ones are frequently obscurely given by two or more other quantities that are more conveniently measured but are far less significant from the fundamental point of view.

The lathe tool offers a good example. The nomenclature adopted by the American Standards Association is defined in Fig. 16.15 for a representative lathe tool, and the standard shorthand method of specifying the several dimensions of a tool is illustrated by an example. It will be noted that two quantities are necessary in this case to define the inclination of the tool face to the tool base, the latter plane being perpendicular to the cutting velocity vector. These two quantities are the side rake (α_s) and the back rake (α_b) angles. These two angles together with the side cutting edge angle (C_s) determine the effective rake and inclination angles of the lathe tool, the other angles being of secondary importance.

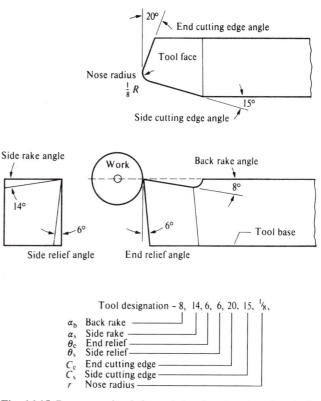

Tool designation – 8, 14, 6, 6, 20, 15, $\frac{1}{8}$,

α_b	Back rake	
α_s	Side rake	
θ_e	End relief	
θ_s	Side relief	
C_e	End cutting edge	
C_s	Side cutting edge	
r	Nose radius	

Fig. 16.15 Representative lathe tool showing American Standards Association nomenclature.

It is of interest to note that British and German engineers thought it preferable to define the position of the tool face relative to the tool base in terms of other pairs of quantities than the back- and side-rake angles adopted in the United States. The German and British systems together with the American system are compared in Fig. 16.16. Here the plane of the paper is taken parallel to the tool base and the lines on the figure show the directions in which the various rake angles are measured. In the German system the back- and side-rake angles are measured parallel and perpendicular to the cutting edge, respectively, rather than along and perpendicular to the axis of the tool shank as in the American system. The tool face is located according to the British standard by specifying the maximum rake angle (see Fig. 16.16). All this illustrates how complex things can be without standardization.

Tool Geometry

Most practical three-dimensional tools are found to have two edges that cut simultaneously. Probably the most common tool of this type is the ordinary lathe

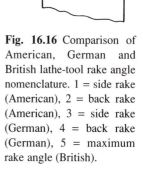

Fig. 16.16 Comparison of American, German and British lathe-tool rake angle nomenclature. 1 = side rake (American), 2 = back rake (American), 3 = side rake (German), 4 = back rake (German), 5 = maximum rake angle (British).

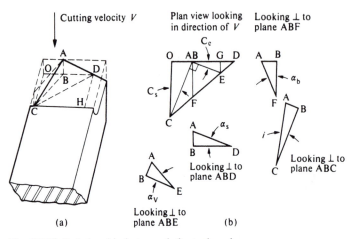

Fig. 16.17 Relationship between lathe tool angles.

tool. Here the primary cutting edge is the side cutting edge while the secondary edge is the end cutting edge. Since the depth of cut is usually much greater than the feed per revolution, what happens along the secondary cutting edge is relatively negligible compared with what happens along the primary cutting edge for cuts of normal aspect ratio. If the nose radius is small compared with the depth of cut, the side cutting edge may be considered to be a simple oblique cutting tool.

In Fig. 16.17a, a lathe tool is shown without clearance or end cutting edge angles for simplicity. In this figure plane OBDC is drawn perpendicular to the velocity vector V, the curvature of which may be ignored if the depth of cut is small compared with the work diameter. Line AB is drawn parallel to V. In Fig. 16.17b a plan view of the construction shown in Fig. 16.17a is given, and the known angles α_s, α_b, and C_s are indicated on auxiliary diagrams as are the unknown quantities i and α_v.

The inclination angle i may be found in terms of the known angles as follows. Since triangles ODC and BDF are similar:

$$\frac{OB + BD}{BD} = \frac{OC}{BF} \qquad \text{or} \qquad 1 = \frac{OC}{BF} - \frac{OB}{BD} \qquad (16.30)$$

and

$$\tan i = \frac{AB}{BC} = \frac{AB}{BC}\left(\frac{OC}{BF} - \frac{OB}{BD}\right) \qquad (16.31)$$

but

$$\frac{AB}{BF} = \tan \alpha_b$$

$$\frac{OC}{BC} = \cos C_s$$

$$\frac{AB}{BD} = \tan \alpha_s$$

$$\frac{OB}{BC} = \sin C_s$$

hence

$$\tan i = \tan \alpha_b \cos C_s - \tan \alpha_s \sin C_s \qquad (16.32)$$

The velocity rake angle α_v may be similarly found by working with similar triangles GDE and BDF:

$$\frac{BD - BG}{BD} = \frac{GE}{BF} \quad \text{or} \quad 1 = \frac{BG}{BD} + \frac{GE}{BF} \tag{16.33}$$

and

$$\tan \alpha_v = \frac{AB}{BE} = \frac{AB}{BE}\left(\frac{BG}{BD} + \frac{GE}{BF}\right) \tag{16.34}$$

but

$$\frac{AB}{BD} = \tan \alpha_s$$

$$\frac{BG}{BE} = \cos C_s$$

$$\frac{AB}{BF} = \tan \alpha_b$$

$$\frac{GE}{BE} = \sin C_s$$

hence

$$\tan \alpha_v = \tan \alpha_s \cos C_s + \tan \alpha_b \sin C_s \tag{16.35}$$

Figure 16.18 is a nomograph for graphically solving Eqs. (16.32) and (16.35) for inclination angle (i) and velocity rake angle (α_v). For the tool of Fig. 16.15 the dotted lines in the nomograph give

$$i = 4.1°$$

$$\alpha_v = 15.6°$$

From Eq. (16.1), α_n is found to be 15.5°, and if Stabler's rule is assumed to hold, the dashed lines on Fig. 16.18 give $\alpha_e = 15.7°$. Thus, to a first approximation the relatively complex tool of Fig. 16.15 is seen to be equivalent to a simple oblique tool having a 4° inclination angle and a 16° effective rake angle.

Representative Tool Dimensions

From the foregoing discussion of the lathe tool, it is evident that there are a number of geometrical quantities that may be varied as well as several operating variables. Many of the more important considerations pertaining to the choice of these quantities will now be considered.

Due to the interdependence of the many variables influencing the performance of a lathe tool, it is not possible to state exact rules pertaining to the choice of operating conditions. One must be content with a presentation of good average practice and the consequence of changing each variable from the recommended value. An additional reason why it is not possible to precisely specify optimum operating variables is that the best choice of variables depends upon whether tool life or

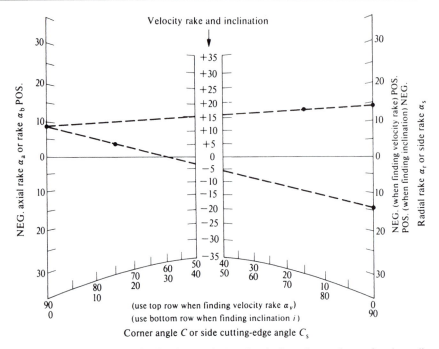

Fig. 16.18 Nomograph for finding inclination angle i and velocity rake angle α_v for three-dimensional cutting tools. (based on Kronenberg, 1966)

finish is of major interest. The method of arriving at optimum operating conditions in practice is to use general recommendations such as those presented here as a starting point and then proceed to improve performance by trial. Of course, a general appreciation of the relative importance of each of the many operating variables is invaluable in reducing the effort involved in approaching the optimum.

Representative values of rake and relief angles for turning tools are given in Table 16.2. Values of α_n, i, and α_e are also given in this table for a range of side cutting edge angles. The recommended values of rake angle are seen to be such that α_n generally decreases as the strength of the metal cut increases. This decrease in α_n provides a greater wedge angle at the tool point to support a greater load. When very hard materials are machined or when intermittent cuts must be taken, the values of α_s and α_b should be reduced to values below those given in Table 16.2 to provide a greater normal rake angle (α_n) to absorb the shock. Under severe shock conditions it is frequently desirable to use negative values of rake angle with carbide-tipped tools. Due to the greater ability of HSS tools to absorb shock, it is preferable to provide them with larger rake angles (on the average $\alpha_e = 10$ to $15°$) than is desirable with carbide tools (on the average $\alpha_e = 5°$) in order to take advantage of the decreased cutting energy that accompanies a shift to a higher effective rake angle.

The rake angles recommended for brass in Table 16.2 are unusually low for a material of this hardness and strength. The reason for this lies in the fact that tools of high rake angle tend to dig into brass and exhibit a gouging action. The resultant force directions for high- and low-rake-angle tools are generally as shown in Fig. 16.19, the vertical components of force in the two cases being in opposite directions. The large-rake-angle case represents a condition of instability with regard to depth of cut, the deflection resulting from the vertical component of force tending to cause an increase in depth of cut which in turn causes a further increase in vertical force, etc. Cutting under

TABLE 16.2 Representative Turning Tool Angles (degrees)

Work Material	α_S Side Rake	α_b Back Rake	Side and End Relief	For $C_s = 0°$ α_n	i	α_e	For $C_s = 15°$ α_n	i	α_e	For $C_s = 30°$ α_n	i	α_e
High speed steel tools												
Aluminum and magnesium alloys	20	20	10	18.9	20	23.7	23.4	14.5	25.8	26.3	7.6	26.9
Copper	15	15	10	14.5	15	17.5	17.9	10.8	19.3	20.0	5.6	20.4
Brass and bronze	0	0	8	0	0	0	0	0	0	0	0	0
Cast iron	10	5	8	9.9	5	10.3	10.9	2.3	11.0	11.1	−0.7	11.1
Steel (C < 0.3)	15	15	8	14.5	15	17.5	17.9	10.8	19.3	20.0	5.6	20.4
Steel (C > 0.3)	12	12	8	11.8	12	13.8	14.5	8.6	15.5	16.2	4.5	16.4
Stainless steels	10	15	8	9.7	15	13.0	13.2	12.0	15.1	15.8	8.2	16.6
Carbide tools												
Aluminum and magnesium alloys	10	0	8	10	0	10	9.7	−2.6	9.8	8.6	−5.0	9.1
Copper	15	0	7	15	0	15	14.5	−4.0	14.7	13.0	−7.7	13.8
Brass and bronze	4	0	7	4	0	4	3.9	−1.0	3.9	3.5	−2.0	3.6
Cast iron	8	0	6	8	0	8	7.8	−2.1	7.9	7.0	−4.1	7.2
All steels	6	0	6	6	0	6	5.8	−1.6	5.9	5.2	−3.0	5.4

All positive value of inclination angle (i) will cause the chip to be directed away from the workpiece, while a negative value of i will direct the chip toward the workpiece.

the high-rake-angle conditions of Fig. 16.19 may thus lead to gouging. Rake angles for which the resultant force vector is directed below the direction of cut are generally to be avoided, particularly when the machine and tool are not very rigid. In the case of brass, the cutting conditions are such that a force vector directed below the direction of cut persists at relatively low values of rake angle and thus the values of α_e to be used for brass are unusually small. Tools that are used to cut aluminum and copper, which have similar hardness and strength to brass, do not exhibit a tendency to dig in; and hence aluminum and copper are cut with larger values of α_e.

The relief angles given in Table 16.2 are seen to decrease slightly as the hardness of the metal cut increases. In cutting very hard materials, the relief angle should be as low as possible; and when the feed is also reduced, relief angles as low as 5° may be successfully used. In cutting very soft metals or materials with low elastic moduli at very high rates of feed, it is sometimes necessary to increase relief angles to as high as 12°. It is preferable to make the end- and side-relief angles the same in the interest of simplicity and to make it easier to blend the nose radius into the flanks of the tool. On carbide tools a secondary relief angle of from 10° to 15° is frequently provided.

The role of the side cutting edge angle (C_s) in influencing the inclination angle (i) and the velocity rake angle (α_v) may be seen in the nomograph of Fig. 16.18. If C_s is zero, the back rake angle becomes the inclination angle (i.e., $\alpha_b = i$)

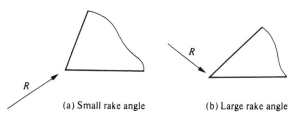

(a) Small rake angle (b) Large rake angle

Fig. 16.19 Comparison of directions of resultant force for tools with (a) small and (b) large rake angles.

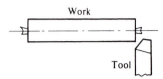

Fig. 16.20 Manner in which initial contact is displaced from tool tip if tool is provided with side cutting edge angle.

and the side rake angle equals the velocity rake angle (i.e., $\alpha_s = \alpha_v$) which in turn is essentially the effective rake angle (α_e). As the side cutting edge angle (C_s) is increased, α_s plays an increasing role in determining the value of i just as α_b plays an increasing role with α_s in determining α_e. In Table 16.2 it is seen that C_s plays a small part in influencing α_n, i, and α_e for carbide tools in the range of C_s extending from 0 to 30°. Table 16.2 on the other hand shows C_s to have a significant influence on these values for HSS tools over the same range of C_s.

Examination of the values of Table 16.2 reveals that a large value of C_s makes it possible to have a large value of α_e while at the same time the corresponding value of α_n is unusually small. This will enable a cut to be made at low power with a tool of large included angle and hence provide improved strength. It would thus appear that tools should be ground with large values of side cutting edge angle (C_s), and this would be the case were it not for the fact that the tendency for a tool to chatter increases with C_s. In order to avoid chatter it is customary to limit C_s to about 15°, although values as high as 45° may be used to advantage if the system is unusually rigid.

Unless it is necessary to turn to a radial shoulder, it is advisable to have a side cutting edge angle other than zero. The shock is thus taken at a point away from the tool tip as the tool engages the work (Fig. 16.20). When machining very rough forgings having a ragged surface, it is particularly important to provide a sufficient side cutting edge angle to protect the tool point. Carbide tools are very susceptible to damage under such conditions. A side cutting edge angle is also useful in releasing the stored elastic energy less abruptly, thus reducing the danger of edge chipping as the tool completes a cut.

Like a relief angle, the end cutting edge angle (C_e) should be as small as possible in the interest of increased tool strength. The normal range of values for C_e is 5° to 15°. The lower values can only be used with very rigid systems since any increase in radial force (due to there not being sufficient end-clearance) will tend to cause tool chatter. Values in the vicinity of 15° may be required at high feeds, particularly when the depth of cut is increased while the feed is engaged.

MILLING

The milling process is distinguished by a tool with one or more teeth which rotate about a fixed axis while the workpiece is fed into the tool. The process is further distinguished from other metal cutting operations by the chips which are produced. These are generally short, discontinuous segments. It should be emphasized that this type of discontinuous chip formation is a result of geometry alone whereas in the turning operation, discontinuous chips are usually due to fracture of the chip. The undeformed chip thickness in most milling operations varies from one end of the chip to the other. It will be shown later that depending upon the particular process the maximum chip thickness may occur at either end or near the middle of the chip. The regular chip discontinuity associated with milling leads to non-steady-state cyclic conditions of force and temperature. As a tooth engages the work, it receives a strong shock followed by a varying force. At the same time the tool chip is relatively cool as it enters the work, is heated as the chip is formed, and cools while awaiting the next engagement. The entering shock is detrimental to tool life, while the cooling period is usually beneficial, unless the tool material is sensitive to thermal shock.

The cyclic variation of force can provide the necessary energy to excite a natural mode of vibration in any part of the machine. Such vibrations can, and frequently do, lead to poor surface finish and decreased tool life.

One of the greatest advantages of the milling operation is that small, light tools can easily be rotated at high speeds to produce flat or curved surfaces on workpieces of a wide variety of sizes and shapes.

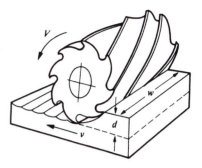

Fig. 16.21 Plane-milling operation.

Types of Milling

Important types of milling are plane-, end- or face-milling. The basic plane-milling cutter used to produce flat surfaces has cutting edges on the periphery which are parallel with the axis of rotation. The perform-ance of this cutter is usually enhanced if the teeth are arranged in helical fashion rather than parallel to the axis of rotation (Fig. 16.21). Use of helical teeth not only provides smoother cutting (in the same way that helical gears run smoother than spur gears) but chip forma-tion is also improved.

An end-milling cutter (Fig. 16.22) has teeth on one end as well as on the periphery. In use, its axis of rotation is perpendicular to the surface produced while, in plane-milling, the cutter axis is parallel to the finished surface. The end-milling cutter is used for producing flat surfaces and profiles—frequently simultaneously. The term end-milling cutter is misleading because in normal operation cutting is done on the periphery rather than by the teeth on the end of the cutter.

One could continue to amplify the number of types of milling cutters almost indefinitely. These would include side-milling cutters, milling saws as well as angle, fly, T-slot, and formed milling cutters. However, all of these are adaptations of the basic types considered above.

Milling cutters are generally made of high speed steel or have cemented carbide teeth. It is common practice with large milling cutters to use inserted teeth which may be removed for grinding and replacement. Smaller cutters are frequently equipped with carbide tips that are brazed in place. Figure 16.23 shows the accepted nomenclature for a plane- or helical-milling cutter. The analysis of this chapter may be applied directly with the aid of Table 16.3.

The normal rake angle is readily computed from Eq. (16.1), while the nomograph of Fig. 16.8 will give the effective rake angle α_e. The helix or inclination angle may be determined by rolling the cutter on a sheet of carbon paper that rests on white paper.

As an example, consider a plane-milling cutter with the following angles:

- radial rake $(\alpha_r) = 10°$
- helix angle $(i) = 25°$

From Eq. (16.1) the normal rake angle (α_n) is found to be 9.1° while the effective rake angle (α_e) is found to be 18° by use of Fig. 16.8.

Nomenclature for an end- or face-milling cutter is shown in Fig. 16.24. The inclination angle (i) and effective rake angle (α_e) may be found if the face-milling cutter angles are replaced by their equivalent turning tool angles given

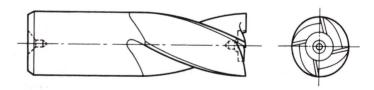

Fig. 16.22 Standard end-milling cutter with straight shank.

TABLE 16.3 Equivalent Turning Tool Angles for Plane-Milling Cutter

Plane-Milling Cutter	Equivalent Three-Dimensional Tool
Radial rake	Velocity rake, α_v
Helix angle	Inclination angle, i

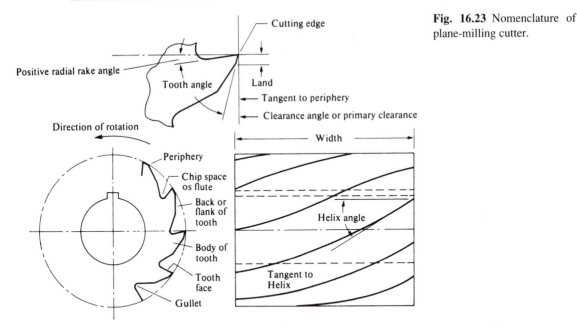

Fig. 16.23 Nomenclature of plane-milling cutter.

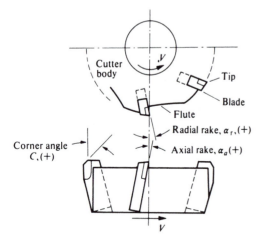

Fig. 16.24 Face-milling cutter with inserted blades, showing principal angles.

in Table 16.4. With these substitutions, the previous analysis of a turning tool enables the desired quantities to be found.

As an example, consider the following case:

- axial rake angle $(\alpha_a) = 10° = \alpha_b$
- radial rake angle $(\alpha_r) = -10° = \alpha_s$
- corner angle $(C) = 30° = C_s$

By following steps similar to those employed for a turning tool, the pertinent values are found to be

- normal rake angle $(\alpha_n) = -3.6°$
- inclination angle $(i) = 13.5°$
- effective rake angle $(\alpha_e) = 0.8°$

As for any cutting tool, sufficient clearance must be provided to prevent the tool from rubbing on the finished surface. In milling, clearance angles must be held to a

TABLE 16.4 Equivalent Turning Tool Angles for Face-Milling Cutter

Face-Milling Cutter	Turning Tool
Radial rake, α_r	Side rake, α_s
Axial rake, α_a	Back rake, α_b
Corner angle, C	Side cutting edge angle, C_s

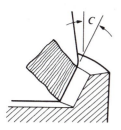

Fig. 16.25 Action of face-milling cutter showing manner in which inclined peripheral cutting edge produces chip. (i = inclination angle; n = number of teeth in cutter.)

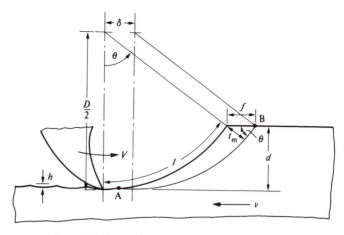

Fig. 16.26 Approximate shape of plane-milling chip.

minimum in order that sufficient metal be available at the tool point to withstand the relatively severe shock pertaining. Figure 16.25 shows the manner in which the inclined cutting edge of a face-milling cutter produces a chip.

Chip Geometry

The following are three geometrical quantities of interest in the milling operation (see Fig. 16.26):

1. maximum undeformed chip thickness (t_m) which corresponds to the feed rate in a turning operation
2. undeformed chip length (l) which gives an indication of the time of contact per tooth
3. peak-to-valley distance (h) of the scallops which are left on the finished surface when plain milling

Unlike turning where the depth of cut and feed can be directly set on the machine, in milling, these important quantities must be computed from the operating variables. Before determining the desired quantities, one must specify the direction of motion of tool and workpiece. When these two velocities have the same direction, the process is called "climb" or "down" milling; however when the two velocities are in opposite directions, a "conventional" or "up" milling process is involved. Figure 16.27 clearly shows the difference between these two types of milling. It is readily seen that with otherwise similar conditions, up-milling gives a slightly smaller t_m, a larger l, and smaller h than does down-milling, because of the smaller average radius of curvature of the chip when down-milling.

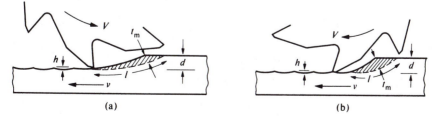

(a)   (b)

Fig. 16.27 Difference between (a) up-milling and (b) down-milling.

The actual shape of an undeformed plane-milling chip is rather complex as the cutting edge traverses a trochoidal path. Equations for milling chip geometry have been developed by Martellotti (1941) but since these derivations are quite tedious, only the final results are given here (see Fig. 16.26 for meaning of symbols):

$$t_{m} = \left[f \left\{ \frac{(D/d) - 1}{(D/2d)^2 (1 \pm v/V)^2 \mp vD/Vd} \right\}^{1/2} \right] \cos i \qquad (16.36)$$

$$l = \frac{D}{2} \theta \pm d \frac{v}{V} \{ (D/d) - 1 \}^{1/2} \qquad (16.37)$$

$$h = \frac{f}{4(D/f) \pm (8n/\pi)} \qquad (16.38)$$

When two signs are given, the upper one refers to up-milling while the lower one refers to down-milling.

If the cutting speed (V) is much greater than (v) (i.e., $v/V \ll 1$), as it always is in practice, a simplified solution can be obtained. Figure 16.27 shows the trochoidal chip shape replaced by a circular arc which is permissible when $V \gg v$. The quantities (δ) and (θ) are as defined above. From this figure, it follows that

$$t_{m} = [f \sin \theta] \cos i \qquad (16.39)$$

$$l = \frac{D}{2} \theta \pm \frac{f}{2} \qquad (16.40)$$

$$h = \frac{f^2}{4D} \qquad (16.41)$$

When $d/D \ll 1$, which is generally the case, these quantities may be further simplified as follows:

$$t_{m} = [2f \sqrt{(d/D)}] \cos i \qquad (16.42)$$

$$l = (Dd)^{1/2} \pm \frac{v}{2nN} \qquad (16.43)$$

$$h = \frac{f^2}{4D} \qquad (16.44)$$

where N is the r.p.m. of the cutter, n is the number of teeth, and f is the feed per tooth.

As an example of values obtained from the foregoing equations, consider a plane-milling cut in which

$$N = 200 \text{ r.p.m.}$$

$$v = 12 \text{ i.p.m. } (305 \text{ mm min}^{-1})$$

$$n = 10 \text{ teeth}$$

$$D = 4 \text{ in } (102 \text{ mm})$$

$$d = \tfrac{1}{4} \text{ in } (6.35 \text{ mm})$$

$$i = 0°$$

TABLE 16.5 Comparison of Exact and Approximate Equations

Method	t_m in[†]		l in		$h (\times 10^6)$ in	
	Up	Down	Up	Down	Up	Down
Martellotti's analysis						
Eqs. (16.36) to (16.38)	0.00290	0.00291	1.017	1.007	2.225	2.270
If $v/V \ll 1$						
Eqs. (16.39) to (16.41)	0.00291	0.00291	1.015	1.009	2.250	2.250
If $v/V \ll 1$ and $d/D \ll 1$						
Eqs. (16.42) to (16.44)	0.0030	0.0030	1.003	1.003	2.250	2.250

[†] Metric values omitted since the purpose of this table is only to compare values obtained from the approximate and exact equations.

For these values, we obtain

$$f = \frac{12}{(200)(10)} = 0.006 \text{ in/tooth (0.15 mm/tooth)}$$

$$\theta = \cos^{-1}\left(1 - \frac{0.5}{4}\right) = 29° = 0.506 \text{ radians}$$

The values of t_m, l, and h obtained from three foregoing approximations are given in Table 16.5 where it is apparent that the approximate methods are good in the practical range.

For face- or end-milling the chip geometry is easily approximated. From Fig. 16.28 it is seen that if the axis of the cutter intersects the workpiece, the maximum chip thickness (t_m) is just equal to the feed per tooth:

$$t_m = \frac{v}{Nn} = f \tag{16.45}$$

The chip length is then approximated by

$$l = \frac{D}{2}\Psi \tag{16.46}$$

where

$$\Psi = \sin^{-1}\left(\frac{2m}{D}\right) + \sin^{-1}\left\{\frac{2(w-m)}{D}\right\} \tag{16.47}$$

If a corner angle (C) exists, the value of t_m should be modified as follows:

$$t_m = \frac{v}{Nn}\cos C \tag{16.48}$$

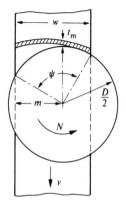

Fig. 16.28 Approximate shape of face-milling chip.

Practical Considerations

The effect of *rake angle* upon tool life, surface finish, and power consumed is the same with a milling cutter as it is with a simple two-dimensional tool. However, the shock loadings to which milling

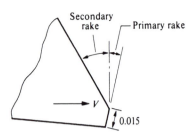

Fig. 16.29 Composite rake angle milling tooth.

cutters are subjected limit the maximum useful rake angle. This is particularly true with carbide-tipped milling cutters. Negative rake angles are frequently used in carbide cutters to provide a greater included angle for increased strength and shock resistance. However, use of negative rake angles is accompanied by an increase in the energy consumed per unit volume of metal cut and the temperature at the tool point.

A composite rake angle tooth has been found to combine some of the better qualities of both positive and negative rake angle tools. The composite rake angle tool shown in Fig. 16.29 was tested by Armitage and Schmidt (1957) with the results shown in Table 16.6. These data show that the addition of a positive secondary rake angle reduces the cutting force (F_P) although it is still higher than when a positive rake angle is used alone. The greatest advantage of this type of tool occurs after a small amount of tool wear. The cutting force (F_P) for the composite rake angle tool is lower than that for either the purely positive rake angle tool or the purely negative rake angle tool.

Naturally, no hard and fast rules can be given for milling-cutter rake angles, but Table 16.7 gives representative values for several materials. When three-dimensional theory is applied to Table 16.7, the values given in Table 16.8 are obtained.

TABLE 16.6 Cutting Force Values (F_P) for Composite Rake Angle Tools[†]

Rake Angles (α) Degrees	Cutting Force (F_P)	
	lb	N
+30	64	285
+15	82	365
−12	115	512
−12 primary +15 secondary	96	427
−12 primary +30 secondary	76	328

[†] After Armitage and Schmidt, 1957.

TABLE 16.7 Representative Rake Angles for Face-Milling Cutters

	Tool Materials					
	HSS		Cast Alloys		Cemented Carbides	
Work Material	α_r	α_s	α_r	α_s	α_r	α_s
Steel and cast iron	12	12	6	6	0	0
Steel and cast iron[†]	−5	25	−5	25	−10	10
Brass and bronze	10	10	6	6	4	4
Aluminum alloys	25	25	12	12	15	15

[†] Cutting with a 45° corner angle; in all other cases corner angle is 0°.

TABLE 16.8 Values of Angles i, α_n, and α_e for the Tools of Table 16.7

Work Material	HSS			Cast Alloys			Cemented Carbides		
	i	α_n	α_e	i	α_n	α_e	i	α_n	α_e
Steel and cast iron	12.0	11.8	13.8	6.0	6.0	6.5	0	0	0
Steel and cast iron[†]	21.4	14.0	19.4	21.4	14.0	19.4	14.1	0	3.4
Brass and bronze	10	9.9	11.4	6.0	6.0	6.5	4.0	4.0	4.3
Aluminum alloys	25	22.9	29.9	12.0	11.8	13.8	15.0	14.5	17.6

[†] Cutting with a 45° corner angle; in all other cases corner angle = 0°.

Clearance angles should be as small as possible consistent with good surface finish. Representative values are 5° when cutting steel and 10° when machining aluminum or similar alloys. Frequently, a secondary clearance angle is provided behind the primary clearance angle to reduce the amount of grinding when sharpening.

The cutter speed in f.p.m. (m min⁻¹) which will give reasonable tool life is of course a strong function of geometry, but, for well designed cutters properly used, the maximum speed is found to depend largely upon the hardness of the metal cut and the tool material. Figure 16.30 gives representative cutting speeds as a function of workpiece hardness and material.

Normally the feed per tooth (δ) is considered an important quantity and is specified in milling operations. However, as previously mentioned, a more meaningful quantity, but one which must be computed, is the maximum chip thickness (t_m). Representative values of δ are given in Table 16.9 to illustrate the order of magnitude of this variable in practice.

In face-milling, initial contact should not occur at the cutting edge of the tooth in order to avoid chipping of the edge. In order to ensure this, the cutter should be larger than the work. A good working rule is to make the cutter diameter at least $1\frac{1}{2}$ times the work width.

In order to cause chips to flow away from the cutter body in face-milling, it is desirable that the teeth be designed to have a substantial positive inclination. A large corner angle (C) combined with a large positive inclination angle will usually provide long tool life. The large corner angle distributes the wear over a long cutting edge, and hence a given volume of metal may be removed with a minimum wear per unit of edge length.

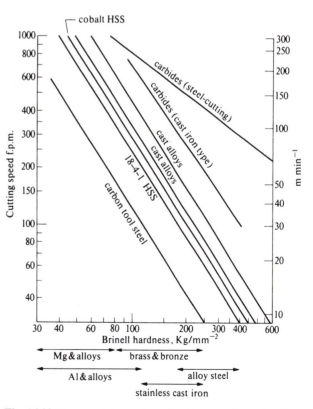

Fig. 16.30 Representative values of cutting speed as a function of work hardness and tool material.

TABLE 16.9 Representative Values of Feed Per Tooth (δ) for Milling Cutters, i.p.r. (mm/rev)

Material	HSS	Carbide and Cast Alloy
Stainless steel	0.005 (0.13)	0.008 (0.20)
Steels	0.005–0.010 (0.13–0.25)	0.010–0.015 (0.25–0.38)
Cast iron	0.010–0.015 (0.25–0.38)	0.010–0.20 (0.25–0.50)
Brass	0.008–0.020 (0.13–0.50)	0.010–0.020 (0.25–0.50)
Al and Mg alloys	0.02 (0.50)	0.02 (0.50)

It is particularly important in milling operations that the work be rigidly supported and that the machine be in good mechanical condition. Cutters should be mounted on large arbors and the arbor supported close to the cutter to further ensure a rigid system. Flywheels should be used whenever possible and should be mounted as close to the cutter as is feasible.

When milling with carbide cutters, it is best to cut steel and cast iron dry using compressed air to blow the chips away. Under such conditions the use of a water-base cutting fluid will actually give significantly shorter tool life than when cutting dry. Aluminum on the other hand should be cut with a copious supply of a water-base cutting fluid.

DRILLING

The drill is a tool designed to produce holes in metal parts quickly and easily. While very precise work can be done with a drill, it is a roughing operation and the primary items of interest are usually long life and high penetration rate. Very frequently drilling is a preliminary operation to reaming, boring or grinding where final finishing and sizing take place. In addition to the common two-flute twist drill, there are several other types in use including flat drills; drills with one, three or four flutes; and core, shell, and spade drills. The discussion here will be limited to the two-flute twist drill since this type adequately illustrates the nature of the problems associated with drills and their analysis.

Nomenclature

The twist drill is a complex tool that usually has two cutting edges designed to produce identical chips. There are four major actions taking place at the point of a drill:

1. a small hole is pierced by the rotating web, in much the same way a tube is pierced by a mandrel in the production of seamless tubing

2. chips are formed by rotating cutting edges (lips)

3. chips are conveyed out of the hole by a screw conveyor in the form of the helical flutes provided in the drill

4. the drill is guided in the hole already produced by the margins

The geometry that is ordinarily provided on a drill (Fig. 16.31) represents a compromise of several conflicting requirements, which include the following:

1. a small web to reduce thrust on the drill but a large web for greater resistance to chipping and greater torsional rigidity

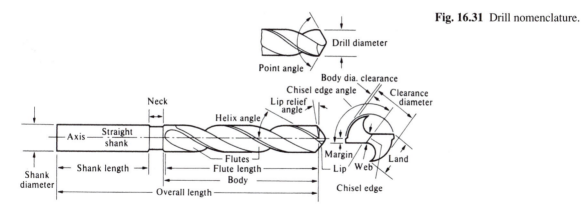

Fig. 16.31 Drill nomenclature.

2. large flutes to provide a larger space for chip transport but small flutes in the interest of torsional rigidity

3. an increase in the helix angle to more quickly remove chips but a decrease in helix angle in the interest of greater strength of cutting edges

Less obvious compromises are associated with the choice of the geometric parameters that influence the effective rake and relief angles.

The more important drill quantities from the analytical point of view are

1. helix angle (δ)

2. point angle $(2p)$

3. web thickness (w)

4. clearance angle (θ)

5. drill diameter (d)

The helix angle varies with the radius (r) to any particular point on the cutting edge. A helix angle without specification refers to the helix angle at the circumference of the drill. The pitch length of the helix (L) is constant for all points along the cutting edge, and the helix angle at any point may therefore be determined by use of the following equation:

$$\delta = \tan^{-1}\left(\frac{2\pi r}{L}\right) \tag{16.49}$$

For example, a 0.750 in (19 mm) drill having a helix angle of 32° is found to have a pitch length (L) of 3.77 in (95.8 mm). This pitch length may then be used to find the helix angle at the point on the drill where $r = 0.250$ in (6.35 mm). When this is done, the helix angle is found to be 22.6°. The helix angle may be most easily determined by rolling a drill across a piece of carbon paper that rests on a sheet of white paper.

The point angle, web thickness, and diameter of a drill may be readily determined by direct measurement.

The clearance angle (θ) at the periphery of a drill may be measured by wrapping a sheet of paper around the drill and marking it as shown in Fig. 16.32. The clearance angle at the center of the drill will be considerably greater than this value. Depending upon the shape of the point that is ground on the drill, the increase in clearance angle from periphery to center may be as much as

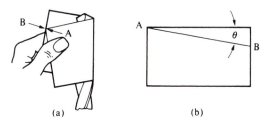

Fig. 16.32 Method of measuring clearance angle θ.

100%. The matter of clearance angles for drills presents a special problem in that a drill is fed downward upon the cut surface. The feed angle (ε) that is generated by any point on the cutting edge at radius r may be obtained from the following expression in terms of the feed per revolution of the drill (f):

$$\varepsilon = \tan^{-1}\left(\frac{f}{2\pi r}\right) \qquad (16.50)$$

The clearance angle (θ) at any radius must provide this much clearance before there is anything left to take care of elastic recovery. The drill clearance that corresponds to that of a conventional tool is ($\theta - \varepsilon$). The quantity ε is seen to increase as the point of the drill is approached.

In Fig. 16.33, values of helix angle (δ), clearance angle (θ), and feed angle (ε) are given for different points on the cutting edge of a representative $\frac{9}{16}$ in (14.33 mm) diameter drill operating under conventional conditions. The effective clearance angle ($\theta - \varepsilon$) is seen to be particularly ample at the center of the drill, and the feed angle (ε) is seen to play a rather insignificant role. The feed angle becomes important only on larger drills and drills that are fed at abnormally high rates.

The quantity (f), the feed rate in inches per revolution (mm/rev), is the feed for two cutting edges; therefore, $f/2$ (for a two-flute drill) corresponds to the feed of a lathe tool. The relief provided at the outside of most drills will be from 8° to 12° (larger values for soft metals).

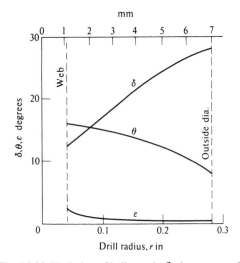

Fig. 16.33 Variation of helix angle δ, clearance angle θ, and feed angle ε, across the cutting edge of a representative $\frac{9}{16}$ in (14.3 mm) diameter drill. Helix angle, 28°; clearance angle, 8°; point angle, 118°; chisel edge angle, 120°; web thickness, 0.08 in (2.03 mm); feed rate, 0.009 i.p.r. (0.23 mm/rev).

TABLE 16.10 Equivalent Angles for Lathe Tool and Drill

Lathe Tool	Drill
Side rake angle, α_s	$\sin^{-1}(w/2r)$
Back rake angle, α_b	helix angle, δ
Side cutting edge angle, C_s	$\frac{1}{2}$ point angle, p

Note: w = web thickness; r = radius to point in question.

Analysis

The analysis of a drill point is much more involved than that for other tools inasmuch as the quantities of interest vary with radial position across the cutting edge. Table 16.10 lists angles that are approximately equivalent for a lathe tool and a drill. The first two drill items in Table 16.10 vary with radius (r) across the cutting edge. By use of the equivalence noted in Table 16.10, the variation of normal rake angle (α_n), effective rake angle (α_e), and inclination angle (i) with radius r may be obtained. For example, the values of Fig. 16.34 were obtained (Shaw, 1954) for a standard drill of 0.750 in (19.05 mm) diameter.

The inclination angle (i) is seen to increase as the center of the drill is approached, while the normal rake angle (α_n) decreases and even assumes large negative values. However, the effective rake angle (α_e) does not become negative as a consequence of the large inclination angles near the drill point. The action of the drill near the web might be described as a slicing action due to the large inclinations obtaining. The support at the cutting edge, which varies inversely with the magnitude of α_n is seen to be greatest near the point of the drill. This is fortunate since the action of a drill at its very center does not involve cutting but might better be described as extrusion.

The manner in which the metal directly under the chisel edge of the drill is extruded is shown in Fig. 16.35. In producing this specimen, a pilot hole the size of the web thickness of a larger drill was made in a brass block and filled with lead. When the larger drill was suddenly stopped, the situation shown in Fig. 16.35 resulted, where two distinct types of chips are evident—the one produced by cutting and the other by a combination of cutting and extrusion.

From the foregoing analysis it is evident that the helix angle (δ), web thickness (w), and point angle ($2p$) are the three quantities of greatest significance in determining drill performance. The first of these is controlled by the drill manufacturer while the latter two quantities can be altered to a certain extent by the user. The effect upon α_n and α_e of a variation of each of the quantities δ, p, and w may be different at the periphery and at the point of a drill. In Table 16.11 the influence of changes in δ, p, and w upon the significant drill angles are given for positions near the periphery and also near the chisel edge. In this table a plus sign signifies an increase, a minus sign a decrease, while zero indicates an insignificant change in the angle in question.

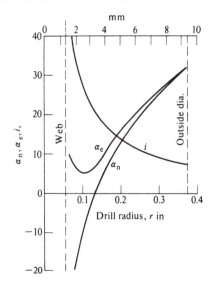

Fig. 16.34 Variation of inclination angle i, normal rake angle α_n, and effective rake angle α_e, across cutting edge of representative 0.750 in (19 mm) drill. Helix angle, 32°; point angle, 118°; web thickness, 0.110 in (2.79 mm).

Fig. 16.35 Comparison of the chips produced by the lips of a drill point and the chisel edge.

TABLE 16.11 Influence of a Change of Drill Dimensions on Drill Angles

| | For Points Near the Drill Periphery | | | | | | For Points Near the Chisel Edge | | | | | |
For increase of	i	C_s	α_b	α_s	α_n	α_e	i	C_s	α_b	α_s	α_n	α_e
Web thickness (w)	+	0	0	−	0	0	+	−	0	−	−	0
Point angle ($2p$)	0	+	0	−	0	0	+	+	0	0	+	+
Helix angle (δ)	0	0	+	+	+	+	0	0	+	0	+	+

TABLE 16.12 Representative Changes in Drill Point Angles Due to Changes in Web Thickness (w) Only[†]

Web Thickness w, in (mm)	Radius r, in (mm)	Degrees					
		i	C_s	α_b	α_s	α_n	α_e
0.05 (1.27)	0.375 (9.53)	3.2	58.9	32.0	17.3	33.0	33.1
0.110 (2.79)	0.375 (9.53)	7.2	58.7	32.0	13.2	31.8	32.3
0.150 (3.81)	0.375 (9.53)	9.9	58.5	32.0	10.2	30.6	31.5
0.050 (1.27)	0.100 (2.54)	12.4	58.1	9.5	−8.9	2.9	5.5
0.110 (2.79)	0.100 (2.54)	28.1	54.3	9.5	−28.2	−10.2	5.5
0.150 (3.81)	0.100 (2.54)	40.0	47.9	9.5	−44.3	−21.3	12.0

[†] Other conditions are the same as those for Fig. 16.34.

TABLE 16.13 Representative Changes in Drill Point Angles Due to Changes in Point Angle ($2p$) Only[†]

Point Angle $2p$ degrees	Radius r, in (mm)	Degrees					
		i	C_s	α_b	α_s	α_n	α_e
100	0.375 (9.53)	6.3	49.5	32.0	21.1	34.9	35.5
118	0.375 (9.53)	7.2	58.7	32.0	13.2	31.8	32.3
140	0.375 (9.53)	7.9	69.7	32.0	4.8	29.8	30.5
100	0.100 (2.54)	24.9	44.9	9.5	−26.1	−11.5	1.0
118	0.100 (2.54)	28.1	54.3	9.5	−28.2	−10.2	5.5
140	0.100 (2.54)	31.1	66.5	9.5	−29.9	−4.3	12.5

[†] Other conditions are the same as those for Fig. 16.34.

The orders of magnitude of the changes in Table 16.11 are illustrated by the numerical values of Tables 16.12 to 16.14. These quantities were computed for the numerical example involving a 0.750 in (19 mm) drill given previously. From this analysis the following observations may be made regarding the influence of changes of w, p, and δ:

TABLE 16.14 Representative Changes in Drill Point Angles Due to Changes in Helix Angle (δ) Only†

Helix Angle δ degrees	Radius r, in (mm)	Degrees					
		i	C_s	α_b	α_s	α_n	α_e
20	0.375 (9.53)	7.2	58.7	20	4.2	17.4	18.0
32	0.375 (9.53)	7.2	58.7	32	13.2	31.8	32.3
40	0.375 (9.53)	7.2	58.7	40	19.9	43.7	44.5
20	0.100 (2.54)	28.1	54.3	5.6	−30.4	−13.3	2.5
32	0.100 (2.54)	28.1	54.3	9.5	−28.2	−10.2	5.5
40	0.100 (2.54)	28.1	54.3	12.7	−26.4	−5.3	9.0

† Other conditions are the same as those for Fig. 16.34.

1. An increase of web thickness (w) has an insignificant influence at the periphery of the drill. The main effect of an increase in web thickness is a strengthening of the cutting edge near the point (as indicated by a decrease in α_n) and a smaller increase in effective rake angle. Larger values of web thickness would thus be indicated when strong materials must be drilled, for under such conditions failure at the drill point is most common. It would appear advisable to increase w for any drill that tends to fail at the center.

2. An increase of point angle also has a much smaller influence at the periphery than at the center of a drill. When p is increased, the effective rake angle is increased near the drill center while the normal rake angle is increased a corresponding amount. This provides freer cutting at the point of the drill but less support for the cutting edge. The smaller change at the periphery of the drill leads to a slightly stronger edge but slightly higher forces. A small point angle would be indicated for the drilling of relatively weak metals such as aluminum alloys where support of the cutting edge can be sacrificed in the interest of decreased cutting forces at the periphery. While a drill usually tends to fail at the periphery when operating in a soft metal, failures at the center of the drill are most common when drilling hard metals and the results of Table 16.13 would indicate that a large point angle would want to be used for a strong metal.

3. Unlike the two previous quantities the helix angle (δ) has its greatest effect near the periphery of the drill. An increase in helix angle causes a large increase in both α_n and α_e at the circumference of the drill and a smaller increase in both quantities near the center. Thus, increased helix angle results in freer cutting but correspondingly less support all across the cutting edge but particularly at the drill periphery. A large helix angle with small point angle and thin web would be indicated for weak materials and the reverse for strong metals such as the high-temperature alloys and stainless steels.

Drill Dimensions

The dimensions of the standard drill point are as follows:

- helix angle = 28° to 32°
- point angle = 118°
- clearance angle = 8° to 12°

The shape of the milled flutes are such that the cutting edge is straight when the proper point angle is ground on the drill. When very hard and strong metals are to be drilled, it is customary to use a high point angle (approaching 140°) and a smaller clearance angle (as low as 5°). In such a case it would probably be advisable to use a lower helix angle to obtain greater support at the cutting edge, but drills of other than standard helix angle are not easily obtained.

For drilling hard alloy steels, the following angles are frequently used:

- helix angle = 28° to 32°
- point angle = 135°
- clearance angle = 6° to 9°

For drilling very soft metals point angles as low as 100° and clearance angles as high as 12° are not uncommon.

The chisel edge angle is a rough measure of the clearance angle of the drill. With the standard 118° point an 8° clearance angle should result in a chisel edge angle of 120° while a 12° clearance angle corresponds to a 135° chisel edge angle. Similarly for the 135° point angle, the chisel edge angle would be 115° for a 6° clearance and 125° for a 9° clearance angle.

The web thickness of the average drill represents a decreasing percentage of the drill diameter as the size of the drill increases. This is illustrated in Table 16.15 where recommended web thicknesses are given. The web of a drill, which is the metal that connects the flutes, is generally made to increase as the shank of the drill is increased in the interest of greater rigidity. This increase in web thickness may be as much as 50%; therefore, when a drill is sharpened a number of times, it may be found to have too large a web thickness. Web thinning is therefore practiced, but a better solution is for manufacturers to provide the useful portion of the drill with a constant web thickness. If a drill has a web that is thicker than necessary, the thrust force will be needlessly high; and if such a drill is used at excessive feed rates, a crack may develop extending up the web of the drill. In such instances a decrease in feed or web thickness is indicated. The clearance at the web should also be checked in cases where the web tends to split since insufficient clearance can also cause this difficulty.

It is customary to provide a drill with decreasing diameter toward the shank (0.0005 to 0.001 in in^{-1}) in order to provide clearance of the drill body in the finished hole. When a drill is sharpened a number of times, it must therefore be expected that the drill will cut undersize. The drill is normally supplied with margins for purposes of guiding. These margins are without circumferential clearance and should be as small as possible in the interest of decreased friction. The margin of a one-inch drill is normally about 0.1 in (2.54 mm).

In the interest of good drill performance and life it is extremely important that both lips of a drill do equal work. The length of the two cutting edges should be checked for equality as well as the height of the two lips. An asymmetrical drill point operating at normal feed rate will produce inaccurate holes and will last a relatively short time since one lip will be overloaded.

Operating Conditions

It is customary to feed drills at a rate dependent on the size of the drill but independent of material, the speed of the drill being adjusted to the particular

TABLE 16.15 Recommended Web Thickness Values

Drill Diameter		Web Thickness
in	mm	% of Diameter
$\frac{1}{8}$	(3.18)	20
$\frac{1}{4}$	(6.3)	17
$\frac{1}{2}$	(12.5)	14
1	(25.4)	12
>1	(>25.4)	11

metal cut. The reason for this is that the flutes of a drill can handle only a certain chip volume and this volume is about proportional to the diameter of the drill. It is therefore common practice to feed drills at a rate that is proportional to the drill diameter in accordance with the relation:

$$f = \frac{d}{65} \tag{16.51}$$

where f is the feed rate and d is the drill diameter. A one-inch diameter drill would thus be fed at about 0.015 i.p.r. (0.38 mm/rev). When holes that are longer than $3d$ are drilled, the feed should be reduced.

Recommended speeds for HSS drills are given in Table 16.16. Carbon steel drills should be operated at speeds of 50% those given in Table 16.16.

While the best combination of speed and feed may differ somewhat from the values recommended above due to differences in relative chip volume, material structure, cutting fluid effectiveness, depth of hole, and condition of drill and machine, the foregoing values are of use as a general guide. If the outer corner of the drill is found to wear prematurely, too high a speed is indicated, while if the cutting edges are found to chip, too great a feed or too much clearance on the drill point may be the cause.

It is customary to drill all of the metals of Table 16.16 with a copious supply of an oil or soluble cutting fluid with the exception of cast iron and magnesium alloys which are normally cut in air. If deep holes are drilled, it may be necessary to supply the fluid under high pressure (10 to 20 atm) in order to lubricate the cutting point and help convey the chips along the flutes of the drill. In such applications an oil-hole drill that is provided with a hole extending down the drill is useful.

Drilling Forces

Although an ordinary twist drill operates in essentially the same way as a single point tool, it is extremely complex geometrically. Analyses of the mechanics of a twist drill are to be found in Oxford (1955). The following quantities influence the torque M and thrust T to which a drill is subjected in producing a hole:

- work material and structure
- drill diameter (d), in
- helix angle (θ), degrees
- length of chisel edge (c) = [web thickness]/[cosine (chisel-edge angle)] = w/cos (120) = 1.15 w, approx.
- point angle ($2p$), degrees
- number of cutting edges (n)
- feed (f), i.p.r. (mm/rev)
- cutting fluid
- drill sharpness

From the treatment of cutting-tool forces in Chapter 3, we should not expect the drilling speed (V), the drill clearance, or small changes in flute shape to be important. The flute geometry behind the cutting edge mainly influences the conveyance of

TABLE 16.16   Recommended Speeds for HSS Drills

Material	f.p.m.	m min^{-1}
Soft cast iron	150	45.7
Medium hard cast iron	80	24.4
Mild steel	90	27.4
Alloy steel	60	18.3
Tool steel	50	15.2
Brass and bronze	100–300	30.5–91.5
Copper	150	45.9
Aluminum and magnesium alloys	300–400	91.5–122

TABLE 16.17 Variables of Primary Importance to Torque

Quantity	Symbol	Dimensions
Drill torque	M	FL
Feed per revolution	f	L
Drill diameter	d	L
Length of chisel edge	c	L
Specific cutting energy of work	u	FL^{-2}

chips and assumes an important role only when chips tend to jam in the flutes. Thus flute-shape details for a normally functioning drill may be ignored.

Useful empirical equations relating torque and thrust with feed and drill diameter have been developed by Boston and Oxford (1930) and Kronenberg (1935). A more analytical approach has been presented by Shaw and Oxford (1957) which will be summarized here.

Three basic actions are involved in the performance of a two-flute twist drill:

1. cutting by the two lips

2. cutting action at the web

3. extrusion by the web

Only the first two of these contribute significantly to the torque (M) while all three are important to the thrust (T).

Variables of primary important to drill torque are listed in Table 16.17. Before dimensional analysis:

$$M = \Psi_1(u, c, f, d) \tag{16.52}$$

After dimensional analysis:

$$\frac{M}{d^3 u} = \Psi_2\left(\frac{f}{d}, \frac{c}{d}\right) \tag{16.53}$$

In the absence of a size effect (increase in u with volume deformed at one time) the torque should be expected to vary as the projected area of the drill or as d^2. Thus, in the absence of a size effect

$$\frac{M}{d^3 u} = \frac{f}{d} \Psi_3\left(\frac{c}{d}\right) \tag{16.54}$$

Shaw and Oxford (1957) have shown that just as the specific cutting energy (u) varies inversely with $t^{0.2}$ in ordinary two-dimensional orthogonal cutting, in drilling, u varies inversely with $(fd)^{0.2}$. Therefore, when drilling a given work material having a Brinell hardness H_B

$$u \sim \frac{H_B}{(fd)^{0.2}} \tag{16.55}$$

Combining Eqs. (16.54) and (16.55):

$$\frac{M}{d^3 H_B} = \frac{f^{0.8}}{d^{1.2}} \Psi_3\left(\frac{c}{d}\right) \tag{16.56}$$

A large number of experiments (about five hundred holes) were run on heat-treated chrome–nickel steel (AISI 3245) with sharp drills cutting dry with

- workpiece hardness (H_B) = 200 kg mm^{-2}
- drilling speed 20 f.p.m. (6.1 m min^{-1})
- $c/d \cong 0.18$
- drills of different diameter ($\frac{1}{4}$ to 1 in = 6.35 to 25.4 mm)

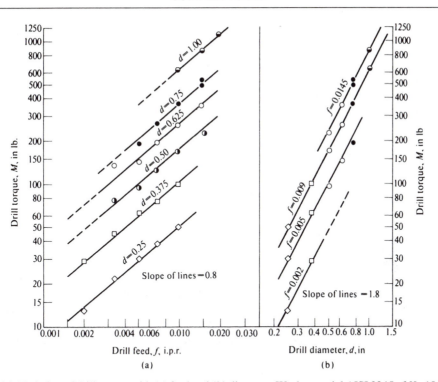

Fig. 16.36 Variation of drill torque with (a) feed and (b) diameter. Work material AISI 3245 of H_B 196–207; drill speed, 20 f.p.m. (6.1 m min^{-1}); cutting fluid, none.

Figure 16.36a shows measured torque (M) plotted against feed (f) while Fig. 16.36b shows measured torque (M) plotted against drill diameter (d). The curves have slopes of 0.8 and 1.8, respectively, in agreement with Eq. (16.56). The relation between drill torque (M) and specific drilling energy (u) may be readily shown to be

$$u = \frac{8M}{fd^2} \tag{16.57}$$

Figure 16.37 shows u obtained from Eq. (16.57) for measured values of M plotted against the product (fd) on log–log coordinates. The experimental points are seen to lie on a straight line in excellent agreement with Eq. (16.55). For these tests Eq. (16.56) may be written

$$\frac{M}{d^3 H_B} = 0.087 \frac{f^{0.8}}{d^{1.2}} \tag{16.58}$$

if H_B is in p.s.i. (1420 times H_B in kg mm^{-2}) and all other quantities are in inch and inch-pound units. While this equation is for the conventional value of c/d of 0.18, further studies have shown that drill torque is relatively insensitive to changes in c/d. Therefore Eq. (16.58) may be used to estimate drilling torque (M) for any material having a Brinell hardness H_B. A similar equation in SI units is

$$\frac{M}{d^3 H_B} = 0.0031 \frac{f^{0.8}}{d^{1.2}} \tag{16.58a}$$

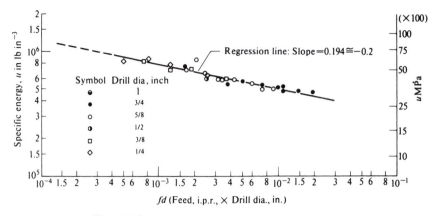

Fig. 16.37 Plot of u against fd for data of Fig. 16.36.

The nondimensional quantity for thrust (T) corresponding to ($M/d^3 H_B$) for torque will be ($T/d^2 H_B$). Since in orthogonal cutting the thrust component of force (F_Q) is usually proportional to the cutting component ($F_P \cong 2F_Q$), it is reasonable to consider the cutting part of the thrust force to be proportional to the torque component. This leads to the following relation for drill thrust where the second term on the right (Ψ_4) is the contribution to thrust due to "extrusion" at the web of the drill:

$$\frac{T}{d^2 H_B} = K_1 \frac{f^{0.8}}{d^{1.2}} + \Psi_4 \left(\frac{c}{d} \right) \tag{16.59}$$

where K_1 is a constant for a given work material and drill, and Ψ_4 is some function of (c/d).

The component of thrust due to extrusion at the web (T_e) should be

$$T_e = \frac{u \pi c^2}{4} \tag{16.60}$$

Since there should be negligible size effect for "extrusion" at the web $u \sim H_B$ and hence

$$\frac{T_e}{d^2 H_B} \sim \left(\frac{c}{d} \right)^2 \tag{16.61}$$

from which it follows that

$$\Psi_4 \left(\frac{c}{d} \right) = K_2 \left(\frac{c}{d} \right)^2 \tag{16.62}$$

where K_2 is a constant for a given work material and drill.

Values of thrust parameter ($T/d^2 H_B$) based on measured values of drill thrust T are shown plotted against ($f^{0.8}/d^{1.2}$) in Fig. 16.38 for the previously discussed tests on chrome–nickel steel (AISI 3245). Here it is seen that the regression line is in excellent agreement with Eq. (16.59). Since the value of c/d for all these drills was approximately 0.18, it follows from the regression equation that

$$\frac{T}{d^2 H_B} = 0.195 \frac{f^{0.8}}{d^{1.2}} + 0.68 \left(\frac{c}{d} \right)^2 \tag{16.63}$$

This equation may be used to estimate the drilling thrust for any material having a Brinell hardness H_B. It should be again noted that in this equation H_B is in p.s.i. (1420 times H_B in kg mm^{-2}) and all other quantities are in inches and inch-pound units. A similar equation in SI units is

$$\frac{T}{d^3 H_B} = 6.962 \frac{f^{0.8}}{d^{1.2}} + 0.68 \left(\frac{c}{d}\right)^2 \qquad (16.63a)$$

where

M = drill torque in N m

d = drill diameter in mm

H_B = Brinell hardness in kg mm^{-2}

f = drill feed in mm/rev

T = drill thrust in N

When the helix angle was varied over the wide range from 18° to 30°, the results shown in Fig. 16.39 were obtained. This shows that an increase in helix angle causes a relatively small decrease in both T and M. However, it should be noted that this is the case where no chip jamming in the flutes occurred. If chip jamming is encountered, both thrust and torque will rise greatly; and under such conditions, helix angle can have a profound influence on promoting chip flow up the flutes. Use of a thin black porous oxide impregnated with oil on the flutes can then play a very important role.

When tests on a wide variety of metals were run, results similar to those shown in Figs. 16.36 and 16.38 were obtained except that the heights of the curves reflected changes in H_B of the work material. This suggests that Eqs. (16.58) and (16.63) may

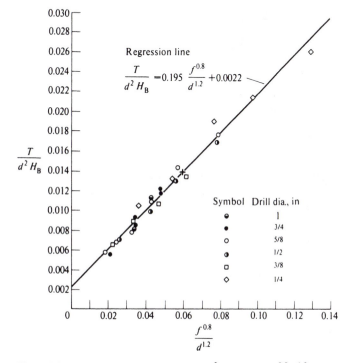

Fig. 16.38 Nondimensional plot of $T/d^2 J_B$ against $f^{0.8}/d^{1.2}$ when drilling AISI 3245 steel.

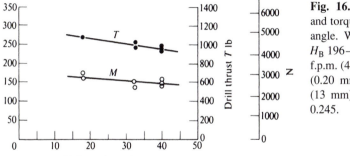

Fig. 16.39 Variation of drill thrust and torque for drills of different helix angle. Work material AISI 3245 of H_B 196–207 kg mm^{-2}; drill speed, 16 f.p.m. (4.9 m min^{-1}); feed, 0.008 i.p.r. (0.20 mm/rev); drill diameter, $\frac{1}{2}$ in (13 mm); cutting fluid, none; c/d = 0.245.

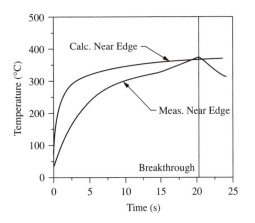

Fig. 16.40 Comparison of measured and calculated temperatures when drilling with a two-fluted drill at a spindle speed of 1000 r.p.m. and a down-feed rate of 0.0059 i.p.r. (0.147 mm/rev). (after Agapiou and Stephenson, 1994)

be used to estimate M and T with differences in workpiece properties being taken care of by use of appropriate values of Brinell hardness (H_B).

The most common causes of drill failure are breakage for small drills [$< \frac{1}{16}$ in (1.59 mm) diameter] and wear for large drills. A drill will usually break as a result of one or the other of two conditions: too great a thrust load or too great a torque.

Drill Temperatures

Agapiou and Stephenson (1994) have published an analytical/experimental study of temperatures associated with drilling. An analytical portion was presented in greater detail in Stephenson and Agapiou (1992). It was found that helix and drill-point angles have the greatest influence on temperature. When the helix angle is increased, this increases the effective rake angle and hence reduces the tool-tip temperature slightly. When the point angle is increased over 10°, this shortens the length of the cutting edge and hence increases the temperature. Chip–tool thermocouple and buried thermocouple results were found to be in reasonable agreement with the analytical results.

Measured temperatures were obtained using chromel-alumel wires 0.010 in (0.25 mm) in diameter running through the oil holes of carbide drills 0.82 in (20.93 mm) in diameter. The wires were welded to the tool flank very close to the cutting edge. A comparison of measured and calculated temperatures is shown in Fig. 16.40. The work material was gray cast iron having a hardness of H_B 170–200 and no coolant was used.

Low-Frequency Down-Feed Oscillation

Toews et al. (1998) studied the effect of low-frequency (< 200 Hz) modulation of the workpiece in the down-feed direction in a drilling operation. Changes in thrust and torque were found to depend on the amplitude of modulation and the ratio of modulation frequency of the part (f_p) to rotational frequency of the drill (f_d). When f_p/f_d was an odd integer, there were large dynamic components of thrust and torque and a reduction in the mean down-feed force. When the dynamic amplitude was greater than half the feed/rev, chips were broken.

Microholes

Odom (2001) has discussed the production of very small holes (0.0005–0.020 in = 0.012–0.51 mm diameter). While there are several competing methods in addition to drilling [e.g., laser, waterjet, electro discharge machining (EDM), and etching], he concludes that drilling is sometimes the best, especially when the material is easy to machine.

CIRCULAR MILLING

Toenshoff (2001) has described a circular milling process that enables holes of reamed quality of different diameter to be produced by a single tool. A drilled hole is enlarged by moving a milling cutter smaller in diameter than the drilled hole along a helical path. This results in a saving in the number of drills involved when a number of holes of different diameter are required. It also results in a saving in tool changing time. The final hole is produced by intermittent cutting which results in better chip control, lower heat generation, and better fluid penetration. Hence, more accurate holes are produced with less tool wear.

ROTARY CUTTING TOOL

Consideration of cutting tool life suggests that the following conditions of tool performance would be highly desirable when attempting to cut metal at a high rate of speed:

1. a large effective rake angle insofar as the shear process is concerned but a small actual rake angle from the standpoint of strength and heat flow

2. positive means for carrying the fluid to the tool point at high cutting speeds as in the case of a journal bearing

3. the possibility of increasing chip velocity without necessitating a corresponding increase in cutting speed

The first item could be accomplished by means of inclination of the tool relative to the work velocity vector. The second two items can be achieved by providing a sidewise motion of the tool relative to the work. Indeed, if a tool which has zero inclination is given a sidewise component of motion V_r (Fig. 16.41a) as well as the customary work velocity V_w, there will be a kinematically induced inclination angle i.

Instead of observing the operation from a fixed point in space as in Fig. 16.41a, the process may be observed from a fixed point on the cutting tool. Such a picture may be obtained by simply translating everything in Fig. 16.41a to the left with a velocity V_r. This will bring the tool to rest, will cause the workpiece to move in an inclined direction with a velocity V, and will cause the chip to appear to flow off to the left rather than to the right, as shown in Fig. 16.41b.

The velocity of the chip across the tool face will be equal to V_{Cr} which is the vector sum of velocities V_C and V_r. The solid lines in Fig. 16.41b represent the cutting process that is kinematically equivalent to that in Fig. 16.41a. This equivalent process is seen to be that of cutting with a simple inclined cutting tool. Thus one of the significant features of a cutting tool with sidewise motion is to provide what is equivalent to an inclined cutting edge, the effective inclination angle i being given by

$$\tan i = \frac{V_r}{V_w} \tag{16.64}$$

The practical way of providing the sidewise motion of the tool is to use a continuous cutting edge in the form of the rotary tool shown in Fig. 16.42. Here, a particular portion of the cutting edge

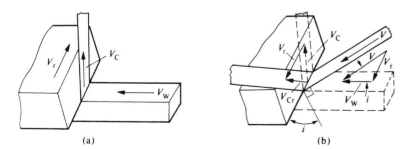

(a)   (b)

Fig. 16.41 Rotary cutting process (a) as observed from fixed point in space and (b) from fixed point on tool.

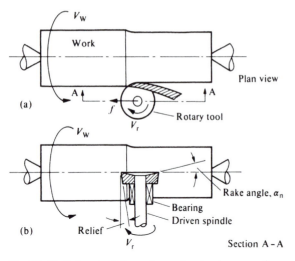

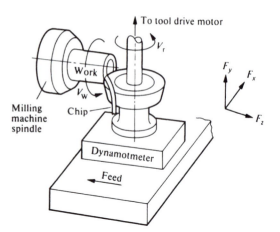

Fig. 16.42 Rotary tool shown in use in a turning operation.

Fig. 16.43 Schematic diagram showing relative arrangement of workpiece and motor-driven rotary tool.

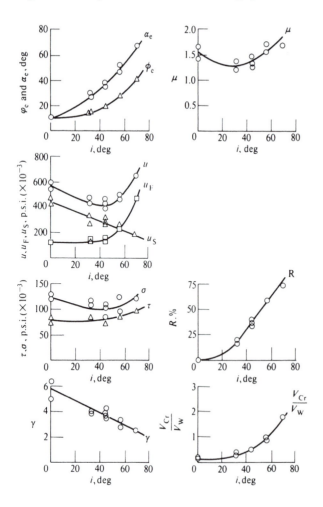

is in operation for a very brief period, which is followed by a much longer rest period during which the thermal energy associated with cutting has ample opportunity to be dissipated to the bulk of the cutter. In this respect the rotary tool resembles a multi-toothed tool such as a milling cutter.

A detailed analysis of the power-driven rotary tool has been given by Shaw et al. (1952). Experimental tests were performed on a rotary tool as shown in Fig. 16.43. Representative data are shown in Fig. 16.44 where the principal cutting variables are shown plotted versus the kinematically induced inclination angle. It is evident that the angle of inclination (i) is the predominant variable that controls performance. The following general observations may be made regarding this motor driven rotary tool:

1. The effective rake angle α_e is seen to increase as the angle of inclination is increased by increasing the tool velocity V_r. It is evident that effective rake angles of over 60° may be obtained with a rotary tool having an actual normal rake angle of but 10°.

Fig. 16.44 Variation of principal cutting variables with angle of inclination for rotary tool with 10° rake angle.

2. The effective shear angle ϕ_e is seen to increase as the effective rake angle increases, from a value of about 10° for the stationary tool to a value of over 40° when the angle of inclination was about 70°.

3. The shear stress on the shear plane is seen to increase with i, while the shear strain decreases. This same paradoxical behavior was also observed for conventional oblique cutting tools.

4. The coefficient of friction on the tool face is seen at first to decrease with increased inclination and then to increase. The minimum value of μ appears to occur at a value of i of about 30°. This result is also in agreement with the friction characteristic of the previous investigation of a conventional oblique tool, where μ was observed to decrease for values of i up to 30°.

5. The total energy per unit volume (u) has a minimum value at an inclination angle of about 45°. This minimum value is seen to be due to a tendency for the shear energy per unit volume u_S to decrease with increased i (due predominantly to the decrease in γ with increased α_e), while the friction energy per unit volume u_F exhibits the opposite trend (due to the tendency for the coefficient of friction to increase with increased α_e). The friction energy begins to increase at a rapid rate with increased inclination at a value of i of about 50°.

6. The quantity R is the ratio of the power required to drive the tool to the total power in the rotary-tool process. For the tool investigated here ($\alpha_n = 10°$) about one-third of the total power goes to drive the tool, while the remaining two-thirds is consumed in driving the workpiece, when the angle of inclination is about 45°. From the aforementioned minimum points in the curves of u and μ, it would appear that the optimum angle of inclination is somewhat under 45°. Therefore a rotary tool should be operated with a surface speed V_r just under that of work speed, V_W, and the capacity of the tool-drive motor should then be about one-half that of the main spindle motor used to drive the workpiece.

The chip–tool interface temperatures of stationary and rotating tools were determined by the chip–tool thermocouple technique. The results of this study are given in Fig. 16.45. Here it is seen that the temperature is a minimum at an angle of inclination of about 40°. This value corresponds approximately to those for minimum coefficient of friction and specific energy (Fig. 16.44). It may also be seen in Fig. 16.45 that a rotary tool operating at optimum speed may have a temperature that is as much as 400 °F lower than that for an equivalent stationary tool. This is a very significant decrease in temperature.

If a rotary tool is mounted so that its axis is inclined to the vertical (with respect to Fig. 16.42b), no drive motor is required since the tool will then be driven by the chip (Shaw et al., 1963). This corresponds to a negative side rake angle that is equal to the side setting angle if the normal rake of the tool is zero (i.e., if the tool face is perpendicular to the tool axis). As illustrated in Fig. 16.46, as the side setting angle of the rotary tool is increased, the rotational speed of the tool increases, and cutting forces and temperatures drop. Optimum performance is obtained at an amazingly high setting angle of about −70°. This corresponds to a tool axis inclined only 20° above the horizontal in Fig. 16.42b (in effect a side rake angle of −70°). At this steep side-setting angle, the speed of the chip just equals the surface

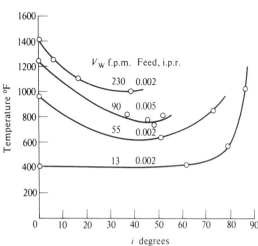

Fig. 16.45 Representative cutting temperatures for rotary tool. Work material SAE 1015 steel; tool material, 18–4–1 high speed steel; tool diameter, 2.5 in (63.5 mm); rake angle $\alpha_n = 10°$.

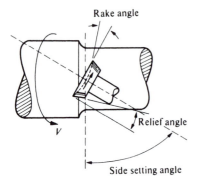

Fig. 16.46 Self-driven rotary tool showing side setting angle.

rotational speed of the tool. There is then no slip between chip and tool, and this is believed responsible for optimum performance occurring at such a high setting angle.

The self-driven rotary tool holds considerable promise for high-speed machining operations since it

1. distributes wear over a long length of cutting edge
2. provides intermittent cooling by heat flowing into the bulk of the tool
3. provides a high effective rake angle while maintaining a relatively massive tool wedge angle of high strength and thermal capacity
4. enables a more effective cutting-fluid action since the moving tool can be more effectively cooled when out of contact with the moving tool and the rotary tool will carry fluid into the chip–tool interface to provide better lubrication

After the self-driven rotary tool was introduced in 1952 (Shaw et al., 1952) and the power-driven rotary tool was introduced in 1963 (Shaw et al., 1963), it took a considerable time before these cutting techniques were adopted by others. By the end of the twentieth century, at least the following publications had appeared: Venuvinod et al. (1983a, 1983b), Chen (1992), Amarego et al. (1994), and Kishawy (1998).

Kishawy et al. (2001) have discussed use of self-propelled rotary tools for dry face-milling. Figure 16.47 shows the cutter used which had a total of eight cartridges. Each cartridge contained a circular insert 1 in (25 mm) in diameter. The cutter diameter was 9 in (0.18 m) in diameter. All inserts were ceramic with chamfered cutting edges and a 0° clearence angle. Rotary tests were performed at a cutting speed of 4500 f.p.m. (1250 m/min), a feed of 0.0018 i.p.m. (3.18 mm/min), and a depth

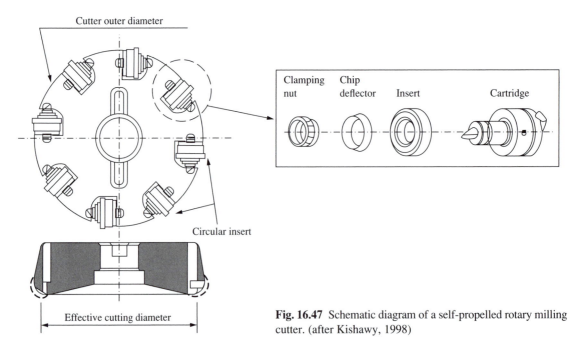

Fig. 16.47 Schematic diagram of a self-propelled rotary milling cutter. (after Kishawy, 1998)

of cut of 0.016 in (0.4 mm). The work material was annealed nodular cast iron (4% Si and 1% Mo), having a hardness of H_B 194–245 (for a 3000 kg load), in the form of automotive exhaust manifold castings. Tests were also performed with ceramic and coated carbide stationary tools with different edge preparations. The rotary tools operated at a much lower tool temperature because the active cutting edge was out of contact with the work most of the time. The wear was uniformly distributed over the circumferance of the rotating contacts and there was no evidence of surface cracks or crater wear as there was with the nonrotating tools. The cutting forces were also lower and the surface finish better when cutting with the rotary inserts than for conventional nonrotary tool cutting with the same parameters.

Kishawy (2002) has discussed the use of self-propelled rotary carbide and TiN coated carbide tools in dry hard turning of heat treated AISI 4340 steel (H_{RC} 54–56). The inserts were 1 in (25.4 mm) in diameter and mounted obliquely to the cutting speed (V). The spinning speed of the inserts was measured using an optical tachometer. The speed of an insert was found to increase linearly with the cutting speed and to be essentially independent of the insert material (carbide, coated carbide or ceramic) as shown in Fig. 16.48a. The insert speed was also found to vary linearly with cutting speed but to be rather strongly influenced by the rake angle on the insert (Fig. 16.48b). The fraction of the total heat generated at the insert–chip interface going to the chip (R_2) with the ratio of tool velocity to work velocity (V_{tool}/V) is shown in Fig. 16.49 for two values of cutting speed (V). An increase of V_{tool}/V decreases R up to a critical speed and then causes an increase. Also, increasing the cutting speed increases the amount of heat carried away by the chip. The critical speed ratio occurs when the time for one revolution of the insert is too small to effectively cool the cutting edge before returning to the cutting zone. One of the main advantages of using rotary tools in hard turning is that coated and noncoated carbide inserts are not practical for use in hard turning unless using a rotory tool, and much more expensive PCBN inserts must be used.

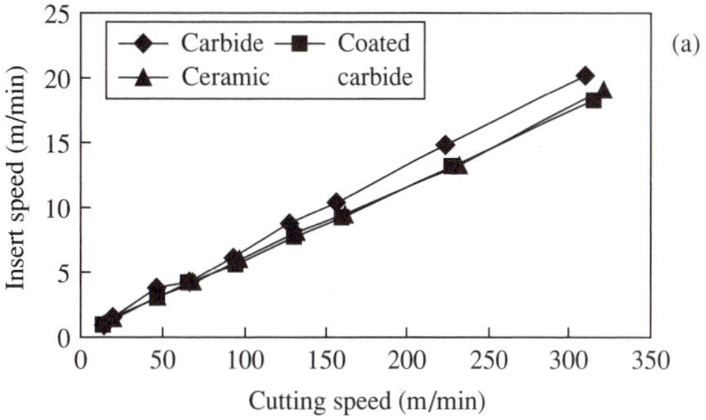

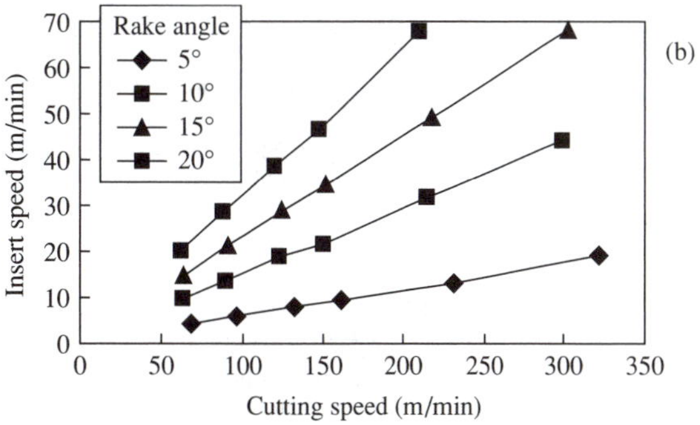

Fig. 16.48 Variation of insert spinning speed versus cutting speed. (a) For three insert materials. (b) For different insert rake angles. (after Kishawy, 2002)

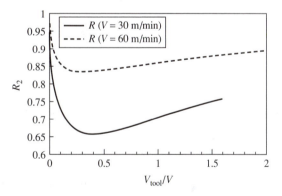

Fig. 16.49 Variation of heat partition ratio between chip and insert (R_2) and ratio of tool velocity (V_{tool}) and cutting velocity (V) for two different cutting speeds. (after Kishawy, 2002)

Figure 16.50 shows the distribution of temperature along the rake face of the tool with and without rotation of the tool. Tool rotation not only reduces the maximum temperature but moves the it closer to the tool tip. When using a rotary tool crater wear does not occur due to the reduction in peak temperature with rotation. Tool wear is distributed uniformly along the circumference of

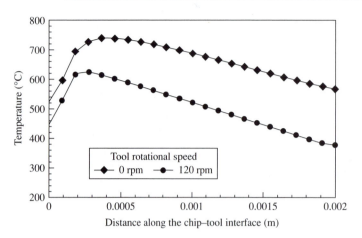

Fig. 16.50 Tool temperature variation with distance along the tool face from the tool tip for nonspinning and spinning inserts. (after Kishawy, 2002)

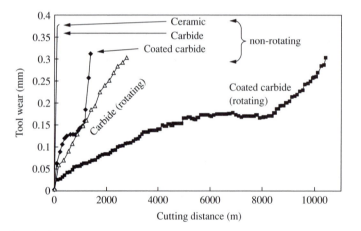

Fig. 16.51 Variation of tool wear-land length versus cutting distance for three tool materials and for nonrotating and rotating inserts. $V = 394$ f.p.m. (120 m/min); $f = 0.008$ i.p.r. (0.2 mm/r); b (DOC) = 0.004 in (0.1 mm). (after Kishawy, 2002)

a rotary tool. Figure 16.51 shows the variation of wear-land wear with cutting distance for carbide, ceramic, and coated carbide tools, with and without rotation for the same cutting conditions. In many cases when sawtooth chips were formed in conventional hard turning (as discussed in Chapter 22), continuous chips of essentially uniform thickness were produced under the same cutting conditions with a self-propelled rotary tool.

Destefani (2002) discusses the use of rotating inserts in milling and boring cutters for use in the production of automotive engine blocks. The cutters employ needle bearings to provide constraint in the rotary and thrust directions with relatively low friction.

ELLIPTICAL VIBRATION CUTTING (EVC)

Moriwaki and coworkers (Shamoto and Moriwaki, 1994; Moriwaki and Shamoto, 1995) have studied a cutting process in which the tool is moved in an elliptical path as shown in Fig. 16.52 in order to reduce tool friction. This causes the tool to cyclically reverse the frictional direction, which results in a reduction of chip thickness and cutting force (F_p). By setting the maximum vibrating speed in the cutting direction higher than the normal cutting speed, the tool is out of contact with the chip during each cycle of vibration. While cutting, the tool is moving in the chip flow direction and moves opposite to the chip flow direction when out of contact with the chip. This results in a virtual lubrication effect that increases the shear angle and reduces the cutting force (F_P).

Figure 16.53 shows the principle of the ultrasonic elliptical vibrator. The cylindrical tool holder is elastically bent cyclically relative to two sets of support that act as nodal points of vibration. The vibrator is vibrated at 20 kHz by two piezoelectric (PZT) plates which are driven by two sinusoidal voltages having a phase shift between them. This causes vibration of the tool holder in an elliptical mode. Figure 16.54 shows the measuring system employed. The vibration system developed enables elliptical vibration with an amplitude less than 4 μm. The maximum vibrating speed is 131 f.p.m. (40 r.p.m.). This needs to be increased since it must be greater than the cutting speed which is greater than this for most practical applications.

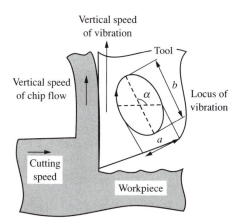

Fig. 16.52 Principle of elliptical vibration cutting (EVC). (after Moriwaki and Shamoto, 1995)

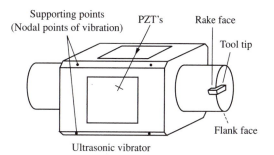

Fig. 16.53 Ultrasonic elliptical vibrator. (after Moriwaki and Shamoto, 1995)

Figure 16.55 shows the quasi-orthogonal mode of cutting employed in the experimental evaluation of the method. Cutting conditions were as follows for ultrasonic elliptical vibration cutting:

- work material: oxygen free copper
- tool material: HSS
- rake angle: 0°
- clearance angle: 13°
- depth of cut: 0.0002 in (5 μm)
- rotational speed of work: 16.4 f.p.m. (5 m/min)
- vibrational frequency: 20 kHz
- amplitude a: 8 μm
- amplitude b: 11 μm
- inclination angle (α in Fig. 16.52): 110°

These results were compared with ordinary conventional cutting and conventional vibration cutting where vibration is mainly in the cutting direction. In this latter case:

- amplitude a: 0 in (0 μm)
- amplitude b: 0.0006 in (16 μm)
- inclination angle: 10°

Figure 16.56 shows mean principal (F_P) and thrust (F_Q) cutting forces for the interrupted cutting tests described above.

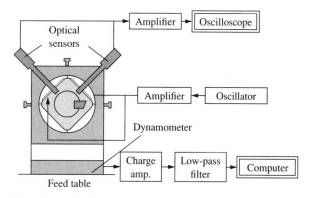

Fig. 16.54 Driving and measuring system for ultrasonic elliptical vibration cutting (EVC). (after Moriwaki and Shamoto, 1995)

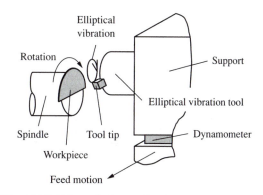

Fig. 16.55 Schematic for quasi-orthogonal interrupted cutting experiments. (after Moriwaki and Shamoto, 1995)

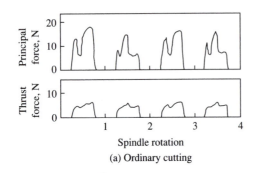

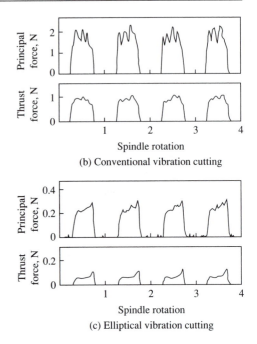

Fig. 16.56 Cutting force component F_P and F_Q for three indicated types of experiments. (after Moriwaki and Shamoto, 1995)

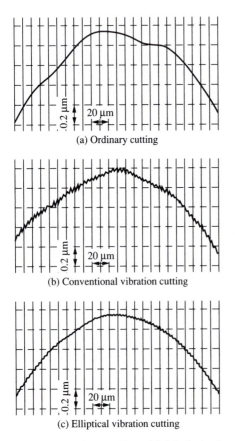

Fig. 16.57 Surface profiles and finish obtained for three indicated types of experiments. (after Moriwaki and Shamoto, 1995)

The chip thickness for conventional vibration cutting (Fig. 16.56b) was about the same as for ordinary cutting (Fig. 16.56a) but the average cutting force (F_P) was reduced by a factor of $\frac{1}{6}$ (the same as the reduction of actual cutting time).

The chip thickness for EVC was significantly lower than for ordinary cutting and the shear angle for EVC was about 51°. The principal cutting force (F_P) was $\frac{1}{42}$ the force for ordinary cutting while the thrust force (F_Q) was $\frac{1}{76}$ the corresponding ordinary cutting force, and the actual cutting time was only $\frac{1}{4}$ of the conventional value.

The width of the EVC chips was equal to the width of the work piece (0.008 in = 200 μm) while that for the other two cases was greater due to burr formation (0.012 in = 300 μm). Thus EVC suppresses burr formation. Figure 16.57 shows surface profiles for surfaces produced by the three types of cutting. The finish for conventional vibration cutting is much poorer than that for the other two cases but the EVC finish is not quite as good as that for ordinary cutting. However, the contour shape of the finished surface is best for the EVC surface.

The experimental apparatus employed in Morawaki and Shamoto (1995) is greatly improved over that employed in Shamoto and Morawaki (1994). Most important, the maximum oscillating frequency employed in the 1994 paper was only 6 Hz while that employed in the 1995 paper was 20 Hz. It would appear that if the power of the oscillator can be greatly increased so that the size of cut can be substantially increased, EVC will be a very useful technique for general turning purposes. Otherwise, the method will still represent a very important contribution to the area of nanochip machining where the concept apparently had its beginning.

The Oxley Machining Theory discussed in Chapter 8 for predicting machining forces has been extended to boring (an internal continuous operation) and to reaming (an internal intermittant operation) in McKendrick et al. (2001). This approach to predicting cutting performance should be applied with caution, however, since the large extrapolation of strain and the large difference in effective size between the materials tests employed to obtain the needed material properties and the size of the deformation zone in chip formation are subject to question for the reasons cited in Chapters 8 and 20.

REFERENCES

Agapiou, J. S., and Stephenson, D. A. (1994). *J. of Eng. for Ind., Trans. ASME* **116**, 54–60.

Amarago, E. J. A., Karri, V., and Smith, A. J. R. (1994). *Int. J. Machine Tools and Mfg.* **34/6**, 785–815.

Armitage, J. B., and Schmidt, A. O. (1957). *Tool Engr.* **27**, 36.

Boston, O. W., and Oxford, C. J., Sr. (1930). *Trans. Am. Soc. Mech. Engrs.* **52**, 5.

Chen, P. (1992). *Annals of CIRP* **41/1**, 59–62.

Destefani, J. (2002). *Mfg. Eng. (SME)* **1** (Jan.), 52–57.

Kishawy, H. A. (1998). Ph.D. dissertation. McMaster University, Hamilton, Ontario, Canada.

Kishawy, H. A. (2002). *Annals of CIRP* **51/1**.

Kishawy, H. A., and Gerber, A. G. (2001). *ASME MED* **23312**, 1–8.

Kronenberg, M. (1935). *Machinery* **45**, 661.

Kronenberg, M. (1966). *Machining Science and Application.* Pergamon Press, Oxford.

Martellotti, M. E. (1941). *Trans. Am. Soc. Mech. Engrs.* **63**, 677.

McKendrick, I. R., Tungka, J., Arsecularatne, J. A., and Mathew, P. (2001). *Machining Sc. and Technology* **53/3**, 375–391.

Merchant, M. E. (1944). *Trans. Am. Soc. Mech. Engrs.* **66**, A–168.

Moriwaki, T., and Shamoto, E. (1995). *Annals of CIRP* **44/1**, 31–34.

Odom, B. (2001). *Mfg. Eng.* (Feb.), 88–102.

Oxford, C. J., Jr. (1955). *Trans. Am. Soc. Mech. Engrs.* **77**, 103.

Shamoto, E., and Moriwaki, T. (1994). *Annals of CIRP* **43/1**, 35–38.

Shaw, M. C. (1954). *Metal Cutting Principles*, 3rd ed. Technology Press, MIT, Cambridge, Mass.

Shaw, M. C., Cook, N. H., and Smith, P. A. (1952). *Trans. Am. Soc. Mech. Engrs.* **74**, 1055.

Shaw, M. C., and Oxford, C. J., Jr. (1957). *Trans. Am. Soc. Mech. Engrs.* **79**, 139.

Shaw, M. C., Smith, P. A., and Cook, N. H. (1952). *Trans. Am. Soc. Mech. Engrs.* **74**, 1065.

Shaw, M. C., Smith, P. A., Cook, N. H., and Loewen, E. G. (1963). Canadian Patent No. 665,459.

Stabler, G. V. (1951). *Proc. Instn. Mech. Engrs.* **165**, 14.

Stephenson, D. A., and Agapiou, J. S. (1992). *J. Mach. Tools and Mfg.* **32**, 521–538.

Toenshoff, H. K. (2001). *Minutes of the CIRP Sc. Tech. Committee*, Paris, February 25, 2001.

Toews, W. D., Compton, W. D., and Chandrasekar, S. (1998). *Prec. Eng.* **22**, 1–9.

Usui, E., Hirota, A., and Masuko, M. (1978a). *Trans. ASME* **100**, 222–228.

Usui, E., and Hirota, A. (1978b). *Trans. ASME* **100**, 229–235.

Usui, E., Shiakashi, T., and Kitagawa, T. (1978c). *Trans. ASME* **100**, 236–243.

Venuvinod, P. K., and Rubenstein, C. (1983a). *Annals of CIRP* **32/1**, 53–58.

Venuvinod, P. K., Lau, W. S., and Reddy, P. N. (1983b). *Annals of CIRP* **32/1**, 59–64.

17 SURFACE INTEGRITY

Table 14.1 shows a turning operation from the point of view of a system. Here a number of inputs are listed together with important internal items and outputs. The outputs are most important since they represent the end-goal of the operation. The integrity of the machined surface is frequently one of the important outputs.

Surface integrity is a term that involves several considerations: surface finish and freedom from cracks, chemical change, thermal damage (burn, transformation, and overtempering), and adverse (tensile) residual stress. The first of these (surface finish) is by far the most important for finishing operations. The others are mainly a concern relative to ground surfaces. Where surfaces finish is relatively important, a finishing operation is often but not always involved.

MEASUREMENT AND SPECIFICATION OF SURFACE FINISH

In engineering design and production, it is important to be able to specify the degree of surface roughness desired. Before 1930 this was done by use of tactual standards. This involved the use of a series of specimens that had different finishes. The man in the shop used these specimens by running his fingernail first across a standard tactual surface and then across the surface he was producing. When the two surfaces were felt to have the same roughness, the workpiece was considered to be smooth enough. In applications where surface finish is important to the functioning of a device such as in face seals, ball bearings, gears, cam surfaces, or journals, it is found that performance varies linearly as the surface roughness varies logarithmically. This means that for there to be a twofold improvement in performance, there must be a tenfold reduction in the mean peak-to-valley roughness Because of this, a set of tactual standards will have peak-to-valley roughness values that stand in the following ratios: 1:2:4:8:16:32:64:128:256:512:1024 instead of being linearly spaced.

Surface finish is specified on a drawing as shown in Fig. 17.1. If a smoother surface is required than one of 64, a finish of 32 would be specified instead of one between 32 and 64.

In the early 1930s surface roughness was made quantitative by the appearance of stylus instruments that traversed a surface to be measured with a diamond stylus, using transducers to convert the vertical and horizontal motions of the diamond into recorded traces. In the United States, the vertical motion of the stylus was analyzed automatically and a meter read the square root of the mean

Fig. 17.1 Method of indicating the surface finish required on a machine part.

of the vertical motions squared. This so-called root-mean-square (RMS) value in micro-inches (μin) was specified on drawings as shown in Fig. 17.1 in terms of the progression 1, 2, 4, 8, 16, 32, 64, 128 μin.

On the continent of Europe and in Japan, a similar development took place but the characteristic maximum peak-to-valley roughness (R_t) is used in place of the RMS value. In Great Britain, the center line average (CLA) value of the surface is used in place of the RMS value. This is now also the standard in the United States and is called arithmetic average (AA) roughness. The CLA or AA roughness (R_a) is obtained by measuring the mean deviation of the peaks from the center line of a trace, the center line being established as the line above and below which there is an equal area between the center line and the surface trace. There is little difference between the CLA and RMS values for a given surface. (RMS = 0.90 AA for a sine wave.) The only virtue of the RMS value is that it is easily measured using an AC voltmeter. However, the CLA value is more readily interpreted physically. Modern tracer instruments have the capability of giving the AA value as a meter reading and the shape of the surface as a magnified tracing. The vertical magnification of the tracing is usually arranged to be five to 100 times as great as in the horizontal direction since practical surfaces have asperities that are very widely spaced relative to their differences in height. Figure 17.2 is a typical stylus trace of a relatively smooth surface (AA = 10 μin = 0.25 μm).

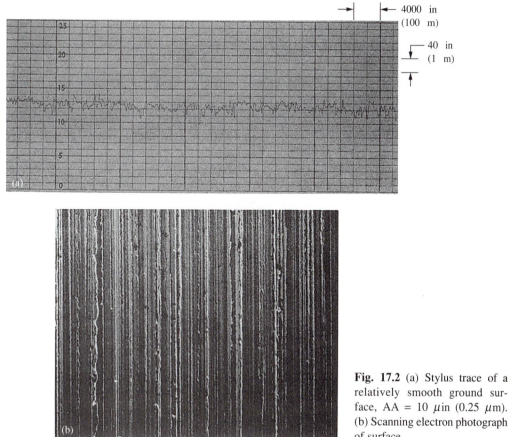

Fig. 17.2 (a) Stylus trace of a relatively smooth ground surface, AA = 10 μin (0.25 μm). (b) Scanning electron photograph of surface.

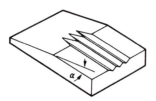

Fig. 17.3 Tapered section.

The peak-to-valley roughness for a ground surface is usually about five times the AA value. However, other ratios pertain for other finishing methods. For example, this ratio is closer to 10:1 for lapped surfaces.

The radius of curvature of the standard diamond stylus is 0.0005 in (12.5 μm) although in research, instruments with stylii having a radius as small as 0.0001 in (2.5 μm) are used.

In order to obtain the true profile of a surface without the influence of the curvature of the tracer, a taper section of the surface may be used. After plating the surface with nickel to protect it, a new flat surface is ground on the nickel which makes a small angle with the original surface as shown in Fig. 17.3. The interface between the steel and nickel may then be studied.

A taper section magnifies the peak-to-valley distance and makes it possible to observe the profile of a surface under study at different magnifications in the horizontal and vertical directions. By making the inclination of the two surfaces differ by $a = 2°17'$, a magnification of CSC $(2°17')$ = 25 times is obtained. If the polished new surface is viewed with a microscope at 100×, then the horizontal magnification of the nickel surface will be 100× while the vertical magnification is 2500×. The taper section method has been discussed in Nelson (1940, 1941), Bowden and Moore (1947), and Moore (1948).

Uchida et al. (1979) have developed a two-dimensional automated method of measuring surface roughness by a light sectioning method. A slit beam is projected on the surface, and the reflected beam passes through a microscope and onto a television camera connected to a minicomputer.

Mitsui and Sato (1974) have described an in-process sensor for measurement of external roughness by use of a laser beam. A photo diode array is used to detect the small displacement of the laser beam by surface roughness. Roughness in the circumferential direction of 80 μin (2 μm) amplitude and 0.002 μin (50 μm) wavelength could be measured at a cutting speed of 180 f.p.m. (500 mm/s).

Table 17.1 gives values of surface finish R_t and R_a for a variety of surfaces of different roughness.

In addition to surface roughness, surfaces may exhibit waviness. The waviness of a surface is the vertical distance between peaks and valleys of relatively long-wavelength irregularities while the corresponding peak-to-valley height for short-wavelength irregularities is surface roughness. The maximum wavelength for surface roughness is usually taken to be 0.03 in (0.75 mm). By filtering

TABLE 17.1 Representative Surface Roughness Values for Different Surfaces

Application	R_a		R_t	
	μin	μm	μin	μm
Castings, brick, fracture surfaces	400	10	1500	38
Rough machined surfaces	250	6.3	1000	25
Medium machined surfaces	125	3.2	600	15
Most noncontact and static surfaces	63	1.6	300	8
Gear teeth, pressed fits	32	0.8	150	4
Heavy loaded gear teeth, splines	16	0.4	80	2
Hydraulic and engine cylinders	8	0.2	50	1.3
Ball bearing races	4	0.1	30	0.8
Gaging surfaces	2	0.05	20	0.5
Gage blocks, ball bearing, ball surfaces	1	0.025	10	0.25

the output from a stylus-type tracer instrument, irregularities having a wavelength greater than 0.03 in may be removed, and it is possible to measure surface roughness without interference due to waviness. Figure 17.4 illustrates the difference between surface roughness and waviness. Figure 17.4a is a flat surface with roughness while Fig. 17.4b is a wavy surface with roughness.

Finish-turning is characterized by relatively small depths of cut and light feeds. The depth of cut will normally be less than 0.06 in (1.5 mm) and the feed less than 0.006 i.p.r. (0.15 mm/rev). The items of major interest in finish-machining are

1. dimensional accuracy
2. surface finish
3. tool life

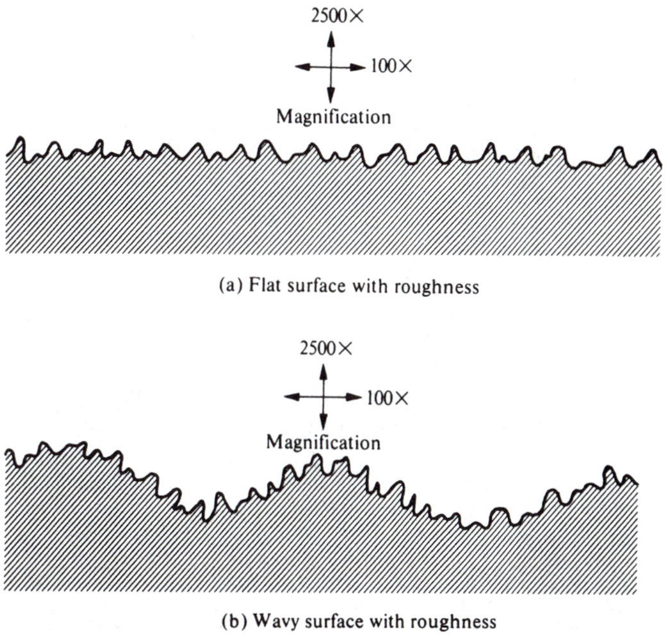

(a) Flat surface with roughness

(b) Wavy surface with roughness

Fig. 17.4 Flat and wavy surfaces with surface roughness.

Dimensional accuracy is mainly a matter of avoiding errors in longitudinal and circumferential form. Errors of longitudinal form result from static deflection of the spindle and workpiece under cutting forces and thermally induced stresses in the machine. Errors in circumferential form result from run-out of the spindle and from vibration of the tool or workpiece. The accuracy achieved in any particular case mainly depends upon the stiffness and stability of the machine tool. Representative studies of errors in form have been presented by Schuler (1957), and the dynamic aspects of machine-tool performance are to be found in many places, for example, Hahn (1953); Doi and Kato (1956); Tobias and Fishwick (1958); Danek, Polacek, Sparek, and Tlusty (1962); Tobias (1965); and Peters (1963).

There are two distinct types of finish encountered in a turning operation:

1. the finish produced by the primary cutting edge
2. the finish produced by the secondary cutting edge

The first of these normally pertains to surface broaching and form-turning, while the second is characteristic of conventional turning using a tool having a nose radius.

The range of roughness values to be expected in finish-machining is as follows:

- turning: $R_a = 32–64$ μin (0.75–1.5 μm)
- broaching: $R_a = 4–16$ μin (0.1–0.4 μm)

The corresponding values of R_t will be about five times these values of R_a. (A very good turned surface will have a finish corresponding to $R_t = 4$ μm or 160 μin.)

Primary Cutting Edge Finish

Knowledge of the finish produced by the primary cutting edge is not only of direct importance in processes where the resultant surface is generated by the major cutting edge (i.e., form-cutting with

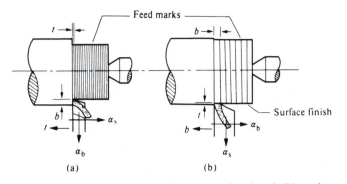

Fig. 17.5 Schematic diagram of (a) conventional and (b) analog cutting processes with an interchange of feed and depth of cut and back (α_b) and side (α_s) rake angles. (after Shaw and Crowell, 1965)

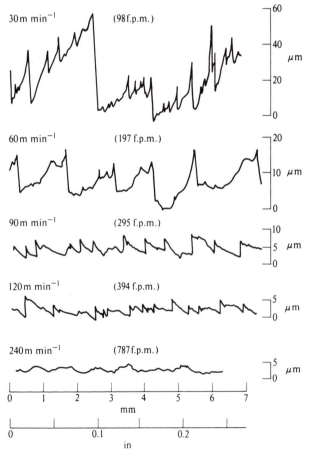

radial feed), but it is also of importance in interpreting the wear on the clearance face of turning tools employing a nose radius. This type of finish may be conveniently studied at low speed using a planing arrangement, or at high cutting speeds by cutting the end of a tube under orthogonal conditions using a doubly overhanging tool and an axial feed. When the work material is not available in the form of a tube, the analog tool shown in Fig. 17.5b is convenient. In the analog arrangement the feed and depth of cut are interchanged along with the side and back rake angles. By use of the analog tool a helical tool path is generated which has a sufficient width (0.1 in, 2.5 mm) to enable surface roughness to be measured with a tracer instrument without interference from the feed marks. Cutting-force values obtained from an analog tool are found to be identical with those from the prototype tool.

Surface profiles produced by the major cutting edge of a tool are shown in Fig. 17.6, and comparable values produced by a secondary cutting edge machining the same material are shown in Fig. 17.7.

The profiles of Fig. 17.6 were obtained by tracing in the circumferential direction while those of Fig. 17.7 were obtained by tracing in the axial direction.

The finish produced is seen to be strongly dependent on the cutting speed (V). The finish for low speeds is very poor, particularly that produced by the primary cutting edge. Much less roughness is obtained at higher speeds, particularly in the case of surfaces generated by the primary cutting edge. The finish obtained by use of a secondary cutting edge is seen

Fig. 17.6 Surface finish produced by a primary cutting edge at different cutting speeds. Work material, normalized AISI 1045 steel; tool material, tungsten carbide; tool geometry, rake and clearance angles = 6°; depth of cut = 0.1 in (2.5 mm); feed = 0.008 i.p.r. (0.2 mm/rev); cutting fluid, none. (after Shaw and Crowell, 1965)

to approach that corresponding to the feed marks at a high cutting speed (V).

Fracture Roughness

When cutting steel at very low speeds, a special type of roughness is frequently observed due to subsurface fracture. Figure 17.8 shows the origin of this type of roughness. When the metal cut is in the vicinity of room temperature, the chip is frequently found to fracture periodically. When it fractures, the crack is frequently found to follow approximately a path of maximum shear stress. Figure 17.9 shows the elastic stress pattern for a concentrated load applied to a plane surface. In this diagram the circles (such as E′) represent points of constant shear stress magnitude while the curved lines (such as OG) give the direction of maximum shear stress. Curved line OG in Fig. 17.9 has a direction similar to line AB in Fig. 17.8.

When the cracks run completely across the chip, the surface has a Moire appearance (Fig. 17.8c) consisting of alternate dull and shiny areas. The shiny areas are those which are actually cut while the dull (gray) areas are the uncut regions of the finished surface. As the speed (temperature) is increased, the cracks may not go all the way across the chip. Such chips will appear to be continuous, but careful examination of the finished surface reveals many small gray craters as shown in Fig. 17.8d. These craters are due to the same subsurface fracture. Above a certain cutting speed, a subsurface pattern will no longer be found and this source of surface roughness disappears.

The low-speed fracture pattern depends upon the number of points of stress concentration in the steel. A resulfurized steel will give a fracture pattern of shorter pitch and likewise of smaller depth. Thus, a resulfurized steel would tend to look like Fig. 17.8c if the nonresulfurized steel looked like Fig. 17.8b. The surface finish obtained with a resulfurized steel is far better than that obtained for a nonresulfurized steel when

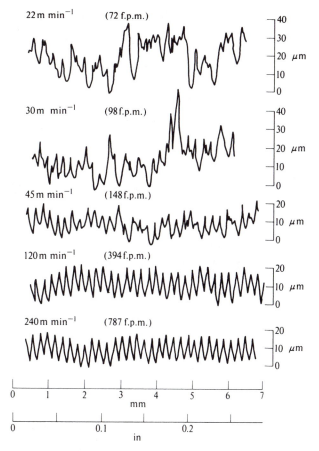

Fig. 17.7 Surface profiles for conventionally turned surfaces. Cutting conditions same as Fig. 17.6 except tool had 15° side- and end-cutting edge angles and the nose radius was 0.032 in (0.8 mm). (after Shaw and Crowell, 1965)

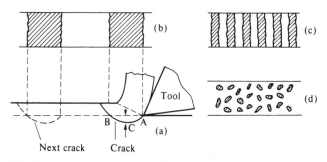

Fig. 17.8 Schematic diagram showing cracks which form at tool tip at low cutting speeds (< 50 f.p.m. or 15 m min⁻¹). Diagrams (b), (c), and (d) are plan views of machined surfaces showing different patterns of fracture-induced roughness. (after Shaw et al., 1962)

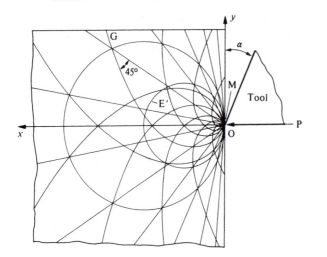

Fig. 17.9 Elastic stress pattern existing at the beginning of a cut. Circles represent lines of constant shear stress magnitude and curved lines are lines of constant shear stress direction. (after Ernst and Martellotti, 1935)

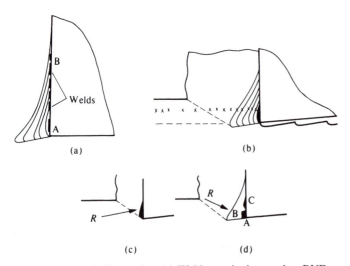

Fig. 17.10 The built-up edge. (a) Welds required to anchor BUE to tool face. (b) Origin of the layered structure and growth and decay of BUE. (c) Magnitude and direction of resultant cutting force (R) for small BUE. (d) Magnitude and direction of resultant cutting force (R) for large BUE. (after Shaw and Sanghani, 1963)

cutting in the low-speed region where subsurface cracks are apt to form.

Built-Up Edge

As the cutting speed is increased, the friction between chip and tool will increase; and when this becomes large enough to cause a shear fracture in the vicinity of the tool face, a built-up edge (BUE) will form (Fig. 17.10). There is no BUE at very low cutting speeds ($V = 1$ f.p.m. or 0.3 m min^{-1}) since the temperature on the face of the chip is then not sufficient to cause the chip surface to behave in a ductile manner. With an increase in cutting speed, the chip metal in contact with the chip face becomes ductile, and the resulting plastic flow causes strong welds to form between chip and tool. The additional plastic flow on the chip face causes strain hardening and a further increase in the force tending to anchor the chip to the tool. When the bonding force between chip and tool exceeds the shear strength of the metal in the main body of the chip, at some particularly weak point near the tool face, the BUE forms.

The BUE is a dynamic body that has a layered structure (Shaw and Sanghani, 1963). When a point of particularly high stress concentration approaches the outer face of the BUE, a new shear surface forms and the BUE grows in size. The origin of the layered structure and the gradual growth of the BUE due to a statistical array of imperfections in the work material is shown diagramatically in Fig. 17.10b. As the BUE grows forward, it will usually also grow downward, causing the finished surface to be undercut. The BUE causes an increase in the rake angle which in turn causes a decrease in the magnitude of the resultant force on the tool and a clockwise rotation of the resultant force vector. For a small BUE the direction of the resultant force is such as to put the BUE into compression and to make it stable. When the BUE becomes large, the resultant force loads the BUE as a cantilever, and eventually the moment at the base of the BUE becomes sufficient to pry it loose. The BUE then passes off partly with the chip and partly on the finished surface. The gradual growth and rapid decay of the size of the BUE causes a sawtoothed finished surface (Fig. 17.10b), which is characteristic of the BUE component of surface roughness.

Influence of Sulfur, Lead, and Cold Work

The BUE has a major influence on the cutting forces encountered at intermediate cutting speeds. Figure 17.11 shows representative cutting force results for analog tools using five steels from the same heat but containing different amounts of manganese sulfide. The solid curve is for a non-resulfurized steel. The tangential or power component of cutting force is designated F_P and F_Q is the

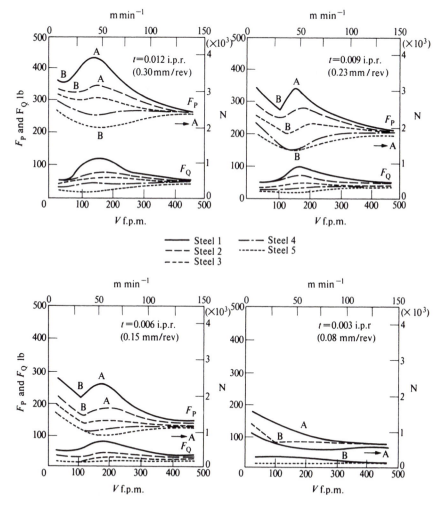

Fig. 17.11 Variation of cutting forces F_P (tangential) and F_Q (radial) for an analog-tool cutting hot-rolled steels of different sulfur content. All materials are plain-carbon steels and 0.08% carbon content, 1% manganese, and the following sulfur contents: steel 1, 0.033%; steel 2, 0.11%; steel 3, 0.18%; steel 4, 0.26%; steel 5, 0.37%. Rake angle, 10°; clearance angle, 5°; nose radius, 0.005 in (0.125 mm); width of cut, 0.10 in (2.5 mm); undeformed chip thickness = t (variable); cutting fluid, none; tool material, M–2 HSS. (after Shaw, Usui, and Smith, 1961)

force component corresponding to the feed force, but which in this case is directed radially since an analog tool is used. Force component F_P is large at very low speeds since no BUE is then present to alter the rake angle. It falls to a minimum at B where the BUE has its maximum size and influence on the effective rake angle. The BUE decreases in size from B to A and there is essentially no BUE at speeds to the left of point A. With increase in cutting speed, the BUE is at first small since the anchoring force between chip and tool is then small. As this anchoring force becomes larger, the maximum size of BUE grows. At higher speeds (temperatures) the material on the tool face begins to soften and the anchoring force decreases. The drop in cutting force to the right of point A is due to the thermal softening of the material on the tool face and the attendant decrease in tool-face friction. The asymptotic level of force in the vicinity of 500 f.p.m. (175 m min^{-1}) is a result of the surface of the chip being essentially liquid-like.

The BUE has been extensively studied by Nakayama (1957) who cites the following ways of eliminating or reducing the size of the BUE:

1. increase the cutting speed
2. make metal that is less ductile (i.e., cold-work or heat-treatment); brittle metals do not form a BUE
3. increase the rake angle
4. use a fluid (but only at low speeds)

Item 1 may be generalized when it is realized that an increase in speed is effective because it causes the chip temperature to reach a critical value. An increase in feed, use of a negative rake angle (which increases tool-face temperature), or use of hot-machining technique will enable the BUE to be eliminated at a lower critical cutting speed. Nakayama found the tool material had no influence on the critical cutting speed. He also found the appearance of the tool-face side of the chip to offer a convenient means for determining when a BUE was present. In the absence of a BUE the chip face was extremely smooth but when a BUE was present it contained evidence of surface welding.

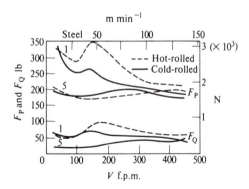

Fig. 17.12 Comparison of cutting-force values for hot-rolled and cold-drawn [two drafts of $\frac{1}{8}$ in (3.18 mm) each from a diameter of $4\frac{1}{2}$ in (114.3 mm)] steels. Cutting conditions are the same as in Fig. 17.11 except for the underformed chip thickness (t) which was 0.009 i.p.r. (0.23 mm/rev). (after Shaw, Usui, and Smith, 1961)

As sulfur is added to the steel, the speed corresponding to a maximum size of BUE (B) is seen to shift to the right as does maximum point A (point where the BUE has disappeared completely). For a feed of 0.012 i.p.r. (0.3 mm/rev), the BUE is seen to disappear at a speed of about 150 f.p.m. (50 m min^{-1}) for the steel with no added sulfur. However, a BUE is still present at 450 f.p.m. (150 m min^{-1}) when cutting the steel of maximum sulfur content. The role that cold-rolling plays on the BUE and hence the cutting forces is reflected in the results of Fig. 17.12 which are for the non-resulfurized steel cut at a feed of 0.009 i.p.r. (0.23 mm/rev).

The variation in the arithmetic average (centerline average) roughness (R_a) is given in Fig. 17.13. These results are for the same hot-rolled steel tests given in Fig. 17.11 and hence are also analog tool results. Sulfur is seen to greatly improve surface finish at low cutting speeds but to cause a slightly poorer finish at high speeds (400 f.p.m. or 130 m min^{-1}), due to the fact that sulfur tends to cause an unusually stable BUE. The main role of manganese sulfide in steel is to provide a large number of stress-raisers of approximately equal

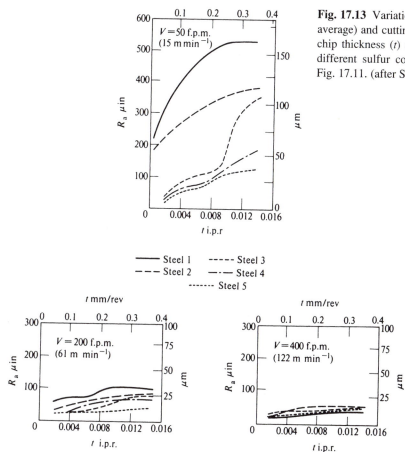

Fig. 17.13 Variation of arithmetic average (centerline average) and cutting roughness (R_a) with undeformed chip thickness (t) and cutting speed (V) for steels of different sulfur content. Cutting conditions same as Fig. 17.11. (after Shaw, Usui, and Smith, 1961)

strength. This in turn causes a more homogeneous BUE consisting of a large number of thin layers. Such a BUE tends to be more stable and not to fluctuate in size so wildly in the large BUE region. It also tends to remain on the tool to very much higher cutting speeds.

Cold rolling improves the finish obtained with a non-resulfurized steel (Fig. 17.12); however, the improvement in finish due to cold-working is not as marked for resulfurized steel cut at light feed.

The best tool finish obtained with an analog tool (12–14 μin or 0.30–0.35 μm) is remarkably independent of undeformed chip thickness (feed) and is obtained at high speeds (> 300 f.p.m. or 100 m min^{-1}) and with a cold-drawn non-resulfurized steel. Sulfur has its greatest influence in improving finish at low and intermediate cutting speeds. For example, at a cutting speed of 200 f.p.m. (61 m min^{-1}) and a feed of 0.003 i.p.r. (0.075 mm/rev), the analog tool finish for the cold-drawn resulfurized steel (5C) was 20 μin (0.5 μm) compared with 35 μin (0.9 μm) for the non-resulfurized cold-drawn steel. At a speed of 50 f.p.m. (15 m min^{-1}) and the same feed, the corresponding values of finish were 15 μin (0.4 μm) for the 5C steel and 350 μin (8 μm) for the 1C steel. Inasmuch as a steel must usually operate over a wide speed range, the best overall performance will be obtained with a cold-drawn resulfurized steel. An analog tool finish R_a of 15–20 μin (0.4–0.5 μm) may be obtained with such a steel over the entire practical range of speeds and feeds (20–800 f.p.m. or 6–244 m min^{-1}; 0.001–0.012 i.p.r. or 0.025–0.3 mm/rev).

Additions of lead to steel are found to improve surface finish, and lead is relatively more effective at light feeds (*t*) which is the reverse of the trend with sulfur additions. Also, lead does not increase the stability of the BUE at high speeds, and unlike sulfur, lead gives better finish at all speeds. Cold-drawing improves the finish obtained with a leaded steel, as in the case of a resulfurized steel.

One of the results of adding sulfur or lead to steel is to cause the chip to curl more and to decrease the chip–tool contact length. This in turn lowers cutting forces and the tendency to form a large BUE at low speeds. The chip–tool contact length may be reduced artificially, by grinding away the tool-face to produce a smaller chip–tool contact length. When this is done a large part of the benefit of sulfur relative to surface finish is lost and a non-resulfurized steel machined using a cut-away tool will give a fairly good finish even at low cutting speeds. However, the poorer finish obtained with sulfur at high cutting speeds becomes even poorer when a cut-away tool is used. A reduced-contact-length tool also gives poorer finish at high speeds when machining a leaded steel (Shaw and Usui, 1962).

At low cutting speeds best finish is obtained when a resulfurized steel is cut using a reduced-contact-length tool having a chip–tool contact length of 3*t* (where *t* is the undeformed chip thickness). For a non-resulfurized steel of low-carbon content the best chip–tool contact length appears to be about 2*t*. From the point of view of finish a reduced-contact-length tool should be used at low cutting speeds but not at high speeds.

Experimental Results: Normalized AISI 1045 Steel

A better understanding of primary cutting edge roughness may be obtained by simultaneously recording surface roughness, cutting forces, tool–chip interface temperature (by tool–chip thermo-couple measurements), and chip formation for a wide range of cutting speeds (V). This is done in the following discussion for normalized AISI 1045 steel.

The following characteristic cutting speeds (V) may be identified when cutting forces, surface roughness values, and chip and workpiece surfaces are examined:

1. V_1 = speed at which tool-face side of chip first becomes shiny
2. V_2 = speed at which the principal cutting force (F) is a maximum
3. V_3 = speed at which cracks in the finished surface disappear
4. V_4 = speed at which the peak-to-valley roughness (R_t) reaches a constant value
5. V_5 = speed at which the arithmetic average (center line average) roughness (R_a) reaches a constant value
6. V_6 = speed at which the work surface first becomes shiny

By surface temperature measurements it is suggested that all of these changes are associated with the attainment of characteristic temperatures on the tool face and tool flank.

Test bars were first bored to produce tubes of 0.12 in (3 mm) wall thickness in order to enable orthogonal cuts to be taken. It was thus a simple matter to obtain samples of surfaces produced by the primary cutting edge by quickly extracting the tool from the work. Actual surfaces were saved for subsequent study by parting a small ring of metal from the workpiece. The tool was held in a two-component dynamometer having a capacity of 4000 lb (1800 kg), connected to a two-channel recorder having a frequency response of 120 Hz.

Carbide tools (P–10) were used. This is a crater and edge wear-resistant grade having a density of 12.42, a hardness of H_{RA} 92, and a transverse rupture strength of 250,000 lb in^{-2} (176 kg mm^{-2}). All tools were of the clamped insert type and had rake and relief angles of 5°. The inserts were triangular in shape (0.37 in i.d. or 9.5 mm i.d.) and were 0.126 in (3.2 mm) thick.

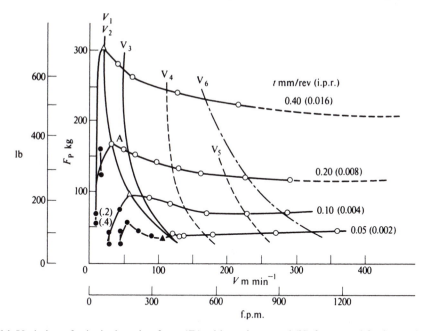

Fig. 17.14 Variation of principal cutting force (F_P) with cutting speed (V) for several feeds (t) when cutting in air. Work material, AISI 1045 steel. (after Nakayama, Shaw, and Brewer, 1966)

Values of the principal or power component of cutting force (F_P) are shown plotted against cutting speed (V) in Fig. 17.14 for several different feeds (t). The appearance of the tool-face side of the chip provides a reproducible indication of whether a large built-up edge (BUE) is present (Nakayama, 1957), and the data points of Fig. 17.14 are coded to indicate the nature of the BUE that was present. The V_2 (maximum cutting force) and V_1 (lowest speed to produce a shiny chip) values were found to be identical as shown in Fig. 17.14.

Figures 17.15 and 17.16 show plots of peak-to-valley roughness (R_t) and arithmetic average roughness (R_a) versus cutting speeds (V) for different values of feed (t). These curves are seen to fall rapidly with increase in speed, reaching a constant asymptotic value of speed V_4 (for R_t) and V_5 (for R_a). Values of V_4 and V_5, which may

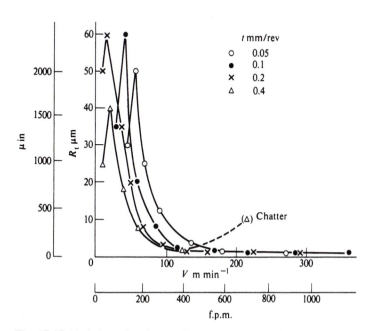

Fig. 17.15 Variation of peak-to-valley roughness (R_t) with cutting speed (V). (after Nakayama, Shaw, and Brewer, 1966)

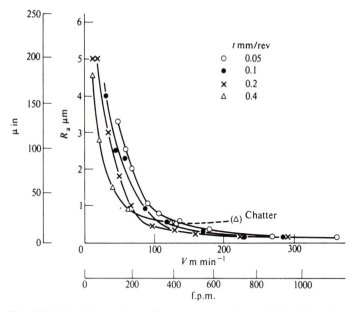

Fig. 17.16 Variation of centerline average roughness (R_a) with cutting speed. (after Nakayama, Shaw, and Brewer, 1966)

only be approximately estimated are shown in Fig. 17.14. The V_5 curve is not extended to a feed of 0.4 mm/rev since chatter occurred with this feed in the vicinity of speed V_5.

The line marked V_6 in Fig. 17.14 shows the speed at which burnishing of the finished surface is first observed at each feed. A smooth burnished chip surface was not produced until a speed of 32 m min^{-1} was reached. For speeds below this value, a BUE was present and a continuous layer of flowed metal was not present on the side of the chip adjacent to the tool face. As the speed was increased above V_1, the thickness of the surface layer decreased as the temperature gradient became steeper. Finally at a speed of 289 m min^{-1}, the burnished layer on the face of the chip became very thin, probably due to the time of contact between chip and tool being very short.

Measured values of chip–tool interface temperature are given in Fig. 17.17. Values are not shown for the 0.4 mm/rev feed since the chips caused a short circuit. Lines corresponding to the V_1 to V_6 curves of Fig. 17.14 are labelled $\bar{\theta}_1$ to $\bar{\theta}_6$.

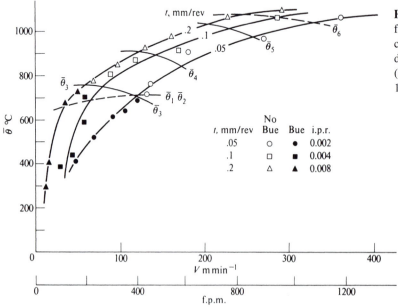

Fig. 17.17 Values of chip–tool interface temperature measured by the chip–tool thermocouple method, for different cutting speeds and feeds. (after Nakayama, Shaw, Brewer, 1966)

The values of F_P in Fig. 17.14 are seen to reach an approximately constant value at high values of speed. This constant value of F_P may be used to compute the energy per unit volume (u). Values of u obtained in this way from Fig. 17.14 are shown plotted in Fig. 17.18. The slope of this curve is approximately -0.2 in accordance with the relationship that usually holds in turning ($u \sim t^{-0.2}$).

Finally, values of $\bar{\theta}$ from Fig. 17.17 are shown plotted against the quantity $u(Vt)^{1/2}$ for each experimental point in Fig. 17.19. A line corresponding to θ versus $u(Vt)^{1/2}$ is found to pass close to all experimental points except those obtained when a BUE was present. In Fig. 17.17, the curves are seen to break and fall off very rapidly below the $\bar{\theta}_1$ line corresponding to the appearance of the BUE. The presence of a BUE decreased the thermoelectric EMF generated and thus suggests a temperature that is too low. Therefore, these points associated with a BUE in Fig. 17.19 should be ignored.

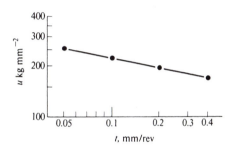

Fig. 17.18 Variation in energy per unit volume (u) with feed (t). (after Nakayama, Shaw, Brewer, 1966)

Side Flow When Machining Hard Material

Kishawy and Elbestawi (1998, 2001) have discussed surface roughness due to side flow when dry-machining hard steel using PCBN tools. D–2 tool steel (1.5 $^w/_o$ C, 12 $^w/_o$ Cr) having a Rockwell C hardness was machined using tools having a rake angle of 0° and a clearence of 6°. The major source of surface roughness was due to side flow that produced burrs on feed mark ridges. Side flow increased with increase in temperature and pressure of the work material at the tool tip. Side flow was found to depend on the nose radius of the tool, tool wear, and cutting speed and feed as previously reported by Bresseler et al. (1997). Two surface characteristics associated with the hard chrom-carbide particles in the work material were surface microcracks and surface cavities.

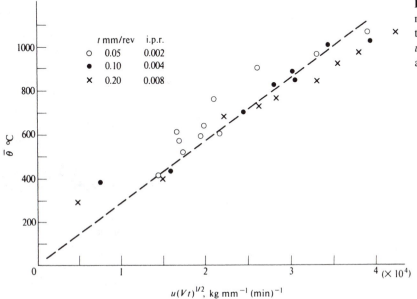

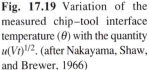

Fig. 17.19 Variation of the measured chip–tool interface temperature (θ) with the quantity $u(Vt)^{1/2}$. (after Nakayama, Shaw, and Brewer, 1966)

The number of microcracks decreased with increase in cutting temperature (i.e., > cutting speed) due to increased ductility. The surface cavities were due to pull-out of chrom-carbide particles during chip formation. While the microcracks are undesirable since they reduce fatigue resistance of the surface, the surface cavities are considered tribologically beneficial, since they provide small pockets that store lubricant.

Higher cutting speeds were found to give lower residual tensile stresses in the surface and also lower peak compressive stresses beneath the machined surfaces. At all speeds an increase in nose radius gave lower tensile stresses in the machined surface. An increase in nose radius, when machining hardened steel, gives a lower cutting temperature in the machined surface since the thermal energy generated is distributed over a larger area over which heat is conducted away from the surface being generated.

A thin white layer of untempered martensite—about 39 μin (1 μm) thick—was found in surfaces produced by worn tools and with chamfered cutting edges, but no white layer was obtained with sharp tools. The thickness of the white layer was found to decrease with increase in cutting speed with a sharp tool. At a high cutting speed less of the heat flows into the work, most of it going to the chip.

Discussion

The entire range of surface finishes produced by a primary cutting edge appear to depend on the temperatures pertaining on the face and flank of the cutting tool.

At very low tool-face temperatures, the junctions on the tool face come apart where they went together and no built-up edge is usually formed. At moderate speeds the temperature on the tool face becomes sufficiently high so that strong bonds are established, and the metal in the chip adjacent to the tool face then flows in a direction close to that corresponding to the direction of maximum shear stress. A BUE is thus established. Pieces of the BUE periodically leave with the chip and with the workpiece, and the face of the chip as well as the workpiece surface are very rough. Values of cutting force will usually be low when a BUE is present due to the large effective rake angle associated with the BUE.

As the speed is further increased, a point is reached where the face of the chip becomes shiny. This is speed V_1. At the same speed (V_2) the cutting forces show a maximum. The surface roughness is still quite high at this speed but has begun to decrease with speed. It is believed that V_1 and V_2 correspond to a mean tool-face temperature which equals the strain recrystallization temperature of the chip ($T_H = 0.5$). In Fig. 17.17, $\bar{\theta}_1$ and $\bar{\theta}_2$ are between 650 and 700 °C. These somewhat-approximate mean temperatures are just below the equilibrium temperature in the iron–carbon diagram (Fig. 15.1) where austenite is first formed (727 °C). As the speed rises slightly above V_1, the α–γ phase transformation will be reached on the tool face. The fact that the $\bar{\theta}_1$ temperature rises with an increase in V_1 (i.e., for decreased feed) may be explained by the fact that recrystallization is a time–temperature reaction. At a higher cutting speed (V_1) and smaller feed (t), the chip is in contact with the tool for a shorter time and hence there is less time for recrystallization to occur. Since recrystallization and the α–γ transformation will occur more rapidly the higher the temperature, and hence the farther the material is removed from equilibrium, the critical temperatures should become higher as the feed decreases and as V_1 increases (i.e., as the tool–chip contact time decreases).

Ostermann (1960) and Opitz and Gappisch (1962) have clearly identified the presence of an austenitic layer on the surfaces of chips produced at high speeds, and they have discussed this phenomenon in terms of tool wear.

At speeds above V_1 a zone of secondary plastic flow is evident on the tool-face side of the chip. This zone is at first thick and then becomes thinner as the speed is increased. The thickness of the zone corresponds to the layer that is partially or completely transformed, and hence softened. The face of the chip is highly burnished and shiny whenever a secondary shear zone is present.

The cutting forces will usually drop as the secondary shear zone becomes thinner. The BUE is probably rapidly lost in going from speed V_2 to V_3. At speeds to the left of V_3 cracks are found in the finished surface which are probably due to the presence of a small, stable, residual BUE. The BUE is probably missing completely to the right of V_3.

It is well known that austenite deforms by grain boundary rearrangement and has a viscous character. Creep is a result. Since grain boundary rearrangement is a more time-dependent mechanism than slip, the deformation of austenite is more strain-rate dependent than the deformation of ferrite. The deformation of austenite may be better approximated by a liquid model, where the motion of holes is analogous to grain boundary rearrangement with regard to strain-rate dependence.

Trent (1963) has suggested that the secondary shear zone often appears as though it were a liquid when sheared.

Schaller (1964) has assumed that the secondary shear zone on the surface of the chip behaves as though it were a viscous liquid that obeys Newton's law of viscous shear.

When the temperature on the tool face has reached the temperature $(\bar{\theta}_1, \bar{\theta}_2)$ at which the BUE disappears and the face of the chip becomes burnished due to the appearance of a secondary shear zone, the temperature on the tool flank will be far below this temperature. Asperities on the newly cut surface will contact asperities on the tool flank, and the real area of contact (A_R) on the clearance face will be far less than the apparent area (A). The high points of actual contact will be separated by extensive regions where there is no contact.

The mechanism of "microchip" formation (Chapter 10) is shown in Fig. 10.1 when a soft metal slides across a harder one. In Fig. 10.1a two asperities are about to make contact, while Fig. 10.1b shows the asperities shortly after contact. The soft metal is seen to deform and cause the lower (soft) surface to rise on the left. With subsequent motion in Fig. 10.1c a shear plane develops and what resembles a chip is formed which flows upward and to the right. The "chip" will tend to force the surfaces apart; and if the system supporting the surfaces is flexible, they will separate and the "chip" will grow to considerable size. In such a case the shear angle ϕ will be relatively large. With a stiff system (cutting tool), the surfaces will separate much less than in Cocks's (1962, 1964, 1965) case (mostly only locally), and the shear angle ϕ will be small. As a consequence the strain in the "microchip" will be large, and the particle thus generated will be highly strain hardened.

As the "chip" grows larger, the separating force will increase, and eventually the local deflection of the surfaces will allow the "microchip" to flow past the hard asperity on the tool as shown in Fig. 10.1e. Or, occasionally the strength of the asperity on the tool face may be exceeded, and a wear particle may be removed from the tool surface as shown in Fig. 10.1d. In this manner a large number of hard particles will be formed on the finished surface which will constitute the main component of roughness between speeds V_1 and V_3 or V_4.

As the speed is further increased, the metal in the finished surface will gradually soften and asperities in the surface will recede downward into the surface on the workpiece. Fewer smaller microchips will be developed. Finally, at speed V_6 the metal in the finished surface will reach the strain recrystallization temperature and then the transformation temperature (727 °C), and a burnished workpiece surface will be formed just as a burnished chip surface is formed at speeds above V_1. At speeds above V_6 the finished surface will possess a layer of secondary shear just as the chip face does above V_1.

The real area of contact on the tool flank abruptly changes from a small value relative to the apparent area to one that approaches the apparent area at speeds in the vicinity of V_4 and V_5. When this occurs, it is to be expected there will be a sudden increase in the rate of tool flank wear.

Adhesion between the tool flank and the workpiece surface is unusually strong in the case of soft (grade $60 = 60,000$ lb in^{-2} or 400 MPa) nodular iron. Trigger et al. (1952) have observed a ribbon-like flow from the clearance face of the tool in such cases, and similar ribbon-like flows have been observed by Ham et al. (1961, 1962, 1964). It is believed that these ribbons are an extreme example of the "microchip" formation described here in which the individual chips are so numerous, they join up to form a small continuous auxiliary chip.

The number of "microchips" that are formed in a given instance will depend upon many factors including the following:

1. the roughness of the tool flank surface and the extent of the wear land
2. the inherent roughness of the freshly generated surface
3. the hardness and flow characteristics of the workpiece material and the response of these quantities to temperature changes
4. the compatibility as measured by the relative tendency for strong bonds to form when clean, dry tool and workpiece surfaces come into contact
5. the atmosphere, lubricant, or contaminant that may be present to prevent weld formation between chip and tool asperities upon contact

This investigation of the surface produced by a primary cutting edge has shown that a sequence of events which resemble those occurring on the tool face at low speeds may also occur on the tool flank at high speeds. A "microchip" that plays the same role on the tool flank as the BUE does on the tool face is responsible for the best finish not being obtained immediately when the BUE is lost and the chip face becomes shiny.

Secondary Cutting Edge Finish

All of the results presented thus far are for the type of roughness obtained in a cutting operation in which there is no tool nose and hence no feed marks. In a conventional turning operation the finish left on the bar is produced by a secondary cutting edge that is separated from the primary cutting edge by a nose radius (Fig. 17.20). Use of a secondary cutting edge to generate the finished surface introduces several complications:

1. Ridges corresponding to the geometry of the tool at its nose and having a pitch equal to the axial feed rate are left behind on the finished surface.
2. The undeformed chip thickness goes gradually to zero at the secondary cutting edge, and this causes uncertainty in the geometry of the cut at the trailing edge, since for a given edge sharpness there is minimum undeformed thickness that will be removed.
3. A concentration of wear occurs at both free surfaces of the cut. The groove thus formed on the end-cutting edge of the tool acts as a forming tool and leaves behind a highly cold-worked ridge on the surface that not only contributes to the roughness but further complicates matters by producing additional grooves on the end-cutting edge as it is recut on subsequent revolutions.
4. The metal at the trailing edge of the tool is subjected to unusually high normal stress and will flow to the side to relieve this stress. This in turn produces a furrow that contributes to the roughness, particularly in the case of a soft, ductile metal.

In addition to these special roughness components, BUE roughness, roughness due to an imperfect cutting edge, and roughness due to tool vibration will also be present.

Surfaces generated by a secondary cutting edge are thus far more complex than those produced by a primary cutting edge.

Surface roughness will generally have different values in different directions, and the direction normally employed is that corresponding to the maximum roughness. For a surface produced by a primary cutting edge, this will be in the direction of cut (parallel to the cutting velocity vector). Consequently all roughness values presented above are for the circumferential direction. For surfaces generated by a secondary edge, the roughness will be greatest in a direction perpendicular to the cutting velocity vector (V), and roughness values should be obtained using a stylus traversed in an axial direction.

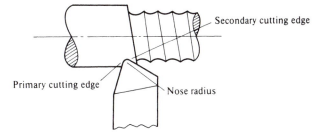

Fig. 17.20 Diagrammatic sketch of cutting operation showing primary and secondary cutting edges and characteristic waveform left on the finished surface.

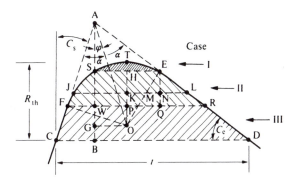

Fig. 17.21 Plan view of cutting tool showing the three cases to be considered in computing the theoretical peak-to-valley roughness (R_{th}) corresponding to any value of feed (t).

GEOMETRICAL CONTRIBUTION TO ROUGHNESS

The component of surface roughness due to tool nose geometry may be readily calculated. Figure 17.20 shows a plan view of a conventional turning operation with ridges left behind on the finished surface. Figure 17.21 shows an enlarged view of the tool tip which is defined in terms of three quantities:

1. nose radius ($r = $ OT in Fig. 17.21)
2. end cutting edge angle (C_e)
3. side cutting edge angle (C_s)

When the nose radius is large and the feed is very small, the surface will be generated by the nose radius alone (Fig. 17.22). This will be referred to as case I and it is evident that

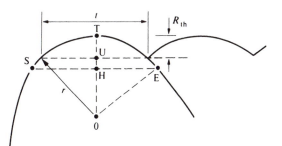

Fig. 17.22 Enlarged view for case I of Fig. 17.21.

$$R_{th} = OT - OU = r - (r^2 - t^2/4)^{1/2} \cong \frac{t^2}{8r} \qquad (17.1)$$

or in nondimensional form

$$\frac{R_{th}}{r} = \frac{1}{8}\left(\frac{t}{r}\right)^2 \qquad (17.2)$$

From Fig. 17.22 it is evident that this equation will hold as long as $t/2$ is less than EH or

$$0 < (t/r) < 2 \sin C_e \qquad (17.3)$$

For larger values of feed both r and C_e must be considered in computing R_{th} and this will be called case II. If t is equal to JL (Fig. 17.21), this corresponds to case II and

$$t = JK + HE + NL$$

$$= [r^2 - (r - R_{th})^2]^{1/2} + r \sin C_e + r\left(\frac{R_{th}}{r} - 1 + \cos C_e\right) \cot C_e \qquad (17.4)$$

$$\frac{t}{r} = \left[2\frac{R_{th}}{r} - \left(\frac{R_{th}}{r}\right)^2\right]^{1/2} + \sin C_e + \left(\frac{R_{th}}{r} - 1 + \cos C_e\right) \cot C_e$$

This equation will hold as long as t lies between length SE ($= 2r \sin C_e$) and length FR,

$$FR = r[\cos C_s + \sin C_e + (\cos C_e - \sin C_s) \cot C_e]$$

that is, as long as

$$2 \sin C_e < t/r < [\cos C_s + \sin C_e + (\cos C_e - \sin C_s) \cot C_e] \qquad (17.5)$$

For still larger values of t, C_s in addition to r and C_e must be considered and this will be called case III. If t is equal to CD, this corresponds to case III and from Fig. 17.21

$$R_{th} = AB - AG + r$$

$$= \frac{t}{\tan C_s + \cot C_e} - \frac{r \cos\left(45 - \dfrac{C_e}{2} - \dfrac{C_s}{2}\right)}{\sin\left(45 - \dfrac{C_e}{2} + \dfrac{C_s}{2}\right)} + r$$

or in nondimensional form

$$\frac{R_{th}}{r} = \frac{t/r}{\tan C_s + \cot C_e} - \frac{\cos\left(45 - \dfrac{C_e}{2} - \dfrac{C_s}{2}\right)}{\sin\left(45 - \dfrac{C_e}{2} + \dfrac{C_s}{2}\right)} + 1 \qquad (17.6)$$

which holds when

$$t/r > \cos C_s + \sin C_e + (\cos C_e - \sin C_s) \cot C_e \qquad (17.7)$$

If the nose radius is very small compared with t, then the last two terms of Eq. (17.6) become negligible compared with the first and

$$\frac{R_{th}}{r} = \frac{t/r}{\tan C_s + \cot C_e}, \qquad \left(\frac{t}{r} \gg 1\right) \qquad (17.8)$$

for all values of t.

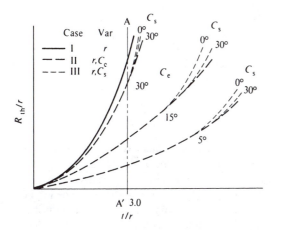

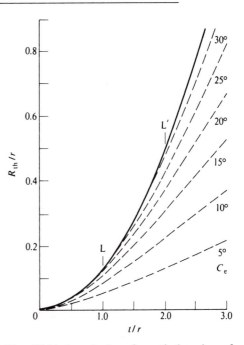

Fig. 17.23 Schematic diagram showing relation between R_{th}/r and t/r for all three cases. (after Shaw and Crowell, 1965)

Fig. 17.24 Actual plot of practical region of Fig. 17.23 to the left of line A'–A. (after Shaw and Crowell, 1965)

Equations (17.2), (17.4), and (17.6) may be used to produce a nondimensional plot as shown schematically in Fig. 17.23, which enables R_{th} to be found for any combination of t, r, C_e, and C_s. However, only the part of Fig. 17.23 to the left of A'–A is of interest in finish machining; therefore, only this part of the diagram is shown to scale in Fig. 17.24. The solid line in Fig. 17.24 represents the solution for case I where the entire surface left behind is generated by the nose radius r. The limiting value of t/r for case I for a side cutting edge angle of 30° is given by Eq. (17.3).

$$\frac{t}{r} = 2 \sin 30° = 1.0$$

and this point is labelled L in Fig. 17.24. However, the case II line for $C_e = 30°$ does not deviate appreciably from the case I line until t/r is about twice this value (point L' in Fig. 17.24). This is true for all other cases and the case I equation may be used to a good approximation as long as $t/r < 4 \sin C_e$, instead of $2 \sin C_e$ as stated in Eq. (17.3). Case III solutions will generally not be needed in finish-machining.

The need for Fig. 17.24 may be illustrated by a practical example. Figure 17.25 shows results (solid lines) obtained by Moll (1939) when using sharp tools of different nose radius r to machine a steel at different values of feed t. The theoretical curves (assuming case I to hold) are shown by the dashed lines. These are seen to lie above the observed curves for large values of feed (i.e., t/r). Actually, case II pertains for large values of t/r, and if Fig. 17.24 were used to plot the theoretical curves instead of Eq. (17.2), the theoretical curves would then lie below the observed curves for all values of t/r. Opitz and Moll (1941) defined a cutting efficiency (C) in terms of the ratio of measured roughness (R_t) to theoretical roughness (R_{th}) as follows:

$$C = R_t/R_{th} \qquad (17.9)$$

This quantity may be considered as a measure of the closeness of approach to ideal finish-machining which will coresspond to $C = 1.0$.

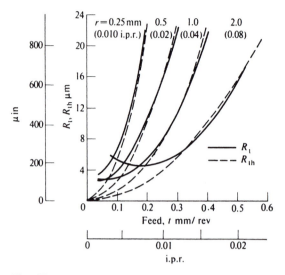

Fig. 17.25 Comparison of measured (R_t) and computed (R_{th}) values of peak-to-valley roughness for tools of different nose radius r used at different feeds. (after Moll, 1939)

The rise in R_t with decrease in t below 0.008 i.p.r. (0.2 mm/rev) will be discussed later.

Martellotti (1941) has derived the following expression for the peak-to-valley roughness (R_t) in a milling operation:

$$R_t = \frac{f^2}{8[r \pm (fn/\pi)]} \qquad (17.10)$$

where f is the feed per tooth, r is the cutter radius, and n is the number of teeth in the cutter. The plus sign pertains to up-milling while the negative sign is for down-milling. This equation was found to be in good agreement with measured values by Martellotti if the run-out of the spindle was held to a very low value.

REPRESENTATIVE FINISH RESULTS

Representative values of peak-to-valley roughness (R_t) and arithmetic average roughness (R_a) are given in Fig. 17.26 for normalized AISI 1045 steel machined using sharp tungsten carbide tools having the following geometry (except as noted in the figures):

- back rake angle, 0°
- side rake angle, 6°
- end relief angle, 6°
- side relief angle, 6°
- end cutting edge angle, 30°
- side cutting edge angle, 0°
- nose radius, 0.4 mm

When tool geometry is specified in the figures, the values are listed in the above order.

Figure 17.26 shows values of R_t plotted against cutting speed V for different values of feed t. The finish is seen to improve rapidly with increased speed and then to level off. The speed at which the level value is reached is seen to increase as the feed decreases which indicates that the improvement is due primarily to loss of the BUE, since the BUE will disappear at a given temperature and this temperature will be obtained at

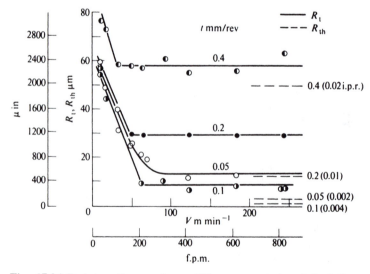

Fig. 17.26 Peak-to-valley roughness (R_t) versus speed and feed for a normalized AISI 1045 steel machined without fluid using sharp carbide tools of geometry 0°, 6°, 6°, 6°, 30°, 0°, 0.4 mm. The theoretical values of R_t (R_{th}) computed from Eq. (17.1) are shown by the dotted lines. (after Shaw and Crowell, 1965)

a higher speed as the feed is decreased. It should be noted that the best finish does not correspond to the lowest feed 0.002 i.p.r. (0.05 mm/rev) but instead corresponds to a feed of 0.004 i.p.r. (0.1 mm/rev). This is due in part to the fact that the cutting energy per unit volume (u) and hence the mean stress on the tool face increases rapidly as the feed falls below 0.004 i.p.r. (0.1 mm/rev). This in turn will cause more plastic side flow from the edge of the cut along the secondary cutting edge. The furrow that is thus formed will add to the discrepancy between the measured R_t and theoretical R_{th} values of surface roughness. The theoretical values of roughness R_{th} shown in Fig. 17.26 were computed using Eq. (17.1), since in this case ($C_e = 30°$) all values correspond to case I. The percentage discrepancy between the observed and theoretical values is seen to increase markedly with decrease in feed. This could be due to the influence of the size effect (increase in energy per unit volume with decrease in t) upon the mean stress on the tool face and hence on the tendency toward side flow at the edge of the cut.

Figure 17.27 shows values of arithmetic average roughness (R_a) corresponding to the peak-to-valley values (R_t) of Fig. 17.26.

Figure 17.28 indicates the importance of nose radius r on the finish produced. The theoretical values of roughness (R_{th}) are about half the actual values (R_t) except for large values of nose radius where chatter is probably responsible for the additional discrepancy. Equation (17.1) was again used to compute values of R_{th} since case I pertains.

Figure 17.29 shows that the depth of cut has a negligible influence on surface roughness.

Figure 17.30 compares theoretical and actual values of surface roughness for different values of feed. The finish R_t is again seen to improve with decrease in t down to a value of 0.004 i.p.r. (0.1 mm/rev),

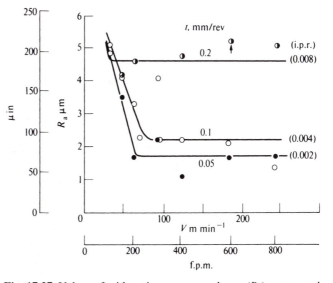

Fig. 17.27 Values of arithmetic average roughness (R_a) corresponding to the peak-to-valley values (R_t) in Fig. 17.26. (after Shaw and Crowell, 1965)

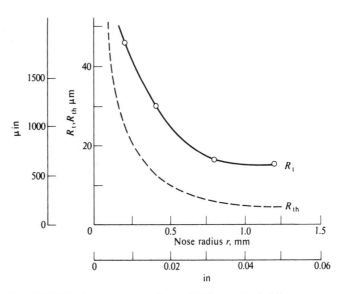

Fig. 17.28 Peak-to-valley roughness (R_t) for tools of different nose radius r when cutting the steel of Fig. 17.25 without a fluid. Tool geometry, 0°, 6°, 6°, 6°, 30°, 0°, var; tool material, carbide; depth of cut, 0.10 in (2.5 mm); feed, 0.008 i.p.r. (0.2 mm/rev); cutting speed, 800 f.p.m. (240 m min⁻¹). (after Shaw and Crowell, 1965)

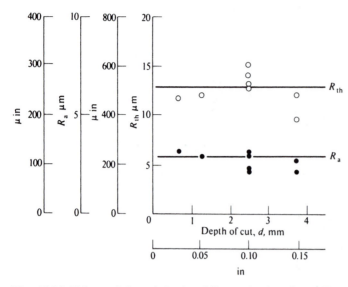

Fig. 17.29 Values of R_t and R_a for different depths of cut. Tool geometry, 0°, 6°, 6°, 6°, 15°, 15°, 0.020 in (0.5 mm); feed = 0.008 i.p.r. (0.2 mm/rev); speed = 394 f.p.m. (120 m min⁻¹); work material, same as in Fig. 17.26; cutting fluid, none. (after Shaw and Crowell, 1965)

when a reversal in the trend takes place. Figure 17.31 shows that the surface finish is not influenced in an important way when the side rake angle is changed from +6° to −6°. The slight increase in R_t with decrease in rake angle is probably due to an increase in the stress on the tool face and consequently an increase in side flow along the secondary cutting edge.

EDGE FINISHING

Burrs that form on edges of finished parts are frequently a concern, and their removal is often a costly extra finishing operation. There are a number of methods of avoiding burr formation and also of effectively removing them when they cannot be avoided. This is a detailed topic too lengthy to discuss here. However, Gillespie (1999) has compiled a handbook that covers this special subject in detail. It is suggested this be consulted when edge finishing is a problem.

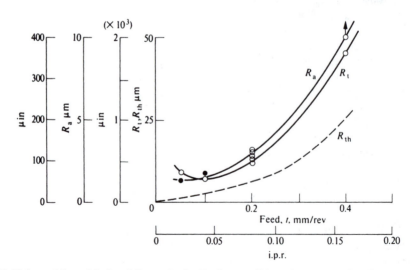

Fig. 17.30 Values of R_t and R_a for different feeds. Cutting conditions the same as for Fig. 17.28 except for the depth of cut, 0.10 in (2.5 mm). (after Shaw and Crowell, 1965)

MINIMUM UNDEFORMED CHIP THICKNESS

Sokolowski (1955) has suggested that there is a minimum undeformed chip thickness below which a chip will not be formed. When this occurs, rubbing takes place instead. This minimum undeformed chip thickness should depend upon the radius of the cutting edge (ρ), the cutting speed, and the stiffness of the system. Measurements made by Sokolowski indicated that for a cutting edge radius of about 0.0005 in (12 μm) and a cutting speed of 650 f.p.m. (210 m min^{-1}), the smallest cut that could be taken was about 0.00016 in (4 μm). Applying this idea to the secondary cutting edge of a turning tool, it is found that a small triangular

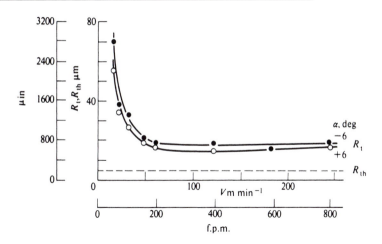

Fig. 17.31 Variation of roughness values (R_t) with rake angle of tool. Work material, normalized AISI 1045 steel; tool geometry, 0°, var., 6°, 6°, 15°, 15°, 0.03 in (0.8 mm); tool material, cemented carbide; cutting fluid, none; feed, 0.01 i.p.r. (0.2 mm/rev); depth of cut, 0.10 in (2.5 mm). (after Shaw and Crowell, 1965)

portion of the material that should be removed will be left behind (Fig. 17.32a). The portion left behind has been analyzed by Bramertz (1960, 1961) and called a Spanzipfel. Figure 17.32b shows an enlarged view of the Spanzipfel used by Brammertz to derive the following expression for the theoretical surface roughness:

$$R'_{th} = \frac{t^2}{8r} + \frac{t_m}{2}\left(1 + \frac{rt_m}{2}\right) \qquad (17.11)$$

The second term in this equation represents the contribution of the Spanzipfel.

Pahlitzsch and Semmler (1960, 1961, 1962) have applied this concept to some data obtained in the fine turning of AISI 1045 steel using newly sharpened ceramic tools. The minimum undeformed chip thickness was taken to be 40×10^{-6} in (1 μm), and all tests were taken after six seconds of cutting at a speed of 1200 f.p.m. (400 m min^{-1}). The degree of agreement between theory and experiment is seen to be very good in Fig. 17.33. The absolute values for R_t and R'_{th} are not only in good agreement, but the theoretical curve is seen to rise for feeds smaller than 0.002 i.p.r. (0.05 mm/rev).

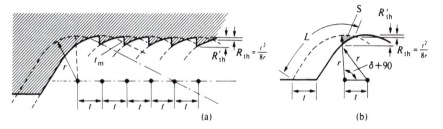

Fig. 17.32 Material left beind on finished surface due to the existence of a minimum undeformed chip thickness (t_m). The theoretical roughness, in the absence of a minimum undeformed chip thickness is $R_{th} = t^2/8r$ in this case (case I). (a) Shows relation between R_{th} and R'_{th}. (b) Shows Spanzipfel (S) shaded.

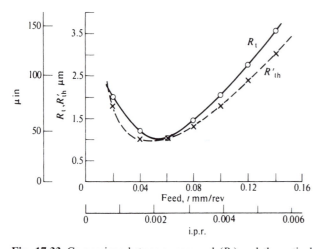

Fig. 17.33 Comparison between measured (R_t) and theoretical (R'_{th}) values of surface roughness for sharp ceramic tools operating at 1200 f.p.m. (400 m min^{-1}) when cutting AISI 1045 steel dry at various values of feed (t). Nose radius = 0.04 in (1 mm); cutting time = 0.1 min. (after Pahlitzsch and Semmler, 1960)

Actually, agreement as good as that shown in Fig. 17.33 should not be expected since the Spanzipfel will be plastically deformed and made smaller as it comes in contact with the clearance surface of the tool. In addition to the surface roughness components already considered, we should expect small contributions to R_t due to vibration and plastic side flow. The latter is apt to be most significant at very small values of feed and could be partly responsible for the rise in the curve for feeds less than 0.002 i.p.r. (0.05 mm/rev) as previously noted.

GROOVE WEAR

Equation (17.11) will hold for a relatively short time, since it only applies to sharp tools. Grooves are very frequently formed at the edges of the cut (A and B in Fig. 17.34) on both the primary and secondary cutting edges. There will usually be a single groove at A, but there may be several grooves at B spaced at a distance equal to the feed. The number of grooves increases with decrease in the end cutting edge angle and with an increase in time.

It is tempting to explain the presence of the grooves on the secondary cutting edge in terms of the Spanzipfel, since this will be a hard body with just the right shape and spacing to account for the grooves. However, there will be no Spanzipfel on the primary cutting edge, and a groove is also found here (often larger than the ones on the secondary cutting edge). This does not mean to say that the Spanzipfel has nothing to do with the groove, but that it is probably not the sole cause for the groove.

There has been much speculation concerning the origin of the concentrated wear which gives rise to the grooves found at the edges of a chip. Hovenkamp and Van Emden (1953) have attributed the action to a work-hardened layer which gives rise to a concentration of stress at the chip edges. Pekelharing and Schuermann (1955) suggested that the grooves on the secondary cutting edge result from the abrasive action of the cusps on the machined surface. Albrecht (1956) has suggested the lateral escape of built-up edge material as the cause. Leyensetter (1956) believes that embedded particles of tool material removed on the previous revolution act as cutting edges to produce the grooves. Lambert (1961) has suggested an increase in hardness of the edge of the chip due to oxidation, and Solaja (1958) has suggested that the groove might result from thermal fatigue. Solaja also suggests that several of these causes might act simultaneously. As discussed in Chapter 11, Shaw et al. (1966) have suggested a difference between plane stress and plane strain as a contributing cause.

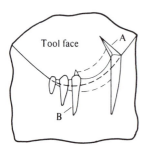

Fig. 17.34 Worn cutting tool showing large groove on primary cutting edge at A and smaller grooves on secondary cutting edge at B.

Groove formation has been studied by Pekelharing (1960) and Lambert (1961) at the Delft Technical University where it has been found that groove formation occurs at a decreasing rate (with respect to helical distance cut) at cutting speeds below 600 f.p.m. (200 m min^{-1}). At higher speeds (temperatures) the rate appears to be constant. This suggests that for rapid groove formation to occur, the metal in the chip must be above the strain recrystallization temperature separating the hot-working region from the cold-working region. It also suggests that the groove is not primarily due to the side flow of BUE material, since we should then expect the rate of groove formation to increase with decrease in speed due to the increasing tendency toward BUE formation with decrease in speed.

SIDE FLOW

The metal left behind on the secondary cutting edge is subjected to sufficiently high pressure to cause the metal to flow to the side as shown in Fig. 17.35. Figure 17.35a shows the profile left behind in the absence of side flow, or "squeezing" as Pekelharing calls this action, while Fig. 17.35b shows the profile with side flow. The peak-to-valley distance is seen to be larger when side flow is present.

Sata (1964) has studied the influence of side flow on finish and has found that this component of roughness will be zero for a brittle material such as free-cutting brass but may contribute up to 6 μm (240 μin) to the roughness (R_t) when an alloy steel is machined. Sata suggests that the side flow contribution to roughness decreases with an increase in cutting speed.

TOOL LIFE

Criteria for tool life in finish-machining have been discussed by Bickel (1954) and Ciragan (1955) who have shown that the arithmetic average roughness (R_a) may be used in finish machining in the same way as the size of the wear land on the clearance face (w) is used in conventional tool-life studies. The R_a versus cutting time (T_c) curve was found to have the same general shape as the conventional wear land (w versus T_c) curve starting with a rapid increase in R_a for the first 3 to 10 min and followed by a more gradual increase in R_a with T_c. The change in workpiece diameter with time was not found to be a sensitive measure of tool wear.

Semmler (1962) has presented tool-life results based on both wear land (w) and peak-to-valley roughness (R_t) criteria. Typical results for a tool cutting AISI 1045 steel are presented in Fig. 17.36 for a feed of 0.0016 i.p.r. (0.04 mm/rev). The speeds corresponding to maximum tool life are seen to be much lower

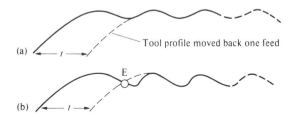

Fig. 17.35 Profile generated by tool point. (a) In absence of side flow. (b) With side flow. (after Pekelharing, 1960)

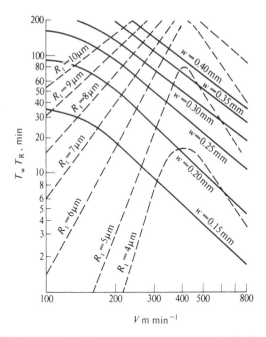

Fig. 17.36 Tool-life curves based on wear-land criterion (solid curves) and surface-roughness criterion (dashed lines). Cutting conditions: work material, AISI 1045 steel; tool material, tungsten carbide; feed, 0.0016 i.p.r. (0.04 mm/rev); depth of cut, 0.120 in (0.3 mm); side rake angle, 5°; back rake angle, 0°; clearance angle, 5°; side cutting edge angle, 30°; nose radius, 0.024 in (0.6 mm); cutting fluid, none. (after Semmler, 1962)

when wear land (w) is the criterion than when surface roughness (R_t) is the criterion, and the tool-life curves based on roughness are seen to be much steeper than those based on land wear. This is probably due to the fact that the BUE has a greater influence on surface roughness than it does on land wear.

The controlling criterion will be the one giving the lower tool life at the operating speed V. The cutting speed should be chosen so that the total machining cost per part is a minimum (Chapter 19).

RECOMMENDED PRACTICE

Recommended conditions for finish machining depend upon the following conditions:

1. rigidity and stability of tool–work–machine system
2. limitations on speed, feed, or depth of cut
3. whether a new or worn tool is considered

If an extremely stiff system is to be used, then the contribution to roughness due to vibration will be small, and a set of conditions that is independent of the tendency to cause chatter may be used. Otherwise, the cutting conditions may have to be shifted to decrease the tendency to chatter.

In finish-machining it is not only important that the cutting temperature be sufficiently high that the BUE be eliminated, but that the layer of material adjacent to the tool face be thermally softened to provide low friction and a high shear angle. The cutting temperature will depend upon the hardness and ductility of the work material and upon the cutting speed, feed, and depth of cut. Finishing cuts are usually of such small feed (0.002 i.p.r. or 0.05 mm/rev) and small depth of cut (0.03 in or 0.75 mm) that a material such as normalized AISI 1045 steel must be machined at a very high speed in order to develop a sufficiently high cutting temperature. Figure 17.36 reveals that cost-optimum speeds are in excess of 1000 f.p.m. (300 m min^{-1}) when AISI 1045 steel is cut at a feed of 0.0016 i.p.r. (0.04 mm/rev). Such speeds are difficult to obtain when bars of ordinary size (1 in or 25 mm diameter) are cut.

The chip equivalent concept first introduced by Woxén (1931) is a convenient means for determining the influence tool geometry has on the relative cutting temperature. The chip equivalent is

$$CE = (L/bt) \qquad (17.12)$$

Where L is the active length of the cutting edge [see L in Fig. 17.32b], b is the depth of cut, and t is feed. It may be shown that

$$L = \frac{b - r(1 - \sin C_s)}{\cos C_s} + \frac{90 - C_s}{180} \pi r + \frac{t}{2} \qquad (17.13)$$

where

$$b = \text{depth of cut}$$
$$r = \text{nose radius}$$
$$C_s = \text{side cutting edge angle}$$
$$t = \text{feed}$$

Woxén showed that when a given material is cut at a specific speed, the temperature will vary inversely with the chip equivalent. Thus, the cutting temperature will be increased with a decrease in chip equivalent, and in cases (most practical instances) where it is difficult to cut at a sufficiently high speed (r.p.m. or stability limitations of machine), it is advisable to use a short chip equivalent. This is obtained by use of a zero side cutting edge angle C_s and a moderate nose radius r. This is despite the fact that the theoretical finish improves significantly with increase in nose radius.

In addition to the thermal reason for using a small chip equivalent, tool stability provides a second important reason. The tendency toward chatter will increase as length L (Fig. 17.32b) is increased. Thus, a system that has limited stability for light cuts (most practical cases) will give far better results when L is reduced by use of a zero value of C_s and a moderate nose radius.

The side and back rake angles are additional variables that influence the cutting temperature since they influence the effective rake angle (α_e) measured in the plane of chip flow. The cutting temperature will increase when α_e is decreased by a decrease of side rake angle or as the back rake angle approaches zero. Thus, in the interest of a higher cutting temperature, a side rake angle of 0° will usually give better results in finish-machining than the more common rake angle of +10° normally used for medium-sized cuts. A back rake angle of 0° is also found to be optimum in the finish-machining of steel.

A 0° end cutting edge angle (C_e) is best geometrically as provided in the Kolosov design in Russia, where the tool is supplied with a secondary cutting edge that is parallel to the workpiece axis and slightly greater in length than the feed. In practice the Kolosov tool fails to give satisfactory finish due to the increased tendency toward chatter and the rapid development of grooves on the secondary cutting edge. When chatter and tool wear are taken into account the best end cutting edge angle (C_e) is found to be about 20°. The optimum clearance angle is governed by similar groove formation considerations and should be about 10° instead of the 6° angle normally used when taking heavier cuts.

The hardness of the workpiece plays an important role in finish-machining. If the work is too soft, side flow will be too great and it will take too high a speed to get rid of the BUE and to produce a thermally softened layer on the tool face. The finish-machining performance of low-carbon steels can be improved by cold drawing, and the performance of higher-carbon steels (AISI 1045 for example) may be improved by heat treatment. Semmler (1962) suggests an ultimate strength of 85,000 p.s.i. (60 kg mm^{-2}) as being optimum for AISI 1045 steel.

The ductility of the work material is also important in finish-machining. Free-machining additives such as sulfur and lead decrease chip ductility and consequently greatly improve finish by providing a smaller BUE and a smaller tendency toward side flow.

The sharpness of the cutting edge is important in finish-machining when the tool is new, and it also influences the way in which the tool subsequently wears. Carbide tools that are ground and lapped with very fine diamond wheels give up to 20% better finish than those subjected to ordinary grinding conditions (Ciragan, 1955). While there is no initial difference in the finish-machining performance between carbide and ceramic tools, it is found that worn carbide tools give better finish than worn ceramic tools. This is because the same initial cutting edge radius ends up as a much larger radius in a worn ceramic tool, particularly in the case of tools of coarser grain size.

Schuler (1957) has shown that a chemically active cutting fluid can reduce surface roughness by a factor of 4 or more at low speeds. However, the effectiveness of a fluid falls off rapidly at proper cutting speeds and in addition may cause an acceleration of tool wear due to the formation of a series of cracks that run perpendicular to the cutting edge (so called "comb" cracks).

EXAMPLE

For cutting a steel such as AISI 1045, the following recommendations may be made for best finish and reasonable life:

- cutting conditions
  speed, 1000 f.p.m. (300 m min^{-1})
  feed, 0.002 in (0.05 mm/rev)
  depth of cut, 0.030 in (0.75 mm)
- tool geometry
  side rake angle, 0° (increase for lower speed)
  back rake angle, 0°
  end cutting edge angle, 20°
  side cutting edge angle, 0°
  end relief, 10°
  side relief, 10°
  nose radius, 0.02 in (0.5 mm)
  edge radius, 0.0001 in (0.003 mm) (diamond-lapped)
- work material, cold drawn with sulfur
- fluid, none

This should produce a finish of approximately R_a = 32 μin (0.8 μm) or R_t = 160 μin (4 μm) if the system is reasonably stable.

SYSTEM STABILITY

The machining system can deviate from desired geometry in two general ways—static deflection leading to inaccuracy of the machined part or dynamic instability leading to periodic errors such as waviness and roughness.

Static Deflection

Static deflection may arise due to lack of stiffness of the machine structures including bearings, deflection of work or tool, differential thermal expansion, or dimensional instability of materials due to changes in residual stresses or due to a density change accompanying a structural transformation. Static deflection due to bearing elasticity may be reduced by use of preloaded nonlinear rolling contact bearings (see, for example, Shaw and Macks, 1949) or may be completely eliminated by use of externally pressurized bearings having variable external resistances with feedback. Such active externally pressurized bearings were first introduced in 1963 (Mayer and Shaw, 1963).

Dimensional accuracy of machine tools may be improved by increasing the precision of components or by correcting for residual inaccuracies that are not convenient to eliminate. For example, lead-screw precision may be improved by use of an error-compensating nut. Similarly, cumulative errors involved in the ruling of large spectroscopic gratings cannot be reduced to a satisfactory level due to differential thermal expansion even when the machine is housed in an air-conditioned room having a temperature variation of only a few degrees and the machine is completely immersed in an oil bath with temperature deviation of only a small fraction of a degree during the ruling process. To counter this difficulty, the spacing of each pair of lines ruled is measured and any deviation from the required value immediately compensated.

Practically all metals contain balanced (equal plus-minus) residual elastic stresses which will cause a change in external dimensions if relaxed. An interesting example of the potential difficulty was experienced by a manufacturer of long cylindrical rolls used in the paper industry. It was found that these rolls which were very straight when leaving the plant were unacceptable relative to straightness after being transported by truck over relatively rough roads. This was traced to the relaxation of residual elastic stresses during shipment. A useful solution was to provide a mechanical stress relieving treat-

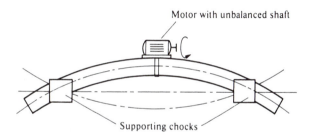

Fig. 17.37 Method of removing residual stress in bar by repeated flexing.

ment before the final cylindrical grinding operation. This was done by supporting the rolls at the nodal points for the first vibrational mode and then inducing a forced vibration of sufficient amplitude to bring the maximum stress induced just to the yield-point. The cyclic force was provided by a small unbalanced mass attached to the shaft of a small electric motor clamped to the center of the freely supported beam (Fig. 17.37). When the roll was shipped after the final finish-grinding operation, it was supported at the same points as when it was subjected to the mechanical stress-relieving treatment. The roll then arrived at its destination as straight as it had been on leaving the plant since the few additional cycles it experienced during the shipment caused negligible additional relaxation of elastic residual stresses.

It is reported that large cast iron castings may be similarly "seasoned" by storing them outdoors for a long time (a year or more) where their temperature changes cyclically. This similarly reduces the level of residual stress before final machining due to cyclic differential thermal expansion.

Dynamic Instability

Dynamic instability in which there is cyclic relative motion between tool and work is of two types—forced vibration and self-excited vibration.

A forced vibration results when a cyclically varying external load has a frequency that is close to one of the natural frequencies of the machining system (tool–work–machine). When this is the case, slightly more energy is absorbed per vibrational cycle than is given back, and the net energy per cycle is available to increase the vibrational amplitude. The vibrational amplitude will increase until the energy dissipated per cycle (as heat = damping) just equals the net energy absorbed by the system per cycle. If damping were constant with amplitude, the amplitude would grow without limit. However, in practical situations, damping increases with amplitude, and a steady state vibrational amplitude is quickly achieved.

Vibrational amplitude due to a forced vibration may be reduced by

1. elimination of the cyclic exciting forces
2. avoiding the coincidence of frequency of exciting force and natural frequency
3. increased stiffness
4. increased damping

The role of the first three of the above items is obvious. In modern times, the fourth item has received considerable attention in machine tool design through modal analysis in which the natural frequencies of a complex structure may be accurately determined relative to different modes of vibrational motion. Analog and digital computers have made such analysis possible.

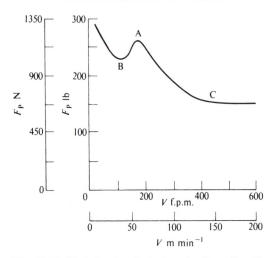

Fig. 17.38 Variation in principal cutting force F_P with cutting speed V. Work material, hot-rolled AISI 1008 steel; tool material, M–2 HSS; feed, 0.006 i.p.r. (0.15 mm/rev); rake angle, 10°; clearance angle, 5°; nose radius, 0.005 in (0.13 mm); depth of cut, 0.1 in (2.54 mm); cutting fluid, none.

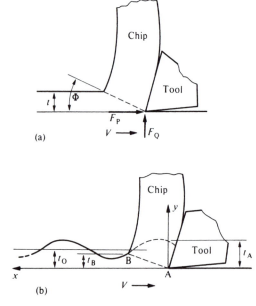

Fig. 17.39 Metal cutting. (a) With constant undeformed chip thickness t. (b) With variable undeformed chip thickness.

Rivin (2000) has presented a very detailed review of the design and performance of the interface between the cutting tool and the machine tool (i.e., the tool holder). This detail is often of importance relative to accuracy, vibration of the tool, and hence surface finish and tool life. Two hundred references are cited from ten countries.

Self-Excited Vibrations

A self-excited vibration results when the internal energy of the system varies in such a way during a vibrational cycle that more energy is stored than is released during a single cycle.

The first source of self-excited vibration to be identified was that due to the decrease of the main (power) component of cutting force (F_P) with cutting speed (V) at high values of V (Arnold, 1946). When a tool vibrates in the F_P direction, the relative velocity between chip and tool will then vary in such a way that a net amount of energy will be stored in the tool per cycle. However, since F_P decreases with V only at relatively high values of V, this mechanism of self-excitation gives rise to high vibrational frequencies such as those associated with a tool (~ 2000 Hz). To induce such a self-excited vibration, the tool need only be bumped or encounter a hard spot in the work material so that it executes a single cycle. The amplitude will then grow until the damping energy per cycle becomes equal to the net energy stored per cycle. Figure 17.38 shows a typical F_P versus V curve capable of inducing self-excited vibration.

In addition to the source of self-excited vibration identified by Arnold, there are undoubtedly several others that pertain at lower vibrational frequencies. One of these discussed by Shaw and Sanghani (1962) is associated with the variation in cutting energy with variable undeformed chip thickness t.

Variable Undeformed Chip Thickness

Figure 17.39a shows orthogonal cutting with constant undeformed chip thickness t while Fig. 17.39b is for an orthogonal cutting situation with cyclically varying t. The case of Fig. 17.39b is capable of causing a self-excited vibration in the F_Q direction since more energy is stored during the part of the cycle when F_Q is increasing than is released by the system when F_Q is decreasing (Shaw and Hölken, 1957). The amplitude of the resulting vibration will depend upon the stiffness of the system and the degree of damping.

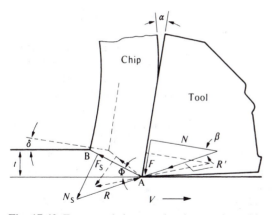

Fig. 17.40 Forces and shear angle when cutting with constant undeformed chip thickness (solid lines) and with increasing undeformed chip thickness (dotted lines).

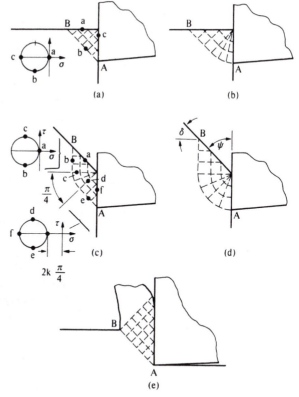

Fig. 17.41 Slip line fields. (after Shaw and Sanghani, 1962)

Figure 17.40 shows the situation for a cut where t is constant (solid lines) as well as one for which the free surface is inclined at an angle (δ) to the velocity vector V (dotted lines). The importance of such a change in direction of the free surface may be visualized by considering some simple two-dimensional slip line fields.

Figure 17.41a shows a frictionless tool engaging the corner of a workpiece. The slip lines (lines of maximum shear stress) will be inclined to both the free surface and the tool face at 45°. This becomes evident when the Mohr's circle diagram (insert) is considered. Corresponding points are similarly labeled in both diagrams. If the material is homogeneous and does not strain harden, it should deform plastically in shear along the lines of maximum shear stress.

If limit-friction (maximum friction possible without subsurface flow) is present on the tool face, then the slip line field will change to that shown in Fig. 17.41b, where the slip lines still intersect the free surface at 45°.

Figures 17.41c and 17.41d are for cases where the free surface makes an angle other than 90° to the tool face. In both of these (zero tool-face friction for Fig. 17.41c and with limit-friction in Fig. 17.41d), the slip lines still make an angle of 45° to the free surface. The slip line field for a fully developed chip is shown in Fig. 17.41e in the absence of tool-face friction. The slip lines are again seen to intersect the free surface at an angle of 45°.

A real cutting process will differ from that shown in Fig. 17.41e in several important ways (Fig. 17.42):

1. The rake angle α will generally not be 0°.

2. Tool-face friction will not be 0.

3. Additional plastic flow will occur along the tool face due to the closing of fracture points and additional burnishing action (as shown, respectively, from A to C and from C to D in Fig. 17.42).

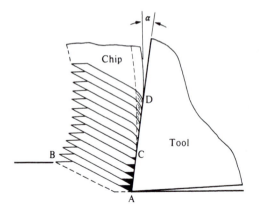

Fig. 17.42 Revised Piispanen model with secondary plastic flow along tool face. (after Shaw and Sanghani, 1962)

The surface of maximum shear stress will not be plane when tool-face friction is present (Figs. 17.41b and 17.41d) or when the angle between the tool face and the free surface is other than 90° (Fig. 17.41c). In Fig. 17.41b tool-face friction is seen to give rise to a low shear angle (0°) near the tool face (A) but this changes to a high value of 45° at the free surface (B).

However, a fully developed Piispanen model cannot be constructed with a shear surface that is concave upward (as shown in Figs. 17.41b, 17.41c, and 17.41d), since this requires the chip to penetrate the tool face. While the shear surface may be as shown in Fig. 17.41b at the very beginning of a cut, the chip soon curls into the tool face and the added pressure developed forces the shear surface to become plane. Hence, under equilibrium conditions the shear surface for a fully developed chip will be a plane (AB in Fig. 17.42) which in general will not correspond to a direction of maximum shear stress. At the tool tip (A) the shear angle will be greater than that corresponding to the direction of maximum shear stress, while at the free surface (B) the shear angle will be less than that corresponding to the plane of maximum shear stress. Thus, the actual shear direction will be between the initial maximum shear stress directions for its two ends.

From Figs. 17.41a through 17.41d it is evident that friction on the tool face and the inclination of the free surface are tending to change the resultant shear angle (ϕ) in the following manner:

1. An increase in tool-face friction will decrease the shear angle.

2. A decrease in the angle between the tool face and the free surface (ψ in Fig. 17.41d) will increase the shear angle.

At the beginning of a cut, the shear direction will always intersect the free surface at the same angle (45°) regardless of the inclination of the free surface (δ) and regardless of the state of friction on the tool face. An equivalent way of expressing this is that the shear angle at the free surface at the beginning of a cut will be

$$\phi_\delta = \phi_0 + \delta \qquad (17.14)$$

where ϕ_0 is the shear angle at the free surface when δ is zero and δ is positive as shown in Fig. 17.41d. The influence of tool-face friction on the direction of shear is completely independent of the influence of the direction of the free surface on the direction of shear at the beginning of a cut.

For a fully developed chip the angle between the shear surface and the free surface is forced to be other than 45° due to the compatability requirement that the shear surface be plane. Angle ϕ_0 is no longer independent of the state of friction on the tool face and in general will not be 45°. It seems reasonable to assume that Eq. (17.14) will also hold to a first order of approximation for a fully developed chip. This is equivalent to saying that the free surface and tool-face friction effects on ϕ are separable and that to a first order of approximation the tool-face friction contribution to ϕ will remain unchanged as δ is varied. Thus, ϕ_0 in Eq. (17.14) will be considered a function of tool-face friction only and independent of changes in δ.

As in most engineering situations where it is expedient to assume an independence of variables, the error associated with the approximation should decrease as the magnitude of δ decreases.

Experimental Verification

In order to check the foregoing observations regarding effects due to changes in δ, some elementary experiments were conducted. The workpiece was moved past a stationary HSS tool (planer fashion) at low speed (1 i.p.m. = 0.03 m min^{-1}). The workpiece was mounted in a dynamometer capable of recording forces F_P and F_Q. The AISI 1020 steel specimen was prepared as shown in Fig. 17.43 with plane faces rather than a sinusoidal surface, as a matter of simplification. The angle of 5° was selected as follows. A workpiece being turned in a lathe at 200 f.p.m. and vibrating at a frequency of 200 c.p.s. will have chatter marks on its surface having a wavelength of $(200 \times 12/60)/200 = 0.200$ in (5 mm). If the amplitude of vibration is 0.01 in (0.25 mm), the corresponding approximate value of δ will be

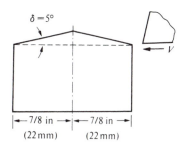

Fig. 17.43 Single-cycle specimen for tests at variable undeformed chip thickness.

$$\delta = \tan^{-1} \frac{0.01}{0.100} \cong 5°$$

A representative result is shown in Fig. 17.44 where the dotted curve shows the variation in undeformed chip thickness at the tool tip (t') with time and the solid curve shows the corresponding cutting-force trace (F_p). The rake angle α was 10° and the maximum value of t was 0.077 in (1.93 mm). Cuts were taken using CCl$_4$ as a cutting fluid in order to minimize effects due to a built-up edge. It is evident that the maximum force occurs before the maximum value of t' is encountered at the tool tip. This suggests that the value of undeformed chip thickness of importance is that at the upper end of the shear plane (B in Fig. 17.42) rather than that at the tool tip (A in Fig. 17.42). This should have been recognized immediately since the length of the shear plane is the quantity of importance and this depends on t_B and not on $t_A = t'$ (Fig. 17.39b) for the situation where the tool moves in a straight line and the free surface is wavy.

It was decided to repeat the previous experiments under conditions that would enable forces to be referred to t_B rather than $t_A = t'$. At the same time it was decided to use specimens (Fig. 17.45) having several cycles present on the surface and to perform the cuts such that (t_{min}) was 0.025 in rather than 0. The four peaks and valleys employed were numbered as shown in Fig. 17.45.

A representative force trace (F_P) is shown in Fig. 17.46. The first two cycles were usually irregular as equilibrium was established, but the last two cycles were always similar and reproducible and hence the ones considered in detail. A marking device with which the recorder was equipped enabled the point-in-time to be recorded when the upper end of the shear plane (B in Fig. 17.42) just reached a peak or a valley on the surface. These points are shown at the top of Fig. 17.46 and represent the locations of the tool tip when the top of the shear plane just reaches a peak or a valley.

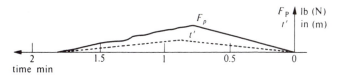

Fig. 17.44 Representative cutting-force trace for specimen of Fig. 17.43. (after Shaw and Sanghani, 1962)

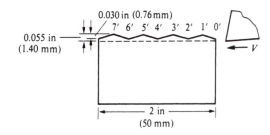

Fig. 17.45 Multi-cycle specimen for tests at variable undeformed chip thickness. (after Shaw and Sanghani, 1962)

Location of tool tip when shear plane reaches points 1',2',3' ... in fig 17.45

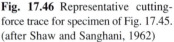

Fig. 17.46 Representative cutting-force trace for specimen of Fig. 17.45. (after Shaw and Sanghani, 1962)

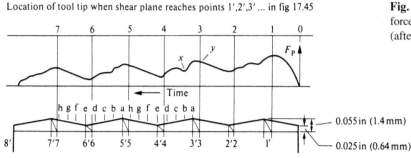

With this information at hand it is possible to draw the surface wave to scale beneath the force plot as shown in the lower part of Fig. 17.46. From this construction we may compute the mean values of shear angle when t is increasing as well as when t is decreasing. Such calculations are given in Table 17.2 where the values in the second column were obtained by converting the measured distances between 0 and 1, 0 and 2, etc., at the top of Fig. 17.46 to equivalent distances on the specimen. The mean value of the shear angle is thus found to be 45.8° when t is increasing and 28.8° when t is decreasing.

The value of δ for the specimen of Fig. 17.45 is $\tan^{-1}(30/250) = 7°$. According to Eq. (17.14) this should give rise to an increase in ϕ of approximately 7° when δ is +7° and a like decrease when

TABLE 17.2 Mean Shear Angle Values When Cutting Downward and Upward

				Shear Angle (degrees)	
Distance	**in ($\times 10^3$)**	**Distance**	**in ($\times 10^3$)**	**Cutting Downward**	**Cutting Upward**
0–1	196	1–1'	$250 - 196 = 54$	$\phi_1 = \tan^{-1}\left(\dfrac{55}{54}\right) = 45.6$	
0–2	458	2–2'	$500 - 458 = 42$		$\phi_3 = \tan^{-1}\left(\dfrac{25}{42}\right) = 30.8$
0–3	700	3–3'	$750 - 700 = 50$	$\phi_3 = \tan^{-1}\left(\dfrac{55}{50}\right) = 47.7$	
0–4	955	4–4'	$1000 - 955 = 45$		$\phi_4 = \tan^{-1}\left(\dfrac{25}{45}\right) = 29.0$
0–5	1190	5–5'	$1250 - 1190 = 60$	$\phi_5 = \tan^{-1}\left(\dfrac{55}{60}\right) = 42.2$	
0–6	1450	6–6'	$1500 - 1450 = 50$		$\phi_6 = \tan^{-1}\left(\dfrac{25}{50}\right) = 26.5$
0–7	1700	7–7'	$1750 - 1700 = 50$	$\phi_7 = \tan^{-1}\left(\dfrac{55}{50}\right) = 47.7$	
				Ave. 45.8°	Ave. 28.8°

Note: 1 in = 25.4 mm.

TABLE 17.3 Values of F_P for Different Values of t_B When t_B Is Decreasing and Increasing

Position of Intersection of Shear Plane with Free Surface (Fig. 17.46)	$t_B = t$ at Shear Plane Intersection in ($\times 10^3$)	F_P lb For Cycle 3 to 5	For Cycle 5 to 7	Ave.
a	55	840	740	790
b	47.5	800	660	730
c	40	540	480	510
d	32.5	560	440	500
e	25	400	240	320
f	32.5	380	280	330
g	40	500	400	450
h	47.5	620	520	570

Note: 1 in = 25.4 mm, 1 lb = 4.45 N.

δ is $-7°$. Half the difference between the two measured values of ϕ (45.8 − 28.8/2 = 8.5°) is to be compared with the value of 7° predicted by Eq. (17.14). Equation (17.14) is thus found to be in reasonably good agreement with experiment, even when the value of δ is quite large.

The upper and lower portions of Fig. 17.46 when used together enable the value of F_P to be determined when the value of (t_B) has any value between the maximum (0.055 in = 1.4 mm) and the minimum (0.025 in = 0.64 mm). Values of F_P for different values of t_B, when t_B is decreasing and increasing, are recorded in Table 17.3 for two cycles of cutting.

The first value of F_P in Table 17.3 (i.e., 840) is that at point y, which is the value of F_P when the top of the shear plane is at point a. Sometimes the chip tends to break as at point x in Fig. 17.46, and this in turn will have a local influence on the force curve.

It is clearly evident from Table 17.3 that F_P is less for the same value of t_B when t_B is increasing than when t_B is decreasing.

The F_Q trace paralleled that for F_P in every instance, and everything that has been said regarding F_P holds for F_Q as well.

Kobayashi and Shabaik (1964) have verified the results of Shaw and Sanghani (1963) by studying orthodonal chip formation with varying undeformed chip thickness for Al 1112 cold-rolled steel at low cutting speed (0.625 i.p.m. = 1.56 mm min⁻¹). Tests were performed for both a tool moving along a triangular wavy surface in a straight line and where a straight surface was cut by a tool moving along a triangularly wavy path. It was found that in both cases the effective shear angle depends upon the state of stress at the point where the shear plane intersects the free surface.

Discontinuous Chips

When relatively brittle materials are machined such as cast iron or β-brass, the chips break periodically in the manner shown in Fig. 17.47, which was drawn from selected frames of a motion picture. The material is seen to fracture cyclically in shear along a line that strongly resembles slip line AB in Fig. 17.42.

Discontinuous cutting is thus seen to involve cutting with a large positive value of δ. Immediately after shear fracture has occurred, a new chip develops, and as it does, δ is decreased due to a humping-up of the material. As the chip grows, it rolls into the tool face, and the resulting normal stress causes

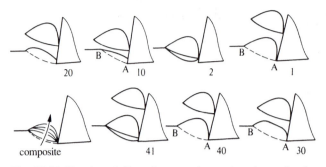

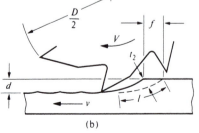

Fig. 17.47 Sketches of discontinuous orthoganal cutting action for β-brass. Numbers under sketches are movie-frame numbers. Depth of cut, 0.007 in (0.18 mm); cutting speed, 0.5 i.p.m. (13 mm min⁻¹); rake angle, 15°; width of cut, 0.161 in (4.1 mm).

subsequent slip to occur on a surface that is less concave upward and farther removed from the surface of maximum shear stress. In the case of a ductile metal, this action will continue until the shear surface becomes perfectly plane. With a brittle material, however, a shear stress sufficiently high to cause a shear failure will be reached before the shear surface becomes plane. The metal does not fracture along the last active shear plane on which shear flow was occurring but along the surface of maximum shear stress.

It is well known that discontinuous cutting is desirable from the point of view of cutting forces. While this was originally believed due to a reduction in tool-face friction associated with the periodic renewal of an adsorbed film on the face of the tool, it now appears that the high positive values of δ that are obtained in discontinuous cutting are chiefly responsible for the relatively low values of F_P in discontinuous chip formation.

Milling

Chips of variable t are characteristic of the plane-milling process. In up-milling (or conventional-milling), the thickness of the chip varies from zero to a maximum value t_1 (Fig. 17.48a), while in down-milling (or climb-milling), the undeformed chip thickness goes from t_2 to zero (Fig. 17.48b).

Obvious advantages of the down-milling process are

1. a shorter length of contact (l) for the same feed per tooth (f) and depth of cut (d)
2. absence of rubbing at the beginning of the cut

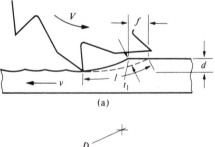

(a)

(b)

Fig. 17.48 Cutter action. (a) Up-milling. (b) Down-milling.

Because of these factors down-milling frequently gives better results than up-milling if the machine has sufficient rigidity to withstand the greater impact associated with the more abrupt tooth engagement in down-milling.

However, the extent of improvement is often not as great as expected and this could be explained in terms of the lower forces associated with conventional-milling. For the same value of t the cutting force will be greater in down-milling than in up-milling, since δ is negative for down-milling but positive for up-milling.

The approximate value of δ will be

$$\delta = \tan^{-1}\left(\frac{t_m}{l}\right) \qquad (17.15)$$

where t_m = maximum undeformed chip thickness.
If $v < V$

$$\delta = \frac{2f}{D}\cos i \qquad (17.16)$$

where f = feed/tooth.

Regenerative Chatter

The foregoing discussion has been concerned with a tool moving in a straight line while cutting a wavy surface, which may be called case I (Fig. 17.49a). When a tool cuts a smooth specimen and vibrates in the y (F_Q) direction, a wavy surface will be generated as shown in Fig. 17.49b. This will be referred to as case II. The effective value of t (with regard to cutting forces) for case II is that at the tool point (t_A), since it is this that determines the length of the shear plane.

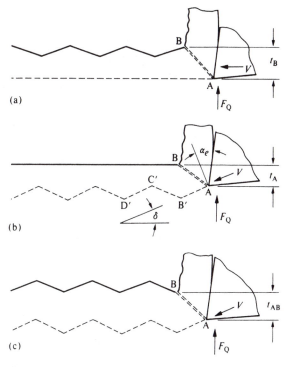

When the tool cuts from A to B′, its resultant velocity (V) will be directed as shown in Fig. 17.49b, and the effective rake angle (α_e) should be measured from a reference line that is perpendicular to this velocity vector. As the tool goes from A to B′, the depth of cut (t) will increase, and δ will be positive.

Thus, there are two major differences between cases I and II:

1. For case I there is no change in rake angle, but for case II the effective rake angle is increased by an amount δ when t is increasing, and vice versa.

2. The effective value of t for case I is t_B while that for case II is t_A.

Fig. 17.49 Cutting with variable undeformed chip thickness. (a) Case I: straight tool motion and wave on free surface. (b) Case II: vibrating tool and smooth free surface. (c) Case III: vibrating tool and wave on free surface.

Due to a change in inclination of the free surface, cutting forces will be less (and the shear angle greater) when the effective value of t is increasing for both cases I and II. In addition, for case II, the increase in rake angle that occurs when t is increasing will cause an additional decrease in cutting forces (and increase in shear angle). Similarly when t decreases, the change in rake angle for case II will cause the cutting force to increase to an even higher value than in case I.

The difference in F_Q, for a given value of t, when t is increasing and decreasing, is significant in connection with self-excited vibrations. Energy is stored elastically in a vibrating tool when the force on the tool (F_Q) is in the same direction as the motion of the tool, and vice versa. When the tool of Fig. 17.49b cuts from A to B′, energy is being released from the tool since tool force and tool motion are in opposite directions. Similarly as the tool goes from B′ to C′, energy is stored since F_Q and tool motion are then in the same direction.

If the energy stored exceeds the energy released over a cycle, then the amplitude of vibration will increase until the damping energy per cycle just equals the net energy stored per cycle. Since force F_Q will be larger when cutting from B′ to C′ (energy stored) than from C′ to D′ (energy released), we have the condition necessary for a self-excited vibration to occur. It should be noted that the foregoing argument is independent of whether the tool moves relative to the work, the work moves relative to the tool, or both.

In a turning operation the surface is recut after each revolution and the wavy surface previously cut will act through feedback to induce additional vibration. Such a case is shown in Fig. 17.49c.

The effective value of t in this instance is the vertical distance between A and B (t_{AB}). Cutting forces will be less when t_{AB} is increasing than when t_{AB} is decreasing. The net energy stored per cycle will depend upon the phase relation between the old and new waves, which will have the same frequency (the natural frequency of the tool–work system). The resulting phase angle will be that which gives the greatest stored energy per cycle (180°), and the resulting energy per cycle will be approximately double the value for the case of Fig. 17.49b. The feedback chatter just described is termed regenerative chatter.

Other Sources of Self-Excitation

Another source of self-excitation energy that may result in low-frequency chatter is due to the unequal growth and decay rates with undeformed chip thickness (t) associated with a BUE. This dynamic characteristic of the BUE is capable of producing a self-excited vibration of a cutting system (Shaw and Sanghani, 1963). It is also likely that the periodic growth and decay of deformed chip thickness (t_C) might similarly contribute to self-excited vibration and there are undoubtedly several other as yet unidentified mechanisms that provide the unsymmetrical energy transfer required for self excitation, thus making this a very complex aspect of metal cutting.

Chatter Threshold

Rahman (1988) has presented a method to reproducably and precisely determine the chatter threshold of a machining operation. This involves the in-process measurement of the workpiece in the horizontal direction. In automated machining systems, it is important to detect the onset of chatter due to tool wear. In process planning, it is also necessary to know the exact chatter-free width of cut for a given machine and set of machining conditions. A chatter threshold may be established by observation of workpiece deflection, cutting forces, or cutting noise level. In this study, it was found that for turning operations the horizontal deflection of the workpiece is the most accurate method of detecting the onset of chatter. Reasons for this conclusion are given based on a sound application of experimental and analytical results.

STRUCTURED SURFACES

Structured surfaces are designed to give a desired performance or appearance. Bryan (1966) and DeVries et al. (1976) have discussed the relationship between surface finish and adhesion, friction, and wear. Evans and Bryan (1999) have presented an extensive review of structured surfaces with an emphasis on definitions. A wide range of applications are considered, including

- optical lenses and reflectors
- mechanical contact control
- friction and wear
- hydrodynamics
- abrasives
- tools (files, burrs, etc.)
- adhesion
- thermal applications
- biological applications (including implants)

The principal production techniques are also considered. These include

- cutting (diamond, laser, and electron beam techniques)
- knurling by forming and embossing
- lithography (photo-resist masking and etching)
- replication by embossing, molding, or casting
- electrodeposition

Structured surfaces provide an important addition to the design tools available to provide improved product performance. De Chiffre et al. (2000) have presented a thorough review of surface texture. Many methods of specification and measurement have evolved, and for the most part these are covered by ISO standards. Figure 17.50 depicts the relation between method of generation, function, and characterization of surface textures.

Since structured surfaces are frequently three-dimensional, with significantly different depth-to-width ratios than for surface finish, different measurement techniques are required (Lonardo et al., 1996; Whitehouse, 1994). This aspect of the subject is also discussed by Evans and Bryan (1999).

Suh and Saka (1987) have investigated friction and wear results for a copper pin sliding on a copper disk at speeds ranging from 3.94 to 8.20 f.p.m. (1.2 to 2.5 m/min) under an axial load on the pin of 0.44 lb (0.2 kg), in the absence of a lubricant. Figure 17.51a shows the coefficient of friction versus sliding distance for such a test when the disk was flat and its surface was uninterrupted. Figure 17.50b shows test results for the same conditions except the disk had a pattern of depressions etched away to a depth of 0.002 in (50 μm). This was done by photolithographic masking. Figure 17.52a shows a scanning electron micrograph of a freshly prepared flat disk, while Fig. 17.52b is for a similar disk with etched depressions. It is evident in Fig. 17.51 that while the coefficient of friction starts out the same (about 0.2) with and without the pattern of depressions, it soon rises and becomes erratic before settling down to a value of about 0.79 for the flat uninterrupted disk. Figure 17.53 shows surfaces of a disk after

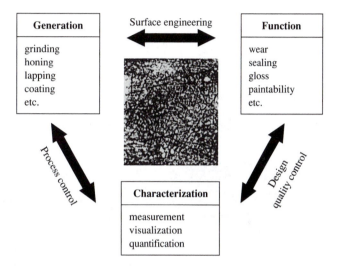

Fig. 17.50 Three important factors related to surface texture and their method of generation, function and characterization. (after De Chiffre et al., 1999)

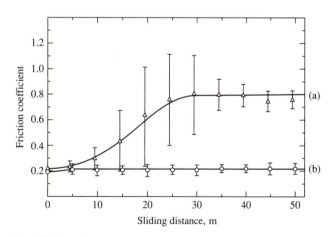

Fig. 17.51 Friction coefficient versus sliding distance for a copper pin sliding on a copper disk. (a) For a flat uninterrupted disk. (b) For a disk with a pattern of small square depressions. (after Suh and Saka, 1987)

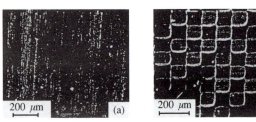

Fig. 17.52 Scanning electron micrograph of unused flat uninterrupted disk at (a) and for a disk with pattern of small square depressions at (b). (after Suh and Saka, 1987)

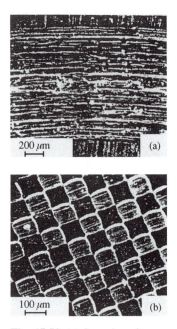

Fig. 17.53 (a) Scanning electron micrograph of worn disk with uninterrupted surface. (b) Scanning electron micrograph of disk with square depressions. Both disks were subjected to identical sliding conditions for a sliding distance of 164 ft (50 m). (after Suh and Saka, 1987)

164 ft (50 m) of sliding contact (a) without depressions to accommodate wear particles and (b) with depressions. There is considerable evidence of plowing in Fig. 17.53a, but not in Fig. 17.53b. Figures 17.52 and 17.53 clearly indicate that what Suh and Saka call engineered surfaces can have an important influence on friction and wear. It is suggested by Suh and Saka that wear particles play an important role in friction and wear, and if provision is taken to remove them from action by means of a pattern of depressions, positive results may be achieved.

RESIDUAL STRESS

Residual stress in the surface of a part may have an important influence on performance. The effect on fatigue life is one of the most important of these consequences. In general fatigue life is increased when the surface stress is compressive, but decreased when the surface stress is tensile. The principal sources of residual stress are mechanical, chemical, or structural. Shot peening of the surface of a part is an important mechanical way of inducing a useful anti-fatigue residual stress. In this case, hard spherical particles impact the surface to produce a large number of overlapping craters. As in the case of a Brinell hardness test, the plastically deformed material around and below each crater is left in a state of residual compressive stress. An important chemical means of inducing a compressive residual surface stress is carburization. In this case, carbon or nitrogen is diffused into the surface at elevated temperature, and after heat treatment the surface is left in a hard state of compression.

Early work concerning residual surface stresses produced in machining was presented by Henriksen (1951), Barish and Schoech (1970), and Okushima and Kakino (1971).

Machined surfaces may be left in a state of tensile or compressive stress depending on the hardness of the workpiece and the machining conditions. Liu and Barish (1982) have studied the patterns of residual stress left in surfaces of a ductile low-carbon steel when cut under a wide range of conditions. The residual stress was estimated by a change in the radius of curvature of a specimen as thin layers of the machined surface were removed in steps of controlled etching. Figure 17.54 shows a specimen resting on two razor blades 2 in (50.8 mm) apart. By measuring the vertical distance between the lower surface of the specimen (to ± 50 μin $= \pm 1.25$ μm) and the supports and also the thickness of the specimen after each etching step, the stress in the layer removed may be estimated by application of principles of elasticity. The thin specimen is masked on all sides except the machined surface and etched with a solution of 14% hydrofluoric acid, 6% nitric acid, and 80% water.

The profile of residual stress is obtained by plotting the residual stress in each layer removed versus its depth below the surface. Figure 17.55 shows the residual stress profiles for a sharp tool machining a ductile low-carbon steel at different speeds and depths of cut. Here it is evident that a

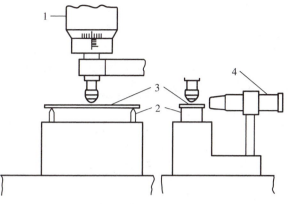

Fig. 17.54 Method of measuring change in curvature of machined surface as layers are removed from the surface by etching: 1 = micrometer, 2 = razor blades, 3 = machined surface oriented with the long direction in the cutting direction or perpendicular to the cutting direction, 4 = microscope. (after Liu and Barish, 1982)

small value of depth of cut does not necessarily give a low value of residual stress. Figure 17.56 shows corresponding results for a worn tool having a flank wear land of 0.010 in (0.25 mm). Similar results were obtained for stresses perpendicular to the cutting direction. In all cases the residual stresses were tensile for low-carbon steel. Higher tensile stresses were obtained at low speeds and low depths of cut. Comparing Figs. 17.55 and 17.56 it is evident that the stress at the surface is only slightly higher for a dull tool but decreases less rapidly with depth below the surface. Edge treatment was found to give results similar to a wear land—somewhat lower tensile stress at the surface and a deeper penetration of tensile stress below the surface.

Figure 17.57 explains the origin of the tensile residual stress obtained when a ductile material is machined. An element *mn* at the level of the finished surface will be subjected to compressive stress before reaching the cutting edge. In the case of a ductile material, a large amount of strain energy will be stored in element *mn* and upon passing the cutting edge the stored energy will be released. If this strain energy is sufficiently high (large undeformed chip thickness and dull tool) and rapidly released (high cutting speed), the expansion of *mn* as it passes the cutting edge may over-

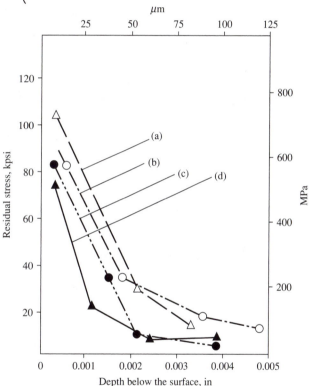

Fig. 17.55 Change in residual stress in the cutting direction with depth below the surface when a ductile low-carbon steel is machined under orthogonal conditions with a sharp tool under the following conditions [speed (m/s), depth of cut (mm)]: (a) 1.53, 0.127; (b) 1.53, 0.254; (c) 4.53, 0.254; (d) 4.53, 0.127. (after Liu and Barash, 1982)

shoot, resulting in a tensile residual stress. The tensile stress should decrease with depth below the surface, but less rapidly for a dull tool (Fig. 17.58a) or one with a small clearance angle (Fig. 17.58b) or small negative clearance edge preparation to prevent chipping of the cutting edge (Fig. 17.58c).

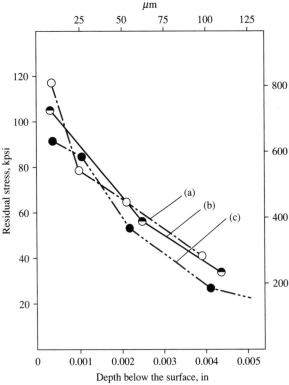

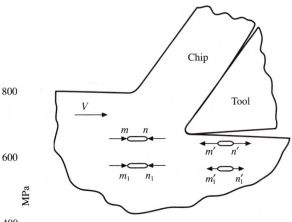

Fig. 17.57 Loading and unloading as metal passes under the tool tip. (after Liu and Barash, 1982)

Fig. 17.56 Change in residual stress in the cutting direction with depth below the surface when a ductile low-carbon steel is machined under orthogonal conditions with a dull tool having a wear land = 0.010 in (0.254 mm). Cutting conditions were as follows [speed (m/s), depth of cut (mm)]: (a) 1.53, 0.254; (b) 2.52, 0.254; (c) 4.63, 0.254. (after Liu and Barash, 1982)

Matsumoto et al. (1984) have studied the residual stress patterns obtained when AISI 4340 specimens having a wide range of hardness were turned as shown in Fig. 17.59 using an Al_2O_3/TiC ceramic tool. In this study, residual stresses were measured using x-ray diffraction. Stress patterns were again obtained by etching the machined surfaces in steps but this time using a solution of 10% hydrofluoric acid, 15% hydrochloric acid, 30% nitric acid, and 45% water. Residual stress profiles were obtained in both the machining direction and the direction perpendicular to the machining direction. Hardness values ranged from H_{RC} 29 to H_{RC} 57. Steady state chips were obtained below H_{RC} 45. Above this hardness, chips that were termed

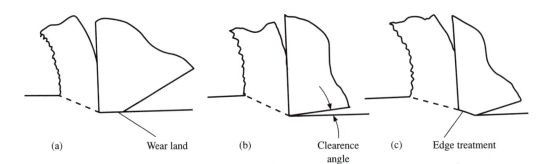

Fig. 17.58 (a) Worn tool with wear land. (b) Tool with small clearance angle. (c) Tool with small negative rake edge treatment to prevent chipping.

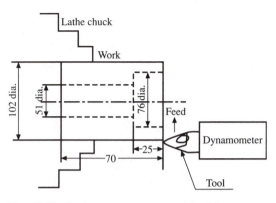

Fig. 17.59 Cutting arrangement used by Matsumoto et al. (1984). Dimensions are given in millimeters (mm).

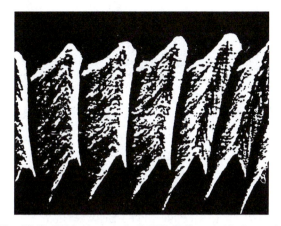

Fig. 17.60 SEM micrograph of AISI 4340 steel chip for workpiece hardness of H_{RC} 55. (after Matsumoto et al., 1984)

"segmental" were obtained. Figure 17.60 is a photomicrograph of a chip from an AISI 4340 workpiece having a hardness of H_{RC} 55.

Figure 17.61 gives the residual stress patterns in the cutting direction with a sharp tool for four values of hardness. The surface residual stress is seen to be tensile for low values of hardness but to change to compressive above a hardness of about H_{RC} 45 where chip formation changes

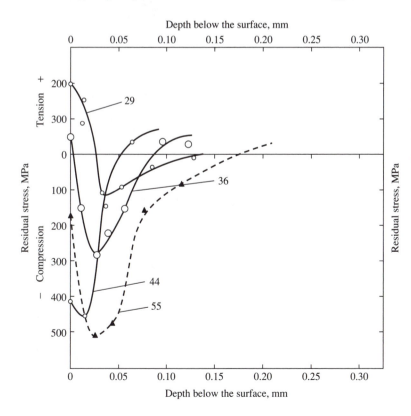

Fig. 17.61 Effect of hardness of workpiece on residual stress distribution in the cutting direction for a speed of 300 f.p.m. (91.47 m/min), a depth of cut of 0.006 in (0.15 mm), and a feed of 0.035 in (0.89 mm). The number associated with each curve is the Rockwell hardness on the C scale (H_{RC}). (after Matsumoto et al., 1984)

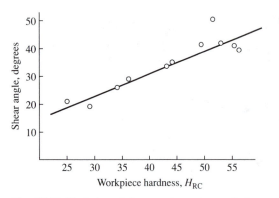

Fig. 17.62 Variation of shear angle versus workpiece hardness for tests of Figs. 17.60 and 17.61. (after Matsumoto et al., 1984)

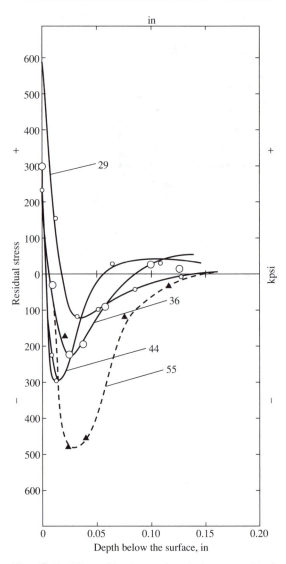

Fig. 17.63 Effect of hardness of workpiece on residual stress distribution in direction perpendicular to the cutting direction for the same cutting conditions as for Fig. 17.60. (after Matsumoto et al., 1984)

from steady state to cyclic. As the hardness of the workpiece increases, the shear angle increases as shown in Fig. 17.62, reaching a value of 45° or higher when the hardness exceeds the value where chip formation ceases to be steady state (about H_{RC} 45). As the shear angle increases, with increase in workpiece hardness, the chip thickness decreases and the depth of the residual stress pattern increases.

Figure 17.63 shows residual stress patterns for AISI 4340 steel for cutting conditions that are the same as in Fig. 17.61, but these stress patterns are for stresses perpendicular to the cutting direction. These results are similar to Fig. 17.61 except that higher tensile stresses at the surface are obtained for low values hardness and lower values of compressive stress are obtained for high values of hardness.

Subsurface compressive stresses for hard workpieces (Figs. 17.61 and 17.63) are due to the high tool-face temperatures, low tool-face friction, and high normal stress on the tool face pertaining. The formation of sawtooth chips discussed in Chapter 22 gives rise to cutting ratios greater than one for very hard work materials. This has been shown to be due to lengthening of the chip due to extrusion as it moves up the tool face under conditions of low friction and high normal stress. Chip lengthening of this sort due to extrusion of the chip should leave the finished surface with a residual compressive stress.

When a hard material is machined at high speed, temperatures are often high enough for a nonetching white layer to be present in photomicrographs of a chip. The origin and significance of this detail is discussed in Chapter 22. From the standpoint of surface integrity it is important that a white layer not be present in the finished surface unless a finishing operation that completely removes this defect is performed.

REFERENCES

Albrecht, P. (1956). *Microtecnic* **10**, 45.

Arnold, R. N. (1946). *Proc. Inst. Mech. Engrs.* **154**, 261.

Barish, M. M., and Schoech, W. J. (1970). *Proc. MTDR Conf.* Pergamon Press, Oxford.

Bickel, E. (1954). *Microtecnic* **8**, 279.

Bowden, F. P., and Moore, A. J. W. (1947). *Inst. of Metals Symposium on Internal Stresses in Metals and Alloys*, pp. 131–137.

Brammertz, P. H. (1960). Dissertation, T. H. Aachen.

Brammertz, P. H. (1961). *Ind. Anz.* **83**, 525.

Bresseler, B., El-Wardany, T. I., and Elbestawi, M. A. (1997). *First French and German Conf. on High Speed Machining*, pp. 196–206.

Bryan, J. B. (1966). CIRP Technical Committee Report.

Ciragan, O. (1955). Dissertation, E. T. H. Zurich.

Cocks, M. (1962). *J. Appl. Phys.* **33**, 2152.

Cocks, M. (1964). *J. Appl. Phys.* **35**, 1807.

Cocks, M. (1965). *Wear* **8**, 85.

Danek, O., Polacek, M., Sparek, L., and Tlusty, J. (1962). *Selbsterregte Schwingungen an Werkzeugmaschinen*. V.E.B. Verlag Technik, East Berlin.

De Chiffre, L., Lonardo, P., Trumphold, H., Lucca, D. A., Goch, G., Brown, C. A., Raja, J., and Hansen, H. N. (2000). *Annals of CIRP* **49/2**, 635–652.

DeVries, W., Field, M., and Kahles, J. (1976). *Annals of CIRP* **25/2**, 569–573.

Doi, S., and Kato, S. (1956). *Trans. Am. Soc. Mech. Engrs.* **78**, 1127.

Ernst, H., and Martelloti, M. (1935). *Mech. Eng.* **57**, 487.

Evans, C. J., and Byran, J. (1999). *Annals of CIRP* **46/1**, 403–440.

Gillespie, L. K. (1999). *Deburring and Edge Finishing*. SME, Dearborn, Mich.

Hahn, R. S. (1953). *Trans. Am. Soc. Mech. Engrs.* **75**, 1073.

Ham, I., Hitomi, K., and Thuering, G. L. (1961). *Trans. Am. Soc. Mech. Engrs.* **83**, 142.

Ham, I., Hitomi, K., and Theuring, G. L. (1962). *Trans. Am. Soc. Mech. Engrs.* **83**, 282.

Ham, I., Hitomi, K., and Theuring, G. L. (1964). *Trans. Am. Soc. Mech. Engrs.* **86**, 141.

Henriksen, E. K. (1951). *Trans ASME* **73**, 461.

Hovenkamp, L. H., and Van Emden, E. (1953). *Microtecnic* **7**, 3, 117.

Kishawy, H. A., and Elbestawi, M. A. (1998). *Int. J. of Machine Tools and Mfg.* **39**, 1017–1030.

Kishawy, H. A., and Elbestawi, M. A. (2001). *Proc. Inst. Mech. Engrs.*, p. 215B.

Kobayashi, S., and Shabaik, A. (1964). Paper 64-Prod 8, ASME Conference and Exposition, Detroit, Mich.

Lambert, H. J. (1961). *Annals of CIRP* **10**, 246.

Leyensetter, W. (1956). *VDI Zeitschrift* **98**, 957.

Liu, C. R., and Barish, M. M. (1982). *J. Eng. for Ind., Trans ASME* **104**, 257–264.

Lonardo, P. M., Trumbold, H., and De Chiffre, L. (1996). *Annals of CIRP*, **45/2**, 589–598.

Martellotti, M. E. (1941). *Trans. Am. Soc. Mech. Engrs.* **63**, 8, 677.

Matsumoto, Y., Barash, M. M., and Liu, C. R. (1984). *ASME PED* **12**, 193–204.

Mayer, J. E., and Shaw, M. C. (1963). *Trans. Am. Soc. Mech. Engrs.* **85**, 2.

Mitsui, K., and Sato, H. (1974). *Seisan-Kenkyu* **26/8**, 304 (in Japanese).

Moll, H. (1939). Dissertation, T. H. Aachen.

Moore, A. J. W. (1948). *Metallurgia* (June), 71–74.

Nakayama, K. (1957). *Bull. Fac. Engng. Yokohama Natn. Univ.* **6**.

Nakayama, K., Shaw, M. C., and Brewer, R. C. (1966). *Annals of CIRP* **14**, 211.

Nelson, H. R. (1940). *Conference on Friction and Surface Finish at MIT*, p. 217.

Nelson, H. R. (1941). *Amer. Mach.* **85**, 743.

Okushima, K., and Kakino, Y. (1971). *Annals of CIRP* **20/1**.

Opitz, H., and Gappisch, M. (1962). *Int. J. Mach. Tool Des. Res.* **2**, 43.

Opitz, H., and Moll, H. (1941). *Herstellung Hochwertiger Drehflachen*. Ber Betriebswiss. Arb., VDI, Verlag, Berlin.

Ostermann, G. (1960). Dissertation, T. H. Aachen.

Pahlitzsch, G., and Semmler, D. (1960). *Z. fur Wirtschaftlich Fertigung* **55**, 242.

Pahlitzsch, G., and Semmler, D. (1961). *Z. fur Wirtschaftlich Fertigung* **56**, 148.

Pahlitzsch, G., and Semmler, D. (1962). *Z. fur Wirtschaftliche Fertigung* **57**, 45.

Pekelharing, A. J. (1960). *Microtecnic* **14**, 2.

Pekelharing, A. J., and Schuermann, R. A. (1955). *Werkstattstech u. Machenbau*, **45**, 59.

Peters, J. (1963). In *International Research in Production Engineering*. ASME, New York, p. 486.

Rahman, M. (1988). *Trans ASME, J. of Eng. for Ind.* **110**, 44–50.

Rivin, E. I. (2000). *Annals of CIRP* **49/2**, 591–634.

Sata, T. (1964). *Annals of CIRP* **12**, 190.

Schaller, E. (1964). Dissertation, T. H. Aachen.

Schuler, H. (1957). Dissertation, T. H. Aachen.

Semmler, D. (1962). Dissertation, T. H. Braunschweig.

Shaw, M. C., and Crowell, J. A. (1965). *Annals of CIRP* **13**, 5.

Shaw, M. C., and Hölken, W. (1957). *Ind. Anz.* **63**, 959.

Shaw, M. C., and Macks, E. F. (1949). *Analysis and Lubrication of Bearings*. McGraw-Hill, New York.

Shaw, M. C., and Sanghani, S. R. (1962). *Proc. Int. Inst. Prod. Engng. Res.* **10**, 340.

Shaw, M. C., and Sanghani, S. R. (1963). *Annals of CIRP* **11**, 59.

Shaw, M. C., Thurman, A. L., and Ahlgren, H. J. (1966). *J. Engng. Ind.* **88**, 142.

Shaw, M. C., and Usui, E. (1962). *Trans. Am. Soc. Mech. Engrs.* **84**, 89.

Shaw, M. C., Usui, E., and Smith, P. A. (1961). *Trans. Am. Soc. Mech. Engrs.* **83**, 181.

Sokolowski, A. P. (1955). *Prazision in der Metallbearbeitung*. VEB Verlag Technik, Berlin.

Solaja, V. (1958). *Wear* **2**, 1, 40.

Suh, N. P., and Saka, N. (1987). *Annals of CIRP* **46/1**, 403–408.

Tobias, S. A. (1965). *Machine Tool Vibration*. Wiley, New York.

Tobias, S. A., and Fishwick, W. (1958). *Trans. Am. Soc. Mech. Engrs.* **80**, 1079.

Trent, E. M. (1963). *J. Iron Steel Inst.* **201**.

Trigger, K. J., Zylstra, L. B., and Chao, B. T. (1952). *Trans. Am. Soc. Mech. Engrs.* **74**, 1017.

Uchida, S., Sato, H., and O-honi, M. (1979). *Annals of CIRP* **28/1**, 419–423.

Whitehouse, D. J. (1994). *Handbook of Surface Metrology*. Inst. of Physics, Bristol, England.

Woxén, R. (1931). *Tekn. Tidskr.* **61**, 43.

18 CHIP CONTROL

An important practical problem concerns the form of chips produced in machining since this has important implications relative to

1. personal safety
2. possible damage to equipment and product
3. handling and disposal of swarf after machining
4. cutting forces, temperatures, and tool life

Naturally brittle materials such as gray cast iron tend to form chips that range from small segments to particles. On the other hand, ductile metals tend to form long, continuous, ribbon-shaped chips that are dangerous and difficult to handle. When machining at high speed, hot chips with sharp, jagged edges represent a hazard to the operator. Such long continuous chips tend to form tangled "nests" that may interfere with the proper functioning of the machine. Under such conditions, it is necessary for the operator to stop the machine periodically to remove the snarled chip "nests" with a hook. This represents a decrease in productivity. If an attempt is made to remove snarled chips while the machine is running, this represents a hazard to the operator.

A "nest" of continuous chips has a very low apparent density making disposal difficult. Lang (1974) has introduced a chip packing ratio (R) where

$$R = \frac{\text{volume of chips}}{\text{equivalent volume of uncut metal}}$$

In practice, R may range from 3 to 10 (or more) but satisfactory chip control calls for a value of R of about 4. To achieve this, chips must break periodically.

Chip curvature plays an important role relative to chip breaking as does the brittleness of the work material and the thickness of the chip (t_C). As chip thickness decreases, chips are more ductile and chip breaking becomes more difficult. A given work material that has been substantially strained during chip formation will have a critical strain at fracture (ε_f) which varies as the ratio t_C/R_C where t_C is the deformed chip thickness and R_C is the radius of curvature of the chip. If the natural chip curvature is not sufficiently small to cause fracture for a given feed rate t it becomes necessary to increases either t_C or chip curl to provide periodic chip fracture.

Chip curl may be increased by use of a so-called chip breaker which is really a chip curler. There are two general types of chip breakers—an inclined obstruction clamped to the tool face

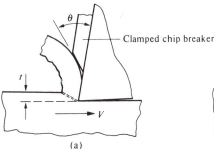

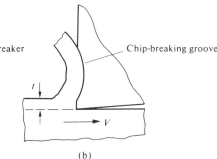

Fig. 18.1 (a) Clamped chip breaker. (b) Chip-breaking groove.

Clamped chip breaker

Chip-breaking groove

(a) (b)

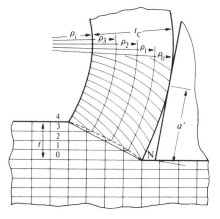

Fig. 18.2 Deformation of rectangular grid during chip formation. (after Nakayama, 1963)

(Fig. 18.1a) or a groove ground or molded into the tool face parallel to the major cutting edge (Fig. 18.1b). If a BUE is present at the tip of a tool, this will cause the chip to curl away from the tool face. Nakayama (1963) has shown that the secondary shear zone which usually develops along the tool face in the absence of a BUE can also cause the chip to curl. The material in the secondary shear zone gradually increases in speed as it travels along the tool face, and as a consequence of flow, continuity becomes thinner. This action is equivalent to a stationary nose of decreasing thickness that plays the same role as a BUE in causing chip curl. Figure 18.2 shows a deformed set of grid lines that are rectangular before reaching the shear plane. Horizontal lines become circular arcs with a common center. By continuity, the deformed, initially horizontal lines have a closer spacing as the tool face is approached and this gives a shear plane that is curved slightly convex upward. Vertical lines are also convex upward. The chip surface nearest the tool face will thus be a circular arc according to Nakayama (1963). The region labelled N (for nose) will be either a BUE or the aforementioned equivalent secondary shear zone. Thus, for volume continuity, the chip will be "born curled" since the shear plane will be slightly curved if a BUE or equivalent secondary shear zone is present at the tip of the tool which is practically always the case. The idea that chips are formed curved has also been suggested by several others (Ernst and Merchant, 1941; Henriksen, 1951; Hahn, 1953; Spaans, 1971), and this appears to be the case except for the rare situation where the shear plane and tool face are perpendicular and there is no secondary shear zone or BUE.

CHIP FLOW DIRECTION AND CHIP TYPES

In pure orthogonal cutting, the chip moves across the tool face in a direction perpendicular to the cutting edge. However, there are two important situations which cause the chip flow direction to be directed otherwise. When the end of a small-diameter tube is turned (Fig. 18.3a), the cutting velocity will be substantially smaller at the inside diameter than at the outside diameter. Since the cutting ratio along the cutting edge will be approximately constant, this will cause the chip to curl to the side as shown in Fig. 18.3a. If, on the other hand, the cutting velocity is essentially constant for all points along the cutting edge but there is an inclination angle i, then the chip will also flow to the side, the direction and magnitude of the side flow being given very approximately by Stabler's rule (Chapter 16). Figure 18.3b is an example of cutting with an inclination angle.

The two sources of side flow mentioned above together with the tendency for the chip to curl away from the tool face give rise to a number of different chip types (Fig. 18.4). The chip type of Fig. 18.4c (also Fig. 18.3a) is periodically broken when the chip encounters the freshly machined surface. This may cause undesirable surface damage in finish machining operations. The chip types in Figs. 18.4d and 18.4e are best broken when an up-curling chip strikes the surface about to be machined. If side flow is sufficient, the up-curling chip may miss the old surface but encounter the clearance face of the tool giving rise to chips of the type in Fig. 18.4f. If there is still greater side flow, chips of the type in Figs. 18.4h, 18.4i, or 18.4j type may be formed.

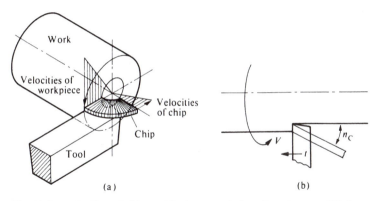

Fig. 18.3 (a) Curling of chip to side due to variation of cutting speed V along cutting edge. (b) Plan view of turning operation showing alteration in chip flow direction (n_C) due to inclination angle i (zero side rake angle, positive back rake angle $\alpha_b = i$). (after Spaans, 1971)

In addition to these types, several others are observed that represent combinations of those shown in Figs. 18.4a–18.4j. For example, the types in Figs. 18.4d–18.4g may form under conditions where the inclination angle i is not zero or where V is not constant across the cutting edge or both. Figure 18.4k is such an example which represents a combination of the actions of the types in Figs. 18.4c and 18.4d. Also, chips may break periodically but every fracture may not be complete. This gives rise to connected chip segments such as the connected C-type segments shown in Fig. 18.4k. The most desirable chip types from all points of view are individual (unconnected) C- or ear-type chips: the types in Figs. 18.4d, 18.4e, 18.4f, and 18.4k. While the type in Fig. 18.4g may yield a satisfactory density ratio R, this type with two or more turns per coil will give rise to tool forces that are higher than necessary due to over-curling.

CHIP BREAKER PERFORMANCE

If chips do not break naturally, then a chip curler (Fig. 18.1) must be used, or the feed must be increased, or the work must be made more brittle. Most operating variables other than feed have a small or negligible influence on chip-breaking tendency other than through their influence on deformed chip thickness t_C. The most effective way of changing t_C of course is to change the undeformed chip thickness (t = feed). For example, since cutting speed V has a relatively small influence on cutting ratio (in the absence of a BUE) and hence on t_C for constant feed, this variable (V) is relatively unimportant relative to chip breaking. It might also be mentioned that an increase in cutting speed has two opposing effects relative to chip fracture:

- an increase in shear rate, which renders the chip more brittle
- an increase in temperature, which makes it less brittle

An increase of side cutting edge angle (C_s) gives a lower value of t_C for the same feed and hence the tendency for chips to break decreases with increase in C_s. If the side cutting edge angle is too high there is an increased tendency to form helical chips instead of the more desirable C- or ear-type chips. The depth of cut and work diameter have little influence on cutting ratio (t_C/t) and hence

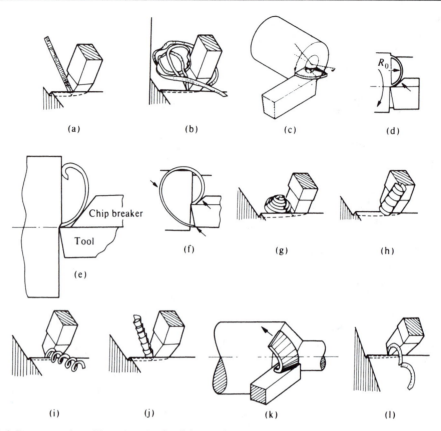

Fig. 18.4 Representative chip types. (a) Straight continuous ribbon ($i = 0$, $n_C = 0$). (b) Wandering continuous ribbon ($i = 0$, n_C variable). (c) Washer-type chip ($i = 0$, $V =$ variable along cutting edge—no upward curl). (d) C-type chip ($i \cong 0$, encounters as-yet-uncut surface). (e) Short ear-type chip ($i \cong 0$, encounters as-yet-uncut surface). (f) Long ear-type chip ($i \cong 0$, encounters tool-flank surface). (g) Coil-type chip ($i = 0$, $V =$ variable along cutting edge plus upward curl). (h) Tubular helix (i positive plus curl, large depth of cut b). (i) Spring-type chip (i positive plus curl, small depth of cut b). (j) Conical helix (combination of types c and h—common for drill). (k) Combination of types c and d ($i \neq 0$). (l) Connected C-type chips. (after Spaans, 1971)

negligible influence on chip breaking. The only exception involves the turning of small-diameter work at large depths of cut where a washer-type action will tend to enter the picture (Fig. 18.4c). Use of an effective cutting fluid will normally decrease contact length between chip and tool (decrease chip curl radius) and this will have a positive effect on chip breaking. Variation of rake angle (α) has a small effect on chip breaking. The deformed chip thickness (t_C) will increase with decrease in α and this will promote chip breaking (a small effect).

Work material of the same nominal chemistry may be made more brittle by heat treatment (hardening) or by cold-working the material. Nakayama (1962) has shown that the carbon content of plain carbon steels has a negligible influence on chip breaking. Addition of very small amounts of free-machining additives (MnS, Pb, Te, Bi, graphite, etc.) will usually improve a material's tendency to form broken chips.

Nakayama (1962) has determined the influence of a limited range of work-material chemistry on chip-breaking tendency using a clamped chip breaker (Fig. 18.1a) with $\theta = 45°$. The order of decreasing ease of chip breaking was found to be as follows:

1. Cr–Mo steel
2. carbon steel (all carbon contents)
3. stainless steel

RATIONAL APPROACH TO CHIP CONTROL

One approach to chip-breaker design is the database approach where a predetermined chart gives empirical chip-breaking results for a given work material. The first charts of this sort were proposed by Henriksen (1951). Nakayama (1962) has suggested a chart of the type shown in Fig. 18.5. This gives combinations of speed (V) and feed (t) that will give ear-type chips for a given work material (AISI 1045 steel, in Fig. 18.5) for points that plot above the curve and continuous (unbroken) chips for combinations of V and t that plot below the curve. Kaldor et al. (1979) have suggested a map indicating different chip types when depth of cut (b) is plotted against feed (t) for a given machining system. A number of additional chip-breaking studies are available in the literature prior to 1980: Henriksen, 1954a, 1954b; TenHorn and Schürmann, 1954; Reinhold, 1959, 1962; Okushima et al., 1960; Trim and Boothroyd, 1968; Jones, 1973; Reinhart and Boothroyd, 1973; Kahng and Koegler, 1977; Gertler, 1978, to mention but a few.

While empirical charts are useful, it often occurs that results for a given work material are not available and some experimentation is required. Both Nakayama (1962) and Spaans (1971) have suggested rational procedures to be followed in such situations.

Nakayama (1962) has suggested the following procedure for a clamped chip breaker. A trial cut is first made under the desired machining conditions and the value of deformed chip thickness (t_C) determined. The

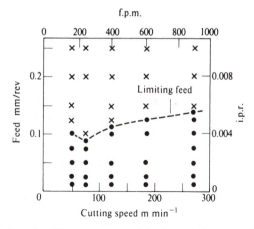

Fig. 18.5 Effects of speed and feed on chip breaking of AISI 1045 steel using clamped carbide chip breaker. $\theta = 45°$; carbide tool geometry, 0, 0, 7, 7, 8, 0, 0.5 mm; depth of cut, 0.08 in (2.0 mm); cutting fluid, none; × = broken chip, ● = continuous chip. (after Nakayama, 1962)

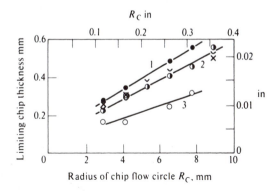

Fig. 18.6 Relation between chip flow radius (R_C) and limiting value of chip thickness (t_C) for different work materials. Curve 1 = Cr–Mo steel. Curve 2 = plain carbon steels of all C contents. Curve 3 = stainless steel. Machining conditions same as for Fig. 18.6. (after Nakayama, 1962)

limiting value of chip curl radius R_C is then obtained from the appropriate empirical curve of Fig. 18.6. This value is reduced by 20% to ensure chip breaking. To limit shock on contact, angle θ on the front face of the chip breaker should not be too steep; and $\theta = 45°$ is recommended. The value of offset distance b may be estimated from the following equation associated with Fig. 18.7:

$$b = \frac{R_C}{\cot \theta/2} \qquad (18.1)$$

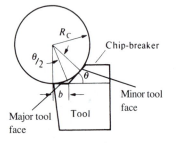

Fig. 18.7 Relation between chip flow radius (R_C) and offset (b) of clamped chip breaker.

For a value of θ of 45°, and R_C from Fig. 18.6, the required offset distance b is found to be

$$b = \frac{0.8R_C}{\cot(22.5°)} = 0.33R_C$$

If the work material of interest is not given in Fig. 18.6, as will usually be the case, an estimate may be had by comparing the properties of the new material with the few that are given and interpolating, if necessary.

Spaans (1971) has suggested a useful means of rating the breakability of a work material. He reasoned that values of the strain at fracture (ε_f) required to produce desirable ear-type chips for several materials would correlate with the ease of chip fracture. The resulting ordering of materials based on the strain at fracture (ε_f) he calls breakability. After the chip breaker has been adjusted to give the desired ear-type chip, the following quantities are measured on a representative chip:

1. chip thickness (t_C)
2. chip curl radius (R_C)
3. distance AB (see insert Fig. 18.8)

A nondimensional value of AB (AB*) is then found as follows:

$$AB^* = AB/R_C \tag{18.2}$$

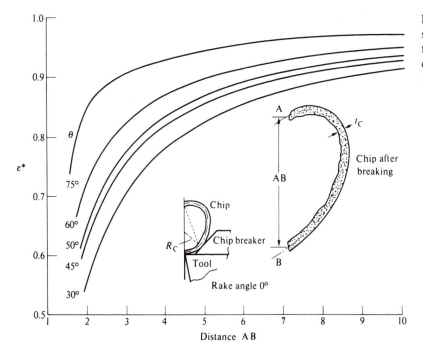

Fig. 18.8 Variation of nondimensional strain in ear-type chip at fracture (ε_f^*) with nondimensional chip size AB*. (after Spaans, 1971)

and the corresponding value of ε_f^* is read from Fig. 18.8 where

$$\varepsilon_f^* = \frac{2R_C}{t_C} \varepsilon_f \tag{18.3}$$

From this latter equation ε_f may be estimated.

CHIP BREAKING GROOVES

As already mentioned, chip breaking may be induced by use of chip-curling grooves. Originally these were ground into the surface of brazed carbide tools. Chip-curling grooves are now being molded into the rake surfaces of carbide inserts before sintering. This makes it possible to use grooves that would be difficult or impossible to grind. An example of this type is shown in Fig. 18.9. The grooves in this triangular insert are of variable chordal width. This is to extend the range of feed over which chip breaking occurs. The proportions of the groove were chosen to give a low increase in tool forces due to chip breaking. This turned out to be a groove of circular shape so proportioned that the depth of the groove was about 10% of the chordal width.

The insert shown in Fig. 18.9 produces a chip of variable cross-section (Fig. 18.10) since metal is forced into the groove parallel to the secondary cutting edge. This is important in preventing partially broken chips held together along the edge farthest from the workpiece axis. Due to the complex state of stress at the nose radius, chips tend to fracture on this edge first. This relieves the stress in the chip and prevents rupture from occurring all the way across the chip. Chip segments are thus joined together by unbroken material on the edge farthest from the nose of the tool (Fig. 18.11). This limits the range of feed over which a chip breaker will give completely broken chips. By thickening the edge of the chip adjacent to the nose of the tool, fracture is postponed and more energy is stored in the chip to cause complete chip fracture once a crack does appear. Thus, the groove segment parallel to the minor cutting edge is important as well as that parallel to the major cutting edge. The role of the latter is to promote upward chip curl while that of the former is to prevent premature chip fracture leading to connected chip segments.

When a chip is being formed, the shear zone is fully plastic as a result of energy provided by the spindle. It then takes a very small additional force to change the upward or sidewise curl of a chip as Pekelharing (1963) has indicated. Figure 18.12 shows a lead chip being formed as a small

Fig. 18.9 Triangular carbide tool insert with chip-breaking grooves molded into the rake face. (after Kaldor et al., 1979)

Fig. 18.10 Cross-section of chips produced by tool of Fig. 18.9. (after Kaldor et al., 1979)

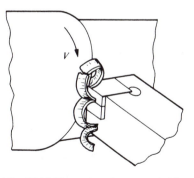

Fig. 18.11 Formation of connected chip segments due to relief of stress due to critical fracture on edge of chip nearest nose of tool. (after Kaldor et al., 1979)

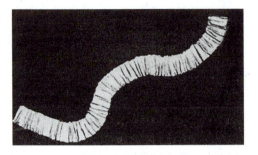

Fig. 18.12 S-shaped chip produced by alternating small external forces while chip is being produced. (after Pekelharing, 1963)

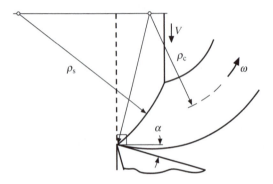

Fig. 18.13 Relation between radius of curvature of shear plane (ρ_s) and radius of curvature of chip (ρ_c).

variable force is applied to the side of the chip. The sidewise force required to cause the variable curl shown is a small fraction of the cutting force as long as the force is applied while cutting is proceeding. However, the force required to change the shape of the chip after cutting stops is many times greater than when cutting is taking place. This is important relative to chip curling. Otherwise the additional force on the tool due to chip curling would be excessive leading to an unacceptable temperature rise and excessive tool wear.

When an external force is applied to a chip, its curl changes due to a change in curvature of the shear zone ($\Delta\rho_s$) giving rise to a much larger change in curvature of the chip ($\Delta\rho_c$) as indicated in Fig. 18.13 ($\rho_s > \rho_c$).

Owen (1994) has suggested there is no universal chip control geometry. As near-net-shape technology advances, thinner chips are involved and chip control becomes a more challenging problem. Also, the choice of tool geometry and operating parameters becomes ever more difficult as the number of choices available increases. Despite the availability of software to help solve the problem, this can only be expected to provide a reasonable starting point to be refined on the job or in the research laboratory in the case of a large company.

The ductility of the material involved is generally the most important parameter relative to chip control, followed by feed, depth of cut, and cutting speed. A sharper rake control groove in the rake face of the tool is needed for a gummy work material such as a low-carbon steel. An increase in toughness of the work material enables an increased rake angle to be used. However, this puts more emphasis on the chip-breaking design that is employed.

CHIP CURLING/BREAKING ANALYSIS

Chip curling/breaking has received ever-increasing research attention since the mid-1980s. Reasons for this are due to the increased importance of

- workpiece safety
- product quality
- higher removal rates
- fully automated production

The breaking of chips into small manageable pieces is very difficult to predict. The most widely used chip control technique is still to induce chip curl that causes periodic fracture due to the development of excessive strain in the chip as it curls into an obstacle (usually the workpiece or tool holder depending on the curling magnitude and direction).

At present, many groove designs are commercially available from even a single manufacturer of inserts. Any one design gives good results for a limited range of machining conditions. The principal conditions influencing the performance of a given chip-breaker design are

- operating conditions (cutting speed, feed, depth of cut, and fluid)
- work material (hardness, strength, brittleness, etc.)

Figure 18.14 shows the many factors that influence chip breaking. Figure 18.15 shows several two-dimensional cutting tool insert alterations that tend to influence chip control and chip breaking. Two three-dimensional chip control groove modifications where the width of the groove varies along its length are shown in Fig. 18.16a, and a more complex three-dimensional insert is shown in Fig. 18.16b with a variety of chip control details molded into the face of the insert. This particular insert is for machining steel at small depths of cut and moderate feeds. Softer, more gummy materials require relatively deep grooves and a back wall closer to the cutting edge to induce greater curl. Small bumps are frequently molded into chip-curling grooves to reduce the chip–tool contact area in order to decrease tool-face friction. Edge preparation (Figs. 18.15a and 18.15c) is frequently employed for roughing cuts.

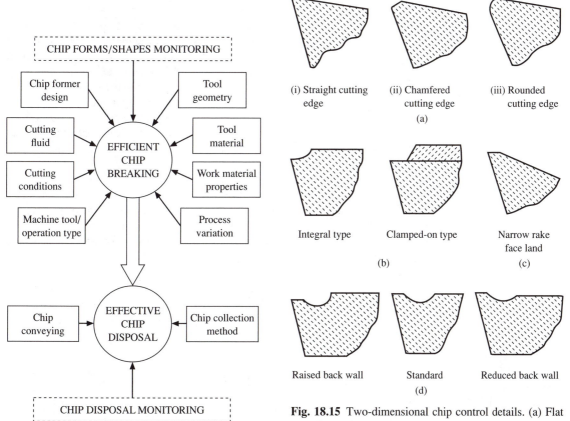

(i) Straight cutting edge (ii) Chamfered cutting edge (iii) Rounded cutting edge
(a)

Integral type Clamped-on type Narrow rake face land
(b) (c)

Raised back wall Standard Reduced back wall
(d)

Fig. 18.15 Two-dimensional chip control details. (a) Flat rake face with altered cutting edge. (b) Changing rake angles. (c) Restricted contact. (d) Conventional chip grooves. (after Jawahir and van Lutterveld, 1993)

CHIP FORMS/SHAPES MONITORING

Chip former design

Tool geometry

Cutting fluid

Tool material

EFFICIENT CHIP BREAKING

Cutting conditions

Work material properties

Machine tool/ operation type

Process variation

EFFECTIVE CHIP DISPOSAL

Chip conveying

Chip collection method

CHIP DISPOSAL MONITORING

Fig. 18.14 The total chip control system. (after Jawahir and van Lutterveld, 1993)

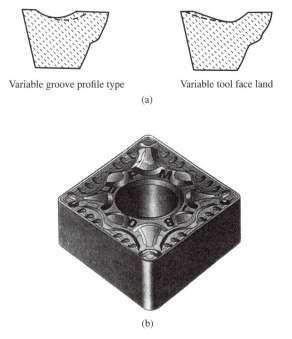

Variable groove profile type   Variable tool face land

(a)

(b)

Fig. 18.16 Three-dimensional chip control grooves. (a) Variable groove width and types. (after Jawahir and van Lutterveld, 1993) (b) Complex molded chip control grooves in surface of insert. (after Koelsch, 1996)

Figure 18.17, based on high-speed camera pictures, illustrates chip breaking associated with an up-curling groove of the type shown in Fig. 18.8. The chip initially penetrates the groove giving rise to a relatively small up-curl radius (r_{ui}) as shown in Fig. 18.17a. When the chip encounters the uncut work surface, a force F_u develops a moment which induces a strain sufficient to fracture the chip as shown in Fig. 18.17b. With further cutting a new cycle begins (Fig. 18.17c).

While chip breakability is not usually listed as a machinability factor such as tool life, surface finish, and production rate, it should be since it is of equal importance. De Chiffre (1988) has presented a useful study of the effect of restricted contact (RC) tools and cutting fluids on chip curvature. Figure 18.18 shows a restricted contact (RC) tool of the step type having a restricted contact length (L_o) subjected to an increasing undeformed chip thickness (t). In Fig. 18.18a, t is so small that the actual contact length (L) is less than L_o. This results in shear (τ) and normal (σ) tool-face stresses that decrease substantially as the chip moves up the face of the tool. This results in substantial chip curl. Figure 18.18b is for an undeformed chip thickness (t) where $L = L_o$ and there is then practically no chip curvature and essentially constant τ and σ along the tool face. Figure 18.18c shows an undeformed chip thickness (t) sufficiently large that negative chip curl occurs resulting in large secondary contact on the relieved region of the tool face. In this case $L > L_o$ and the stresses on length L_o are constant.

Figure 18.19a shows an RC tool of the secondary rake type. For such an RC tool, chip-curl radius (ρ) will vary with t as shown. As the chip–tool contact length approaches the secondary rake length (L') with increase in t, the chip radius (ρ) will approach infinity as shown in insert a. However, as L exceeds L' with further increase in t, as in insert b, a stable BUE forms and chip curl will again increase (i.e., ρ will decrease with increase in t). Also shown in Fig. 18.19 (insert c) is the variation of ρ with t for a tool without a secondary tool face (natural tool).

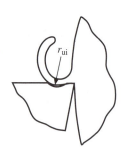

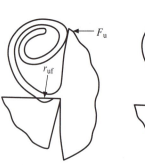

(a)   (b)   (c)

Fig. 18.17 Development of chip up-curl radius and fracture. (after Jawahir and van Lutterveld, 1993)

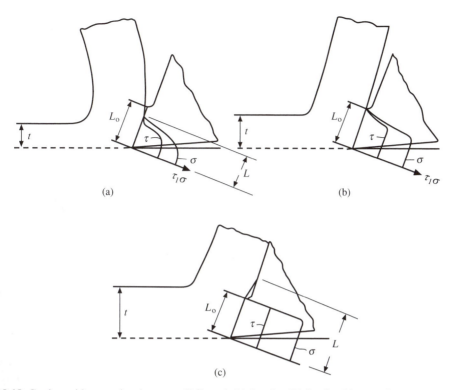

Fig. 18.18 Cutting with a restricted contact (RC) tool: (a) $L < L_o$, (b) $L = L_o$, (c) secondary contact at large feeds. (after De Chiffre, 1988)

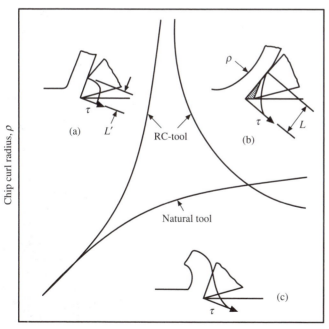

Fig. 18.19 Observed behavior of chip curl using RC tools. (after De Chiffre, 1988)

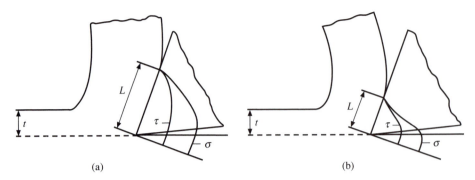

Fig. 18.20 Effect of lubrication while cutting. (a) Dry cutting. (b) Cutting under good conditions of lubrication. (after De Chiffre, 1988)

It was also found by De Chiffre (1988) that application of an effective cutting fluid will change the stress distribution along the tool face, the chip–tool contact length (*L*) and hence chip curl as shown in Fig. 18.20. This is a result of microcrack contamination that prevents rewelding at low speeds (Chapter 13) and by cooling the back of the chip at high speeds. Both of these effects tend to cause a steeper decrease in stresses as the chip moves up the tool face resulting in greater chip curl, shorter contact length (*L*), and lower forces and temperatures.

The foregoing results highlighting improvements associated with RC tools and cutting fluids by controlling the chip–tool contact length have been verified by numerous orthogonal cutting tests on electrolytic copper and an aluminum alloy performed in De Chiffre's laboratory. From the foregoing discussion of Fig. 18.18, it is evident that the significance of the chip–tool contact (*L*) depends upon the value of (*t*) involved, and De Chiffre has proposed the use of a nondimensional quantity $\eta = L/t$ as a measure of the tendency for chip curl to occur. He calls η the contact length factor.

Nakayama has extended the geometrical study of chip shapes and has identified and explained all possible chip types (Nakayama, 1972; Nakayama and Ogawa, 1978; Nakayama, 1984; Nakayama and Arai, 1992). All possible forms of steady state chips are identified in the 1992 paper and found to be functions of

- the radius of upward curvature
- the radius of sideward curvature
- the chip flow angle relative to the line of tool–chip separation

The latter of these quantities (v_c) is indicated in Fig. 18.21 which shows the line of tool–chip separation relative to the cutting edge and the direction of chip flow for a case where the tool–chip separation line is straight. A straight tool–chip separation line is not always obtained particularly when the rake face is not flat. These analytical studies of Nakayama are very valuable in explaining what is observed, but their complexity suggests that the design of chip control grooves in the tool face is probably best approached experimentally using the theory to guide the experimental approach to an optimum solution.

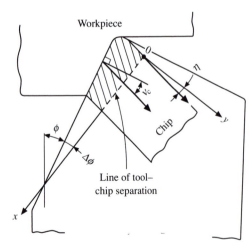

Fig. 18.21 Relation between chip flow direction and straight line of tool–chip separation (v_c). (after Nakayama and Arai, 1992)

Jawahir and van Luttervelt (1993) have presented a thorough review of a wide range of topics associated with metal cutting in general and chip control in particular. This review is accompanied by an extensive literature base (over 400 entries) by Jawahir (1993). The importance of chip–tool interface friction emphasized by De Chiffre (1988) relative to tool-face temperatures and broken chip size and shape is also considered in Jawahir and van Luttervelt (1993).

Figure 18.22 shows broken chips produced by identical cutting conditions except for the coatings applied to the carbide inserts. The coating for the chip in Fig. 18.22a had an outer layer of PVD-deposited TiN plus an inner layer of CVD-deposited TiC and TiCN. These inserts had a gold color and produced brown C-shaped chips because of the relatively low tool-face friction and temperature pertaining. The Fig. 18.22b chips were produced with an insert having an outer coating of CVD-deposited Al_2O_3 and inner layers of CVD TiC and TiCN. These inserts were black in color and produced blue, tightly curled chips because of the high tool-face friction and temperature involved.

In general an outer CVD coating gives a relatively rough, brittle surface and requires edge honing to prevent edge chipping. This edge rounding is one reason for the higher chip temperatures obtained. Also, an outer TiN layer gives lower friction and temperature than an Al_2O_3 outer layer as Fig. 18.22 clearly indicates.

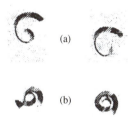

Fig. 18.22 Chips produced by identical cutting conditions for AISI 1018 steel except for outer layer of coating on insert. (a) PVD outer layer of TiN (brown chips with low friction). (b) CVD outer layer of Al_2O_3 (blue chips with high friction). (after Jawahir and van Lutterveld, 1993)

Jawahir et al. (1994) have discussed the effects of chip flow on tool wear of grooved tools. Change of groove geometry due to wear is often a more important tool-life consideration than flank wear, grooving, or other wear factors. Figure 18.23a shows wear concentrated at the back wall of a groove which normally occurs for low-feed conditions (tool-face wear land is large relative to the feed). Figure 18.23b is for a case where the feed is large relative to the tool-face wear land. This causes the chip to curl considerably and not to contact the back wall of the groove giving rise to considerable wear of the inner wall of the groove and the nose of the tool. This also gives rise to deterioration of surface finish. Figure 18.23c shows a better, more uniform groove wear situation where the major wear occurs simultaneously at the front and back walls and eventually also along the bottom of the groove. This later combination wear condition leads to a longer, more predictable tool life. In any of these cases, when sufficient wear has occurred, chip breaking may cease, introducing an additional criterion for tool life.

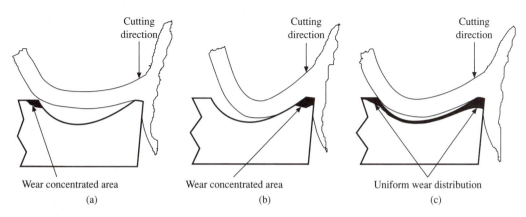

Fig. 18.23 Different tool-wear situations for grooved tools. (a) Back wall wear. (b) Front wall wear. (c) Balanced wear. (after Jawahir et al., 1994)

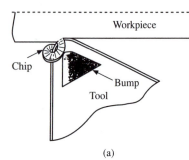

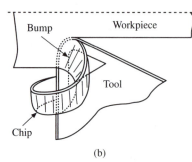

(a)

(b)

Fig. 18.24 Performance of a molded-in bump in the tool face near the tool tip. (a) Small depth of cut with large side curl. (b) Same tool and cutting conditions but with larger depth of cut giving combination of side-curl and up-curl with the latter dominating. (after Jawahir et al., 1994)

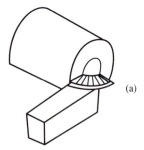

(a)

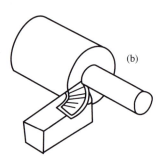

(b)

Fig. 18.25 Othogonal cutting with chip curl (a) of a tube and (b) of a bar. (after De Chiffre, 1998)

Jawahir et al. (1994) have also considered the role of bumps on the tool face near the tool tip on chip curling/breaking characteristics of tools with molded tool faces similar to Fig. 18.16b. Figure 18.24b shows the same bump as that in Fig. 18.24a operating with a larger depth of cut causing more up-curl. In both cases the chip is broken when it contacts the flank face of the insert away from the cutting edge. The swarf will consist of small C-shaped segments in the low depth of cut case and larger equally manageable C-shaped segments in the larger depth of cut case.

Fang and Jawahir (1993) have studied the effects of tool wear on chip breakability in metal cutting. While most chip curl/breaking studies have been conducted using sharp, unworn tools, chip breakability changes with cutting time at a rate that depends on work material and cutting conditions. Due to the complexity of the chip breakability problem, it is assumed, based on many experiments, that chip breakability depends upon the following factors with the indicated weighting factors:

- size of chip (curl radius, length, etc.), 60%
- shape of chip (helical, spiral, straight, etc.), 25%
- difficulty/ease of chip formation, 15%

Each of these factors is difficult to define precisely, and hence a fuzzy logic approach is employed in which chip breakability is expressed in terms of a chip breakability factor ranging from 0 to 1 (from 0 for nonbreakability to 1 for complete breakability) for the types of chips found in most turning operations.

De Chiffre (1998) has discussed the influence of chip–tool contact length and lubrication on chip curvature. When a tube is cut under orthogonal conditions as shown in Fig. 18.25a, the chip curls inward due to the difference of the length of cut from OD to ID. Inward curl is obtained with or without use of a lubricant. However, when a bar is cut, as shown in Fig. 18.25b, the chip will curl outward due to the flow constraint that is present at the inner diameter. This causes the chip to be thinner and longer at the ID giving rise to an outward curl. When an effective lubricant is applied, the chip–tool contact length at the OD is reduced resulting in less resistance to chip flow and a decrease in chip curl.

When a tool having a variable restricted contact length along the cutting edge of a shaping tool makes an orthogonal cut, the chip will curve toward the region of shorter contact length when cutting dry. Application of an effective lubricant will again result in less chip curl due to the smaller effect of the fluid in reducing friction on the side having the smaller chip–tool contract length.

It was suggested that hot machining will give a similar effect as good lubrication in reducing chip curl in both the cases of Fig. 18.25 and when cutting with a variable restricted contact along the cutting edge.

Jawahir and Balaji (2000) have presented an extensive review of the state of chip curling/breaking technology with special emphasis on three-dimensional chip flow associated with complex grooved tools.

Batzer and Sutherland (2001) have presented a detailed study of semi-spiral chip formation in orthogonal cutting before and after the chip encounters an obstruction. Figure 18.26a shows chip formation before the chip contacts an obstruction having an inclination γ. The radius of curvature of the chip is constant up to this point of contact. After contact a moment on the shear plane is present, as first suggested by Pekelharing in 1963. This moment causes a linear change in radius of curvature with further chip flow as suggested by Cook et al. in 1963. Figure 18.26b shows chip formation after contact with an obstruction. If the chip is not too ductile, it will fracture when the moment on the shear plane is sufficiently high giving rise to easily handled ear-type chips as shown in Fig. 18.26b. Batzer and Sutherland (2001) have extended the results of Pekelharing and Cook et al. analytically and have presented experimental results that generally verify their analysis of semi-spiral chip formation.

McClain et al. (2000) have presented an FEM analysis of the effect of increased width and depth of chip-curling grooves upon the shear strain at the critical

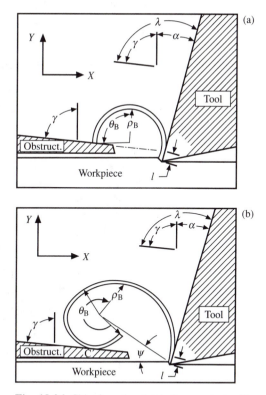

Fig. 18.26 Chip fomation (a) before contact with an obstruction and (b) after contact with an obstruction. (after Batzer and Sutherland, 2001)

point of chip fracture. This is at the free surface end of the shear plane as Nakayama has indicated (Nakayama, 1962; Nakayama, 1984). The McClain et al. analysis was based upon the following assumptions that

- the coefficient of friction for the well-lubricated tool–chip contact area was 0.1
- an arbitrary node separation distance was assumed
- an assumed strain hardening power law for AISI 1112 steel was employed
- strain rate and thermal softening were ignored
- young's modulus for the tool was assumed to be 4 times that of the work
- a perfectly sharp tool having a +30° rake angle and a cutting speed of 52 f.p.m. (263 mm/s) was assumed
- the groove extended to the cutting edge with no front land
- the cutting conditions and performance results were taken from Huang and Black (1996)

Despite the large number of arbitrary assumptions and limited range of this study, it is mentioned here because of the lack of such analytical studies devoted to chip curl/breaking. However, a great deal more experimental work and FEM analysis based on fewer assumptions would be required before FEM analysis could be considered to be a useful approach to the selection of optimum chip-curling groove parameters. A more effective approach appears to be the empirical chip-breaking chart approach first suggested by Henriksen (1951), extended by Nakayama (1962) and applied using the automated chart method described in the next section that was developed by De Chiffre and his students at the Technical University of Denmark.

AUTOMATED PRODUCTION OF CHIP CONTROL CHARTS

Andreasen and De Chiffre (1993) describe an automatic system for chip-breaking detection in turning that monitors the feed component of the cutting force during cutting. The feed force spectrum is filtered and an average of 20 instant values is obtained by a signal analyzer. This is accomplished in about 800 ms. Figure 18.27 shows average RMS feed force values with and without chip breaking. A chip-breaking algorithm is used to automatically check for chip breaking. This involves identification of a broad peak consisting of five spectrum lines.

A series of straight turning experiments was conducted on bars of carbon steel, a NiCrMo heat-treated alloy steel, and an austenitic stainless steel on two different CNC lathes. These tests used standard triangular coated inserts having various chip-breaker designs. Most experiments were conducted dry, but a few were repeated using a water-base cutting fluid. Chips were collected and classified in accordance with 1977 ISO standard 3685 (Annex G) shown in Fig. 18.28.

Representative chip-breaking diagrams derived from these tests are given in Fig. 18.29. The lines running across these diagrams separate combinations of depth of cut (d) and feed (f) that give satisfactory and unsatisfactory chip breaking. The thin lines were derived by reference to Fig. 18.28 while the thick lines were obtained by automatic detection. The small space between the lines represents the degree of uncertainty of the automatic system. The relatively few, automatically determined uncertain combinations of d and f are shown by dark triangular points.

Andreasen and De Chiffre (1998) presented a review and extension of their 1993 paper. In the automatic process, peak-based values were found best for steels while mean-based detection proved best for aluminum alloys.

Evaluation of different chip-breaking systems was expressed in terms of a hit rate (HR) value where

$$HR \% = \frac{\text{number of correct predictions}}{\text{total number of trials}} \times 100$$

Figure 18.30 shows peak-based HR values for different chip types obtained when turning five different materials under the ranges of conditions given in Fig. 18.31. From Fig. 18.30 it is evident that the automatic chip-breaking system depends strongly on the type of chip involved as follows:

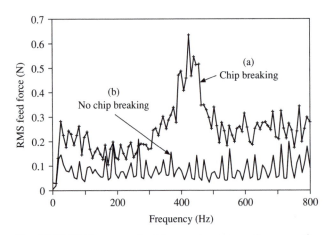

Fig. 18.27 RMS average feed force values versus frequency for (a) feed (f) = 0.2 mm/r to produce chip breaking and (b) feed (f) = 0.06 mm/r without chip breaking. Material, AISI 1035 steel; cutting speed, V = 250 m/min. (after Andreasen and De Chiffre, 1993)

Type	Ave. HR %
Snarled chips	98
Short arcs	90
Short spirals	83
Long spirals	38

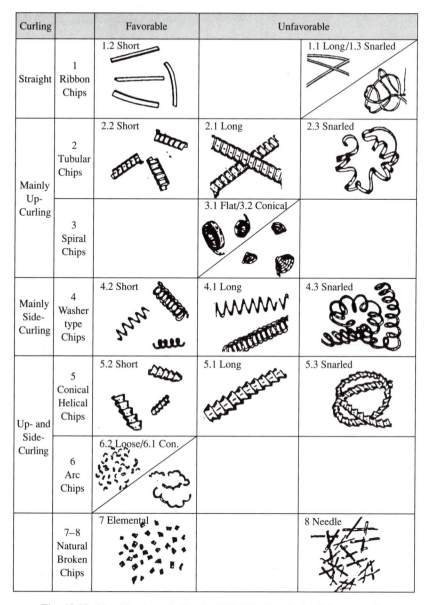

Curling		Favorable	Unfavorable	
Straight	1 Ribbon Chips	1.2 Short		1.1 Long/1.3 Snarled
Mainly Up- Curling	2 Tubular Chips	2.2 Short	2.1 Long	2.3 Snarled
	3 Spiral Chips		3.1 Flat/3.2 Conical	
Mainly Side- Curling	4 Washer type Chips	4.2 Short	4.1 Long	4.3 Snarled
Up- and Side- Curling	5 Conical Helical Chips	5.2 Short	5.1 Long	5.3 Snarled
	6 Arc Chips	6.2 Loose/6.1 Con.		
	7–8 Natural Broken Chips	7 Elemental		8 Needle

Fig. 18.28 Classification of chips in 1977 ISO Standard 3685 (Annex G).

From this it is evident that the automatic system is most reliable for detecting snarled chips and gives very poor results for the aluminum alloy.

Figure 18.32 gives results for tests similar to Fig. 18.30 but based on mean instead of maximum values. From this it is clear that the aluminum alloy gives much better results for mean-based values than for peak-based values. Figure 18.33 shows HR values for different chip types when AISI 1045 steel is turned dry and with a coolant. There was relatively little difference, but the dry tests gave slightly higher HR values. From the foregoing discussion, it appears that chip control charts offer a promising future particularly when generated automatically, and this topic deserves much more study.

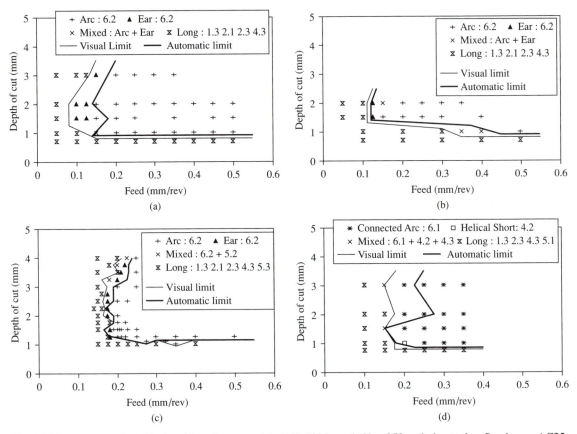

Fig. 18.29 Representative chip breaking diagrams: (a) AISI 1035 steel, $V = 250$ m/min, tool = Sumitomo AC25; (b) AISI 1045 steel, $V = 250$ m/min, tool = Sumitomo AC10; (c) Ni–Cr–Mo steel, $V = 200$ m/min, tool = Widia TZ15; (d) Stainless steel, $V = 18$ m/min, tool = Seco TP15-MF2. The numbers in the inserts refer to those in Fig. 18.28. (after Andreasen and De Chiffre, 1993)

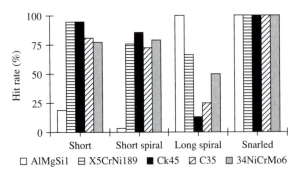

Fig. 18.30 Hit rate for peak-based crack detection for different chip types and workpiece materials (all 925 tests in Fig. 18.31). (after Andreasen and De Chiffre, 1993)

VARIABLE FEED

Since the magnitude of t_C and hence of undeformed chip thickness (t = feed rate) plays such an important role in chip breaking, it is not surprising that chip control may be improved by use of a periodically varying feed rate. When this is done, the chip will tend to break when the feed is high even though it may not break when cutting at the mean feed rate. This method of chip control will, of course, lead to a visible pattern on the surface of the work that is far greater than the change in surface roughness would suggest. However, since the appearance of a machined surface is often equated

Workpiece material	Carbon steel (C35) (AISI 1035)	Carbon steel (Ck45) (AISI 1045)	Heat treated alloyed steel (34Ni CrMo6)	Auste nitic stain less (X5Cr Ni189)	Alumi nium (AlMg Si1)
Inserts	Nine different commercial inserts				
Cutting speed (m/min)	100–300	60–390	150–200	90–180	300–700
Depth of cut (mm)	0.75–3	0.2–3	1–4	0.6–3	0.5–3
Feed (mm/rev)	0.05–0.5	0.05–0.6	0.15–0.4	0.06–0.6	0.05–0.3
No of tests	86	325	83	190	241

Fig. 18.31 Cutting conditions for system evaluation. (after Andreasen and De Chiffre, 1993)

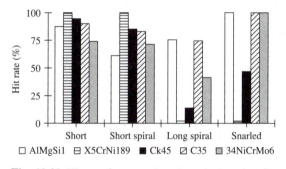

Fig. 18.32 Hit rate for mean-based crack detection for different chip types and workpiece materials (all 925 tests in Fig. 18.31). (after Andreasen and De Chiffre, 1993)

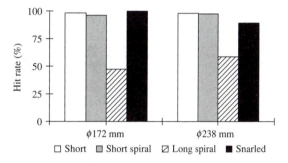

Fig. 18.33 Hit rate for different work diameters based on 324 tests (some dry and some with coolant). Speed = 150 m/min; work material = AISI 1045 steel. (after Andreasen and De Chiffre, 1993)

to surface quality in the mind, the variable feed solution to chip control is often not acceptable even though the resulting finished surface may be equal or even superior functionally to one produced at a constant feed rate.

PULSATING FLUID JET

Gettleman (1991) has suggested that one way of preventing long stringy chips (sometimes referred to as *bird nests*) when cutting ductile metals is to apply a mixture of a water-base cutting fluid and CO_2 gas. The CO_2 is introduced into the cutting fluid just before it reaches the tools and is directed along the tool face in a pattern of pulses. This not only periodically breaks the chip but extracts heat from the tool. Because of the high pressure (up to 10,000 p.s.i.) of the fluid/CO_2 jet, it is extremely important that the machining area be enclosed to contain the mist that is generated.

REFERENCES

Andreasen, J. L., and De Chiffre, L. (1993). *Annals of CIRP* **42/1**, 45.

Andreasen, J. L., and De Chiffre, L. (1998). *Annals of CIRP* **47/1**, 35.

Batzer, S. A., and Sutherland, J. W. (2001). *Mach. Sc. and Technology* **5/1**, 63–76.

Cook, N. H., Jahaveri, P., and Nayak, N. (1963). *J. of Eng. for Ind., Trans. ASME* **85**, 374–380.

Da, Z. J., Sadler, J. P., and Jawahir, I. S. (1998). *Trans. NAMRI/SME* **29**, 129–134.

De Chiffre, L. (1988). *Proc. 6th International Colloquoium at Tech. Akad. Esslingen* **2**, 5–1 to 5–9.

De Chiffre, L. (1998). *Comments at Group C Meeting of CIRP in Paris*, Jan. 1998.

Ernst, H., and Merchant, M. E. (1941). In *Surface Treatment of Metals.* Am. Soc. for Metals, Metals Park, Ohio, pp. 299–378.

Fang, X. D., and Jawahir, I. S. (1993). *Wear* **160**, 243–252.

Gertler, S. (1978). *Cut Tool Eng.* **29**, 11.

Gettleman, K. M. (1991). *Modern Machine Shop* (Sept.), 55–61.

Hahn, R. S. (1953). *Trans. Am. Soc. Mech. Engrs.* **75**, 581.

Henriksen, E. K. (1951). *Trans. Am. Soc. Mech. Engrs.* **73**, 461.

Henriksen, E. K. (1954a). *Steel* **134**, 93.

Henriksen, E. K. (1954b). *Am. Mach, N.Y.* **98**, 118.

Huang, J. M., and Black, J. T. (1996). *J. Mfg. Sc. and Eng., Trans. ASME* **118**, 545–554.

ISO Standard 3685, Annex G, 1977.

Jawahir, I. S. (1993). *Annals of CIRP* **42/2**, 686–693.

Jawahir, I. S., and Balaji, A. K. (2000). *J. Mach. Sc. and Tech.* **4/3**, 399–443.

Jawahir, I. S., Dillon, O. W., Balaji, A. K., Redetzky, M., and Fang, N. (1998). *J. Mach. Sc. and Tech.* **2**, 253–276.

Jawahir, I. S., Ghosh, R., Fang, X. D., and Li, P. X. (1994). *Wear* **184**, 145–154.

Jawahir, I. S., and van Lutterveld, C. A. (1993). *Annals of CIRP* **42/2**, 659–685.

Jones, D. G. (1973). Soc. Manuf. Engrs. Tech. Paper 73–215.

Kahng, C. H., and Koegler, W. C. (1977). *Man. Engng. Trans.* **6**.

Kaldor, S., Ber, A., and Lenz, E. (1979). *J. Engng. Ind.* **101**, 241.

Koelsch, J. R. (1996). *Mfg. Eng. (SME)* (Jan.), 71–76.

Lang, M. (1974). Dissertation, T. H. Munchen.

McClain, B., Team, W., Maldonado, G. I., and Fang, X. D. (2000). *J. Mach. Sc. and Tech.* **4/2**, 305–316.

Nakayama, K. (1962). *Bull. Jap. Soc. Mech. Engrs.* **5**, 192.

Nakayama, K. (1963). *Bull. Jap. Soc. Precis. Eng.* **1**, 25.

Nakayama, K. (1972). *J. of JSPE* **38**, 592–598.

Nakayama, K. (1984). *Bull. Japan Soc. of Prec. Engrs.* **18/2**, 97.

Nakayama, K., and Arai, M. (1992). *Annals of CIRP* **41/1**, 71–74.

Nakayama, K., and Ogawa, M. (1978). *Annals of CIRP* **27/1**, 17–2.

Okushima, K., Hoshi, T., and Fujinawa, T. (1960). *Bull Jap. Soc. Mech. Engrs.* **3**, 199.

Owen, J. V. (1994). *Mfg. Eng. (SME)* (July), 57–62.

Pekelharing, A. J. (1963). *Annals of CIRP* **12/3**, 143–147.

Rahman, W. K. A. M., Seah, P., and Zhang, X. D. (1995). *Int. J. Mach. Tools and Mfg.* **35/7**, 1015–1031.

Reinhart, L. E., and Boothroyd, G. (1973). In *Proc. First N. Am. Metalworking Res. Conf.* McMaster University, Hamilton, Ontario, p. 13.

Reinhold, R. (1959). *Fertigungstechnik u. Betreib* **9**, 161.

Reinhold, R. (1962). *Fertigungstechnik u. Betrieb* **12**, 475.

Spaans, C. (1971). Dissertation, Delft Technical University, The Netherlands.

TenHorn, B. L., and Schürmann, R. A. (1954). *Tool Engineer* **32**, 37.

Trim, A. R., and Boothroyd, G. (1968). *Int. J. Prod. Res.* **6**, 227.

Wang, L. (1996). Ph.D. dissertation. University of Kentucky, Lexington.

19 OPTIMIZATION

There are generally a number of ways of producing a specific part, some of which are more efficient than others. When a new design appears, the initial methods of producing the desired parts are usually not the most efficient. With experience, the processing operations are optimized or more efficiently carried out. In any optimization procedure, it is important to identify the output of chief importance and to establish a figure of merit against which different processing paths may be rated. In the area of manufacturing engineering, the most common optimization variables are specific cost (cost per part) and production rate (parts produced per unit time). The latter should normally be expected to pertain only in a time of national emergency or if a production bottleneck develops due to lack of machine capacity. However, in practice, maximum production rate often takes precedence over minimum cost per part as the optimization objective.

LEVELS OF OPTIMIZATION

There are several levels of process optimization including the following:

1. improved details relative to present methods of production
2. new conventional production paths
3. adoption of a radically new production path

Generally, optimization levels 2 and 3 are progressively less feasible. If the present equipment available in a plant is reasonably efficient and can meet production requirements, it often cannot be justified economically to replace the old equipment even though the new equipment is more efficient. This, of course, is nature's way of softening the blow associated with loss of production capabilities due to war or natural disaster. Very often, the new equipment has a much larger production capacity than the old, and a new machine may have more production capacity than a potential user requires. There is then the prospect of having the new machine lie idle part of the time or to sell excess time on the machine or excess products produced by the machine in order that the cost of meeting primary production requirements not be excessive. Another constraint associated with the adoption of new, more efficient machinery is the probability the new machinery is more sophisticated to the point it may not be consistent with the levels of skill of available operators or maintenance personnel. Another important deterrent to the adoption of new, more efficient manufacturing facilities is the fact that a new generation of machines generally involves a larger capital investment and if, as is usual, there is a shortage of capital available, a manufacturer may be prevented from replacing old machinery by new even though the new machinery can be justified on the basis of a reduced production cost per part including the true machine cost (taking into account possible unused capacity, greater maintenance cost, and the greater unit cost of a more sophisticated operator or programmer).

Level 2 optimization is best pursued by periodically brainstorming major production operations with input from internal or external consultants or both. The decision to try a new conventional processing route should be based upon a sound economic analysis of alternatives with current production costs as a basis for comparison.

The situation associated with level 3 optimization is still more difficult. It will usually take a minimum of five years to bring a new manufacturing concept through the many development steps associated with its progression from a laboratory feasibility study through the pilot plant stage and finally into full-scale production. Each of these steps involves unforeseen hazards and problems that require new thinking and frequently innovative solutions. All of this takes time, careful planning, and a well-coordinated team effort. Therefore, level 3 process optimization involves important constraints of time, availability of development personnel, and risk, in addition to the constraints associated with level 2 optimization. This does not mean that optimization activity at levels 2 and 3 are less important than level 1. The wise manufacturer will periodically review the manufacturing program to be sure a new approach is not being overlooked. Such periodic review should involve a quantitative consideration of alternatives. It is important to emphasize the need for periodic review (say annually) since the options available to the manufacturing engineer are constantly changing at an ever-increasing rate.

Since the situation most frequently encountered involves optimization at level 1, the remainder of the discussion here will be concerned with optimization at this level. However, the basic approach to be used at all levels is the same:

1. Identify the appropriate optimization objective.
2. Quantitatively consider processing options available within existing constraints.
3. Adjust processing options to correspond as closely as possible to optimum values.

CONSTRAINTS

Just as there are several constraints for process optimization at level 2 (capital available, unit machine size and production capacity, level of skill of personnel, etc.) and level 3 (time, risk, etc.), there are a number of constraints for level 1 optimization. These include

1. tool material strength and wear resistance
2. range of speeds, power, and stability of machines available
3. finish, surface integrity, and dimensional accuracy required, etc.

The problem is to adjust the machine consistent with constraints so that production is carried out as close to optimum conditions as possible.

MACHINING ECONOMICS

Some of the most important problems of the workshop involve the choice of cutting speeds and feeds, tool geometry, tool and work materials, cutting fluids, and machine tools themselves. There appear to be only three basic considerations associated with all of these decisions. First of all, the chosen cutting conditions must be capable of producing parts that meet the required specifications of size, shape, and finish. Second, the required production schedules must be met; and thirdly, the parts should be produced at the lowest possible cost. In the large-scale mass production of parts, sufficient equipment can often be procured to meet production requirements under a wide range of conditions, and the important problem might then become to choose the operating conditions so that quality requirements are just met and the total cost per part is a minimum.

Since the cost of the tools that are used on a given operation represents a major component of the total machining cost, tool life plays a major role in the choice of optimum machining conditions. The several different criteria for tool life which must be adopted under different practical situations have been discussed in Chapter 11; and it has been shown that if the range of cutting speed V or feed t is not great, the tool life (T, min) defined in any of the several possible ways may be represented as follows:

$$\text{For constant feed: } VT^n = C \tag{19.1}$$

where n and C are constants depending on tool and work material, tool geometry, etc.

In several sections that follow, answers to the following questions will be discussed:

1. What is the most satisfactory machinability index for use in rating the ease of machining a number of different workpiece materials, and what is a reasonable approximation to this quantity?

2. On what basis should a water-base cutting fluid be chosen from a group of fluids of different cost?

3. What is the amount one is justified in paying for a stepless speed changer on a lathe?

4. Which of two tools which cost different amounts and have different wear characteristics should be chosen?

5. Which of two work materials having different machinability characteristics and cost should be used in a given instance?

6. Is the clearance which gives maximum tool life always the one which should be used in practice?

In discussing these questions, it will be assumed in the interest of brevity that the feed is fixed by considerations of finish, machine tool power, or maximum allowable cutting force for the tool holder; and only the choice of the otpimum cutting speed will be considered. The analyses might, however, be readily extended to include the choice of optimum feed rate.

Machining Costs

Gilbert (1950) first approached machining economics in a rational way and the following discussion of principles involved is based on his approach.

If cutting is done at very high speed, the machine and labor cost will be low but the tool cost will be high. The opposite is true if cutting is done at very low speed. There is an optimum cutting speed for any machining operation where the total machining cost per part is a minimum.

The costs to be included in such an analysis are the following:

1. direct labor and machine cost (xT_c), where x is the value of the machine and operator with overhead (cents/min) and T_c is the machining time per part,

$$T_c = \frac{\pi dl}{12Vt} = \frac{L}{V} \tag{19.2}$$

where

$$d = \text{work diameter, in (mm)}$$
$$l = \text{axial length of work to be cut, in (mm)}$$
$$t = \text{feed, i.p.r. (mm/rev)}$$
$$V = \text{cutting speed f.p.m. (m min}^{-1})$$
$$L = \text{helical length machined per part, ft (m)}$$

[When SI units are used, the 12 in Eq. (19.2) is replaced by 1000.]

2. tool changing cost per part $(xT_d(T_c/T))$, where T_d is the downtime in minutes to change a tool

3. tool cost per part $(y(T_c/T))$, where y is the mean value of a single cutting edge

The total cost per part $(\cent)$ is then

$$\cent = xT_c + xT_d\frac{T_c}{T} + y\frac{T_c}{T} \qquad (19.3)$$

(work changing cost is not included since it is independent of V).

The quantity $\cent$ is a minimum when $\partial\cent/\partial V$ is zero, and this may be found after T_c and T are expressed as functions of V using Eqs. (19.1) and (19.2). In this way, it is found that $\cent$ is a minimum when tool life (T) has a particular value which we will call $T^{\#}$:

$$T^{\#} = \left(\frac{xT_d + y}{x}\right)\left(\frac{1}{n} - 1\right) = R\left(\frac{1}{n} - 1\right) \qquad (19.4)$$

where R (minutes) is called the cost ratio and n is the exponent of the tool-life equation [Eq. (19.1)]. The corresponding cutting speed is then

$$V^{\#} = \frac{C}{(T^{\#})^n} \qquad (19.5)$$

If this value is put back into the cost equation [Eq. (19.3)], we obtain the minimum machining cost per part:

$$\cent^{\#} = \frac{x}{1 - n}T_c^{\#} \qquad (19.6)$$

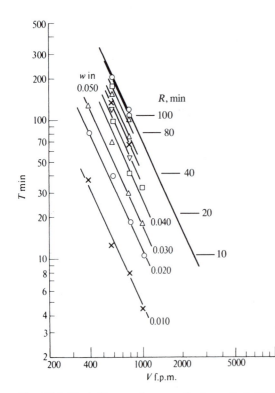

Fig. 19.1 Tool life–speed curves for hot-rolled AISI 1020 steel cut using a carbide tool having the following American Standards Association (ASA) geometry: –7, –7, 7, 7, 15, 15, $\frac{1}{32}$ in (0.75 mm). Depth of cut, 0.100 in. (2.5 mm); feed, 0.0104 i.p.r. (0.26 mm/rev); fluid, tap water applied at $1\frac{1}{2}$ g.p.m. at 60 °F (5.5 l min^{-1} at 17 °C).

Representative tool life–speed data are shown in Fig. 19.1 for a low-carbon steel machined with a carbide insert. Two commonly used tool-life criteria are considered here:

1. total destruction of the tool (heavy line)

2. lines of constant wear land on the clearance face of the tool (w)

The lines of constant R shown in this figure are a convenient addition to such a plot for they immediately enable the cost optimum values ($T^\#$ and $V^\#$) to be determined. The line corresponding to $R = 10$ min was obtained by measuring the value of n (0.48 in the case of Fig. 19.1) and substituting the proper values into Eq. (19.4) to obtain

$$T^\# = 10\left(\frac{1}{0.48} - 1\right) = 10.8 \text{ min}$$

The line corresponding to $R = 10$ is thus drawn opposite $T = 11$ min. The corresponding value of $V^\#$ (speed for lowest total machining cost per part) for a tool operated to total destruction is seen to be over 2000 f.p.m. (610 m min^{-1}).

The cost values for the lathe and throw-away tool used in obtaining the data of Fig. 19.1 were as follows:

$$x \ = \text{cost of operator, lathe, and overhead} = 10\cancel{c}/\text{min}$$

$$T_d = \text{tool changing and adjusting time} = 3 \text{ min}$$

$$y \ = \text{value of single cutting edge} = 60 \text{ cents}$$

The corresponding value of R is 9 min and hence, for minimum cost per part, this tool should be operated at a speed considerably in excess of 2000 f.p.m. to give a tool life of about 9 min. This speed may be beyond the range of currently available lathes and indicates that the maximum available speeds for lathes are presently too low, even for throw-away carbide tools, not to mention ceramic tools.

Lines of constant R may be readily applied to tool life T versus speed V plots even when the curves are not linear. In the latter case the only difference is that the slope n is no longer a constant but different values must be used from point to point. The R scales previously shown on Figs. 15.19 and 15.20 represent a useful addition to these plots.

It is frequently convenient to be able to rate workpiece materials in relative order of ease of machining or to express the potential of one tool material relative to another. Taylor (1907) measured the relative performance of his high speed steel tools in terms of the speeds needed to give a 20 min tool life.

$$M_{20} = \frac{V_{20,1}}{V_{20,2}} \tag{19.7}$$

The machinability index in most common use is that corresponding to the ratio of speeds to give a 60 min tool life or

$$M_{60} = \frac{V_{60,1}}{V_{60,2}} \tag{19.8}$$

Since we are normally most interested in producing parts at minimum cost, a better machinability index for these same materials would be

$$M^\# = \frac{\cancel{c}_2^\#}{\cancel{c}_1^\#} \tag{19.9}$$

where $\cancel{c}_1^\#$ is the minimum cost per part for a given tool, work material, or fluid designated (1), and $\cancel{c}_2^\#$ is the corresponding minimum cost per part for the second tool, work material, or cutting fluid, etc. It is of interest to know the difference which exists between these machinability indices.

From Eq. (19.6)

$$M^\# = \frac{(xT_c^\#/1 - n)_2}{(xT_c^\#/1 - n)_1}$$ (19.10)

but, since

$$\frac{(T_c^\#)_2}{(T_c^\#)_1} = \frac{V_1^\#}{V_2^\#}$$

and $x_2 = x_1$

$$M^\# = \frac{C_1(T_2^\#)^{n_2}(1 - n_1)}{C_2(T_1^\#)^{n_1}(1 - n_2)}$$ (19.11)

Similarly, the machinability index based on a given tool life (M_T) may be written

$$M_T = \frac{V_{T,1}}{V_{T,2}} = \frac{C_1}{C_2} T^{n_2 - n_1}$$ (19.12)

If the value of T in Eq. (19.12) is taken as the mean between $T_1^\#$ and $T_2^\#$, then

$$T = \frac{T_1^\# + T_2^\#}{2} = \frac{R}{2}\left[\frac{1 - n_1}{n_1} + \frac{1 - n_2}{n_2}\right]$$ (19.13)

since $R_1 = R_2$. The value of the ratio between the commonly used index based on speeds (M_T) and that based on minimum cost per part $(M^\#)$ will be

$$\frac{M_T}{M^\#} = \left(\frac{1 - n_1}{n_1} + \frac{1 - n_2}{n_2}\right)^{(n_2 - n_1)} \frac{[2(1 - n_1/n_1)]^{n_1}}{[2(1 - n_2/n_2)]^{n_2}} \left(\frac{1 - n_2}{1 - n_1}\right)$$ (19.14)

This ratio is equal to 1 whenever n_1 is equal to n_2. The ratio will be close to 1 whenever n_1 and n_2 do not differ by more than 0.2. However, whenever n_1 and n_2 differ widely, then $(M_T/M^\#)$ is appreciably different from 1, and the two machinability criteria are not equivalent.

From the foregoing discussion, it is clear that there is little difference between the easily computed machinability index based on speed and the less easily computed (although more important) index based on costs, provided the slopes of the tool-life curves (n) for the conditions compared do not differ very much. Since the tool life T on which the speed index M_T is based should correspond to a good average of economic tool-life values $T^\#$ for a variety of applications, it would appear that $T = 60$ min in Eq. (19.12) is a reasonable value and perhaps a better choice for present-day conditions than Taylor's original value of 20 min. It is surprising to find that the form of Taylor's original machinability index gives values that are so very close to those obtained by use of the more fundamental machinability index based on cost $(M^\#)$.

Water-Base Cutting Fluids

The many fluids available may be divided into two general classes—oil-base fluids and water-base fluids. The oil-base fluids are generally acknowledged to be the best boundary lubricants, while the water-base fluids are capable of the greatest cooling action. At low cutting speeds, the production of good finish is the chief problem, and boundary lubrication of the tool-chip interface is of major importance; whereas, at high cutting speeds, tool life is more important and cooling the cutting tool becomes more necessary.

The objective of this discussion is to present a rational method for studying the performance of a variety of water-base cutting fluids based upon machining costs. Answers are to be sought to such questions as: Is the cost of a cutting fluid of importance if it really provides superior performance? And how much additional cost for a cutting fluid is justified in a given instance?

After a number of preliminary tests, a disposable carbide tool was found which gave very good results, and this tool was adopted for this study. The characteristics of this tool and the cutting conditions used follow:

- carbide type, steel cutting grade C–5
- carbide size, $\frac{3}{8}$ in $\times \frac{3}{8}$ in $\times 1\frac{1}{2}$ in (9.53 $\times$ 9.53 $\times$ 38 mm), rectangular parallelopiped
- tool angles

    back rake angle, $-7°$

    side rake angle, $-7°$

    end relief angle, $7°$

    side relief angle, $7°$

    end cutting edge angle, $15°$

    side cutting edge angle, $15°$

    nose radius, $\frac{1}{32}$ in (0.79 mm)
- feed, 0.0104 i.p.r. (0.26 mm/rev)
- depth of cut, 0.100 in (2.5 mm)
- cutting speed, variable
- rate of fluid flow, 1.50 g.p.m. (5.7 l min^{-1})

Each tool insert had eight cutting edges, and when these had been used, the carbide insert was discarded. The work material used was annealed AISI 4340 steel (H_B 185). It was thought best to run all tools to total destruction so that a comparison could be made between curves of constant wear land and total destruction. This required that a large amount of material be cut, and some tests involved as many as 16 hours of cutting time.

Tool-life results for several water-base cutting fluids are given in Fig. 19.2. In these tests the data have been confined to the practical turning region where built-up edge (BUE) is negligible. Both lines of constant wear land and total destruction are included. In each case lines of constant R have also been given, and the value of R corresponding to the tool used in this study ($R = 11.5$ min) is shown by a dotted line.

Three examples will be used to compare the performance of these fluids. First, it will be assumed that we are interested in a roughing operation for which it is permissible to use tools to total destruction. The optimum cutting speed $V^{\#}$ and tool life $T^{\#}$ in this case corresponds to the point where $R = 11.5$ crosses the total destruction line for each fluid (Fig. 19.2). The values of $T^{\#}$, $V^{\#}$, and n for each of the fluids tested are given in Table 19.1.

The minimum cost per unit part $\cent^{\#}$ is given by Eq. (19.6) from which it is evident that [see Eq. (19.2)]

$$\cent^{\#} = (xL)\frac{1}{V^{\#}(1 - n)} \qquad (19.15)$$

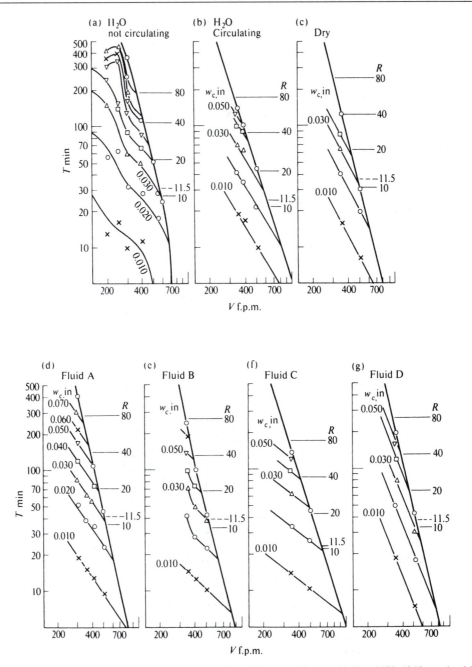

Fig. 19.2 Tool life–speed curves for different cutting fluids used in machining AISI 4340 steel with a carbide tool with the following ASA geometry: −7, −7, 7, 7, 15, $\frac{1}{32}$ in (0.79 mm). Depth of cut, 0.100 in (2.5 mm); feed, 0.0104 i.p.r. (0.26 mm/rev). Rate of fluid flow, $1\frac{1}{2}$ g.p.m. (5.7 l min⁻¹). (a) Noncirculating water at 60 °F (16 °C). (b) Circulating water at 90 °F (32 °C). (c) Dry tool. (d) Cutting fluid A, 2.5% in water. (e) Cutting fluid B, 2.5% in water. (f) Cutting fluid C, 2.5% in water. (g) Cutting fluid D, 2.5% in water. (after Shaw, 1959)

TABLE 19.1 Cutting Fluid Test Results for Total Destruction

Fluid	n	$V^{\#}$ f.p.m.	m min^{-1}	$T^{\#}$ min	$\mathcal{c}^{\#}/xL$ (f.p.m.)$^{-1}$	(m/min)$^{-1}$	Relative Cost
Water, noncirculating 60 °F (16 °C)	0.27	570	174	31.0	0.00240	7.87×10^{-3}	0.88
Water, circulating 90 °F (32 °C)	0.32	610	186	24.5	0.00241	7.91×10^{-3}	0.88
Cutting fluid A, 2.5%	0.22	500	152	40.8	0.00256	8.40×10^{-3}	0.94
Cutting fluid B, 2.5%	0.23	515	157	38.5	0.00252	8.27×10^{-3}	0.92
Cutting fluid C, 2.5%	0.32	610	186	24.5	0.00241	7.91×10^{-3}	0.88
Cutting fluid D, 2.5%	0.23	510	155	38.5	0.00255	8.37×10^{-3}	0.93
Dry tool	0.24	480	146	36.4	0.00274	8.99×10^{-3}	1.00

Note: R = 11.5.

Since the quantity xL is constant for any operation, the optimum cost per part $\mathcal{c}^{\#}$ will be proportional to $(1/V^{\#}(1 - n) = \mathcal{c}^{\#}/xl)$ for any given cutting fluid, and values of this quantity are included in Table 19.1. The values of relative cost given in the last column of Table 19.1 are based on a cost per part in dry cutting of unity.

From this study it is evident that all water-base cutting fluids give better results than dry cutting. The poorest fluid showed a decrease in cost per part over dry cutting of about 6% while the best fluid showed a decrease in cost per part of 12% based on dry cutting. The fact that none of the commercial water-base cutting fluids was better than water indicates that the major role of the fluid in this speed range (500–600 f.p.m. or 152–183 m min^{-1}) is one of cooling. Water should be expected to be the best coolant and the performance of the water-base fluids should be expected to decrease from that of water as their tendency to form a thick oily deposit on metal surfaces increases. Such a thick oily film will tend to decrease the heat transfer coefficient between metal and fluid. The four water-base cutting fluids tested might be rated in the following order of decreasing effectiveness for use in the speed range 500–600 f.p.m. (152 to 183 m min^{-1}):

Fluid	% Improvement Over Dry Cutting
C	12 (circulating water 12%)
B	8
D	7
A	6

A second comparison may be made for the case where a wear land of 0.040 in (1 mm) cannot be exceeded for reasons of, let us say, surface finish. The best operating point in this case ($V^{\#}$ and $T^{\#}$) will be the point where the $w = 0.040$ in (1 mm) curve intersects the curve of total destruction. Such values of $T^{\#}$ and $V^{\#}$ are given in Table 19.2.

The equation which in this case corresponds to Eq. (19.15) may be readily shown to be

$$\mathcal{c}^{\#} = (xL)\frac{T^{\#} + R}{T^{\#}V^{\#}} \qquad (19.16)$$

TABLE 19.2 Values of $T^{\#}$ and $V^{\#}$ for Constant Wear Land of 0.040 in (1 mm)

Fluid		$V^{\#}$		$T^{\#}$ min	$\phi^{\#}/xL$		Relative Cost
		f.p.m.	m min^{-1}		(f.p.m.)$^{-1}$	(m/min)$^{-1}$	
Water, noncirculating	60 °F (16 °C)	525	160	40	0.00246	8.07×10^{-3}	0.87
Water, circulating	90 °F (32 °C)	410	125	88	0.00276	9.06×10^{-3}	0.97
Cutting fluid A,	2.5%	460	140	60	0.00259	8.50×10^{-3}	0.91
Cutting fluid B,	2.5%	450	137	65	0.00262	8.60×10^{-3}	0.92
Cutting fluid C,	2.5%	420	128	80	0.00272	8.92×10^{-3}	0.96
Cutting fluid D,	2.5%	440	134	72	0.00264	8.66×10^{-3}	0.93
Dry tool		420	128	60	0.00284	9.32×10^{-3}	1.00

Note: R = 11.5.

where the quantity xL is a constant for any given operation. Hence, the minimum cost per part ($\phi^{\#}$) with any cutting fluid will now be directly proportional to the quantity

$$\left(\frac{T^{\#} + R}{V^{\#}T^{\#}} = \frac{\phi^{\#}}{xL} \right)$$

given in Table 19.2 for a value of $R = 11.5$.

The optimum speed is now seen to range from 420 to 460 f.p.m. (128 to 140 m min^{-1}), and for this speed range water and water-like fluids are the least effective. The relative ratings of fluids for the 0.040 in (1 mm) wear-land condition are now as follows in decreasing order of effectiveness:

Fluid	% Improvement Over Dry Cutting
A	9
B	8
D	7
C	4 (circulating water = 3%)

Apparently lubrication begins to be important in this speed range and fluid A that was poorest at the high-speed range of 500–600 f.p.m. (157–183 m min^{-1}) (Table 19.1), because of its poorer heat transfer, now appears to be the most effective fluid for the 420–460 f.p.m. (128–140 m min^{-1}) speed range. The range of effectiveness of the commercial fluids is still seen to be about 2:1, but the relative order of effectiveness is entirely changed.

Still another comparison of the effectiveness of these cutting fluids can be made for a case where the tooling is relatively expensive. Let us consider the specific case where tools are run to just short of total destruction and the value of R has been increased by a factor of 4 (i.e., to a value of 46). The quantities of interest for this example are given in Table 19.3.

TABLE 19.3 Optimum Cost Values for Expensive Tool for Total Destruction

Fluid	n	$V^{\#}$ f.p.m.	m min^{-1}	$T^{\#}$ min	$\cancel{c}^{\#}/xL$ (f.p.m.)$^{-1}$	(m/min)$^{-1}$	Relative Cost
Water, noncirculating 60 °F (16 °C)	0.27	390	119	124	0.00352	11.55×10^{-3}	0.92
Water, circulating 90 °F (32 °C)	0.32	390	119	98	0.00376	12.34×10^{-3}	0.98
Cutting fluid A, 2.5%	0.22	370	113	163	0.00346	11.35×10^{-3}	0.91
Cutting fluid B, 2.5%	0.23	375	114	154	0.00347	11.38×10^{-3}	0.91
Cutting fluid C, 2.5%	0.32	390	119	98	0.00377	13.37×10^{-3}	0.99
Cutting fluid D, 2.5%	0.23	370	113	154	0.00351	11.52×10^{-3}	0.92
Dry tool	0.24	345	105	146	0.00382	12.53×10^{-3}	1.00

Note: R = 46.

The optimum speed is now still lower, ranging from 345 to 390 f.p.m. (105 to 110 m min^{-1}). The relative ratings of the fluids are as follows:

Fluid	% Improvement Over Dry Cutting
A	9
B	9
D	8
C	1 (circulating water = 2%)

Note that water and (water-like) fluid C are now seen to be not only the poorest fluids of the group but to be practically without effect.

The foregoing analysis of water-base cutting fluids has clearly shown that two distinct actions are present which are not completely compatible—a lubrication action and a cooling action. For the AISI 4340 steel–carbide–tool combination used in these experiments, cooling seems to be of negligible importance for speeds of 300 f.p.m. (91 m min^{-1}) and below, while lubrication assumes an increasing role of importance. On the other hand, at speeds of 600 f.p.m. (183 m min^{-1}) and above, only cooling seems to be beneficial and any lubricating properties a fluid may have appear to be detrimental. Of the fluids tested, it would appear that fluids A and B have the best lubricating characteristics, while water and fluid C have about the same cooling characteristics. For cutting at speeds of 300–500 f.p.m. (91–152 m min^{-1}), fluids A or B should be used, but for speeds above 500 f.p.m. (152 m min^{-1}), fluid C would be distinctly better.

The quantity $\cancel{c}^{\#}/xL$ is not only useful in deciding which fluid of a group is best, but it can also be used to determine how much one is justified in spending for a particular cutting fluid. This is most effectively illustrated by an example. Assume that a part is to be produced under the following conditions:

x = value of machine and labor = 10 cents/min ($6 per hour)

d = diameter of part = 2 in (50 mm)

l = axial length of part = 2 in (50 mm)

t = feed = 0.010 i.p.r. (0.25 mm/rev)

The quantity in parentheses in Eq. (19.15) is then

$$xL = \frac{x\pi dl}{12t} = \frac{(10)\pi(2)(2)}{12(0.010)} = 1050$$

From Table 19.1 the value of cutting fluid A to this operation (assuming tools are used to total destruction) is clearly

$$\cent_{\text{dry}}^{\#} - \cent_{\text{A}}^{\#} = (1050)\left(\frac{\cent^{\#}}{xL_{\text{dry}}}\right) - \left(\frac{\cent^{\#}}{xL_{\text{A}}}\right) = 0.19 \text{ cents per part}$$

If this fluid is used as a $2\frac{1}{2}\%$ solution and the concentrate costs g cents per gallon, then the cost of the solution will be $(0.025)g$ cents/gallon. If N parts can be made per gallon of solution (including drag-out with chips, evaporation, replacement, etc.), then the cost of fluid per part will be $(0.025\ g/N)$ cents. The use of cutting fluid A is justified on a cost basis if

$$\cent_{\text{dry}}^{\#} - \cent_{\text{A}}^{\#} > \frac{0.025g}{N}$$

In the above example, fluid A costs \$2.00 per gallon; hence, for the fluid to be justified on a cost basis

$$N > \frac{\cent_{\text{dry}}^{\#} - \cent_{\text{A}}^{\#}}{(0.025)(200)} = \frac{0.19}{(0.025)(200)} = 0.04$$

The number of parts that may be produced per gallon (3.78l) of solution will greatly exceed 0.04; hence, it would appear that the fluid is certainly justified on a cost basis.

The optimum total cost per part including machining costs and fluid costs will be

$$\left(\cent^{\#} + \frac{cg}{N}\right)$$

in any particular case where c is the concentration of the active ingredient that is used. The fluid of a group for which this quantity is a minimum is the most economical one to use and hence normally should be the one used if other conditions are equivalent. The cost of the fluid per part (cg/N) will normally be completely negligible compared with the machining cost per part; hence, that fluid which gives the smallest machining cost per part is normally the most economical regardless of the cost of the fluid concentrate. Thus, it is clear that if a particular fluid really does a superior job and the machining cost per part $(\cent^{\#})$ is reduced any measurable amount by using it, this water-base fluid should be used regardless of whether the concentrate costs one, ten, or twenty dollars per gallon.

The optimum selection of speeds and feeds available on a machine tool is considered in Shaw (1965). The commonly used geometric progression is found to be optimum based on tool life, surface finish, and minimum manufacturing cost over the life of the machine tool. A general method of determining the optimum number of intermediate speeds or feeds for a specific machine tool is derived and illustrated.

Stepless Speed Change

Most lathes are constructed with a number of finite speeds which differ one from the next by a constant ratio $(1 + K)$, where K is usually from 0.2 to 0.4. A stepless speed control can be supplied at a greater machine tool cost and it is of interest to know how much additional cost is justified for this feature.

If, in a given instance, the speed corresponding to minimum cost per part ($¢^{\#}$) is $V^{\#}$, then the maximum discrepancy in speed due to the lack of a stepless speed changer will be ($\pm KV^{\#}/2$), while the average discrepancy will be ($\pm KV^{\#}/4$). The $+$ or $-$ sign pertains to whether the actual speed employed is greater or less than $V^{\#}$. The ideal cost per part when $V^{\#}$ can be realized will be

$$¢^{\#} = xT_c^{\#}\left(1 + \frac{R}{T^{\#}}\right) \tag{19.17}$$

The cost per part with the average discrepancy present will be

$$¢_1 = \frac{xT_c^{\#}}{1 \pm K/4}\left[1 + \frac{R}{T^{\#}}\left(1 \pm \frac{K}{4}\right)^{1/n}\right] \tag{19.18}$$

The average additional cost per part due to the lack of a stepless speed changer will be

$$¢_1 - ¢^{\#} = xT_c^{\#}\left\{\frac{\mp K/4}{1 \pm K/4} + \frac{R}{T^{\#}}\left[\left(1 \pm \frac{K}{4}\right)^{(1/n-1)} - 1\right]\right\} \tag{19.19}$$

If Z parts are to be made during the entire life of the machine, then the total saving associated with the stepless speed control is

$$S = Z(¢_1 - ¢^{\#}) = ZxT_c^{\#}\left\{\frac{\mp K/4}{1 \pm K/4} + \left(\frac{n}{1 - n}\right)\left[\left(1 \pm \frac{K}{4}\right)^{(1/n-1)} - 1\right]\right\} \tag{19.20}$$

The quantity (ZT_c) is the total time the machine will be making useful parts. If the use factor for the machine is F (i.e., F is the percentage of the nominal time the machine is in use that it will be cutting) and its total useful life is M min, then (ZT_c) may be replaced by (FM).

In evaluating Eq. (19.20), the average between the solutions with the plus and minus signs should be taken and hence

$$\bar{S} = \frac{FMx}{Z}\left\{\frac{2(K/4)^2}{1 - (K/4)^2} + \left(\frac{n}{1 - n}\right)\left[\left(1 + \frac{K}{4}\right)^{(1/n-1)} + \left(1 - \frac{K}{4}\right)^{(1/n-1)} - 2\right]\right\}$$

$$= FMx\left\{\left(\frac{K}{2}\right)^2\left[\frac{1}{1 - (K/4)^2} + \frac{1 - 2n}{2n}\right]\right\} = FMx\psi(K, n) \tag{19.21}$$

which is obtained by expanding the

$$\left(1 + \frac{K}{4}\right)^{(1/n-1)} \qquad \text{and} \qquad \left(1 - \frac{K}{4}\right)^{(1/n-1)}$$

terms into a binomial series and noting that $(K/4)^2 \ll 1$;

- use factor (F) = 0.8
- life of machine (M) = 2 shifts for five days per week for 5 years = 1.2×16^6 min
- value of machine, labor, and overhead (x) = 10 cents/min
- $K = 0.3$
- $n = 0.1, 0.2, 0.3, 0.5$

TABLE 19.4 Savings with Stepless Speed Changer

n	$\psi(K, n)$	$S, \$$
0.1	0.0281	2700
0.2	0.0140	1350
0.3	0.0094	900
0.5	0.0056	540

The results of the calculation for the total savings associated with the stepless speed change ($\bar{S}$) are given in Table 19.4.

The amount one is justified in paying for a stepless speed control on a representative lathe is seen to be about $1500 when no interest is charged on the investment and nothing is allowed for maintenance. Let the worth of the invested capital be taken at 10% per year and let 5% of the initial investment be allowed for maintenance per year. If the amount we are justified in spending for the speed changer at the beginning is x dollars, then the net investment at the end of the first year will be

$$x(1.1) + 0.05x - 300$$

where 300 is the amount of benefit derived from the speed changer during the first year. At the end of the second year the net investment in the speed changer is

$$[x(1.1) + 0.05x - 300]1.1 + 0.05x - 300 = (1.1)^2 x + (0.05x - 300)(1.1 + 1)$$

At the end of the fifth year the net investment is:

$$(1.1)^5 x + (0.05x - 300)[(1.1)^4 + (1.1)^3 + (1.1)^2 + 1.1 + 1]$$

which will be zero for the correct value of x.

Solving for x we obtain

$$x = \$953$$

Thus, for this example, the justified expenditure for the speed changer at the time of purchase is $953. This value depends on the expected return on the investment which may be three or more times the value (10%) assumed in the above example. Further discussion of the economics of stepless machine tool drives, particularly for facing operations, is to be found in Shaw (1960).

Tool and Work Materials

In the past the improvements associated with changes in tool and work materials have usually been studied in terms of tool life. While tool life is the proper criterion in such cases only when the costs of the two processes are the same, a comparison based on cost per part is always proper. A cost analysis for two ceramic tools has been presented by Shaw (1957) in which it was found that the tool which gave the poorest life but which cost the least was the one to be used since it resulted in the lowest cost per part. There are many problems of this sort that arise from day to day in the workshop but only one is considered here to illustrate rational procedures which can be followed in such cases.

A problem that frequently arises is that of deciding which of two steels, that are equivalent with regard to end-product, should be used, when the machinability and the cost of the two steels differ. The two steels are designated A and B. The cost per pound for these steels is M_A and M_B (where $M_B > M_A$), respectively, while the machinability equations [Eq. (19.1)] for the steels are

$$VT^n = C_A$$
$$VT^n = C_B$$

where $C_B > C_A$ and the values of n are assumed to be approximately the same.

From Eq. (19.6), the difference in minimum cost per part will be

$$\cancel{c}_A^\# - \cancel{c}_B^\# = \frac{x}{1-n}[T_{c,A}^\# - T_{c,B}^\#] \qquad (19.22)$$

By use of Eq. (19.2)

$$\cancel{c}_A^\# - \cancel{c}_B^\# = \frac{\pi dlx}{12t(1-n)}\left(\frac{1}{V_A^\#} - \frac{1}{V_B^\#}\right) \qquad (19.23)$$

The difference in cost of the workpiece material will be $W(M_B - M_A)$ where W is the weight of the workpiece before machining. In order that use of the new work material be justified on a cost basis,

$$W(M_B - M_A) < \cancel{c}_A^\# - \cancel{c}_B^\# \qquad (19.24)$$

or

$$M_B - M_A < \frac{\pi dlx}{12t(1-n)W}\left(\frac{1}{V_A^\#} - \frac{1}{V_B^\#}\right) \qquad (19.25)$$

As an example consider the following case:

- diameter to be cut $(d) = 2$ in (50 mm)
- axial length of cut $(l) = 4$ in (100 mm)
- value of machine, operator, and overhead $(x) = 10$ cents/min
- weight of workpiece $(W) = 4$ lb (1.82 kg)
- $R = 20$ min
- $n = 0.2$
- $t = 0.01$ i.p.r. (0.25 mm/rev)

If the present material A costs 15 cents per pound while the new material B costs 17 cents per pound, then

$$\frac{1}{V_A^\#} - \frac{1}{V_B^\#} > \frac{12(0.01)(0.8)4}{\pi(2)(4)(10)}(2) = 0.00304$$

This means that if the cost-optimum cutting speed for the present material $(V_A^\#)$ is 100 f.p.m. (30.5 m min^{-1}), then

$$V_B^\# > \frac{1}{1/100 - 0.00304} = 144$$

in order that the proposed substitution be justified on a cost basis.

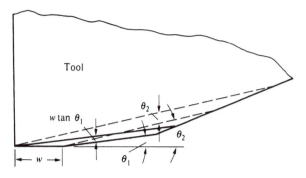

Fig. 19.3 Material to be removed when resharpening tools of different relief angle θ.

Relief Angle

There are two criteria on which the relief angle for a tool may be chosen:

1. minimum cost per part

2. close dimensional tolerance

The volume rate of wear on the clearance face of a tool is experimentally found to increase linearly with cutting time or the helical length of the surface machined. If the tool life T corresponds to a certain value of wear land, then it follows that

$$T \sim \tan \theta \qquad (19.26)$$

where θ is the relief angle (see Fig. 19.3). This means that the volume available for wear to give a certain wear land w should be as large as possible. However, when the relief angle is increased, there is less metal at the cutting edge to support the load and to conduct heat away. In practice the maximum tool life based on a 0.03 in (0.75 mm) wear land is found to correspond to a relief angle of about 12°.

However, this does not mean that such a large clearance angle should always be used in practice, and it becomes evident why this is so when costs are considered. The mean cost of a tool and the cost of reconditioning a tool often represent an appreciable portion of the total cost for machining a part. The amount of metal that must be removed when a tool is reconditioned is (Fig. 19.3)

$$x = w \tan \theta \qquad (19.27)$$

The larger the layer x that must be removed to recondition the tool, the more grinding time will be required and the fewer the number of regrinds that will be possible before the tool must be replaced. It is thus evident that, except in the case of the throw-away tool, the relief angle giving minimum cost does not correspond to that giving maximum tool life, but will be a somewhat smaller relief angle. This explains why tool life can usually be increased by increasing the relief angles beyond values in general use in workshops but why it is frequently not wise to increase these relief angles even though an increase in tool life results.

From Fig. 19.3, it is evident that whenever dimensional accuracy is of major concern, both w and θ must be as small as possible. Thus, broaches (where dimensional accuracy is extremely important) are fitted with tools of very small relief angle (1° to 3°) and are resharpened when the wear land is but a few thousandths of an inch. On the other hand, rough-turning tools (where minimum cost is of major interest) are fitted with relief angles of from 5° to 10° and the wear land is usually measured in hundredths of an inch.

Tools Cutting Simultaneously

When two identical tools take identical cuts at the same time on the same machine involving a single operator, the entire operation will be completed in half the time (Fig. 19.4) and this alters the economic balance. The cost of machine, operator, and overhead per minute (x) will be the same as when a single tool is used, but the value of the downtime to change tools (T_d) and the cost of the new tools ($2y$) will be double the value for the single-tool case. The cost ratio for the two-tool case will thus be

$$R_2 = \frac{xT_d + y}{x} = 2R_1 \qquad (19.28)$$

where R_1 is the value of R for the single-tool case. From Eq. (19.4), the cost optimum tool life (min) for the two-tool case will be

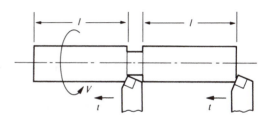

Fig. 19.4 Turning with two identical tools.

$$T_2^{\#} = \left(\frac{1}{n} - 1 \right) R_2 = 2T_1^{\#} \qquad (19.29)$$

where $T_1^{\#}$ is the cost-optimum tool life for the single-tool case.

It is thus evident that as the total tool and tool changing cost increases relative to machine and labor cost per minute (x), the optimum tool life increases and the machine should be operated at a lower speed.

An extreme multi-tool situation is the multispindle automatic screw machine. In this case several tools operate simultaneously on each of several spindles. After the slowest tool has completed its cut, the spindles are indexed, and a completed part is obtained after a given spindle has been indexed through all stations.

All spindles operate at the same r.p.m. and an important problem is to determine the cost-optimum r.p.m. There is one tool that requires the greatest time to complete its cut and this tool determines the cycle time (time to complete a single part = cutting time for this critical tool times the number of spindles). If all tools are changed at the same time (this need not be the case), then a good approximation to this complex optimization problem is to compute the R value for each tool and then determine the optimum tool life ($T^{\#}$) as though the critical tool (one determining cycle time) is acting alone with an R value equal to the sum of the R values of all tools. The cost optimum tool life ($T^{\#}$) may thus be estimated by substituting the sum of the R values for all tools for R in Eq. (19.4) and using the value of n for the critical tool.

In actual practice, tools are normally changed in groups rather than all at the same time, but this does not change the fact that a multispindle automatic screw machine is one having an unusually high R value. It is, therefore, to be expected that the cost-optimum operating speed for the critical tool will be significantly lower than for an equivalent single-tool situation. It is, therefore, to be expected that multi-tool operations or operations having unusually expensive tooling (such as in broaching) should be performed at lower speeds than single-tool operations involving inexpensive tooling (such as single-point turning). However, due to the natural tendency to integrate all experience into a single intuitive reaction, it is often found that simple single-tool turning operations are performed at too low a speed while multi-tool operations involving relatively expensive tooling are operated at too high a speed.

OPTIMUM PRODUCTION RATE

An alternative optimization procedure involves the selection of cutting speed for maximum rate of production. This involves writing an expression for the total machining time including cutting time and tool- and work-changing times, and then differentiating this with respect to cutting speed and equating to zero.

The total machining time will be

$$\Sigma T = T_{\mathrm{c}} + T_{\mathrm{d}}\frac{T_{\mathrm{c}}}{T} + T_{\mathrm{w}} \tag{19.30}$$

where

T_{c} = cutting time given by Eq. (19.2)

T_{d} = downtime to change and reset tool

T = tool life, min

T_{w} = work-changing time (assumed independent of V)

Substituting for T_{c} and T from Eqs. (19.1) and (19.2), taking the partial derivative relative to V (t constant), and equating to zero give the following value for production-rate-optimum tool life:

$$T^{\#\#} = T_{\mathrm{d}}\left(\frac{1}{n} - 1\right) \tag{19.31}$$

When this equation is compared with Eq. (19.4):

$$\frac{T^{\#\#}}{T^{\#}} = \frac{T_{\mathrm{d}}}{R} = \frac{T_{\mathrm{d}}}{T_{\mathrm{d}} + y/x} \tag{19.32}$$

it is evident that the optimum tool life based on rate of production will be less than that based on minimum cost per part.

The cutting speed corresponding to maximum production rate will be

$$V^{\#\#} = \frac{C}{(T^{\#\#})^{n}} \tag{19.33}$$

Figure 19.5 shows diagrammatically the relationship between the optimum cutting speeds for minimum cost ($V^{\#}$) and maximum production rate ($V^{\#\#}$).

ADAPTIVE CONTROL

Databases Versus Adaptive Control

Level 1 optimization is most widely carried out in industry by use of a database approach. Past experience of the company is stored in a computer and used to estimate the unit production cost when using a variety of tool materials, tool geometries, feeds, speeds, etc. By comparing the unit costs for each combination, it is possible for the production engineer to estimate the optimum set of conditions for a job and to specify these conditions on the production order going to the operator with the parts to be processed and the expendible tools to be used. The difficulty with this approach is that the number of variables that must be specified to characterize a

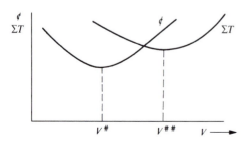

Fig. 19.5 Relation of cost-optimum cutting speed ($V^{\#}$) and production-rate-optimum cutting speed ($V^{\#\#}$).

job completely is too large to be used in practice; and when only the most important variables are considered, the result is too approximate. Also, the inputs to the process such as workpiece hardness, workshop temperature, or even operator fatigue are not static but vary with time. The problem of predicting the optimum set of conditions for a given machining operation is similar to the problem of predicting the weather. For any one month, the average of all weather predictions is apt to be fairly accurate even though the day-to-day predictions are relatively inaccurate. In the case of database-based predictions, the mean performance for a number of jobs will similarly be far closer to what is predicted than the performance for individual jobs will be. A database approach, although approximate, usually represents a good starting point. To do better, some form of adaptive control must be used in which performance is monitored and the machine adjusted according to the results obtained.

Adaptive control may be classified as follows:

1. *Primary adaptive control versus secondary.* In primary adaptive control, the actual optimization variable is measured or calculated and the machine adjusted stepwise to approach optimization. In secondary adaptive control, secondary variables such as rate of tool wear, cutting temperature, cutting forces or power, chatter, finish, or accuracy are monitored and the machine adjusted in the direction of the optimum based on an assumed relationship between primary and secondary variables.

2. *Automatic versus manual adaptive control.* In automatic adaptive control (AAC), a set of sensors continuously measures performance and a computer combines these inputs with assumed weighting factors and adjusts the machine to approach optimum performance after each set of inputs. In manual adaptive control (MAC), the operator is part of the control loop. He replaces the required group of sensors and also serves as the interface between an inexpensive programmable computer and the machine tool. Manual adaptive control is a particularly attractive alternative when capital funds are in short supply.

Automatic Adaptive Control (AAC)

Practically all applications of AAC are of the secondary type. For example, it is well established that in a turning operation, the feed should be as high as possible consistent with the maximum force on the cutting tool to cause tool fracture and the surface finish required. In some instances, the hardness of the workpiece will vary significantly from point to point on a given workpiece and from one workpiece to another. When this is the case, the cost per part and the production rate can be improved by holding the cutting speed constant and adjusting the feed to the highest value consistent with a maximum allowable cutting force that is safe relative to tool fracture and a maximum allowable amplitude of chatter vibration relative to the finish required. If either the allowable force or amplitude of chatter is exceeded, the feed is automatically decreased until either the force or vibrational amplitude decreases to what is thought to be the acceptable level. In this case, both cutting force and vibrational amplitude are secondary variables related in a complex way to the items of primary interest—tool fracture and surface finish. Since there is no known way of sensing when a tool is about to fail in brittle fracture, one must be content with monitoring a secondary variable—cutting force. Likewise, tool chatter is only one component of surface roughness and since there are no workshop-proven in-process sensors for surface roughness, one must be content to measure a secondary variable (vibrational amplitude) that is related in a complex way to the main variable of interest (surface roughness).

The automatic adaptive control of this example is useful even though far from perfect. It is useful in that the rate of material removal will be greater than if the feed rate were to be set at such a level that constant values of tool force or vibrational amplitude not be exceeded regardless of the

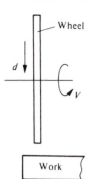

Fig. 19.6 Abrasive cut-off operation.

local workpiece hardness, presence of surface scale, or workpiece-to-workpiece variation in hardness or structure. It is less than perfect in that there is a wide dispersion of tool-to-tool fracture strength and the fact than an interaction exists between tool-fracture tendency and surface finish produced due to a change in workpiece hardness. When workpiece hardness increases, the cutting force increases, calling for a decrease in feed. However, an increase in workpiece hardness leads to a better finish (aside from chatter-induced roughness) calling for an increase in feed.

The adaptive control system just described automatically senses secondary variables: cutting force and vibrational amplitude, related in a complex way to the primary variables of interest (tool-facture tendency and surface finish). It may, therefore, be classified as an example of secondary automatic adaptive control (SAAC).

The only known example of primary automatic adaptive control (PAAC) involves the control of an abrasive cutoff machine (Gall, 1969). Figure 19.6 shows schematically the abrasive cut-off operation. A high-speed grinding wheel ($V = 12,000$ f.p.m. or 3600 m min^{-1}) is fed vertically downward at a feed rate d. The variable of interest in this case will usually be the total cost per cut which involves the following components:

1. cost of machine, labor, and overhead (cents/cut)
2. value of abrasive consumed per cut (cents/cut)
3. work-changing cost per cut (cents/cut)

The optimum (minimum) cost per cut ($\cent$#) is obtained when operating at the highest wheel speed (V) possible consistent with safety and machine capability. However, an optimum down-feed (d#) exists that usually lies in the practical range of down-feeds available. This optimum value of d depends on workpiece size, shape, and properties, several grinding-wheel characteristics, as well as environmental and operator influences in such a complex way that it is very difficult to predict the cost-optimum value of d, based on the general accumulated past experience.

A more practical approach is to sense the change in wheel-diameter after each cut and introduce this into a computer along with the down-feed pertaining and several fixed quantities [wheel speed, mean value of grinding wheel (cents/vol), value of machine, operator, and overhead (cents/min), and certain workpiece and grinding-wheel dimensions]. The computer calculates the cost per cut, takes the previous value from storage and automatically changes the down-feed by a small increment (Δd) upward or downward depending on the sign of Δd and whether the new cost per cut is greater or less than the previous one. In this way the machine may be adjusted to begin operation at the best estimate of d based on previous experience but will automatically adjust d to correspond to the cost-optimum value d# after a few trials. By monitoring each cut, the machine will seek a new cost-optimum down-feed d# as soon as any one of a large number of variables changes sufficiently to require a new optimum. This includes such obvious variables as workpiece hardness and grinding-wheel homogeneity as well as less obvious ones such as workshop temperature and operator fatigue. If the machine is manually serviced, the work-changing time should be included in the time per cut. The optimization will then include any increase in work-changing cost associated with an increase in rate of production and operator fatigue.

The above is an example of primary automatic adaptive control (PAAC). While the procedure may appear complicated, the hardware involved is relatively inexpensive since it employs readily available analog elements assembled into an inexpensive dedicated computer. The only sensor required (measurement of wheel diameter) is relatively simple in this case. Versatility is provided by allowing quantities such as cost of machine, cost of operator (x, cents/min), and mean value of consumable wheel volume (y, cents/vol) to be dialed in at the beginning of each job. Details on this development are to be found in Gall (1969).

Schulz and Spath (1997) have discussed the difficult problem of integrating operator experience into numerical control (NC) manufacturing, which cannot be performed on the workshop floor. This involves use of a new manufacturing-feature-oriented method of describing different machining operations. This is designed to introduce operator experence into the manufacturing process without the distraction involved when this is attempted on the shop floor.

Bouzakis et al. (1994) have discussed the problem of integrating computer-aided technology into the production process. It is suggested that this be realized by

- creating a database covering typical machined parts including tool details
- employing algorithms that cover optimum cutting conditions
- applying computer-aided tolerance estimation

Manual Adaptive Control (MAC)

The difficult elements of automatic adaptive control are the sensors, interface between sensors and computer, and the cost of the computer. The appearance of small, inexpensive, programmable computers makes it possible to expand the concept of an incremental approach to optimum operation of a machine tool by having the machine operator perform the function of the required group of sensors as well as the machine–computer interface. This type of adaptive control is termed manual adaptive control (MAC) to distinguish it from the more sophisticated automatic adaptive control (AAC) and has been the subject of several articles (Shaw and Komanduri, 1977; Shaw et al., 1978).

A trend in industry is to reduce the input of the operator by removing him from the decision-making process and by depending less on his operating skills. This has tended to reduce the role of the operator to that of a watchman leading to boredom and aggravating the job-satisfaction problem.

The MAC approach to adaptor control appears to be useful until sensor technology and the cost of control equipment drops to the point where primary automatic adaptive control can be a practical reality.

It should be emphasized at the outset that in manual adaptive control, the operator uses the computer at the machine on the shop floor. Of course, it is possible for an engineer to use a programmable computer to help analyze data collected in the workshop but that is not what MAC is about. The basic idea is to provide the operator with a tool programmed by an engineer which enables the operator to analyze data as it is generated by the machine and to make wise decisions for change based on this analysis. The operator is thus enabled to make important calculations he does not have time to make even when he possesses the mathematical skills required. The system is designed to work as though the engineer were continuously at the side of the operator to analyze results coming from the machine and to make the changes in operating conditions necessary to optimize the process. The programmable computer merely enables the engineer to perform these activities indirectly *in absentia*.

In order to focus attention on the principles involved, a very simple example will be considered—that of turning a workpiece in a lathe by a single-point tool. However, detailed programs have been prepared and applied in industry to manufacturing operations involving multispindle automatic screw machines, boring, milling, sawing, and a variety of grinding situations. For turning on a lathe, the quantity to be optimized will usually be cost per part (to be minimized) or production rate in parts per hour (to be maximized). In both cases, the feed should be as large as possible consistent with constraints (e.g., power available, tool fracture, finish). In this case there is an optimum cutting speed that will usually be in the practical range of values and the problem is to operate as close to this speed as possible.

When MAC is applied to a straight single-tool turning situation, the operator uses the program prepared previously by an engineer. This program contains approximate values of machine, operator, and overhead cost (x) and the mean cost of a cutting edge (y) for the corresponding machine operator and tool to be used. The program should be delivered to the operator with the shop order, tools, and parts to be machined.

The operator decides when the tool should be changed which may be because of any one of a combination of reasons (e.g., tool breakage, tool wear, excessive cutting temperature, chatter, poor finish, lack of dimensional accuracy). After each tool is replaced, the operator keys in two numbers:

1. time elapsed from last tool change
2. number of parts machined

A key is then pressed that causes the computer to determine the mean cost per part and subtract this value from the corresponding value taken from storage for the previous tool. If this difference is positive, the operator increases the spindle speed by a small increment (say 5% or the next available speed on the machine if the speed is not continuously variable) and repeats the procedure. If the number that comes up is negative, the operator decreases the speed by a corresponding amount. In this way, the machine is adjusted to reach the optimum cutting speed ($V^{\#}$) in a few trials. If workpiece hardness changes, or any other of a large number of items that will affect tool life, the machine will automatically be guided to a new optimum value of cutting speed.

In each application of MAC, the operation must be analyzed to determine whether one or more operating variables may be optimized and then a program written and tested in the workshop. This requires a certain amount of skill and has proven to be an excellent vehicle for training students to approach manufacturing problems in a rational way.

MAC was successfully applied to a centerless grinding machine used to grind cylindrical track pins in a tracked vehicle manufacturing plant. A preliminary consideration of the process indicated that there should be an optimum production rate for minimum cost. At a high rate of production, the cost of wheel dressing (including wheel and diamond cost and downtime during dressing) is excessive while at a lower rate of production, labor cost is too high. The program was written to provide total cost per part when the time between dressing operations and the number of parts produced were entered into the computer. Production rate was changed by changing the r.p.m. of the regulating wheel. The initial setting for the first part tested was ground at the feed rate given by the database programmed into the company's central computer. This was based on actual performance values for the machine in use that had been accumulated over a fairly long period of time. After the first wheel-dressing cycle, the computer called for an incremental increase in feed rate. This was repeated again and again until the optimum value was reached. At this point the total cost for this particular grinding operation had been cut approximately in half. The next job that was put on this machine was similarly treated. This time the computer led the operator downward in speed and finally the optimum cost was appreciably less than the starting value suggested by the database.

These results are typical of those to be expected from a database approach. While the average production rate will be about right for a variety of jobs, the predicted value for each individual case may be appreciably in error—some values being too low and some too high. More details concerning this case may be found in Stelson et al. (1977).

The MAC system is also of value when used in connection with a preprogrammed numerically controlled machine. In one example, a large sophisticated NC machining center was programmed to produce parts based on previous experience. However, this machine was provided with an override capability that enabled the operator to change the production rate without changing the proportional time involved with each detail. When MAC was used to adjust the cycle time for minimum cost per part, an appreciable saving in total cost per part was realized. This illustrates that MAC can play a valuable role when used to fine-tune and monitor a preprogrammed sophisticated NC machine.

The MAC system has also proven to be of use in evaluating different tool materials on a given job. In such a case, it is important that conditions be optimized for each material before comparing the resulting cost per part. Details on how this may be done and an example of where production rate is optimized instead of cost per part is to be found in Fucich et al. (1978). An additional reference discussing and applying MAC is to be found in Shaw (1988). Shaw and Avery (1979) have also presented a report that discusses the improvement of productivity by application of MAC.

Role of Capital in Productivity

It should be noted that throughout the discussion of optimization presented here, both labor and capital cost were included. This is consistent with the concept that productivity is not alone a matter of the output per man-hour but should take into account the capital involved in increasing output per hour. As the cost of more sophisticated machinery increases, as interest rates rise, and as capital available to industry for investment decreases, it becomes ever more important to be sure that available capital is wisely used.

The big problem relative to process optimization is not in knowing what to do but in applying what is already known. There are many reasons why it is difficult to implement in the workshop concepts that are fully developed and available in the literature. The single most important reason is a lack of broadly trained, innovative, manufacturing engineers.

TOOL CONDITION MONITORING

Byrne et al. (1995) have presented an overview of tool condition monitoring (TCM). Important topics discussed are sensor systems, signal and information processing, optimization, application of multiple sensor systems, and increased reliability of sensor systems.

Computer numerical control (CNC) machining offers special opportunities for optimization of manufacturing processes. Tounsi and Elbestawi (2001) have presented an example of how improved productivity in three-axis milling may be achieved by feed-rate scheduling. This involves what the authors call an optimized feed schedule strategy (OFSS). By scheduling the feed rate, the variation of cutting force may be greatly reduced leading to reduced cutting time and the prevention of tool breakage by limiting the maximum force encountered in the machining cycle.

REFERENCES

Bouzakis, K. D., Efstathion, K., and Paraskevopoulou, R. (1994). *Man. Systems* **23/2**, 137–143.
Byrne, G., Dornfeld, D., Inasaki, I., Ketteler, G., Koenig, W., and Teti, R. (1995). *Annals of CIRP*, **44/2**, 541–567.
Fucich, L., Obermiller, K., Shuster, A., and Shaw, M. C. (1978). *Proc. 6th N. Am. Metal Working Res. Conf.* Soc. Manuf. Engrs. Dearborn, Mich., p. 372.

Gall, D. (1969). *Annals of CIRP* **17**, 359.

Gilbert, W. W. (1950). *Machining Theory and Practice*. Am. Soc. for Metals, Metals Park, Ohio, p. 465.

Schulz, H., and Spath, D. (1997). *Annals of CIRP* **46/1**, 415–418.

Shaw, M. C. (1957). *Ind. Anz.* **79**, 847.

Shaw, M. C. (1959). *Microtecnic* **13**, 103.

Shaw, M. C. (1960). *Int. J. Mech. Sci.* **1**, 89.

Shaw, M. C. (1965). *Int. J. Mach. Tool Des. and Res.* **5**, 25–34.

Shaw, M. C. (1988). *Proc. of CIRP 20th Internat. Seminar on Manf. Systems at Tbilisi, USSR* **18/3**, 199–202.

Shaw, M. C., and Avery, J. P. (1979). *Proc. of 5th World Congress on Theory of Machines and Mechanisms* **1**, 777–780.

Shaw, M. C., Avery, J. P., and Frankman, J. (1978). *Annals of CIRP* **27/1**, 425–428.

Shaw, M. C., and Komanduri, R. (1977). *American Machinist* **121**, 81.

Stelson, T. S., Komanduri, R., Shaw, M. C. (1977). In *Proc. 4th Int. Prod. Eng. Conf.* (Tokyo). Taylor and Francis, London.

Taylor, F. W. (1907). *Trans. Am. Soc. Mech. Engrs.* **28**, 33.

Tounsi, N., and Elbestawi, M. A. (2001). *Machining Sc. and Tech.* **5/3**, 293–414.

20 MODELING OF CHIP FORMATION

This chapter considers the history and difficulties of *predicting* metal cutting chip formation. Only the simplest type of chip will be considered—the steady state continuous chip that is widely encountered when a ductile metal is cut under orthogonal conditions (Fig. 1.2). In order to accomplish this, we must be able to predict two outputs—the shear angle (ϕ) and the mean flow stress on the shear plane (τ_s).

Before about 1940, process planning of metal cutting performance was based on making modifications of experimental results that had been obtained previously in the workshop. This has been referred to as a craft approach. There was essentially no attempt to relate performance to details of mechanics such as stress, strain, strain rate, localized microfracture, and temperature of the chip. Then, beginning in the late 1930s and continuing through the 1940s and 1950s, cutting performance was explained in terms of basic principles of mechanics (the scientific approach). Piispanen (1937) and Merchant (1945a, 1945b) were initially responsible for the major contributions to this aspect of modeling which have been covered in Chapters 3, 8, and 9.

The scientific approach was followed by the compilation of data banks that gave operating conditions for a wide variety of cutting conditions. Most of these data banks were sponsored by large companies on a proprietary basis. However, a very large one that was government sponsored and available to the public gave rise to the *Machining Data Handbook* (1980).

Another important attempt to predict metal cutting performance was to extrapolate material test results (conventional compression, torsion, and indentation results) to the very large values of strain, strain rate, and temperature pertaining in metal cutting chip formation. A valuable presentation of this effort is to be found in Oxley (1989). Oxley used experimental flow fields to model an appropriate slip line field (SLF). Experimental flow fields were obtained from magnified motion pictures of grids on the side of a chip, side flow being prevented by a glass plate attached to the side of the workpiece. Experimental flow fields were also obtained from explosion-activated quick-stop photos of grids at the center of a split workpiece. An experimentally based SLF was then altered until boundary conditions of velocity and force were satisfied. This SLF was then used to determine the shear angle, cutting forces, power, and temperature using a value of flow stress for the work material that was constant and independent of hydrostatic stress. A computer was used to construct slip line fields consistent with the experimental flow fields. Initially, modeling was for low cutting speeds where the effect of strain rate and temperature on flow stress could be ignored. Later, strain hardening and strain rate effects were introduced that gave more realistic results for higher cutting speeds.

THE FINITE ELEMENT METHOD (FEM)

FEM analysis employs a number of finite points (nodal points) covering the domain of a function to be evaluated. The subdomains within these nodal points are called finite elements. Thus, the entire domain is a collection of elements connected at their boundaries without gaps. Each element is identified by its nodes. Instead of seeking a solution covering the whole space, an approximate solution is obtained just for the nodes. Governing equations associated with geometry, velocity, and forces are applied to the nodes, and then all equations are solved numerically. An approximate solution is obtained by interpolation between the nodes.

Since FEM analysis involves the simultaneous solution of a large number of equations, it was not viable until digital computers and suitable software became available in the 1960s. At first, application of FEM analysis was to stress analysis and metal forming problems. The method was first applied to metal cutting chip formation problems in the 1970s. Details of FEM are given in Zienkiewicz (1989). Applications of FEM to metal cutting chip formation are very well covered in Childs et al. (2000).

In applying the FEM to numerical modeling, the following approaches have usually been taken:

- where the mesh is attached to the material (Lagrangian formulation = LF)
- where the mesh is fixed in space (Eulerian formulation = EF)

The LF is usually applied to the solution of solid mechanics problems. The EF is usually applied to problems where a control volume is involved (e.g., fluid mechanics, rolling and extrusion of metals). Both of these formulations have been applied to metal cutting chip formation. When considering the initiation of chip formation from initial indentation of the tool, the shape of the chip is not known and LF is employed. In cases where the steady state shape of the chip is known or assumed, EF is more efficient.

In applying FEM to metal cutting chip formation, two important problems involved are

- the fact that new surface is being formed
- the fact that extremely large strains and change of shape are involved

The first of these is called the separation problem. Several methods used to solve the separation problem are

- separation when the material reaches a certain distance from the tool tip (Usui and Shirakashi, 1982)
- critical strain at node nearest the tool tip (Carroll and Strenkowski, 1988)
- ductile fracture criteria (Iwata et al., 1984)
- strain energy density from a tensile test (Lin and Lin, 1992)
- a critical damage value (Cerretti et al., 1996)
- a fracture mechanics approach (Zhang and Bagchi, 1994)

When determination of a steady state chip shape is involved, rather than modeling chip formation from initial tool–work contact, a method devised by Shirakashi and Usui (1976) is convenient. This involves assuming a chip shape and then finding the actual shape by iteration until the assumed and calculated flow fields correspond. This is called the iterative convergence method (ICM). The flow characteristics are extrapolated to metal cutting strains from Hopkinson bar tests, and the tool–chip friction characteristics are based on split tool experiments.

The separation problem may be avoided by adopting a continuous flow approach. As the tool advances, the mesh is distorted and it must be regenerated (Sekhhon and Chenot, 1993; Marusich and Ortiz, 1995; Madhavan et al., 1993). A continuous flow approach requires frequent remeshing resulting in high computational cost. Any approach to the separation problem must avoid use of a gross crack running outward from the tool tip since such a crack is not observed experimentally.

A more recent FEM for use in studies of the metal cutting problem is one called the Arbitrary Lagrangian Eulerian (ALE) method. This is well suited to problems involving very large strains that have characteristics of both solid and fluid flow. In this approach the nodal points of the finite element mesh are neither attached to the work material or fixed in space. The mesh points may have an arbitrary motion best suited to the problem. A detailed description of the AEL approach is presented in Movahhedy et al. (2000).

In all FEM applications, the von Mises yield criterion is employed, and this represents an analytical weakness when the strains involved are very large as in metal cutting chip formation. This will be discussed in the following portions of this chapter.

SHEAR-ANGLE PREDICTION

In Chapter 8, attempts to predict the shear angle are considered in detail. The conclusion reached is that Eq. (8.11) represents the best approach where C is a constant (for a given material and set of machining conditions) approximately equal to the slope of the curve of shear stress (τ_s) versus compressive stress (σ_s) on the shear plane.

Figures 8.9 and 8.10 illustrate how use of the Merchant empirical machining "constant" C that gives rise to Eq. (8.11) gives values of ϕ in reasonably good agreement with experimentally measured values. While Figs. 8.9 and 8.10 are only for two medium carbon steels, equally good results were obtained for several other materials resulting in values of C given in Table 8.3. Thus, it is generally found that shear stress on the shear plane (τ_s) increases with normal stress on the shear plane (σ_s) in metal cutting. This is not a result peculiar to the conditions of Figs. 8.9 and 8.10 as has been sometimes suggested (e.g., Pugh, 1958; Oxley and Welsh, 1967). Table 8.3 shows the increase in τ_s with σ_s to be a common occurrence when a number of polycrystalline materials are subjected to metal cutting involving continuous chip formation.

Hydrostatic stress plays no role in the plastic flow of metals if they have no porosity. Yielding then occurs when the von Mises criterion reaches a critical value. Merchant (1945b) has indicated that Barrett (1943) found that for single crystal metals τ_s is independent of σ_s. Merchant suggests this is related to the presence of grain boundaries (i.e., structural defects) in polycrystalline metals. Merchant also found that τ_s is independent of σ_s when plastics such as celluloid are cut.

In general, if a small amount of compressibility is involved, yielding will occur when the von Mises criterion reaches a certain value. However, based on the results of Figs. 8.9 and 8.10 and Table 8.3, the role of compressive stress (σ_s) on shear stress on the shear plane (τ_s) in metal cutting is substantial. The fact there is no outward sign of voids or porosity in steady state chip formation of a ductile metal and yet there is a substantial influence of normal stress on shear stress on the shear plane represents an interesting paradox. Merchant also points out that Nobel laureate Bridgman (1952) found the torsional shear stress at a shear strain of one varied as shown in Fig. 20.1 when specimens of hollow drill rod were subjected to different levels of axial compressive stress. This also clearly shows an increase in shear stress at a given strain with normal stress on the shear plane.

It is interesting to note that Piispanen (1937) had assumed that shear stress on the shear plane would increase with

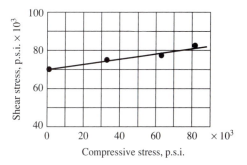

Fig. 20.1 Shear stress versus compressive stress at strain = 1 for drill rod. (after Bridgman, 1943)

normal stress and had incorporated this into his graphical treatment. This was unknown to Merchant in writing his 1945 papers, since a translation of Piispanen's paper from Finnish into English became available only between the time when Merchant's papers were written and when they were published. It is not uncommon that important discoveries are made quite independently at different locations at about the same time. A case in point is discovery of the dislocation quite independently by Orowan, Polanyi, and Taylor all in 1934. Apparently there is a right time for an important discovery to be made.

ROLE OF MICROCRACKS

The influence of σ_s on τ_s in the metal cutting of polycrystalline metals was suggested by Shaw (1980) to be due to the formation and rewelding of localized microcracks as they move across the shear plane. The role that grain boundaries play (Merchant, 1945b) in this connection is undoubtedly due to their acting as points of stress concentration along with second-phase particles and impurities in the generation of microcracks. Since these microcracks do not extend to the side surface of a chip, they would not be visable during chip formation and hence not suspected of being there.

In Chapter 9, the tests of Walker and Shaw (1969) shown in Fig. 9.7 provide additional evidence for the appearance of microcracks above a certain shear strain depending on the axial compressive stress on the shear plane.

In Chapter 13, an interesting shear test (Fig. 13.15) designed to explain why CCl_4 is so effective at low cutting speeds is shown. A piece of copper was prepared as shown in Fig. 13.15. The piece that extends upward and appears to be a chip is not a chip, but a piece of undeformed material left there when the specimen was prepared. A vertical flat tool was then placed precisely opposite the free surface as shown in Fig. 13.15 and fed horizontally. Horizontal (F_P) and vertical (F_Q) forces were recorded as the shear test proceeded. It was expected that the vertical piece would fall free from the lower material after the vertical region had been displaced a small percentage of its length. However, it went well beyond the original extent of the shear plane and was still firmly attached to the base. This represents a huge shear strain since the shear deformation was confined to a narrow band. When a single drop of CCl_4 was placed at A in Fig. 13.15 before conducting the shear test, the protrusion could be moved only a fraction of the displacement in air before gross fracture occurred on the shear plane. Figure 13.17 shows photomicrographs for tests with and without CCl_4. It is apparent that CCl_4 is much more effective than air in preventing microcracks from rewelding.

Sawtooth chip formation for hard steel discussed in Chapter 22 is another example of the role microcracks play. In this case gross cracks periodically form at the free surface and run down along the shear plane until sufficient compressive stress is encountered to cause the gross crack to change to a collection of isolated microcracks.

FLUID-LIKE FLOW IN METAL CUTTING CHIP FORMATION

At the General Assembly of the International Institution for Production Engineering Research (CIRP) in 1952, an interesting paper was presented by Eugène (1952). Figure 20.2 shows the apparatus he used. Water was pumped into baffled chamber A, which removed eddies, and then was caused to flow under gravity past a simulated tool at B. Powdered bakelite was introduced at C to make the streamlines visible as the fluid flowed past the "tool." The photographs taken by the camera at D were remarkably similar to quick-stop photomicrographs of actual chips. It was thought by this author at the time that any similarity between fluid flow and plastic flow of a solid was not to be expected. That was long before it was clear that the only logical explanation for the results of Bridgman and Merchant involved microfracture (Shaw, 1980).

At the General Assembly of CIRP 47 years later, a paper was presented that again suggested that metal cutting might be modeled by a fluid (Kwon et al., 1999). However, this paper was concerned with ultraprecision machining (depths of cut < 4 μm), and potential flow analysis was employed instead of the experimental approach taken by Eugène.

It is interesting to note that chemists relate the flow of liquids to the migration of vacancies (voids) just as physicists relate the plastic flow of solid metals to the migration of dislocations. Henry

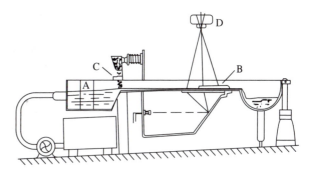

Fig. 20.2 Apparatus used by Eugène (1952) to photograph flow paths past tools having different rake angles.

Eyring and coworkers (Eyring et al., 1958; Eyring and Ree, 1961; Eyring and Jhon, 1969) have studied the marked changes in volume, entropy, and fluidity that occur when a solid melts. For example, a 12% increase in volume accompanies melting of argon suggesting the removal of every eighth molecule as a vacancy upon melting. This is consistent with x-ray diffraction of *liquid* argon that showed good short-range order but poor long-range order. The relative ease of diffusion of these vacancies accounts for the increased fluidity that accompanies melting. A random distribution of vacancies is also consistent with the increase in entropy observed on melting. Eyring's theory of fluid flow was initially termed the hole theory of fluid flow, but later it was changed to *the significant structure theory* (Eyring and Jhon, 1969).

According to this theory the vacancies in a liquid move through a sea of molecules. Eyring's theory of liquid flow is mentioned here since it explains why the flow of a liquid approximates the flow of metal past a tool in *chip formation*. In this case the microcracks (voids) move through a sea of crystalline solid.

KECECIOGLU'S CONTRIBUTIONS

Kececioglu has presented several papers that illustrate the difficulty of predicting the mean shear stress on the shear plane in steady state chip formation (Kececioglu, 1958a, 1958b, 1958c, 1960). The last of these builds on the others and is most important relative to the modeling being discussed here. Based on a large number of dry cutting experiments on AISI 1015 steel tubing having a hardness of 118 Brinell and using a steel cutting grade of carbide, Kececioglu concludes that the mean shear stress on the shear plane (τ_s) depends upon the following variables:

- mean normal stress on the shear plane (σ_s)
- the shear volume of the shear zone (e)
- the mean strain rate in the shear plane ($\dot{\gamma}$)
- the mean temperature of the shear plane ($\bar{\theta}_S$)
- the degree of strain hardening in the work before cutting

If this were not complicated enough, it is clear from Kececioglu's experimental results that these items do not act independently (i.e., the influence of a high normal stress σ_s on τ_s depends upon the combination of the other variables pertaining). This is an important result since it means that in general it is not possible to extrapolate materials test values of τ_s to vastly different cutting conditions (i.e., to metal cutting values of γ, $\dot{\gamma}$, σ_s, $\bar{\theta}_S$, and e).

It is interesting to note that Kececioglu suggests that the specific energy should be related to the shear plane area (e) instead of t as in Eq. (3.36). The shear plane area is

$$e = (bt/\sin\phi)\Delta y \text{ (in}^3 \text{ or mm}^3) \tag{20.1}$$

This appears to be a useful suggestion. In both cases the inverse relation between u and t or e is due to the greater chance of encountering a stress-reducing defect as t or e increases. The use of e instead of t is a more general, although more complex, way of expressing the "size effect."

The range of values covered by Kececioglu in his orthogonal experiments on AISI 1015 steel were as follows:

- rake angle, $-10°$ to $+37°$
- undeformed chip thickness (t), 0.004 to 0.012 in (0.2 to 0.3 mm)
- cutting speed (V), 126 to 746 f.p.m. (38.4 to 227.4 m/min)
- inclination angle (i), 0° to 35°

This resulted in the following wide range of dependent variables:

- width of shear zone (Δy), 0.002 to 0.007 in (0.10 to 0.18 mm)
- rate of strain ($\dot{\gamma}$), 20,000 to 40,000 s^{-1}
- mean shear stress on shear plane (τ_s), 62,000 to 84,000 p.s.i. (427.5 to 579.2 MPa)
- mean normal stress on shear plane (σ_s), 1000 to 12,000 p.s.i. (6.9 to 83 MPa)
- mean shear plane temperature ($\bar{\theta}_S$), 410 to 840 °F (210 to 449 °C)

Figure 20.3a shows considerable scatter in the variation of τ_s with σ_s due to the fact that the effect of changing one independent variable on τ_s depends upon a combination of the other variables as well. Figure 20.3b shows the variation of the mean value of τ_s for a group of points in the vicinity of the value of σ_s plotted. This indicates on the average that τ_s increases with an increase in σ_s. This is consistent with the view expressed in Chapter 9 that under the very high strains involved in metal cutting (a mean value of about 3.5 in Kececioglu's experiments), localized microcracks are likely to form on the shear plane; and an increase in normal stress should decrease their number and give rise to an increase in shear stress on the shear plane.

Figures 20.4a and 20.4b show similar results for the mean shear stress on the shear plane versus the volume of the shear zone (e). This shows clearly

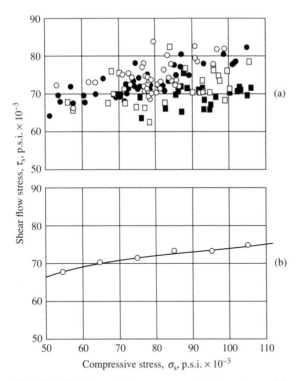

Fig. 20.3 Variation of shear stress on the shear plane with normal stress on the shear plane. (a) For 43 tests involving a wide range of rake angles, cutting speeds, undeformed chip thicknesses, and inclination angles. (b) Mean values of shear stress for values of normal stress indicated. (after Kececioglu, 1960)

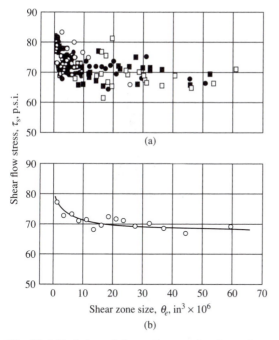

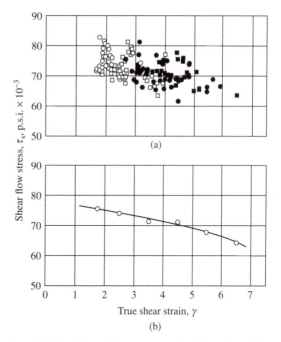

Fig. 20.4 Variation of shear stress on the shear plane with shear zone volume. (a) For 43 tests involving a wide range of rake angles, cutting speeds, undeformed chip thicknesses, and inclination angles. (b) Mean values of shear stress on the shear plane for values of e close to the values of e indicated. (after Kececioglu, 1960)

Fig. 20.5 Variation of shear stress on the shear plane with shear strain on the shear plane. (a) For 43 tests involving a wide range of rake angles, cutting speeds, undeformed chip thicknesses, and inclination angles. (b) Mean values of shear stress on the shear plane for values of γ close to the values of γ indicated. (after Kececioglu, 1960)

that in general a decrease in the shear zone volume causes an increase in the shear stress on the shear plane particularly for shear volumes below 10^{-5} in³.

Figure 20.5 shows results similar to Figs. 20.3 and 20.4 for the mean shear stress on the shear plane (τ_s) versus the shear strain (γ). This yields the unusual result that the shear stress decreases with increase in strain which is not consistent with ordinary materials test results that involve strain hardening. The reason for this paradox is due to the fact that in addition to strain (γ) several other variables are involved ($\dot{\gamma}$, $\bar{\theta}_S$, σ_s, e) and the net effect is a decrease in τ_s with the mean value of γ.

ZHANG AND BAGCHI

Zhang and Bagchi (1994) have presented a valuable analysis of the chip separation problem in FEM in terms of microfracture mechanics. This is based on the fact that ductile metals fail in three steps: nucleation, growth, and coalescence of microvoids that initiate at points of stress concentration (Anderson, 1991). Figure 20.6 illustrates how the three steps lead to gross fracture in shear. Figure 20.7 shows a random distribution of defects (points of stress concentration) in a ductile metal chip. As the work material approaches a stationary tool, defects along the x-axis in the cutting direction are subjected to an increasing stress. This leads to nucleation of a void at A in Fig. 20.7a, growth as the void moves to B in Fig. 20.7b, and coalescence with the tool as shown in Fig. 20.7c.

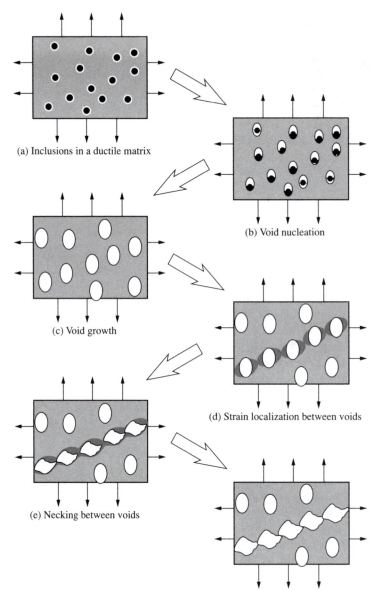

(a) Inclusions in a ductile matrix

(b) Void nucleation

(c) Void growth

(d) Strain localization between voids

(e) Necking between voids

(f) Void coalescence and fracture

Fig. 20.6 Void nucleation, growth, and coalescence in ductile metals. (after Anderson, 1991)

    Zhang and Bagchi (1994) suggest that voids approaching the tool tip grow but do not coalesce before reaching the tool tip. This is important since coalescence would lead to a gross crack extending in front of the tool tip, and this is never observed experimentally even when hard brittle materials are machined (Chapter 22). Zhang and Bagchi (1994) have applied the continuum model for the void nucleation of Argon et al. (1975) and the void growth of Rice and Tracey (1969). It is assumed that separation occurs when the leading void reaches its maximum size.

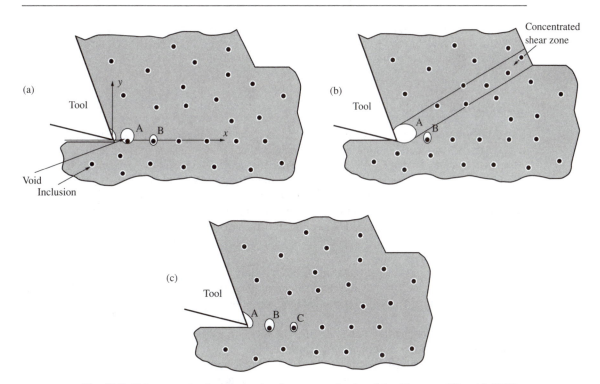

Fig. 20.7 Chip separation based on microfracture mechanics. (after Zhang and Bagchi, 1994)

The theory presented by Zhang and Bagchi based on the presence of points of stress concentration, formation of microvoids, void growth, and void coalescence within the tool tip offers a very reasonable explanation for the flow separation problem involved when FEM is applied to metal cutting chip formation. The energy involved in the generation of new surface area associated with flow separation at the tool tip is negligible (Chapter 3) and hence need not be considered in any

FEM analysis for cutting forces. The flow separation problem is a special application of the role microvoids (in ductile materials = microcracks in highly worked materials) play in metal cutting chip formation.

Figure 20.8 shows a diagram equivalent to Fig. 20.7 including action along the shear plane. Here the material in the shear plane is subjected to very large strains, and any points of stress concentration should be expected to give rise to microcracks instead of microvoids. In the presence of relatively high normal stress on the shear plane and the absence of a contaminating film, such as CCl_4 vapor, these microcracks will reweld after moving a relatively short distance. Also, due to the very high strains associated with the shear

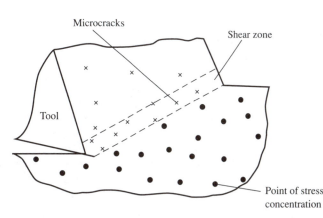

Fig. 20.8 Chip formation based on transport of microcracks across the shear plane by formation, displacement, and rewelding.

plane, a considerably higher density of microcracks will be involved than the density of microvoids involved in the undeformed work material of Fig. 20.7. Details of the microcrack theory of steady state chip formation are to be found in Shaw (1980).

Concluding Remarks

Based on the discussion presented, it is concluded that microcracks usually play an important role in steady state metal cutting chip formation because of the unusually large strains involved. Also, there is considerable experimental evidence that normal stress on the shear plane has a substantial influence on shear stress on the shear plane. Both of these rule out use of the von Mises criterion in FEM modeling except as a rough first approximation. Future studies should be devoted toward methods designed to include microfracture mechanics into the constitutive relation employed in FEM analysis of steady state chip formation with a ductile metal. For the time being, "prediction" of the shear angle appears to involve use of Eq. (8.11) that employs the empirical constant C.

Although FEM has been successfully used in metal forming analysis (Kobayashi et al., 1989), the strains involved are considerably lower than in metal cutting chip formation. This enables the von Mises criterion to be used in metal forming since microcracks are not involved in the purely plastic flow associated with forming. FEM analysis is also an effective way of estimating temperature fields associated with chip formation provided the energy input is derived from experimental cutting forces and the von Mises criterion is not involved.

Monitoring of cutting operations offers an alternative approach to improved performance. This consists of measuring forces, torque, acoustic emission, tool and workpiece displacement, cutting temperatures and sound during processing, and tool wear and surface finish out of process. This is sometimes referred to as an intelligent manufacturing approach. In some cases two or more signals are combined to give a result that improves performance by either discrete or continuous changes. A review of monitoring technology is to be found in Byrne et al. (1995).

There have been attempts to employ computer-aided tool selection (COATS) in computer-aided process planning (CAPP). However, this is a very difficult problem. In general, it has not been possible to completely replace human experts in the workshop with their heuristic knowledge by computer-based expert systems that employ fuzzy logic (Zimmerman, 1991) and other techniques that handle vagueness and constraints.

While great progress has been made during the past half century in *understanding* the metal cutting process, relatively little progress has been made in *predicting* performance.

In FEM chip formation modeling, the flow stress employed is usually based upon an empirical equation that contains three or more adjustable constants that reflect the effect of strain, strain rate, temperature, etc. These empirical equations are often based on Hopkinson bar impact experiments that yield flow stress values for a given material at a range of strain rates and temperatures, but at shear strains less than one. Strains involved in chip formation are several times higher than one and are obtained by linear extrapolation. Shirakashi et al. (1988) give an excellent account of the procedure for ferrous metals and Usui and Shirakashi (1982) for nonferrous metals. Values from an empirical equation are then used with the von Mises criterion to carry out the FEM analysis. By using appropiate constants in the empirical equation, predicted and experimental values may be found to be in reasonable agreement. However, all this should be considered to yield an approximate result since the role of microcracks has been ignored.

Since 1995, a working committee of the CIRP, with expert members throughout the world, has spent considerable effort studying modeling of the metal cutting process. The objective is to develop a collection of models that will *predict* performance for a wide variety of metal cutting outputs such as

- chip type
- cutting forces
- tool life including tool fracture and tool wear
- tool-face friction
- tool-face temperature distribution
- surface integrity including finish
- workpiece accuracy

This is to cover a variety of metals and operations such as

- turning
- milling
- drilling, boring, and reaming
- production of complex shapes

The aim is that all this may be achieved by 2050.

While this is a very tall order, it may be possible considering metal cutting progress that was made during the second half of the twentieth century. However, a detail that increases the difficulty of achieving success is that operating conditions are not static, but are continuously changing. A few changes that have occurred during the second half of the twentieth century include

- new work materials (titanium alloys, etc.)
- new cutting tool materials [steel cutting grade carbides, ceramics, cubic boron nitride (CBN), and polycrystalline diamond (PCD)]
- new tool geometries including chip control grooving
- more efficient use of expensive materials
- tool coatings and improved cutting fluids
- high-speed machining
- machining of hard materials to eliminate finish grinding
- new machine tool concepts (numerical control, machining centers, etc.)
- increased emphasis on safety and the environment

In May of 1998, an International Workshop on Modeling of Machining Operations was organized by CIRP and held in Atlanta, Georgia. This was organized into four discussion groups:

- analytical modeling based on scientific principles
- computation modeling (computer-aided modeling including finite element methods (FEMs)
- empirical modeling based on experimental results
- integrated modeling (combinations of models)

The main problem encountered with the first two items is lack of constitutive equations that are suitable for metal cutting applications. It is apparent that a uniquely different approach is required than that which is applicable to forming, materials testing, and other processes involving classical plastic flow.

A very detailed progress report covering this CIRP endeavor has been published (van Luttervelt et al., 1998). The entire effort was divided into the following categories:

- industrial promotion and application
- fundamentals of modeling
- computational mechanics
- compilation of a modeling literature database

All of these areas are covered, plus a very extensive general discussion of the problems involved in metal cutting modeling. The main industrial finding was a general lack of interest by industry unless a product (software) is made available that enables application with a minimum learning period. At the time the progress report was published (August 1998), there were 3500 items in the modeling literature database available from Professor P. K. Venuvinod of the City University of Hong Kong.

REFERENCES

Anderson, T. L. (1991). *Fracture Mechanics*. CRC Press, Boca Raton, Fla.

Argon, A. S., Im, J., and Safoglu, R. (1975). *Metallurgical Trans.* **6a**, 825.

Barrett, C. S. (1943). *Structure of Metals*. McGraw-Hill, New York, pp. 295–296.

Blazynski, T. Z., and Cole, J. M. (1960). *Proc. Instn. Mech. Engrs.* **174**, 757.

Bridgman, P. W. (1943). *J. Appl. Phys.* **14**, 273–283.

Bridgman, P. W. (1952). *Studies in Large Plastic Flow and Facture*. McGraw-Hill, New York.

Byrne, G., Dornfeld, D., Inasaki, I., Ketteler, G., Koenig, W., and Teti, R. (1995). *Annals of CIRP* **44/2**, 541–567.

Carroll, J. T., and Strenkowski, J. S. (1988). *Internat. J. Mech. Sc.* **30**, 899–920.

Cerretti, J. E., Fallbehmer, P., Wu, W. T., and Altan, T. J. (1996). *Mat. Proc. Tech.* **59**, 169–181.

Childs, T. H. C., Maekawa, K., Obikawa, T., and Yamane, Y. (2000). *Metal Machining*. Arnold Pub. Co., England.

CIRP International Workshop on Modeling Machining Operations. (1998). Edited by A. K. Jawahar, R. Balaji, and R. Stevenson. University of Kentucky Publishing Services, Lexington.

Drucker, D. C. (1949). *J. Appl Phys.* **20**, 1013.

Ernst, H. J., and Merchant, M. E. (1941). *Trans. Am. Soc. Metals* **29**, 299.

Eugène, F. (1952). *Annals of CIRP* **52/1**, 13–17.

Eyring, H., and Jhon, M. S. (1969). *Significant Theory of Liquids*. Wiley, New York.

Eyring, H., and Ree, T. (1961). *Proc. Nat. Acad. of Sc.* **47**, 526–537.

Eyring, H., Ree, T., and Hirai, N. (1958). *Proc. Nat. Acad. of Sc.* **44**, 683.

Heidenreich, R., and Shockley, W. (1948). *Report on Strength of Solids, Phys. Soc. of London*, p. 57.

Iwata, K., Osakada, A., and Terasaka, T. (1984). *J. Eng. Mat. Tech.* **106**, 132–128.

Kececioglu, D. (1958a). *Trans. ASME* **80**, 149–157.

Kececioglu, D. (1958b). *Trans. ASME* **80**, 158–168.

Kececioglu, D. (1958c). *Trans. ASME* **80**, 541–546.

Kececioglu, D. (1960). *J. Eng. for Ind., Trans. ASME* **82**, 79–86.

Kobayashi, S., Oh, S., and Altan, T. (1989). *Metal Forming and the Finite Element Method*. Oxford University Press, New York.

Komanduri, R., and Brown, R. H. (1967). *Metals Materials* **95**, 308.

Kwon, K. B., Cho, D. W., Lee, S. J. and Chu, C. N. (1999). *Annals of CIRP* **47/1**, 43–46.

Lankford, G., and Cohen, M. (1969). *Trans. ASM* **62**, 623.

Lin, Z. C., and Lin, S. F. (1992). *ASME J. Eng. Mat. Tech.* **114**, 218–226.

Machining Data Handbook, 3rd ed. (1980). Metcut Research Associates Inc., Cincinatti, Ohio.

Madhaven, V., Chandrasekar, S., and Farris, T. N. (1993). In *Materials Issues in Machining III and the Physics of Machining Process*. Minerals, Metals, and Materials Soc., pp. 187–209.

Marusich, T. O., and Ortiz, M. (1995). *Int. J. Num. Meth. Eng.* **38**, 3675–3694.

Merchant, M. E. (1945a). *J. Appl. Phys.* **16**, 267–275.

Merchant, M. E. (1945b). *J. Appl. Phys.* **16**, 318–324.

Merchant, M. E. (1950). In *Machining Theory and Practice*. ASM, pp. 5–44.

Movahhedy, M. R., Gadala, M. S., and Altintas, Y. (2000). *Mech. Sc. and Tech.* **4**, 15–42.

Ortiz, M. (1995). *Int. J. Num. Meth. Eng.* **38**, 3675–3694.

Oxley, P. L. B. (1989). *Mechanics of Machining*. Ellis Horwood, Ltd., England.

Oxley, P. L. B., and Welsh, M. J. B. (1967). *Trans. ASME* **89**, 549–555.

Piispanen, V. (1937). *Eripaines Teknillisesla Aikakausehdesla* (Finland) **27**, 315.

Piispanen, V. (1948). *J. Appl. Phys.* **19**, 876.

Pugh, H. L. D. (1958). *Proc. of Conf. on Technology of Manufacturing*. The Inst. of Mech. Engrs., London.

Rice, J. R., and Tracey, D. M. (1969). *J. of Mechanics and Physics of Solids* **27**, 201–217.

Sekhhon, G. S., and Chenot, J. L. (1993). *Eng. Comput.* **10**, 31–48.

Shaw, M. C. (1950). *J. Appl. Phys.* **21**, 599.

Shaw, M. C. (1980). *Int. J. of Mech. Sc.* **22**, 673–686.

Shirakashi, T., Maekawa, K., and Usui, E. (1988). *Bull. of Soc. of Prec. Engrs.* **17**(3), 161–166.

Shirakashi, T., and Usui, E. (1976). *Proc. Internat. Conf. on Prod. Eng., Tokyo*, pp. 535–540.

Usui, E., Gujral, A., and Shaw, M. C. (1961). *Int. J. Mach. Tools and Res.* (London) **1**, 187–197.

Usui, E., and Shirakashi, T. (1982). *ASME PED* **7**, 13–35.

van Luttervelt, C. A., Childs, T. H. C., Jawahir, I. S., Klocke, F., and Venuvinod, P. K. (1998). *Annals of CIRP* **47/2**.

Walker, T. J. (1967). Ph.D. dissertation. Carnegie Mellon University, Pittsburgh, Pa.

Walker, T. J., and Shaw, M. C. (1969). *Advances in Machine Tool Design and Research*. Pergamon Press, Oxford, pp. 241–252.

Zhang, B., and Bagchi, A. (1994). *J. Eng. for Ind., Trans. ASME* **116**, 289–297.

Zienkiewicz, O. C. (1989). *The Finite Element Method*, 4th ed. McGraw-Hill, New York.

Zimmerman, H. J. (1991). *Fuzzy Set Theory and Applications*. Kluwer, Boston.

21 WAVY CHIP FORMATION

Before 1940 only three basic types of chips were identified and studied in detail. Ernst (1938) described the three following types:

- the discontinuous or segmental chip (type 1)
- the continuous steady state chip (type 2)
- the continuous chip with built-up edge (type 3)

During the 1940s the type 2 chip received a great deal of attention mainly by Merchant (1945, 1950) and Piispanen (1948). This has been discussed in Chapters 8 and 9.

Masuko (1953) suggested that the tip of even a sharp tool will have a radius that gives rise to an indentation force. This detail associated with steady state (type 2) chip formation has been discussed with Figure 10.14.

EARLY STUDY OF WAVY CHIP FORMATION

Bickel (1954) used high-frequency flash photography to study chip formation at higher cutting speeds than could be used with ordinary motion picture cameras operating at 16 frames/s. Speeds ranging from 65 to 460 f.p.m. (20 to 140 m/min) were studied. It was found that in addition to the three types of chips mentioned above, chips of cyclic thickness were often obtained that suggested a regular fluctuation of shear angle, chip speed, and cutting forces. Bickel also found that under certain conditions the chip adheres locally to the tool face, and when adhesion occurs, chip thickness increases. Bickel suggested that in addition to the steady state chip that had been receiving so much attention in the United States, cyclic chips of variable thickness were also worthy of study. Chips that were cyclic including discontinuous type chips were originally called segmental chips.

Landberg (1956) also found that under certain conditions segmental chips were formed. He concluded that in general the chip formation process was responsible for this type of chip and not the stability of the system (tool stiffness or that of the machine tool). However, when the segment frequency coincided with a natural frequency of the system, a large amplitude of vibration resulted leading to a rough surface. Segmental chips were considered to be detrimental in general, but they could be useful in connection with chip control.

Rice and coworkers (1961, 1962, 1971) also obtained chips that varied cyclically in thickness. Chips of this type were also studied by Albrecht (1961) who observed that wavy chips were associated with a cyclic variation in the shear angle due to some unidentified cause.

NOMENCLATURE

At this point it is important to introduce a few definitions of cyclic chip types. There are three types of "segmental" chips, which will be designated as follows:

- discontinuous chips
- wavy chips (continuous cyclic chips having a smooth wavy outer surface)
- sawtooth chips (continuous cyclic chips having uniformly spaced sharp points along the outer surface)

Discontinuous chip formation is discussed in connection with Figs. 15.10 through 15.18 and Fig. 17.47, wavy chip formation is discussed in this chapter, and sawtooth chip formation will be discussed in Chapter 22.

OECD-SPONSORED STUDIES

During the 1960s, an extensive study of metal cutting performance sponsored by the Organization for Economic Cooperation and Development (OECD) was conducted in laboratories in Europe and the United States by members of CIRP. Chip formation, cutting force, surface finish, and tool-life studies were made on chrom–moly and nickel–chrome alloy steels having a 0.4 $^w/_o$ carbon content. Six different structures were used for each alloy. The results of the OECD/CIRP study were presented at a conference held in Paris (*OECD Seminar*, 1966).

After this seminar an investigation of a detail not covered in the OECD seminar was continued at Carnegie Mellon University. This had to do with the occurrence of wavy chips when machining the OECD steels. These results were covered in detail in an unpublished report (Noh and Shaw, 1968) which is abstracted below. The cutting conditions were as follows:

- type of cut, orthogonal cuts on tubes of 4 in (102 mm) diameter and 0.125 in (3 mm) wall thickness (width of cut)
- work material, Cr–Mo and Ni–Cr steels of 0.4 $^w/_o$ carbon content having 6 types of heat-treated structure (hardness) each
- tool material, P–10 carbide
- feed, 0.008 i.p.r. (0.20 mm/rev)
- fluid, none
- back rake angle, variable
- side cutting edge angle, 0°
- side rake angle and clearance angle, 1.5°

The different heat treatments and hardness levels involved are given in Table 21.1.
The following results were obtained:

1. Wavy chips were obtained only for structure (hardness) levels 3 and 4 in Table 21.1, corresponding to a hardness range of H_B 225 to H_B 270.
2. Wavy chips were obtained for cutting speeds from 50 to 105 f.p.m. (5 to 32 m/min).
3. Wave pitch was 0.040 in (1 mm) at 740 f.p.m. (20 m/min).
4. Tool stiffness (overhang) was not important.

TABLE 21.1 Hardness Values of OECD Steels Used in the Wavy Chip Study of Noh and Shaw (1968)

No.	Hardness		Heat Treatment		
	H_{RB}	H_B	Quench	Tempering	Structure
1	78	330	850 °C	560 °C 1 hr	Tempered Martensite
2	74	300	850 °C	575 °C 1 hr	Tempered Martensite
3	72	270	850 °C	660 °C 1 hr	Tempered Martensite
4	69	225	850 °C	690 °C 9 hr	Sphr. Cementite in αFe
5	65	195	Isothermal Trans.	660 °C, 20 min	Lam. Pearlite in αFe
6	63	180	Isothermal Trans.	690 °C, 9 hr	Sphr. Pearlite in αFe

5. Wavy chips were obtained at all feeds down to 0.0008 in/rev (0.02 m/rev).

6. Wavy chips were obtained for rake angles of +5° and −5° but not for +15°.

7. Wavy chips were obtained for workpiece temperatures of −30 °C to 200 °C but not for 300 °C (Fig. 21.1).

8. The ratio of minimum to maximum wavy chip thickness was approximately 0.5 for a cutting speed of 104 f.p.m. (30 m/min) and a feed of 0.004 in/rev (0.10 mm/rev).

9. When cutting with oil, no wavy chips were observed, but when cutting with water or when cutting dry, wavy chips were obtained (Fig. 21.2).

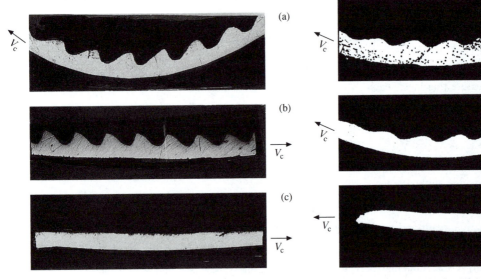

Fig. 21.1 Wavy chips obtained for different workpiece temperatures: (a) −30 °C, (b) 200 °C, and (c) 300 °C. Cutting conditions: cutting speed, 98 f.p.m. (30 m/min); feed, 0.008 i.p.r. (0.2 mm/rev); rake angle, +5°; cutting fluid, dry; tool material, P–10 carbide. (after Noh and Shaw, 1968)

Fig. 21.2 Wavy chips obtained with different cutting fluids: (a) dry, (b) water, and (c) oil. Cutting conditions: cutting speed, 98 f.p.m. (30 m/min); feed, 0.008 i.p.r. (0.2 mm/rev); rake angle, +5°. (after Noh and Shaw, 1968)

10. When different tool materials were used (ceramic or P–10 carbide), the results were essentially the same (Fig. 21.3).

11. When starting a cut from rest, it took about 10 s for a wavy chip to appear.

12. When cutting a tube with the upper half removed (interrupted cutting), a wavy chip did not form even after cutting for 20 s or more.

13. The frequency of wave formation for a speed of 104 f.p.m. (30 m/min), a feed of 0.004 i.p.r. (0.10 mm/rev), and a rake angle of +5° was 165 Hz; but for a rake angle of −5°, the frequency was 330 Hz.

From the above results wavy chip formation appears to be related to the nature of energy dissipation on the shear plane and on the tool face.

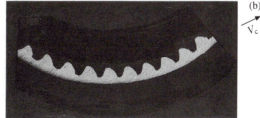

Fig. 21.3 Wavy chips obtained with different tool materials: (a) P–10 tungsten carbide and (b) white ceramic. Cutting conditions: cutting speed, 98 f.p.m. (30 m/min); feed, 0.004 i.p.r. (0.1 mm/rev); rake angle, −5°. (after Noh and Shaw, 1969)

CONTRIBUTIONS OF BROWN AND KOMANDURI

In 1969, R. H. Brown of Monash University in Melbourne, Australia, spent a sabbatical at Carnegie Mellon University. This was when the work on wavy chip formation covered in Noh and Shaw (1968) was being continued. Upon returning to Monash University, Professor Brown assigned a study of wavy chip formation to R. Komanduri as his Ph.D. thesis problem. Komanduri defended an excellent thesis in 1972 (Komanduri, 1972).

The material used throughout his study was a relatively soft low-carbon (AISI 1015) steel having a hardness of H_B 175. Wavy chips were obtained when machining the end of a tube in air (dry) under orthogonal conditions with a high speed steel tool having a rake angle of −5° and a clearance angle of 1.7°. A speed of 180 f.p.m. (55 m/min) and a feed of 0.0096 i.p.r. (0.24 mm/rev) were used.

The main driving force responsible for wavy chip formation was identified as a negative stress-strain characteristic on the shear plane (i.e., negative strain hardening) due to the formation, transport, and rewelding of microcracks at the very high strains pertaining. These microcracks give rise to a cyclic variation in shear angle, cutting forces, chip thickness, and chip speed. This is best illustrated by a few figures from Komanduri and Brown (1981).

Figure 21.4 shows two quick-stop wavy chip photomicrographs machined with a −5° rake angle carbide tool under orthogonal conditions. The photomicrograph in Fig. 21.4a shows the chip just after its thickness reaches the minimum value and is beginning to increase. The shear angle in Fig. 21.4a is seen to be higher than that in Fig. 21.4b where the chip thickness is beginning to decrease.

Figure 21.5 shows scanning electron microscope (SEM) photomicrographs at the midpoint on the shear plane in Fig. 21.4b. The photomicrograph in Fig. 21.5a shows many white pearlite colonies aligned along the shear plane in a ferrite matrix. The photomicrograph in Fig. 21.5b at a higher magnification shows black microcracks that have formed in some of the white pearlite bands shown in Fig. 21.5a.

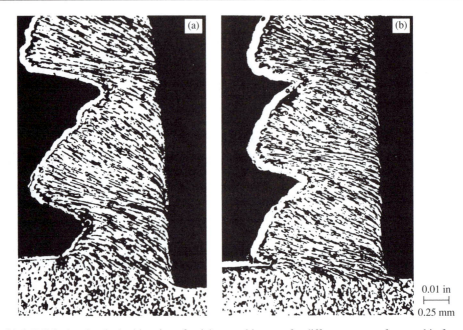

Fig. 21.4 Polished and etched midsection of quick-stop chip roots for different stages of wavy chip formation. (a) High shear angle with chip increasing in thickness. (b) Low shear angle with chip thickness just beginning to decrease. (after Komanduri and Brown, 1981)

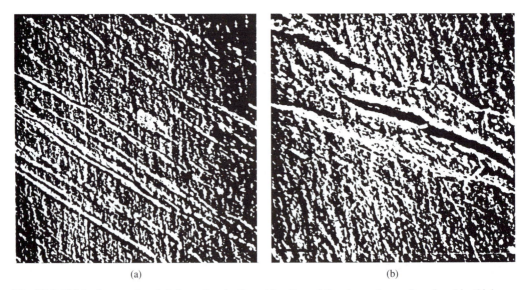

Fig. 21.5 SEM micrographs of deformation in the midsection of the shear plane when the chip thickness is a maximum and the shear angle is a minimum in Fig. 21.4b. (a) Pearlite colonies (white bands) in ferrite matrix aligned in the shear plane. (b) Area of (a) at higher magnification showing micirocracks (black bands) in pearlite colonies.

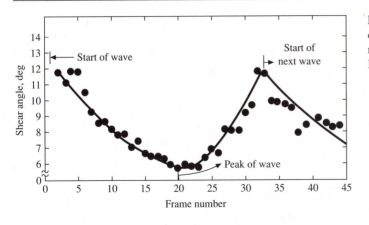

Fig. 21.6 Variation in shear angle in wavy chip formation determined by high-speed motion pictures. (after Komanduri and Brown, 1981)

Figure 21.6 shows the variation in shear angle with frame number (time) while Fig. 21.7 shows the variation of cutting forces F_P (in the cutting direction) and F_Q (in the feed direction) with frame number (time). Figures 21.6, 21.7, and 21.8 are all consistent with Fig. 21.4.

In addition to changes in chip formation associated with the shear plane, cyclic variations in friction on the tool face also play a role. Komanduri and Brown (1981) refer to a stick-slip behavior on the shear plane. However, this involves stick-slip of individual asperities and not of the chip as a whole. The variation in chip velocity with time shown in Fig. 21.8 indicates that gross stick-slip is not involved but only a cyclic variation in speed. The fluctuation in chip speed on the tool face will also have an influence on the shape of a wavy chip. For example, a symmetrical variation in chip velocity with time as that shown in Fig. 21.8 should be expected to yield a wavy chip that is symmetrical such as those in Figs. 21.2a and 21.2b.

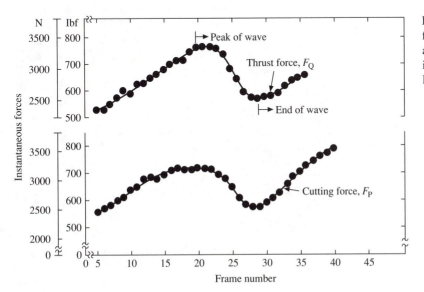

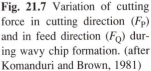

Fig. 21.7 Variation of cutting force in cutting direction (F_P) and in feed direction (F_Q) during wavy chip formation. (after Komanduri and Brown, 1981)

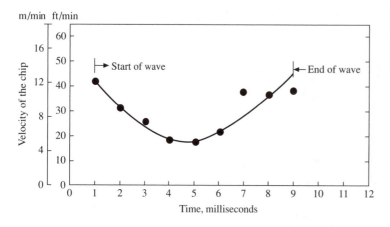

Fig. 21.8 Variation of chip velocity in wavy chip formation determined from high-speed motion pictures. (after Komanduri and Brown, 1981)

ENERGY CONSIDERATIONS

At the end of Chapter 17 regenerative chatter is discussed from an energy point of view. Figure 17.49 identifies three possible situations that may lead to regenerative chatter based on a cyclic change in undeformed chip thickness (t). When the effective value of t is increasing, cutting forces will be less than when t is decreasing. This will result in more elastic energy being stored in the system during the part of a cycle when t is decreasing than is released when t is increasing. The difference in stored energy will be available to induce chatter, and the amplitude will increase until damping energy per cycle equals the net stored energy for a cycle of vibration.

The three situations considered in Fig. 17.49 are for cases where there is no change in shear angle and the net stored energy is due to a cyclic change in the length of shear plane and whether it is increasing or decreasing with time. In the case of wavy chip formation, there is a cyclic change in shear angle that gives rise to a cyclic change in chip thickness (Figs. 21.4 and 21.6). In the wavy chip case, elastic energy is stored when the shear angle decreases and the cutting force increases. Elastic energy is released when the shear angle increases and the cutting forces decrease. The stored energy is greater than the energy released per cycle, and the difference gives rise to a cyclic change in chip thickness and shear angle. The difference in chip thickness increases until the net energy stored per cycle equals the damping energy per cycle.

Details that determine occurrence and shape of a wavy chip are very complex involving ductility of the work material, microfracture and rewelding on the shear plane, and lubrication and friction on the tool face. It appears that at this time prediction of the occurrence and wave shape in wavy chip formation are goals too complex to achieve. However, an understanding of what occurs is of value. Also, at present it is not possible to predict which of the several types of chips will be obtained for a given combination of machining parameters. This is best done by performing a relatively simple cutting experiment. Once the type of chip involved has been identified, the results of experience with that type of chip may be employed to achieve improved performance.

REFERENCES

Albrecht, P. (1961). *Trans ASME* **84**, 405.
Bickel, E. (1954). *Annals of CIRP* **54/1**, 90–91.
Ernst, H. (1938). In *Machining of Metals*. ASM, pp. 1–34.

Komanduri, R. (1972). Ph.D. thesis. Monash University, Melbourne, Australia.

Komanduri, R., and Brown, R. H. (1972). *Metals and Materials* **6**, 531.

Komanduri, R., and Brown, R. H. (1981). *J. of Eng. for Ind., Trans. ASME* **103**, 33.

Landberg, P. (1956). *Annals of CIRP* **56/1**, 219.

Masuko, M. E. (1953). *Trans. Jap. Soc. Mech. Engrg.* **19**, 32.

Merchant, M. E. (1945). *J. Appl. Phys.* **16**, 267–295, 318–324.

Merchant, M. E. (1950). *Metal Cutting Research in Machining Theory and Practice.* ASM, pp. 5–44.

Noh, A., and Shaw, M. C. (1968). Unpublished report. Carnegie Mellon University, Pittsburgh, Pa.

OECD Seminar on Metal Cutting. (1966). Published by OECD, Paris.

Piispanen, V. (1948). *J. Appl. Phys.* **19**, 876–881.

Rice, W. B. (1961). *Eng. J., Eng. Inst. of Canada* **44**, 41.

Rice, W. B., Salmon, R., and Russell, L. T. (1962). *Eng. J. of Canada* **45**, 59.

Rice, W. B., Sharma, C. R., and Salmon, R. (1971). *Annals of CIRP* **19**, 545.

Shaw, M. C. (1980). *Int. J. of Mech. Sc.* **22**, 673–686.

22 SAWTOOTH CHIP FORMATION

Sawtooth chips have been identified as one of the basic types of chips in Chapter 1. In this chapter the mechanics of this type of chip will be considered together with associated characteristics and applications. As in all areas of production engineering, it is important to identify the key element in an operation and to give far less attention to secondary effects that tend to cloud the issue. The important objective of this chapter is to establish the key action involved in sawtooth chip formation, which is fracture, and to then develop the mechanics involved as far as it is possible to proceed at this time. Applications to the hard machining of steel as well as other materials will be considered. The key requirement for sawtooth chip formation is brittle behavior which may have several causes:

- high hardness due to heat treatment
- high strain rate (high speed)
- high degree of strain hardening before machining
- fracture-inducing chemistry or structure

THE SAWTOOTH CHIP

Figure 22.1 shows a comparison of chips produced when turning a titanium alloy and steel under the same conditions (Shaw et al., 1954). This study was made when titanium was first being considered as a structural material because of its high strength-to-weight ratio and resistance to corrosion. Figure 22.1a is a two-dimensional titanium chip produced at relatively high speed and undeformed chip thickness (feed), while Fig. 22.1b shows the same material turned with the same tool at a lower speed and feed. As speed and feed are reduced, chip formation changes from the sawtooth type to the steady state type. Chips of the type shown in Fig. 22.1a are often called segmental chips, but a more descriptive term introduced by Nakayama (1974) that emphasizes the characteristic pointed nature is *sawtooth chip*, and this will be used here. It is also an objective of this chapter to distinguish the sawtooth chip from the entirely different wavy chip or discontinuous chip, all of which are often called segmental chips.

For comparison, AISI 1045 steel chips are shown in Figs. 22.1c and 22.1d when turned under the same conditions as Figs. 22.1a and 22.1b, respectively. The steel chips are seen to be of the steady state type in both cases.

At the time these first sawtooth chips were found, the concept of adiabatic shear (material continuing to shear on an initial plane due to thermal softening when there is not time for the thermal

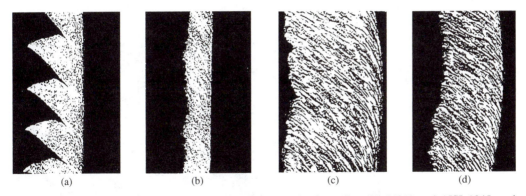

(a)   (b)   (c)   (d)

Fig. 22.1 Comparison of chips produced when turning a titanium alloy (Ti-140A) and AISI 1045 steel with identical carbide tools having a rake angle of +5 [work, cutting speed (*V*), feed (*t*)]: (a) T140A, 150 f.p.m. (45.7 m/min), 0.0104 i.p.r. (0.26 mm/rev); (b) Ti-140A, 100 f.p.m. (30.5 m/min), 0.0052 i.p.r. (0.13 mm/rev); (c) AISI 1045 steel, 150 f.p.m. (45.7 m/min), 0.0104 i.p.r. (0.26 mm/rev); and (d) AISI 1045 steel, 100 f.p.m. (30.5 m/min), 0.0053 i.p.r. (0.13 mm/rev). (after Shaw et al., 1954)

energy generated to dissipate) was in the forefront of ballistic studies of projectile penetration (Zener, 1948). It was suggested by Shaw et al. (1954) that the root cause of sawtooth chip formation might be adiabatic shear, and this view has continued to the present. It is an objective of this chapter to demonstrate that adiabatic shear is not the root basis for sawtooth chip formation and to develop an alternate view that is supported by substantial experimental evidence.

CYCLIC CRACK APPROACH TO SAWTOOTH CHIP FORMATION

In the early 1970s, Nakayama found that sawtooth chips were produced when highly cold-worked (60% reduction) 40/60 (Zn/Cu) brass was turned under orthogonal conditions. Figure 22.2 shows a quick-stop chip produced under the following cutting conditions:

- cutting speed (*V*): 2.95 i.p.m. (0.075 m/min)
- undeformed chip thickness (*t*): 0.0064 in (0.16 mm)
- rake angle (*α*): −15°

Subsequently, Nakayama (1974) suggested a new theory of sawtooth chip formation based on many experiments on highly cold-worked (brittle) brass cut at very low speeds. At point C in Fig. 22.3a metal in the free surface begins to rise and assumes a direction CD parallel to the resultant force (*R*) on the tool face. A shear crack initiates at D and runs from the surface downward along shear plane DO toward the tool point at O. With further advance of the tool (Fig. 22.3b), the chip glides outward like a friction slider along the cracked surface until the next crack forms at D′ (Fig. 22.3c), and a new cycle begins. The principle stresses at D are shown in the inset in Fig. 22.3a. A shear crack will initiate at the free surface at D where the crack arresting normal stress is zero and will proceed downward along the shear plane toward the tool tip (O). Initially the crack will be continuous across the width of the chip (called a gross crack = GC) but will

Fig. 22.2 Sawtooth chip produced when turning highly cold-worked brass at very low speed and with negative rake tool. (after Nakayama, 1972)

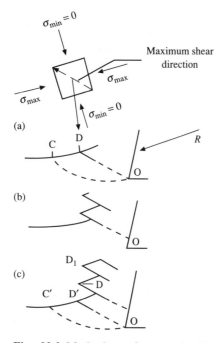

Fig. 22.3 Mechanism of sawtooth chip formation. (after Nakayama, 1974)

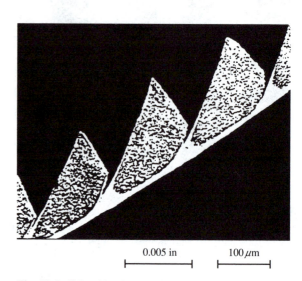

Fig. 22.4 Chip of hard (H_{RC} 62) case carburized steel produced by turning under the following conditions: α, −7 deg; V, 338 f.p.m. (103 m/min); feed and depth of cut, 0.011 i.p.r. (0.28 mm/rev); and nose radius at tool tip, 0.125 in (318 mm). (after Vyas and Shaw, 1999)

become discontinuous as higher crack-arresting normal stresses are encountered. These disconnected localized cracks will be called microcracks (MC), and the microcracked region of the shear plane is represented by dashed lines in Fig. 22.3.

Figure 22.2 is a case where periodic gross cracks extend nearly all the way from the free surface to the tool tip. The depth of "teeth" (the distance each segment slides outward) is controlled by the cracking frequency. The outward glide occurs only until the next crack forms.

Nakayama's suggestion that sawtooth chip formation is initiated by periodic crack formation rather than the periodic formation of adiabatic concentrated shear bands is very important relative to a basic understanding of this process. On the one hand, the root cause of sawtooth chip formation is periodic fracture, while on the other (the adiabatic shear theory), it is of thermal origin. The fact that sawtooth chips form at extremely low cutting speeds (low temperature) where periodic cracks may be readily seen to initiate at the free surface and proceed downward to the tool tip provides strong evidence that periodic fracture is the root cause of sawtooth chip formation and not adiabatic shear. Further evidence will be discussed later.

MECHANICS OF SAWTOOTH CHIP FORMATION

Figure 22.4 shows a photomicrograph of a polished and etched chip of very hard case-carburized steel turned with a negative rake polycrystalline cubic boron nitride (PCBN) tool under hard turning conditions designed to give a surface finish comparable to that obtained in fine grinding.

Figure 22.5a shows the chip of Fig. 22.4 oriented along the negative rake tool face as a free body, and just below, the tool is shown in the process of making a cut. Just as Fig. 22.3a is a snapshot of a sawtooth chip at the instant when a crack forms at D, Fig. 22.5 is a snapshot of a sawtooth chip an instant after crack formation, where the segment just formed has slid outward a small distance DC_1.

Following Nakayama's suggestion, the equal and oppositely directed forces R and R' should be parallel to CD. Forces R and R' are shown resolved parallel and perpendicular to the tool face (F and N, respectively) and parallel and perpendicular to the shear plane (F_S and N_S, respectively). In this instance the tool-face friction force F is small while there is a very significant "secondary" shear zone (the portion of the chip between the bottom of the "teeth" and tool face T). The gross cracked region of the chip (GC) extends from C_2 to D_1 while the microcracked region (MC) extends from D_1 to T. The hodograph for the GC region is given in Fig. 22.5b.

The cutting ratio (r) for this chip may be found by dividing the undeformed chip thickness (t) by the mean chip thickness ($\bar{t}_c$). However, the composite surface $C_1D_1 + C_2D_2 + C_3D_3 +$ etc. was found to correspond to the equivalent length of uncut surface on the work. This was demonstrated by coating the original surface of

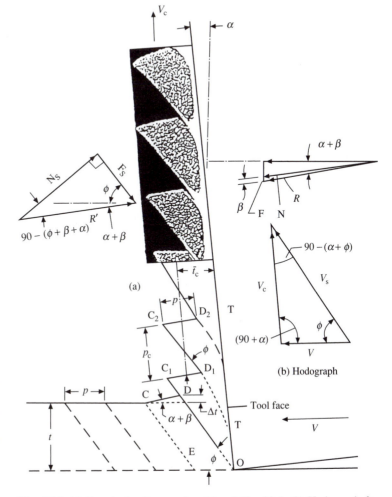

Fig. 22.5 (a) Free body diagram for chip of Fig. 22.4. (b) Hodograph for gross cracked region of chip only. (after Vyas and Shaw, 1999)

the work with soot and then producing a replica of the surface of the chip by pressing a soft plastic material into the back of the chip. Valleys on the chip become peaks on the replica with slopes CD coated with soot. Microscopic measurements on the replica revealed that mean length CD corresponds to the mean distance between cracks on the work (p). Therefore, a convenient method of finding r is to divide the mean tooth pitch (p_c) by the mean value of CD ($= p$). Thus, the chip length ratio is

$$r = p_c/p = V_c/V \qquad (22.1)$$

There is a small tooth-to-tooth variation of p_c and p for the chip of Fig. 22.4. The mean value of p_c/p was found to be

$$\overline{p_c/p} = 1.59 \pm 0.16 \ (1.59 \pm 10\%) = V_c/V$$

The cutting ratio (r) will usually be greater than one when hard steel sawtooth chips are produced by a zero- or negative-rake-angle tool operating at a practical speed. The cutting ratios for the chips of Fig. 22.1 were as follows:

$$\text{Fig. 22.1a: } r = 1.22$$
$$\text{Fig. 22.1b: } r = 0.74$$
$$\text{Fig. 22.1c: } r = 0.49$$
$$\text{Fig. 22.1d: } r = 0.42$$

Only the sawtooth chip had a value of $r > 1$.

SAWTOOTH SPECIFIC ENERGY ANALYSIS

The total specific energy (u) for a sawtooth chip has the following components that are significant:

$$u = u_{GC} + u_{MC} + u_F \tag{22.2}$$

where

u_{GC} = specific energy for the gross cracked region of the chip

u_{MC} = specific energy for the microcracked region of the chip

u_F = specific energy associated with friction between tool and chip

From Fig. 22.5a:

$$u = \frac{R \cos (\alpha + \beta)V}{Vbt} = \left(\frac{R}{bt}\right) \cos (\alpha + \beta) \tag{22.3}$$

$$u_{GC} = \frac{\mu_s N_S V_S}{Vbt} = [\mu_s \cos (\phi + \alpha + \beta)]\left(\frac{V_S}{V}\right)\left(\frac{R'}{bt}\right) \tag{22.4}$$

where μ_s is the coefficient of sliding fiction along the GC surfaces

$$u_F = \left(\frac{R}{bt \sin \beta}\right)\left(\frac{V_c}{V}\right) \tag{22.5}$$

In the above analysis, and that which follows, α is the absolute value of the rake angle. For the chip of Fig. 22.4,

$$\alpha = -7°$$
$$r = 1.59$$
$$\beta \cong 3°$$

From the hodograph for the GC region (Fig. 22.5b),

$$r = \frac{\sin \phi}{\cos (\alpha + \phi)} \tag{22.6}$$

and hence,

$$\phi = 52.9°$$

Also, from Fig. 22.5b,

$$\frac{V_S}{V} = \frac{\cos \alpha}{\cos (\alpha + \phi)} = \frac{\cos 7}{\cos 59.9} = 1.98$$

From Eqs. (22.3), (22.4), and (22.5),

$$\frac{u_{GC}}{u} = \left[\frac{\mu_s \sin (\phi + \alpha + \beta)}{\cos (\alpha + \beta)} \right] \left(\frac{V_S}{V} \right)$$

$$= \left(\frac{\sin 62.9}{\cos 10} \right) (1.98) = 1.79 \mu_s$$

$$\frac{u_F}{u} = \frac{\sin \beta}{\cos (\alpha + \beta) \left(\dfrac{V_c}{V} \right)}$$

$$= \left(\frac{\sin 3}{\cos 10} \right) (1.59) = 0.085$$

The value of u_{MC} cannot be obtained as in the case of u_{GC} because of the complexity of deformation in the MC sector. The value of u_{MC} must therefore be obtained by difference:

$$u_{MC} = u - u_F - u_{GC} \qquad (22.7)$$

The maximum possible value of μ_s will be 0.51, since if μ_s is greater than this u_{MC} would be negative [from Eq. (22.7)], and this is not possible. It is expected that μ_s will be quite high since it involves two identical unlubricated nascent surfaces in sliding contact. A reasonable value for μ_s appears to be 0.4, but this is only an estimate. Assuming μ_s is 0.4, then the energy distribution for the chip of Fig. 22.4 will be approximately as follows:

$$\frac{u_F}{u} = 9\%$$

$$\frac{u_{GC}}{u} = (0.4)(1.79) = 72\%$$

$$\frac{u_{MC}}{u} = 19\%$$

CRACKING FREQUENCY

The cracking frequency (CF) may be found as follows:

$$CF = V_c/p_c = rV/p_c \qquad (22.8)$$

where p_c is the mean measured value from a chip (0.0065 in = 0.165 mm for the chip of Fig. 22.4). Therefore, for the chip of Fig. 22.4,

$$CF = rV/p_c = \frac{[(1.59)(338)(12/60)]}{(0.0065)} = 16,500 \text{ Hz}$$

This approaches the upper limit of the audio-frequency range.

The very inhomogeneous strain in the MC region of Fig. 22.5 will give high temperatures in the bands of concentrated shear along the microcracked extensions of the gross cracks, for high cutting speeds. It is only these bands that involve adiabatic shear, and when present as in Fig. 22.4, they follow in time the formation of gross cracks. For ferrous alloys, the temperature in these concentrated shear bands may exceed the transformation temperature where ferrite (α-iron) changes to austenite (γ-iron). Evidence of this transformation is found in the nonetching white bands in Fig. 22.4 and other similar photomicrographs of sawtooth chips of ferrous alloys machined at high speed. These white areas have been identified as a combination of untempered martensite and austenite, as shown in experiments discussed later in this chapter. However, before rapid cooling and during chip formation, these bands will be γ-iron at high temperature which is a relatively soft material that offers little resistance to plastic deformation. It is the presence of high temperature γ-iron along the tool face that gives such low values of u_F and u_{MC} (estimated) for the chip of Fig. 22.4.

As previously mentioned, the distance one segment slides relative to its neighbor during one cycle (D_1C_2 in Fig. 22.5a) will depend upon pitch (p). When $p_c > p$ ($r > 1$), this is a result of the compressive stress on the material in the MC zone being sufficient to cause elongation of the MC region. Material in the GC region is carried along with the MC material, resulting in r for the entire chip being > 1.

CUTTING FORCES

Cutting forces that fluctuate at a frequency over 10,000 cycles per second are very difficult to measure. This is because the relatively large mass between the transducers and the point of cutting in a conventional piezoelectric dynamometer limits the frequency response to about 1500 Hz in a turning operation. However, an estimate of the percentage change in the cutting forces and the frequency of force fluctuation may be obtained by mounting a very small wire resistance strain gage close to the tool tip and recording the cyclic change in potential across the gage in millivolts (Sampath and Shaw, 1983; Ramaraj et al., 1988; Shaw and Ramaraj, 1989). Figure 22.6a shows the location of a very small strain gage [gage length = 0.03 in (0.75 mm)] mounted close to the cutting edge of the polycrystalline CBN tool used to produce the chip shown in Fig. 22.4, while Fig. 22.6b shows a representative output from this strain gage. The strain gage was connected to a bridge circuit and an Ellis BAM-1 bridge amplifier having a frequency response to 50,000 Hz was used.

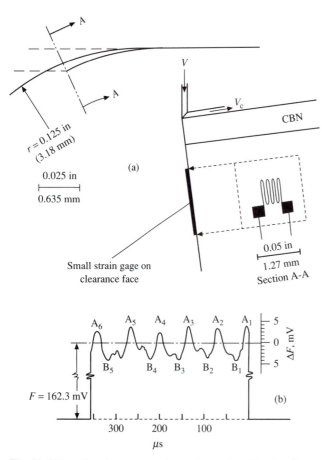

Fig. 22.6 Use of strain gage to estimate fluctuation of cutting forces in sawtooth chip formation. (a) Location of strain gage. (b) Output of strain gage in mV versus time in μs. (after Vyas and Shaw, 1999)

The output from the strain gage consists of a series of peaks (A) and valleys (B) which constitute a series of points of maximum and minimum force, respectively. It is not possible to assign values of force to these points since a single active gage was used to measure a force having two components neither of which is aligned with the strain gage. However, it is reasonable to assume that the fluctuation of forces F_P and F_Q will be proportional to the pattern of Fig. 22.6b. The mean values at the maximum (A) and minimum (B) points in Fig. 22.6b suggest that a range of force fluctuation relative to the mean ($\Delta F/\bar{F}$) is as follows:

$$\frac{\Delta F}{\bar{F}} = 0.06 \text{ (i.e., 6\%)}$$

This is a relatively small fluctuation that is primarily due to the release of stored elastic energy when each crack occurs at the free surface. Points A in Fig. 22.6b should correspond to the force pertaining just before crack initiation. Therefore the mean cracking frequency ($\bar{C}\bar{F}$) should correspond to the reciprocal of the mean time elapsed between points A in Fig. 22.6b.

The time in microseconds (μs) between the several peaks in Fig. 22.6b are as follows:

Peaks	Time (μs)
A_1 to A_2	65
A_2 to A_3	65
A_3 to A_4	63
A_4 to A_5	67
A_5 to A_6	72
	Average = 66.4 μs

Thus,

$$\bar{C}\bar{F} = \frac{1}{66.4 \times 10^{-6}} = 15,000 \text{ Hz}$$

This is in excellent agreement with the value previously obtained (16,500 Hz) considering the different approximations made in the two cases.

DETAILS

The conditions that control sawtooth chip formation include the following:

- properties of work material (primarily hardness and brittleness)
- undeformed chip thickness (t)
- cutting speed (V)
- rake angle (α)

A decrease in the shear fracture strain of a material plays an important role in the tendency for sawtooth chips to form, as does the depth of the GC zone and the spacing of cyclic cracks (p on the work surface or p_c on the chip). Increased brittleness due to chemical composition, heat treatment, strain hardening, or residual stress is important.

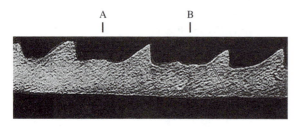

Fig. 22.7 Chip in transition between sawtooth and steady state types. All cutting conditions were identical with Fig. 2.4 except for workpiece hardness which was H_{RC} 48 instead of H_{RC} 62. (after Vyas and Shaw, 1999)

Fig. 22.8 Photomicrograph of chip of case carburized 8620 steel with back surface protected by nonetching white layer of electroless nickel having a hardness of H_{RC} 5. (after Vyas and Shaw, 1999)

The crack spacing (p) is found to increase with an increase in the undeformed chip thickness (t). The depth of gross crack penetration also increases as (t) becomes larger. As (t) is decreased, there is a point below which sawtooth chip formation ceases and steady state chips are formed. Figures 22.1a and 22.1b illustrate this point. For large values of removal rate, gross cracks may penetrate all the way to the tool tip so that there is no MC zone and hence no region of adiabatic shear (as in Fig. 22.2).

The cutting speed has a mixed influence on sawtooth chip formation. An increase in speed gives rise to an increase in strain rate which increases brittleness and hence sawtooth chip formation. On the other hand, an increase in V increases the temperature of the chip which decreases brittleness. The net effect of an increase in V is usually a modest increase in the tendency for sawtooth chip formation.

The tendency toward sawtooth chip formation increases as the rake angle in the direction of the chip flow becomes more negative. The values of p on the work and p_c on the chip are remarkably constant for a given set of machining conditions.

Figure 22.7 is a photomicrograph of a chip of case carburized 8620 steel having a hardness of H_{RC} 48. The cutting conditions for this chip were identical to those of Fig. 22.4 except for the hardness of the work material (H_{RC} 48 versus H_{RC} 62). Figure 22.7 is a chip in transition between steady state and sawtooth chip formation. "Teeth" are missing at points A and B but are present at other points in accordance with the expected cyclic pattern for a sawtooth chip. At points A and B there is evidence of some upward rise, but no evidence of crack initiation. The fact that cracks develop at some points but not at others and for the deviation of the spacing of cracks (p) from point to point on the work in this figure as well as in Fig. 22.4 is due to inhomogeneity of the work material. However, the effect of this lack of homogeneity is a minor one; otherwise, sawtooth chips would not display the remarkable cyclic behavior they do.

Figure 22.8 shows a photomicrograph of a chip of case carburized AISI 8620 steel of H_{RC} 62 hardness turned under the following conditions:

- speed (V): 126 f.p.m. (38 m/min)
- feed (f): 0.005 i.p.r. (0.13 mm/rev)
- depth of cut (b): 0.005 in (0.13 mm)
- rake angle (α): $-7°$

This chip was produced with a tool having a nose radius of 0.125 in (3.2 mm). Two details frequently observed in sawtooth chips are evident in this figure. The geometry at the tips of the "teeth" is variable, and a gap frequently develops at points D_1, D_2, etc. in Fig. 22.5a.

The variable cracking patterns at the tooth tips is due to inhomogeneity of the work material that leads to variation in the manner in which the cracks initiate. For a perfectly homogeneous material, a crack should form in the maximum shear direction at the free surface [i.e., $45 - (\alpha + \beta)$ degrees from the horizontal]. It should then soon shift to the direction of the shear plane ($\phi°$ from the horizontal). Due to inhomogeneity at the surface of the work material, the crack initiates in a variety of ways as evident in Fig. 22.8, before taking the direction of the shear plane.

The gaps that develop at D_1, D_2, etc. (Fig. 22.5a) are due to the elongation of material in the MC region. The mean gap size will increase as the cutting ratio exceeds one. There should be no gaps for a homogeneous material having a cutting ratio of one or less.

Elbestawi et al. (1996) have also adopted the view that the root cause of sawtooth chip formation is cyclic crack formation. They further suggest that defects in the free surface play a role relative to the pitch of crack formation (p). While their defect-related fracture mechanics approach will influence the variable way cracks initiate, as illustrated in Fig. 22.8, it will have little influence on the main direction taken by a crack. This depends on the value of the cutting ratio (r) as given by Eq. (22.6).

Figure 22.9 shows two quick-stop photomicrographs of sawtooth chips in the process of being cut from a Ti–6Al–4V alloy under the following conditions:

- speed (V): 172 f.p.m. (52 m/min)
- feed (f) = (t): 0.007 i.p.r. (0.012 mm/rev)
- depth of cut (b): 0.100 in (2.54 mm)
- rake angle (α): $-7°$
- tool material: tungsten carbide

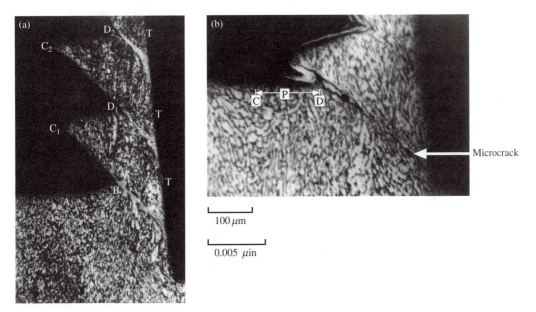

Fig. 22.9 Quick-stop photomicrographs of Ti–6Al–4V alloy (a) halfway between cyclic cracks and (b) shortly after formation of a crack showing the extent of the CG and MC regions. (after Vyas and Shaw, 1999)

Figure 22.9a is for a chip about halfway between cracks. The upward rise begins at A and is about half completed at B. The distance between cracks on the work (p) is approximately C_2D_2. The distance D_2T is less than D_1T and D_1T is less than BT, as it should be since the cutting ratio (r) is greater than 1. The cutting ratio in this case is approximately

$$r = \frac{C_1C_2}{C_2D_2} \cong 1.18$$

Figure 22.9a suggests that the equilibrium chip shape is reached after 3 or 4 "teeth" have passed up the tool face. There is a slight indication of adiabatic shear in the MC region in Fig. 22.9a.

Figure 22.9b shows a quick-stop photomicrograph obtained under the same cutting conditions as Fig. 22.9a. However, this time a crack has just occurred. The tool has advanced a small distance into the work, and the last segment formed has slid a short distance along the shear plane since the last crack occurred. The gross crack appears to have penetrated about one-half the distance along the shear plane in this case. A single microcrack is evident in the shear plane in the MC region.

ADIABATIC SHEAR THEORY

The adiabatic shear theory suggests that the root cause of sawtooth chip formation is a catastrophic thermoplastic instability where the decrease in flow stress due to thermal softening associated with an increase in strain more than offsets the associated strain hardening. Proponents of this theory refer to this as chip segmentation (a term that lumps together wavy, discontinuous, and sawtooth chip formation). The model employed to explain sawtooth chip formation is similar to that used to explain discontinuous chip formation (Fig. 22.10). Here it is assumed that for some reason a thermoplastic instability occurs along dashed line (5) extending from the tool tip curving upward toward the free surface. As the tool advances, the material above (5) is extruded into the shape (2, 3, 4, 5), line (5) becoming surfaces (3) and (4). At the same time, inclined surface (1) becomes surface (2). The process then starts over again. It is stated that adiabatic shear surface (5) rolls down on the tool face and remains static there until the next cycle begins, a stick-slip motion up the tool surface being postulated. According to this model the final chip shape is completed after one cycle.

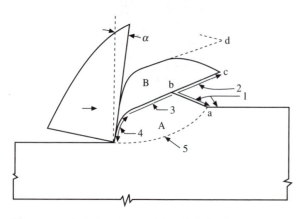

Fig. 22.10 Model used to explain adiabatic shear theory of sawtooth chip formation. (after Recht, 1985)

COMPARISON OF THE TWO THEORIES

The quick-stop photomicrographs of Fig. 22.9 are useful in comparing the two theories. First of all, a thermally initiated process should be expected to have its origin where the temperature is a maximum which is at the tool tip. This is in agreement with the model of Fig. 22.10 but not with reality (Fig. 22.9). The crack in Fig. 22.9b clearly runs from the free surface toward the tool tip along a relatively straight shear plane. A shear crack should be expected to initiate near a point of maximum shear stress where the compressive stress is a minimum (i.e., at the free surface). Finite element analysis of stresses along the shear plane in steady state orthogonal cutting (assuming a von Mises constitutive relation as a first approximation) has revealed a substantial increase in normal stress in progressing along the shear plane from the free surface to the tool tip (Sampath and Shaw, 1983). As higher normal stresses are encountered as a shear crack progresses downward from the free surface toward the tool tip, a continuous gross crack may gradually be converted into a discontinuous microcracked region. This is seen to be the case in Fig. 22.9. It should be noted that a titanium alloy was employed in Fig. 22.9 which should favor the adiabatic shear theory due to its low thermal conductivity and specific heat.

There is no evident reason why the material should extrude into a sharp point in the model of Fig. 22.10. The reason for the sharp points actually observed in sawtooth chips is apparent in the crack initiation theory. The fact that workpiece hardness (brittleness) is so important relative to the onset of sawtooth chip formation (Fig. 22.7) further supports the crack initiation theory.

The adiabatic shear model of Fig. 22.10 shows concentrated shear bands that are parallel to the tool face initiating at the tool tip. The photomicrograph of Fig. 22.9b shows an initially straight shear plane that gradually bends downwards, becoming parallel to the tool face only after the chip has moved a considerable distance up the tool face.

A number of papers suggest adiabatic shear as the origin of sawtooth chip formation. A few of these are Recht (1964, 1985), Komanduri et al. (1984), Koenig et al. (1984), Zhen-Bin and Komanduri (1995), Sheikh-Ahmed and Bailey (1997), and Davies et al. (1996, 1997). The most recently proposed model based on the adiabatic shear theory is given in Fig. 22.11. In this model the adiabatic shear band runs from the tool tip A in Fig. 22.11a along a straight line to the free surface. As the chip moves forward, the concentrated shear band just formed rolls down onto the tool face as block (1) glides along two adjacent shear bands. While this model explains the sharp point, it is not in agreement with Fig. 22.9b in that the adiabatic shear band shown from C to D in Fig. 22.11b is not found experimentally in Fig. 22.4 or Fig. 22.8.

Even when surfaces CB in Fig. 22.11 are carefully protected with a hard material to prevent alteration or loss during polishing, no evidence of adiabatic has been found on such surfaces with an electron microscope at very high magnification. While there is no apparent reason the concentrated shear bands bend down and approach the tool face in the model of Fig. 22.11, the reason for this is apparent in the shear crack theory.

The fact that workpiece hardness (brittleness) is so important relative to the onset of sawtooth chip formation further supports the crack initiation theory.

It must be concluded that the root cause of sawtooth chip formation is periodic fracture at the free surface of the work and not adiabatic shear as was originally suggested as a possibility (Shaw et al., 1954). When adiabatic shear is involved, it occurs in the microcracked region that is an extension of a gross crack region. Thus, the onset of sawtooth chip formation cannot be predicted by thermal analysis alone as has been assumed so many times in the past. A very complex combination of thermal and fracture analysis is involved, and hence prediction is not a possibility at this time. However, the next best thing to prediction is understanding.

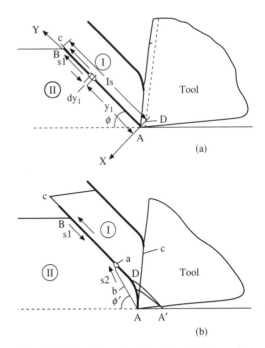

Fig. 22.11 Model used by Zhen-Bin and Komanduri for thermal analysis based on adiabatic shear theory. (after Zhen-Bin and Komanduri, 1995)

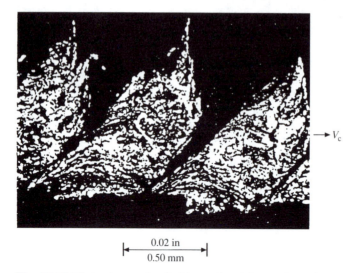

$\longrightarrow V_c$

$$\vdash\!\!\!-\!\!\!-\ \underset{0.50\ mm}{0.02\ in}\ -\!\!\!-\!\!\!\dashv$$

Fig. 22.12 Photomicrograph of chip produced in rough cutting a moderately hard rolling mill roll. (after Binns, 1961)

In looking back over the route taken to the present in the development of sawtooth chip mechanics, the following quotation of the philosopher of science Karl Popper comes to mind: "We hate the very idea that we may be mistaken and we cling dogmatically to our conjectures. However, it is actually good to discover that your ideas are false because once you eliminate false ideas, what's left may be the truth."

Hard Turning of Steel

At present the most important application of sawtooth chip formation is the hard turning of steel. With the application of superhard cutting tool materials (ceramics and PCBN), it is possible to finish-turn work materials such as hardened alloy steel and case carburized steel after heat treatment rather than by grinding.

The hard turning of steel appears to have started in the 1950s when J. Binns found that difficult-to-machine hard steel mill rolls could be finish-machined at unusually high removal rates using ceramic tools, provided the entire system was extremely rigid and high horsepower was available. Binns (1961) gives a number of examples, and Fig. 22.12 shows a photomicrograph of a chip produced under the following conditions:

- work material: high-tensile-strength cast steel
- hardness: H_{RC} 36, H_B 330
- tool: Ceramic
- rake angle: $-5°$
- lead angle: $60°$
- speed (V); 190 f.p.m. (60 m/min)
- feed (f): 0.035 i.p.r. (0.9 mm/rev)
- depth of cut: 1 in (25.4 mm)
- undeformed chip thickness: 0.0175 in (0.44 mm)
- width of cut: 2 in (50.8 mm)

This is clearly a sawtooth chip having a velocity ratio of about 1.3 ($\phi = 49°$) and a friction angle (β) of about $11°$ ($\mu_F = 0.19$) Gross cracks penetrate all the way to the tool face and there is no evidence of any adiabatic shear. It was found that when cutting under these conditions, the power consumed was only about $\frac{1}{3}$ that required when noncyclic flow type chips were produced. The importance of this is that it enables a higher removal rate for the same power available. However, this requires a system of unusual stiffness since otherwise the fluctuating force associated with sawtooth chip formation may cause chatter and hence poor finish. The cracking frequency for the chip of Fig. 22.12 will be approximately as follows [Eq. (22.8)]:

$$CF = \frac{rV}{p_c} = \frac{(1.3)(190)(12/60) \text{ i.p.s.}}{0.0225 \text{ in}} = 2200 \text{ Hz}$$

This is within the audible range and could excite vibration, and hence a very rigid system is essential.

Binns (1961) has taken advantage of the sound emitted to adjust the speed and feed for maximum removal rate for a given power. For cuts such as that of Fig. 22.12, he found that a crisp sizzling sound is preferable to a lower frequency growl. Binn's machines were constructed with a base of wrought steel, a minimum number of parts (joints) and gears, and drive motors of up to 500 hp (375 kW). Two rigidity tests were employed:

(a) the ability to stop a heavy cut in progress and to restart without tool breakage or a surface defect

(b) the ability of sawtooth chips to form without segments flying off

Binns (1961) has also found that there is an optimum combination of speed and feed that gives the maximum removal rate for a given situation. For example, when machining a roll material having a hardness H_{RC} 49 (H_B 460) with a 1.25 in (32 mm) diameter ceramic insert, the optimum combination of speed and feed was 200 f.p.m. (61 m/min) and 0.035 i.p.r. (0.9 m/rev), respectively, with a depth of cut of 0.20 in (5.1 mm). The tool life under these conditions was 120 min. When operating at the optimum removal rate, tool life will usually be short enough that the downtime to change tools is very important. Binns has solved this problem by designing tool changing arrangements that require downtime for a tool change that is a fraction of a minute.

The tool-face side of the chip of Fig. 22.12 is unusually rough. This is because there is no MC region in this case. The upward glide of each section between cracks produces a stepped lower surface on the chip. If there is not sufficient normal stress between chip and tool, only the high points on the stepped surface of the chip will be burnished.

It should be noted, however, that the roughness on the back of the chip is not indicative of the finish produced on the work. The upward glide of adjacent blocks that occur in the chip will have little influence on the surface generated at the tool point. In addition, it is necessary that a land be provided at the tool tip to prevent chipping of the cutting edge, and this provides additional burnishing action on the finished surface to improve finish. Binns reports that surface roughness values as low as $R_a = 50$ μin (1.25 μm) may be obtained when rough turning steel mill rolls.

Use of a large diameter ceramic cylindrical tool, inclined to provide a negative rake angle, has been found to be an effective way of cutting under conditions to give sawtooth chips in roll turning. This enables a high removal rate as well as a good surface finish. In one example (Binns, 1984), a 2 in diameter cylinder was used to cut a cast steel roll (H_{RC} 32, H_B 302) at a depth of cut of 0.5 in (12.7 mm) and a feed (f) of 0.031 i.p.r. (0.79 mm/rev) at 500 f.p.m. (152 m/min). In this case the theoretical peak-to-valley roughness (R_{th}) for this very high removal rate will be only

$$R_{th} = \frac{f^2}{8r} = \frac{(0.031)^2}{8(1)} = 120 \text{ } \mu\text{in (3 } \mu\text{m)}$$

There also appears to be great promise for machining hardened steel parts (rolling contact bearing components, gears, cams, etc.) using cubic boron nitride tools operating under sawtooth chip forming conditions. If the speed and feed are sufficiently high and the nose radius of the tool (r) is also sufficiently high, sawtooth chips having concentrated shear bands of γ-iron will be obtained.

This allows a relatively high removal rate to be achieved at a relatively low power level as well as a finish that does not require further improvement by a costly final grinding operation.

Other early hard turning references include Herzog (1975), Hodgson and Trendler (1981), Lindberg and Lindstrom (1983), Matsumoto et al. (1984), Schwarzhofer and Kaelin (1986), and Koenig et al. (1984).

Lindberg and Lindstrom (1983) studied sawtooth chip formation of AISI 1035 steel and found that sawtooth chips were not formed even at very high speeds if the undeformed chip thickness (feed) had a low value. For example, sawtooth chips were formed at a frequency of about 14,000 Hz at a cutting speed of 490 f.p.m. (150 m/min), a feed of 0.012 i.p.r. (0.315 mm/rev), and a depth of cut of 0.080 in (2 mm); but continuous chips were formed at the same speed and depth of cut when the feed was reduced to 0.004 i.p.r. (0.100 mm/rev). Since the highest natural frequency of any of the components of the tool–work–machine system is only about 1000 Hz, it follows that the stiffness of the system should have no influence on the frequency of segment formation. This has been found to be so for all cases of sawtooth chip formation (Komanduri et al., 1984).

In hard turning of steel, the objective is to remove the required layer of hard material as rapidly as possible. It has been found advantageous to use a tool having a large nose radius (r) that is an order of magnitude larger than the feed, and a depth of cut about equal to the feed.

If all cutting is on the nose radius, the depth of the feed marks remaining on the finished surface will be small and independent of the depth of cut.

Several papers concerning the hard turning of steel have been presented by Kishawy and Elbestawi (1997a, 1997b, 1999, 2001). In the first three of these studies, they used ceramic inserts (70% Al_2O_3 to 30% TiC) to turn AISI 550 case carburized steels having a hardness of H_{RC} 60 under a wide range of cutting conditions (speeds, feeds, depths of cut). Sawtooth chips were obtained in all cases except when the feed was less than the nose radius. For the fourth study, they employed PCBN inserts having different edge preparation and cutting D-2 tool steel. Based on all these studies it was concluded that the basic origin of sawtooth chip formation was due to the periodic formation of cracks at the free surface of the chip, in agreement with the results of Vyas and Shaw (1997, 1999, 2000).

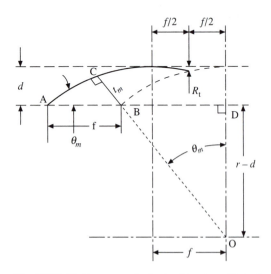

Fig. 22.13 Cutting geometry for hard turning with tool having a relatively large nose radius (not to scale).

SURFACE FINISH

Figure 22.13 is a typical hard turning arrangement where a nose radius (r_o) of 0.118 in (3 mm) is making a cut at a feed rate of 0.005 i.p.r. (125 μm/rev). The scallop left behind on the finished surface will give a theoretical peak-to-valley roughness (R_t) of $f^2/8r$ (independent of depth of cut). To a good approximation, the theoretical arithmetic average roughness (R_{th}) will be $f^2/32r_o$ [Eq. (17.2)].

For this example,

$$R_a = (125 \times 10^{-6})^2[(32)(0.003)] = 6.4 \ \mu\text{in} \ (0.16 \ \mu\text{m})$$

The following relationships represent good approximations for the area of cut (A_{cut}), the width of cut (b), and the mean undeformed chip thickness ($\bar{t}$). These quantities may be obtained by reference to Fig. 22.13.

$$A_{cut} = fd \qquad (22.9)$$

$$b = r\theta_m + \frac{f}{2} \ (\theta_m \text{ in radians}) \tag{22.10}$$

$$\frac{\cos \theta_m \ (r - d)}{r_\mathrm{o} - f \sin \theta} \tag{22.11}$$

$$\bar{t} = \frac{A_{\mathrm{cut}}}{b} \tag{22.12}$$

For the above example (with $d = f$),

$$A_{\mathrm{cut}} = fd = 24.18 \ \mu\mathrm{in}^2 \ (0.0156 \ \mathrm{mm}^2)$$
$$\theta_m = 20°, \text{ from Eq. (22.11) by iteration}$$
$$b = 0.044 \ \mathrm{in} \ (1.11 \ \mathrm{mm})$$
$$\bar{t} = 560 \ \mu\mathrm{in} \ (14 \ \mu\mathrm{m})$$

The quantity that determines whether plane strain pertains is $b/\bar{t}$, and this should be greater than about 10, for plane strain to hold. For this example, $b/\bar{t} = 79$. Hence plane strain exists even though $d \cong f$.

A series of tests was performed on case carburized steels of different hardnesses (H_{RC}) at different cutting speeds (V) and different values of f and d using a PCBN tool having a nose radius (r) of 0.118 in (3 mm). These results are given in Table 22.1 where

F_P = power component of cutting force perpendicular to the paper in Fig. 22.13

u = specific cutting energy = $F_\mathrm{P}/A_{\mathrm{cut}}$ in lb in^{-3} (Nm m^{-3}) (22.13)

$h.p.$ = horsepower at cutting tool = $F_\mathrm{P}V/(33,000)$ (22.14)

The ratio of $b/\bar{t}$ in the tests of Table 22.1 has a minimum value of 18. This means that plane strain will pertain. The values of $\bar{t}$ in Table 22.1 range from 0.0002 to 0.0017 in (5 to 42 μm), which are lower than values pertaining in ordinary finish turning.

The mean values of specific energy (u) for the tests of Table 22.1 are shown plotted in Fig. 22.14 versus the mean undeformed chip thickness together with bars that show the range of variation of u for rather wide ranges of cutting speed or workpiece hardness. The mean values of u are seen to decrease with increase in mean values of $\bar{t}$ as might be expected from conventional turning experience but only to a value of $\bar{t}$ of about 10^{-3} in. It is not clear why $\bar{u}$ increases above this value.

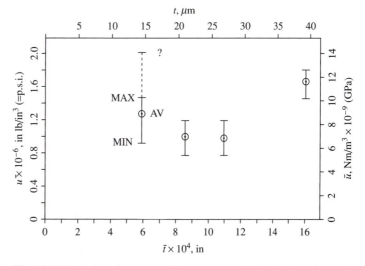

Fig. 22.14 Variation of mean specific energy required for hard turning steel at different speeds, feed, depths of cut, and workpiece hardness using a PCBN insert having a nose radius of 0.118 in (3 mm). (after Shaw and Vyas, 1998)

TABLE 22.1 Representative Hard Turning Results[†]

(a) $f = 0.005$ i.p.r. (0.127 mm/rev); $d = 0.005$ in (0.127 mm); $A_{cut} = 25 \times 10^{-6}$ in^2 (0.016 mm^2); $\theta_m = 15°$; $b = 0.028$ in (0.711 mm); $\bar{t} = 880$ μin (22 μm) $R_t = 26$ μin (0.66 μm); $R_a = 6.6$ μin (0.17 μm); $b/\bar{t} = 32$

V, f.p.m.	H_{RC}	F_P, lb	u, p.s.i.	h.p.
120	62	37.0	1.48×10^6	0.135
120	52	35.5	1.42×10^6	0.129
120	44	32.5	1.30×10^6	0.118
120	28	27.0	1.08×10^6	0.098
255	62	32.0	1.28×10^6	0.247
255	52	30.2	2.08×10^6	0.236
255	44	28.0	1.12×10^6	0.216
255	28	29.0	1.16×10^6	0.224
515	62	33.0	1.32×10^6	0.515
515	52	26.0	1.04×10^6	0.406
515	44	27.0	1.08×10^6	0.421
515	28	24.0	0.96×10^6	0.375
			Average: 1.28×10^6	

(b) $f = 0.005$ i.p.r. (0.127 mm/rev); $d = 0.010$ in (0.254 mm); $A_{cut} = 50 \times 10^{-6}$ in^2 (0.032 mm^2); $\theta_m = 21°$; $b = 0.047$ in (1.19 mm); $\bar{t} = 1064$ μin (27 μm) $R_t = 26$ μin (0.66 μm); $R_a = 6.6$ μin (0.17 μm); $b/\bar{t} = 44$

V, f.p.m.	H_{RC}	F_P, lb	u, p.s.i.	h.p.
120	62	58.0	1.16×10^6	0.211
120	52	51.0	1.02×10^6	0.185
120	44	54.5	1.09×10^6	0.198
120	28	54.5	1.09×10^6	0.198
255	62	54.0	1.08×10^6	0.417
255	52	47.0	0.94×10^6	0.036
255	44	46.0	0.92×10^6	0.355
255	28	47.5	0.95×10^6	0.367
515	62	50.5	1.01×10^6	0.788
515	52	44.0	0.88×10^6	0.687
515	44	47.0	0.94×10^6	0.733
515	28	43.0	0.86×10^6	0.671
			Average: 1.000×10^6	

TABLE 22.1   *(Cont'd)*

(c) f = 0.010 i.p.r. (0.254 mm/rev); d = 0.005 in (0.127 mm); A_{cut} = 50 × 10⁻⁶ in² (0.032 mm²); θ_m = 12°; b = 0.030 in (0.754 mm); $\bar{t}$ = 1684 μin (42 μm); R_t = 105 μin (2.6 μm); R_a = 26 μin (0.66 μm); $b/\bar{t}$ = 18

V, f.p.m.	H_{RC}	F_P, lb	u, p.s.i.	h.p.
120	62	60.0	1.20×10^6	0.218
120	52	52.0	1.04×10^6	0.189
120	44	50.0	1.00×10^6	0.182
120	28	50.0	1.00×10^6	0.182
255	62	57.5	1.15×10^6	0.494
255	52	48.5	0.97×10^6	0.375
255	44	47.0	0.94×10^6	0.363
255	28	50.0	1.00×10^6	0.386
515	62	51.0	1.02×10^6	0.796
515	52	42.0	0.84×10^6	0.655
515	44	50.0	1.00×10^6	0.780
515	28	39.5	0.79×10^6	0.616
			Average: 1.000×10^6	

(d) f = 0.010 i.p.r. (0.254 mm/rev); d = 0.010 in (0.254 mm); A_{cut} = 100 × 10⁻⁶ in² (0.064 mm²); θ_m = 20°; b = 0.046 in (1.168 mm); $\bar{t}$ = 217 μin (5.4 μm) R_t = 106 μin (2.65 μm); R_a = 26 μin (0.66 μm); $b/\bar{t}$ = 212

V, f.p.m.	H_{RC}	F_P, lb	u, p.s.i.	h.p.
120	62	89.0	1.78×10^6	0.324
120	52	86.5	1.73×10^6	0.315
120	44	92.0	1.84×10^6	0.335
120	28	90.0	1.80×10^6	0.327
255	62	82.0	1.64×10^6	0.634
255	52	81.0	1.62×10^6	0.626
255	44	82.0	1.64×10^6	0.637
255	28	81.0	1.62×10^6	0.626
515	62	76.0	1.52×10^6	1.186
515	52	66.5	1.33×10^6	1.038
515	44	76.0	1.52×10^6	1.186
515	28	73.0	1.46×10^6	1.139
			Average: 1.63×10^6	

[†] After Shaw and Vyas, 1998.

The theoretical values of surface finish (R_a) of 6.6 and 2.6 μin (0.17 and 0.65 μm) for the tests of Table 22.1 are within the range of those for fine grinding (8 to 16 μin = 0.20 to 0.40 μm).

For purposes of estimating the required horsepower for hard turning steel of any hardness, the specific energy needed may be found approximately from Fig. 22.14 and the *hp* obtained from the following equation:

$$hp = \frac{ufdV}{33,000} \tag{22.15}$$

It has been demonstrated how the main chip-forming characteristics of a material may be readily identified by chips produced under a range of cutting conditions. These are very simple procedures that do not involve elaborate instrumentation or complicated techniques. Before attempting to generalize or extrapolate cutting results or to perform a statistically designed experiment, it is very important to be sure that chip formation is not shifting from one type to another within the range investigated. Otherwise, useful results are very unlikely due to the proverbial confusion associated with the failure to distinguish between apples and oranges.

A recommended procedure for setting up a hard turning operation follows:

1. Determine the minimum depth of cut required to remove all defects and provide the geometry required.
2. Determine the maximum nose radius that is available and that will give the required shape.
3. Determine the maximum roughness (R_a) that is acceptable.
4. Determine the maximum feed (f) to give R_a.
5. Select the highest cutting speed (V) consistent with the machine and an acceptable tool life.
6. The horsepower required at the tool may be estimated from Eq. (22.15) for the hard turning of steel.

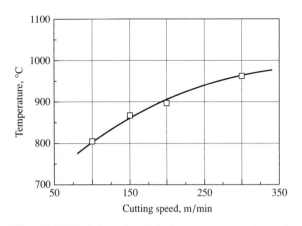

Fig. 22.15 Variation of tool-flank temperature at the tool tip for hard (H_{RC} 60) AISI 52100 ball bearing steel with cutting speed. Feed, 0.004 i.p.r. (0.10 mm/rev); depth of cut, 0.004 in (0.10 mm); rake angle, 5°; tool material, 60 $^v/_o$ CBN in TiN binder. (after Ueda et al., 1999)

Ueda et al. (1999) measured the temperature on the flank face at the tip of a CBN tool using a two-color pyrometer with a fused coupler for several different work materials (see Ueda et al., 1995, for details). The highest temperature was obtained for a hardened ball bearing steel (AISI 52100) and this varied appreciably with cutting speed (Fig. 22.15). The temperature was also quite sensitive to work-piece hardness but far less sensitive to feed and depth of cut than cutting speed or hardness.

It has been established that subsurface residual compressive stress increases the fatigue limit of high-strength steels loaded in tension (Field and Kahles, 1971). Scott et al. (1962) showed that compressive residual stress produced by grinding improves the fatigue life of ball bearings particularly when the depth of the residual stress is increased. Matsumoto et al. (1999) have compared the life of rolling contact bearings finished by hard turning and grinding. They performed fatigue tests on taper roller

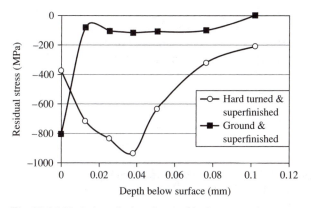

Fig. 22.16 Variation of subsurface residual compressive stress with depth below the surface for hard turned and ground surfaces both superfinished before testing. (after Matsumoto et al., 1999)

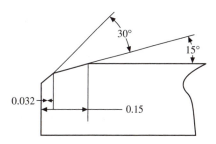

Fig. 22.17 Double chamfer cutting edge treatment used on hard turning tests by Matsumoto et al. (1999).

bearings of case carburized steel (carbon content at surface = 0.8%) with hardness in the range of H_{RC} 58 to H_{RC} 62. The inner ring races were ground or hard turned on a high-precision lathe and then superfinished to a finish of $R_a = 0.1$ μm. Both large (17.7 in o.d. = 450 mm) and small (4.8 in o.d. = 120 mm) bearings were tested. The fatigue life of the hard turned and superfinished bearings was found to be somewhat better than the ground and superfinished ones, particularly for the small bearings. Only about 1 μm was removed in the superfinishing operation and the resulting finish was $R_a = 0.1$ μm.

Figure 22.16 compares circumferential residual stress profiles in the circumferential direction for the two methods of finishing. The hard turned specimens showed a compressive stress pattern extending further beneath the surface, and this was found to be responsible for the better performance. Superfinishing was found to alter the residual stress only near the surface, and this was not found to be as important as the depth of the compressive stress with regard to fatigue life. Further study

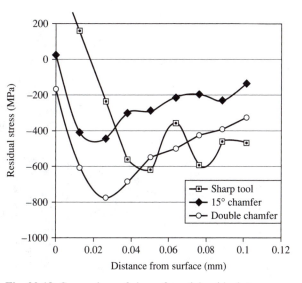

Fig. 22.18 Comparison of circumferential residual stress profiles for a sharp tool, a tool with single 15° chamfer, and a tool with the double chamfer shown in Fig. 22.17. (after Matsumoto et al., 1999)

revealed that a tool having a double chamfer as shown in Fig. 22.17 gave a deeper, more desirable residual stress pattern than a sharp tool or one having only a 15° chamfer (Fig. 22.18). A honed cutting edge (0.02 mm in extent) gave essentially the same result as one ground in accordance with Fig. 22.17 but was more difficult to reproduce accurately. Since hard turning is a finishing operation, attention must be paid to surface integrity which has been discussed in Chapter 17.

MILLING HARD CASE CARBURIZED STEEL

When case carburized AISI 8620 steel [H_{RC} 61 with 0.050 in (1.25 mm) case depth] is subjected to a plane-milling operation under the following conditions, sawtoothed chips very similar to those in Fig. 22.4 are obtained:

- cutting speed: 500 f.p.m. (152.61 m/min)
- depth of cut: 0.010 in (0.25 mm)
- feed: 0.010 i.p.r. (0.25 mm/rev)
- rake angle: −7°
- tool: five PCBN inserts, each 0.500 × 0.188 in (12.5 × 480 mm); nose radius, 0.031 in (0.78 mm); cutter diameter, 3 in (76 mm); 80° × 100° diamond-shaped inserts (100° corner used)
- cutting fluid: air

Figure 22.19 shows a typical sawtooth chip produced under the above conditions. The white layer in the chips was found to be without structure in SEM micrographs at 3500×.

The analysis for milling chips of this type should be the same as for previously discussed hard turning chips such as those in Fig. 22.4.

TITANIUM AND STAINLESS STEEL

Titanium alloys such as Ti–6Al–4V and stainless steel (303S) form cyclic sawtooth chips under many conditions. Chips produced by the quasi-orthogonal turning operation (Fig. 22.20) under a wide range of cutting conditions for these two materials were collected and analyzed.

Cutting results are given in Tables 22.2 and 22.3. All of the chips were cyclic sawtooth types having a spacing p_c of fracture planes in the chip.

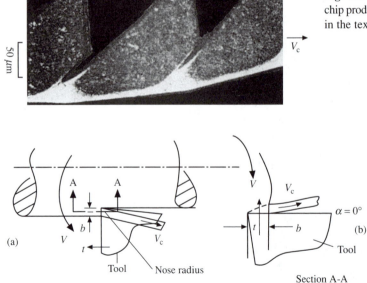

Fig. 22.19 Optical photomicrograph of sawtooth chip produced by face milling under conditions given in the text. (after Shaw and Vyas, 1993)

Fig. 22.20 Relation of (a) quasi-orthogonal cutting to (b) two-dimensional cutting model.

TABLE 22.2 Representative Cutting Data for Ti6Al4V[†]

Speed V, f.p.m.	Feed t, i.p.r.	Rake Angle α, deg	Fracture Pitch on Chip, p_c, in	Specific Energy u, p.s.i.	Citting Ratio, r
50	0.015	+6		242,200	1.10
145	0.015	+6	0.012	213,300	1.21
205	0.015	+6	0.009	195,500	1.11
295	0.015	+6	0.013	186,600	1.23
415	0.015	+6	0.010	151,100	1.41
50	0.015	−5	0.013	266,700	1.90
145	0.015	−5	0.011	224,400	1.44
205	0.015	−5	0.013	222,300	1.27
295	0.015	−5	0.014	222,200	1.45
415	0.015	−5	0.014	168,900	1.52
145	0.005	+6	0.004	280,000	1.18
145	0.011	+6	0.008	200,000	1.13
145	0.015	+6	0.012	213,300	1.24
145	0.022	+6	0.015	212,100	1.24
145	0.033	+6	0.018	149,500	1.81
145	0.005	−5	0.008	333,300	1.06
145	0.011	−5	0.009	200,000	1.17
145	0.015	−5	0.011	224,400	1.44
145	0.022	−5	0.017	174,200	1.68
145	0.033	−5	0.022	197,000	1.68

[†] Tool, steel cutting grade tungsten carbide; side cutting edge angle, 0°; nose radius, 0.031 in; depth of cut, 0.15 in. (after Shaw et al., 1991)

TABLE 22.3 Representative Cutting Data for 303S Stainless Steel[†]

Speed V, f.p.m.	Feed t, i.p.r.	Rake Angle α, deg	Fracture Pitch on Chip, p_c, in	Specific Energy u, p.s.i.	Cutting Ratio, r
50	0.015	+6	0.025	266,700	0.61
145	0.015	+6	0.014	266,700	0.91
205	0.015	+6	0.013	280,000	0.92
295	0.015	+6	0.018	280,000	0.74
415	0.015	+6	0.015	280,000	0.88
50	0.015	−5	0.015	311,100	0.80
145	0.015	−5	0.020	364,400	0.89
205	0.015	−5	0.013	328,900	0.88
295	0.015	−5	0.025	328,900	0.85
415	0.015	−5	0.016	328,900	0.88
205	0.005	+6	0.008	600,000	0.59
205	0.011	+6	0.011	351,500	0.87
205	0.015	+6	0.013	280,000	0.92
205	0.022	+6	0.021	260,600	0.86
205	0.033	+6	0.027	214,100	0.72
205	0.005	−5	0.010	520,000	0.62
205	0.011	−5	0.011	351,500	0.89
205	0.015	−5	0.013	328,900	0.88
205	0.022	−5	0.013	272,700	0.97
205	0.033	−5	0.023	230,300	0.97

[†] Tool, steel cutting grade tungsten carbide; side cutting edge angle, 0°; nose radius, 0.031 in; depth of cut, 0.15 in. (after Shaw et al., 1991)

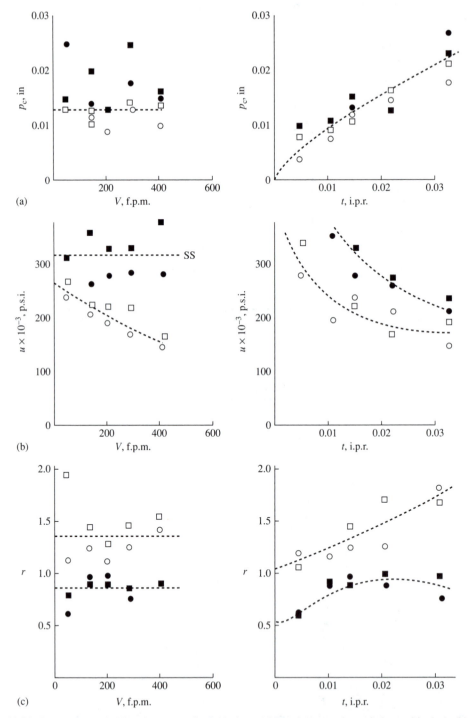

Fig. 22.21 Constant speed (V) and constant feed (t) plots. (a) For fracture plane pitch on chip (p_c). (b) For specific cutting energy (u). (c) For cutting ratio (r). Symbol, material, α: ○, Ti, +6°; □, Ti, −5°; ●, SS, +6°; ■, SS, −5°. (after Shaw et al., 1991)

Figure 22.21 shows data from Tables 22.2 and 22.3 plotted versus speed (V) and feed (undeformed chip thickness, t). The spacing of fracture planes is essentially independent of speed (V) but increases approximately linearly with increase in undeformed chip thickness (t) for both materials. The approximately linear variation of p_c with t (Fig. 22.21a) appears to be associated with a geometrically similar model for all feeds with t acting as a scaling factor as indicated in Fig. 22.22.

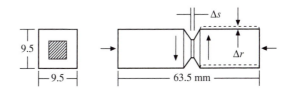

Fig. 22.22 Two-dimensional chips showing relation between fracture pitch on chip (p_c) and feed (t) for geometrically similar formation of chips. (after Shaw et al., 1991)

Figure 22.21b shows the variation of specific cutting energy (u) with V and t. For the stainless steel, u is roughly independent of V but decreases with an increase in t. For the titanium alloy, u decreases with an increase of V apparently due to thermal softening. It is well established that u increases exponentially with decrease in t (the size effect), and this is seen to be the case for both stainless steel and the titanium alloy.

Figure 22.21c shows the variation in cutting ratio (r) with V and t. For the titanium alloy, the

Fig. 22.23 Plane strain shear-compression test ($\Delta x = 0.010$ in).

cutting ratio is essentially independent of V but increases with increase of t having values greater than one in all cases. For 303S stainless steel, the cutting ratio is independent of V but increases with t up to a value of 0.02 i.p.r. (0.15 mm/rev) but having values less than one in all cases.

Simple plane strain-shear tests were also performed on each of the materials to total fracture with different values of axial compressive stress on the shear plane (σ_s). The specimen and loading involved are shown in Fig. 22.23. Details of the test procedure are given in Walker and Shaw (1969).

Figure 22.24a shows plots of shear stress (τ) versus shear strain (γ) for different levels of compressive stress on the shear plane (σ_s) for the Ti–6Al–4V alloy. Figure 22.24b gives similar results

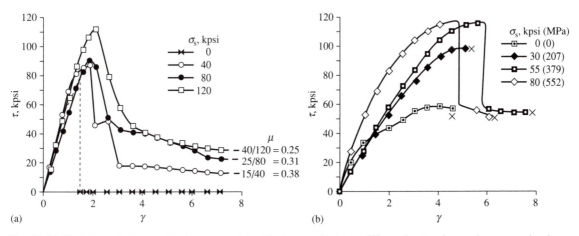

Fig. 22.24 Variation of plane strain shear stress (τ) with shear strain (γ) at different levels of normal stress on the shear plane (σ_s). (a) Ti–6Al–4V alloy. (b) 303S stainless steel. (after Shaw et al., 1991)

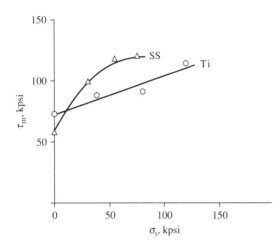

Fig. 22.25 Variation of maximum shear stress at gross fracture (τ_m) versus normal stress on the shear plane in plane strain shear-compression test of Fig. 23.23 for 303S stainless steel and Ti–6Al–4V. (after Shaw et al., 1991)

for the 303S stainless steel. Gross fracture occurs at the point of maximum shear stress (τ_m) or soon thereafter.

Total fracture indicated by x in Fig. 22.24b occurs as soon as a critical shear stress (τ_c) is reached in the case of 303S stainless steel but does not occur with the titanium alloy (Fig. 22.24a). Instead, shear resistance due to adhesive friction persists to very large strains (τ) after gross cracking occurs at (τ_m). The shear stress before gross fracture (τ_m) is seen to increase with increased stress (σ_s) in Fig. 22.25.

There is no white layer with sawtooth chip formation for titanium and stainless steel, since there is no phase transformation. However, bands of very high concentrated strain extending down from periodic gross cracks play the same role as the white layers obtained with hardened steel. In both cases concentrated shear bands begin as a straight shear plane and bend down becoming parallel to the tool face as the chip proceeds up the face of the tool. In the case of most alloy steels and titanium alloys, the concentrated shear bands are extended in compression giving rise to an increase in chip speed as the chip proceeds up the tool face. This gives rise to a cutting ratio (r) greater than one in the case of hard alloy steels and titanium alloys but not for 303S stainless steel (Fig. 22.21c).

WHITE LAYER

Non-etching white regions are usually associated with the hard turning of steel (Fig. 22.4). There are two regions of potential white layer interest:

(**a**) in the chip

(**b**) on the finished surface

Most studies of the white layer have been concerned with item b, where dull tools operating at very high speed have been used to obtain nonetching white material on the finished surface. However, item a is far more important than item b, since a white layer in the chip has a major influence on the chip-forming process (cutting forces, tool life, and surface integrity). On the other hand, a white layer on the finished surface is only of academic interest since any white layer there represents an unacceptable defect in finished machining conducted under unrealistic cutting conditions.

Chips of the type shown in Fig. 22.4 were carefully mounted and locally thinned by polishing and etching. Since it is impractical to make the entire specimen transparent to electrons, a "dimple" was mechanically polished at a selected area of the specimen as shown in Fig. 22.26. The dimpled area was then made electron transparent by ion milling using ions of 3–6 keV in a Gatan dual ion mill (model 600) followed by a precision ion milling system (Gatan model 645 MK4). A Topcon II2B (200 keV) transmission electron microscope was used for analyzing the specimen.

Results of electron microprobe analysis revealed that the bulk of the chip has the same composition as the white layer. The scanning electron microscope showed no microstructural detail in the white layer. Similar results have been reported in studies of white layers of ballistic origin (Timothy, 1987; Wittman et al., 1990). The electron diffraction pattern for the white layer of Fig. 22.4 was obtained, digitized, and analyzed for ring diameters using image-processing software (Saxton et al., 1979). Two phases were identified—untempered martensite and austenite. There was no indication of Hägg carbide as reported by Wittman et al. (1990) in the adiabatic shear bands obtained in ballistic experiments. Further details concerning electron diffraction analysis of the white layer may be found in Vyas (1997).

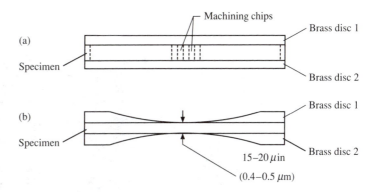

Fig. 22.26 (a) Specimens between protective brass discs. (b) Specimen after dimpling by grinding/polishing. (after Vyas and Shaw, 2000)

The formation of untempered martensite involves the following sequence: heating above the transformation temperature, dissolution, dispersion of carbides, and rapid quenching. The second step is usually a slow process since it involves diffusion. In ordinary heat treatment about one hour at maximum temperature is allowed for dissolution and dispersion of carbon. It is, therefore, somewhat surprising to find untempered martensite in the white layer of a hard turning chip, where the time for transformation to start would be less than or equal to the time for one sawtooth to form. Many experiments have been carried out to determine the transformation rate from α- to γ-iron in rapid heating situations at rates as high as 10^6 C/s (Speich and Szirmae, 1968; Ashby and Easterling, 1984; Lyasotskii and Shtanskii, 1993). However, the very large strain involved and the fact that the carbon is already dispersed in a hard steel should greatly shorten the time required for the second step. It is well established that the transformation from the γ to σ structure is very rapid (of the order of 10^{-7} s) since no diffusion, but merely an atomic shift, is involved. Since the same atomic rearrangement is involved in the α to γ transformation, this should also involve about 10^{-7} s and hence time should not be a factor.

Microhardness measurements of the white layer of the chip of Fig. 22.4 for which the bulk hardness was H_{RC} 62 ranged from H_{RC} 69 to H_{RC} 71. However, it should not be inferred that the white layer would be this hard during chip formation that would lead to enhanced abrasive tool wear. In fact, austenite above the transformation temperature will offer very little resistance to deformation. This should tend to reduce the tool-face friction and the resistance to shear in the MC region and hence lower the power required. This is a plus relative to abrasive tool wear. However, the fact that the existence of a white layer is associated with a high transformation temperature on the tool face coupled with an increased chip speed over the tool face when the cutting ratio is greater than one should lead to an increase in diffusion tool wear.

Chou and Evans (1997) have investigated the white layer that forms on the finished surface of a hard turned AISI 52100 steel (H_{RC} 59 to H_{RC} 61) using an AI-TIC insert having a rake angle of $-30°$ and a wear land on the clearance face.

Akcan et al. (1999) have presented a very detailed study of white layers produced on hard steel surfaces turned under a variety of conditions. It was found that these white layers consisted primarily of martensite differing from that produced by heat treatment by being unusually hard and of submicron grain size. These unusual characteristics were attributed to the very high stresses and strains associated with chip formation and the very rapid rate of cooling. A photomicrograph of an etched taper section of an AISI 52100 surface is shown in Fig. 22.27a, and the same photomicrograph digitized to enhance contrast is shown in Fig. 22.27b. The undulations are feed marks spaced approximately 80 μm along the axis.

Figure 22.28 shows similar processed photomicrographs for AISI 52100 steel machined at different cutting speeds (Fig. 22.28a), for different sized wear lands (Fig. 22.28b), and in a comparison of white layers produced under the same cutting conditions for hard samples of AISI M2,

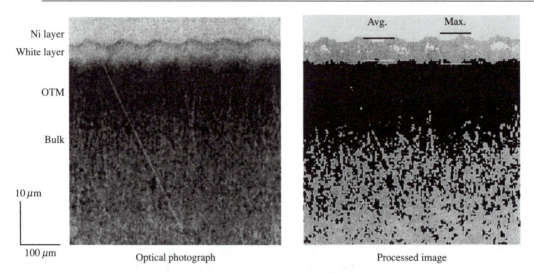

Optical photograph   Processed image

Fig. 22.27 Taper section of hard AISI 52100 steel turned under the following conditions: cutting speed, 442 f.p.m. (150 m/min); depth of cut, 0.008 in (0.2 mm); feed, 0.004 i.p.r. (0.1 mm/rev); wear land, 0.012 in (0.3 mm). (a) Optical photomicrograph. (b) Digitized version of (a). (after Akcan et al., 1999)

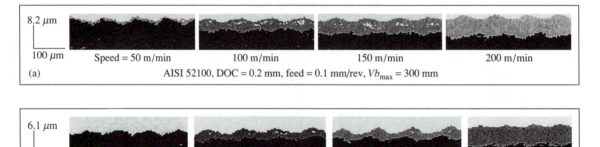

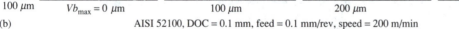

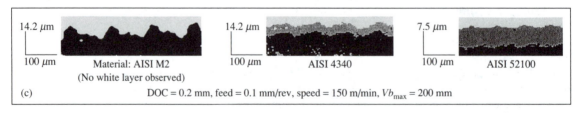

Fig. 22.28 Digitized photomicrographic specimens to show (a) effect of cutting speed on AISI 52100 steel machined with all other conditions the same as in Fig. 22.27, (b) effect of size of wear land on AISI 52100 steel with all other conditions the same as in Fig. 22.27, and (c) effect of type of steel turned under the following conditions: cutting speed, 492 f.p.m. (150 m/min); depth of cut, 0.008 in (0.2 mm); and feed, 0.004 i.p.r. (0.1 mm/rev). (after Akcan et al., 1999)

AISI 4340, and AISI 52100 steels (Fig. 22.28c). The size of the white layer is seen to decrease with increase in the transformation temperature, the transformation temperature for AISI M2, 4340, and 52100 being 1200 °C, 830 °C, and 800 °C, respectively.

In general, a white layer on the finished surface constitutes an unacceptable defect and indicates an excessively high temperature on the clearance face of the tool. Since interest in the present case is concerned with the influence of a white layer in the chip in sawtooth chip formation, only sharp tools were used in further consideration of white layers.

Figure 22.9 shows quick-stop photomicrographs of Ti–6Al–4V chips turned using tungsten carbide tools. While white layers are not formed with a titanium alloy, the strain will be equally inhomogeneous and the regions of concentrated shear will play the same role as a white layer does in the MC region of a hard steel chip.

Figure 22.9a suggests that the temperature increases as the chip moves up the face of the tool. This is because the MC region increases in size and becomes closer and more parallel to the tool face, while the chip speed increases as the chip proceeds up the tool face. This is fortunate since it is important that the maximum temperature be as far as possible from the tool tip in order to avoid a white layer from reaching the finished surface. It is important that the useful role associated with an increase in the white layer on the tool face with increased cutting speed, which reduces tool-face friction, not be counteracted by an increase in the temperature at the tool tip. A temperature increase at the tool tip will limit tool life and may cause the white layer to appear on the finished surface. Two ways of avoiding this are to

1. use a tool material having a low coefficient of conductivity (ceramic instead of PCBN, for example)

2. use a tool-face geometry that causes the maximum temperature to be as far as possible from the tool tip

The fact that the white layer is harder than the bulk of the chip after cooling and that it contains considerable untempered martensite supports the idea that the white layer is a region that has transformed to austenite. However, this does not rule out the possibility that the white layer results from some melting since very rapid cooling of molten material could give rise to a structureless material of high hardness. However, whether the white layer is completely austenite or partially molten during chip formation is not too important since in either case the white layer would offer little resistance to deformation during chip formation and this appears to be the most important characteristic of the white layer. The melting possibility associated with the white layer has been discussed in Shaw and Vyas (1993).

REFERENCES

Akcan, S., Shah, S., Moylon, S. P., Cahhaba, P. N., Chandrasekar, S., and Farris, T. N. (1999). *Proc. ASME International Conf.*

Ashby, M. F., and Easterling, K. E. (1984). *Acta. Metall.* **32/11**, 1935–1948.

Binns, J. (1961). *Iron and Steel Engineer* (Oct.), 1–8.

Binns, J. (1984). *Mech. Eng. (ASME)* (Mar.), 55–59.

Chou, Y. K., and Evans, C. J. (1997). *ASME Mfg. and Tech. MED* **6/2**, 75.

Davies, M. A., Burns, T. J., and Evans, C. J. (1997). *Annals of CIRP* **46/1**, 1–6.

Davies, M. A., Chou, Y., and Evans, C. J. (1996). *Annals of CIRP* **45/1**, 77–82.

Elbestawai, M. A., Srivastava, A. K., and El-Wardany, T. I. (1996). *Annals of CIRP* **45/1**, 71–76.

Field, M., and Kahles, J. E. (1971). *Annals of CIRP* **20/2**, 153–163.

Herzog, D. E. (1975). *Mfg. Eng.* (Oct.), 29.

Hodgson, T., and Trendler, P. H. H. (1981). *Annals of CIRP* **30/1**, 63–66.

Kishawy, H. A., and Elbestawi, M. A. (1997a). *Proc. of Int. MATADOR Conf.*, pp. 253–258.

Kishawy, H. A., and Elbestawi, M. A. (1997b). *ASME Mfg. Sc. and Tech.*, *MED* **6/2**, 13–20.

Kishawy, H. A., and Elbestawi, M. A. (1999). *Proc. Int. Symp. Plasticity*, Cancun, Mexico, pp. 361–364.

Kishawy, H. A., and Elbestawi, M. A. (2001). *Proc. Inst. Mech. Engrs.* **215B**, 755–767.

Koenig, W. A., Klinge, M., and Link, R. (1990). *Annals of CIRP* **39/1**, 61–66.

Koenig, W. A., Komanduri, R., Toenshoff, H. K., and Ackeshott, G. (1984). *Annals of CIRP* **33/2**, 417–427.

Komanduri, R., Schroeder, J. A., Hazra, J., von Turkovich, B. F., and Flom, D. G. (1984). *ASME J. of Eng. for Ind.* **104**, 121–131.

Lindberg, B., and Lindstrom, R. (1983). *Annals of CIRP* **32/1**, 17–20.

Lyasotskii, I. V., and Shtanskii, D. V. (1993). *Physics of Metals and Metollography* **75/1**, 77–82.

Matsumoto, Y., Barish, M. M., and Liu, C. R. (1984). *ASME PED* **12**, 193–204.

Matsumoto, Y., Hashimoto, F., and Lahoti, G. (1999). *Annals of CIRP* **48/1**, 59.

Nakayama, K. (1972). Private communication.

Nakayama, K. (1974). *Proc. Int. Conf. on Prod. Eng.*, Tokyo, pp. 572–577.

Ramaraj, T. C., Santhanam, S., and Shaw, M. C. (1988). *ASME J. of Eng. for Ind.* **110**, 333–338.

Recht, R. F. (1964). *ASME J. Appl. Mech.* **39**, 189–193.

Recht, R. F. (1985). *ASME J. of Eng. for Ind.* **107**, 309–315.

Sampath, W. S., and Shaw, M. C. (1983). *Proc. 11th N. Amer. Metal Working Research Conf.* SME, Dearborn, Mich., pp. 281–285.

Saxton, W. O., Pitt, T. J., and Horner, M. (1979). *Ultramicroscopy* **4**, 343–354.

Schwarzhofer, R. P., and Kaelin, A. (1986). *Annals of CIRP* **32/1**, 45–50.

Scott, R. L., Kopple, R. K., and Miller, M. H. (1962). *Proc. Sym. on Rolling Contact Fatigue.* Elsevier, pp. 301–316.

Shaw, M. C., Dirke, S. O., Smith, P. A., Cook, N. H., Loewen, E. G., and Yang, C. T. (1954). *Machining Titanium.* MIT Report to U.S. Air Force.

Shaw, M. C., Janakiram, M., and Vyas, A. (1991). *NSF Grantees Conf.*, Austin, Tex. SME, Dearborn, Mich., pp. 359–366.

Shaw, M. C., and Ramaraj, T. C. (1989). *Annals of CIRP* **38/1**, 59–63.

Shaw, M. C., and Vyas, A. (1993). *Annals of CIRP* **42/1**, 29–33.

Shaw, M. C., and Vyas, A. (1998). *N. Amer. Metal Working Research Conf.*, Atlanta, Ga. SME, Dearborn, Mich.

Sheikh-Ahmad, J., and Bailey, J. A. (1997). *J. of Mfg. Sc. and Eng.* **119**, 307.

Speich, G. R., and Szirmae, A. (1968). *Trans. ASME* **245**, 1063–1074.

Timothy, S. P. (1987). *Acta. Metall.* **35/2**, 301–306.

Ueda, T., Al Huda, M., Yamada, K., and Nakayama, K. (1999). *Annals of CIRP* **48/1**, 63–66.

Ueda, T., Sato, M., Sugita, T., and Nakayama, K. (1995). *Annals of CIRP* **44/1**, 325–328.

Vyas, A. (1997). Ph.D. dissertation. Arizona State University, Tempe.

Vyas, A., and Shaw, M. C. (1999). *J. of Mfg. Sc. and Eng.* **121**, 163–172.

Vyas, A., and Shaw, M. C. (2000). *Mach. Sc. and Tech.* **4/1**, 169–175.

Walker, T. J., and Shaw, M. C. (1969). *Proc. 10th Int. Mach. Tool Design and Research Conf.* Pergamon Press, Oxford, p. 291.

Wittman, C. C., Meyers, M. A., and Pak, H. (1990). *Met. Trans.* **21/4**, 707–716.

Zener, C. (1948). *The Micromechanism of Fracture* in Fracture of Metals. ASM, pp. 3–31.

Zhen-Bin, H., and Komanduri, R. (1995). *Annals of CIRP* **44/1**, 69–73.

23 PRECISION ENGINEERING

Precision engineering involves

- precise measurement of size, shape, and finish (metrology)
- production and testing of parts for engineering products
- nanoprocessing of products

The last of these items involves the removal of particles that approach a nanometer ($0.04 \, \mu in = 10^{-9}$ m) in size. Whereas conventional (macro) processing involves chips and areas of deformation that are sufficiently large that conventional plasticity and fracture may be employed, nanoprocessing concerns particles that approach atomic dimensions in size, this requires a relatively new deformation technology called molecular dynamics (MD), discussed later in this chapter.

Taniguchi (1993) has presented a history of the progressive increase of machining accuracy during the twentieth century. In Fig. 23.1 the ordinate represents total machining accuracy which is the sum of systematic error (tool positioning error, etc.) and random error (due to defects such as backlash, etc. of processing equipment). This estimated by extrapolation that at the beginning of 2000, the attainable level of processing accuracy would reach the nanometer level. Taniguchi (1992) has also discussed the potential of a wide variety of noncutting removal processes from the point of view of their possible role in nanoprocessing. Taniguchi et al. (1989) have published a treatise on manufacturing involving various energy sources which has important implications for nanoprocessing of materials.

Precision engineering is an area of technology that has been gaining in importance in recent years. The term nanotechnology was first introduced by N. Taniguchi at the International Conference on Production Engineering in Tokyo in 1974. This was a meeting sponsored by the Japan Society for Precision Engineering (JSPE) and CIRP. Nanotechnology is the basis for ultra-precision processing and involves the limit of measurement of length which is the spacing of atoms (approximately $0.3 \, nm = 0.3 \times 10^{-9}$ m).

PRECISION ENGINEERING SOCIETIES

JSPE was founded in 1947. This was preceded by the Association of Precision Machinery that had been founded by T. Aoki of the University of Tokyo in 1933. The initial emphasis of JSPE was on metrological instruments, precision machine tools, and production of precision mechanical

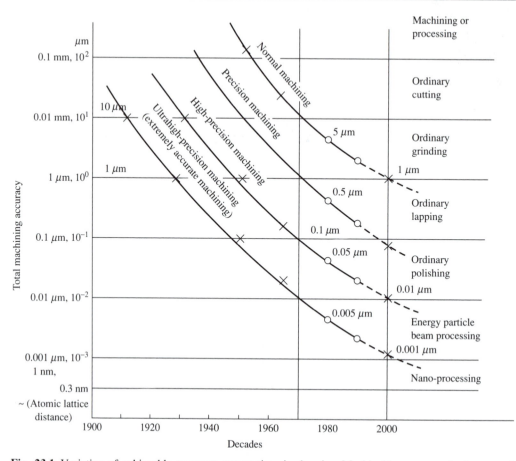

Fig. 23.1 Variation of achievable accuracy versus time in decades. Machinable accuracy is the sum of systematic error and random error. (after Taniguchi, 1993)

elements. However, with time JSPE has extended its activities to include new developments as they emerge such as information processing, CAD/CAM systems, and, most recently, nanoprocessing.

The American Society of Precision Engineers (ASPE) was founded in 1986 and also covers a wide range of subjects.

The European Union Society for Engineering and Nanotechnology (EUSPEN) was founded in 1998.

Beginning in 1977 the journal *Precision Engineering* published articles related to high-accuracy engineering from around the world. Recently it became the *Journal of the International Societies for Precision Engineering and Nanotechnology* (JSPE, ASPE, and EUSPEN).

Corbett et al. (2000) describe nanotechnology as an activity for developing new, smaller, faster, and cheaper products by application of nanoscience—the science of behavior approaching the atomic level.

Nanotechnology is of particular interest in the optical and electronic industries and involves single-point diamond turning (SPDT) as well as ultra-precision diamond grinding (UPDG). These two processes are similar in that chips of unusually small size are involved. Both are capable of producing surfaces with mirror finish without polishing, using specially designed machine tools of high rigidity with air bearing spindles. Yoshioka et al. (1985) have discussed the requirements and design philosophy for such machines. Details concerning precision turning machines are reviewed in Aronson (1994).

In this chapter, SPDT will be considered. UPDG of very brittle materials, such as glass, is discussed in Shaw (1996).

The major problem with both SPDT and UPDG is the appearance of subsurface defects usually in the form of microcracks. UPDG is usually performed on very hard brittle materials, such as glasses and ceramics, while SPDT is often performed on very soft ductile metals, such as pure copper and aluminium.

CONSEQUENCES OF SMALL CHIP SIZE

A major consequence of small chip size is that the normally ductile metal in SPDT behaves as though it were as brittle as glass when the volume deformed is limited to a very small size. Herring and Galt (1952) found that very small whiskers produced when a normally ductile metal was deposited electrolytically at a very small rate were essentially defect free. When these whiskers were tested in bending, they were found to be completely brittle (yield stress > fracture stress), to have unusually high elastic fracture strain (> 0.1), and to have a fracture stress approaching the theoretical strength of the material (shear modulus/n, where n is in the range 2π to 10, depending on the approximations involved in the estimate). This clearly demonstrates that when the size of the deformation zone approaches the mean defect structure spacing, normally ductile materials behave as though they were very brittle.

Another consequence of the small undeformed chip thicknesses involved in SPDT and UPDG is that the chip-forming model shifts from one involving concentrated shear (Fig. 1.2) to microextrusion (Fig. 23.2). When the undeformed chip thickness (t) becomes less than the radius (r) at the tool or grit tip, the effective rake angle of the tool has such a large negative value that Fig. 23.2 replaces Fig. 1.2 (Shaw, 1972). In the case of Fig. 23.2, a relatively large volume of material has to be brought to the fully plastic state in order for a relatively small amount of material to escape as a chip. This is one of the reasons for the exponential increase in specific energy with a decrease in undeformed chip thickness (the size effect) in UPDG and SPDT.

A further, more important increase in specific energy with a decrease in the undeformed chip thickness is due to a decrease in the probability of encountering a stress-reducing dislocation or a defect capable of inducing dislocations as the deformation zone decreases in volume. Lucca and Seo (1994) have measured the depth of subsurface deformation (δ) versus undeformed chip thickness (t) and have obtained the results shown in Fig. 23.3. This is surprising in that it indicates that δ is

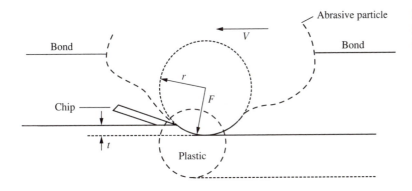

Fig. 23.2 The mechanism of chip formation at a single abrasive particle in fine grinding. (after Shaw, 1972)

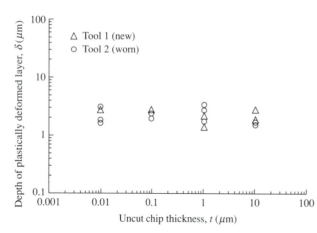

Fig. 23.3 The measured depth of plastically deformed layer (δ) versus the uncut chip thickness (t) after machining a Te–Cu alloy with a sharp and a worn diamond tool. (after Lucca and Seo, 1994)

essentially independent of t. The ratio δ/t is about 200 at $t = 0.01$ μm, but two orders of magnitude smaller at $t = 1.0$ μm. The mean defect spacing for this material is about 1 μm. Thus, in order that a dislocation or dislocation-generating defect be encountered for plastic flow to occur at $t = 0.01$ μm would require the elastic stress field to penetrate very deeply and to have a much higher mean value than when t is 1 μm or greater.

SINGLE-POINT DIAMOND TURNING (SPDT)

The important role of tool tip sharpness in SPDT is illustrated in Fig. 23.4. Both tools A and B are relatively sharp but A is sharper than B. Tool sharpness is seen to be relatively unimportant relative to cutting force per unit width cut (F'_P) but extremely important relative to feed force per unit width (F'_Q). The less sharp tool B has values of F'_Q that are more than an order of magnitude greater than those for the very sharp tool A and exhibits a pronounced size effect. The difference in behavior of these two tools is due to a difference in the radius at the tip of the tool. The tip radius for tool B is large enough relative to the undeformed chip thickness t so that the removal model for grinding (Fig. 23.2) holds, while that for tool A is small enough so that the metal cutting model (Fig. 1.2) is approached. Thus, when cuts are made having values of undeformed chip thickness that are smaller than the tool tip radius, the removal process in SPDT approaches that of fine grinding.

Ikawa et al. (1987) suggest that with care, it is possible to polish diamond tools to a roughness of about 1 nm and a sharpness (tip radius) of about 10 nm. It should be possible, therefore, to remove material by SPDT with an undeformed chip thickness of about 50 nm and still have the concentrated shear model as the one pertaining.

Chips produced by sharp tool A are shown in Fig. 23.5a, and chips produced by less sharp tool B in Fig. 23.5b. The chips shown in Fig. 23.5a resemble those in ordinary metal cutting, with one side smooth and the other side rough. These clearly conform to the concentrated shear metal cutting model of Fig. 1.2. The chips in Fig. 23.5b have an entirely different appearance and suggest formation by the extrusion-like mechanism of grinding shown in Fig. 23.2.

Figure 23.6 shows a long continuous chip obtained when diamond face turning electrodeposited Cu with an undeformed chip thickness of 1 nm. This was produced by a very sharp tool and resembles the chips of Fig. 23.5a, except that there are

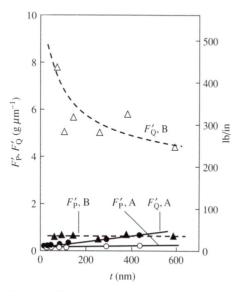

Fig. 23.4 The variation of the cutting forces per width cut, F'_P and F'_Q with the undeformed chip thickness t for tool A which is sharper (has a smaller radius of curvature at the cutting edge) than tool B. (after Ikawa et al., 1987)

(a)   (b)

Fig. 23.5 SEM micrographs of chips produced at 4 μin ($t = 100$ nm) (a) for sharp tool A and (b) for less sharp tool B of Fig. 23.4. (after Ikawa et al., 1987)

many more microcracks in the surface, having a pitch of 50–100 nm. There is obviously a great deal of fracture involved for the chip of Fig. 23.6. The continuity of the chip is due to one or a combination of the following:

- individual microcracks do not extend far across the *width* of the chip
- individual microcracks do not extend all the way across the *thickness* of the chip
- microcracks reweld as the chip proceeds up the tool face

The greater amount of fracture involved in Fig. 23.6 compared with Fig. 23.5a is undoubtedly due to the presence of more stress-concentrating defects in electrodeposited copper than in wrought copper and also the smaller undeformed chip thickness for Fig. 23.6. Both Figs. 23.5a and 23.6 clearly indicate that micro-fracture and the defect structure of the metal cut play important roles in ultra-precision machining. It would, therefore, appear that any attempt mathematically to model this process that does not include normal or hydrostatic stress in the constitutive relationship to take fracture into account will be inadequate. It also appears that any attempt to model ultra-precision machining by molecular dynamic simulation should include a defect structure.

Lucca and Seo (1994) have studied the effect of undeformed chip thickness (t) and tool tip radius (ρ) on the cutting forces and the depth of the deformed layer in SPDT of elecrodeposited copper and a 0.5 $^w/_o$ Te–Cu alloy. Tool tip radii were measured with an atomic force microscope (AFM) using standard carboxylate spheres of 519 ± 10 nm diameter to estimate the radius of the cantilever tip of the microscope, which was found to be about 30 nm [see Lucca and Seo (1994) for details of the basic method]. A piezoelectric dynamometer was used to measure cutting forces F_P and F_Q, and the taper section technique was used to estimate the depth of deformation (δ) after cutting. The nominal rake angle of the tool was $0°$ and the clearance angle was $5°$. The undeformed chip thickness values used were from 10 to 10^4 nm in orthogonal

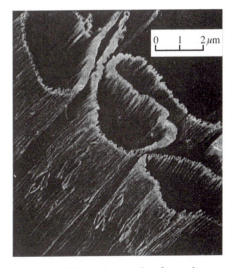

Fig. 23.6 SEM micrograph of an electro-deposited Cu chip produced at 0.04 μin ($t = 1$ nm) in an ultra-precision turning operation. (after Donaldson et al., 1987)

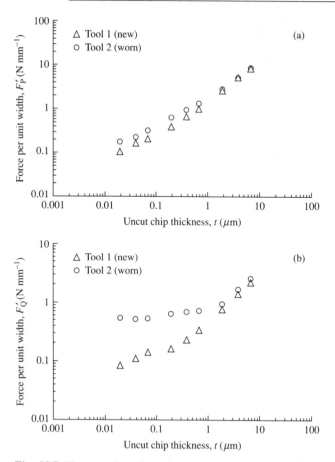

Fig. 23.7 The variation of the forces per unit width (a) F_P' (in cutting direction) and (b) F_Q' (normal to cutting direction) with undeformed chip thickness (t) for a new diamond tool (1) and a used tool (2) when cutting a 0.5 $^w/_o$ Te–Cu alloy. (after Lucca and Seo, 1994)

cutting. The surfaces of the cut specimens were carefully protected to prevent edge rounding when being taper-sectioned at an inclination of 5°. The depth of the deformed layer was made visible by etching with an aqueous solution of $Fe(NO_3)_3.9H_2O + HCl$ for several seconds.

The cutting tool tip radius was found to be 75 ± 15 nm for the new tool and 105 nm for a used tool. The variation of unit force component in the cutting direction (F_P') versus undeformed chip thickness (t) is shown in Fig. 23.7a and for the orthogonal direction (F_Q') in Fig. 23.7b. Approximate values of forces F_P', F_Q', and the ratio F_P'/F_Q' are given in Table 23.1 for the new and used tools for a wide range of undeformed chip thicknesses (t). It is seen that the ratio F_P'/F_Q' is greater than one for the sharp tool, as it should be when cutting is predominantly in the concentrated shear mode. However, this ratio is less than one (except for the very light cut at 10 nm) for the used tool. A value less than one is to be expected when cutting is predominantly in the microextrusion mode. The unusually high value of F_P'/F_Q' for the dull tool operating at small t is undoubtedly due to most of the energy being associated with friction without removal for such a light cut with a dull tool.

Ueda and Manabe (1992) have observed microcutting chip formation of an amorphous metal ($Fe_{78}B_{13}Si_9$) by *in-situ* cutting in an SEM at extremely low cutting speeds (15–850 μm min^{-1}, or 600–34,000 μin min^{-1}). A typical chip is shown in Fig. 23.8, where a

TABLE 23.1 *Approximate Values of Force Components F_P' and F_Q' and the Ratio F_P'/F_Q' for SPDT of a Te–Cu Alloy with New and Used Tools†*

Item	New Tool at t of			Used Tool at t of		
	10 nm	**10^3 nm**	**10^4 nm**	**10 nm**	**10^3 nm**	**10^4 nm**
F_P' (N mm^{-1})	0.1	1.0	10.0	0.1	1.0	10.0
F_Q' (N mm^{-1})	0.06	0.6	3.0	0.6	0.6	3.0
F_P'/F_Q'	1.66	1.66	3.33	6	0.6	0.3

† After Lucca and Seo, 1994.

card-like structure is clearly seen. The authors report that the concentrated shear bands which were less than 0.03 μm (12 μin) in width showed no evidence of gross fracture at very high magnification. The spacing of the "cards" (S) was found to scale with the undeformed chip thickness t, as shown in Fig. 23.9.

The spacing of the shear bands (S) in Fig. 23.8 is approximately half the undeformed chip thickness t and the cutting ratio is approximately one-third of t. This chip is unusual in that the side of the chip contacting the tool face (A in Fig. 23.8) is not burnished. Because of the extremely low cutting speed in these tests (0.15 i.p.m. or 370 μm/min), temperature effects will be nil. Further discussion of SPDT of amorphous metal (metallic glass) is given in the section on molecular dynamics (MD) at the end of this chapter.

Since the geometry of the tool tip plays such an important role, the wear of diamond tools is extremely important in SPDT. Diamond tool wear depends on structural anisotropy, imperfections present (influencing strength and microchipping), the affinity for the metal being cut (influencing adhesive wear), and the role of temperature and

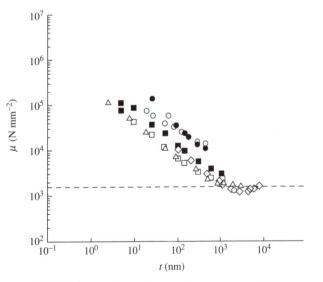

Fig. 23.8 Top and side views of an in-situ SEM chip of amorphous metal cut as follows: $V = 0.015$ i.p.m. (370 μm min^{-1}); rake angle = 20°; t = 200 μin (5 μm). (after Ueda and Manabe, 1992)

atmosphere (influencing weight loss of diamond due to oxidation and graphitization). The questions of optimum crystal orientation and identification of defects in diamonds are complex issues that have been pursued in the literature (Ikawa et al., 1987).

A number of workers have observed a pronounced increase in specific energy with a decrease in undeformed chip thickness t, as shown in Fig. 23.10, when diamond turning copper under ultraprecision conditions. The approximate specific energy for machining copper in the conventional finish machining regime ($t \approx 0.005$ in or 0.13 mm) is shown by the dashed line in Fig. 23.10.

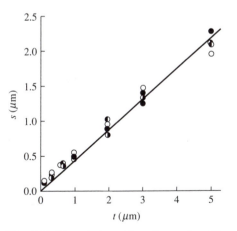

Fig. 23.9 The relation of lamella spacing (S) and undeformed chip thickness (t) for cutting speeds ranging from 600–34,000 μin min^{-1} (15 μm min^{-1} to 850 μm min^{-1}) and a rake angle of 20°. (after Ueda and Manabe, 1992)

Fig. 23.10 The variation in the specific energy u with the undeformed chip thickness t in diamond micromachining of copper, from several independent investigators. (after Ikawa et al., 1991)

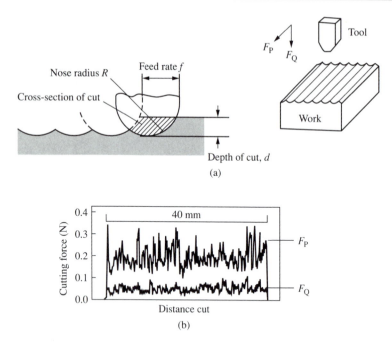

Fig. 23.11 (a) An ultra-precision slow speed ($V = 1180$ f.p.m. = 6 m s^{-1}) cutting test arrangement. (b) Fluctuations in cutting forces when cutting polycrystalline aluminum [mean grain size, 0.026 in (640 μm); H_V 56] with $V = 1180$ f.p.m. (6 m s^{-1}), $f = 0.0016$ in per stroke (40 μm), $d = 240$ μin (6 μm), $R = 0.020$ in (0.5 mm). (after Moronuki et al., 1994)

Moronuki et al. (1994) have performed slow-speed ultra-precision cutting tests on several materials, using the cutting arrangement shown in Fig. 23.11a. The cutting forces shown in Fig. 23.11b were obtained when machining pure polycrystalline aluminum under the conditions indicated in the figure caption. The cutting force F_P varied from a minimum value of about 0.022 lb (0.1 N) to a maximum value of about 0.067 lb (0.3 N), with a peak of varying magnitude occurring at each millimeter of travel. Force F_Q was only about one-quarter of F_P and showed a similar threefold fluctuation for each millimeter of tool travel. The authors attribute the rather large fluctuations in the cutting forces to grain-to-grain anisotropy and suggest that crystalline anisotropy plays an important role when the undeformed chip thickness approaches mean crystal size.

Wang et al. (1994a) have studied the effect of crystal orientation in single-point diamond turning (SPDT) of copper. It was found that tools having a (100) crystal orientation on both rake and flank faces had better wear resistance than other crystal orientations. Liang et al. (1994) have used finite element analysis to verify the influence of grain boundaries and crystal orientation on microcutting mechanics.

Komanduri et al. (1999) have studied the effects of crystal orientation when single-crystal aluminum is machined under nanometric cutting conditions. It was found that depending on the crystal orientation and direction of cutting, different modes of plastic flow occurred giving rise to different amounts of subsurface flow and the direction of shear flow in front of the tool. It was found that under single-crystal nanometric cutting conditions, planes of maximum shear strain (shear planes) assumed directions ranging from the cutting direction to normal to the cutting direction (i.e., shear angles from 0° to 90°).

Wang et al. (1994b) have studied the machining conditions on the surface finish of copper when turned with a diamond tool. It was found that cutting speed had no effect on surface roughness. Feed rate had a negligible effect on surface roughness for values less than 10 μm/rev, but roughness increased linearly with feed above this value as shown in Fig. 23.12.

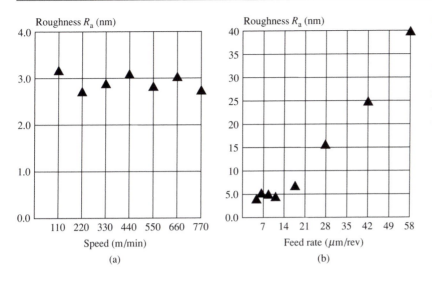

Fig. 23.12 Variation in R_a roughness with (a) cutting speed and (b) feed rate for SPDT of copper. Cutting conditions: rake angle, $-2°$; clearance angle, $2°$; nose radius, 10 mm; depth of cut, 1 μm; fluid, isopropanol. (after Wang et al., 1994b)

SPDT TOOL WEAR

Paul et al. (1996) have studied the chemical aspects of tool wear in SPDT for a wide range of pure materials. Carbon atoms that are pulled from the tool under conditions giving strong adhesion between chip and tool react with oxygen to give CO or CO_2 or with the work to give a carbide. Table 23.2 lists elements in increasing order of melting temperature together with the number of unpaired d-shell electrons capable of forming strong bonds with cubic carbon. It also indicates whether the material is reasonably turnable with a diamond tool (Y) or not (N). Tests were performed with single-crystal diamond tools having a $0°$ rake angle, a clearance angle of $6°-8°$, and a 0.020 in (0.508 mm) nose radius. The cutting fluid employed was a light mineral oil. Facing cuts had a depth of 2 μm and a feed rate of 2 μm/rev (except for Si, which was 20 μm/rev).

The melting temperature and the number of unpaired electrons in the d-shell correlate approximately with the Y's and N's except for the four elements between the bars. The elements having a low melting temperature and zero unpaired electrons in the d-shell are in general diamond turnable. Plastics were very diamond machinable but steel and Ti were not, steel being somewhat more resistant to wear than pure iron and Ti somewhat more resistant to wear than pure iron (probably due to the greater number of d-shell electrons for Fe than for Ti).

Nickel is a special case. While pure nickel is not diamond turnable, electroless deposited nickel (eNi) is diamond machinable.

Donaldson et al. (1985) have studied SPDT of electrolesss nickel (eNi) with concentrations of phosphorous varying from 1.8 to 13%. Facing cuts were made on 2 in diameter copper discs with 0.003 in of eNi deposited on the flat surfaces. The single-crystal diamond tools used had a rake angle of $0°$, a rotational speed of 1000 r.p.m. (V_{max} = 524 f.p.m.), a feed rate of 100 μin/rev and a cutting fluid of light mineral oil. No wear was detected for samples containing above 11% P. FEM analysis was also used to study the effect of tool-face friction (f) on chip formation. Figure 23.13 shows that as the tool-face friction increases, chip curl decreases. It was also found that cutting forces increased and the depth of penetration of residual stress also increases with increase in tool-face friction.

Use of an eNi coating on a material that is not diamond machinable (such as ferrous alloys) enables the final finishing operation to be performed by SPDT. This enables the unusually excellent

TABLE 23.2   Data for Elements with Known Turning Properties[†]

Element		Melting Point, °C	No. of Unpaired d-Shell Electrons	D Turnable? (Yes/No)
In	Indium	157	0	Y
Sn	Tin	232	0	Y
Pb	Lead	373	0	Y
Zn	Zinc	420	0	Y
Pu	Plutonium	640	0	Y
Mg	Magnesium	649	0	Y
Al	Aluminum	660	0	Y
Ge	Germanium	937	0	Y
Ag	Silver	962	0	Y
Au	Gold	1064	0	Y
Cu	Copper	1083	0	Y
U	Uranium	1132	1	N
Mn	Manganese	1244	5	N
Be	Beryllium	1277	0	Y
Si	Silicon	1410	0	Y
Ni	Nickel	1453	2	N
Co	Cobalt	1495	3	N
Fe	Iron	1535	4	N
Ti	Titanium	1660	2	N
Cr	Chromium	1857	5	N
V	Vanadium	1890	3	N
Rh	Rhodium	1966	2	N
Ru	Ruthenium	2310	3	N
Nb	Niobium	2468	4	N
Mo	Molybdenum	2617	5	N
Ta	Tantalum	2996	3	N
Re	Rhenium	3180	5	N
W	Tungsten	3410	4	N

[†] After Paul et al., 1996.

finish and precision associated with diamond machining to be achieved with an excellent tool life. It was found that the phosphorous content of 13% and a heat treatment of 200 °C for two hours gave optimum tool life. A hard crystalline phase (N_3P) forms in nickle containing phosphorous. The amount of the crystalline phase depends on heat treatment. Heating to 200 °C gives a more amorphous material while heating to 400 °C gives a more crystalline form with a higher tool wear rate. Phosphorous is beneficial in tying up unpaired d-shell electrons in nickel, and this reduces diamond tool wear. However, too much phosphorous or too low a heat treating temperature gives rise to too high a crystalline content that increases diamond tool wear. This accounts for the optimum phosphorous content and heat treating temperature.

TOOL-FACE TEMPERATURE IN SPDT

Ueda et al. (1999) have measured temperatures on the rake face of single-crystal diamond tools when cutting aluminum and copper. Infrared rays radiated from the chip–tool contact area and through the transparent diamond tool were measured using an infrared radiation pyrometer. The infrared rays radiating through the tool were conducted by a chalcogenide fiber optic to a two-color detector that divides the radiation, sending one branch to an InSb detector and the other to a HgCdTe detector. The temperature distribution on the tool face and the maximum temperature were determined numerically using an FEM program. This is similar to the system described in Chapter 12 in connection with Figs. 12.7 through 12.13.

The experimental two-color pyrometer setup employed is shown in Fig. 23.14. Figure 23.15 shows the temperature distribution measured with the two-color pyrometer when aluminum was cut at 1700 f.p.m. (518 m/min) under other cutting conditions listed in Table 23.3. Similar results were obtained for a range of cutting speeds for both Al and Cu. At the same time, cutting forces, shear angles, and cutting ratios were measured for all of these tests; and from these, measured values of maximum cutting temperatures were calculated using the FEM analysis of Childs et al. (1988). Figure 23.15 is a representative

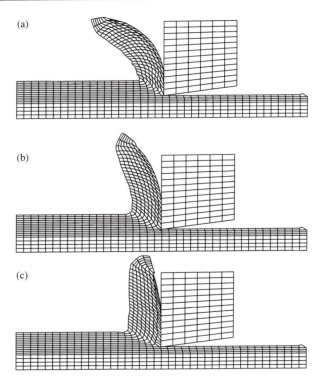

(a)

(b)

(c)

Fig. 23.13 Effect of coefficient of friction (f) on the rake face in single point turning: (a) $f = 0$, (b) $f = 0.1$, and (c) $f = 0.3$. (after Donaldson et al., 1985)

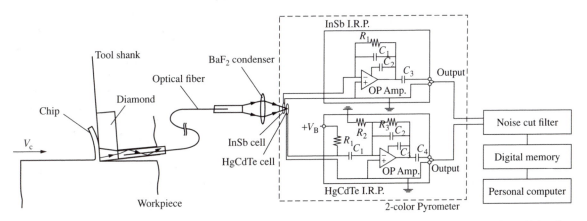

Fig. 23.14 Experimental setup used to measure chip–tool contact temperatures on the tool face. (after Ueda and Manabe, 1992)

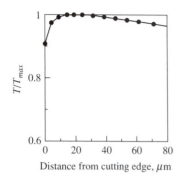

Fig. 23.15 Variation of the measured ratio of local temperature on the tool face to the maximum temperature for aluminum machined at 1700 f.p.m. (518 m/min) and other conditions given in Table 23.3. (after Ueda et al., 1999)

TABLE 23.3 Cutting Conditions

Cutting tool	Single crystal diamond
Rake angle	$-5°$
Clearance angle	$5°$
Workpiece	Aluminum, Copper
Depth of cut	10 μm (400 μin)
Width of cut	1 mm (0.040 in)
Cutting speed	400–900 m/min (1312–2950 f.p.m.)
Dry cutting	

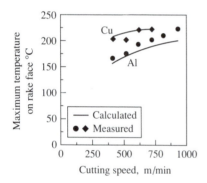

Fig. 23.16 Variation of maximum tool-face temperature with cutting speed for copper and aluminum under conditions given in Table 23.3. (after Ueda et al., 1999)

example of the change of the ratio of radiation pyrometer values of tool-face temperature for an aluminum workpiece. Figure 23.16 shows the variation of maximum tool-face temperature versus cutting speed for Al and Cu for values calculated from experimental results (thin lines) and from measured values from the radiation pyrometer apparatus (solid points). The maximum temperature values for copper are slightly higher than those for aluminum, and the maximum temperature values are seen to increase slightly with cutting speed for both materials.

ENERGY DISSIPATED IN NANOSCALE CUTTING

Inamura et al. (1993) have presented a study of the mechanics and energy dissipation in nanoscale cutting. The material considered was monocrystalline copper being cut with a diamond tool having a rake angle of 6.3°, a relief angle of 5°, and a very sharp cutting edge. The depth of cut was 1 nm. It was found that cutting proceeded cyclically, building as the work deformed elastically and dropping when plastic flow occurred. The extent of plastic deformation above (region I) and below (region II) the cutting plane is shown in Fig. 23.17a, and a pie graph of the distribution of the total energy consumed is shown in Fig 23.17b. It is seen that plastic flow extends considerably in front of the chip and below the tool. There is no evidence of a distinct shear plane in Fig. 23.17a.

A detailed description of the MD-FEM analysis of these results is presented in the section on molecular dynamics at the end of this chapter. This involved considering the process from both a continuum and an atomic perspective, providing energy conservation and displacement consistency between the two models.

Inamura et al. (1994) have extended the preceding paper with greater emphasis on stress and strain in nanoscale cutting. It was found in this later study that the workpiece is subjected to concentrated compression and shear strain, but at the chip–tool interface the strain is tensile suggesting that this is due to an adhesive force that develops between the tool and the chip. The stress distribution also suggested that instead of a concentrated shear zone as in macrochip formation, the region in front of the tool was subjected to a complex pattern of severe compressive and shear strain giving rise to a type of buckling deformation.

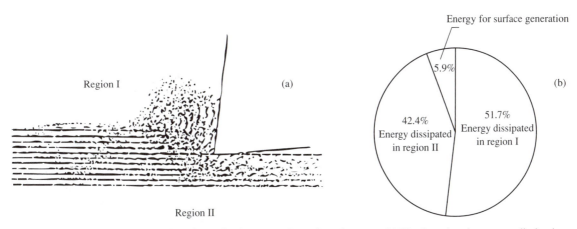

Fig. 23.17 (a) Distribution of plastic deformation in nanoscale cutting of copper. (b) Pie chart showing energy dissipation. (after Inamura et al., 1993)

SPDT of Hard Materials

High-precision cutting and grinding of optical components of glass offers the possibility of obtaining surfaces of excellent finish and accuracy without the need for time-consuming polishing. However, the problem with this approach is that unless great care is taken, the finished surface will not be crack free.

Fujita and Shibata (1993) have performed single-groove cutting tests on a glass cylinder using 90° diamond cones having a negative rake angle of 2°. In order to obtain a groove free of cracks, it was necessary for the normal force on the diamond to be < 0.007 lb (0.03 N), for the radius at the tip of the diamond to be < 200 μin (5 μm), and for the depth of cut to be < 8 μin (0.2 μm). The critical depth of cut to avoid subsurface cracking was found to increase with cutting speed to 1640 f.p.m. (500 m min^{-1}) and then to remain constant. This could be explained in terms of an increase in temperature with speed, causing the glass to be less brittle.

Takahashi et al. (1993) have also performed single-point diamond cutting tests on glass in the fly cutting mode, as shown in Fig. 23.18. The tools were Vickers indenters with values of tip radii of 80 and 20 μin (2 μm and 0.5 μm). At a cutting speed of 49,200 f.p.m. (250 m s^{-1}), wavy marks were found on each side of the groove when examined with a scanning tunneling microscope. These marks had a pitch of 80 μin (2 μm). This corresponds to a time between peaks of 8 ns and a frequency of 124×10^6 Hz, thus ruling out the possibility of a vibrational origin. Similar grooves were found with both values of tip radius, but were more pronounced with the 80 μin (2 μm) radius than with the

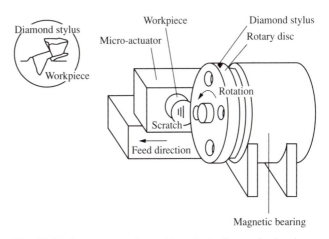

Fig. 23.18 An apparatus for making single-diamond grit micro-fly milling cuts on a glass workpiece at speeds up to 50,000 f.p.m. (250 ms^{-1}). (after Takahashi et al., 1993)

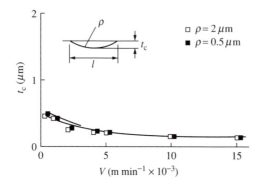

Fig. 23.19 The variation of the critical depth of cut t_c with the cutting speed V for Vickers diamond tools with two different tip radii (R) when cutting BK7 glass $t_c = \rho^2 - (\rho - l^2/4)^{1/2}$. (after Takahashi et al., 1993)

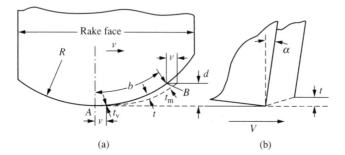

Fig. 23.20 The cutting arrangement used by Blake (1988) to determine the maximum critical undeformed chip thickness in an SPDT operation (face turning) of polycrystalline Ge. (a) Front view in V direction. (b) Side view. Typical values employed were as follows: V, up to 2814 f.p.m. (14.3 m s^{-1}); $v = 100$ μin (2.5 μm); $d = 100$ μin (2.5 μm); $R = 0.125$ in (3.125 μm); $\alpha = -10°$; $b = 5000$ μin (125 μm); $t_m = 0.4$ μin (0.1 μm); clearance angle, $-6°$; fluid, distilled water; estimated tool-tip radius, 0.4 μin (10 nm). Under these conditions, the R_a surface roughness was found to be 3.2 μin (80 nm), and the critical undeformed chip thickness to avoid subsurface damage was about 1.8 μin (45 nm). (after Blake, 1988)

20 μin (0.5 μm) radius. Etched grooves had a molten appearance, suggesting that they were of thermal origin. The critical maximum depth of cut at which subsurface cracks appeared (t_c) is shown plotted in Fig. 23.19 together with the method of deriving t_c from measured values of groove width l and tip radius ρ. The critical depth of cut decreases with increase in speed to 32,800 f.p.m. (10,000 m min^{-1}) and then remains constant at about 8 μin (0.2 μm) with further increase in speed.

While the critical depths of cut (t_c) found by Fujita and Shibata (1993) and Takahashi et al. (1993) were approximately the same, the former found t_c to increase with cutting speed, while the latter found t_c to decrease with cutting speed. This could be due to Fujita and Shibata employing continuous cutting at constant depth, while Takahashi et al. employed intermittent cutting at a varying depth of cut.

Although SPDT is best adapted to relatively ductile materials (Cu, Al, Pb, etc.), because of the rapid loss of cutting edge sharpness with harder materials, Blake (1988) has applied this process to diamond turning of two very brittle materials—single-crystal germanium and silicon. It was found that surfaces of excellent finish that were free of surface pits could be produced by use of a gem-grade diamond tool having a large ($\frac{1}{8}$ in = 3.18 mm) nose radius in the feed direction, as shown in Fig. 23.20. The tool-tip radius at the cutting edge was estimated to be 10 nm. The actual operation was face turning with a tool having a large nose radius, that was fed slowly toward the center of the work, which had a circumferential velocity V at the tool. The kinematics were similar to a phonograph (record = work and needle = tool) with the tool fed inward at a fixed rate. The width of cut b was about 50 times the feed per revolution v. The undeformed chip thickness was not constant as in orthogonal cutting but varied from 0 at A to a maximum (t_m) at B (Fig. 23.20a).

The tool was quickly raised out of the cut after equilibrium was established so that the cut shoulder could be examined for finish and defects. The defects were in the form of pits that increased in depth and density as t increased up the shoulder. It was found that there was a value of undeformed chip thickness (t_c) below which no pits were found. In addition, the value of t_c was so small over the feed per revolution distance that there was appreciable burnishing, leading to very good finish ($R_a \approx 3.2$ μin or 80 nm). This was called ductile regime machining, even though practically all of the material removed involved very brittle behavior. Figure 23.21 is an SEM of germanium chips produced under "ductile regime" cutting, which clearly indicates appreciable fracture associated with their formation.

The most important results of Blake's study are as follows:

- By use of a large nose radius and a small feed, the bulk of the material may be removed without regard for fracture as long as t at the feed distance (t_v in Fig. 23.20) is less than t_c.

- There is a critical value of undeformed chip thickness below which no subsurface cracks are obtained. Thus, t_c under a standard set of conditions (V, v, R, α, and fluid) may be used to characterize a material in SPDT.

- The depth of cut (d) is of negligible importance but, instead, undeformed chip thickness (t) is all important.

Fig. 23.21 SEM micrograph of chips produced under the conditions or Fig. 23.20. (after Blake, 1988)

- Values of t_c increase with a decrease (more negative) in the rake angle and with a decrease in the clearance angle.

- The critical undeformed chip thickness is greater for Si than for Ge, but Ge gives the better finish.

- Crystal orientation is important, t_c being lowest for the (110), cutting direction.

The poorer finish obtained with silicon was attributed to microchipping of the cutting edge of the diamond. It appears that SPDT is best suited to ultra-precision machining of relatively soft materials (Cu, Al, etc.), because it requires a tool having a very small tip radius which will relatively quickly wear when cutting hard materials, even with a diamond tool. It appears that UPDG offers greater promise than SPDT for precision finishing of hard, brittle materials such as glass and ceramics.

The fact that ductile regime cutting is really a result of the deformation volume being small relative to the defect spacing is further illustrated by the observation that microhardness impressions made in marble (a very brittle material) will show no signs of fracture provided the load on the indenter (and hence the deformation volume) is sufficiently small as shown in Fig. 5.35.

Further discussion of the true nature of ductile regime behavior (size effect) is given in Shaw et al. (1990). Figure 23.22 shows the variation of resisting shear stress over a very wide range of deformation volume as measured by the equivalent undeformed chip thickness (t).

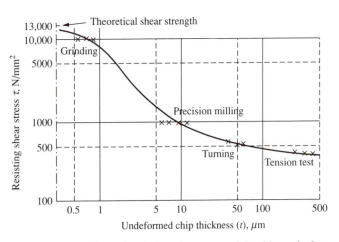

Fig. 23.22 Variation of resisting shear stress (τ) with equivalent undeformed chip thickness (t) for SAE 1112 steel. (after Backer et al., 1952; altered by Taniguchi, 1993)

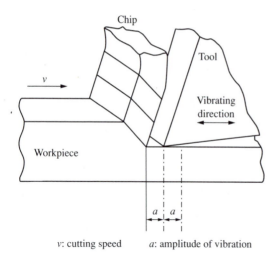

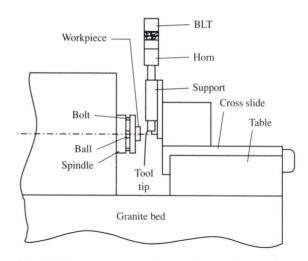

Fig. 23.23 Arrangement used in initial ultrasonic vibrating tool studies. (after Moriwaki and Shamoto, 1991)

Fig. 23.24 Arrangement used to cut independent grooves with ultrasonically vibrating diamond tool. (after Moriwaki and Shamoto, 1992)

VIBRATING TOOLS

Moriwaki and Shamoto (1991) found that diamond tool wear when cutting steel was greatly reduced when the cutting tool was vibrated in the cutting direction with an ultrasonic frequency of 40 kHz as shown in Fig. 23.23. Because of the arrangement employed, there was also a vibration in the cutting width direction with an amplitude about $\frac{1}{15}$ that in the cutting direction at the same frequency. When face-turning free-machining 303 stainless steel (+0.15% Se) under the following conditions, an optical quality mirror of stainless steel having a surface finish of $R_t = 0.026$ μm was obtained:

- cutting speed (V): 4.2 to 1.4 m/min
- rake angle (α): 0°
- relief angle: 7°
- nose radius: 0.8 mm
- depth of cut: 2.3 μm
- cutting fluid: light mineral oil

Surfaces having a finish of less than 0.1 μm (R_t) were obtained after cutting a distance of 1600 m. This clearly shows the potential of UPDT using a vibrating tool.

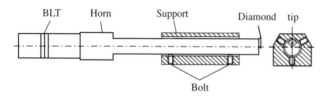

Fig. 23.25 Detail of ultrasonically vibrated tool holder. (after Moriwaki and Shamoto, 1992)

Moriwaki and Shamoto (1992) next improved their apparatus as shown in Fig. 23.24. This consisted of a bolted langevin transducer (BLT) attached to the bar holding the single-crystal diamond tool as shown in Fig. 23.25. The stepped horn magnified the amplitude of vibration by a factor of three. The tool vibrated at a frequency of 40 kHz in the axial direction.

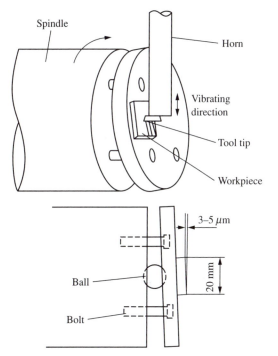

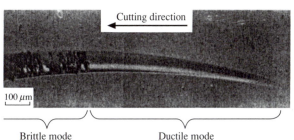

Fig. 23.27 Photomicrograph of a groove cut with gradually increasing depth of cut. The ductile/brittle transition is clearly evident, thus enabling the critical transition depth of cut to be determined. (after Moriwaki and Shamoto, 1995)

Fig. 23.26 Method of providing taper groove cutting by inclining workpiece surface to surface of spindle. (after Moriwaka and Shamoto, 1992)

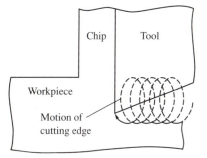

Fig. 23.28 Tool vibrating in helical mode.

This apparatus was used to cut grooves of gradually increasing and decreasing depth by mounting the surface of the workpiece at a small angle as shown in Fig. 23.26. This was done by placing a steel ball between two clamping screws. In this way the inclination of the normal to the workpiece surface and the axis of the spindle could be adjusted to be a very small value. Figure 23.27 shows a portion of a groove going from a ductile mode to a brittle mode as the depth of cut went from zero to the critical ductile/brittle value. The material tested was soda lime glass. By use of a very rigid machine tool, a sharp diamond tool (small ρ), a rake angle of 0°, a clearance angle of 7°, and a nose radius of 0.8 mm, it was possible to obtain a surface finish of $R_a = 1$ nm provided the width and depth of cut were sufficiently small to provide ductile mode cutting. It is also important that the amplitude of tool vibration in the cutting direction be sufficiently large relative to the cutting speed to enable the chip and tool to leave contact. When the tool is moving in the same direction as the chip, the flow of the chip from the work is aided. If the relative motions between chip and tool periodically oppose each other, while in contact, this reduces tool-face friction. Both of these effects tend to reduce chip thickness, increase shear angle, decrease the rate of tool wear, and improve surface finish.

Moriwaki and Shamoto (1995) next developed a design causing the tool to vibrate ultrasonically in a *helical* pattern as shown in Fig. 23.28. The tool holder was supported at nodal points for the first mode of bending. Piezoelectric ultrasonic plates (PZTs) were cemented to flat surfaces on the tool holder as shown in Fig. 23.29. The tool was caused to resonate at a frequency of 20 kHz. Sinusoidal voltages were applied to each PZT with a phase shift that enabled any elliptical shape to be developed with an amplitude of 4 μm and a vibrating speed of 131 f.p.m. (40 m/min). Since the

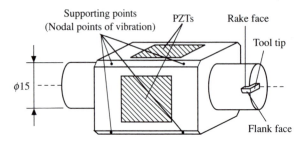

Fig. 23.29 Ultrasonic elliptical vibrating tool holder. (after Moriwaki and Shamoto, 1995)

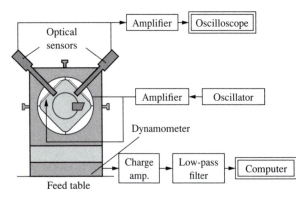

Fig. 23.30 Driving and measuring system for ultrasonic elliptical vibration cutting. (after Moriwaki and Shamoto, 1995)

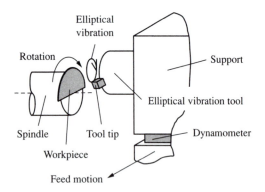

Fig. 23.31 Schematic illustration of quasi-orthogonal cutting experiments. (after Moriwaki and Shamoto, 1995)

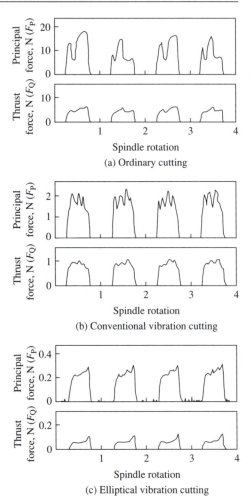

(a) Ordinary cutting

(b) Conventional vibration cutting

(c) Elliptical vibration cutting

Fig. 23.32 Principal (F_P) and thrust (F_Q) force components for three cutting modes. (after Moriwaki and Shamoto, 1995)

cutting speed must be less than the vibrating speed for the elliptical motion to be effective, the maximum cutting speed is 131 f.p.m. (40 m/min) for this design. Figure 23.30 shows the driving and measuring system employed which included a piezoelectric dynamometer.

Interrupted orthogonal cutting tests were performed as shown in Fig. 23.31. Figure 23.32 shows cutting force traces for three types of cutting:

(a) ordinary cutting (no tool vibration)

(b) conventional vibration cutting (Moriwaki and Shamoto, 1991)

(c) elliptical ultrasonic vibrating tool (Moriwaki and Shamotto, 1995)

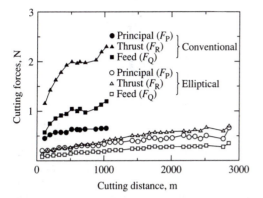

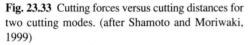

Fig. 23.33 Cutting forces versus cutting distances for two cutting modes. (after Shamoto and Moriwaki, 1999)

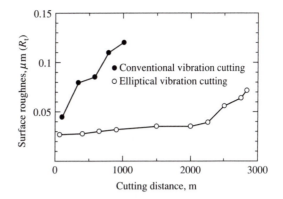

Fig. 23.34 Surface roughness versus cutting distances for two cutting modes. (after Shamoto and Moriwaki, 1999)

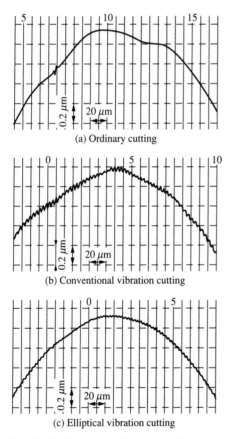

(a) Ordinary cutting

(b) Conventional vibration cutting

(c) Elliptical vibration cutting

Fig. 23.35 Surface profiles obtained by three cutting methods. (after Shamoto and Moriwaki, 1999)

The chip thicknesses were about the same for case a ($t_c = 20 \ \mu$m) and case b ($t_c = 19 \ \mu$m) but much smaller for case c ($t_c = 4 \ \mu$m). Figure 23.33 shows the variation of cutting forces with distance cut for the three types of cutting, while Fig. 23.34 shows the change in surface roughness with distance cut. Figure 23.35 shows the deviation in the shape and finish for the three types of cutting.

Shamoto and Moriwaki (1999) describe the more compact elliptical vibration tool holder shown in Fig. 23.36a. Figure 23.36b shows the bending pattern

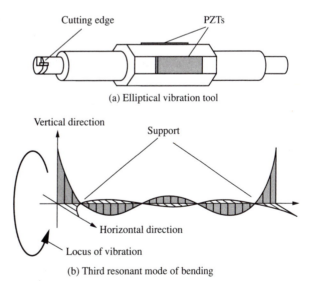

(a) Elliptical vibration tool

(b) Third resonant mode of bending

Fig. 23.36 Improved elliptical vibration tool holder. (after Shamoto and Moriwaki, 1999)

for this tool holder. This vibrator is resonant at 20 kHz. Additional results on ultrasonic vibration cutting is to be found in Ma (1999).

Schmuetz et al. (2001) have studied subsurface deformation of vibration cutting of oxygen-free high-conductivity (OFHC) copper. Diamond tools were vibrated in the cutting direction at a frequency of 20 kHz and an amplitude of 60 μin (1.5 μm) in two-dimensional orthogonal turning operations with a width of cut of 0.080 in (2 mm) and a range of depths of cut from 400 to 2000 μin (10 to 50 μm). The rake angle for vibration cutting was found to increase by a factor of 2 to 2.5 times compared with corresponding cuts without tool vibration. Plastic deformation of the cut surface was found to extend to a depth of about half the depth of cut for cutting with and without tool vibration. Strain hardening of the finished surfaces was studied by nanoindentation, and the indentation was found to increase linearly with the applied normal load on the indenter. It was suggested that grain boundaries act as barriers to plastic deformation and that the depth of plastic deformation was at least one layer of grains deep (2000 μin = 60 μm) for the diamond turned OFHC copper considered.

McGeough (2001) has edited a book that covers several of the important aspects of micro-machining by a number of experts in the field. Topics covered include

- measurement techniques
- molecular dynamics in microcutting
- micromachining by grinding and with diamond cutting tools
- application of several special materials
- other removal operations (electrodischarge, laser, electrochemical, ion beam, electron beam, high-resolution lithography)

TESTS WITH VARIABLE DEPTH OF CUT

Brinksmeier et al. (1994) have also studied the behavior of a diamond tool as it moves across a surface that is inclined slightly to the path of the tool, as shown in Fig. 23.37. Many workpiece materials were studied. This reference covers results obtained with workpieces of electroless nickel (eNi) and oxygen-free high-conductivity copper (OFHC–Cu). The eNi had a phosphorus content of 11% and was deposited on an OFHC–Cu substrate with a 12 μm thickness. The tool was sharp (10 nm edge radius), had a 0° rake angle, a nose radius of up to 6 mm, and a cutting speed of 5 mm/s. There was considerable side flow with OFHC–Cu but essentially none with eNi. The edges of the copper chips were unusually rough probably due to the increasing depth of cut and the effect of grain boundaries.

In Brinksmeier et al. (1995) results were presented for inclined plunge cutting for monocrystalline brittle materials (silicon and germanium). Diamond tools were used having rake angles of 0° and −45°, a nose radius of 6 mm, an edge radius of 40 nm, and a cutting speed of

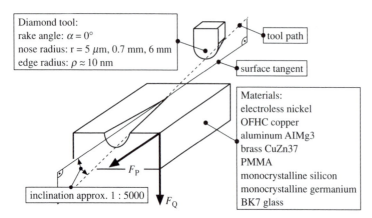

Fig. 23.37 Principle of plunge-cut machining. (after Brinksmeier et al., 1994)

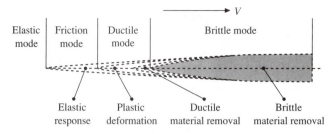

Fig. 23.38 Regimes of behavior in plunge-cut machining. (after Brinksmeier et al., 1995)

20 mm/min. The following modes of performance were tested for an increasing depth of cut (Fig. 23.38):

- elastic mode (no trace)
- friction mode (no groove, slight trace)
- ductile mode (groove without cracks)
- brittle behavior (transverse cracks)

Figure 23.39 gives transition values and a thrust force (F_Q) trace versus cutting distance for a silicon workpiece.

Figure 23.40 gives cutting (F_P) and tranverse (F_Q) forces for the ductile-brittle transition for a range of conditions for both silicon and germanium. The transverse force (F_Q) is larger than the cutting force (F_P) in all cases, and germanium is seen to be much more brittle than silicon. Comparison of results for rake angles of 0° and −45° for germanium indicated greater brittleness for 0° than for −45°.

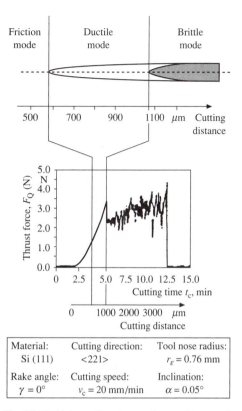

Material: Si (111)	Cutting direction: <221>	Tool nose radius: $r_\varepsilon = 0.76$ mm
Rake angle: $\gamma = 0°$	Cutting speed: $v_c = 20$ mm/min	Inclination: $\alpha = 0.05°$

Fig. 23.39 Values of cutting mode transitions and a thrust force trace (F_Q) for a silicon workpiece. (after Brinksmeier et al., 1995)

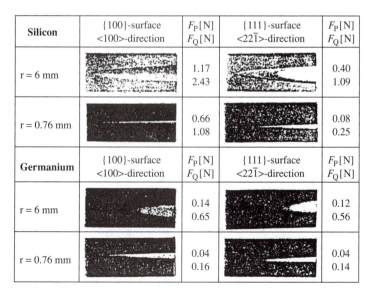

Silicon	{100}-surface <100>-direction	F_P[N] F_Q[N]	{111}-surface <22$\bar{1}$>-direction	F_P[N] F_Q[N]
r = 6 mm		1.17 2.43		0.40 1.09
r = 0.76 mm		0.66 1.08		0.08 0.25
Germanium	{100}-surface <100>-direction	F_P[N] F_Q[N]	{111}-surface <22$\bar{1}$>-direction	F_P[N] F_Q[N]
r = 6 mm		0.14 0.65		0.12 0.56
r = 0.76 mm		0.04 0.16		0.04 0.14

Fig. 23.40 Representative cutting force values at start of brittle mode cutting for diamond tools having a −45° rake angle when cutting single crystal silicon and germanium having different crystal orientations. (after Brinksmeier et al., 1995)

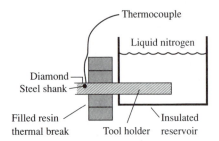

Fig. 23.41 Cryogenic tool holder used for low workpiece temperature tests. (after Evans and Bryan, 1991)

SPDT with Low Workpiece Temperature

Paul et al. (1996) studied tool wear in SPDT at ambient and liquid nitrogen temperatures for SPDT of stainless steel. The same result was obtained at both temperatures for pure iron. However, while the wear rate was higher for steel than for iron at ambient temperature, the reverse was true at liquid nitrogen temperature. This may be explained by the fact that at ambient temperature, steel will give a higher temperature rise than pure iron resulting in more graphitation. However, at low temperature, the diffusion of carbon into iron is of greatest importance favoring a greater diffusion rate into pure iron than into steel.

Evans and Bryan (1991) have also suggested that tool wear in diamond turning of stainless steel should be greatly reduced at very low workpiece temperatures. This should be the case since the reaction of carbon and iron at elevated temperatures plays an important role, and reduction of the tool-face temperature should exponentially reduce the reaction rate and hence the wear rate. Using the arrangement shown in Fig. 23.41, the wear rate in diamond turning of stainless steel was reduced considerably.

SPDT of Optical Glass

Fang and Venkatest (1998) have studied SPDT of a silicon optical glass. Taper-cutting of grooves was again employed, and the conclusion was reached that in order to attain a finish of $R_a = 1$ nm on a surface initially with $R_a = 24$ nm by SPDT without polishing requires ductile cutting (i.e., the depth of cut (d) and feed (f) must be below a critical volume based on the fact that the probability of encountering a crack-causing defect decreases with the decrease of deformation volume, frequently referred to as the size effect). It is also important that the cutting force component in the depth of cut direction (F_Q) be large relative to the force in the cutting direction (F_P). This is strongly influenced by the effective rake angle (α_e) and the nose radius of the tool (R), not to be confused with the sharpness of the cutting edge (ρ). Figure 23.42 shows three situations:

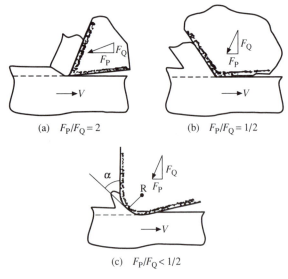

Fig. 23.42 Three SPDT tool situations for finishing optical glass without lapping and polishing. (after Fang and Venkatesh, 1998)

(a) a sharp tool (small ρ), $R = 0$ and a positive rake angle (α)

(b) a sharp tool (small ρ), $R = 0$ and a negative rake angle (α)

(c) a tool with a large nose radius (R) relative to the depth of cut and a 0° rake angle

This figure shows representative values of forces F_P, F_Q, and F_P/F_Q. In order to promote ductile cutting without needlessly decreasing d and f (which is costly relative to productivity), Fig. 23.42c is to be preferred. This gives rise to a large hydrostatic compressive stress that prevents crack formation with a relatively small cutting force (F_P). A small F_P is important since this consumes less energy giving rise to a lower wear rate. To obtain the same finish with a rake angle (α) of $-25°$ as when $\alpha = 0$ requires a reduction in the cutting speed of about 10×. For example, a 1 nm finish (R_a) may be achieved at about 90 m/min while a tool with a rake angle of $-25°$ would require a speed of about 9 m/min.

It is recommended that for SPDT of single-crystal or polycrystalline silicon, a single-crystal diamond tool be used with a rake angle of $0°$, a nose radius (R) of 0.5 mm, a depth of cut of 1 μm, a feed of 0.4 mm/rev, and a cutting speed of about 10 m/min. This should convert an initial surface having an R_a roughness of 25 nm ($R_t \approx 1.4$ nm) to one having an R_a of about 1 nm ($R_t \approx 10$ nm).

While it is possible to produce a somewhat better finish by SPDT on optical glass, the process is relatively slow, and the usual technique of fine grinding, lapping, and polishing may be less expensive except in mass production operations.

MOLECULAR DYNAMICS

In the late 1980s there was increasing interest in finishing copper and aluminum surfaces by micro-machining using SPDT instead of grinding and polishing. This called for a better understanding of chip formation at very small depths of cut where atomic behavior was involved instead of only plastic flow as in conventional machining. This required the development of a new branch of science called molecular dynamics (MD).

MD involves the interaction of a large number of atoms as deformation occurs on an atomic scale. A potential energy function is used that approximates the interaction of individual atoms in accordance with Newtonian mechanics. The major problem initially encountered was the large amount of computer time required to solve an enormous number of simultaneous equations. This initially required a super main frame computer.

With time, a number of approximations evolved that greatly reduced the computing requirements. It was found profitable to consider the atoms in the surface of a cutting tool to be rigid as well as those a reasonable distance from the action. The active atoms were arranged according to the crystal system involved (e.g., FCC for Cu and Al, BCC for Fe-based materials, and tetragonal for diamond). No defect structure was assumed which still represents an important limitation for modeling real materials. In cutting applications the cutting speed has to be unusually high; otherwise, computing time becomes enormous. The tool must be moved into the work in time steps that are smaller than the period of atomic vibration. These time steps are usually 10 to 15 fs (1 fs = 10^{-9} s). Molecular statics (MS) is a technique for reducing computing time where only the motion of atoms for which the resultant force is zero is considered. A joint investigation involving MD in cutting was undertaken by Lawrence Livermore National Laboratory (LLNL) in the United States and the Precision Engineering Department at Osaka University in Japan. A molecular dynamics modeling video applied to indentation and metal cutting was first presented by Stowers at the CIRP General Assembly in Berlin in August 1990. Then, Ikawa et al. (1991b) presented the first report from the Osaka group concerning an atomistic approach to nanochip formation. This was followed by a further publication concerning this subject (Stowers et al., 1991).

Shimada et al. (1993) described experimental and MD simulation results of nanometric chip formation that were in relatively good agreement. It was demonstrated that with a sharp diamond tool and a high-grade machine tool, it was possible to produce precise cuts as small as 1 nm.

Figure 23.43 shows values of F_Q (width cut) (open points) measured by a piezoelectric dynamometer having a resolution of 1 gram force used in cutting oxygen-free high-conductivity copper

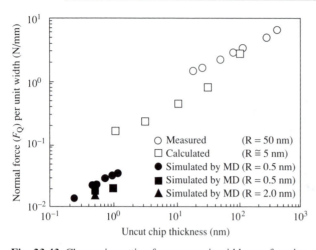

Fig. 23.43 Change in cutting force per unit width as a function of uncut chip thickness (t) (R = nose radius; MD = molecular dynamics). (after Shimada et al., 1993)

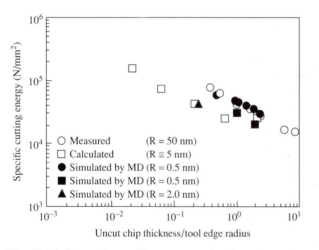

Fig. 23.44 Change in specific work as a function of nominal uncut chip thickness normalized by cutting edge radius (ρ). (after Shimada et al., 1993)

(OFHC–Cu) with an uncut chip thickness (t) ranging from 20 to 440 nm. Other cutting conditions were as follows:

- tool: diamond, rake angle $\alpha = 0°$, clearance angle = $-7°$, estimated radius (ρ) = 20 nm
- work: polycrystalline OFHC–Cu
- width of cut: 6–19 μm
- work temperature: 293 K

The simulated results for different values of nose radius (R) are indicated by solid points.

Cutting forces were found to increase with increase in tool nose radius (R) probably due to an increase in burnishing/plowing action.

MD simulation is useful for estimating cutting forces that are smaller than those conveniently measured by a dynamometer. Figure 23.43 is an example.

Figure 23.44 shows the variation of specific energy (cutting energy per unit volume) with increase in the uncut chip thickness (t) divided by the radius of the cutting edge (ρ). This size effect is probably due to the same reason pertaining in conventional machining, that is, due to a decrease in the probability of encountering a defect with decrease in the volume involved.

When the distribution of kinetic energy in an atomic model is expressed in terms of equivalent temperature, a wide temperature distribution is obtained as shown in Fig. 23.45. This represents an entirely different temperature distribution than is found in macromachining.

Whereas the transfer of heat in metals is controlled by the transport of electrons, in MD energy is transferred by lattice vibration. This gives rise to a lower thermal gradient in MD than is found experimentally. This requires an adjustment (scaling factor) that takes this difference in thermal transport into account, making thermal analysis such as that leading to Fig. 23.45 quite complex.

Shimada et al. (1993) have used MD to investigate the maximum depth of cut that is possible and the ultimate surface finish and depth of deformed layer that is possible for a diamond tool cutting copper and aluminum. Table 23.4 summarizes the results for two values of cutting edge radius (ρ). Figure 23.46 shows the extent of atomic disturbance for an uncut chip thickness of 0.5 nm for one of the conditions in Table 23.4 ($\rho = 5$ nm). Dislocations are seen to extend to a greater depth for Al than for Cu, and the depth of the deformed layer is greater for Al than for Cu. The surface finish is also seen to be poorer for Al than for Cu which was also found to be the case experimentally.

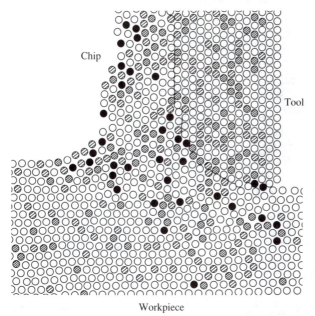

Fig. 23.45 Distribution of kinetic energy in the model as expressed in equivalent temperature. Cutting speed: 39,360 f.p.m. (200 m/s); uncut chip thickness: 0.02 μin (0.5 nm); tool edge radius: 0.08 μin (2.0 nm). The equivalent temperature of individual atoms is expressed as follows: $\bigcirc$ < 500 K; $\oslash$ 500–1000 K; $\obslash$ 1500–2000 K; $\bullet$ > 2000 K. (after Shimada et al., 1993)

TABLE 23.4 Effect of Cutting Parameters on the Surface Quality (with a Cutting Speed of 200 m/s)[†]

Work Material	Cutting Edge Radius (nm)	Uncut Chip Thickness (nm)	Surface Roughness (nm)	Depth of Deformed Layer (nm)	Depth of Dislocation (nm)
Cu	5	0.5	0.5	4.5	5.3
Cu	10	0.6	0.7	4.1	7.6
Al	5	0.5	0.8	6.5	9.7
Al	10	1.2	1.4	9.7	≥ 10

[†] After Shimata et al., 1993.

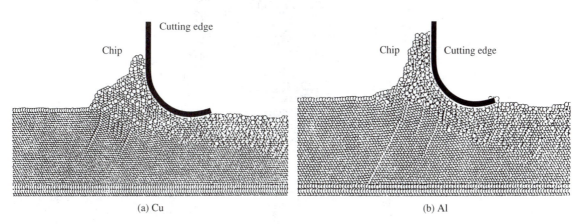

(a) Cu   (b) Al

Fig. 23.46 Roughness and deformed layer of work surface when microcutting (a) copper and (b) aluminum. Cutting speed: 3936 f.p.m. (200 m/s); cutting edge radius: 0.2 μin (5 nm); uncut chip thickness: 0.02 μin (0.5 nm). (after Shimada et al., 1993)

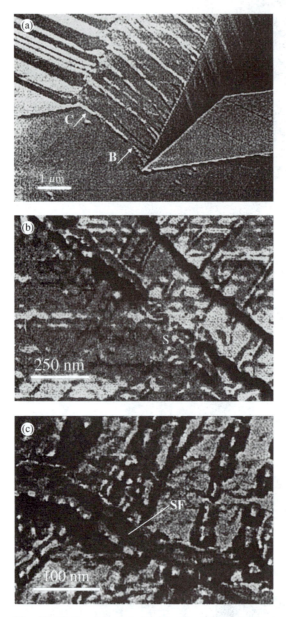

Fig. 23.47 Concentrated shear bands produced when a noncrystalline amorphous material is machined in an atomic force microscope (AFM) at low speed. Figure 23.47b is a magnified view of region B in Fig. 23.47a, while Fig. 23.47c is a magnified view of region C in Fig. 23.47a. S in Fig. 23.47b indicates a region of concentrated shear without gross fracture while SF in Fig. 23.47c shows discontinuous regions of fracture perpendicular to the main shear plane direction. (after Ueda et al., 1999)

Shimada et al. (1995) have studied (by MD simulation) the brittle-ductile transition by microindentation and micromachining of defect-free surfaces. From the results presented it may be concluded that any material regardless of its macroscale brittleness can be machined or indented in a ductile mode if the undeformed chip thickness or depth of indentation is sufficiently small (a nanometer or less).

Inamura et al. (1994) studied the transformation of an atomic model to the corresponding continuum (macro) model in a nanoscale setting. This involves periodically adjusting the potential energy within a given volume in the atomic model to be approximately equal to the strain energy in the same volume in the continuum model.

Ueda et al. (1999) have continued their previously discussed study of atomic scale chip formation (Ueda and Manabe, 1992) by cutting an amorphous metal (metallic glass = Fe_{78}, B_{11}, Si_9) in an atomic force microscope (AFM) and using MD-FEM computer simulation. The amorphous metal was used to study the microcutting behavior of a material that is noncrystalline and has no slip system, grain boundaries, dislocations, or stacking faults.

Orthogonal cuts were made in the AFM with a single-crystal natural diamond under the following conditions:

- rake angle: 30°
- nose radius: 30 nm
- undeformed chip thickness: 5 nm to 5 μm
- width of cut: 0.02 mm
- cutting speed: 0.01 to 10 μm/s

The chips were found to have a lamellar structure with a spacing of concentrated shear zones ranging in width from 0.6 to 1.0 μm. The spacing of concentrated shear zones increased with increase in undeformed chip thickness but was essentially independent of cutting speed. Figure 23.47 shows concentrated shear bands in a typical cut made in

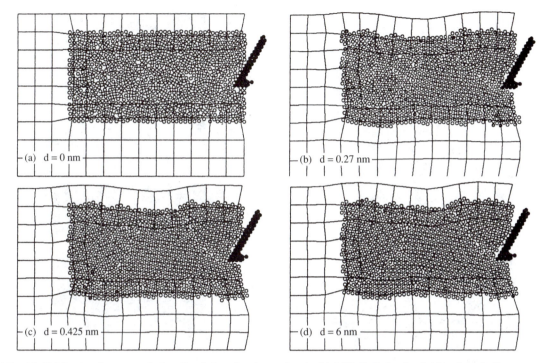

Fig. 23.48 The development of a localized shear band associated with MD analysis of a rigid tool gradually penetrating an amorphous metal. (after Ueda et al., 1999)

the AFM. Since it is surprising to find chips having a card-like structure in a noncrystalline amorphous material, it was decided to perform a MD-FEM study in search of an explanation.

Two grids were employed: an outer one for the rigid plastic FEM analysis and an inner one for the molecular dynamics (MD) analysis that involved 984 atoms. The region where these two treatments come together was adjusted to be consistent after each 30 steps of tool advance of 0.05 nm/step. Figure 23.48 shows results for four different displacements (d) of the rigid 30°-rake-angle tool ranging from 0 to 6 nm for a 4.5 nm depth of cut. Localized shear is seen to begin in Fig. 23.48b with a shear angle of about 45°. The chip segment that began in Fig. 23.48b continues to develop at Fig. 23.48c and is completed at Fig. 23.48d. More detailed MD views of the development of a periodic plane of concentrated shear are shown in Fig. 23.49. This shear plane develops as a result of cooperative group movement of atoms that involve formation and coalescing of voids.

This study shows that even when cutting a noncrystalline material at the atomic level, periodic planes of concentrated shear will develop that are similar except for size to those pertaining in ordinary cutting of a polycrystalline material.

Komanduri and Raff (1999b) have presented a very thorough review of the application of MD to the nanocutting process (including 121 references). A discussion of the advantages and limitations of the process is included together with the influence of a wide range of processing parameters (rake angle, α; edge radius, ρ; nose radius, R; and crystal orientation).

Komanduri et al. (2000) have employed MD to study atomic scale friction in indentation/ scratching tests when an indenter having a 45° rake angle (90° included angle) was drawn across a single-crystal aluminum surface. Scratch indentation depths ranging from 0 to 0.8 nm were used. Figure 23.50 shows MD simulation plots for a scratch depth of 0.4 nm. The ratio of cutting (horizontal)

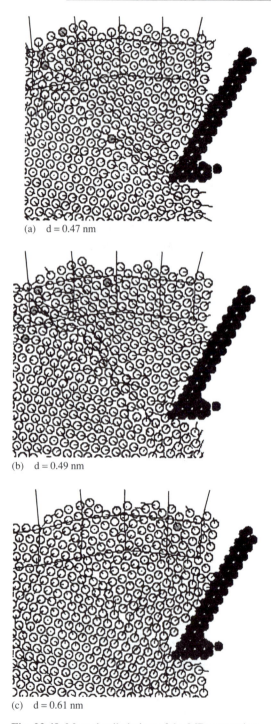

(a)   d = 0.47 nm

(b)   d = 0.49 nm

(c)   d = 0.61 nm

Fig. 23.49 More detailed view of the MD generation of a shear band in an amorphous metal. (after Ueda et al., 1999)

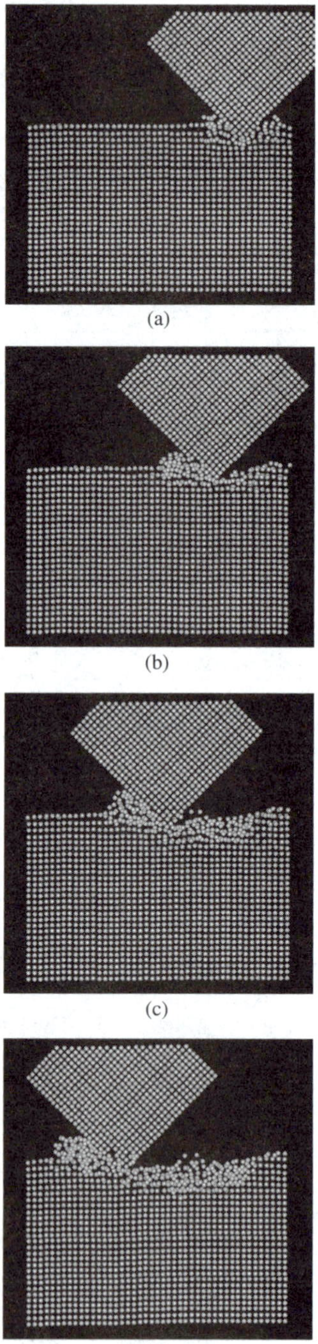

(a)

(b)

(c)

(d)

Fig. 23.50 MD simulation plots for a clean 90° included angle indenter moving across a single crystal aluminum surface at an indentation depth 0.01×19^{-9} in (0.4 nm). (after Ueda et al., 1999)

to thrust (vertical) forces (coefficients of friction) was found to be 0.6 for all depths. The fact that both the indentation and scratch hardness values increased with decreasing depth strongly suggests a size effect. The friction coefficient dropped below 0.6 only when the penetration dropped to zero and pure sliding pertained. The surfaces involved were very clean and the presence of a lubricant would undoubtedly lower the friction coefficient.

REFERENCES

Aronson, R. B. (1994). *Mfg. Eng.* **113**, 63.

Backer, W. R., Marshall, E. R., and Shaw, M. C. (1952). *Trans ASME* **74**, 61–72.

Blake, P. N. (1988). Ph.D. dissertation. North Carolina State University, Raleigh.

Blake, P. N., and Scattergood, R. O. (1988). *J. Am. Ceram. Soc.* **73/4**, 949–959.

Brinksmeier, W., Preuss, W., and Riemer, O. (1994). *Proc. 3rd Int. Conf. on Ultraprecision in Mfg. Eng.*, Aaken, Germany.

Brinksmeier, W., Preuss, W., and Riemer, O. (1995). *Proc. 8th Int. Precision Eng. Seminar*, Compiegne, France.

Childs, T. H. C., Maekawa, K., and Maulik, P. (1988). *Material Science and Tech.* **4/11**, 1006–1019.

Corbett, J., McKeown, P. A., Peggs, P. N., and Whatmore, R. (2000). *Annals of CIRP* **49/2**, 523–546.

Donaldson, R. R., Syn, C. K., Taylor, J. S., Ikawa, N., and Shimada, S. (1987). *UCRL 97606*.

Donaldson, R. R., Syn, C. K., Taylor, J. S., and Riddle, R. A. (1985). *Chip Science Report in Fabrication Technology. Lawrence Livermore National Laboratory*, pp. 69–85.

Evans, C., and Bryan, J. B. (1991). *Annals of CIRP* **40/1**, 571–575.

Fang, F. Z., and Venkatesh, V. F. (1998). *Annals of CIRP* **47/1**, 45–49.

Fujita, S., and Shibata, J. (1993). *J. Japan Soc. Prec. Engrs.* **27**, 138.

Herring, C., and Galt, J. (1952). *Phys. Rev.* **85**, 1060.

Ikawa, N., Shimada, S., and Morooka, H. (1987). *Bull. Japan Soc. Prec. Engrs.* **21**, 233.

Ikawa, N., Donaldson, R. R., Komanduri, R., Koenig, W., McKeon, P. A., and Moriwaki, T. (1991a). *Annals of CIRP* **40(2)**, 587.

Ikawa, N., Shimada, J., Tanaka, H., and Ohmori, G. (1991b). *Annals of CIRP* **40/1**, 551–554.

Inamura, T., Takezawa, N., and Kumaki, Y. (1993). *Annals of CIRP* **42/1**, 79–82.

Inamura, T., Takezawa, N., Kumaki, Y., and Sata, T. (1994). *Annals of CIRP* **43/1**, 47–50.

Komanduri, R., Chandrasekaran, N., and Raff, L. (1999). *Annals of CIRP* **48/1**, 69–72.

Komanduri, R., Chandrasekaran, N., and Raff, L. (2000). *Physical Review B* (The Amer. Phys. Soc.) **61/20**, 14007–14019.

Komanduri, R., and Raff, L. (1999a). In *Proc. of U.S.-Japan Symposium on Finishing of Advanced Materials*, Oklahoma State University, pp. 153–207.

Komanduri, R., and Raff, L. (1999b). "New Developments in the Finishing of Advanced Materials," *U.S.-Japan Symposium* sponsored by the U.S. National Science Foundation, Oklahoma State University, pp. 153–208.

Liang, Y., Moronuki, N., and Furukawa, Y. (1994). *Prec. Engng.* **16**, 132.

Lucca, D. A., and Seo, Y. W. (1994). *Annals of CIRP* **43**(1), 43.

Ma, C. (1999). Dr. Eng. dissertation. Kobe University.

McGeough, J. A. (2001). *Micromachining of Engineering Materials*. Marcel Dekker, New York.

Moriwaki, T., and Shamoto, E. (1991). *Annals of CIRP* **40/1**, 559–562.

Moriwaki, T., and Shamoto, E. (1992). *Annals of CIRP* **41/1**, 141–144.

Moriwaki, T., and Shamoto, E. (1995). *Annals of CIRP* **44/1**, 31.

Moronuki, N., Liang, Y., and Furukawa, Y. (1994). *Prec. Engng.* **16**, 124.

Paul, E., Evans, C. J., Mangameilli, A., McGlaufin, M. L., and Polvani, R. S. (1996). *Prec. Eng.* **18/11**, 4–19.

Schmuetz, J., Brinksneier, E., and Bischoff, E. (2001). *Prec. Eng.* **25**, 218–223.

Shamoto, E., and Moriwaki, T. (1999). *Annals of CIRP* **48/1**, 441–444.

Shaw, M. C. (1972). *Mech. Chem. Engng. Trans., Inst. Engrs. (Australia)* **MC8**, 73.

Shaw, M. C. (1996). *Principles of Abrasive Processing*. Clarendon Press, Oxford.

Shaw, M. C., Shafer, H. G., and Adler, M. (1990). *Am. Soc. Prec. Eng. Annual Meeting*. Rochester, N.Y.

Shimada, S., Ikawa, N., Inamura, T., Takezawa, N., Ohmori, H., and Sata, T. (1995). *Annals of CIRP* **44/1**, 523–526.

Shimada, S., Ikawa, N., Tanaka, H., Ohmori, G., and Uchikoshi, J. (1993). *Annals of CIRP* **42/1**, 91–94.

Stowers, I. F., Belak, J. F., Lucca, D. A., Komanduri, R., Moriwaki, T., Okuda, K., Ikawa, N., Shimada, S., Tanaka, H., Dow, T. A., and Dresher, J. D. (1991). *Proc. of Amer. Soc. Prec. Eng. Annual Meeting*.

Takahashi, M., Ueda, S., and Kurobe, T. (1993). *Japan Soc. Prec. Engng.* **27**, 140.

Taniguchi, N. (1992). *Int. J. Japan Soc. Prec. Engrs.* **27**, 14.

Taniguchi, N., Ikeda, M., Miyamoto, I., and Miyazaki, T. (1989). *Energy-Beam Processing of Materials*. Clarendon Press, Oxford.

Taniguchi, T. (1993). *ASPE Distinguished Lecture*. Seattle, Wash.

Ueda, K., Fu, H., and Manabe, K. (1999). *Machining Science and Technology* **3/1**, 61–75.

Ueda, K., and Manabe, K. (1992). *Annals of CIRP* **41**(1), 129.

Wang, Z. Y., Sahay, C., and Rajurkar, K. P. (1994a). SME-NAMRI publication MR 94–144.

Wang, Z. Y., Sahay, C., and Rajurkar, K. P. (1994b). *Proc. S. M. Wu Symposium on Mfg. Sc. at NW University*. ASM, Detroit, **1**, 98–100.

Yoshioka, J., Hashimoto, F., Miyashita, M., Kanai, A., Abo, T., and Daito, M. (1985). *ASME PED* **16**, 209.

24 UNUSUAL APPLICATIONS OF THE METAL CUTTING PROCESS

The primary objective in metal cutting is to remove unnecessary material to produce a useful part, and previous chapters of this book have discussed a number of problems associated with this objective. However, there are a number of applications where the item of primary interest may be quite different such as

- the production of chemically active nascent surfaces
- the absorption of unwanted energy
- production of thin ribbon-like strips
- new methods of producing electrical conductors
- cut-welding
- the production of metal fibers for reinforcing concrete, etc.
- the production of nanocrystalline structures

ACTIVE SURFACES

Reactions involving metals and organic compounds have been used in chemical synthesis for a long time. For example, zinc was reacted with organic (carbonaceous) compounds by E. Frankland in 1849. V. Grignard was awarded the Nobel Prize in 1912 for his contributions to chemistry. Grignard reactions are among the most versatile available to the organic chemist. Practically any compound may be produced by a Grignard reaction. However, use of these reactions is limited because they are difficult to initiate and control and are extremely dangerous and labor intensive. Danger is associated with the highly combustible and explosive ethyl ether used as a catalyst/solvent. High cost is associated with the fact Grignard reactions are normally carried out as a batch rather than a continuous flow process. A challenging design problem is to devise a Grignard process that is controllable, safe, and economical. In the paragraphs below the conventional Grignard operation is first described followed by description of an improved technique called mechanical activation.

The Grignard process involves two steps. First an organic halide (organic hydrocarbon containing chlorine, bromine, or iodine) is reacted with magnesium. This is normally done by placing magnesium chips in a vessel and adding the halide and ethyl ether. A condenser is then attached, and the mixture is refluxed (boiled/condensed) in the absence of air and moisture as shown in Fig. 24.1a until reaction begins. There is usually an induction period during which no reaction

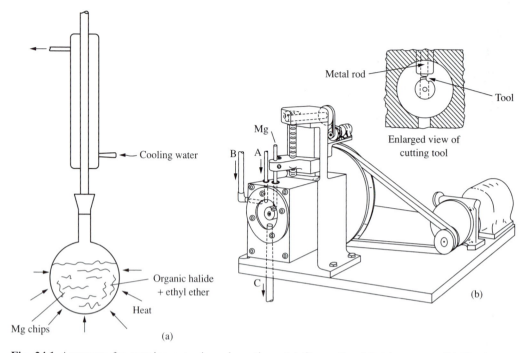

Fig. 24.1 Apparatus for carrying out grignard reactions. (a) Conventional batch process. (b) Continuous process involving mechanical activation. (after Shaw, 1945, 1947, 1948)

occurs. However, once reaction begins, the source of heat must be quickly removed and an ice pack quickly applied to reduce the rate of reaction to a safe level. This is because these reactions are exothermic (release heat) and autocatalytic (the rate of reaction varies exponentially with the amount of product produced). If the reaction once started is not brought under control, the hazardous ether vapor may explode with serious consequences.

The following equation illustrates the first step in a Grignard synthesis:

$$
\underset{\text{amylbromide}}{
\begin{array}{c}
\text{H} \ \ \text{H} \ \ \text{H} \ \ \text{H} \ \ \text{H} \\
| \quad | \quad | \quad | \quad | \\
\text{H} - \text{C} - \text{C} - \text{C} - \text{C} - \text{C} - \text{Br} \\
| \quad | \quad | \quad | \quad | \\
\text{H} \ \ \text{H} \ \ \text{H} \ \ \text{H} \ \ \text{H}
\end{array}}
\ + \ \text{Mg} \
\xrightarrow[\text{Heat}]{\text{Ether}}
\underset{\text{amylmagnesiumbromide}}{
\begin{array}{c}
\text{H} \ \ \text{H} \ \ \text{H} \ \ \text{H} \ \ \text{H} \\
| \quad | \quad | \quad | \quad | \\
\text{H} - \text{C} - \text{C} - \text{C} - \text{C} - \text{C} - \text{MgBr} \\
| \quad | \quad | \quad | \quad | \\
\text{H} \ \ \text{H} \ \ \text{H} \ \ \text{H} \ \ \text{H}
\end{array}}
\quad (24.1)
$$

The second step involves reacting the Grignard reagent with one of a wide variety of compounds to obtain the desired final product. For example, if formaldehyde and then water is reacted with the Grignard reagent in Eq. (24.1), the final product will be the normal primary alcohol hexanol:

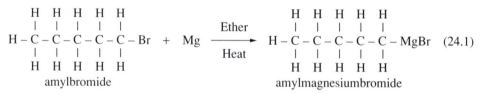

Equation (24.1) may be generalized as follows:

$$R\text{-}X + Mg \xrightarrow[\text{Heat}]{\text{Ethyl Ether}} R\text{-}Mg\text{-}X \qquad (24.3)$$

where R represents any aromatic or aliphatic hydrocarbon and X is a halide (Cl, Br, or I). Similarly the second reaction may involve one of a wide variety of secondary items. For example, water alone will lead to a hydrocarbon, $CO_2 + H_2O$ will lead to an organic acid, etc.

An improved procedure for carrying out a Grignard reaction is illustrated in Fig. 24.1b. Here a magnesium rod is converted into small chips under the surface of the liquid reactants (halide and ethyl ether) as shown in the insert in Fig. 24.1b. Liquids A and B are introduced continuously into the sealed reaction chamber along with dry nitrogen to exclude air and water vapor. First stage reaction products are extracted continuously at C. Small chips, having an undeformed thickness of about 0.001 in (25 μm), are generated under the surface of the liquid reactants and begin to react immediately as they fall toward the outlet at C.

There are three reasons why a reaction begins immediately:

- high pressure (pressures up to the hardness of the metal cut are present on the surface of the chip as it is generated)
- high temperature (surface temperatures approaching the melting point of the metal being cut will be present at the chip–tool interface)
- highly reactive surfaces are continuously generated

When a new surface is generated, there is an excess of electrons in the new surface. For new surfaces generated in vacuum, these excess electrons are expelled to the atmosphere as the surface comes to equilibrium and are called exoelectrons. For new surfaces generated in air, an oxide will normally form. For new surfaces generated under the surface of a liquid, an appropriate reaction will occur induced by the presence of excess free electrons. Collectively the three items mentioned above (high temperature, high pressure, and enhanced surface reactivity) are referred to as mechanical activation.

This method of organometallic synthesis is preferable to conventional synthesis for the following reasons:

- safety (a very small quantity of material is reacting at any one time, and even this is contained in a steel reaction chamber)
- control (if the rate of reaction becomes too great, the cutting speed need only be reduced by reducing the rpm of the cutter)
- continuity (the conventional batch arrangement is replaced by a continuous one; the main feature involved is converting *chemical science* into *chemical engineering*)

When properly set up, inputs at A and B are in equilibrium with the output at C.

When the two steps of a Grignard synthesis are combined, this is called a Barbier reaction. Such a procedure which is normally not possible in a conventional batch procedure is frequently made possible by mechanical activation, and this constitutes an added advantage.

Additional discussion of this topic is presented in Shaw (1945, 1947, 1948).

Diamond Synthesis

A method of converting graphite to diamond that involves metal cutting (grinding) has been described in Komanduri and Shaw (1974). Diamond grits in grinding wheels are frequently coated with nickel (about 55% of the weight of a diamond grit) in resin-bonded wheels used in wet

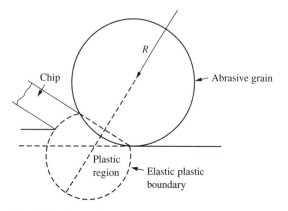

Fig. 24.2 Mechanism of chip formation in fine grinding. (after Shaw, 1972)

grinding operations. This increases the volume of a grit by 15%, and the increase in diameter provides a larger area that improves bonding. The metal coating will also promote heat transfer away from the wheel as well as protect the grits against loss due to mechanical shock or thermal fatigue.

The synthesis of diamonds from graphite in the size range < 0.004 to 0.040 in (< 0.1 to 1 mm) requires an ultra-autoclave capable of operating at pressures and temperatures in the vicinity of 100,000 atm and 2000 °K as previously described in Chapter 14. In addition, a molten metal solvent/catalyst is required so that the conversion of graphite to diamond may be achieved in a relatively short time (minutes). Catalysts employed include chromium, manganese, and tantalum plus all elements of group VIII of the periodic table, but nickel is used most frequently. The shaded region of Fig. 14.10 shows the range of combinations of pressure and temperature used for diamond synthesis when nickel is the catalyst.

Marshall and Shaw (1952) have shown that in surface grinding, where small chips are formed, the specific energy of chip formation is about 10×10^6 in lb in^{-3} (7000 kg/mm^2). One reason for this value being so high is that a large volume of material is deformed plastically relative to the volume of material actually removed. Figure 24.2 shows the mechanism of chip formation that pertains in fine grinding. The entire region within the elastic-plastic boundary will be plastically deformed as in an indentation hardness test while only the material in the chip is removed (Shaw and DeSalvo, 1972). This gives rise to an unusually large mean pressure between the abrasive particle and the chip which is estimated to be about $\frac{1}{5}$ that of the specific energy or about 2×10^6 p.s.i. (130,000 atm).

In addition, surface temperatures in grinding have been shown to be well above the melting point of the metal ground (Mayer and Shaw, 1957), even though there is no sign of melting in the ground surface. The reason for this anomaly is that melting is a process involving considerable structural change which requires time to occur. The time at temperature in grinding is of such short duration that melting does not have time to occur even though the surface temperature is substantially above the usual (equilibrium) melting value.

It is estimated that when grinding steel, surface temperatures of the order of 3000 °C are obtained with grit–chip contact pressures as high as 2×10^6 p.s.i. (130 katm). This condition exceeds the range of operation employed in diamond synthesis (see Fig. 14.11). It is to be expected that under such extreme conditions, the diamond nuclei produced will be extremely small, closely spaced, and spontaneously formed.

The fact that a very thin layer of nickel catalyst is involved in regular diamond synthesis is suggested by the following statement from Bovenkerk et al. (1959): "The actual transformation from carbon to diamond occurs across a very thin film (of nickel) about 0.1 mm [0.004 in] thick, which separates the carbon from the diamond. Thus, the transformation is almost 'direct' but the catalyst is essential." For the controlled growth of individual diamond crystals, Bovenkerk (1961) suggested the use of alternate layers of nickel and carbon in the reaction chamber.

In order to test the feasibility of generating diamond nuclei at the tip of an abrasive grit, the following experiment was performed: A tungsten carbide sphere, 0.060 in (1.5 mm) in diameter, was coated with alternate layers of nickel and graphite. The tungsten carbide sphere was first cleaned by sputter etching, and then three thin layers were deposited in the order nickel–carbon–nickel using a sequential sputtering apparatus.

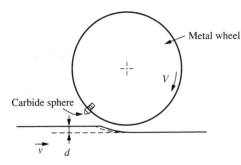

Fig. 24.3 Single-point cutting arrangement. (after Komanduri and Shaw, 1974)

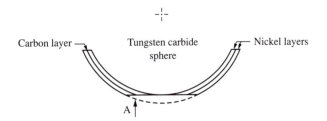
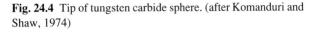

Fig. 24.4 Tip of tungsten carbide sphere. (after Komanduri and Shaw, 1974)

The coated tungsten carbide sphere was then cemented into a hollow head screw which in turn was threaded radially into the surface of an 8 in (20 cm) diameter aluminum disc. The disc was mounted on the spindle of a surface grinding machine in place of the usual grinding wheel and used to cut a groove in a small piece of hardened ball bearing steel (AISI 52100 steel of hardness H_{RC} 61) as shown schematically in Fig. 24.3. The speed of the disc (V) was in the range of conventional grinding (6000 f.p.m. = 30 m/s) while the table speed (v) and depth of cut (d) were adjusted to provide individual chips that were comparable in size to those in ordinary surface grinding (V = 6 i.p.m. or 2.54 mm/s, and d = 0.001 in or 25 μm). The length of groove produced was 0.5 in (12.7 mm), which corresponds to 250 individual cuts.

The tungsten carbide sphere acted like an abrasive grit in the surface of a grinding wheel and produced individual chips that were projected into the atmosphere as sparks.

After cutting, a small flat was ground on the tip of the tungsten carbide sphere as shown in Fig. 24.4 at A. When the worn surface of the composite coating was examined in a scanning electron microscope for evidence of diamond nuclei from direction A in Fig. 24.4, many small nucleation sites about a micron in diameter were evident at the inner nickel layer (Fig. 24.5). In the synthesis of diamond, the actual transformation from carbon to diamond was reported to occur in a very thin layer of nickel that separates the carbon from the diamond (Bovenkerk et al., 1959). The nuclei observed in Fig. 24.5b were produced in the thin nickel layer and are believed to be diamond nuclei generated under the extreme conditions of temperature and pressure pertaining during cutting.

The Debye–Scherrer x-ray diffraction technique was employed in an attempt to verify that the nuclei evident in Fig. 24.5 were in fact diamond. The sample was held stationary in the camera with the x-ray beam aimed at the wear scar. Tungsten carbide, titanium carbide, and nickel patterns were clearly evident. In addition, one spot (corresponding to the strongest diamond line) was also found, possibly indicating the presence of diamond. The small concentration of nuclei makes it very difficult to obtain certain verification of their structure.

The practical implications of the possibility of generating diamond nuclei at the tip of an abrasive grit appear to lie in at least three areas:

1. to reduce the rate of wear of diamond grinding wheels

2. as a source of diamond seeds for epitaxial growth of diamond by vapor phase deposition of diamond

3. as a convenient method of studying the formation of new high-temperature, high-pressure stable phases of materials (e.g., CBN)

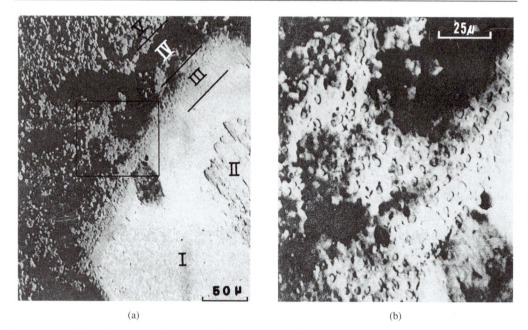

(a)   (b)

Fig. 24.5 (a) Scanning electron micrograph of tungsten carbide specimen abraded at high speed (temperature). I = tungsten carbide sphere, II = wear scar, III = inner nickel layer, IV = carbon layer, V = outer nickel layer. (b) Portion of inner nickel layer at higher magnification showing many hard nuclei. (after Komanduri and Shaw, 1974)

It is believed that the thin layer of nickel that smears over the surface of a worn coated diamond can act to prevent the diamond surface from being converted to graphite. Or, some of the graphite formed by degradation of diamond during wear may be reconverted to diamond if nickel is present. The presence of nickel would thus alter the rate of conversion of diamond to graphite. If this is so, then it should be advantageous to provide a source of graphite.

Hydrogen Generation

Uehara et al. (2000) have presented a study in which small amounts of hydrogen gas were generated when aluminum or an aluminum alloy was cut beneath the surface of water at room temperature. Bubbles were evident as long as cutting occurred but ceased as soon as cutting stopped. The initial aim of the study was concerned with behavior of cutting fluids. However, when gas bubbles were observed and these were identified to be H_2, interest shifted to the possibility of metal cutting being a source of hydrogen production for use in automotive fuel cells.

Slow-speed orthogonal chip formation was performed under the following conditions:

- tool material: HSS
- rake angle: 0°
- width of cut: 0.160 in (4 mm)
- depth of cut: 0.004 and 0.008 in (0.1 and 0.2 mm)
- cutting speed: 1.2 i.p.m. (30 mm/min)
- work materials: pure aluminum, a Sic–Al composite (Duralcan), and AISI 1045 steel

Figure 24.6 is a schematic of results obtained with a video camera having a macroscopic lens, when machining the aluminum or Duralcan in water. Bubbles were found whenever a freshly formed surface was involved (1) along the finished surface, (2) along the smooth side of the chip, (3) along the back of the chip, and (4) along the edge of the chip. The bubbles at (3) and (4) were associated with microcrack roughness in these regions. When Duralcan was machined in oil or when steel was machined in water, no bubbles were observed. When bubbles were present, the gas was collected in a test tube and identified as H_2.

In considering possible reactions that might occur, the following equations were considered where ΔG is the change of Gibb's free energy for each reaction:

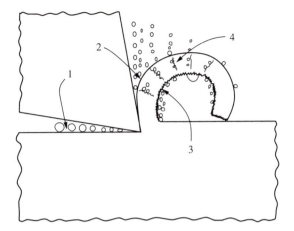

Fig. 24.6 Schematic of bubble formation when cutting Al in water. (after Uehara et al., 2000)

$$\textbf{1.} \quad 2Al + 3H_2O = Al_2O_3 + 3H_2 \qquad \Delta G = -435.2 \text{ kJ/mol} \qquad (24.4)$$

$$\textbf{2.} \quad 2Al + 4H_2O = 2AlOOH + 3H_2 \qquad \Delta G = -449.2 \text{ kJ/mol} \qquad (24.5)$$

$$\textbf{3.} \quad 2Al + 4H_2O = 2AlOOH + 3H_2 \qquad \Delta G = -436.3 \text{ kJ/mol} \qquad (24.6)$$

$$\textbf{4.} \quad 2Al + 6H_2O = 2Al(OH)_3 + 3H_2 \qquad \Delta G = -444.1 \text{ kJ/mol} \qquad (24.7)$$

The sign and magnitude of ΔG for all of these possibilities is so close that none may be ruled out. Also, all are equally efficient in releasing the main product of interest, H_2. In all cases the surface will be covered by a water-impervious layer of alumina (case 1), hydroxide (case 4), or oxyhydroxide (cases 2 and 3), all of which will terminate the generation of H_2 as soon as cutting is stopped.

For the cutting or grinding of aluminum or an aluminum alloy in water to be a practical source of hydrogen, means must be found for the reaction once started during cutting to continue until all of the chip has been consumed. If this can be achieved, then this technique would be of interest as a source of hydogen to be used as an alternative fuel for vehicles. According to any of the previous equations, 9 kg of aluminum would produce 1 kg of H_2 when the efficiency is 100%. It is estimated that 1 kg of H_2 would fuel a compact car for a distance of 125 km. A corresponding conventional car would require about 10 liters of gasoline to travel the same distance (125 km).

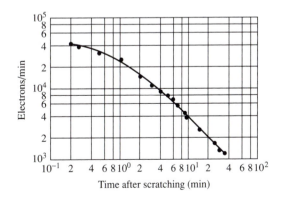

Fig. 24.7 Variation of rate of electron emission with time after scratching an aluminum surface. (after Meyer, 1967)

Curved Platelets

When a new metal surface is generated in high vacuum, a large number of free electrons are emitted and the surface ions move apart. This emission of free electrons when a new metallic surface is formed is known as the Kramer Effect (Kramer, 1950; Grunberg, 1958). Figure 24.7 shows

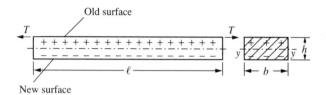

Fig. 24.8 Freshly generated platelet with surface tension force acting on outer (old) surface only. (after Komanduri and Shaw, 1975)

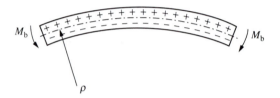

Fig. 24.9 Elastic beam subjected to pure bending. (after Komanduri and Shaw, 1975)

the change in electron emission with time when aluminum is deeply scratched in air to form a new surface. It has been found that the emission of electrons is accelerated by irradiation with x-rays, ultraviolet light, or visible light. The reason equilibrium is attained so slowly in the case of Fig. 24.7 is due to the process being diffusion controlled.

The excess electron density of a freshly generated surface of a thick specimen will be made consistent with its ion spacing by the emission of electrons from the new surface (Kramer Effect). In the case of a very thin platelet (1 μm or less), an alternative way of achieving equilibrium is for the platelet to curl with the new surface of the platelet on the concave side. This causes the ions on this side to be more closely spaced, making it unnecessary for electrons to be emitted from this newly formed surface. Figure 24.8 shows the situation for a newly formed platelet before equilibrium is established. The old surface will be subjected to surface tension (T) relative to the rest of the platelet. This will give rise to a constant bending moment (M_b) along the length of the platelet:

$$M_b = \frac{Tbh}{2} \tag{24.8}$$

According to Crandall and Dahl (1959) a linerally elastic beam that is subjected to pure bending (Fig. 24.9) will assume a radius of curvature (ρ) as follows:

$$\frac{1}{\rho} = \frac{M_b}{EI_{yy}} \tag{24.9}$$

Where E is Young's modulus of elasticity and I_{yy} is the moment of inertia of the cross-section of the beam about the neutral axis (yy). For the beam of Fig. 24.8,

$$I_{yy} = \tfrac{1}{12} bh^3 \tag{24.10}$$

and hence

$$\rho = \frac{Ebh^3}{12Tb(h/2)} = \frac{Eh^2}{6T} \tag{24.11}$$

For a platelet that is wide relative to its length ρ will be the radius of a sphere instead of a cylinder.

Thus, a first consequence of the formation of a newly formed platelet being out of equilibrium relative to electron density is a bending moment tending to make the platelet curl. A second consequence is that the curved platelet will be highly polarized, being positively charged on its outer (convex) surface and negatively charged on its concave surface. As a result, a curved platelet will tend to form layered spherical or cylindrical structures. For example, Fig. 24.10 shows a spherical particle formed from small individual platelets, each covering a small area of a sphere. The reason curling may be favored over electron emission (Kramer Effect) for chips in the form of very thin platelets is apparently due to curling being practically an instantaneous elastic effect whereas electron emission is a much slower diffusion-controlled process.

ABSORPTION OF UNWANTED ENERGY

Auto safety has been receiving more and more attention. Concern is not only with the welfare of the driver but with the reduction in property damage associated with front- and rear-end collisions. Some time ago, the latter aspect was brought into sharp focus by the offer of an insurance company to reduce insurance rates 20% on cars capable of withstanding a front- or rear-end crash at a speed of 5 mph without damage.

An acceptable solution should provide a simple, inexpensive yet predictable and reliable means of absorbing energy associated with the crash, without subjecting the frame or bumper of the car to too high a force (P) or passengers to too high an acceleration (a). As in Fig. 24.11 the solution can be represented by a "black box" inserted between the bumper and car

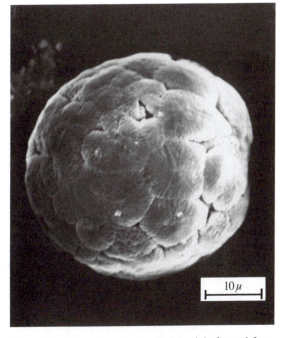

Fig. 24.10 Illustration of spherical particle formed from small individual platelets, each covering a small area of a sphere. (after Komanduri and Shaw, 1975)

frame. The mechanism in the black box must absorb the kinetic energy of the vehicle by allowing the bumper to move a distance ℓ, while the resisting force is limited to a value P.

Equating the kinetic energy of the vehicle of weight W and velocity V to the work done by mean force P acting through a distance ℓ:

$$\frac{WV^2}{2g} = P\ell \tag{24.12}$$

where g is the acceleration due to gravity. From Newton's second law:

$$P = \left(\frac{W}{g}\right)a \tag{24.13}$$

Combining these equations:

$$\ell = \frac{V^2}{2a} \tag{24.14}$$

If acceleration a is limited to 3g (the highest safe acceleration for a passenger without a seat belt), then for $V = 5$ mph:

$$P = 3W$$

$$\ell = \frac{V^2}{6g} = 3.33 \text{ in}$$

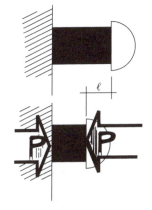

Fig. 24.11 Energy-absorbing black box located between bumper and frame of car.

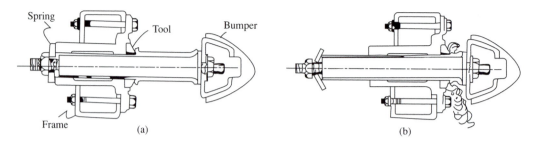

Fig. 24.12 Metal cutting energy-absorbing bumper unit. (a) Before impact. (b) After impact. (after Shaw, 1972b)

Thus, for a 4000 lb car the force on the frame and bumper would be 12,000 lb and a minimum of 3.33 in of travel would be required between bumper and car frame. These values appear to be reasonable and may be taken as the design requirements.

The least expensive way of protecting electrical circuits is by means of a fuse consisting of a wire which burns through when the current exceeds a certain safe limit. By analogy, the black box might be developed as an inexpensive, expendable item that allows motion when the resisting force on the bumper reaches a value P, and in so moving, absorbs the unwanted energy. This type of energy-absorbing device may be thought of as a "mechanical fuse."

The energy absorbed should be irreversibly converted into heat. This rules out springs which return the energy stored to the car after the crash, causing a reverse acceleration which subjects passengers to whiplash.

When a metal is plastically deformed, practically all of the energy involved ends up as heat. While there are many ways of plastically deforming materials, such as bending, denting, shearing, drawing, extruding, or cutting, not all are equally efficient relative to energy absorption.

The cutting process represents a very efficient use of metal for energy absorption. It is also easy to predict the forces and energy involved. For mild steel, about 300,000 in lb are required to produce a cubic inch of chips, and this value is essentially independent of the speed of the operation.

Use of the cutting process to absorb energy is not initially attractive since it normally involves a tool, guideways, power transmission, and a complex system of tool control. However, the energy-absorbing bumper design of Fig. 24.12 is a simple means of absorbing the unwanted energy. Before impact, the nut at the left is tightened against the spring, and the assembly is capable of withstanding normal towing and bumperjack loads. Upon impact, the surface of the rod is cut after a critical force P develops. Equating the energy absorbed to the cutting energy gives

$$P\ell = \ell(\pi dt)u \qquad (24.15)$$

where d is the diameter of the tool, t is the radial depth of cut, and u is the specific cutting energy (energy per unit volume cut).

Then, solving for t:

$$t = P/(\pi du) \qquad (24.16)$$

For the 4000 lb car previously considered, the required force is 12,000 lb. When cutting mild steel in this way, u will be about 0.3×10^6 in lb in^{-3}. Thus, for a tool of 1 in diameter and for two points of attachment on the bumper, the wall thickness (t) required is

$$t = \frac{(12{,}000/2)}{\pi(1)(0.3 \times 10^6)} = 0.0064 \text{ in}$$

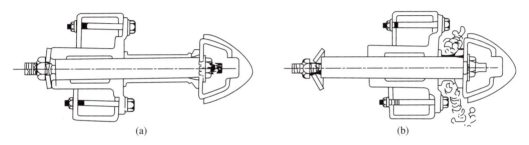

Fig. 24.13 Expected action of metal cutting unit when entire tube is cut. (a) Before impact. (b) After impact. (after Shaw, 1972b)

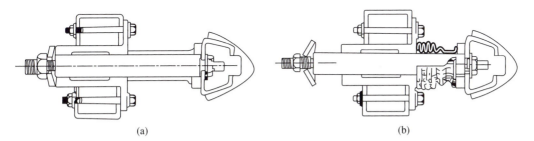

Fig. 24.14 Actual performance of unit shown in Fig. 24.13. (a) Before impact. (b) After impact. (after Shaw, 1972b)

Fig. 24.12b shows the device after the tool has cut 3.33 in of steel and the car has been brought to rest. The rods are axially unrestrained and will now rattle in the guide bearings, thus warning the operator that the "mechanical fuse" should be replaced. To restore the bumper to its original condition, it is only necessary to remove two nuts, pull the entire bumper and guide rods out of the guide bearings, and replace the inexpensive expendable steel tubes. This could be done at any service station using color-coded tubes of different diameters for cars of different weight.

The design is simple, inexpensive, easily predictable, and reliable and does not cause whiplash since over 95% of the energy ends up as heat in the chips. The tool is a simple carbon-steel washer, and the unit operates equally well over a wide range of velocities.

The only disadvantage of this design is that *all* of the tube is not converted into chips. An attempt was, therefore, made to have the tool slide directly on the steel guide rods with a small clearance (Fig. 24.13a). However, instead of cutting chips as shown in Fig. 24.13b, the tube buckled, accordion fashion (Fig. 24.14). This result was completely unexpected.

It is interesting to note that the unique high specific energy involved in metal cutting at first appeared to offer the best solution. However, when it was found necessary to cut only part of the tube, whereas the entire tube could be pleated under the stabilizing influence of the mandrel, this changed the picture. There were then two solutions: metal cutting of part of the tube and mandrel-controlled pleating of the entire cross-section of the tube. The second of these solutions was not anticipated at first since the specific energy per unit volume in forming (including pleating) is considerably less than in cutting.

A more complete discussion of this subject including mechanics of the pleating alternative, nontechnical obstacles of applying either of these solutions, and other applications of the absorption of unwanted energy is available in Shaw (1972b).

CHIP PULLING

Finnie and Wollak (1963) employed the orthogonal cutting process with a tensile force applied to the chip leaving the tool to obtain a better understanding of the plastic behavior of materials at the very high strains and strain rates that pertain in metal cutting. Figure 24.15a shows the method used while Fig. 24.15b defines the pulling angle (η), the rake angle (α), and the shear angle (ϕ). When $\eta > \alpha$ the chip is lifted from the tool face, reducing the chip-tool contact length and the cutting forces but increasing the shear angle and cutting ratio (r).

Orthogonal cutting tests were performed on pure aluminum with and without chip pulling forces at different angles. Cutting forces F_P and F_Q were measured using a two-component dynamometer. Mean shear stresses and shear strains were determined by considering the chip as a free body. The shear force on the shear plane (F_S) will be as follows when a pulling force (P) is acting:

$$F_S = F_Q \cos \phi - F_P \sin \phi - P \sin (\eta - \phi) \tag{24.17}$$

The average shear stress (τ), shear strain (γ), and ϕ were then obtained by the methods given in Chapter 3.

Compression tests were also made on cylinders of the same aluminum that were 0.75 in (1.91 mm) in diameter and 1 in (2.5 mm) long, and the effect of barrelling was removed periodically. The effective shear stress (τ_e) and effective shear strain (γ_e) for the compression tests were assumed to be

$$\tau_e = \sigma/2 \tag{24.18}$$

$$\gamma_e = \sinh 3\varepsilon/2 \tag{24.19}$$

where σ and ε are the mean shear stress and shear strain, respectively, in the compression test. Figure 24.16a shows a comparison between shear-stress shear-strain results at 68 °F for a compression

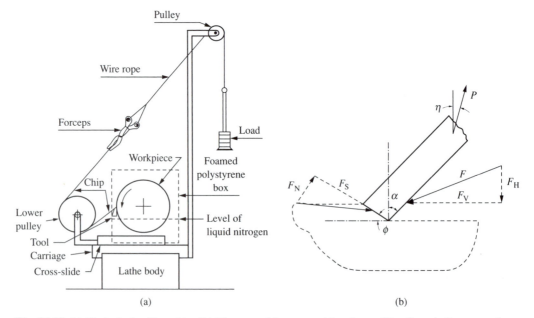

(a)   (b)

Fig. 24.15 (a) Method of pulling chip. (b) Diagram of forces on chip when pulling force is P at an angle η. (after Finnie and Wolak, 1963)

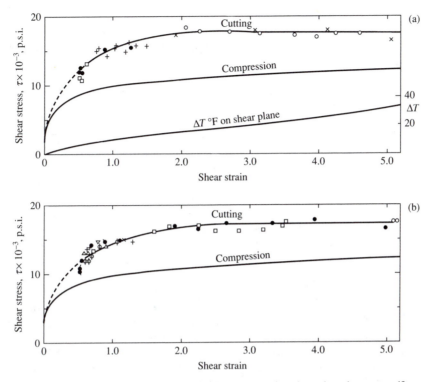

Fig. 24.16 (a) Cutting and compression tests at 68 °F. Compression data plotted as $\tau = \sigma/2$ versus shear strain = sinh $3\varepsilon/2$. Estimated temperature rise (F) on shear plane is shown in lower part of Fig. 24.16a. Rake angles and pulling angles corresponding to data points in cutting are $\circ$: $\alpha = 45°$, $\eta = 45°$; $\times$: $\alpha = 45°$, $\eta = 23°$; $+$: $\alpha = 60°$, $\eta = 60°$; $\bullet$: $\alpha = 60°$, $\eta = 25°$; $\square$: $\alpha = 60°$, $\eta = 16°$. (b) Additional data to that in (a) compared with the same compression data. Rake angles and pulling angles corresponding to data points in cutting are $\bullet$: $\alpha = 45°$, no pulling; $\circ$: $\alpha = 45°$, $\eta = 0°$; $\square$: $\alpha = 45°$, $\eta = -15°$; $+$: $\alpha = 60°$, $\eta = 45°$; $\triangledown$: $\alpha = 60°$, $\eta = 30°$; $\blacklozenge$: $\alpha = 60°$, $\eta = 15°$, $\otimes$: $\alpha = 60°$, $\eta = -15°$; $\times$: $\alpha = 60°$, $\eta = -30°$; $\circ$: $\alpha = 60°$, $\eta = -45°$; ϕ: $\alpha = 60°$, $\eta = -60°$. (after Finnie and Wolak, 1963)

test and cutting tests with and without chip pulling. Figure 24.16b shows the same compression results with additional cutting data, with and without chip pulling. It is evident in both Figs. 24.16a and 24.16b that the shear stress ceases to rise appreciably with strains above about 2. However, the most important result is that the shear stress at very high strain (~ 5.0) is about 35% greater in the cutting tests than it is in a compression test. The following possible explanations for this difference were considered without reaching a definitive conclusion:

- change in shear zone temperature
- normal stress on the shear plane in cutting
- rubbing forces on finished surface
- size effect (a smaller specimen in cutting)
- high strain rate in cutting
- concentration and interaction of dislocations

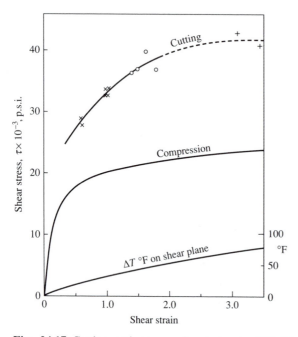

Fig. 24.17 Cutting and compression tests at −320 °F. Compressive test data plotted as $\sigma/2$ versus sinh $3\varepsilon/2$. Estimated temperature rise (°F) on shear plane during cutting is shown in lower part of figure. Rake angles corresponding to data points are × = 60°, ○ = 45°, + = 30°. (after Finnie and Wolak, 1963)

Both cutting and compression tests were also performed at very low temperature (~ 320 °F). These results plotted under the same conditions as for Fig. 24.16 are given in Fig. 24.17. It is evident that at a very low temperature the difference between cutting and compression results are very much greater than at ambient temperature (shear stress in cutting is 2 times shear stress in compression).

When the chip pulling angle η is very large (e.g., 90°), a periodic instability occurs resulting in kinks in the chip as shown in Fig. 24.18.

The main objective of the chip pulling technique presented in Finnie and Wolak (1963) was to obtain a wide range of shear strains with but two tools (i.e., 2 rake angles) making it possible to use metal cutting as a large strain materials test. From Fig. 24.16 it is seen that but two rake angles are sufficient to give a shear-stress–shear-strain curve covering a very wide range of shear strain. However, before this may be applied, a better understanding of the large difference between the effective stress–strain curves for cutting and compression must be available. Based on experimental results discussed in Chapter 20, it appears that all of the reasons mentioned above may play a role.

Nakayama (1964) also performed chip pulling experiments but in this case to better understand the chip curling mechanism. His experimental apparatus is shown in Fig. 24.19. The chip is

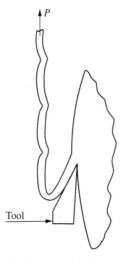

Fig. 24.18 Drawing of kinks in chip pulled at $\eta = 90°$. (after Finnie and Wolak, 1963)

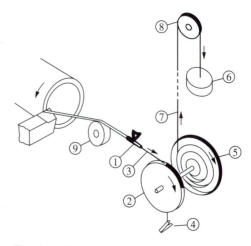

Fig. 24.19 Experimental arrangement used by Nakayama (1964).

attached to a steel wire (3) at (1) which, in turn, is attached to the pulley at (2). When pedal (4) is depressed, a brake is released and a stepped pulley (5) begins to rotate, being driven by the falling weight (6). During four rotations of (5), the force on (3) varies from W to $2W$ to $3W$ to $4W$ where W is the weight at (6). The change in pulling direction is obtained by adjusting the guide roller (9).

The workpiece was a copper tube of 1.9 in (48 mm) diameter and 0.080 in (2 mm) wall thickness. This was cut at a speed of 279 f.p.m. (85 m/min) and a feed of 0.004 i.p.r. (0.1 mm/rev) using HSS tools having rake angles of 10° and 30°. The tensile force was varied between 0 and 13.2 lb (0 and 6 kg) in steps of 3.3 lbs (1.5 kg) for the 10°-rake-angle tool and between 0 and 8.8 lb (0 and 4.0 kg) in steps of 2.2 lb (1 kg) for the 30°-rake-angle tool. Force components F_P and F_Q were measured with a dynamometer. Results are given in Fig. 24.20.

Figure 24.21 shows the moment $M = Pd$ introduced when a chip pulling force P is employed at a pulling angle η. Bending moment M complicates determination of d since stress along the shear plane is then no longer uniform.

Nakayama (1964) has shown that inclusion of M and P gives analytical results that are in good qualitative agreement with the experimental results which indicate that with chip pulling

- chip tool contact length ℓ_c decreases
- forces F_P, F_Q, F, and chip thickness (t_c) decrease
- shear angle (ϕ) and chip curl increase

The force diagrams used when chip pulling force P is included is shown in Fig. 24.22.

It is interesting to note that there are optimum values of η for which cutting forces are a minimum. The work of chip pulling is negligible compared with the total work of chip removal. The main effect of chip pulling is to reduce the chip–tool contact length which in turn reduces friction on the tool face, cutting forces F_P and F_Q and chip thickness. The bending moment introduced by chip pulling increases the thickness of the shear zone, thus making analysis based on a concentrated shear plane approximate.

The main objective of the Nakayama chip pulling investigation was to study the effect on chip curl. Chip pulling enables a much larger effect on chip curl than use of chip curling geometry on the tool face. A second effect of chip pulling and possibly the more important one is introduction of moment M. This gives rise to an increase in normal stress on the shear plane which in turn causes a less concentrated shear zone, fewer (if any) microcracks, and a more ductile chip. A more ductile chip would be of importance for cases where the chip is the product of interest as in the production of wire by cutting.

THIN RIBBON-LIKE STRIPS

Magnesium foil has been produced for use in the atomic energy field by hot machining a billet of magnesium (Alfille et al., 1963). It was not possible to produce thin magnesium strips by rolling due to the brittleness of magnesium because of its hexagonal structure. The magnesium foil is spiraled on a fuel element and oxidized in place to produce an effective neutron shield.

Stainless steel ribbon approximately 4 in (100 mm) wide and 0.003 in (0.08 mm) thick has also been produced in the automotive industry by turning but without heating the work material.

In both of these cases a cylindrical billet is peeled by a tool that is fed radially as shown in Fig. 24.23. The chip is removed under light tension and with a pulling angle that decreases the chip–tool contact length, and hence the shear strain. This results in a product (chip) that is much more ductile.

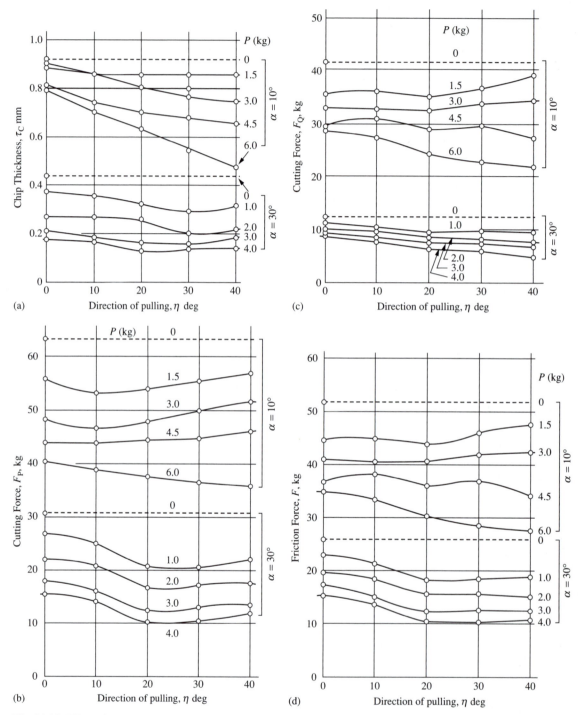

Fig. 24.20 Effect of rake angle (α), chip pulling force (P), and pulling direction (η) on (a) chip thickness (t_c), (b) cutting force (F_P), (c) cutting force (F_Q), and (d) friction force on the tool face (F). Work material: copper tube 1.2 in (48 mm) o.d. and 0.080 in (2 mm) wall thickness. HSS tools: cutting speed, 279 f.p.m. (85 m/min) and feed of 0.004 i.p.r. (0.1 mm/rev). (after Nakayama, 1964)

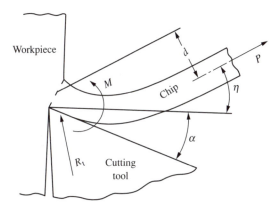

Fig. 24.21 Bending moment M induced with pulling force P acting in direction η. (after Nakayamo, 1964)

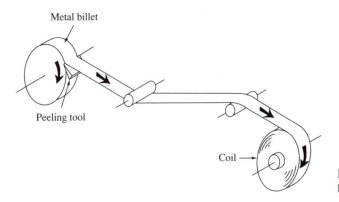

Fig. 24.23 Method of producing metal foil by billet peeling.

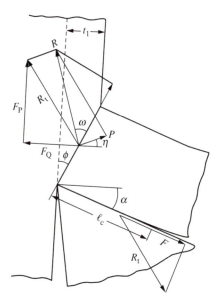

Fig. 24.22 Forces acting on chip including pulling force P.

WIRE PRODUCTION BY METAL CUTTING

Practically all wire commercially produced today is made by drawing material through successively smaller circular dies. This is an old, highly developed procedure. Its main disadvantage lies in the high cost of capital equipment and the large inventory of dies required.

The large number of dies arises in the requirement that the reduction in area produced by any one die is limited by the force that may be applied to the wire without rupture. In the case of copper wire, the reduction in area is usually limited to about 21% (i.e., a reduction in diameter of 11%).

There have been a number of attempts to develop methods of producing wire by means other than drawing. Most of these involved extrusion where the metal is pushed through a die instead of being pulled through. This enables a larger reduction in area to be used per die and hence requires fewer dies.

One of these processes has been described by Green (1972). This is called the continuous extrusion forming (CONFORM) process and is capable of converting 0.3125 in (7.94 mm) diameter wire into a wire of 0.040 in (1.02 mm) diameter by use of a single die. The required extrusion pressure is obtained by dry friction developed between a grooved wheel and the feed stock. Another extrusion process that has been proposed is called extolling (Avitzur, 1975).

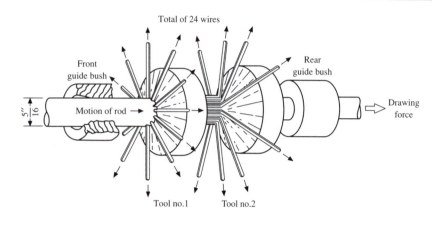

Fig. 24.24 Schematic diagram of proposed machine for making 24 wires each 0.045 in (1.14 mm) square from 0.3125 in (7.94 mm) diameter feed stock. (after Shaw and Hoshi, 1976)

Shaw and Hoshi (1976) have proposed the method of producing wire shown in Fig. 24.24. This involves a pair of shaving tools which remove the outer layer of material. The incoming rod is guided by a bushing to the first tool which produces 12 wires each 0.045 in (1.14 mm) wide. The radial depth of cut is 0.018 in (0.457 mm) resulting in a square product 0.045 × 0.045 in (1.14 × 1.14 mm) assuming a cutting ratio of 0.4. There is a sidewise expansion of 12% which causes the undeformed chip width of 0.040 in to increase to 0.045 in.

The grooved rod proceeds to the second tool which removes the remaining material to produce 12 additional square wires. The 24 wires thus produced may be fed directly to a stranding unit if a square section meets product requirements. Otherwise, the individual wires may be spooled pending further processing or fed directly to single dies which convert the square wire into round wire with little change in diameter.

An important distinguishing characteristic of wire production by the metal cutting process is that the billet is deformed only locally instead of across its entire cross-section as in extrusion or drawing. This means that the forces associated with the process will be relatively small and, as a consequence, the machine need be less massive and less costly.

The following factors have been found to be important in order to produce good wire by cutting copper:

1. a high positive rake angle

2. a sharp cutting edge

3. low friction between tool and chip

4. a high cutting speed

5. highly cold-worked metal of homogeneous structure

6. the chip should be pulled from work at an appropriate angle

7. an elevated shear plane temperature

A convenient measure of performance is the chip length ratio:

$$r = \frac{\text{length of chip}}{\text{undeformed chip length}}$$

The larger this ratio, the better the quality of wire produced. The chip length ratio (r) should not be less than 0.35 in the case of copper wire production and preferably 0.5 or greater. The chip length ratio increases with decrease in friction between tool and work.

In order to develop the necessary data to check the feasibility of the design shown in Fig. 24.24 a series of tests was performed on a lathe using single-point tools that simulate the actual situation.

Figure 24.25 shows three different cutting arrangements that were investigated. Cutting mode A employs a tool cutting a helical groove that simulates the situation of the first tool in Fig. 24.24. Cutting mode B involves removal of the land produced in mode A and represents the situation of the second tool of Fig. 24.24. Cutting mode C represents conventional turning. These modes of cutting represent different conditions of side constraint. Both sides of the chip-forming area are constrained in the case of mode A, one side is constrained in mode C, and there is no contraint in the case of mode B.

Chips produced were wound on a 5 in (127 mm) drum that provided a tensile force to the chip of about 1 lb (4.45 N) directed along the rake face of the tool. Tensile forces of up to 60 lb (267 N) were applied by falling weights attached tangentially to the drum.

All chips produced were collected and their shape, strength, and freedom from cracks were studied.

Figure 24.26 shows the effect of cutting speed and workpiece hardness on the chip length ratio, tensile strength of chip, and specific cutting energy. In all these tests, cutting mode C was used with a feed of 0.50 i.p.r. (1.27 mm/rev), a depth of cut of 0.020 in (0.508 mm), and a rake angle of +20°. The solid points indicate chips having visible surface cracks while the open points correspond to chips without cracks. It is evident that a chip length ratio of at least 0.35 is required to obtain crack-free chips. It is also evident that annealed material gives poor chips and that higher cutting speeds give chips of higher quality at lower values of specific energy.

Variation of cutting characteristics for mode C with variation in pulling force (P) is shown in Fig. 24.27a, and variation in pulling angle (η) is shown in Fig. 24.27b.

A study of the effect of side constraint revealed an increase in chip quality and chip

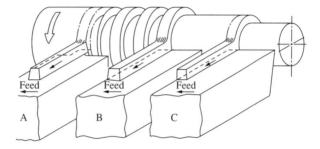

Fig. 24.25 Three cutting modes tested to evaluate the influence of side constraint on the chip. Mode A involves tool producing a chip constrained on both edges. Mode B involves tool producing an unconstrained chip. Mode C involves tool producing a conventional chip that is constrained on one edge only. (after Shaw and Hoshi, 1976)

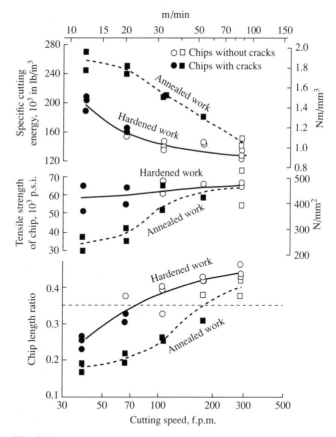

Fig. 24.26 Variation of chip length ratio, tensile strength of chip, and specific cutting energy with cutting speed and workpiece hardness. Cutting conditions: conventional turning (mode C, Fig. 24.25); rake angle, +20°; depth of cut, 0.020 in (0.508 mm); feed (width of cut), 0.050 i.p.r. (1.27 mm/rev); work hardness, hard (H_{RB} 60) and annealed (H_{RB} 23). (after Shaw and Hoshi, 1976)

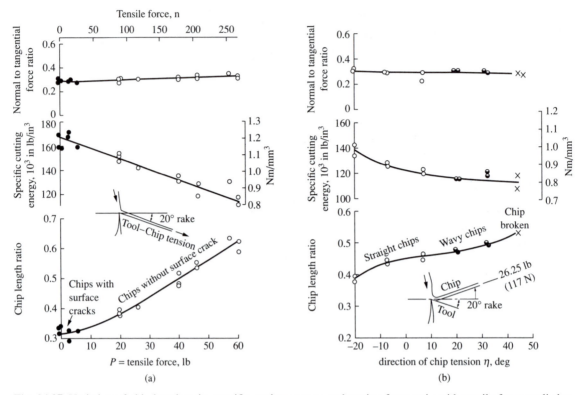

Fig. 24.27 Variation of chip length ratio, specific cutting energy, and cutting force ratio with tensile force applied to chip. Cutting conditions: cutting speed, 67 f.p.m. (20.4 m/min); cutting mode, C in Fig. 24.25; depth of cut, 0.020 in (0.508 mm); feed, 0.050 i.p.r. (1.27 mm/rev); work hardness, H_{RB} 60. (a) $\eta = 20°$ and variable P. (b) $P = 26.25$ lb (117 N) and variable η. (after Shaw and Hoshi, 1976)

length ratio with increase in side constraint. Tests in which the rake angle of the tool was changed indicated an improvement in chip quality with increase in rake angle.

An increase in tensile force is seen to result in better quality chips of greater length, produced at lower values of specific energy. The study of the direction of chip tension indicated the optimum value to be about 0° (horizontal).

The values obtained in this study are in excellent agreement with those published by Nakayama (1964) who studied the effect of tension applied to copper chips to provide a better understanding of the mechanics of cutting rather than to produce a new product. Professor Nakayama found that a tensile force applied to a chip had a synergistic effect upon the total expended energy in that a reduction in the cutting energy as great as 40 times the pulling energy was observed. The main effect of tension appears to be to reduce tool-face friction which, in turn, has an amplified influence on the shear angle and hence on the cutting forces. The single-point simulation tests suggest a rake angle of 20° for the first grooved tool in Fig. 24.24 and a somewhat larger rake angle (30°) for the second tool. The increase in rake angle should offset the difference in restraint for tools 1 and 2.

A cutting speed of 200 f.p.m. (60.9 m/min) should be satisfactory with a high speed steel tool but could be as high as 500 f.p.m. (152 m/min) with carbide tools. At such a speed and with the rake angles indicated there should be no difficulty achieving a chip length ratio of 0.4 or higher. It appears safe to assume a mean specific cutting energy of 140,000 in lb in^{-3} (1.00 Nm/mm^3) when cold-worked copper is cut under the foregoing conditions. This would require a total force of 2480 lb applied to the 0.277 in (7.02 mm) diameter rod leaving the system. Such a force would produce a tensile stress of 41,150 p.s.i. in the rod at exit. This is sufficiently below the minimum expected value of tensile strength for fully worked copper (60,000 p.s.i.) to prevent rupture. The 0.045×0.045 in (1.13×1.13 mm) wire size (Fig. 24.24) apparently represents close to an upper limit for this arrangement. The production of finer wires should provide no difficulty.

The power required for the situation shown in Fig. 24.24 when operating at a cutting speed of 500 f.p.m. (152 m/min) would be 37.6 h.p. (28.02 kW). The output would be 24 wires each traveling at 200 (0.4×500) f.p.m. (60.9 m/min) or a total wire production of 4800 f.p.m. (1460 m/min). This is about half the output of a Number One wire drawing machine which, however, requires 400 h.p. (298.4 kW).

For materials less ductile than copper, it should be useful to heat the chip by DC electrical hot machining. Special cutting fluids may also be useful. Wires as fine as 0.010 in (0.254 mm) diameter may be produced continuously by cutting and as small as 0.0001 in (0.0254 m) diameter if long continuous lengths are not a prime requirement. Steel wire about 0.001 in (0.03 mm) in diameter with a length to diameter ratio of about 10 have been mixed into concrete (with a steel concentration of 1 or 2 $^{w}/_{o}$) for use in areas subjected to high surface stress. These small wires play the same role in limiting crack propagation as SiC whiskers do in carbide cutting tests. Use of steel whiskers in concrete was introduced by Professor Romualdi at Carnegie Mellon University in the 1950s.

Wire production by turning offers an unusually versatile process involving inexpensive equipment that is easy to set up which should be useful for small lot manufacture.

De Chiffre (1976) has described a process termed *extrusion cutting* that is potentially useful for producing flat strips of material. Figure 24.28a shows the process known as side extrusion with an appropriate slip line field. In this case the plunger moving to the left forces plastic material upward and along the inclined surface of the die. Figure 24.28b shows the inclined upward surface replaced by a cutting tool. The same slip line field pertains as in the corresponding side extrusion process shown in Fig. 24.28a. The main effect is to replace the relatively thin shear plane in ordinary cutting by a large deformation zone with a much higher normal stress on the tool face. The product in this case is the chip which will be much less brittle than a comparable ordinary metal cutting chip. The reason for the greater ductility of chips made by extrusion cutting is the lower strains involved and greater normal stress on the deformation zone. Both of these effects enable chip formation in the absence of the discontinuous microcracks involved in ordinary metal cutting discussed in Chapter 20. The absence of microcracks in this type of chip formation provides a product (chips) that is much more valuable than those produced in ordinary metal cutting. Also, the absence of microcracks on the shear plane provides a chip that is equally smooth on both sides. The absence of the rough surface found on the back of an ordinary chip (as in Fig. 6.28) suggests that microfracture is not involved. Hence, ordinary plastic flow analysis involving the von Mises criterion should be applicable unlike the situation for ordinary turning discussed in Chapter 20. De Chiffre (1976) has presented a valuable analytical treatment of extrusion cutting based on slip line field analysis. This was found to be in good agreement with experiments using a modeling wax (Filia) having a stress-strain curve similar to metal. Results in good agreement with theory were also obtained when cutting a tube of free-machining brass.

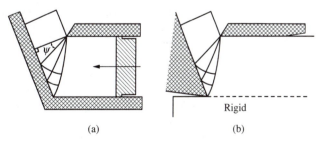

(a)   (b)

Fig. 24.28 Analogy between side extrusion at (a) and extrusion-cutting at (b). (after De Chiffre, 1976)

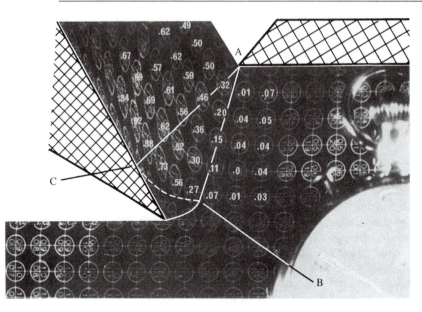

Fig. 24.29 Extrution-cutting of wax with rake angle of 30° and cutting speed of 2.41 i.p.m. (1 mm/s). (after De Chiffre, 1976)

Figure 24.29 shows a typical wax modeling result. The numbers in the circles are measured effective strains. There is essentially no strain below the finished (rigid) surface and in the material to the right of line AB and no further strain to the left of line AC. All of the strain occurs in the wedge ABC.

The important result of this investigation is that extrusion cutting is a valuable way of producing flat conductor cable and other small items involving extrusion. Flat conductor cable was introduced in the space program and offers important advantages over round conductor cable for house wiring and production of wiring harnesses in the automotive industry.

Hoshi and Shaw (1977) have developed a method of producing wire in which cutting and forming are combined into one operation (called cut-forming). A chip is first formed by cutting a billet with a special tool containing a hole into which the chip flows. This hole acts as a die to convert the rectangular chip into a round wire. The total force in this case corresponds to the chip-forming force plus the extrusion force.

Figure 24.30 illustrates the principle of the chip-forming process. The sharp cutting edge produces a rectangular chip with depth of cut (d) and width of cut (f) (feed per revolution). This flows into the hole in the tool which gradually changes the shape by extrusion from a rectangle at entrance to a circular wire of diameter D at exit.

In order to check the feasibility of the concept, a scaled up version of the tool was made from plexiglass. This transparent tool had an orifice diameter of 0.5 in (12.7 mm) and was used to cut modeling clay. The shapes of the cutting edge and die surfaces were altered until a perfectly round piece of modeling clay issued from the orifice. This geometry was then transferred to the much smaller steel tools used in the aluminum trials.

An electric conductor grade (EC) aluminum billet 4 in (100 mm) in diameter and 6 in (150 mm) long was turned using a 10 h.p. engine lathe. The special tool was made of M2 high speed steel heat treated to a hardness of H_{RC} 42–46. The orifice diameter for this tool was 0.089 in (2.26 mm). The as-cast aluminum billet had a hardness of H_B 19.3–21.8 without prior heat treatment or cold work. A second aluminum specimen was cold-worked to a hardness of H_B 24.8–26.5, in order to study the effect of cold work. A commercial drawing compound (Duo Kote X72-25) was applied at the tool-work interface to provide lubrication.

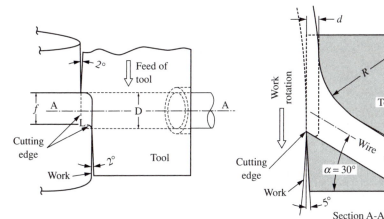

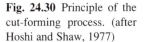

Fig. 24.30 Principle of the cut-forming process. (after Hoshi and Shaw, 1977)

Forces exerted on the tool were measured using a two-component strain gage dynamometer in conjunction with a dual-channel carrier amplifier recorder.

Continuous chips were also produced by turning identical EC aluminum billets using high speed steel single-point tools of conventional design. The back rake angle of these conventional tools was varied from 30° to 50°, while other values were kept constant as follows:

- side rake angle: 0°
- side and end relief angles: 5°
- side cutting edge angle: 0°
- end cutting edge angle: 5°
- nose radius: 0 (sharp)

Wire produced by the cut-forming process as well as some chips produced by conventional cutting were fully annealed [650 °F (343 °C) for 2 hr] and then drawn through conventional circular dies of four successive diameters: 0.071, 0.064, 0.057, and 0.050 in (1.80, 1.63, 1.45, and 1.27 mm). Wire thus produced was examined relative to ultimate tensile strength, percent conductivity relative to the standard value for annealed copper, microstructure, and surface quality.

Figure 24.31 gives tangential and normal force values when round wire is produced by the cut-forming process from

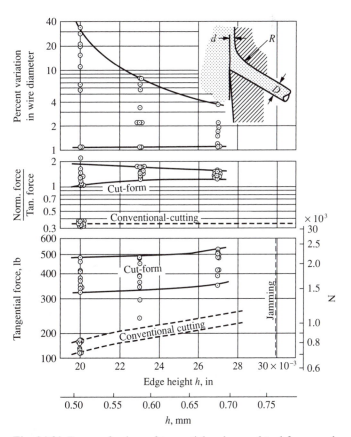

Fig. 24.31 Range of values of tangential and normal tool forces and roundness error when cut-forming as-cast EC grade Al. Material hardness: H_B 19.3–21.8; cutting speed, 90–720 f.p.m. (27–219 m/min); feed (f), 0.071 i.p.r. (1.8 mm/rev); orifice diameter (D), 0.089 in (2.26 mm); rake angle (α), 30°; orifice upper corner radius (R), 0.062 in (1.6 mm). (after Hoshi and Shaw, 1977)

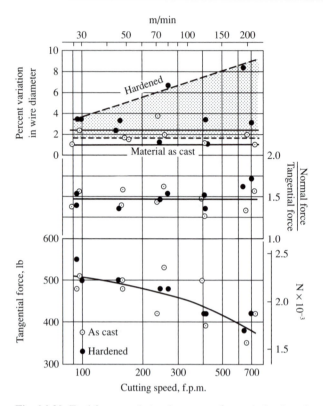

Fig. 24.32 Tool forces and roundness error for strain hardened EC grade aluminum. Material hardness, H_B 24.8–26.5; feed (f) 0.071 i.p.r. (1.8 mm/rev); orifice diameter (D), 0.089 in (2.26 mm); rake angle (α), 30°; orifice upper corner radius (R), 0.062 in (1.6 mm). (after Hoshi and Shaw, 1977)

aluminum in the as-cast condition. These tests were made with different values of d obtained by successively grinding away the front surface of a tool. In all tests the orifice diameter was 0.089 in (2.26 mm), the rake angle was 30°, and the feed was 0.071 i.p.r. (1.8 mm/rev). Cutting speeds ranged from 90 to 720 f.p.m. (27 to 219 m/min).

Values of d greater than 0.030 in (0.76 mm) resulted in jamming due to excessive material input. Extrusion occurred for depths of cut below 0.027 in (0.69 mm). While the power required (proportional to the tangential force) increased with d, the roundness of the product also improved with increased d. The latter quantity may be measured by the percent deviation of wire diameter and is given in the top portion of Fig. 24.31.

Cutting force values for chips produced by conventional turning are also shown in Fig. 24.31. These results are for a rake angle of 30°, feed of 0.071 i.p.r. (1.8 mm/rev), and depth of cut of 0.020 in (1.5 mm). The tangential force required in cut forming is seen to be from 1.4 to 2.8 times that required for conventional chip formation under comparable conditions. In conventional cutting the ratio of normal force to tangential force is about 0.35, while in the cut-forming process it is about 1.5. This calls for a radially rigid setup when cut-forming.

A comparison of results for as-cast (soft) and strain hardened aluminum is given in Fig. 24.32 for different values of cutting speed. Workpiece hardness is seen to be of little significance in cut forming. However, the tangential force decreased appreciably with increased cutting speed. The roundness error was found to be less for the softer work material. This suggests that a soft billet is preferred when cut-forming. This is contrary to experience in wire production by conventional turning where a highly work hardened workpiece is preferred (Shaw and Hoshi, 1976).

Figure 24.33 gives results for approximately square chips produced by conventional turning with the intention of converting them to round wire by subsequent drawing. The higher the rake angle employed, the higher the observed value of cutting ratio (r = length of chip/undeformed chip length). It was found in a previous study of copper wire production by cutting (Shaw and Hoshi, 1976) that r should be 0.35 or greater to produce good-quality wire and the higher the value of r, the better the wire quality. These findings appear to be confirmed by Fig. 24.33 for aluminum. To produce high-quality wire by cutting, a rake angle of at least 45° should be used, and the workpiece should be highly work hardened.

Values of ultimate tensile strength and conductivity of aluminum wire by cut-forming and conventional cutting are given in Table 24.1. All values of conductivity in this table are within expectation for processed high-conductivity aluminum wire.

TABLE 24.1 Average Values of Ultimate Tensile Strength and Electric Conductivity for Aluminum[†]

	Diameter		Ultimate Tensile Strength		Percent Conductivity Relative to Annealed Standard Copper
	in	mm	10^3 p.s.i.	10^7 N/m^2	
1. Wire produced by cut-forming	0.089	2.26	17.10	11.80	61.2
2. No. 1 above, annealed	0.089	2.26	8.28	5.71	63.1
3. No. 2 above drawn to 0.050 in (1.27 mm) diameter	0.050	1.27	15.8	10.89	64.0
4. Wire made by drawing approximately square chips after annealing	0.050	1.27	15.5	10.69	64.8

[†] After Hoshi and Shaw, 1977.

CUT-WELDING

Sherwood and Milner (1969) have shown that fresh surfaces generated in vacuum may be cold welded in vacuum with application of very little pressure.

De Chiffre (1989) has introduced a method of welding that takes advantage of the fact that freshly cut surfaces are free of oxide and other contaminants that prevent welding and in addition have excess electrons that promote binding. Figure 24.34a illustrates the principle involved. A pair of tools each having a sharp cutting edge are placed between two surfaces to be welded. After the materials to be welded are aligned, pressure is applied which completes the weld (Fig. 24.34b).

Cut-welding may be performed in air and without application of heat. If one of the materials is harder than the other, the harder one may serve as one of the cutting tools. Figure 24.35 illustrates this case where a hard cylinder (A) is being cut-welded into a softer plate (B). A hard circular tool (C) is mounted on top of the softer plate (B). The cutting phase of the operation is shown in Fig. 24.35a while the bonding (compression) phase is shown in Fig. 24.35b. In this case the cut-welding operation is completed by a single stroke of the ram.

Bay (1985) has presented an extensive discussion of the importance of surface preparation in cold pressure operations.

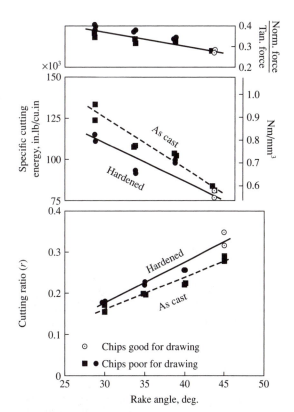

Fig. 24.33 Variation of cutting ratio, specific cutting energy, and force ratio with rake angle and workpiece hardness for conventional cutting. Cutting speed, 270 f.p.m. (82 m/min); feed, 0.050 i.p.r. (1.27 mm/rev); depth of cut, 0.020 in (0.51 mm); cutting fluid, none; work material, same as Figs. 24.31 and 24.32. (after Hoshi and Shaw, 1977)

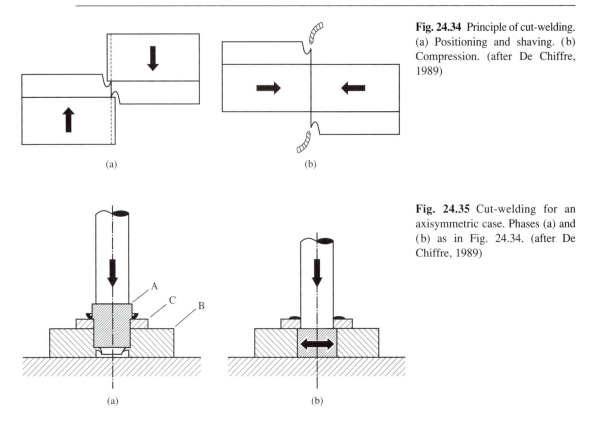

Fig. 24.34 Principle of cut-welding. (a) Positioning and shaving. (b) Compression. (after De Chiffre, 1989)

Fig. 24.35 Cut-welding for an axisymmetric case. Phases (a) and (b) as in Fig. 24.34. (after De Chiffre, 1989)

Cut-welding may be successfully performed with a wide variety of materials including aluminum and titanium which have an unusually strong tendency to oxidize, which normally makes these materials very difficult to cold weld.

Figure 24.36 shows an interrupted cut where a copper slug having a hardness of H_V 110 is enlarging a circular hole in an aluminum plate having a hardness of H_V 40. In this case only one surface (Al) is cut while the surface of copper is only abraded. De Chiffre (1989) has shown the strength of a cut-welded joint to be very much better when both surfaces are machined instead of just one as in Fig. 24.36.

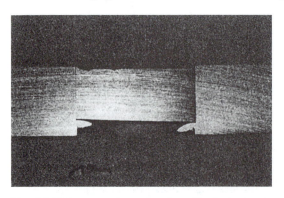

Fig. 24.36 Interrupted cut showing chip formation in cut-welding a copper stud in an aluminum plate. (after De Chiffre, 1989)

FIBER-REENFORCED PRODUCTS

The initial application of fiber-reenforced materials was to the strengthening of concrete by J. P. Romualdi at Carnegie Mellon University in the early 1970s. He employed about 150 lb of steel wire fragments per cubic yard of concrete for this purpose. Initially the thin wire fibers had an aspect ratio of about 60. The wire fibers increased the strength of the concrete in about

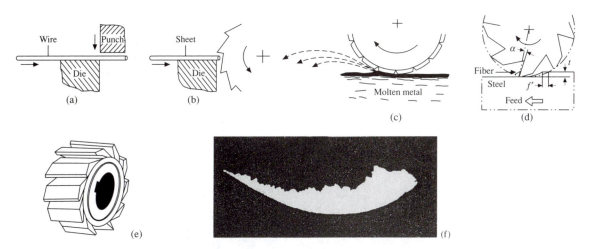

Fig. 24.37 Early methods of producing steel fibers for strengthening concrete against gross fracture. (a) Shearing of drawn steel wire. (b) Shearing of steel sheat. (c) Extraction from molten metal. (d) Down-milling of steel block, 1 in (25 mm) wide using 25 in (100 mm) diameter cutter having an inclination angle of 15° shown in (e) to produce long thin strips as shown in (f).

the same way as carbon fibers increase the strength of ceramic tools as discussed in Chapter 14, with apparently no connection between these two developments. When a microcrack is formed at a point of stress concentration in fiber-reenforced concrete, it runs only a short distance before encountering a wire which changes its direction to one of lower stress, thus preventing the crack from reaching a critical size and leading to gross fracture.

Initially the steel fibers were produced by shearing fine wire as shown in Fig. 24.37a. However, this method of producing thin steel wires was too costly due to the relatively high cost of drawing the wire prior to shearing. The next method of producing thin steel fibers was by shearing steel strip as shown in Fig. 24.37b. While this method had a much higher production rate, the cost of the rolled starting material was still too high.

The next method of fiber production was the melt-extraction process shown in Fig. 24.37c, where a water-cooled serrated disc caused threads of liquid steel to be periodically removed from the molten pool of metal. However, the liquid fibers cooled so rapidly that they required a tempering treatment to restore ductility. This melt-extraction process was developed at the Battelle Memorial Institute in Columbus, Ohio, where methods of applying Romualdi's method of strengthening concrete was being developed and where the basic patent for wire reenforcing concrete was registered under the name WIRAND (for wire and concrete).

Kobayashi (1975) presented an important paper in Japan outlining the virtues of wire-reenforced concrete, which aroused considerable interest in the application and extension of this idea. Yanagisawa and Nakagawa (1980) introduced a very important improvement in the production of steel fiber for reenforcing concrete. An English translation of this article is available in Nakagawa and Yanagisawa (1981). This new method of producing steel fibers employs a special tool holder designed to produce a self-excited vibration in which the tool leaves contact with the work periodically as shown in Fig. 24.38a. The length of a fiber corresponds to the depth of cut, the equivalent diameter of a fiber to the feed per revolution, and the number of fibers produced per second to the frequency of vibration in hertz. The cross-section of a fiber is triangular as shown in Fig. 24.38b, and the appearance of a cluster of fibers is shown in Fig. 24.38c.

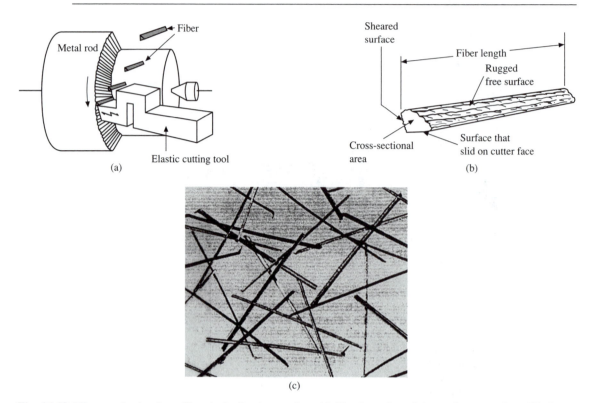

Fig. 24.38 Fiber production by self-excited vibration-cutting. (a) Elastic tool used in turning operation. (b) Cross-section of a chip (fiber). (c) Appearance of group of vibration-cut fibers.

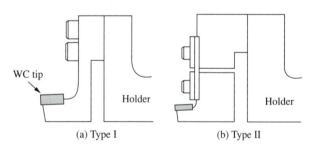

Fig. 24.39 Tool holder arrangements used for producing steel fibers by vibration-cutting.

Figure 24.39 shows the types of tool holders that were used in fiber production by vibration-cutting. Type I was used initially. However, the maximum equivalent fiber diameter produced by this type of tool holder was only 0.00044 in (0.11 mm) due to its high natural frequency. For concrete reenforcement, a larger equivalent steel fiber is required (0.010 in = 0.24 mm). This requires a lower natural frequency. The type II tool holder has a larger vibrating mass and hence a lower natural frequency of vibration. The type II design gives a tool life of about two hours with K10 carbide with coolant when vibration-cutting a free-machining steel. A free-cutting steel is usually used for concrete-reenforcing fiber because of its superior ductility and machinability.

The illustrations in Figs. 24.37–24.39 are taken from a series of papers published by Nakagawa and his coworkers at the Institute of Industrial Science at the University of Tokyo. The milling fiber production process was covered by Japanese patent 4,298,660 granted to Nakagawa in 1981. The vibration-cutting method of producing fibers was covered by U.S. Patent 4,560,622 granted to Tezuka, Nakagawa, and Kobayachi in 1985.

Nakagawa and coworkers have extended the use of fibers of materials other than steel to produce composites for a number of applications other than concrete, including the following:

- for improved brake pads consisting of a nylon base with brass or steel fibers
- for improved grinding wheels consisting of cast-iron fibers and diamond grits in a bonding material
- for production of conductive plastics by incorporating brass fibers in a plastic substrate
- for shielding against electromagnetic interference (EMI) using brass fibers in a plastic to line instrument cases
- for antistatic floor material to prevent dangerous static buildup and spark generation in association with dangerous chemicals (ethyl ether in operating rooms, for example)
- for producing filters by sintering fibers that are more effective than those consisting of compacted powder
- for production of self-lubricating bearings, for example, bearings consisting of 40% graphite incorporated in a matrix of cast-iron or bronze fibers
- for production of reenforced plastics with fibers having a minimum diameter of 0.0002 to 0.0004 in (5 to 10 um) and a minimum length of 0.200 in (5 mm)
- for reenforcing refractory furnace blocks by incorporating 18-8 stainless steel fibers

Nakagawa and his coworkers have made many contributions to fiber-reenforced materials technology and have published many reports on the subject. Two representative examples are Nakagawa and Suzuki (1977) and Nakagawa et al. (1982). For his many contributions to fiber-enhanced materials technology, Nakagawa has been awarded the prestigious Okoshi Memorial Prize, an honor he justly deserves.

NANOCRYSTALLINE STRUCTURES

Brown et al. (2002) have identified the presence of nanocrystalline crystals ranging in size between 100 and 800 nm (4 and 32 μm) in turning chips of a variety of pure metals and alloys. The hardness and strength of these chips was found to be significantly greater than the bulk material before chip formation. It was suggested such chips be ball-milled to produce particles of unusual hardness and strength for use in a variety of composites. This represents an important extension of the previously discussed use of wire-like chips produced by vibration cutting by Nakagawa and Yanagisawa (1981) to improve the strength of concrete. This discovery of nanocrystalline structures in metal cutting chips is consistent with the previously suggested role the unusually large strains involved in metal cutting play in chip formation due to microcrack migration in addition to dislocation migration. It is also consistent with the role microcracks play in connection with the size effect in metal cutting.

REFERENCES

Alfille, L., Ropers, J., and Weill, R. (1963). *Revue du Metallurgie* (Feb.), 169.
Avitzur, B. (1975). SME Tech. Paper MF 75–140.
Bay, N. (1985). D.Sc. dissertation. Tech. University of Denmark.
Bovenkerk, H. P. (1961). U.S. Patent 2,992,900.

Bovenkerk, H. P., Bundy, F. P., Hall, H. T., Strong, H. M., and Wentdorf, R. H. (1959). *Nature* **184**, 1094–1098.

Brown, T. L., Swaminathan, S., Chandrasekar, S., Compton, W. R., King, A. H., and Trumble, K. P. (2002). *J. Mater. Res.* **17/10**, 1–5.

Crandall, S. H., and Dahl, N. C. (1959). *An Introduction to the Mechnics of Solids*. McGraw-Hill, New York.

De Chiffre, L. (1976). *Int. J. Mach. Tool Des. and Res.* **16**, 137–144.

De Chiffre, L. (1989). *Annals of CIRP* **38/1**, 125.

De Chiffre, L. (1990). D.Sc. dissertation. Tech. University of Denmark.

Finnie, I., and Wolak, J. (1963). *ASME J. of Eng. for Ind.* **858**, 351–355.

Green, D. (1972). *J. Inst. of Metals* **100**, 295.

Grunberg, L. (1958). *J. Appl. Phys. (UK)* **9**, 85.

Hoshi, T., and Shaw, M. C. (1977). *Trans. ASME* **99**, 225–229.

Kobayashi, K. (1975). *Concrete Journal* (in Japanese) **13/8**, 21.

Komanduri, R., and Shaw, M. C. (1974). *Nature* **248**, 582–584.

Komanduri, R., and Shaw, M. C. (1975). *Phil. Mag.* **32/4**, 711–724.

Kramer, J. (1950). *Der Metallishce Zustand (Gottingen)*.

Marshall, E. R., and Shaw, M. C. (1952). *Trans. ASME* **74**, 51–59.

Mayer, J. E., and Shaw, M. C. (1957). *J. of Amer. Soc. of Lub. Engrs.* **13/1**, 21–22.

Meyer, K. (1967). *Deutschen Acad. der Wissenshaften zu Berlin*.

Nakagawa, T., and Suzuki, K. (1977). *Annals of CIRP* **25/1**, 55–58.

Nakagawa, T., Suzuki, K., Uematsu, T., and Koyama, H. (1982). *Proc. 23rd Int. MTDR Conf.*

Nakagawa, T., and Yanagisawa, A. (1981). *Proc. Japan–USA Conf. on Composite Materials*, Tokyo, pp. 271–280.

Nakayama, K. (1964). *Bull. Faculty of Eng., Yokohama National University* (13/March).

Shaw, M. C. (1945). British Patent 571,539.

Shaw, M. C. (1947). U.S. Patent 2,416,717.

Shaw, M. C. (1948). *J. Appl. Mech.* **15**, 37.

Shaw, M. C. (1972a). *Trans. I.E. Australia* **MC8**, 73.

Shaw, M. C. (1972b). *Mech. Eng.* **94/4**, 23–29.

Shaw, M. C., and DeSalvo, G. J. (1972). *Metals Eng. Quarterly* **12**, 1.

Shaw, M. C., and Hoshi, T. (1976). *Proc. 16th Int. Conf. on Mach. Tool Design and Res.* Macmillan Press, pp. 459–465.

Sherwood, W. C., and Milner, D. R. (1969). *J. Inst. Met.* **97**, 1.

Uehara, K., Takekshita, H., and Kotaka, H. (2000). *Proc. Int. Conf. on Prec. Eng.*, Singapore.

Vigor, C. W., and Leibring, W. (1973). SAE paper 730, 101.

Yanagisawa, A., and Nakagawa, T. (1980). *Proc. of 6th Symp. on Composite Materials* (in Japanese), pp. 60–63.

INDEX